Characterization and Modelling of Composites, Volume II

Characterization and Modelling of Composites, Volume II

Editor

Stelios K. Georgantzinos

MDPI • Basel • Beijing • Wuhan • Barcelona • Belgrade • Manchester • Tokyo • Cluj • Tianjin

Editor
Stelios K. Georgantzinos
National and Kapodistrian
University of Athens
Greece

Editorial Office
MDPI
St. Alban-Anlage 66
4052 Basel, Switzerland

This is a reprint of articles from the Special Issue published online in the open access journal *Journal of Composites Science* (ISSN 2504-477X) (available at: https://www.mdpi.com/journal/jcs/special_issues/Characterization_Modelling_Composites_Volume_II).

For citation purposes, cite each article independently as indicated on the article page online and as indicated below:

LastName, A.A.; LastName, B.B.; LastName, C.C. Article Title. *Journal Name* **Year**, *Volume Number*, Page Range.

ISBN 978-3-0365-5507-2 (Hbk)
ISBN 978-3-0365-5508-9 (PDF)

Contents

About the Editor

Stelios K. Georgantzinos

Stelios K. Georgantzinos is currently Assistant Professor in the Department of Aerospace Science and Technology at National and Kapodistrian University of Athens. Stelios does research in Mechanical Design, Structural Analysis and Additive Manufacturing with applications in Composite Structures, Machine Elements and Nanotechnology. He has more than 100 publications in referred journals, conference papers, and book chapters, as well as he has edited technical reports, textbooks, and laboratory manuals. His research work is cited more than 1100 times in the open literature. He is a member of the Editorial Board of the Journal Mathematical Problems in Engineering (Hidawi), and Topic Board member of the journal Molecules (MDPI). He has also served as guest editor in Materials (MDPI) and Journal of Composites Science (MDPI). He is a member of the Greek Association of Computational Mechanics (GR.A.C.M.) and the Technical Chamber of Greece. He is also a registered Mechanical and Electrical Engineer in Greece.

Preface to "Characterization and Modelling of Composites, Volume II"

The increasing demands for more durable, lighter, and smarter structures have led to the development of new and advanced composites. Increased strength and simultaneous weight reduction have resulted in energy savings and applications in several manufacturing industries, such as the automotive and aerospace industries as well as in the production of everyday products. Their optimal design and utilization are a process, which requires their characterization and efficient modeling. The papers published in this Special Issue of the *Journal of Composites Science* will give composite engineers and scientists insight into what the existing challenges are in the characterization and modeling for the composites field, and how these challenges are being addressed by the research community. The papers present a balance between academic and industrial research, and clearly reflect the collaborative work that exists between the two communities, in a joint effort to solve the existing problems.

Stelios K. Georgantzinos
Editor

Journal of
Composites Science

MDPI

Editorial

Editorial for the Special Issue on Characterization and Modelling of Composites, Volume II

Stelios K. Georgantzinos

Laboratory for Advanced Materials, Structures, and Digitalization, Department of Aerospace Science and Technology, National and Kapodistrian University of Athens, 34400 Psachna, Greece; sgeor@uoa.gr

Abstract: The increasing demands for more durable, lighter, and smarter structures have led to the development of new and advanced composites. Increased strength and simultaneous weight reduction have resulted in energy savings and applications in several manufacturing industries, such as the automotive and aerospace industries as well as in the production of everyday products. Their optimal design and utilization are a process, which requires their characterization and efficient modeling. The papers published in this Special Issue of the *Journal of Composites Science* will give composite engineers and scientists insight into what the existing challenges are in the characterization and modeling for the composites field, and how these challenges are being addressed by the research community. The papers present a balance between academic and industrial research, and clearly reflect the collaborative work that exists between the two communities, in a joint effort to solve the existing problems.

Keywords: composites; composite structures; nanocomposites; additive manufacturing; characterization; modeling; finite element analysis; simulation; experiments

Citation: Georgantzinos, S.K. Editorial for the Special Issue on Characterization and Modelling of Composites, Volume II. *J. Compos. Sci.* **2022**, *6*, 274. https://doi.org/10.3390/jcs6090274

Received: 13 September 2022
Accepted: 13 September 2022
Published: 17 September 2022

Publisher's Note: MDPI stays neutral with regard to jurisdictional claims in published maps and institutional affiliations.

The last few decades have seen extraordinary progress in the science and technology of composites. Their distinctive characteristics make composites desirable for engineering applications in a wide variety of industrial sectors. New manufacturing methods are driving cost reduction, and emerging areas such as nanocomposites, green composites, 2D/3D textile composites, multifunctional composites, and smart composites have been the focus of much exciting and innovative research activity. This Special Issue is the continuation of the successful first companion Special Issue [1,2]. It contains a snapshot of the research typical of this activity, and includes thirty-two papers on topics such as additive manufacturing of composites, structural and failure analysis, process modeling, fundamental mechanisms at nano- and microscales in both nano- and biocomposites, and non-destructive testing methods for advanced carbon composites.

Fiber-reinforced polymer matrix composites continue to attract scientific and industrial interest since they offer superior strength-, stiffness-, and toughness-to-weight ratios. The research of Youssef et al. [3] characterizes two sets of E-Glass/Epoxy composite skins: stressed and unstressed. The stressed samples were previously installed in an underground power distribution vault and were exposed to fire while the unstressed composite skins were newly fabricated and never-deployed samples. The mechanical, morphological, and elemental composition of the samples were methodically studied using a dynamic mechanical analyzer, a scanning electron microscope (SEM), and an x-ray diffractometer, respectively. Sandwich composite panels consisting of E-glass/Epoxy skin and balsa wood core were originally received, and the balsa wood was removed before any further investigations. Skin-only specimens with dimensions of ~12.5 mm wide, ~70 mm long, and ~6 mm thick were tested in a Dynamic Mechanical Analyzer in a dual-cantilever beam configuration at 5 Hz and 10 Hz from room temperature to 210 °C. Micrographic analysis using the SEM indicated a slight change in morphology due to the fire event but confirmed the effectiveness of the fire-retardant agents in quickly suppressing the fire. Accompanying Fourier transform infrared and energy dispersive X-ray spectroscopy studies corroborated the mechanical and morphological results. Finally, X-ray diffraction showed that the fire event consumed the surface level fire-retardant and the structural attributes of the E-Glass/Epoxy remained mainly intact. The results suggest the panels can continue field deployment, even after short fire incident.

Mechanical properties of fiber-reinforced engineering materials often depend on their local orientation of fibers. Analytical orientation models such as the Folgar Tucker model are widely applied to predict the orientation of suspended non-spherical particles. The accuracy of these models depends on empirical model parameters. Dietemann et al. [4] assess how well analytical orientation models can predict the orientation of suspensions not only consisting of fibers but also of an additional second particle type in the shape of disks, which are varied in size and filling fraction. They mainly focus on the FT model and compare its accuracy to more complex models such as Reduced-Strain Closure model, Moldflow Rotational Diffusion model, and Anisotropic Rotary Diffusion model.

Building owners have become interested in a sustainable and healthy environment, which is a trend favoring ecological materials with outstanding performance. In addition, currently thermal insulation can be considered to be a hot issue for civil engineering that tries to reduce cooling and heating costs and, at the same time, eliminate CO_2 emissions. Ninikas et al. [5] investigate the technical feasibility of manufacturing low density insulation particleboards that were made from two renewable resources, namely hemp fibers (Cannabis sativa) and pine tree bark, which were bonded with a non-toxic methyl cellulose glue as a binder. Four types of panels were made, which consisted of varying mixtures of tree bark and hemp fibers (tree bark to hemp fibers percentages of 90:10, 80:20, 70:30, and 60:40). An additional set of panels was made, consisting only of bark. The results showed that addition of hemp fibers to furnish improved mechanical properties of boards to reach an acceptable level. The thermal conductivity unfavorably increased as hemp content increased, though all values were still within the acceptable range. Based on cluster analysis, board type 70:30 (with 30% hemp content) produced the highest mechanical properties as well as the optimal thermal conductivity value. It is concluded that low density insulation boards can be successfully produced using these waste raw materials.

Recently, a rapid development has been observed concerning drive shafts made of composite materials that, in addition to their low weight, are able to respond effectively to their functional demands. The potential performance improvement of specific mechanical components due to the use of nanomaterials has not been extensively reported yet. Georgantzinos et al. [6] carried out the modal and linear buckling analysis of a laminated composite drive shaft reinforced by multi-walled carbon nanotubes using an analytical approach, as well as the finite element method. Their theoretical model is based on classical laminated theory. The fundamental frequency and the critical buckling torque were determined for different fiber orientation angles. The Halpin–Tsai model was employed to calculate the elastic modulus of composites having randomly oriented nanotubes. The effect of various carbon nanotube volume fractions in the epoxy resin matrix on the material properties of unidirectional composite laminas was also analyzed. The fundamental frequency and the critical buckling torque obtained by the finite element analysis and the analytical method for different fiber orientation angles were in good agreement with each other. The results were verified with data available in the open literature, where possible. For the first time in the literature, the influence of carbo nanotube fillers on various composite drive shaft design parameters such as the fundamental frequency, critical speed, and critical buckling torque of a hybrid fiber-reinforced composite drive shaft was predicted.

During construction works, it is advisable to prevent strong thawing and an increase in the moisture content of the foundations of engineering structures in the summer. Since the density of water and ice differ, due to the difference bulging of the foundation sections can occur when it freezes back in winter. Sivtsev et al. [7] numerically investigated the effect of fiber-reinforced piles on the thermal field of the surrounding soil; that is, the study of the influence of aggregates with high and low thermal-physical properties on the temperature of frozen soils is conducted. Basalt and steel fiber reinforcement were compared. The difficulty of this work is that the inclusions inside piles are too small compared to the pile itself. Therefore, to solve the Stefan problem, a generalized multiscale finite element method (GMsFEM) was used. In the GMsFEM, the usual conforming partition of the domain into a coarse grid was used. It allowed reducing problem size and, consequently, accelerating the calculations. Results of the multiscale solution were compared with fine-scale solution, the accuracy of GMsFEM was investigated, and the optimal solution parameters were defined. Therefore, GMsFEM was shown to be well suited for the designated task. Collation of basalt and steel fiber reinforcement showed a beneficial effect of high thermal conductive material inclusion on freezing of piles in winter.

The demand for high strength and lightweight steel is inevitable for technological, environmental, and economic progress. Umar et al. [8] uses EBSD data of two thermo-mechanically processed medium carbon (C45EC) steel samples to simulate micromechanical deformation and damage behavior. Two samples with 83% and 97% spheroidization degrees are subjected to virtual monotonic quasi-static tensile loading. The ferrite phase is assigned already reported elastic and plastic parameters, while the cementite particles are assigned elastic properties. A phenomenological constitutive material model with critical plastic strain-based ductile damage criterion is implemented in the DAMASK framework for the ferrite matrix. At the global level, the calibrated material model response matches well with experimental results, with up to ~97% accuracy. The simulation results provide essential insight into damage initiation and propagation based on the stress and strain localization due to cementite particle size, distribution, and ferrite grain orientations. In general, it is observed that the ferrite–cementite interface is prone to damage initiation at earlier stages triggered by the cementite particle clustering. Furthermore, it is observed that the crystallographic orientation strongly affects the stress and stress localization and consequently nucleating initial damage.

The mechanical and ring stiffness of glass fiber pipes are the most determining factors for their ability to perform their function, especially in a work environment with difficult and harmful conditions. Usually, these pipes serve in rough underground environments of desert and petroleum fields; therefore, they are subjected to multi-type deterioration and damage agents. In polymers and composite materials, corrosion is identified as the degradation in their properties. Hassan et al. [9] carried out tension and compression tests before and after preconditioning in a corrosive agent for 60 full days to reveal corrosion influences. Moreover, the fracture toughness is measured using a standard single edge notch bending. Ring stiffness of such pipes which, are considered characteristic properties, is numerically evaluated using the extended finite element method before and after preconditioning. The results reported that both tensile and compressive strengths degraded nearly more than 20%. Besides the fracture toughness decrease, the stiffness ring strength is reduced, and the finite element results are in good agreement with the experimental findings.

Components made from multidirectional fiber-reinforced composite laminates experience several distinct damage mechanisms when exposed to fatigue loads. A model that predicts the stiffness degradation in multidirectional reinforced laminates due to off-axis matrix cracks is proposed by Drvoderic et al. [10] and evaluated using data from fatigue experiments. Off-axis cracks are detected in images from the fatigue tests with automated crack detection to compute the crack density of the off-axis cracks which is used as the damage parameter for the degradation model. The purpose of this study is to test the effect of off-axis cracks on laminate stiffness for different laminate configurations. The hypothesis is that off-axis cracks have the same effect on the stiffness of a ply regardless of the acting stress components if the transverse stress is positive. This hypothesis proves to be wrong. The model can predict the stiffness degradation well for laminates with a ply orientation such as the one used for calibration but deviates for plies with different in-plane shear stress. This behavior can be explained by the theory that off-axis cracks develop by two different micro damage modes depending on the level of in-plane shear stress. It is found that besides influencing the initiation and growth of off-axis cracks, the stiffness degradation is also mode dependent.

Lack of cost information is a barrier to acceptance of 3D woven preforms as reinforcements for composite materials, compared with 2D preforms. A parametric, resource-based technical cost model (TCM) was developed by Clarke et al. [11] for 3D woven preforms based on a novel relationship equating manufacturing time and 3D preform complexity. Manufacturing time, and therefore cost, was found to scale with complexity for seventeen bespoke manufactured 3D preforms. Two sub-models were derived for a Weavebird loom and a Jacquard loom. For each loom, there was a strong correlation between preform complexity and manufacturing time. For a large, highly complex preform, the Jacquard loom is more efficient, so preform cost will be much lower than for the Weavebird. Provided production is continuous, learning, either by human agency or an autonomous loom control algorithm, can reduce preform cost for one or both looms to a commercially acceptable level. The TCM cost model framework could incorporate appropriate learning curves with digital twin/multi-variate analysis so that cost per preform of bespoke 3D woven fabrics for customized products with low production rates may be predicted with greater accuracy. A more accurate model could highlight resources such as tooling, labor, and material for targeted cost reduction.

In tape placement process, the laying angle and laying sequence of laminates have proven their significant effects on the mechanical properties of carbon fiber reinforced composite material, specifically, laminates. To optimize these process parameters, an optimization algorithm is developed by Mou et al. [12] based on the principles of genetic algorithms for improving the precision of traditional genetic algorithms and resolving the premature phenomenon in the optimization process. Taking multi-layer symmetrically laid carbon fiber laminates as the research object, this algorithm adopts binary coding to conduct the optimization of process parameters and mechanical analysis with the laying angle as the design variable and the strength ratio R as the response variable. A case study was conducted, and its results were validated by the finite element analyses. The results show that the stresses before and after optimization are 116.0 MPa and 100.9 MPa, respectively, with a decrease in strength ratio by 13.02%. The results comparison indicates that, in the iterative process, the search range is reduced by determining the code and location of important genes, thereby reducing the computational workload by 21.03% in terms of time consumed. Through multiple calculations, it validates that "gene mutation" is an indispensable part of the genetic algorithm in the iterative process.

In industrial applications where contact behavior of materials is characterized, fretting-associated fatigue plays a vital role as a failure agitator. While considering connection, it encounters friction. Biomaterials such as polytetrafluoroethylene (PTFE) and ultra-high-molecular-weight polyethylene (UHMWPE) are renowned for their low coefficient of friction and are utilized in sophisticated functions such as the hip joint cup and other biomedical implants. In addition to the axial stresses, some degree of dynamic bending stress is also developed occasionally in those fretting contacts. Shah et al. [13] investigated the fracture behavior of a polymer PTFE under bending fretting fatigue. Finite element analysis justified the experimental results. A mathematical model is proposed by developing an empirical equation for fracture characterization in polymers such as PTFE. It was found that the bending stiffness exists below the loading point ratio (LPR) 3.0, near the collar section of the specimen. Along with fretting, the bending load forces the specimen to crack in a brittle-ductile mode near the sharp-edged collar where the maximum strain rate, as well as stress, builds up. For a loading point ratio of above 3, a fracture takes place near the fretting pads in a tensile-brittle mode. Strain proportionality factor, k was found as a life optimization parameter under conditional loading. The microscopic analysis revealed that the fracture striation initiates perpendicularly to the fretting load. The fretting fatigue damage characteristic of PTFE may have a new era for the biomedical application of polymer-based composite materials.

Conveyor belts are used in a wide range of applications such as supermarkets, logistic centers, and mining. The conveyor belts in mining are reinforced with steel cables to reach the high strengths required. Frankl et al. [14] introduces a finite element model of a steel cable-reinforced conveyor belt to accurately compute stresses in the splice. In the modelled test rig, the belt runs on two drums and is loaded with a cyclic longitudinal force. An explicit solver is used to efficiently handle the high number of elements and contact conditions. This, however, introduces some issues of dynamics in the model, which are subsequently solved: (a) the longitudinal load is applied with a smooth curve and damping is introduced in the beginning of the simulation, (b) residual stresses are applied in regions of the belt that are initially bent around the drums, and (c) supporting drums are introduced at the start of the simulation to hinder oscillations of the belt at low applied forces. To accurately capture the tensile and bending stiffness of the cables, they are modelled by a combination of solid and beam elements. The results show that numerical artefacts can be reduced to an acceptable extent. In the region of highest stresses, the displacements are additionally mapped onto a submodel with a smaller mesh size. The results show that, for the investigated belt, the local maximum principal stresses significantly increase when this region of highest stresses comes into contact with, and is bent by, the drum. Therefore, it is essential to also consider the belt's bending to predict failure in such applications.

The microstructure-based finite element modeling (MB-FEM) of material representative volume element (RVE) is a widely used tool in the characterization and design of various composites. However, the MB-FEM has a number of deficiencies, e.g., time-consuming in the generation of a workable geometric model, challenge in achieving high volume-fractions of inclusions, and poor quality of finite element mesh. Luo [15] first demonstrate that for particulate composites the particle inclusions have homogeneous distribution and random orientation, and if the ratio of particle characteristic length to RVE size is adequately small, elastic properties characterized from the RVE are independent of particle shape and size. Based on this fact, it is proposed a microstructure-free finite element modeling (MF-FEM) approach to eliminate the deficiencies of the MB-FEM. The MF-FEM first generates a uniform mesh of brick elements for the RVE, and then a number of the elements, with their total volume determined by the desired volume fraction of inclusions, is randomly selected and assigned with the material properties of the inclusions; the rest of the elements are set to have the material properties of the matrix. Numerical comparison showed that the MF-FEM has a similar accuracy as the MB-FEM in the predicted properties. The MF-FEM was validated against experimental data reported in the literature and compared with the widely used micromechanical models. The results show that for a composite with small contrast of phase properties, the MF-FEM has excellent agreement with both the experimental data and the micromechanical models. However, for a composite that has large contrast of phase properties and high volume-fraction of inclusions, there exist significant differences between the MF-FEM and the micromechanical models. The proposed MF-FEM may become a more effective tool than the MB-FEM for material engineers to design novel composites.

In recent years, finite element analysis (FEA) models of different porous scaffold shapes consisting of various materials have been developed to predict the mechanical behavior of the scaffolds and to address the initial goals of 3D printing. Although mechanical properties of polymeric porous scaffolds are determined through FEA, studies on the polymer nanocomposite porous scaffolds are limited. Kakarla et al. [16] carried out FEA with the integration of material designer and representative volume elements (RVE) on a 3D scaffold model to determine the mechanical properties of boron nitride nanotubes (BNNTs)-reinforced gelatin (G) and alginate (A) hydrogel. The maximum stress regions were predicted by FEA stress distribution. Furthermore, the analyzed material model and the boundary conditions showed minor deviation (4%) compared to experimental results. It was noted that the stress regions are detected at the zone close to the pore areas. These results indicated that the model used in this work could be beneficial in FEA studies on 3D-printed porous structures for tissue engineering applications.

Variable stiffness composite laminates can improve the structural performance of composite structures by expanding the design space. Carvalho et al. [17] explores the application of variable stiffness laminated composite structures to maximize the fundamental frequency by optimizing the tow angle. To this end, an optimization framework is developed to design the fiber angle for each layer based on the maximization of the fundamental frequency. It is assumed that the design process includes the manufacturing constraints encountered in the automated fiber placement process and a linear fiber angle variation. The current study improves existing results by considering embedded gap defects within the optimization framework. The plates are assumed symmetric, with clamped and simply supported boundary conditions. The optimal results and a comparison between the non-steered and steered plates with and without gaps are presented. Results show that, although gaps deteriorate the structural performance, fiber steering can still lead to an increase in the fundamental frequency depending on the plate's geometry and boundary conditions.

Sandwich structures benefit from the geometrical stiffening effect due to their high cross-sectional area moment of inertia. Transferred to carbon fiber-reinforced plastic (CFRP) components, the needed amount of carbon fiber (CF) material can be reduced and with it the CO_2 footprint. The combination of a light foam core with continuous fiber-reinforced face sheets is a suitable material combination for lightweight design. Traditionally, CFRP sandwich structures with a foam core are manufactured in a two-step process by combining a prefabricated foam core with fiber-reinforced face sheets. However, in addition to the reduction in the used CFRP material, manufacturing processes with a high efficiency are needed. Behnisch et al. [18] investigated sandwich manufacturing and characterization by using the Direct Sandwich Composite Molding (D-SCM) process for the one-step production of CFRP sandwich structures. The D-SCM process utilizes the resulting foaming pressure during the reactive polyurethane (PUR) foam system expansion for the impregnation of the CF-reinforced face sheets. The results of this work show that the production of sandwich structures with the novel D-SCM process strategy is feasible in one single manufacturing step and achieves good impregnation qualities. The foam density and morphology significantly influence the core shear properties and thus the component behavior under a bending load.

The quadratic function of the original Tsai–Wu failure criterion for transversely isotropic materials is re-examined by Li et al. [19]. According to analytic geometry, two of the troublesome coefficients associated with the interactive terms—one between in-plane direct stresses and one between transverse direct stresses—can be determined based on mathematical and logical considerations. The analysis of the nature of the quadratic failure function in the context of analytic geometry enhances the consistency of the failure criterion based on it. It also reveals useful physical relationships as intrinsic properties of the quadratic failure function. Two clear statements can be drawn as the outcomes of the present investigation. Firstly, to maintain its basic consistency, a failure criterion based on a single quadratic failure function can only accommodate five independent strength properties, viz. the tensile and compressive strengths in the directions along fibers and transverse to fibers, and the in-plane shear strength. Secondly, amongst the three transverse strengths—tensile, compressive and shear—only two are independent.

Thermoset polymers offer great opportunities for mass production of fiber-reinforced composites and are being adopted across a large range of applications within the automotive, aerospace, construction and renewable energy sectors. They are usually chosen for marine engineering applications for their excellent mechanical behavior, including low density and low cost compared to conventional materials. In the marine environment, these materials are confronted by severe conditions, thus there is the necessity to understand their mechanical behavior under critical loads. The high strain rate performance of bonded joints composite under hygrothermal aging has been studied by Lagdani et al. [20]. Initially, the bonded composite specimens were hygrothermal aged with the conditions of 50 °C and 80% in temperature and relative humidity, respectively. After that, gravimetric testing is used to describe the moisture diffusion properties for the adhesively bonded composite samples and exhibit lower weight gain for this material. Then, the in-plane dynamic compression experiments were carried out at different impact pressures ranging from 445 to 1240 s^{-1} using the SHPB (Split Hopkinson Pressure Bar) technique. The experimental results demonstrated that the dynamic behavior varies with the variation of strain rate. Buckling and delamination of fiber are the dominant damage criteria observed in the sample during in-plane compression tests.

Haas et al. [21] tried to enhance and validate a systematic approach for the structural finite element (FE) analysis of thermoplastic impregnated 3D filament winding structures (fiber skeletons). The idealized modeling of geometrically complex fiber skeletons used in previous publications is refined by considering additional characteristic dimensions and investigating their mechanical influence. Moreover, the modeling approach is transferred from the meso- to the macro-level in order to reduce modeling and computational effort. The properties of meso- and macro-level FE models are compared using the example of simple loop specimens. Based on the results, respective application fields are defined. In the next step, the same modeling approach is applied to a more complex, three-dimensional specimen—the inclined loop. For its macro-level FE model, additional material characterization and modeling, as well as enhancements in the modeling of the geometry, are proposed. Together with previously determined effective composite properties of fiber skeletons, these results are validated in experimental tensile tests on inclined loop specimens.

For some categories of aircraft, such as helicopters and tiltrotors, fuel storage systems must satisfy challenging crash resistance requirements to reduce or eliminate the possibility of fuel fires and thus increase the chances of passenger survival. Therefore, for such applications, fuel tanks with high flexibility (bladder) are increasingly used, which can withstand catastrophic events and avoid fuel leakages. The verification of these capabilities must be demonstrated by means of experimental tests, such as the cube drop test (MIL-DTL-27422). To reduce development costs, it is necessary to execute experimental tests with a high confidence of success, and, therefore, it is essential to have reliable and robust numerical analysis methodologies. Cristillo et al. [22] tried to provide a comparison between two explicit FE codes (i.e., Abaqus and Ls-Dyna), which are the most frequently used for such applications according to experimental data in the literature. Both codes offer different material models suitable for simulating the tank structure, and therefore, the most suitable one must be selected by means of a specific trade-off and calibration activity. Both can accurately simulate the complex fluid–structure interaction thanks to the use of the SPH approach, even if the resulting sloshing capabilities are quite different from each other. Additionally, the evolution of the tank's deformed shape highlights some differences, and Abaqus seems to return a more natural and less artificial behavior. For both codes, the error in terms of maximum impact force is less than 5%, but, even in this case, Abaqus can return slightly more accurate results.

Additively manufactured parts play an increasingly important role in structural applications. Fused Layer Modeling (FLM) has gained popularity due to its cost-efficiency and broad choice of materials, among them, short fiber reinforced filaments with high specific stiffness and strength. To design functional FLM parts, adequate material models for simulations are crucial, as these allow for reliable simulation within virtual product development. Witzgall et al. [23] presented a new approach to derive FLM material models for short fiber reinforced parts; it is based on simultaneous fitting of the nine orthotropic constants of a linear elastic material model using six specifically conceived tensile specimen geometries with varying build direction and different extrusion path patterns. The approach is applied to a 15 wt.% short carbon-fiber reinforced PETG filament with own experiments, conducted on a Zwick HTM 5020 servo-hydraulic high-speed testing machine. For validation, the displacement behavior of a geometrically more intricate demonstrator part, printed upright, under bending is predicted using simulation and compared to experimental data. The workflow proves stable and functional in calibration and validation.

The quality of the fiber orientation of injection molding simulations and the transferred fiber orientation content, due to the process–structure coupling, influence the material modeling and thus the prediction of subsequently performed structural dynamics simulations of short-fiber reinforced, thermoplastic components. Existing investigations assume a reliable prediction of the fiber orientation in the injection molding simulation. The influence of the fiber orientation models and used boundary conditions of the process–structure coupling is mainly not investigated. Kriwet and Stommel [24] investigated the influence of the fiber orientation from injection molding simulations on the resulting structural dynamics simulation of short-fiber reinforced thermoplastic components. The Advani–Tucker Equation with phenomenological coefficient tensor is used in a 3- and 2.5-dimensional modeling approach for calculating the fiber orientation. The prediction quality of the simulative fiber orientations is evaluated in comparison to experiments. Depending on the material modeling and validation level, the prediction of the simulated fiber orientation differs in the range between 7.3 and 347.2% averaged deviation significantly. Furthermore, depending on the process–structure coupling and the number of layers over the thickness of the model, the Kullback–Leibner divergence differs in a range between 0.1 and 4.9%. More layers lead to higher fiber orientation content in the model and improved prediction of the structural dynamics simulation. The investigations prove that the influence of the fiber orientation on the structural dynamics simulation is lower than the influence of the material modeling. With a relative average deviation of 2.8% in the frequency and 38.0% in the amplitude of the frequency response function, it can be proven that high deviations between experimental and simulative fiber orientations can lead to a sufficient prediction of the structural dynamics simulation.

Abbasi et al. [25] conducted an FE numerical study to assess the effect of size on the shear resistance of reinforced concrete (RC) beams strengthened in shear with externally bonded carbon fiber-reinforced polymer (EB-CFRP). Although a few experimental studies have been performed, there is still a lack of FE studies that consider the size effect. Experimental tests are time-consuming and costly and cannot capture all the complex and interacting parameters. In recent years, advanced numerical models and constitutive laws have been developed to predict the response of laboratory tests, particularly for issues related to shear resistance of RC beams, namely, the brittle response of concrete in shear and the failure modes of the interface layer between concrete and EB-CFRP (debonding and delamination). Numerical models have progressed in recent years and can now capture the interfacial shear stress along the bond and the strain profile along the fibers and the normalized main diagonal shear cracks. This paper presents the results of a nonlinear FE numerical study on nine RC beams strengthened in shear using EB-CFRP composites that were tested in the laboratory under three series, each containing three sizes of geometrically similar RC beams (small, medium, and large). The results reveal that numerical studies can predict experimental results with good accuracy. They also confirm that the shear strength of concrete and the contribution of CFRP to shear resistance decrease as the size of beams increases.

Lithium-based batteries with improved safety performance are highly desired. At present, most safety hazard is the consequence of the ignition and flammability of organic liquid electrolytes. Dry ceramic-polymer composite electrolytes are attractive for their merits of non-flammability, reduced gas release, and thermal stability, in addition to their mechanical strength and flexibility. Denney and Huang [26] fabricated free-standing solid composite electrolytes made up of polyethylene oxide (PEO), $LiBF_4$ salt, and $Li_{1+x}Al_xGe_{2-x}(PO_4)_3$ (LAGP). This study is focused on analyzing the impacts of LAGP on the thermal decomposition characteristics in the series of PEO/$LiBF_4$/LAGP composite membranes. It is found that the appropriate amount of LAGP can (1) significantly reduce the organic solvent trapped in the polymer network and (2) increase the peak temperature corresponding to the thermal degradation of the PEO/$LiBF_4$ complex. In the presence of LAGP, although the peak temperature related to the degradation of free PEO is reduced, the portion of free PEO, as well as its decomposition rate, is effectively reduced, resulting in slower gas release.

Mechanical properties of fiber-reinforced polymers are sensitive to environmental influences due to the presence of the polymer matrix but inhomogeneous and anisotropic due to the presence of the fibers. Hence, structural analysis with mechanical properties as a function of loading, environment, design, and material condition produces more precise, reliable, and economic structures. Esha et al. [27] developed an analytical model that can predict engineering values as well as non-linear stress–strain curves as a function of six independent parameters for short fiber-reinforced polymers manufactured by injection molding. These parameters are the strain, temperature, humidity, fiber content, fiber orientation, and thickness of the specimen. A three-point test matrix for each independent parameter is used to obtain experimental data. To insert the effect of in-homogenous and anisotropic distribution of fibers in the analytical model, microCT analysis is conducted. Similarly, dynamic mechanical thermal analysis (DMTA) is performed to insert the viscoelastic effect of the material. The least mean square regression method is used to predict empirical formulas. The standard error of regression for the fitting of the model with experimental stress–strain curves is closely controlled below 2% of the stress range. This study provides user-specific material data for simulations with specific material, loading, and environmental conditions.

Spiral steel cables feature complex deformation behavior due to their wound geometry. In applications where the cables are used to reinforce rubber components, modeling the cables is not trivial, because the cable's outer surface must be connected to the surrounding rubber material. There are several options for modeling steel cables using beam and/or solid elements for the cable. So far, no study that lists and evaluates the performance of such approaches can be found in the literature. Pletz et al. [28] investigate such modeling options for a simple seven-wire strand that is regarded as a cable. The setup, parameter calibration, and implementation of the approaches are described. The accuracy of the obtained deformation behavior is assessed for a three-cable specimen using a reference model that features the full geometry of the wires in the three cables. It is shown that a beam approach with anisotropic beam material gives the most accurate stiffness results. The results of the three-cable specimen model indicate that such a complex cable model is quite relevant for the specimen's deformation. However, there is no single approach that is well suited for all applications. The beam with anisotropic material behavior is well suited if the necessary simplifications in modeling the cable–rubber interface can be accepted. The present work thus provides a guide not only for calibrating but also for selecting the cable-modeling approach. It is shown how such modeling approaches can be used in commercial FE software for applications such as conveyor belts.

Recent investigations have highlighted the multi-resolution and high throughput characteristics of the spherical indentation experimental and analysis protocols. Millan-Espitia and Kalidindi [29] further demonstrate the capabilities of these protocols for reliably extracting indentation stress–strain (ISS) responses from the microscale constituents as well as the bulk scale of dual phase materials exhibiting bimodal microstructures. Specifically, we focus on bimodal microstructures produced in an α–β Ti6242 sample. Combining the multi-resolution indentation responses with microstructural statistics gathered from the segmentation of back-scattered electron images from the scanning electron microscope allowed for a critical experimental evaluation of the commonly utilized Rule of Mixtures based composite model for the elastic stiffness and plastic yield strength of the sample. The indentation and image analyses protocols described in this paper offer novel research avenues for the systematic development and critical experimental validation of composite material models.

Due to the high design freedom and weight specific properties carbon fiber reinforced plastics (CFRP) offer significant potential in light-weighting applications, specifically in the automotive sector. The demand for medium to high production quantities with consistent material properties has paved the way for the use of high-pressure resin transfer molding

(HP-RTM). Due to high experimental cost and number of the operational parameters the development of numerical simulations to predict part quality is growing. Despite this, erroneous assumptions and simplifications limit the application of HP-RTM models, specifically with regard to the energy models used to model the heat transfer occurring during infiltration. Sherratt et al. [30] investigate the operating parameters at which the thermal non-equilibrium energy model's increased computational cost and complexity is worth added accuracy. It was found that in nearly all cases, using the thermal non-equilibrium is required to obtain an accurate prediction of the temperature development and resulting final properties within the mold after the infiltration process.

Additive manufacturing or 3D printing has been utilized for parts built of composite materials. Both developing composite materials by adding reinforcement particles and printing such a combination of materials can cause a certain kind of defect and anomaly generation in printed parts or the inappropriate functionality of the 3D printers. Baechle-Clayton et al. review [31] the potential failures and flaws associated with fused deposition modeling (FDM) or fused filament fabrication (FFF) 3D printing technology. The focus is on presenting the failures and flaws that are caused by the operational standpoints and which are based on the many years of experience with current and emerging materials and equipment for the 3D printing of polymers and composites using the FDM/FFF method. This study provides discussions and insights into the potential factors that can cause the failure of 3D printers when producing a part and presents the type and characteristics of potential flaws that can happen in the produced parts. Common defects posed by FDM printing have been discussed, and common nondestructive detection methods to identify these flaws both in-process and after the process is completed are discussed. The discussions on the failures and flaws in machines provides useful information on troubleshooting the process if they happen, and the review on the failures and flaws in parts helps researchers and operators learn about the causes and effects of the flaws in a practical way.

Robust finite element models are utilized for their ability to predict simple to complex mechanical behavior under certain conditions at a very low cost compared to experimental studies, as this reduces the need for physical prototypes while allowing for the optimization of components. Rabiee and Ghasemnejad [32] reviewed various parameters in finite element techniques to simulate the crushing behavior of glass/epoxy tubes with different material models, mesh sizes, failure trigger mechanisms, element formulation, contact definitions, single and various numbers of shells and delamination modeling. Six different modeling approaches, namely, a single-layer approach and a multi-layer approach, were employed with 2, 3, 4, 6, and 12 shells. In experimental studies, 12 plies were used to fabricate a 3 mm wall thickness GFRP specimen, and the numerical results were compared with experimental data. This was achieved by carefully calibrating the values of certain parameters used in defining the above parameters to predict the behavior and energy absorption response of the finite element model against initial failure peak load (stiffness) and the mean crushing force. In each case, the results were compared with each other, including experimental and computational costs. The decision was made from an engineering point of view, which means compromising accuracy for computational efficiency. The aim is to develop an FEM that can predict energy absorption capability with a higher level of accuracy, around 5% error, than the experimental studies.

The benefit of fiber-reinforced composites originates from the interaction between the fiber reinforcement and the matrix. This interplay controls many of its mechanical properties and is of utmost importance to enable its unique performance as a lightweight material. However, measuring the fiber–matrix interphase strength with micromechanical tests, such as the Broutman test, is challenging, due to the many, often unknown boundary conditions. Vogtmann et al. [33] use state-of-the-art, high-resolution X-ray computed microtomography (XRM) as a tool to investigate post mortem the failure mechanisms of single carbon fibers within an epoxy matrix. This was conducted at the example of single carbon fiber Broutman test specimens. The capabilities of today's XRM analysis were shown in comparison to classically obtained light microscopy. A simple finite element model was used to enhance the understanding of the observed fracture patterns. In total, this research reveals the possibilities and limitations of XRM to visualize and assess compression-induced single fiber fracture patterns. Furthermore, comparing two different matrix systems with each other illustrates that the failure mechanisms originate from differences in the fiber–matrix interphases. The carbon fiber seems to fail due to brittleness under compression stress. Observation of the fiber slippage and deformed small fracture pieces between the fragments suggests a nonzero stress state at the fragment ends after fiber failure. Even more, these results demonstrate the usefulness of XRM as an additional tool for the characterization of the fiber–matrix interphase.

Finally, in this Special Issue established by Georgantzinos et al. [34] the concept of six-dimensional (6D) printing as a new branch of additive manufacturing investigating its benefits, advantages as well as possible limitations concerning the design and manufacturing of effective smart structures. The concept of 6D printing, to the author's best knowledge, is introduced for the first time. The new method combines the four-dimensional (4D) and five-dimensional (5D) printing techniques. This means that the printing process is going to use five degrees of freedom for creating the final object while the final produced material component will be a smart/intelligent one (i.e., will be capable of changing its shape or properties due to its interaction with an environmental stimulus). A 6D printed structure can be stronger and more effective than a corresponding 4D printed structure, can be manufactured using less material, can perform movements by being exposed to an external stimulus through an interaction mechanism, and it may learn how to reconfigure itself suitably, based on predictions via mathematical modeling and simulations.

Conflicts of Interest: The author declares no conflict of interest.

References

1. Georgantzinos, S.K. Editorial for the Special Issue on Characterization and Modelling of Composites. *J. Compos. Sci.* **2021**, *5*, 47. [CrossRef]
2. Georgantzinos, S.K. *Characterization and Modelling of Composites*, 1st ed.; MDPI: Basel, Switzerland, 2022.
3. Youssef, G.; Newacheck, S.; Huynh, N.U.; Gamez, C. Multiscale Characterization of E-Glass/Epoxy Composite Exposed to Extreme Environmental Conditions. *J. Compos. Sci.* **2021**, *5*, 80. [CrossRef]
4. Dietemann, B.; Bosna, F.; Kruggel-Emden, H.; Kraft, T.; Bierwisch, C. Assessment of Analytical Orientation Prediction Models for Suspensions Containing Fibers and Spheres. *J. Compos. Sci.* **2021**, *5*, 107. [CrossRef]
5. Ninikas, K.; Mitani, A.; Koutsianitis, D.; Ntalos, G.; Taghiyari, H.R.; Papadopoulos, A.N. Thermal and Mechanical Properties of Green Insulation Composites Made from *Cannabis* and Bark Residues. *J. Compos. Sci.* **2021**, *5*, 132. [CrossRef]
6. Georgantzinos, S.K.; Antoniou, P.A.; Markolefas, S.I. A Multi-Scale Method for Designing Hybrid Fiber-Reinforced Composite Drive Shafts with Carbon Nanotube Inclusions. *J. Compos. Sci.* **2021**, *5*, 157. [CrossRef]
7. Sivtsev, P.V.; Smarzewski, P.; Stepanov, S.P. Numerical Study of Soil-Thawing Effect of Composite Piles Using GMsFEM. *J. Compos. Sci.* **2021**, *5*, 167. [CrossRef]
8. Umar, M.; Qayyum, F.; Farooq, M.U.; Guk, S.; Prahl, U. Qualitative Investigation of Damage Initiation at Meso-Scale in Spheroidized C45EC Steels by Using Crystal Plasticity-Based Numerical Simulations. *J. Compos. Sci.* **2021**, *5*, 222. [CrossRef]
9. Hassan, M.K.; Mohamed, A.F.; Khalil, K.A.; Abdellah, M.Y. Numerical and Experimental Evaluation of Mechanical and Ring Stiffness Properties of Preconditioning Underground Glass Fiber Composite Pipes. *J. Compos. Sci.* **2021**, *5*, 264. [CrossRef]
10. Drvoderic, M.; Pletz, M.; Schuecker, C. Modeling Stiffness Degradation of Fiber-Reinforced Polymers Based on Crack Densities Observed in Off-Axis Plies. *J. Compos. Sci.* **2022**, *6*, 10. [CrossRef]
11. Clarke, J.; McIlhagger, A.; Dixon, D.; Archer, E.; Stewart, G.; Brelsford, R.; Summerscales, J. A Cost Model for 3D Woven Preforms. *J. Compos. Sci.* **2022**, *6*, 18. [CrossRef]
12. Mou, X.; Shen, Z.; Liu, H.; Xv, H.; Xia, X.; Chen, S. FEM-Validated Optimal Design of Laminate Process Parameters Based on Improved Genetic Algorithm. *J. Compos. Sci.* **2022**, *6*, 21. [CrossRef]
13. Shah, Q.M.Z.; Kowser, M.A.; Chowdhury, M.A.; Chani, M.T.S.; Alamry, K.A.; Hossain, N.; Rahman, M.M. Modeling Fracture Formation, Behavior and Mechanics of Polymeric Materials: A Biomedical Implant Perspective. *J. Compos. Sci.* **2022**, *6*, 31. [CrossRef]
14. Frankl, S.M.; Pletz, M.; Wondracek, A.; Schuecker, C. Assessing Failure in Steel Cable-Reinforced Rubber Belts Using Multi-Scale FEM Modelling. *J. Compos. Sci.* **2022**, *6*, 34. [CrossRef]
15. Luo, Y. Microstructure-Free Finite Element Modeling for Elasticity Characterization and Design of Fine-Particulate Composites. *J. Compos. Sci.* **2022**, *6*, 35. [CrossRef]

16. Kakarla, A.B.; Kong, I.; Nukala, S.G.; Kong, W. Mechanical Behaviour Evaluation of Porous Scaffold for Tissue-Engineering Applications Using Finite Element Analysis. *J. Compos. Sci.* **2022**, *6*, 46. [CrossRef]
17. Carvalho, J.; Sohouli, A.; Suleman, A. Fundamental Frequency Optimization of Variable Angle Tow Laminates with Embedded Gap Defects. *J. Compos. Sci.* **2022**, *6*, 64. [CrossRef]
18. Behnisch, F.; Brütsch, J.; Werner, H.O.; Henning, F. The Direct Sandwich Composite Molding (D-SCM) Process: Sandwich Manufacturing and Characterization. *J. Compos. Sci.* **2022**, *6*, 81. [CrossRef]
19. Li, S.; Xu, M.; Sitnikova, E. The Formulation of the Quadratic Failure Criterion for Transversely Isotropic Materials: Mathematical and Logical Considerations. *J. Compos. Sci.* **2022**, *6*, 82. [CrossRef]
20. Lagdani, O.; Tarfaoui, M.; Rouway, M.; Laaouidi, H.; Jamoudi Sbai, S.; Amine Dabachi, M.; Aamir, A.; Nachtane, M. Influence of Moisture Diffusion on the Dynamic Compressive Behavior of Glass/Polyester Composite Joints for Marine Engineering Applications. *J. Compos. Sci.* **2022**, *6*, 94. [CrossRef]
21. Haas, J.; Aberle, D.; Krüger, A.; Beck, B.; Eyerer, P.; Kärger, L.; Henning, F. Systematic Approach for Finite Element Analysis of Thermoplastic Impregnated 3D Filament Winding Structures—Advancements and Validation. *J. Compos. Sci.* **2022**, *6*, 98. [CrossRef]
22. Cristillo, D.; Di Caprio, F.; Pezzella, C.; Paciello, C.; Magistro, S.; Di Palma, L.; Belardo, M. On Numerical Models for Cube Drop Test of Bladder Fuel Tank for Aeronautical Applications. *J. Compos. Sci.* **2022**, *6*, 99. [CrossRef]
23. Witzgall, C.; Völkl, H.; Wartzack, S. Derivation and Validation of Linear Elastic Orthotropic Material Properties for Short Fibre Reinforced FLM Parts. *J. Compos. Sci.* **2022**, *6*, 101. [CrossRef]
24. Kriwet, A.; Stommel, M. The Impact of Fiber Orientation on Structural Dynamics of Short-Fiber Reinforced, Thermoplastic Components—A Comparison of Simulative and Experimental Investigations. *J. Compos. Sci.* **2022**, *6*, 106. [CrossRef]
25. Abbasi, A.; Benzeguir, Z.E.A.; Chaallal, O.; El-Saikaly, G. FE Modelling and Simulation of the Size Effect of RC T-Beams Strengthened in Shear with Externally Bonded FRP Fabrics. *J. Compos. Sci.* **2022**, *6*, 116. [CrossRef]
26. Denney, J.; Huang, H. Thermal Decomposition Characteristics of PEO/$LiBF_4$/LAGP Composite Electrolytes. *J. Compos. Sci.* **2022**, *6*, 117. [CrossRef]
27. Hausmann, J. Development of an Analytical Model to Predict Stress–Strain Curves of Short Fiber-Reinforced Polymers with Six Independent Parameters. *J. Compos. Sci.* **2022**, *6*, 140.
28. Pletz, M.; Frankl, S.M.; Schuecker, C. Efficient Finite Element Modeling of Steel Cables in Reinforced Rubber. *J. Compos. Sci.* **2022**, *6*, 152. [CrossRef]
29. Millan-Espitia, N.; Kalidindi, S.R. Study of a Bimodal α–β Ti Alloy Microstructure Using Multi-Resolution Spherical Indentation Stress-Strain Protocols. *J. Compos. Sci.* **2022**, *6*, 162. [CrossRef]
30. Sherratt, A.; Straatman, A.G.; DeGroot, C.T.; Henning, F. Investigation of a Non-Equilibrium Energy Model for Resin Transfer Molding Simulations. *J. Compos. Sci.* **2022**, *6*, 180. [CrossRef]
31. Baechle-Clayton, M.; Loos, E.; Taheri, M.; Taheri, H. Failures and Flaws in Fused Deposition Modeling (FDM) Additively Manufactured Polymers and Composites. *J. Compos. Sci.* **2022**, *6*, 202. [CrossRef]
32. Rabiee, A.; Ghasemnejad, H. Finite Element Modelling Approach for Progressive Crushing of Composite Tubular Absorbers in LS-DYNA: Review and Findings. *J. Compos. Sci.* **2022**, *6*, 11. [CrossRef]
33. Vogtmann, J.; Klingler, A.; Rief, T.; Gurka, M. 3D X-ray Microscopy as a Tool for in Depth Analysis of the Interfacial Interaction between a Single Carbon Fiber and an Epoxy Matrix after Mechanical Loading. *J. Compos. Sci.* **2021**, *5*, 121. [CrossRef]
34. Georgantzinos, S.K.; Giannopoulos, G.I.; Bakalis, P.A. Additive Manufacturing for Effective Smart Structures: The Idea of 6D Printing. *J. Compos. Sci.* **2021**, *5*, 119. [CrossRef]

Journal of
Composites Science

MDPI

Review

Failures and Flaws in Fused Deposition Modeling (FDM) Additively Manufactured Polymers and Composites

Maggie Baechle-Clayton [1], Elizabeth Loos [2], Mohammad Taheri [3] and Hossein Taheri [1,*]

1 Laboratory for Advanced NonDestructive Testing, In-Situ Monitoring and Evaluation (LANDTIE), Department of Manufacturing Engineering, Georgia Southern University, Statesboro, GA 30458, USA; mb11180@georgiasouthern.edu
2 Department of Chemistry and Biochemistry, Georgia Southern University, Statesboro, GA 30458, USA; el01817@georgiasouthern.edu
3 Department of Mathematics and Statistics, South Dakota University, Bookings, SD 57007, USA; mohammad.taheri@jacks.sdstate.edu
* Correspondence: htaheri@georgiasouthern.edu

Abstract: In this review, the potential failures and flaws associated with fused deposition modeling (FDM) or fused filament fabrication (FFF) 3D printing technology are highlighted. The focus of this article is on presenting the failures and flaws that are caused by the operational standpoints and which are based on the many years of experience with current and emerging materials and equipment for the 3D printing of polymers and composites using the FDM/FFF method. FDM or FFF 3D printing, which is also known as an additive manufacturing (AM) technique, is a material processing and fabrication method where the raw material, usually in the form of filaments, is added layer-by-layer to create a three-dimensional part from a computer designed model. As expected, there are many advantages in terms of material usage, fabrication time, the complexity of the part, and the ease of use in FDM/FFF, which are extensively discussed in many articles. However, to upgrade the application of this technology from public general usage and prototyping to large-scale production use, as well as to be certain about the integrity of the parts even in a prototype, the quality and structural properties of the products become a big concern. This study provides discussions and insights into the potential factors that can cause the failure of 3D printers when producing a part and presents the type and characteristics of potential flaws that can happen in the produced parts. Common defects posed by FDM printing have been discussed, and common nondestructive detection methods to identify these flaws both in-process and after the process is completed are discussed. The discussions on the failures and flaws in machines provides useful information on troubleshooting the process if they happen, and the review on the failures and flaws in parts helps researchers and operators learn about the causes and effects of the flaws in a practical way.

Keywords: fused deposition modeling (FDM); fused filament fabrication (FFF); polylactic acid (PLA) filament; acrylonitrile butadiene styrene (ABS) filament; nondestructive testing (NDT); 3D printing; additive manufacturing (AM)

Citation: Baechle-Clayton, M.; Loos, E.; Taheri, M.; Taheri, H. Failures and Flaws in Fused Deposition Modeling (FDM) Additively Manufactured Polymers and Composites. *J. Compos. Sci.* **2022**, *6*, 202. https://doi.org/10.3390/jcs6070202

Academic Editor: Francesco Tornabene

Received: 26 May 2022
Accepted: 6 July 2022
Published: 8 July 2022

1. Introduction

1.1. Fused Deposition Modeling (FDM) as a Cost-Effective Alternative to Traditional Manufacturing

Fused Deposition Modeling (FDM) or fused filament fabrication (FFF) is the application of additive manufacturing technology that uses a heating chamber to liquify a polymer that is then fed and extruded by a system in the form of a filament [1]. In this paper, the term FDM will be used for the purpose of consistency. Many times, the filaments used in an FDM printer consist of wax and/or a thermoplastic polymer (Figure 1) [2]; however, this technology also offers the possibility of introducing and printing composite materials as well [3,4]. FDM technology is widely used in commercial applications [5]. This is due to the relative strength of the parts produced using FDM applications. One of its largest consumer applications is within aerospace engineering. Such applications produced from polymers created by Stratasys and Orbis are FAA-approved ULTEM 9085 aircraft air ducts [6]. It has a large range of materials and applications that surpass other AM technologies and traditional manufacturing.

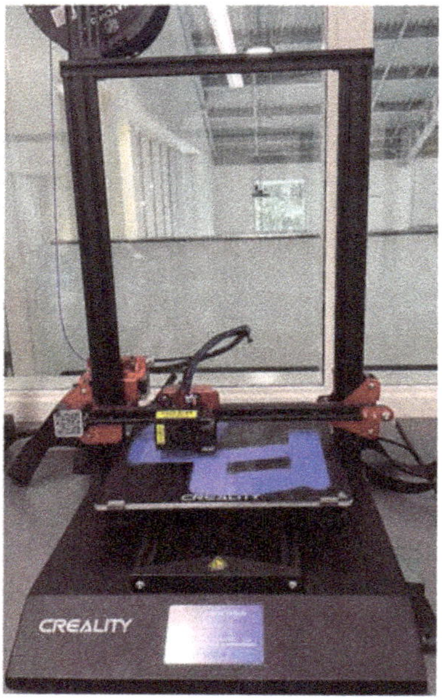

Figure 1. Example of an FDM 3D printer with a polymer filament spool (top-left).

Compared to traditional manufacturing, additive manufacturing via FDM methods is more cost effective. While the initial investment into 3D FDM printing can be substantial, its application in prototyping and part design saves both time and economic resources. On average, a 1 kg roll of Polylactic Acid (PLA) filament is approximately USD 20.00 [7]. In addition to the benefits from the lower cost of raw materials, the reduction in fabrication time and labor, as well as the close-to-zero waist of material due to no subtractive fabrication process, significantly affect the overall cost of part manufacturing compared to traditional manufacturing techniques. The type and size of a 3D printer determine the overall associated cost. An initial investment for a larger industrial-scale FDM printer is the in the range of USD 70,000 to USD 400,000. When considering smaller FDM printers such as a desktop model for mostly prototyping purposes, the initial cost could be in the range of USD 1300 to USD 2500 [8–10]. With the complex geometry available to manufacturers/fabricators that would be unrealistic by traditional means, FDM is the answer. The major issues posed by the increasing application of FDM additive manufacturing in comparison to traditional manufacturing are the ability to find internal defects and the assessment of the quality of the parts.

Additive manufacturing or 3D printing has been also utilized for parts built of composite materials. The combination of two materials as the base for developing composite materials has been developed over the years to achieve enhanced mechanical and material properties such as higher strength-to-weight ratios as well as thermal and/or electrical functionalities [11]. The 3D printing of composite materials comes with its own pros and cons. It provides the capability to print parts with complex geometries and controlled dimensions, but at the same time, it has certain limitations regarding the reinforcement materials that can be added to the base polymer to print a polymer matrix composite [12].

In addition, both developing composite materials by adding reinforcement particles and printing such a combination of materials can cause a certain kind of defect and anomaly generation in printed parts or the inappropriate functionality of the 3D printers [13]. Some of the major types of such defects include warping and geometrical distortion, voids and porosities in the matrix and filament, poor fiber–matrix bonding, and uneven reinforcement distribution in the matrix base polymer. As an example, the homogeneity of the reinforcing and base polymer mixing is a common challenge which can cause significant changes in composite material properties. Related common types of defects and failures of polymer and composite materials and their 3D printing are discussed in the following section.

1.2. Common Defects Posed by FDM Printing

With the large application and rapid growth of additively manufactured materials, the concern for the integrity and safety of the parts and testing methods is paramount. Unlike traditionally manufactured parts, due to the nature of the polymer-based FDM materials, traditional stress/strain tests or defect inspection tests would result in the damage and destruction of the part. Currently, the most common defects are porosity and density changes within the material [14]. This is caused by a various number of issues, but the most common is the environmental and production control of the application of thermal and humidity changes within the fabrication process [15,16].

FDM technology fabricates a part through a heating and cooling cycle of a filament. For PLA, the recommended temperature is between 401 °F +/− 27° [7]. However, the recommended temperature will vary based on the diameter of the filament and the extrusion nozzle of the printer head. The extrusion temperature is crucial in FDM printing. The temperature directly impacts the viscosity and adhesion of the filament, which are deciding factors of the functionality of the print [17]. For parts that are designed for aesthetic purposes, the issue is not a consideration. For pieces fabricated for their resultant functionality within a whole integrated system, this issue is daunting.

1.2.1. Thermal Inconsistencies Affecting Part Fabrication

The application of heat is vital in FDM for both polymer and composite printing. The process of extrusion has to be maintained throughout the printing process. Within FDM technology, the largest issue is heating consistency. There are several different types of FDM printers in circulation during the 21st century. While there is a wide variety of filament types, this paper will focus on the application of acrylonitrile butadiene styrene (ABS) and polylactic acid (PLA). In FDM printers, there are two common applications of heat. The first application of heat is within the thermocouple attached to the printer head extruder. This thermocouple is controlled by the software used by the printer. It runs using cycle heating, meaning that the thermocouple will reach the required temperature, and then the heat application will be removed until it drops to a temperature outside of the tolerance range. This lack of consistency is a reoccurring and substantial issue with FDM applications. With the inconsistent heating/cooling of the polymer filament, the adhesion of the applied layers will differentiate, causing an ununiformed density and consistency between the layers of the part.

FDM printing layer adhesion is greatly affected by the ambient temperature created by the thermocouples connected to the print bed. When in an enclosed system, the thermocouples attached to the underside of the print bed produce and maintain an average temperature. This temperature is commanded by the software of the printer per the filament requirements stated in the slicing software. Fluctuations in the print bed temperature will result in the warping of the base layer of the print. This warping will produce inconsistent prints and failures for the rest of the print. When the base layers warp, the chances of being caught on the printer extrusion head are drastically increased, and the adhesion of the print to the print bed will be jeopardized. If the filament is applied to the print bed at a temperature that increases the viscosity, the print will fail to adhere at a microlevel, therefore reducing the surface tension between the print bed and the print itself. This type of defect is very common in an exposed printer environment, mainly due to the ambient fluctuations of an uncontrolled environment.

When a print bed is heated to a temperature higher than that recommended for the print, the risk of over-melting the print and the reduction of the print bed/part adherence greatly increases. If the filament is not allowed to harden at a gradual pace, the print head will extrude filament that will not adhere to the print bed. This will result in "air printing", or the filament extruding into the air rather than attaching to the print bed. This is a more common issue with printers that have fewer than five thermocouples and an open printing environment [18]. Printers with fewer than five thermocouples have a higher chance of cold spots and inconsistent heating due to overlapping ranges. This is because thermocouples work on an average range of temperatures rather than keeping a constant temperature. The thermocouples will fluctuate within the range, creating that average temperature reported between them. When thermocouples work, they must take into account not just the temperature ranges of the hot end but also the impact of the external environment [19].

1.2.2. Fiber-Related Defects in Composites

However, adding reinforcing materials or fibers to the base polymer has been shown to be beneficial in enhancing material properties in composites, but they can also cause several challenges in terms of defect generation and challenges with enhanced mechanical properties [20]. One of these challenges arises from the feeding material development and usage process. Mixing the fibers with the matrix polymer must be done such that the final composition is as uniform as possible. The uniformity of the composition is not only an essential factor toward the enhanced mechanical properties but also eliminates stress concentration in particular areas or weak regions in the composite material [21]. When 3D printing composite materials, it is expected to have a well-defined fiber orientation in the final part. Several different factors from both the feeding composite material and the manufacturing process may cause a local alignment deviation of fibers. Such localized misalignment of the fibers is called waviness [22]. If a composite part contains waviness flaws, it is more likely to be subjected to failure at or around the waviness regions during the strength test or over the operating life of the structure. Several studies have evaluated the waviness and its influence on mechanical properties and the failure of the composite material [23–25]. As explained earlier, fiber waviness, in addition to a few other factors such as manufacturing deficiency and curved geometries and coordination, can generate a related defect which is generally due to fiber misalignment [26]. Fiber misalignment is particularly important in unidirectional composite materials and is shown to have a significant effect on the mechanical properties of the composite materials [27]. In a fiber-reinforced polymer composite, fibers are considered to be the main load-bearing component of the material. Due to the important role of fibers in the structural integrity of composite materials, fiber breakage has a significant effect on the overall strength and toughness of the composite parts [28]. Since the fiber breakage is a distributed type of defect in composite materials, both the localized and overall effect of these defects must be investigated and assessed for a load-bearing composite part susceptible to fiber breakage flaws [29].

Figure 2 shows the common defects and faults in parts and machines posed by the FDM printing of polymers and composites, which are described in the following sections.

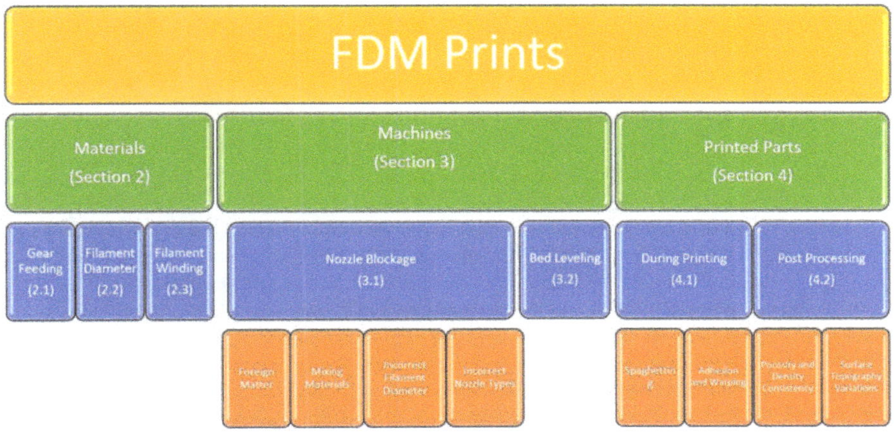

Figure 2. Common defects posed by the FDM printing of polymers and composites.

2. Nondestructive Testing (NDT) Methods

Nondestructive Testing (NDT) Applications are a large asset to Additive Manufacturing Engineering [30–34]. Due to the time-consuming nature and recyclability of most filaments, any traditional integrity and safety testing of parts would render the printed part unusable. This paper aims to review the major NDT techniques for the evaluation of FDM parts.

2.1. Nondestructive Testing in the Process

Temperature and thermal signatures monitoring has been widely used for the in-process quality monitoring of machines and parts. These techniques are commonly being applied through two main methods. The first method is based on temperature measurement with traditional or advanced thermocouple sensors. Thermocouples are used to measure the temperature of critical parts or locations during the 3D printing processes including bed or base plate temperature, nozzle temperature, and filament temperature. If the measurement is conducted on the machine parts such as the nozzle, the goal is usually to monitor the health of the machine for smooth operation. Temperature measurement is also conducted on the parts over the printing process. In this case, the goal is usually to monitor the integrity of the part and correlate the measurement to the potential flaws and defects. As an example, a large temperature gradient or shock usually causes cracking of the printed parts, which propagate even after the printing is completed or over the operational lifetime of the parts. The second thermal-based method for quality monitoring is based on thermal imaging using thermal cameras. Thermal cameras are devices that work based on capturing the thermal emissions from the objects and recording an image or a video on their detector. Based on the type and working mechanism of the thermal cameras, the temporal and spatial resolution of the recorded image and/or video, as well as their speed and resolutions, might be different, which is an important factor in using these devices for the in situ monitoring of 3D printing. Thermal imaging will detect the thermal inconsistencies within the part in real time, communicate with the operator, and allow them to adjust the process accordingly [11,35–38]. This will assist in a more uniform print, thus reducing density and layer adhesion inconsistencies. Depending on the extruded temperature of the filament, the rigidity and porosity are subject to change. Upon the extrusion temperature being above the recommended ranges of the filament, the viscosity of the filament will reduce, resulting in an unstable filament extrusion. This will produce a layer that will spread unnecessarily and fuse incorrectly to the rest of the print. The porosity and thickness of the layer will also be affected. In the instance where the extrusion temperature is lower than that of the recommended temperature, the viscosity of the filament will be higher than the suggested levels. This will result in a thicker and denser layer, causing improper bonding between layers. This porosity can be visualized either through the above-mentioned thermal imaging or even by a laser profilometer [35,39]. The laser profilometer would measure the amount of polymer that fills each layer. If the laser profilometer was used to detect under- or over-filled lower layers, that detection would cause the upper layers to compensate so that the final product would become smooth. This would, of course, depend on the adhesion between the layers. A way to alleviate the inconsistent heating between the thermocouples and the print bed is to use the skirt adhesion type build plate [40]. It was found in a study conducted by Nguyen et al. that the adhesion type was the most significant parameter tested that affected the vibrational response of the print. This change from raft to skirt adhesion has been shown to result in better inter-layer fusion and increased heat consistency, which in turn yields a better polymer print.

Layer-wise quality monitoring over the printing process through imaging techniques can also be done using optical images. In this method, optical cameras with a sufficient resolution and sensitivity are used to capture images on each layer during the fabrication process. Such an image will then be compared to the information from the G-code of the part's model for the automatic detection of the flaws and anomalies in the layer. Optical camera integration usually costs less than thermal cameras and is easier to integrate and operate. Computer vision algorithms are effective techniques for analyzing the images and extracting the information from them.

Shmueli et al. described in their paper that thermal imaging can be used during the 3D printing process, and X-ray scattering can also be used during that time. The X-ray scattering was shown to be indicative of the overall strength of the composite. This indication was given by the measured crystallinity of the polymer layer. Depending on the orientation of the molecules within the layers, the determination of the brittleness was shown to be directly correlated with the overall crystallinity of the composite [37,41]. X-rays are an efficient technique for the inspection of the 3D-printed parts, specifically for the assessment of the internal structure and volumetric flaw detection. Higher initial costs of the equipment and operation, safety concerns, higher technical skills requirements, and the complexity of the integration into the 3D printing machine are the major factors that limit the application of X-rays for in-process quality monitoring. This is even more critical for the 3D printing of polymers and composites since there might be less of a chance for cost justification.

The application of the NDT during the production process of the part would then allow for a layer-by-layer account of the part, since layer-wise enables the in situ monitoring of the process for each individual layer if the proper sensor(s) are integrated into the system.

2.2. Nondestructive Testing after the Process

After the completion of the printing process, the finished parts can be inspected for quality and potential flaws using various NDT techniques depending on the material, geometrical, and surface conditions. Ultrasonic, radiography, thermography, and microwave NDT are among the most applicable inspection techniques for polymer and composite materials. The above-mentioned uniformity issue can be detected using phase array ultrasound technology (PAUT), producing an image for the engineers to view without destroying the part to detect defects within the part [42–44]. PAUT used an array of an ultrasonic transducer to send and receive ultrasonic waves into the part in an electronically programmed sequence. Within the ultrasound, the largest issue is the conduction of the projected frequencies through the air and into the part. Due to the layering texture of the produced parts or their surface condition, it is highly difficult to produce a perfectly flush surface mating [45]. The surface irregularities produce "noise", the detection of frequencies due to air, thus reducing the accuracy of the ultrasonic testing technology. To limit the feedback noise from the irregular surface, the use of media or a controlled AM part is needed. As of now, the largest issue is the absence of a piece that adapts to the changing surface texture of the part and remains solid for the sensor to rest on. To solve this issue, additive manufacturing could be the answer. A controlled AM part would need to fit the surface texture and fill in the ridges and valleys produced by FDM. A promising way to negate those effects would be to use stereolithography (SLA) printing combining a rigid and semi-aqueous resin material [46–48].

The produced part would then act as a conductor between the sensor and the FDM part. A series of control tests using the ultrasound sensor and AM medium need to be conducted to allow for the removal of any potential noise remaining and effectively "zero" the sensor out for the testing application of the FDM part. The rigid resin portion of the mating piece would act as a sturdy, flush surface for the sensor to rest on. This would reduce the amount of noise feedback caused by air and ambient vibrations. The semi-aqueous resin material would then shape and form to the ridges and valleys of the part to test. This would again limit the noise produced by the air and ambient vibrations. To further reduce unwanted feedback, using a jelly medium would further increase the surface contact and produce a physical pathway for the ultrasound waves to penetrate the parts.

The media described above for surface mating techniques would then allow for the flat-bottomed sensors to be used on a larger variety of AM geometry parts. One of the largest concerns of using ultrasound technology is the complex geometry allowance provided by AM production technology, mainly within FDM. These issues are mainly focused on complex internal geometry. However, in using surface mating techniques and applying NDT methods to several faces, the ultrasound readings could effectively produce a 3D image of the part in question. This would then allow the researcher to break the part down to a layer level and detect where any defects occurred when compared to the digital slicing software layer models [49,50].

Radiography testing (RT) is always a reliable NDT technique for the post-process evaluation of the parts. For polymers and composites, RT faces less of a challenge regarding the penetration into these materials, since they are much less dense when compared to the metals and ceramics. Due to this characteristic, RT can penetrate into considerably thick parts of polymers and composites without being limited by the surface condition or geometry of the part. Then, the whole internal structure of the parts can be visually imaged in a 3D image using computed tomography (CT) or just in conventional 2D X-ray images. Similar to other imaging techniques, image processing (in 2D and 3D RT) and image reconstruction techniques (in 3D or CT imaging) are crucial to obtaining accurate measurements and information from the RT. The dependency of the RT images on the angle of projection is a major issue in 2D X-ray images. If a flaw has a very high aspect ratio, e.g., a long line or crack type, then it may be projected very well in an angle perpendicular to the projection plane, but it might be totally missed in the other 90-degree perpendicular view. However, this is a less important issue if a 3D CT image is produced, which takes more time and will have a larger size.

3. Failure and Flaws in Raw Feeding Materials

3.1. Gear Feeding

The gear feeding process consists of two cylinders with a tooth inlay, as shown in Figure 3. They are normally held with a screw spring that provides an adjustable tension for the extruding gears. This tension allows the gears to increase or decrease friction on the filament, which allows for the ease of movement when using different materials (PEEK, ABS, PLA, Carbon Fiber, etc.), as each material has a different hardness and density.

In 3D printing, there are two different feeding styles. The most common is Bowden extruding, where the gears feeding the filament are located on the body of the printer and a tube carries the filament to the thermal coupling hot end. An example of a 3D Filament printer that uses Bowden extruding is in Figure 4. This helps to reduce the possibility of breakage and thermocouple burnout.

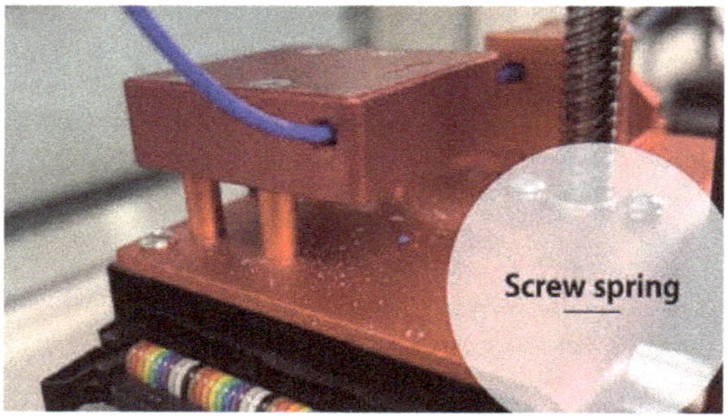

Figure 3. Screw spring that provides tension for the gear feeding process.

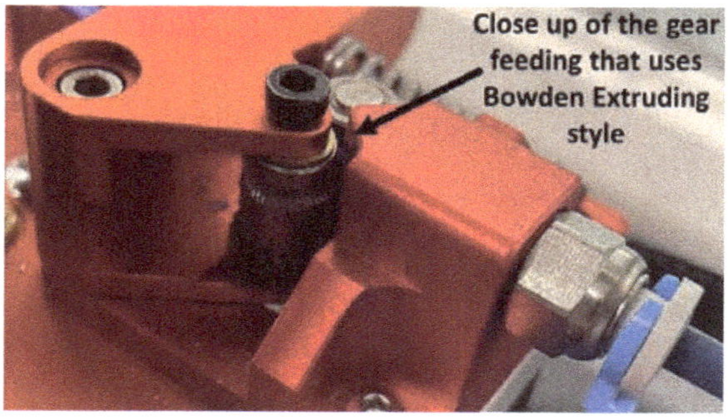

Figure 4. Close-up of the gear feeding that uses the Bowden extruding style.

For direct extruding printers, the extrusion gears are located behind the thermocouple hot ends. This has a lower failure rate due to the lack of tension placed on the filament and the reduced travel time between the gears and the thermocouple. However, it reduces the quality of the print because of the slop produced by the gear motors.

Both extruding systems normally have an eye or a light sensor attached behind the extruding gears to notify the printer when the filament runs out. This then pauses the printing process and allows it to cool down mid-print. When this occurs, the location and progress of the print are saved and allow the print to be continues once a new filament is loaded in. However, due to the location of the notification system being behind the gears, if the filament's diameter has been chewed away, the print will fail. The notification system, being behind the failing location, will show the filament still being processed through the extruder, even though it has faulted. To remove issues such as these, it would be beneficial to relocate the notification system as close to the thermocouple location as possible or install a two-part system. A two-part notification will greatly increase the chance of catching a gear feeding failure quicker, reducing the possibility of a failed print. In both systems, one of the most common failures is gear slippage. There are two possible ways for this to occur.

One cause is moisture content within the filament itself. If the filament is too dry, it becomes brittle and encourages breaking. Due to the feeding pressure and possible twisting of the filament due to the unrolling process, snapping may occur between the gears and the feeding tube of the extrusion system. This causes separation between the two parts of the filament and reduces the chance of the gears catching the material and feeding it into the guide tube. If the gears gain traction on the filament, it has a high chance of pushing the material forward but missing the guiding tube. This then extrudes the filament into the air and out of the system, resulting in a failed print.

The second cause is the shedding of the filament into the gear's grooves or the wearing down of the gear's grooves. Both will result in the slipping of the gears against the filament and cause the diameter to be worn down over time. This will cause the filament to stay stationary within the system and keep any material from extruding from the hot ends. This will again cause the printer to "print air". Since the sensor is behind the gear, the system will detect the material (even if it is stationary) and it will consider the material to be extruding from the hot ends and ruin the print.

3.2. Diameter

As mentioned above, the diameter of the filament plays a crucial role in the success of a printed object. The most common diameter sizes are 3.0 mm and 1.75 mm. The production of these filaments states that, over a certain length of the filament, the average diameter of it is one of those two values stated above. Due to the variations of the diameter within a tolerance range for these filaments, it is possible for the gear systems to lose traction on the filament and stop the material extrusion process (Figure 5).

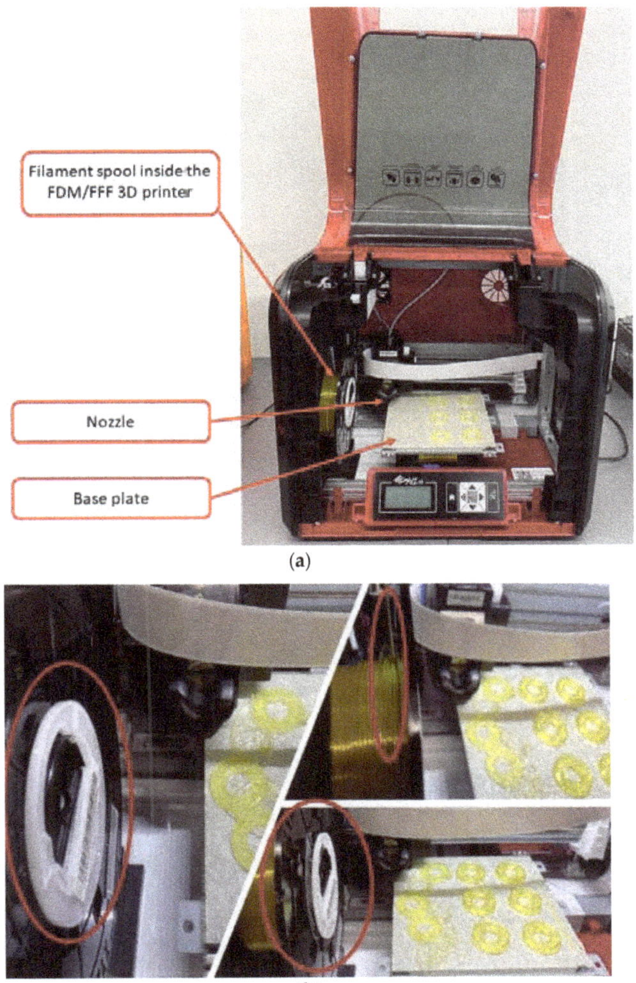

Figure 5. (**a**) Location and arrangement of the filament spool inside the FDM/FFF 3D printer; (**b**) Instability (misalignment) of the filament spool winding (red circles) during the printing process.

3.3. Filament Winding

Filament winding errors occur either during the winding process of the filament spool's manufacturing or due to the failure of the tension while printing.

4. Failure and Flaws in Machines (3D Printers)

4.1. Nozzle Blockage

There are several possibilities for material to cause nozzle blockages while printing. Foreign matter, material mixing, and the filament diameter are some of the most common causes of nozzle blockages (Figure 6).

Figure 6. Nozzle blockage fault in a 3D printing machine.

4.1.1. Foreign Matter Blockage

Focusing less on the machine and more on the environment the printer resides in, there are many different possibilities for material blocking. One of the most prominent is tape. There are several different filament manufacturers that initialize the coiling of the filament using tape to adhere it to the roll. Due to the light sensors mentioned above for the gear feeding, this causes one of two issues. The first issue is that the tape remains adhered to both the filament and the roll. This inadvertently causes the machine to be placed under tension, gear chewing on the filament, and a failed print. The second issue is that the tape dislodges from the roll yet remains adhered to the filament roll. As the gears continue feeding, this tape can stay attached to the filament and result in the clogging of the feeding tube or the extrusion nozzle.

In cases where the tape has clogged the feeding tube, it is necessary to remove and replace the whole tube due to the blockage. If the tape has clogged the extrusion nozzle, it is necessary to conduct a "hot push" of material. This is when the thermocouple of the extruder is heated to the top end of the temperature range of the filament and the filament is manually pushed through. In doing this, the tape is melted down with the filament and forcefully removed from the system. Theoretically, this will then completely remove the foreign material from the extruder.

4.1.2. Mixing Materials

The most commonly used filaments are PLA and ABS. The extrusion tips used for PLA can then be used for ABS after a hot push of the new material. This is due to the lower melting point of PLA. The hot push using ABS effectively removes any remaining PLA from the walls of the extrusion tip and allows for a clean extrusion of the new material. However, once a tip has been used for ABS, it is generally relegated to be used for only that material. This is because residual ABS on the walls of the extrusion tip remain, even after a hot push. This leads to an inconsistent temperature of the hot end and introduces impurities within the extruded filament, therefore ruining the print.

As stated previously in Section 1.1, FDM can be used to produce composites. When mixing those composites, it is very important to know the bonding mechanism, mechanical performance, fatigue behavior, rheological behavior, and thermal properties of the individual components [51–57]. The bonding mechanism between polymers can be affected by many different things. It all comes down to how the chemicals/polymers interact with each other. For example, a negatively charged polymer will not bind to another negatively charged polymer. Therefore, one of them has to have a positive charge for the bond to happen. These charges can be changed based on many different factors such as temperature or pH. These interactions can happen at the nanometer level, as evidenced by a paper published by Arkhurst et al. and Fan et al. [51,52]. Fan et al. go on to discuss how the mechanical properties of the polymers change the correlating properties of the composite. It was concluded by Fan et al. that when the bonding quality increased, so did the fracture strain limit [52]. When reviewing the fatigue behavior of polymeric composites, Shanmugan et al. discovered that this parameter has not been studied extensively and needs to receive further analysis [53]. Rheological behavior is how a polymer actually flows and moves as a liquid. This is extremely important for FDM printing since the polymer is heated and has to be pushed through a nozzle. Once it is cooled, it becomes a solid composite or, in the case of this paper, a PLA-based thermoplastic [55]. Thermoplastic composites are an extremely common type of plastic. These are made by heating a polymer to its melting point and then manipulating it to the desired shape. The consistent heat allows the polymer to stay in liquid form long enough for the composite to be shaped [54,58]. These composites can be used in many different ways such as cladding for walls or plastics that are nonbiodegradble or biodegradable depending on the application. In a study conducted by Park et al., it was found that adding an ABS polymer board to a cement wall curtain significantly increased the stability of the wall when the wind speed was increased to 150% [57]. Biodegradability has always been a hot topic when discussing polymers. In most cases, the mixture of polymers needs to be biodegradable. However, there are several cases in which the composite needs to have a high stability to resist biodegradation. In a study published by Harris et al., it was found that a composite that had a partial biodegrading blend had a better stability than blends that were made with nondegrading polymers [56].

4.1.3. Incorrect Filament Diameter

There are two commonly produced diameter extrusion tips: 3.0 mm is the largest diameter for most hobby printers, while the other diameter is 1.75 mm. When printing, it is imperative to ensure that the correct diameter filament is matched to the extruder tip. While it might not fail, it is generally not recommended to use a 1.75 mm filament in a 3.0 mm extruder. It increases the chance of introducing air to the extruded material and reducing the desired material properties of the filament. Most printer systems are set up to handle one of the two diameters. Common hobby printers are generally set up to use the 1.75 mm filament, and, as such, very few have the ability to plug-and-play a 3.0 mm conversion kit. This is mainly due to the light sensor for the material feeding, as well as the guiding systems used for the filament itself.

4.1.4. Incorrect Nozzle Types

While the printers are focused on one of the two main diameter filaments, there can be an arrangement of extrusion point diameters. This information is required during the slicing software process. This will allow the software to calculate the movement speed of the print head, the cooling time, the layer thickness, and the layer height. It also calculates the tolerances and allows for the part's shrinkage value. Generally, the larger the extrusion point, the more material there is that is used in a print, the higher the temperature needed during the initial extrusion, and the slower the movement speed. This allows for the filament to resolidify before the next layer is produced, while ensuring proper layer adhesion.

One of the main materials for extruder tips is brass. This information must be taken into account when printing due to the heat conductivity of the metal.

4.2. Bed Leveling

As the beginning of a printing process, it is recommended to go through a leveling process. For most slicing software, this is automatically coded into the system. This allows for the technician to ensure that the extruded material will meet the build plate at the proper height. Bed leveling at the initial start of the print will reduce the possibility of failure due to non-adhesion. However, for most non-enclosed sections, it is possible for the bed height and pitch to be adjusted during the print. If there is a pitch within the bed, there is a higher chance of non-adhesion for the extruded material (depending on the diameter of the extrusion nozzle itself).

5. Failure and Flaws in Printed Parts

5.1. Failures and Flaws during the Printing Process

Due to the nature of Additive Manufacturing, flaws and failures can be very costly and time consuming. In material removal processes, most can be remelted and then reproduced. For printing, due to the nature of the processes, any failure within the part or during the process will render the part unusable.

5.1.1. Spaghetting

Spaghetting is a failure that occurs when the material being extruded does not connect or adhere to the build plate but rather adheres to the extrusion nozzles (Figure 7). This is the unofficial term for this failure; however, it is a common issue and wastes large amounts of filament. One factor that causes this failure is foreign matter on the extruder tip. This is normally caused by old filament burning onto the nozzle from past prints. Another possibility is dragging. This is when the extrusion nozzle is not offset to correct for possible height differences in the print bed. The nozzle then drags across the build plate and removes some of the build plate material. This then melts and adheres to the extrusion nozzle and produces a possible adhesion point for the extruding material. This failure causes a large loss of material and failed prints if not rectified quickly. If the failure does occur, removing the material from the extrusion tip increases the chance of saving the print from a full failure.

5.1.2. Adhesion and Warping

The most common failure for adhesion is normally print bed adhesion. Print bed adhesion is the first few layers of the print and assists in keeping the print stationary. If the bed adhesion fails, the print will not progress. To reduce the chance of bed adhesion failure, it is recommended to use a heated and slightly abrasive surface. This will create friction for the filament and increase the chance of adhesion by increasing the surface area contact. Heated beds soften the extruded material, which in turn allows for the adhesion between the rough surface top layer and the first material layer [19,59].

Figure 7. Spaghetting of unsuccessful deposition of the melted filament material.

Layer adhesion is another failure. This is when the extruded material fails to melt and adhere correctly to the previous layer. There are several factors that can cause this type of failure. One of the largest contributing factors is the environment. The ambient temperature of the area surrounding a print can cause the cooling process to alter during printing. If the ambient temperature is over the normal threshold for the print, it will cause the extruded material to cool at a slower rate, which can cause slippage. The exact opposite occurs when the ambient temperature of the printing area is cooler than the required temperature. This will cause the layers to not bond together.

5.2. *Failures and Flaws after Processing*

5.2.1. Porosity and Density Consistency

There are two main material property measurements. Porosity is a measure of the space within a part that is void of material, while density is defined as the measurement of mass per unit of volume. Regarding additive manufacturing, much like casting, trapped bubbles of air can occur during the printing process. Since the material is being extruded, there are instances where a bubble is introduced into the material while the print is occurring. Prints are produced in a linear pattern, which means there are micro-gaps between the layers and walls. These micro-gaps produce porosity in the prints. As such, due to the change in porosity, density is also affected. If the printed component is sectioned into different cross sections, the porosity and density of each varies. Voids cause fatigue, which, as stated earlier, is not a majorly discussed topic. In this article published in 2022, voids' interaction with fatigue behavior is discussed [60]. The impulse excitation technique to find the Young's modulus is a way to remedy this problem [61,62]. Through this process, the natural frequencies of the sample can be quantified, and then the mechanical properties can be calculated. It was found that when the Young's modulus was calculated, an optimization of the geometry of the print could be performed to solve the porosity issues. Optimization of the print can be achieved in composites as well. It was found by Ahmed et al. that changing the printing parameters on the interfacial bond strength of ABS/carbon fiber-reinforced polylactic acid (CF-PLA) also changed the quality of the print [63].

5.2.2. Surface Topography Variations

Due to the layer printing technology used by FDM, the surface topography varies from component to component. As the machine prints, it extrudes semi-melted material onto the print surface or upon the previously laid layers. While this occurs, as stated previously in the paper, air bubbles can become trapped between layers, or the material can break during printing. If this happens, it can cause topography variations in the final component. While most of the errors are within tolerances, there are times where the topography varies widely. This can be due to the print failing and collapsing on itself or to air pockets being introduced on the surface [64]. A solution for this problem is to add various coatings that can increase the composite's mechanical properties including the tensile strength of the composite. A battery of tests can then be done on the finished product, including the proposed methodology by Bernal et al. [61].

6. Discussion

The results from past experiments and research show that the ultrasound technology was highly effective in showing/recording defects and faults in both machined and purposefully produced defects within printed parts. In addition, ultrasound- and acoustic-based techniques showed promising capabilities for detecting acoustic signatures during the printing process and thus become a capable technique for the in situ monitoring of the machine and parts quality. Ultrasonic techniques can provide a rich amount of information about the quality of the part and the condition of the process. They are very precise and efficient in the post-process quality monitoring of the parts. However, the bottleneck of using ultrasonic for in situ monitoring is the speed of the data recording and analysis and the complexity of the data processing. Thermal-based techniques are still an effective method for quality monitoring, specifically in the FDM/FFF process, since the range of temperatures in FDM/FFF can be more easily included in the operational range of many thermal cameras. Future studies via thermal imaging are necessary to evaluate the appropriate characteristics of images, including resolution, speed, and accuracy for both the in-process and post-process assessment of the 3D printed parts' quality.

When reproducing the experiments, the environment should be controlled. The ambient temperature and humidity need to be held constant and at an optimal level for the filament being used. The results will vary if the printer is enclosed versus open to the environment. If the printer is in a confined environment, feedback from the thermal imaging sensor will be expected and will artificially change the results. The testing application of the sensors for ultrasound technology should be repeated in a relatively constant spot to reduce errors and outlier data. These factors are very important practical points when designing a monitoring and/or control system for 3D printers since they affect the complex process of the 3D printing.

When NDT techniques are used for the in situ monitoring of the 3D printing processes, the immediate benefit would be the sustainability of the process, by which the dispensable waste of materials, energy, and production time can be eliminated if a serious flaw or inconsistency is detected in a layer. To reduce the false alarm in a monitoring methodology, in addition to the appropriate sensor integration, accurate and robust data analysis and signal processing are necessary. Machine learning and computer vision techniques can reduce the false alarm and enhance the accuracy of data analysis for this purpose [65].

To further develop the knowledge and understanding of the defects in 3D-printed parts and the potential failures in 3D printing machines, more investigations into both materials and manufacturing processes are necessary. Some of the suggested future work includes specific experimental investigations into the causes of defects in 3D-printed polymers and composites, studies on the optimization of manufacturing processes toward the minimum possibility of defect generation, studies on the cases of machine failures and the implementation of potential feedback controls to avoid such failures, and studies on the process parameters and their influence on faults and failures.

7. Conclusions

Additively manufacturing or 3D printing polymer and composite materials provides many opportunities for the design and fabrication of a variety of parts and components, specifically for parts with complex geometries, prototyping purposes, the fabrication of parts with low production rates, and tools and fixture design. Despite the valuable advantages that the 3D printing of polymers and composite materials such as FDM provides, the structural integrity and quality of the manufactured parts remain a main concern.

Understanding the type and properties of the potential defects and flaws in 3D printed parts, as well as the potential failures in 3D printing machines, as discussed in this paper, is crucial to avoid manufacturing low-quality parts or operating the equipment in inappropriate conditions. The application of nondestructive testing should be conducted in all additively manufactured parts due to their destructible and degradable nature when checked by traditional methods. Due to their quick feedback and accuracy, thermal imaging, X-ray scattering, laser profilometry, and ultrasound are the most applicable nondestructive testing methods. They are able to be conducted in-line and after the processing of the part, and this allows for the early detection of defects as well as quality control.

Author Contributions: Conceptualization, M.B.-C.; methodology, M.B.-C.; investigation, M.B.-C. and E.L.; resources, M.B.-C.; writing—original draft preparation, M.B.-C.; writing—review and editing, H.T. and E.L.; supervision, H.T. and M.T.; project administration, H.T. and M.T.; funding acquisition, M.B.-C. All authors have read and agreed to the published version of the manuscript.

Funding: This research was funded by the Undergraduate Research Award of the Allen E. Paulson College of Engineering and Computing of Georgia Southern University under award number UGRA2020-MB.

Institutional Review Board Statement: Not applicable.

Informed Consent Statement: Not applicable.

Acknowledgments: The authors thank the undergraduate research program and the Dean's office at the Allen E. Paulson College of Engineering and Computing of Georgia Southern University.

Conflicts of Interest: The authors declare no conflict of interest.

References

1. Gibson, I.; Rosen, D.; Stucker, B.; Khorasani, M. *Additive Manufacturing Technologies*; Springer: Cham, Switzerland, 2020.
2. Groover, M.P. *Fundamentals of Modern Manufacturing*; Wiley: Hoboken, NJ, USA, 2022.
3. Zhao, H.; Liu, X.; Zhao, W.; Wang, G.; Liu, B. An Overview of Research on FDM 3D Printing Process of Continuous Fiber Reinforced Composites. In *Journal of Physics: Conference Series*; IOP Publishing: Bristol, UK, 2019; p. 052037.
4. Pervaiz, S.; Qureshi, T.A.; Kashwani, G.; Kannan, S. 3D Printing of Fiber-Reinforced Plastic Composites Using Fused Deposition Modeling: A Status Review. *Materials* **2021**, *14*, 4520. [CrossRef] [PubMed]
5. Singh, R.; Singh, S.; Mankotia, K. Development of ABS Based Wire as Feedstock Filament of FDM for Industrial Applications. *Rapid Prototyp. J.* **2016**, *22*, 300–310.
6. Albakri, M.I.; Sturm, L.D.; Williams, C.B.; Tarazaga, P.A. Impedance-based non-destructive evaluation of additively manufactured parts. *Rapid Prototyp. J.* **2017**, *23*, 589–601. [CrossRef]
7. Matterhackers. Available online: https://www.matterhackers.com/ (accessed on 6 December 2021).
8. Makerbot Replicator. Available online: http://store.makerbot.com/replicator (accessed on 6 December 2021).
9. Makerbot Replicator Mini. Available online: http://store.makerbot.com/replicator-mini (accessed on 6 December 2021).
10. Cubify Cube Store. Available online: http://cubify.com/cube/store.aspx (accessed on 6 December 2021).
11. Wickramasinghe, S.; Do, T.; Tran, P. FDM-Based 3D Printing of Polymer and Associated Composite: A Review on Mechanical Properties, Defects and Treatments. *Polymer* **2020**, *12*, 1529. [CrossRef]
12. Blanco, I. The use of composite materials in 3D printing. *J. Compos. Sci.* **2020**, *4*, 42. [CrossRef]
13. Kalsoom, U.; Nesterenko, P.N.; Paull, B. Recent developments in 3D printable composite materials. *RSC Adv.* **2016**, *6*, 60355–60371. [CrossRef]
14. Triyono, J.; Sukanto, H.; Saputra, R.M.; Smaradhana, D.F. The effect of nozzle hole diameter of 3D printing on porosity and tensile strength parts using polylactic acid material. *Open Eng.* **2020**, *10*, 762–768. [CrossRef]
15. Bhagia, S.; Bornani, K.; Agrawal, R.; Satlewal, A.; Ďurkovič, J.; Lagaňa, R.; Bhagia, M.; Yoo, C.G.; Zhao, X.; Kunc, V.; et al. Critical review of FDM 3D printing of PLA biocomposites filled with biomass resources, characterization, biodegradability, upcycling and opportunities for biorefineries. *Appl. Mater. Today* **2021**, *24*, 101078. [CrossRef]
16. Swetham, T.; Reddy, K.M.M.; Huggi, A.; Kumar, M.N. A Critical Review on of 3D Printing Materials and Details of Materials Used in FDM. *Int. J. Sci. Res. Sci. Eng. Technol.* **2017**, *3*, 353–361.
17. Zhu, Q.; Yu, K.; Li, H.Q.; Zhang, Q.Q.; Tu, D.W. Rapid residual stress prediction and feedback control during fused deposition modeling of PLA. *Int. J. Adv. Manuf. Technol.* **2022**, *118*, 3229–3240. [CrossRef]
18. Kuo, C.C.; Wu, Y.R.; Li, M.H.; Wu, H.W. Minimizing warpage of ABS prototypes built with low-cost fused deposition modeling machine using developed closed-chamber and optimal process parameters. *Int. J. Adv. Manuf. Technol.* **2019**, *101*, 593–602. [CrossRef]

19. Nazan, M.A.; Ramli, F.R.; Alkahari, M.R.; Abdullah, M.A.; Sudin, M.N. An exploration of polymer adhesion on 3D printer bed. *IOP Conf. Ser. Mater. Sci. Eng.* **2017**, *210*, 012062. [CrossRef]
20. Taheri, H. Utilization of Non-destructive Testing (NDT) Methods for Composite Materials Inspection (Phased Array Ultrasonic). Master's Thesis, South Dakota State University, Brookings, SD, USA, 2014.
21. Lin, T.; Jia, D.; He, P.; Wang, M.; Liang, D. Effects of fiber length on mechanical properties and fracture behavior of short carbon fiber reinforced geopolymer matrix composites. *Mater. Sci. Eng. A* **2008**, *497*, 181–185. [CrossRef]
22. Alves, M.; Junior, C.C.; Ha, S. Fiber waviness and its effect on the mechanical performance of fiber reinforced polymer composites: An enhanced review. *Compos. Part A Appl. Sci. Manuf.* **2021**, *149*, 106526. [CrossRef]
23. Senthil, K.; Arockiarajan, A.; Palaninathan, R.; Santhosh, B.; Usha, K. Defects in composite structures: Its effects and prediction methods–A comprehensive review. *Compos. Struct.* **2013**, *106*, 139–149. [CrossRef]
24. Kugler, D.; Moon, T.J. Identification of the most significant processing parameters on the development of fiber waviness in thin laminates. *J. Compos. Mater.* **2002**, *36*, 1451–1479. [CrossRef]
25. Thor, M.; Sause, M.G.; Hinterhölzl, R.M. Mechanisms of origin and classification of out-of-plane fiber waviness in composite materials—A review. *J. Compos. Sci.* **2020**, *4*, 130. [CrossRef]
26. Bednarcyk, B.A.; Aboudi, J.; Arnold, S.M. The effect of general statistical fiber misalignment on predicted damage initiation in composites. *Compos. Part B Eng.* **2014**, *66*, 97–108. [CrossRef]
27. Li, Y.; Stier, B.; Bednarcyk, B.; Simon, J.-W.; Reese, S. The effect of fiber misalignment on the homogenized properties of unidirectional fiber reinforced composites. *Mech. Mater.* **2016**, *92*, 261–274. [CrossRef]
28. Kidangan, R.T.; Krishnamurthy, C.V.; Balasubramaniam, K. Identification of the fiber breakage orientation in carbon fiber reinforced polymer composites using induction thermography. *NDT E Int.* **2021**, *122*, 102498. [CrossRef]
29. Liu, T.; Gao, Y.; Fan, W.; Gao, X.; Ma, J. Predictions of the axial tensile property of the unidirectional composite influenced by microfiber breakage defects. *Text. Res. J.* **2022**, *92*, 15–29. [CrossRef]
30. Taheri, H. Nondestructive Evaluation and In-Situ Monitoring for Metal Additive Manufacturing. Ph.D. Thesis, Iowa State University, Ames, IA, USA, 2018; pp. 61–75.
31. Koester, L.W.; Bond, L.J.; Taheri, H.; Collins, P.C. Nondestructive Evaluation of Additively Manufactured Metallic Parts. *Nondestruct. Eval. Mater.* **2018**, *17*, 544–552. [CrossRef]
32. Koester, L.W.; Bond, L.J.; Taheri, H.; Peter CCollins, P.C. Nondestructive evaluation of additively manufactured metallic parts: In situ and post deposition. In *Additive Manufacturing for the Aerospace Industry*; Elsevier Inc.: Philadelphia, PA, USA, 2018; Chapter 18; pp. 401–417. [CrossRef]
33. Koester, L.W.; Taheri, H.; Bigelow, T.; Collins, P.C.; Bond, L.J. Nondestructive Testing for Metal Parts Fabricated Using Powder-Based Additive Manufacturing. *Mater. Eval.* **2018**, *76*, 514–528.
34. Koester, L.W.; Taheri, H.; Bond, L.J.; Barnard, D.; Gray, J. Additive manufacturing metrology: State of the art and needs assessment. In *AIP Conference Proceedings*; AIP Publishing LLC: Melville, NY, USA, 2016; p. 130001.
35. Borish, M.; Post, B.K.; Roschli, A.; Chesser, P.C.; Love, L.J. Real-Time Defect Correction in Large-Scale Polymer Additive Manufacturing via Thermal Imaging and Laser Profilometer. *Procedia. Manuf.* **2020**, *48*, 625–633. [CrossRef]
36. Sauerbrunn, E.; Chen, Y.; Didion, J.; Yu, M.; Smela, E.; Bruck, H.A. Thermal imaging using polymer nanocomposite temperature sensors. *Phys. Status Solidi. (A)* **2015**, *212*, 2239–2245. [CrossRef]
37. Shmueli, Y.; Jiang, J.; Zhou, Y.; Xue, Y.; Chang, C.; Yuan, G.; Satija, S.K.; Lee, S.; Nam, C.; Kim, T.; et al. Simultaneous In Situ X-ray Scattering and Infrared Imaging of Polymer Extrusion in Additive Manufacturing. *ACS Appl. Polym. Mater.* **2019**, *1*, 1559–1567. [CrossRef]
38. Abdelrahman, M.; Starr, T.L. Quality certification and control of polymer laser sintering: Layerwise temperature monitoring using thermal imaging. *Int. J. Adv. Manuf. Technol.* **2016**, *84*, 831–842. [CrossRef]
39. Giesko, T.; Zbrowski, A.; Czajka, P. Laser Profilometers for Surface Inspection and Profile Measurement. *Probl. Eksploat.* **2007**, 97–108.
40. Nguyen, H.T.; Crittenden, K.; Weiss, L.; Bardaweel, H. Experimental Modal Analysis and Characterization of Additively Manufactured Polymers. *Polymer* **2022**, *14*, 2071. [CrossRef]
41. Liao, Y.; Liu, C.; Coppola, B.; Barra, G.; Di Maio, L.; Incarnato, L.; Lafdi, K. Effect of Porosity and Crystallinity on 3D Printed PLA Properties. *Polymer* **2019**, *11*, 1487. [CrossRef]
42. Taheri, H.; Du, J.; Delfanian, F. Experimental Observation of Phased Array Guided Wave Application in Composite Materials. *Mater. Eval.* **2017**, *75*, 1308–1316.
43. Taheri, H.; Hassen, A.A. Nondestructive Ultrasonic Inspection of Composite Materials: A Comparative Advantage of Phased Array Ultrasonic. *Appl. Sci.* **2019**, *9*, 1628. [CrossRef]
44. Lee, J.; Hasanian, M.; Saboonchi, H.; Baechle, M.; Taheri, H. Ultrasonic evaluation of polymer additively manufactured parts for defect inspection and structural integrity assessment. In *Nondestructive Characterization and Monitoring of Advanced Materials, Aerospace, Civil Infrastructure, and Transportation IX, 4*; SPIE proceeding; SPIE Press: Bellingham, WA, USA, 2020; p. 70.
45. Taheri, H.; Koester, L.; Bigelow, T.; Bond, L.J.; Braconnier, D.; Carcreff, E.; Dao, G.; Caulder, A.; Hassen, A.A. Fast Ultrasonic Imaging with Total Focusing Method (TFM) for Inspection of Additively Manufactured Polymer Composite Component. In Proceedings of the ASNT 27th Annual Research Symposium Proceedings, 10, Houston, TX, USA, 29–31 October 2018; pp. 212–220.
46. Finnes, T. High Definition 3D Printing-Comparing SLA nad FDM Printing Technologies. *J. Undergrad. Res.* **2015**, *13*, 10–26.

47. Kafle, A.; Luis, E.; Silwal, R.; Pan, H.M.; Shrestha, P.L.; Bastola, A.K. 3D/4D Printing of Polymers: Fused Deposition Modelling (FDM), Selective Laser Sintering (SLS), and Stereolithography (SLA). *Polymer* **2021**, *13*, 3101. [CrossRef] [PubMed]
48. Szykiedans, K.; Credo, W. Mechanical Properties of FDM and SLA Low-cost 3-D Prints. *Procedia Eng.* **2016**, *136*, 257–262. [CrossRef]
49. Bond, L.J.; Koester, L.W.; Taheri, H. NDE in-process for metal parts fabricated using powder based additive manufacturing. In *Smart Structures and NDE for Energy Systems and Industry 4.0, 3*; SPIE proceeding; SPIE Press: Bellingham, WA, USA, 2019; p. 1.
50. Taheri, H.; Koester, L.W.; Bigelow, T.A.; Faierson, E.J.; Bond, L.J. In Situ Additive Manufacturing Process Monitoring With an Acoustic Technique: Clustering Performance Evaluation Using K-Means Algorithm. *J. Manuf. Sci. Eng.* **2019**, *141*, 041011. [CrossRef]
51. Arkhurst, B.M.; Seol, J.B.; Lee, Y.S.; Lee, M.; Kim, J.H. Interfacial structure and bonding mechanism of AZ31/carbon-fiber-reinforced plastic composites fabricated by thermal laser joining. *Compos. Part B-Eng.* **2019**, *167*, 71–82. [CrossRef]
52. Fan, C.Z.; Shan, Z.D.; Zou, G.S.; Zhan, L.; Yan, D.D. Interfacial Bonding Mechanism and Mechanical Performance of Continuous Fiber Reinforced Composites in Additive Manufacturing. *Chin. J. Mech. Eng.* **2021**, *34*, 11. [CrossRef]
53. Shanmugam, V.; Das, O.; Babu, K.; Marimuthu, U.; Veerasimman, A.; Johnson, D.J.; Neisiany, R.E.; Hedenqvist, M.S.; Ramakrishna, S.; Berto, F. Fatigue behaviour of FDM-3D printed polymers, polymeric composites and architected cellular materials. *Int. J. Fatigue* **2021**, *143*, 15. [CrossRef]
54. Yang, C.C.; Tian, X.Y.; Liu, T.F.; Cao, Y.; Li, D.C. 3D printing for continuous fiber reinforced thermoplastic composites: Mechanism and performance. *Rapid Prototyp. J.* **2017**, *23*, 209–215. [CrossRef]
55. Arrigo, R.; Frache, A. FDM Printability of PLA Based-Materials: The Key Role of the Rheological Behavior. *Polymer* **2022**, *14*, 1754. [CrossRef] [PubMed]
56. Harris, M.; Mohsin, H.; Potgieter, J.; Ishfaq, K.; Archer, R.; Chen, Q.; De Silva, K.; Le Guen, M.J.; Wilson, R.; Arif, K.M. Partial Biodegradable Blend with High Stability against Biodegradation for Fused Deposition Modeling. *Polymer* **2022**, *14*, 1541. [CrossRef] [PubMed]
57. Park, M.J.; Bae, J.; Ju, Y.K. Structural Behavior of a Composite Curtain Wall Fabricated by the Fused Deposition Modeling 3D Printing Method. *Polymer* **2022**, *14*, 1431. [CrossRef]
58. Sehhat, M.H.; Mahdianikhotbesara, A.; Yadegari, F. Impact of temperature and material variation on mechanical properties of parts fabricated with fused deposition modeling (FDM) additive manufacturing. *Int. J. Adv. Manuf. Technol.* **2022**, *120*, 4791–4801. [CrossRef]
59. Nazan, M.A.; Ramli, F.R.; Alkahari, M.R.; Sudin, M.N.; Abdullah, M.A. Optimization of Warping Deformation in Open Source 3D Printer Using Response Surface Method. In Proceedings of the Mechanical Engineering Research Day, Melaka, Malaysia, 31 March 2016; pp. 71–72.
60. Hassanifard, S.; Behdinan, K. Effects of voids and raster orientations on fatigue life of notched additively manufactured PLA components. *Int. J. Adv. Manuf. Technol.* **2022**, *120*, 6241–6250. [CrossRef]
61. Bernal, J.D.B.; Silva, E.C.N.; Rubio, W.M. Characterization of effective Young's modulus for Fused Deposition Modeling manufactured topology optimization designs. *Int. J. Adv. Manuf. Technol.* **2019**, *103*, 2879–2892. [CrossRef]
62. Ahmad, M.N.; Ishak, M.R.; Mohammad Taha, M.; Mustapha, F.; Leman, Z.; Anak Lukista, D.D.; Irianto; Ghazali, I. Application of Taguchi Method to Optimize the Parameter of Fused Deposition Modeling (FDM) Using Oil Palm Fiber Reinforced Thermoplastic Composites. *Polymer* **2022**, *14*, 2140. [CrossRef]
63. Ahmed, S.W.; Hussain, G.; Altaf, K.; Ali, S.; Alkahtani, M.; Abidi, M.H.; Alzabidi, A. On the Effects of Process Parameters and Optimization of Interlaminate Bond Strength in 3D Printed ABS/CF-PLA Composite. *Polymer* **2020**, *12*, 2155. [CrossRef]
64. Jin, Y.; Wan, Y.; Zhang, B.; Liu, Z. Modeling of the chemical finishing process for polylactic acid parts in fused deposition modeling and investigation of its tensile properties. *J. Mater. Process. Technol.* **2017**, *240*, 233–239. [CrossRef]
65. Taheri, H.; Gonzalez Bocanegra, M.; Taheri, M. Artificial Intelligence, Machine Learning and Smart Technologies for Nondestructive Evaluation. *Sensors* **2022**, *22*, 4055. [CrossRef]

Journal of
Composites Science

MDPI

Review

Finite Element Modelling Approach for Progressive Crushing of Composite Tubular Absorbers in LS-DYNA: Review and Findings

Ali Rabiee and Hessam Ghasemnejad *

Centre for Structures, Assembly and Intelligent Automation, Cranfield University, Cranfield MK43 0AL, UK; ali.rabiee@ruroc.com
* Correspondence: hessam.ghasemnejad@cranfield.ac.uk; Tel.: +44-(0)-1234-754395

Abstract: Robust finite element models are utilised for their ability to predict simple to complex mechanical behaviour under certain conditions at a very low cost compared to experimental studies, as this reduces the need for physical prototypes while allowing for the optimisation of components. In this paper, various parameters in finite element techniques were reviewed to simulate the crushing behaviour of glass/epoxy tubes with different material models, mesh sizes, failure trigger mechanisms, element formulation, contact definitions, single and various numbers of shells and delamination modelling. Six different modelling approaches, namely, a single-layer approach and a multi-layer approach, were employed with 2, 3, 4, 6, and 12 shells. In experimental studies, 12 plies were used to fabricate a 3 mm wall thickness GFRP specimen, and the numerical results were compared with experimental data. This was achieved by carefully calibrating the values of certain parameters used in defining the above parameters to predict the behaviour and energy absorption response of the finite element model against initial failure peak load (stiffness) and the mean crushing force. In each case, the results were compared with each other, including experimental and computational costs. The decision was made from an engineering point of view, which means compromising accuracy for computational efficiency. The aim is to develop an FEM that can predict energy absorption capability with a higher level of accuracy, around 5% error, than the experimental studies.

Keywords: crashworthiness; composites; FEA; progressive crushing; LS-DYNA

Citation: Rabiee, A.; Ghasemnejad, H. Finite Element Modelling Approach for Progressive Crushing of Composite Tubular Absorbers in LS-DYNA: Review and Findings. *J. Compos. Sci.* **2022**, *6*, 11. https://doi.org/10.3390/jcs6010011

Academic Editor: Stelios K. Georgantzinos

Received: 16 November 2021
Accepted: 16 December 2021
Published: 29 December 2021

1. Introduction

Two classes of finite element methods are available: either the implicit or explicit method [1]. The implicit method is widely available and used in a broad range of problems, including nonlinear stress analysis and static. The explicit method is widely used in highly nonlinear stress analysis and dynamics with contact-dominated problems. A car crash, for instance, or metal stamping simulations are applications well suited for the explicit method.

Due to the high cost of conducting experimental studies, there is a need for reliable computational models capable of predicting the crushing response of composite structures. There have been various attempts to develop explicit finite element models (FEMs), with different degrees of precision, for circular tubes [2–6], square tubes [4–13], angle-stiffeners [8], C-channels [8] and hat-stiffeners [14]. The classification of structural FEM can be divided into two groups. The first group is the micro-mechanical one [15–20]. In this group, the finite element models try to simulate the composite crushing phenomenon through a detailed modelling of its micromechanical behaviour. A very fine solid mesh is developed to accurately capture the micro-mechanics matrix crack propagation phenomenon. The computational effort demanded by this kind of model is very high, which makes it unpractical for engineering crash analysis. This approach is used mainly to perform simulations concerning the delamination phenomenon, in which the growth behaviour of a single crack is studied in a very detailed way [7]. The second group is the macro-mechanical one [3].

This type of model provides a macro-mechanical description of the material collapse. It is much more computationally effective, and consequently, it is a suitable choice for engineering crash analysis. However, it is not capable of precisely modelling all the main collapse modes that occur simultaneously during a crush event.

The FE modelling of composite structures can be either shell or solid elements. Solid element models require more computation time and are less widely used compared with shell elements used in the axial crushing of composite structures as mentioned above. A single layer [17,21–24] or multiple layers [3,25] of shell elements can be used to model a laminate. In the single-layer model, this can be modelled as a single layer of shell elements, with each ply being represented by the thickness integration point, also referred to as the integration point in the thickness direction. This kind of model is not capable of modelling the interlaminar collapse modes shown by composites under crushing in an accurate way. However, it is useful if a detailed representation of the failure physics is not required and only load and energy level predictions are required. The main advantages are its simplicity and computational effectiveness, so for that point, they are highly suitable for practical engineering crash analysis. On the other hand, they have a notably lower level of robustness due to the large amount of parameter calibration required to obtain acceptable global results for a given test configuration [7].

The authors of [8] used this configuration for C-channel, angle-stiffeners and the hollow square tube modelling technique. In the multi-layer configuration model, the laminate is modelled by multiple shell elements, with each layer representing either a single ply or a group of plies, and the layers are glued together using an automatic contact definition (surface to surface).

In recent studies, the single-layer configuration model has been used to simulate the behaviour of various composite structures in the case of, for instance, thin-walled square sections. This method is capable of accurately capturing the local buckling and unstable collapse; however, due to the complex failure mechanism of composite structures, this method was ineffective in depicting the progressive failure process [4]. Providing precise input of key material properties and numerical parameters in the material model (e.g., eleven parameters in MAT 54), and defining the contact definition between the test specimen and the impactor and applying an all degrees of freedom constraint on the end of the test specimen, the single-layer model was able to yield good correlation with the experimental load–displacement graph for cross-sections studied in [8]. However, this configuration, as mentioned above, is not appropriate for failure mechanisms and the crushing behaviour, as these are mostly neglected.

The multi-layer modelling technique can be utilised for better capturing the failure process of the tubes undergoing progressive crushing [3–7]. However, in the multi-layer model, the correlation with the experimental load–displacement was not always satisfactory. The composite hollow tube was modelled in LS-DYNA using MAT 54 to analyse its crushing behaviour. The FEA results were satisfactory and agreed with the experimental load–displacement graphs; however, instead of the brittle failure mechanism observed in experimental studies, a significant local buckling of the tube was observed in FEA.

A finite element model was developed by [4]. This model was able to accurately predict the peak load of thin-walled square CFRP tube, although the specific energy absorption was underestimated by 33%. In the FEA simulation, the crushed elements were deleted instead of forming debris, as was observed in the experiment. One of the parameters that contributed to the energy absorption was the debris formation wedged between the fronds of the tube's wall, which was neglected in the simulation. The author noted that this parameter alone affected the SEA value in a significant way. In [5,6], the FEM was developed to capture the crushing behaviour of hollow circular and square tubes and compare the results with the experimental observations. It was noted by the authors that the model was not able to reproduce the axial matrix splits observed experimentally. This resulted in different load–displacement curves, deformation, and failure behaviour when compared with the experiment. Several parameters that influenced the crushing behaviour of square hollow tubes were analysed in [7]. The parameters were element size, number of shell layers, coefficient of friction and interlaminar material properties [26–28]. The author noted that by increasing the number of shells, the main collapse mode was unchanged. The influence of the friction coefficient between the tube and the machine plate and of the element dimension was also studied. The element size was coarse mesh (7 mm) and fine mesh (4 mm). The static friction coefficient values were 0.1, 0.2 and 0.3, and the dynamic friction coefficient was kept constant at 0.65. The increase in the friction coefficient from 0.1 to 0.3 and the use of finer mesh did not significantly change the crush zone morphology in the sense that all three tubes collapsed in the same way. For all three values of friction, the load–displacement curves have almost the same shape. In this sense, the effect of the friction variation was similar to a scaling of the magnitude of the load curves. The model with bigger elements shows a magnitude of force oscillation higher than the one shown by the refined model; however, the average force value is the same.

The authors of [10] combined a series of experimental investigations and examined their effect using finite element analysis and concluded proportional failure mechanisms leading to SEA, which is a result of dominated energy absorption mechanisms of approximately 60% material damage and 30% friction with 10% related to contact parameter [29]. Although a detailed examination of the material behaviour is essential to reach a load adapted lightweight design. Axial finite element modelling of composite structures uses different modelling methods. Various approaches were developed to obtain ideal force–displacement curves that are aligned with experimental data. One approach includes a hybrid mesh of shell and solid elements to capture ideal load–displacement curves [30]. This method is used for modelling of crack propagation with finite cohesion elements [31] that consist of shell elements representing the material and a solid element to represent delamination failure. Alternatively, material 54 in LS-DYNA is used to predict crushing behaviour and the idealised force–displacement graph [32]. One strategy is the SOFT parameter implemented in LS-DYNA to map pre-damage and consequently create a progressive crashfront of CFRP square tubes or composite sections, and the author concluded that Mat_Composite_Damage of LS-DYNA for GFRP box structures was found to be in accordance with experimental results when it was modelled as a double-shell configuration. Some similar conclusions were obtained by other researchers that used multi-shell configuration, which can predict energy absorption and maximum force under crushing [32,33].

Abdallah [34] numerically optimised fibre orientation of glass/phenolic tubes under tensile and compression loadings. Failure parameters in the Mat_54 material card in the LS-DYNA Library are defined by ALPH (shear stress parameter for the nonlinear term), BETA weighting factor for the shear term in tensile fibre mode), FBRT (softening for fibre tensile strength), YCFAC (reduction factor for compressive fibre strength after matrix failure), TFAIL (time step size criteria for element deletion), SOFT (softening reduction factor for material strength in the crashfront element) and EFS (effective failure strain) to improve the accuracy of the final results. Abdallah concluded that the developed numerical models are capable of predicting tensile and compression loading of small and large scale GFRP specimens within 10% discrepancy. This trial-and-error or parametric study approach has also been used in References [34–38]. However, in another study by Xiao [39], axial crushing of braided carbon tubes was modelled using MAT58 in LS-DYNA to predict peak forces within 20% of the experimental values, but the predicted crushing forces and SEA were generally lower than experimental results. This is likely due to a deficiency in MAT_58 in modelling the subsequent unloading response of partially damaged material.

Many studies focused on the contact parameter [29], crack propagation modelling with de-cohesion elements [40] or user-defined material model, which requires extensive experimental investigations [41].

In [42,43], the effect of failure trigger mechanisms on the energy absorption capability of CFRP tubes under axial compression is experimentally and numerically investigated. The conventional approach to introduce a failure mechanism is to apply a 45° chamfer on one end, and the failure could initiate progressively. Alternatively, an attachment of crush-cap can be utilised to initiate the progressive crushing. Two different types of crush-caps were studied, each causing the crushed material to fold either inward or splay outward. The effect of the corner radius of the crush-caps on the peak load, SEA, and crush behaviour was investigated. The author noted that the chamfer failure trigger was most effective at reducing the initial peak load while maintaining a high-sustained crush load and specific energy absorption (SEA). The inward-folding failure trigger approach was not as effective at reducing the initial peak load but was more effective than using a chamfer for maintaining a high-sustained crush load and SEA. However, the modelling technique of the simulation, which was based on a multi-shell configuration with failure a trigger, showed a high level of correlation with the experimental results for both the chamfer and combined failure trigger cases. The simulation was able to predict key deformation characteristics, which were observed during the experimental crushing process, and progression of matrix splits and the direction of splaying of plies.

In the modelling approach, it is necessary to develop models that are simple enough to be employed in practical analysis situations but at the same time capable of providing results with a suitable level of accuracy. At the same time, the approach shall be numerically robust and practical in the model build phase [44–46].

The primary focus of this study is to develop a 'multi-layer' finite element modelling approach to capture the crushing behaviour of composite tubular structures. This approach requires a series of parametric studies to maximise prediction of FEM, i.e., failure trigger mechanism to initiate a progressive crushing, materials card formulation, delamination interface, element formulation, mesh sensitivity and number of shells. This paper illustrates the effect of the named parameters on the predictability of developed FEM compared to experimental studies.

The modelling techniques further investigate the effect of various loading on the crushing behaviour of composite tubes. The experimental and FEA results are compared based on force–displacement curves. The approach of FEM is categorised based on parameters affecting energy absorption capability and the ability of the model to capture real-life crashing and composite material behaviour. Hence, this review paper has covered the essential parameters and calibration procedures to establish the findings.

2. LS-DYNA Software

A commonly used piece of software for crashworthiness application by industry and academics is LS-DYNA, which was developed by LSTC and is suited for highly nonlinear transient dynamic finite element analysis.

LS-DYNA, within the past decade, has added many new features such as new material types, contact algorithms, element formulation, etc. LSTC has gradually expanded to develop a universal tool for most implicit and most vastly used explicit coding for aerospace, automotive, military and construction. LS-DYNA has its own pre-processor called LS-PrePost.

2.1. Material Models

New material models are developed and added to LS-DYNA regularly. Approximately 200 material models are implemented in the software. For unidirectional composite materials, these material models narrow down to the following [47]:

1. Material mode—22: Composite Damage
2. Material model—54 and 55: Enhanced Composite Damage
3. Material model—58: Laminated Composite Fabric
4. Material model—59: Shell/Solid Composite Failure Model

Other material models are adopted to predict composite behaviour, i.e., MAT_161 and MAT_162, where the matrix phase is modelled within the material card and also these material cards can be used to model both unidirectional or bidirectional fibres; however, the concentration of this paper is to isolate material cards from the matrix phase.

Material properties such as shear modulus, elastic modulus, and Poisson's ratio are essential parameters for a material model. Strength properties for failure analysis are also essential to predict material behaviour. These properties are transverse compressive strengths, longitudinal compressive strength, transverse tensile strength and shear strength. Each model has an option to determine the material axes, such as local and global orthotropic material axes. For a given geometry and load, the process of calculation is in three steps,

1. Stress and strain distributions around the stress concentrated areas are calculated,
2. Failure (maximum) load is predicted,
3. Mode of failure is determined.

The analysis consists of two major parts: stress analysis and failure analysis. The most often used material models are described with parametric studies to compare the differences.

2.1.1. MAT_022: Composite Damage Model

This is the first composite failure material model implemented in LS-DYNA, which was proposed by Chang–Chang [48,49]. The keyword for this model is *MAT_COMPOSITE_DAMAGE or *MAT_022. This model can be used in solid and shell elements. By using the user-defined integration rule, the constitutive constants vary through the thickness of the shell.

MAT-022 uses three criteria defined by Chang–Chang and five material parameters to define failure modes.

The three failure criteria used are:

1. A fibre breakage failure mode,
2. A compression failure mode,
3. A matrix failure mode.

The five material parameters are:

1. Shear Strength, $S_{2,}$
2. Traverse tensile strength, $S_{12,}$
3. Transverse compressive strength, $C_{2,}$
4. Longitudinal tensile strength, $S_{1,}$
5. Nonlinear shear stress parameter.

The matrix cracking failure mode is determined from Equations (1) and (2),

$$F_{matrix} = \left(\frac{\sigma_2}{S_2}\right)^2 + \overline{\tau} = 1 \tag{1}$$

$$\overline{\tau} = \frac{\frac{\tau_{12}^2}{2G_{12}} + \frac{3}{4}\alpha\tau_{12}^4}{\frac{S_{12}^2}{2G_{12}} + \frac{3}{4}\alpha S_{12}^4} \tag{2}$$

The failure is assumed when F_{matrix} is less than 1, then E_2, G_{12}, ν_1 and ν_2 are set to zero. The compression failure is determined from Equation (3). Failure is assumed when F_{comp} is less than 1 and E_2, ν_1 and ν_2 are set to zero.

$$F_{comp} = \left(\frac{\sigma_2}{2S_{12}}\right)^2 + \left[\left(\frac{\sigma_2}{2S_{12}}\right) - 1\right]\frac{\sigma_2}{C_2} + \overline{\tau} = 1 \tag{3}$$

The fibre breakage failure mode is determined from Equation (4). Failure is assumed when F_{fibre} is less than 1, then E_1, E_2, G_{12}, ν_1 and ν_2 are set to zero.

$$F_{fibre} = \left(\frac{\sigma_1}{S_1}\right)^2 + \overline{\tau} = 1 \tag{4}$$

where E_1 and E_2 are the longitudinal and transverse elastic moduli, respectively, G_{12} is the shear modulus, and ν_1 and ν_2 are the in-plane Poisson's ratios. C_2 and S_2 are the transverse compressive strength and shear strength, S_1 is the longitudinal tensile strength and S_{12} is the in-plane shear strength, and α is the nonlinear shear stress parameter.

2.1.2. MAT_54-55: Enhanced Composite Damage Model

These material models are improvised versions of the Chang–Chang composite damage model. The keyword for this model is *MAT_ENHANCED_COMPOSITE_DAMAGE or *MAT_054 or *MAT_055. This model is used for thin shells only. When the model is undamaged, the material is assumed to be orthotropic and linear elastic, and when the damage occurs, nonlinearity is introduced into the material. Material 54 is suggested by Chang, which is called the Chang matrix failure criterion, and material 55 is suggested by Tsai-Wu, which is called the Tsai-Wu matrix failure criterion. These two models have very similar formulations.

Material 54 is the same as material 22 but with added compressive fibre failure mode, and it also includes compressive and tensile fibre failure and compressive and tensile matrix failure modes.

The Chang–Chang criterion (MAT_54) is given below,

Tensile fibre ($\sigma_{11} > 0$).

$$\left(\frac{\sigma_{11}}{S_1}\right)^2 + \overline{\tau} = 1 \tag{5}$$

All moduli and Poisson's ratios are set to zero when the tensile fibre failure criteria are met, that is $E_1 = E_2 = G_{12} = \nu_{12} = \nu_{21} = 0$. All the stresses in the elements are reduced to zero, and the element layer has failed. Where E_1 and E_2 are the longitudinal and transverse elastic moduli, respectively, G_{12} is the shear modulus, ν_{12} and ν_{21} are the in-plane Poisson's ratios.

Failure mode for compressive fibre ($\sigma_{11} > 0$),

$$\left(\frac{\sigma_{11}}{S_{12}}\right)^2 = 1 \tag{6}$$

For this mode, $E_1 = \nu_{12} = \nu_{21} = 0$

Failure mode for tensile matrix ($\sigma_{11} > 0$),

$$\left(\frac{\sigma_{22}}{S_2}\right)^2 + \overline{\tau} = 1 \tag{7}$$

For this mode, $E_2 = G_{12} = \nu_{21} = 0$

Failure mode for compressive matrix

$$\left(\frac{\sigma_{22}}{2S_{12}}\right)^2 + \left[\left(\frac{C_2}{2S_{12}}\right) - 1\right]\frac{\sigma_{22}}{C_2} + \overline{\tau} = 1 \tag{8}$$

where $\overline{\tau}$ and other parameters are defined in the above section. For this mode, $E_2 = G_{12} = \nu_{12} = \nu_{21} = 0$. For brittle material, when the failure criterion is met, a reduction factor is applied to reduce the compressive fibre strength, and a softening factor is used to reduce tensile fibre strength.

When the nonlinear shear stress parameter is set to 0, all the above failure criteria reduce to the original failure criteria of Hashin [43].

Material model 55 formulation is very close to the material model 54. It uses the Tsai-Wu failure criteria [50] for compressive and tensile matrix failure modes, which are given in single expression as follows:

$$\frac{\sigma_{22}^2}{C_2 S_2} + \left(\frac{C_2}{S_{12}}\right)^2 + \frac{(C_2 - S_2)\sigma_{22}}{C_2 S_2} = 1 \tag{9}$$

This material model (Mat_055) is similar to the Chang–Chang failure model except for the compressive and tensile matrix failure mode, which is replaced with the above expression and transverse shear parameters in this material model.

In the material input, additional parameters such as effective failure strain and maximum strains are required besides strengths. When the strains values are reached, then the element is deleted. The element is removed when failure occurs in all the composite layers, and these layers are defined through shell thickness integration points.

Elements having the same nodes with the deleted elements become "crashfront" elements. By using a softening reduction factor, the strengths can be reduced and become moduli of the crashfront elements; this fact results in a stable crushing process and, consequently, a sudden release of stress concentration is compensated. To understand the strain parameters, a four-node single-shell element undergoes a tensile load in the direction of the fibre, see Figure 1. The material and strength properties are taken from [51]. SOFT takes into account the softening strength effects on the crashfront elements by scaling down the initial input strength.

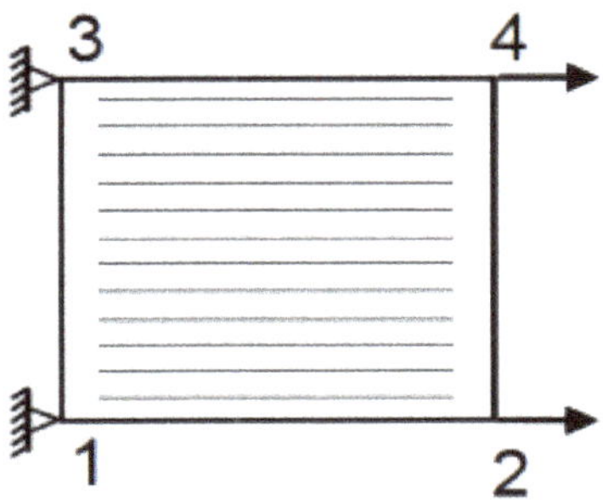

Figure 1. Single 4-node shell element under tension [51].

Initially, the element is loaded at a constant strain rate of 1/s in the fibre direction. The stress in the element increases linearly in the fibre direction up to the maximum value, and all elastic properties and stresses are reduced to zero in 100 time steps, as shown in Figure 2. The maximum strain for fibre tension is set to a default value of 0.

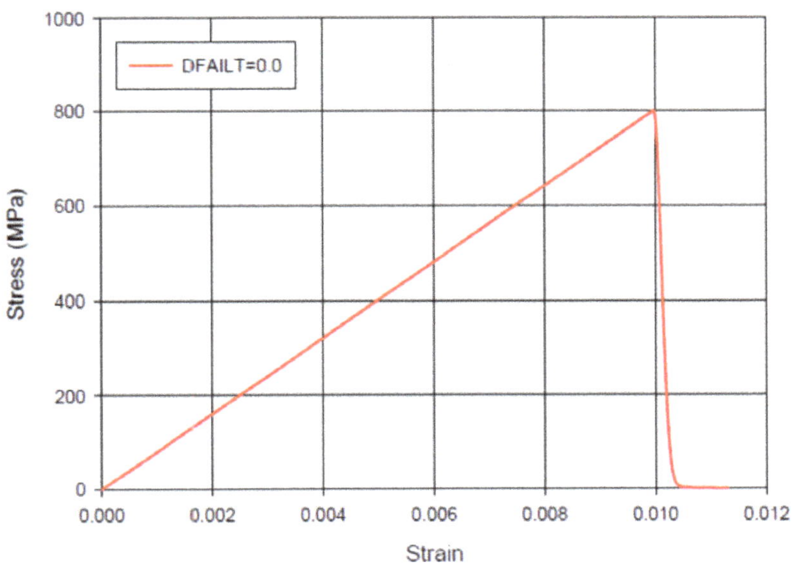

Figure 2. Stress–strain curve in fibre direction under tension, DFAILT = 0.0 [51].

Then, all the elastic properties and stresses are kept constant except the maximum strain for fibre tension, which is set to 0.02. After the value of maximum stress is reached, the elastic constants and the stresses remain constant until the maximum strain value is reached, then the element is deleted, immediately resulting in the stress being reduced to zero, as shown in Figure 3.

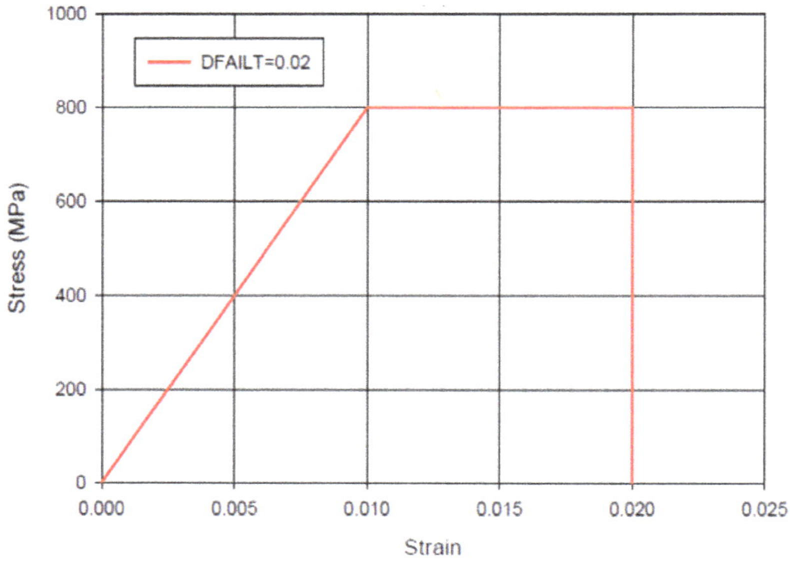

Figure 3. Stress–strain curve in fibre direction under tension, DFAILT = 0.02 [51].

Stresses are kept constant when compressive matrix and fibre criteria are met. In material models 54 and 55, ultimate failure can occur in any four different ways:

1. Chang–Chang failure criterion is satisfied in tensile fibre mode,
2. Maximum fibre tensile strain is met,
3. Maximum effective failure strain is met,
4. Minimum time step is met.

Softening Reduction Factor (SOFT)

A crashfront parameter is used to reduce the crush failure. This parameter is the softening reduction factor for the element material strength that nodes share of the crushed element. By default, this value is set to 1, which means that the element contains 100% of its strength, and when this value drops to 0.5, it indicates that the raw elements that share the same nodes of crashfront elements have only 50% of its original strength.

In numerical parameters, SOFT is considered as one of the influential parameters that could amend the shape of the force–displacement curve to match with the experimental results. For every geometric structure, this parameter value needs to be amended through trial and error to obtain a good agreement with the numerical and experimental force–displacement results. The SOFT parameter can be found in materials 54, 55, 58 and 59. This parameter can be activated by giving a positive value for TFAIL in materials 54 and 55, which is the time step size for element deletion, and by giving a positive value for TSIZE in materials 58 and 59, which is the time step for element deletion. When this time step is reached, the element is deleted. When the degree of curvature of the structure is higher, then the structure is more efficient in crushing by fragmentation. For lower value, the structure gives frond formation. These large fronds are accompanied by long delamination, which results in creating an effective damage length that is inefficient for energy absorption. Francesco Deleo, Wade, Paolo Feraboli [8] simulated the model using material 54 in LS-DYNA and noted that by using the SOFT parameter, the damage length can be changed by this parameter to reduce the material strength of the row of the element that is ahead of the crashfront. Sivarama Krishnamoorthy [52] showed that the crashfront parameter influences the mean crushing force in the force–displacement curve. Therefore, it can be concluded that the SOFT parameter significantly influences the amount of energy absorption. As the value of SOFT increases, more energy is absorbed by the composite structure due to the higher strength of the structure. Other authors claimed the similar conclusions [53–55]

2.1.3. MAT_58: Laminated Composite Fabric Failure Model

Based on the strain-based failure surface, this model can be used for modelling composite materials, which have unidirectional layers, woven fibres and laminates. The keyword for this model is *MAT_LAMINATED_COMPOSITE_FABRIC or *MAT_058. This model is used for shell elements only. This model is implemented in LS-DYNA by Matzenmiller, Lubliner and Taylor [56], which is also called an MLT composite model and is based on a plane stress continuum damage mechanic model.

In material 58, Hashin failure criteria [49] are used with changes for different types of composites. The maximum effective strain is applicable for element layer failure for any different type of composites.

2.1.4. MAT_59: Composite Failure Model

This material model is also called an elastic-plastic material model, which is an enhanced version of Mat_022. It works on the basis of failure surfaces, which are: faceted failure surfaces and ellipsoidal failure surfaces. It will be able to model the progressive material failure due to many failure criterions, which includes longitudinal and transverse directions in tension and compression, respectively, and through-thickness direction in compression and shear.

2.2. Delamination Models

Delamination modelling has several approaches in LS-DYNA. Tiebreak contacts have been vastly used, and it is proven to be a robust contact algorithm and relatively simple. Depending on the model of study, different contacts can be employed to achieve better prediction.

One-way contact types allow for compression loads to be transferred between the slave nodes and the master segments. Tangential loads are also transmitted if relative sliding occurs when contact friction is active. A Coulomb friction formulation is used to transition from static to dynamic friction. This transition requires that the static friction coefficient is larger than the dynamic friction coefficient and a decay coefficient is defined. The one-way contact is used to indicate that only the user-specified slave nodes are checked for penetration of the master segments.

The algorithm ties nodes that are initially in contact by creating a linear spring, and the debonding of the surface initiates when the maximum stress criterion is met, which leads to scaling down of the stress by a linear damage curve until the critical separation is reached and the spring is removed [47].

$$\left(\frac{\sigma_n^2}{NFLS}\right)^2 + \left(\frac{\sigma_s^2}{SFLS}\right)^2 = 1 \quad (10)$$

In which σ_n and σ_s are the normal and shear stresses acting at the interface, and *NFLS*, *SFLS* and *PARAM* are the normal and shear strength of the tie and critical distance, respectively. Once the damage is initiated, the two surfaces begin to separate, and the interfacial stresses are then scaled down as a linear function of the separation distance. *PARAM* is the critical distance at which the failure occurs (i.e., deletion of tiebreak and advancing of delamination) [53].

$$PARAM = \frac{2 \times E_{tie}}{S} \quad (11)$$

where:

$$S = \sqrt{\max{(\sigma_n, 0)}^2 + (\sigma_s^2)^2} \quad (12)$$

Due to the failure of the tiebreak interface, E_{tie} is the energy released. With trial and error procedures, a sensitivity study was conducted of Mode-I and Mode-II to determine their relative effect(s) on the tiebreak failure process. It can be noted that for composite crushing simulations, Mode-I fracture is a dominant mode of failure during the tiebreak failure process. Thus, to simplify the simulations, a pure Mode-I delamination was assumed.

$$G_{IC} = \frac{1}{2}\, \sigma_n\, PARAM \quad (13)$$

In Equation (13), the critical normal separation of the surface is determined, named *PARAM*, based on the energy release rate in Mode-I (G_{IC}) and the critical normal stress.

The laminate of the open cross-sections was modelled as multi-shell configurations of shell elements, with each layer representing a various number of plies. In Section 4.7, the effect of the number of plies in each shell was investigated based on energy absorption capability. Hence, the tiebreak was adopted for each case of study and the tiebreak contact was defined only between these shell layers, rather than between individual plies. However, delamination could occur along any of the plies, and if not all ply interfaces, during specimen crushing, as was observed experimentally [56–66]. To account for the energy dissipated by these additional delamination interfaces, *PARAM* was scaled by the ratio of the number of ply interfaces $n_{delamination}$ to the number of tiebreak interfaces n_{tie} defined as:

$$PARAM' = PARAM \times \frac{n_{delamination}}{n_{tie}} \quad (14)$$

based on the experimental observations, it was assumed that delamination occurred among all plies. The values that were used and calculated in Equations (13) and (14) are listed in Table 1.

Table 1. Tiebreak input parameters [59–66].

E_1 (GPa)	E_2 (GPa)	G_{IC} (kJ/m^2)	G_{IIC} (kJ/m^2)	*NFLS* (MPa)	*SFLS* (MPa)	*PARAM* (mm)	*PARAM'* (mm) (2 Shells)
39 ± 3	11.8 ± 1	1.2	0.9	22.5	57.3	0.025	0.15

2.3. Laminate Stiffness

Laminate stiffness is critical in designing composite structures. In crashworthiness, this can be a critical factor due to the complexity of energy transfer and dissipation through the structure against occupant safety. Four major categories are considered during the composite design approach,

1. Cost,
2. Mass,
3. Stiffness,
4. Geometrical stability.

Stiffness can be calculated from load–deflection graphs:

$$Stiffness = \frac{Total\ Load}{Deflection} \quad (15)$$

Equation (15) can help with experimental studies and obtaining the general performance from a specimen. However, due to the orthotropic behaviour of composite materials, the stiffness of a laminate, either symmetric or antisymmetric, depends on the angle of fibres, materials, and thickness of the plies, which can zero out some elements of the three stiffness matrices [A], [B] and [D]. This is known as ABD Matric and is used to calculate composite laminate stiffness. The stiffness matrix in the principal directions is:

$$[C] = \begin{bmatrix} \frac{E_1}{1-v_{12}v_{21}} & \frac{v_{12}E_2}{1-v_{12}v_{21}} & 0 \\ \frac{v_{21}E_1}{1-v_{12}v_{21}} & \frac{E_2}{1-v_{12}v_{21}} & 0 \\ 0 & 0 & G_{12} \end{bmatrix} \quad (16)$$

These elements may result in reducing or zeroing out the coupling of forces, normal and shear forces, and twisting and bending moments. Hence, a balanced lay-up sequence can reduce or zero bending moments in a laminate, which improves mechanical performance under various loading conditions. It can be proved that the coupling matrix [B] = 0 for symmetric laminates, and hence, the force and moment equations can be decoupled.

$$\begin{bmatrix} N_x \\ N_y \\ N_{xy} \end{bmatrix} = \begin{bmatrix} A_{11} & A_{12} & A_{16} \\ A_{12} & A_{22} & A_{26} \\ A_{16} & A_{26} & A_{66} \end{bmatrix} \begin{bmatrix} \varepsilon_x^0 \\ \varepsilon_y^0 \\ \gamma_{xy}^0 \end{bmatrix} \quad \begin{bmatrix} M_x \\ M_y \\ M_{xy} \end{bmatrix} = \begin{bmatrix} D_{11} & D_{12} & D_{16} \\ D_{12} & D_{22} & D_{26} \\ D_{16} & D_{26} & D_{66} \end{bmatrix} \begin{bmatrix} k_x \\ k_y \\ k_{xy} \end{bmatrix} \quad (17)$$

A laminate is called quasi-isotropic if its extensional stiffness matrix [A] behaves similar to that of an isotropic material. This not only implies $A_{11} = A_{22}$, $A_{16} = A_{26}$, and $A_{66} = (A_{11} - A_{12})/2$ but also that these stiffnesses are independent of the angle of rotation of the laminate. It is called quasi-isotropic and not isotropic because [B] and [D] may not behave as an isotropic material. Examples of quasi-isotropic laminates include [0/±60], [0/±45/90]s, [0/36/72/−18/−54].

3. Simulation Setup

For the simulations, an Explicit FE LS-DYNA code is used with a multi-layered shell configuration to reduce numerical costs. Composite tubes were modelled as multi-layers of Belytschko–Tsay circular shell elements with one integration point in the element plane to represent the direction of the stacking sequence. In double-shell configuration, the GFRP innermost shell has six integration points, with another six integration points being assigned to the outermost shell to represent all twelve UD layers. In a GFRP tube, each individual layer has a thickness of 0.25 mm. The total thickness of both shells is 3 mm. Each fibre orientation is assigned with insertion of an integration point with respect to the stacking sequence used with its associated thickness. The material properties are obtained from [56,59,60]; see Table 2.

Table 2. Material properties of GFRP (TenCate E772) [56–60].

GFRP (TenCate E772)							
E_1 (GPa)	E_2 (GPa)	G_{12} (GPa)	v_{12}	σ_u 0° (MPa)	σ_u 90° (MPa)	τ_s (MPa)	V_f (%)
39 ± 3	11.8 ± 1	3.2 ± 0.5	0.29	836 ± 20	29 ± 2	97 ± 4	58

In shell theory, the thickness of the shell is considered as mid-plane. In double-shell configuration, two shells with radiuses of 37.75 and 39.25 mm to represent the inner and outer shells with lengths of 80 and 77.5 mm were modelled, respectively, using LS-PrePost representing the GFRP tube geometry. Each shell was glued to itself so that the triggering at the top of the shell would not detach during the crushing process, as a separate shell was used. In this triggering approach, two shells were used, one with 2.5 mm in height acting as the trigger (one element size), and the other, depending on whether it was the inner or outer shell, had its representative height assigned. Therefore, the top shell in each FEA case study represents the trigger. The quadrilateral shell element was used with each element size of 2.5 × 2.5 mm. The trigger mechanism was modelled by reducing the first-row thickness of the shell elements to represent the bevel trigger, from 1.5 to

0.05 mm in each shell. A solid element rigid block was modelled to represent the striker. The LS-DYNA material model of Enhanced_composite_damage (Mat_54-55), which is an orthotropic material, with the failure criterion of Chang–Chang was used. This failure criterion is a modification of Hashin's failure criterion for assessing lamina failure. The hourglass was set at 10% [3–7,43,53].

Modelling interlaminar separation or delamination failure (Mode-I) requires either detailed experimental investigation for the cohesive zone or three-dimensional representation that both result in an increase in computational and experimental costs. Delamination failure causes energy absorption, and this can be modelled with a multi-layered shell configuration with a contact card that is capable of a G_{IC}-implemented energy release rate [40–43]. The One_Way_Surface_To_Surface_Tiebreak contact between the two shells is defined, with the inner tube being the master and the outer being the slave.

Tiebreak contacts allow the modelling of interlaminar debonding, which transmits both compressive and tensile forces with optional failure criteria. The tiebreak option enables the detachment of the contact surfaces by creating springs between two surfaces, and after reaching maximum normal stress (*NFLS*) or shear stress (*SFLS*), if the failure parameter, driven by occurring normal and shear stresses, become 1, the contact forces soften linearly until contact distance *PARAM* is reached and the interface failure is completed. Based on the interlaminar utilisation of the contact, the parameters are determined by the mechanical properties of the matrix material. Consequently, shell layers detach when the interlaminar stress exceeds the matrix properties, which are mainly responsible for interlaminar strength. Maximum normal and shear contact stresses for the tiebreak contact is based upon the mechanical properties of the epoxy resin. The maximum contact distance is set to 0.15 mm. Automatic_Node_To_Surface contact was defined for the striker and inner shell, with the striker being the master and the inner shell being the slave. The Automatic_Single_Surface contact algorithm was utilised. This prevents penetration of the crushing tube by its own nodes.

To satisfy quasi-static conditions, it is important that the load is applied in a manner that would yield a minimal inertial effect on the results, and the ratio of the kinetic energy to the internal energy must be reasonably small. Time-scaling was utilised to apply the load at a higher rate to reduce total simulation time. A constant loading rate of 0.65 m/s was applied, and the kinetic energy to the internal energy was less than 10% upon initial contact and less than 5% throughout the remainder of the crushing process.

All bottommost nodes of all shell element layers are constrained in their translational degrees of freedom. The impactor is modelled rigidly with a mass of 108.4 kg and a velocity of 7.022 m/s. Gravity is modelled with an acceleration factor of 9.81 m/s^2. All simulation results are smoothed using an SAE 300 Hz filter [43].

4. Finite Element Modelling

4.1. Delamination Interface

Many researchers have used friction to simulate delamination, e.g., [56–60]. Friction influences the energy absorption capability; however, using friction influences the SEA value, increases friction between the shells, and causes a higher SEA value (see Section 5.3), and this compared with experimental data cannot be considered as a correct FEM. Due to this, a different approach was considered. Tiebreak option 8 was utilised instead of friction to model delamination as this contact card can define the Mode-I and Mode-II energy release rate, which simulates delamination.

The tiebreak contact definition implemented in LS-DYNA allows for the simulation of delamination at the interface between adjacent shell element layers. Tiebreak Option 8 formulations were investigated for this study, namely, tiebreaks with a bilinear traction-separation law. This requires interlaminar normal and shear strengths and a critical distance to interface failure as input parameters to model delamination in crush simulations [10]. However, the optimal critical failure distance parameter selection has not been thoroughly studied in the open literature. The formulation of required input parameters, such as interlaminar normal and shear strengths, fracture toughness under pure Mode-I and Mode-II loading, and interfacial stiffness for normal and shear modes; a description of the model setup using each formulation is explained in Section 4.3.1. To determine the energy rate of Mode-I and Mode-II, DCB and 3ENF tests were carried out.

DCB test determines the Mode-I energy release rate (G_I) for delamination growth. Unidirectional GFRP with resin E722 was used to manufacture 20 by 150 mm specimens with stacking sequence of (−45, 45, 0, 90)s. A thin Teflon film is placed at the mid-plane to avoid bonding and develop a pre-determined crack at 60 mm. The samples are cured at 120 degrees Celsius for an hour and a half under composite vacuum pressure in a sealed bag wrapped in breather cloth and two metal sheets to evenly apply pressure on the laminates. The delaminated end of the specimen is attached to two metal tabs. Load is then applied on the metal tabs to create a crack growth on the specimen at a rate of 2 mm/min (quasi-static). This is carried out by taking the difference in crack lengths. Unidirectional GFRP with resin E722 was used to manufacture 20 by 150 by 6 mm specimens according to ESIS standard with a stacking sequence of (−45, 45, −45, 45, 0, 90, 0, 90, 0, 90, 0, 90)s to carry out three end-notched flexure tests. The input parameters of tiebreak contact option 8 are shown in Table 2.

4.2. Boundary Conditions and Contact Definitions

The loading striker was modelled as a rigid body. The tubes were placed in the Z-direction upright and the loading striker at the chamfered end of the tube. The interaction between the loading striker and the tube was modelled using a node-to-surface contact definition (automatic contact from node to surface). The tiebreak contact definition between the shell layers not only facilitates the simulation of delamination but also prevents layers from penetrating each other after the tiebreak has failed, as the contact definition would remain in effect. In summary, Automatic_Node_To_Surface contact was defined for the striker and inner shell with striker being the master and inner shell being the slave. The Automatic_Single_Surface contact algorithm was utilised. This prevents penetration of the crushing tube by its own nodes, which is due to the sticker's nodes potentially causing disturbance to the model and the inner shell penetrating its own nodes and elements. All bottommost nodes of all shell element layers are constrained in their translational degrees of freedom.

4.3. Material Model

Material models Mat_022, Mat_054-055, Mat_058 and Mat_059 were used to capture an ideal initial peak and mean crushing force against computational costs. These parameters determine the reliability and the ability of these material models in cylindrical composite structures. The initial peak illustrates the stiffness of the material, and the mean crushing force shows the progressive crushing behaviour. In this section, the computational cost is one of the main parameters of case consideration. The time taken for the simulation to converge against the extracted results can be compromised to ideally have a model that converges within a reasonable timeframe and its effect on the extracted results. All bottommost nodes of all shell element layers are constrained in their translational degrees of freedom.

4.3.1. Material Modelling of Mat-045-055

This model allows the user to create a local material coordinate system to specify the orientation of each ply. There are 21 parameters in Mat_54 that need to be specified, 15 of which are physical parameters and 6 are numerical parameters [47]. From the 15 physical parameters, 10 parameters are material constants, the values of which were obtained from [55,59,60], as shown in Table 1. The remaining five physical parameters are the tensile and compressive failure strains (element deletion strains) in the fibre direction (DFAILT and DFAILC), the matrix and shear failure strains (DFAILM and DFAILS), and the effective failure strain (EFS). The six numerical parameters can be adjusted to yield desired material behaviour. Based on an extensive parametric study, it was concluded that of these six parameters, the crashfront element softening parameter (SOFT) is of key importance to this study. This parameter reduces the strength of elements surrounding a damaged or deleted element.

As mentioned, there are five physical parameters (failure strains) and six numerical parameters in MAT54 whose values need to be determined numerically. A comprehensive parametric study was performed to investigate the effect of these parameters on the simulated load–displacement behaviour. It was determined that the physical parameter DFAILC (fibre compression failure strain) had the greatest effect on the value of initial peak load while the numerical parameter SOFT (crashfront element softening parameter) had the greatest effect on the value of sustained crush load, which determined the value of SEA.

Parameters DFAILT, DFAILM and DFAILS (shear failure strains) were found to have a marginal effect on the results and were kept constant at arbitrarily selected values of 0.02, 0.02 and 0.03, respectively. However, increasing the DFAILM value increased computational cost unreasonably (see Section 6.1). It was found that simulations with DFAILC = −0.004 and SOFT = 0.75 yielded the mean crushing force value and displacement behaviour for chamfered tubes that matched very well with experimental data, as shown in Tables 3 and 4.

Table 3. The parametric study shows the effect of DFAILC in MAT54 on the peak load, crush and SEA of the circular tube [66].

DFAILC	Peak Load (Num/Exp%)	Crush Load (Num/Exp%)	SEA (Num/Exp%)
−0.0100	157.9%	159.2%	166.1%
−0.007	67.2%	86.9%	87.2%
−0.006	80.3%	86.9%	87.2%
−0.005	106.2%	86.9%	87.2%
−0.004	102.2%	86.9%	87.2%
−0.003	94.1%	86.9%	87.2%

Table 4. The parametric study shows the effect of SOFT in MAT54 on the peak load, crush and SEA of the circular tube [66].

SOFT	Peak Load (Num/Exp%)	Crush Load (Num/Exp%)	SEA (Num/Exp%)
0.75	102.2%	103.4%	101.1%
0.80	102.2%	116.2%	125.8%
0.85	102.2%	137.6%	138.8%
0.90	102.2%	153.8%	159.4%

A parametric study was conducted to determine the optimal values of the unknown parameters for the multi-layer modelling approach. The resultant values are presented in the above tables. This developed FEM, with the SOFT parameter set to 0.75 (75%) and DFAILC set to −0.004, can produce an accurate prediction of experimental results. The DFAILC value is negative due to compression.

4.3.2. Material Model Results

Figure 4 shows a comparison of the material models in the previous sections. MAT-022 has an initial peak of 141 kN with a mean crush force of 13 kN, and the displacement reaches 70 mm with a computational cost of 14 h. In comparison with the experimental data, this material model is inaccurate at predicting the experimental material behaviour.

MAT_054-055, on the other hand, illustrates an ideal prediction of material prediction, with an initial peak value of 80 kN, mean crushing force of 67 kN, displacement of 33 mm and computational cost of 28 h. In comparison with the experimental data, which included an initial peak of 78 kN and mean crush force of 69 kN, Mat_054-055 was on average 5% off. Mat_058 illustrated that it over-predicts the mean crushing force by 7 kN, although the initial peak value has been improved to 79 kN compared with Mat_054-055. The displacement reduced to 31 mm and the computational cost increased to 36 h. The difference between the two material models lays in the mean crushing force, and Mat_058 over predicts, which results in a reduction in the displacement value, resulting in greater difference to the experimental data, with a 7.5% overall difference. Mat_059, which is a modified version of Mat_022, has shown greater improvement in the prediction of crushing behaviour in composites. However, the initial peak is 10 kN off, the mean crushing force value is underpredicted by 11 kN, and the displacement value is 36 mm. The overall difference is 9%. The computational cost value is 45 h.

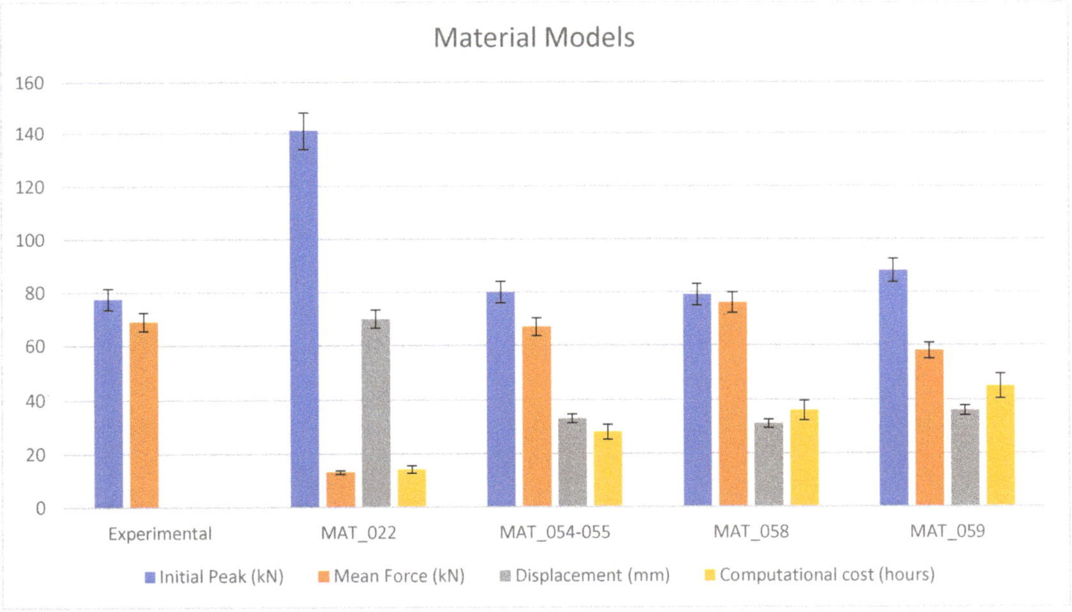

Figure 4. Material model comparison [66].

By considering all four reviewed parameters, Mat_054-055 can predict the material behaviour with respect to the energy absorption capability and having a reasonable computational cost compared with the extracted results, as predicted by [3–7,43,53,59,60].

4.4. Element Formulation

The possibility of using under-integrated elements results in a reduction in computational cost and compromises the accuracy of the prediction. To compare the performances, several relevant element formulations were employed, and the results have been discussed. The relevant chosen element formulations are Hughes–Liu, S/R Hughes–Liu, S/R co-rotational Hughes–Liu, fully integrated and Belytschko–Tsay (default). In consideration of this test, since energy absorption capability is the main concern, the parameters chosen are the initial peak, mean crush force and displacement; however, in numerical studies, the computational cost plays a major role, and therefore, this parameter is also considered.

Figure 5 shows the differences between the chosen element formulations against the reference parameter of experimental data. The result of Hughes–Liu element formulation showed an initial peak of 82 kN, mean crush force value of 90 kN with displacement of 27 mm, and computational cost of 105 h. The total difference in performance value was 13% compared with experimental data. The extracted result from S/R Hughes–Liu showed an initial peak of 81 kN, mean crush force of 65 kN and displacement of 34 mm, and the computational cost was 115 h. The mean difference from the experimental data was 7%.

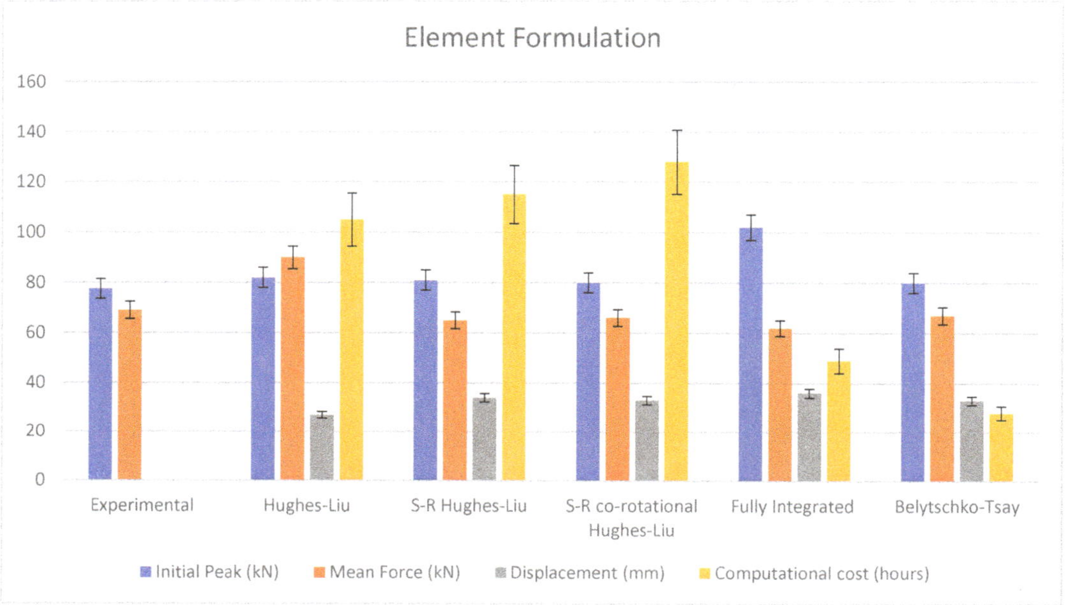

Figure 5. Element formulation comparison [66].

Both element formulations have staggering computational costs with around 4 to 5 days to converge each of the simulations. The results obtained from S/R co-rotational Hughes–Liu showed an initial peak value of 80 kN, mean crush force of 66 kN, displacement of 33 mm and computational cost of 128 h. Fully integrated element formulation, on the other hand, was off by 24 and 7 kN in the initial peak value and mean crush force, respectively. Although the computational cost is much lower compared with mentioned element formulations, both element formulations were off by 6% and 12%, respectively.

The Belytschko–Tsay element formulation, which is the default parameter in LS-DYNA, was rather close to the experimental data with a computational cost of 28 h. The initial peak value was 80 kN, mean force value was 67 kN, and the displacement was 33 mm, with a total mean difference of 5%.

By taking into account all four reviewed parameters, the Belytschko–Tsay element formulation is the cheapest computational cost compared with other element formulations, as supported by other researchers [3–7,43,53,59,60], which leads to the use of this type of element formulation hereafter.

4.5. Mesh Size

In numerical modelling, one of the influential parameters is the mesh size. The mesh sensitivity test is beneficial for establishing a mesh size regarding a specific model to obtain an acceptable accuracy. In a numerical study, compromising acceptable accuracy for computable costs is also relatively important. Six different element sizes were modelled with 5.5, 4.5, 3.5, 2.5, 1.5, 0.5 mm quad elements (see Figure 6).

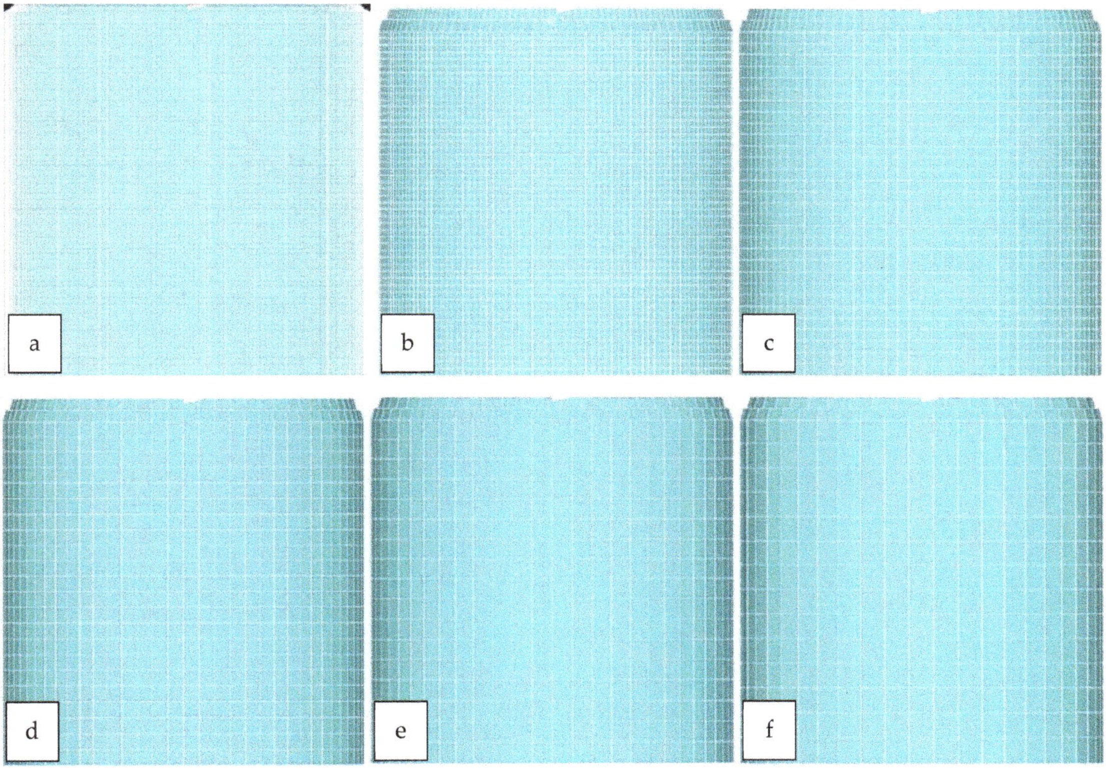

Figure 6. Mesh sensitivity models. (**a**) 0.5, (**b**) 1.5, (**c**) 2.5, (**d**) 3.5, (**e**) 4.5, (**f**) 5.5 mm [66].

From a design point of view, the aim is to achieve the cheapest model in terms of computational cost and predict energy absorption capability with an acceptable accuracy. As the mesh becomes increasingly finer, the computational cost increases dramatically, and relatively higher accuracy is achieved. A balance of the two needs to be chosen in that the energy absorption capability of the model is within an acceptable range and the computational cost is within an acceptable range.

Figure 7 shows the mesh sensitivity comparison of the modelled mesh sizes mentioned in Figure 6. The results illustrate a noticeable fact that the mesh 5.5 mm size is too coarse with very high peak forces and low mean crushing force, with 86 kN difference between the two. This mesh size has the lowest computational cost; however, the result is nearly 40% off from the experimental data. As the mesh size becomes smaller, the accuracy improves. At 4.5 mm, the results improved from the 5.5 mm case study, although the difference is 29% and at 3.5 mm, the difference is 18%. The peak is higher than the experimental study by 8 kN, and the mean crushing force is lower by 7 kN.

At a mesh size of 2.5 mm, the result is in line with experimental data. The difference is 5%, and the computational cost is lower than the mesh size of 1.5 and 0.5 mm by 170% and 280%, respectively. Although the accuracy is 1.5% for both cases, the balanced case to accurately calculate and predict energy absorption is 2.5 mm, which is supported by many studies [43,53,59,60].

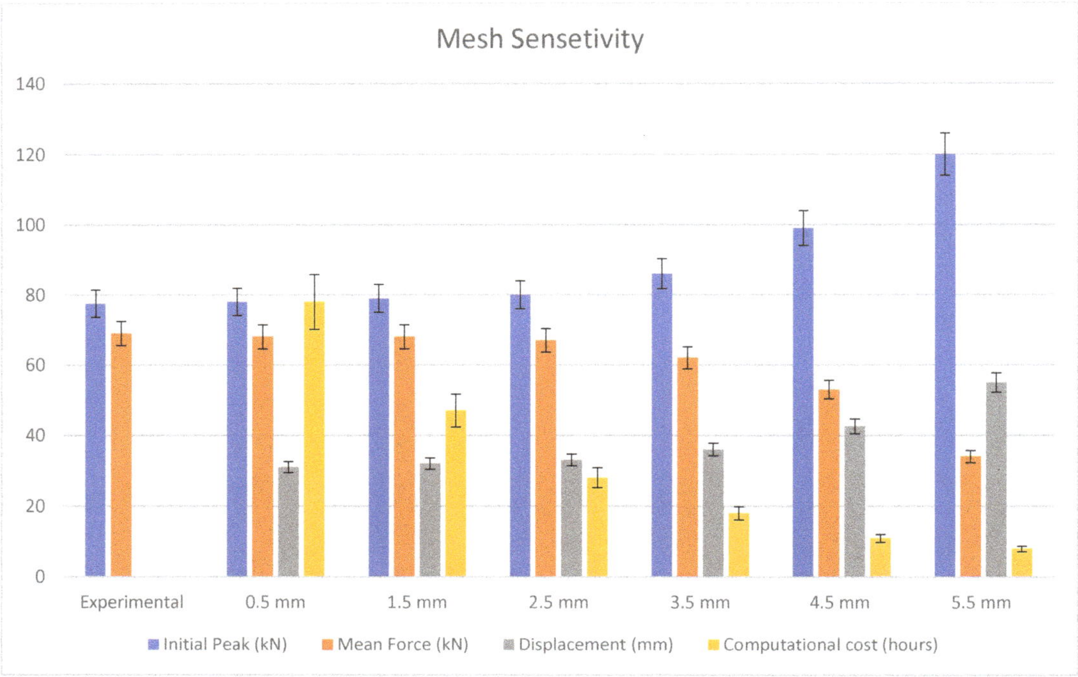

Figure 7. Mesh sensitivity comparison [66].

4.6. Trigger Modelling

In experimental studies, a 45° bevelled trigger mechanism was utilised, and similarly, in numerical studies, a suitable trigger mechanism is needed to initiate the progressive failure that matches the experimental studies. The maximum crush force in the FEA model tends to be overestimated significantly [10,54]. Few approaches were raised to study the effect of different trigger mechanisms on the initial peak value of the load and mean crushing force (see Figure 8). These case studies are, (a) single-shell with no trigger, (b) single-shell inward-chamfer, (c) single-shell outward-chamfer, (d) double-shell level size inward-chamfer, (e) double-shell with 2.5 mm shell size difference inward-chamfer, (f) double-shell with 2.5 mm shell size difference outward-chamfer, (g) double-shell with 2.5 mm shell size difference inward-chamfer with different reduced element sizes, (h) double-shell with 5 mm shell size difference inward-chamfer, (i) double-shell with 2.5 mm shell size difference inward and outward-chamfer.

The results from the case studies are compared in Figures 9 and 10. The aim of this research is to find the optimum finite element modelling case to obtain a prediction of the initial peak force value.

Single-shell configuration with trigger mechanism has shown better peaks and mean crushing force prediction (case b and c) than the one without the trigger mechanism (case a). However, the double-shell configuration has led to better prediction with less than 5% error compared with experimental data. Outward-chamfer in all cases lead to an increase in computational costs with lower mean crushing force. Cases (e) and (f) have shown better prediction of the initial peak values of 80 and 78 kN. The configurations of the two are similar in sectioning and positioning of the shells, with the outer shell being 2.5 mm (one element size) shorter than the inner shell. However, the difference lays in the computational costs, in which the outward-chamfer, as shown in all cases, increases by 5–15 h. Furthermore, the trigger mechanism affects progressive crushing behaviour. The mean crushing force in case (f) has a lower value than the one in case (e).

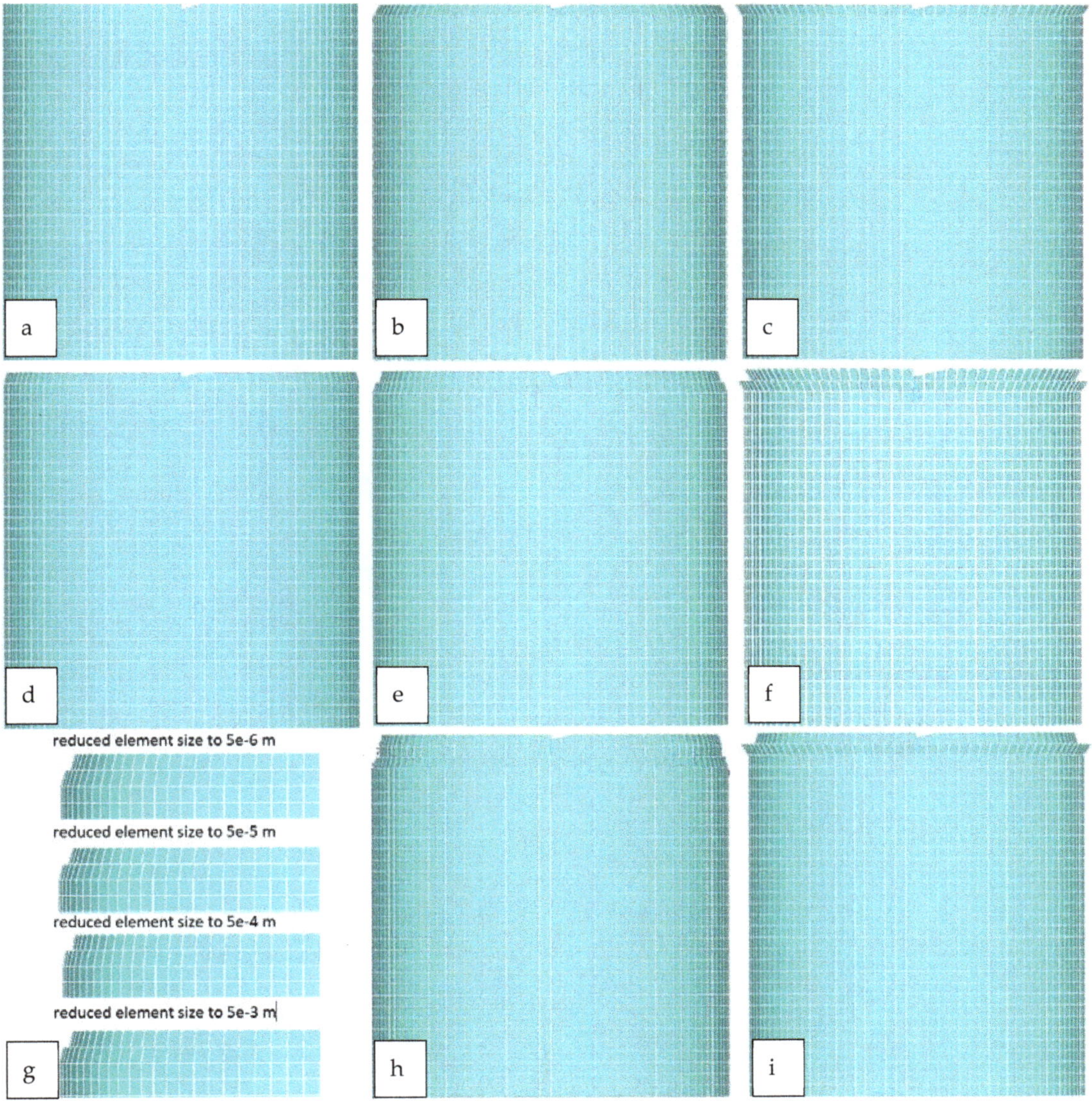

Figure 8. Trigger mechanism modelling cases, (**a**) single-shell no trigger, (**b**) single-shell inward-chamfer, (**c**) single-shell outward-chamfer, (**d**) double-shell level size inward-chamfer, (**e**) double-shell 2.5 mm shell size difference inward-chamfer, (**f**) double-shell 2.5 mm shell size difference outward-chamfer, (**g**) double-shell 2.5 mm shell size difference inward-chamfer different reduced element sizes, (**h**) double-shell 5 mm shell size difference inward-chamfer, (**i**) double-shell 2.5 mm shell size difference inward and outward-chamfer [65].

Case (g) is based on case (e), and the configuration of the trigger is the same. In this trigger mechanism, the element size is reduced to a smaller size, and this acts as the trigger mechanism. This technique is supported by [43,44,59,60]. In Figure 10, the reduced element sizes are compared against their initial peak value and mean crushing force and computational costs. It can be concluded that as the element size becomes thicker, the peak value increases and from 0.05 mm onwards, and this reduction in element size has no effect on initial peak value.

The prevailing case that matches experimental results is a 2.5 mm sectioned double-shell configuration with an inward-chamfering trigger mechanism and a reduced element size of 0.05 mm.

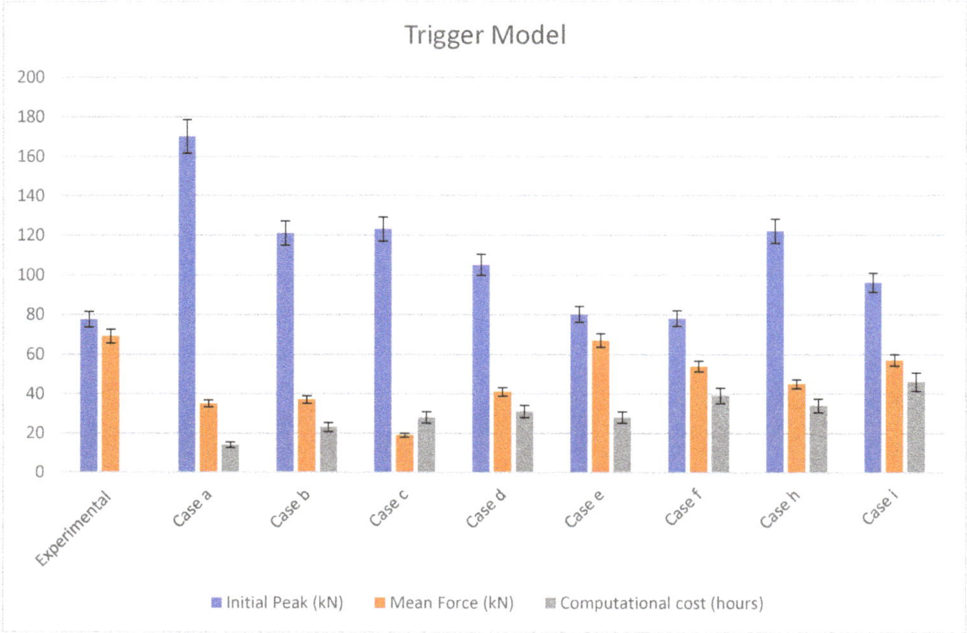

Figure 9. Trigger model comparison [66].

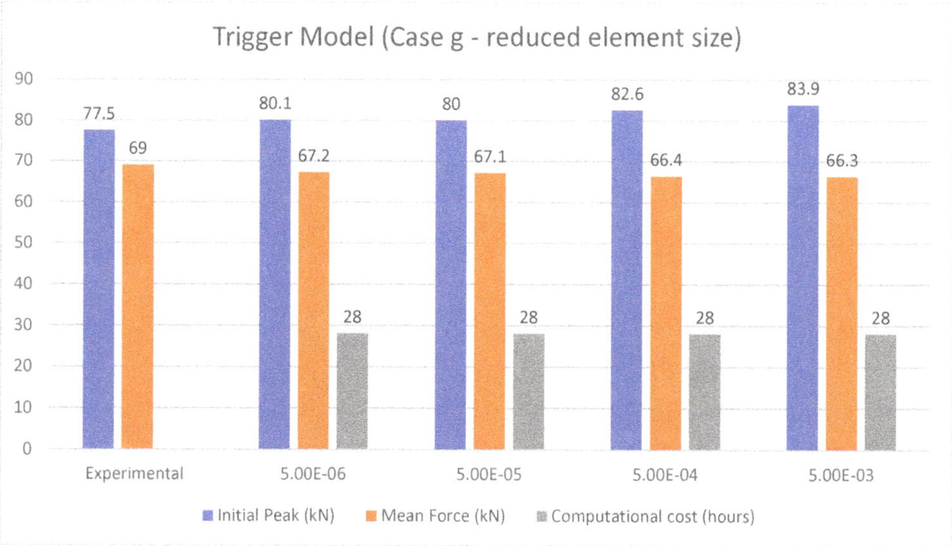

Figure 10. Trigger model case (g) comparison [66].

4.7. Number of Shell(s) Configuration

The effect of the number of shells on the energy absorption prediction of FEM is investigated. The case studies are 1, 2, 3, 4, 6, and 12 shells (see Figure 11).

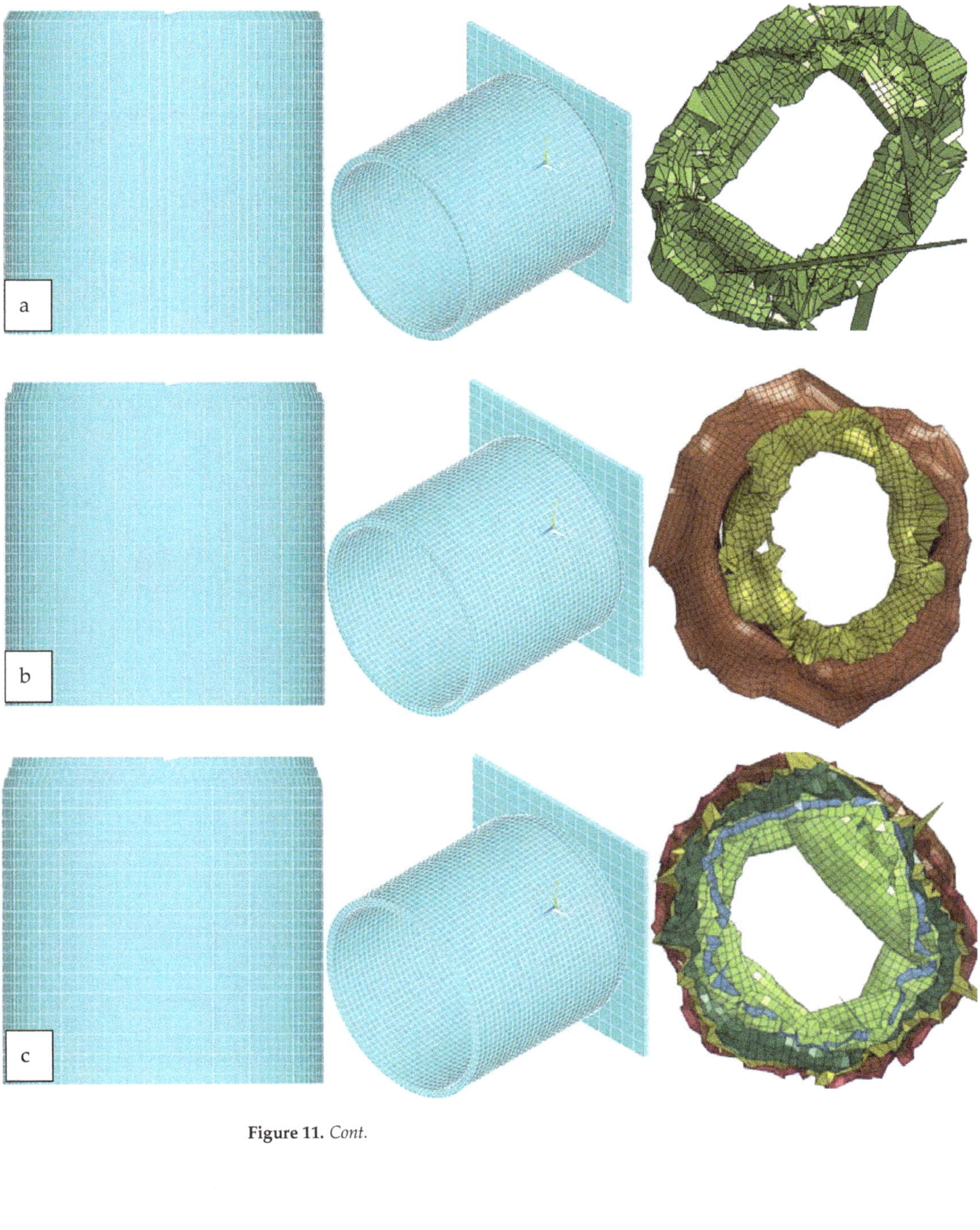

Figure 11. *Cont.*

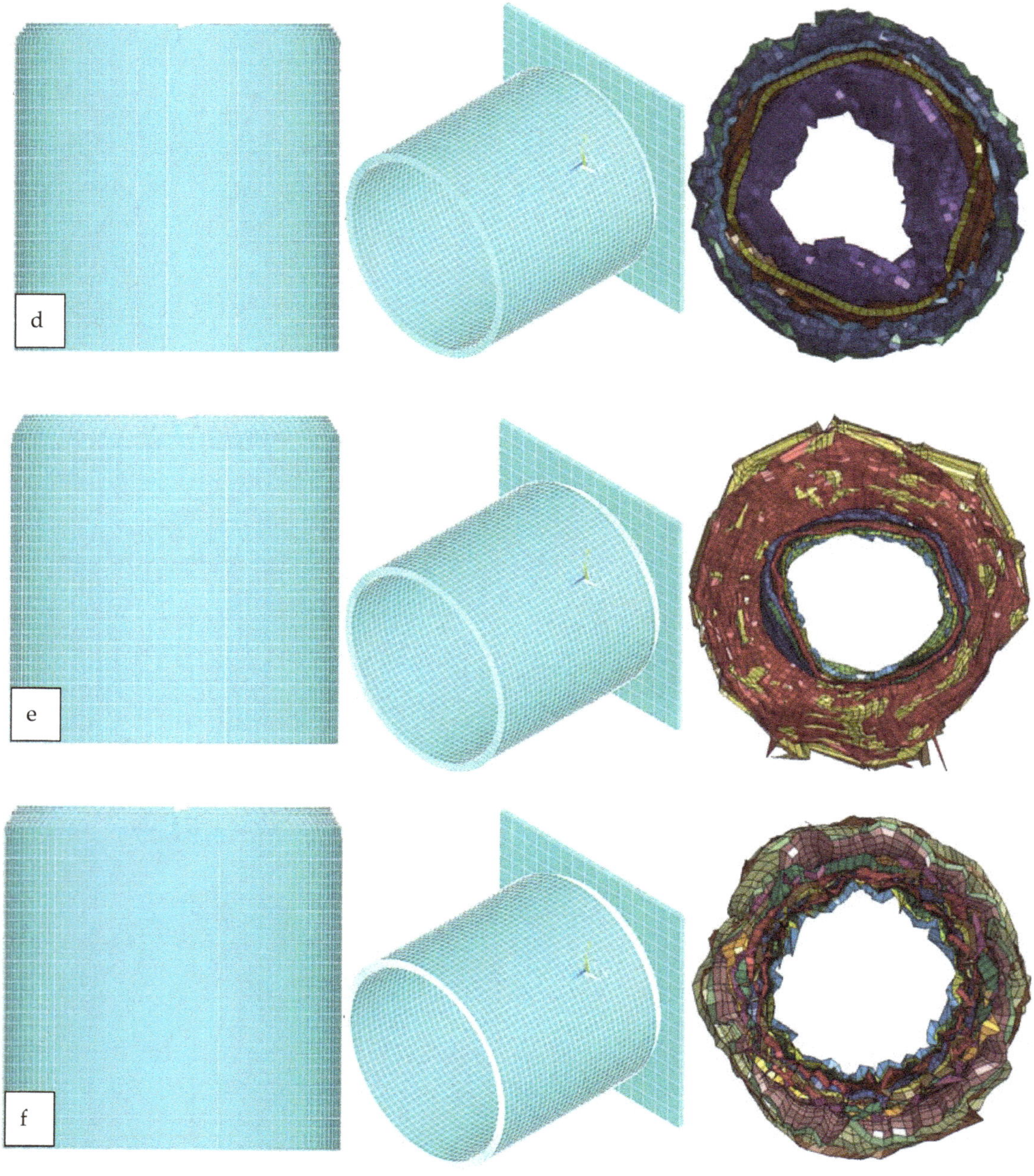

Figure 11. Number of shell configurations, (**a**) 1 shell, (**b**) 2 shells, (**c**) 3 shells, (**d**) 4 shells, (**e**) 6 shells, (**f**) 12 shells [65].

In this study, the crushed morphology is one of the factors that is greatly influenced, as seen in Figure 11. From the crushed morphology point of view, it can be concluded that, as the number of shells increase, the symmetricity and prediction of crushing behaviour improves. Since 12 plies were used in experimental studies, six cases were considered. All cases were subjected to the same trigger mechanism and applied energy.

In Figure 12, the shell configurations are compared. In single-shell configuration, the initial peak and mean crushing force is off by 45%. However, in double-shell configuration, this improves to a 5% difference. The initial peak value is 80 kN with a crushing force of 67 kN. Hereafter, the computational cost is the deciding parameter. Since the crushing behaviour slightly or minimally improves by adding more shells. This improvement is in both the initial peak value and mean crushing force value. However, the computational cost increases rapidly by adding more shells to the model. Using the 12 shell configuration compared with experimental data, it has a 1.5% difference, and using a double-shell configuration, it has a 5% difference. From an engineering point of view, this compromising 3.5% results in solving the problem with 4.5 times less computational costs; one takes 28 h to converge and the other 123 h.

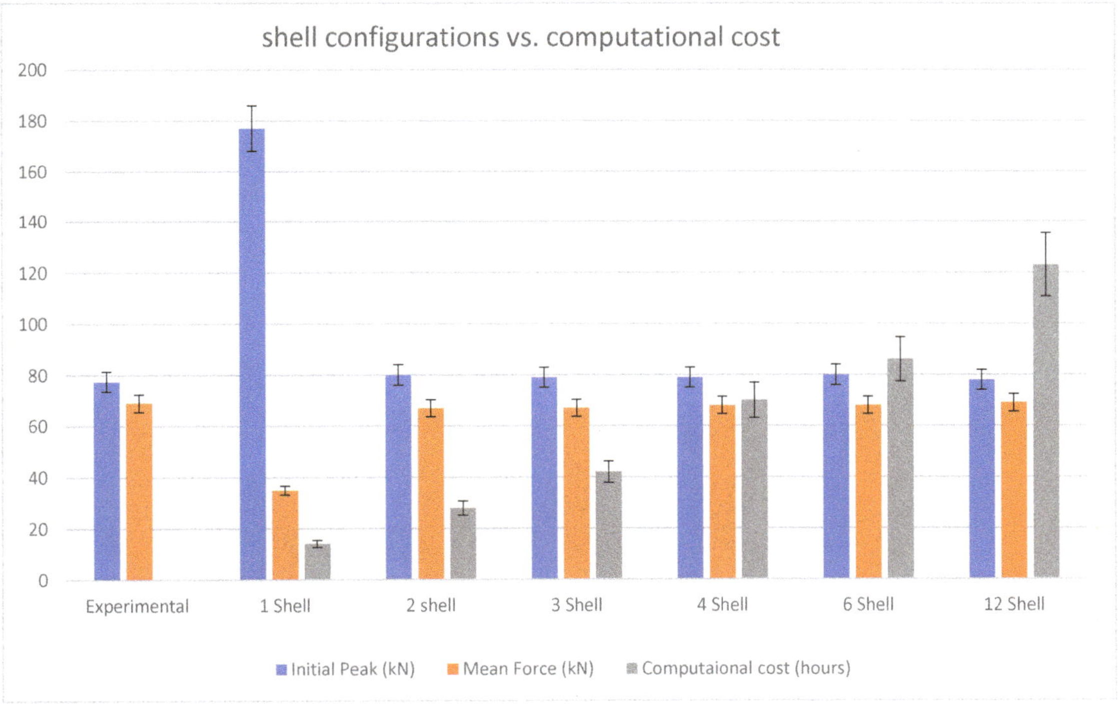

Figure 12. Shell configuration comparison [65].

At this stage, since the main concentration is energy absorption capability, the reliable and cheaper computational cost of double-shell configuration is considered in this study and is also supported by [43,54,59,60].

5. Model Sensitivity to Physical Parameters

A robust finite element model needs to tolerate small variations in modelling parameters and be able to capture the differences. The aim of this section is to carry out a study of stability and sensitivity with respect to the material parameters, delamination, friction and impact loading. The reference model is the one developed earlier in this section and tweaking the parameters by ±10%.

5.1. Material Model

The following parameters from Mat_54-55 in LS-DYNA have been studied: stiffness, compressive strength, strain to failure in compression and strain to failure in tension. The stiffness and strength have an influence on fibre and matrix arrangements, and the strain to failure is a parameter that influences the experimental energy absorbed per unit of crushed volume/mass. The developed model will be compared with the experimental data. The SEA value from the experimental and numerical data should be around the error percentage, which is 5%.

5.1.1. Laminate Stiffness

In Figure 13, the comparison of laminate stiffness is shown with a change of stiffness within ±10%. The SEA, initial peak and average crushing force are the main concentrations of this study compared with the relevant FEM reference. It is noticeable that the computational cost remained the same with minimal change in all cases. The influence of stiffness on crushing behaviour is illustrated in the results. A 10% increase resulted in an increase in the initial peak value by 3 kN, the mean crushing force increased to 71 kN, and the SEA value increased by 4 kJ/kg. The displacement value was affected by a drop of 2 mm. A 10% decrease resulted in a reduction in initial peak value by 2.2 kN, the mean crushing force decreased to 63 kN, and the SEA value dropped by 12 kJ/kg. The displacement value was affected by an increase of 6 mm.

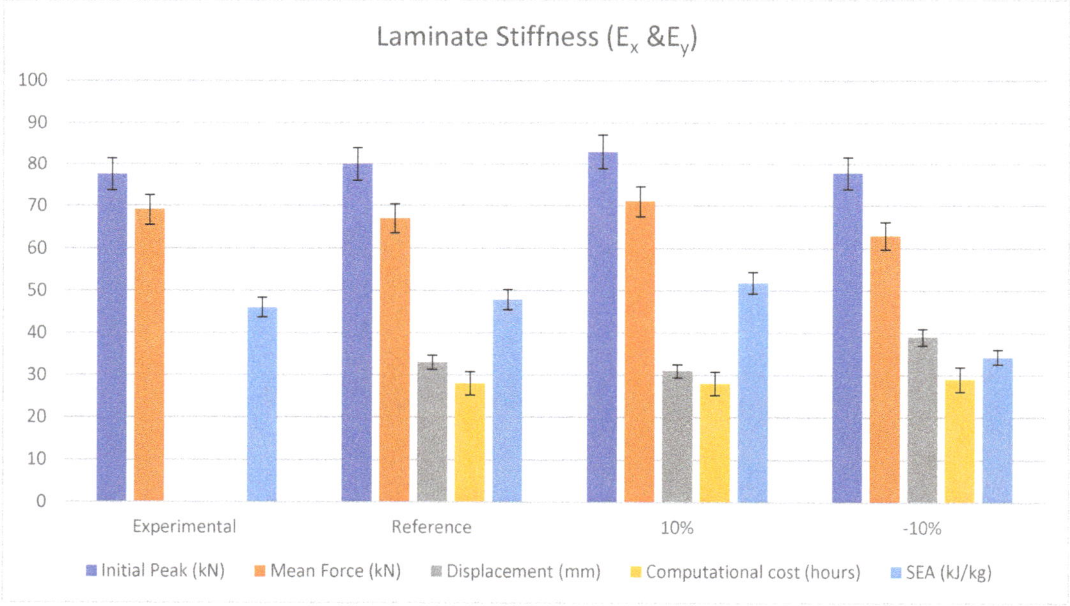

Figure 13. Laminate stiffness comparison [66].

The results showed that an increase in laminate stiffness caused a higher initial peak value and mean crushing force value. Vice versa, the reduction in laminate stiffness caused a reduction in these values, which shows the model is responsive to small changes in the parameters.

5.1.2. Compressive Strength

Regarding energy abruption capability in the Mat_54-55 card, compressive strength plays an important role. Figure 14 shows the comparison between compressive strength with an increase and reduction of ±10%. The SEA, initial peak and average crushing force is the main concentration of this study, compared with reference FEM. The computational cost remained the same with minimal changes in all cases. Increasing compressive strength results in an increase in initial peak value by 6.5 kN, an increase in mean crush force by 4 kN, a reduction in displacement by 3 mm, and an increase in SEA value by 2 kJ/kg. A reduction in compressive strength by 10% results in a reduction in initial peak force value by 3 kN, a reduction in mean crushing force by 1 kN, an increase in displacement by 1 mm, and a reduction in SEA value by 3 kJ/kg.

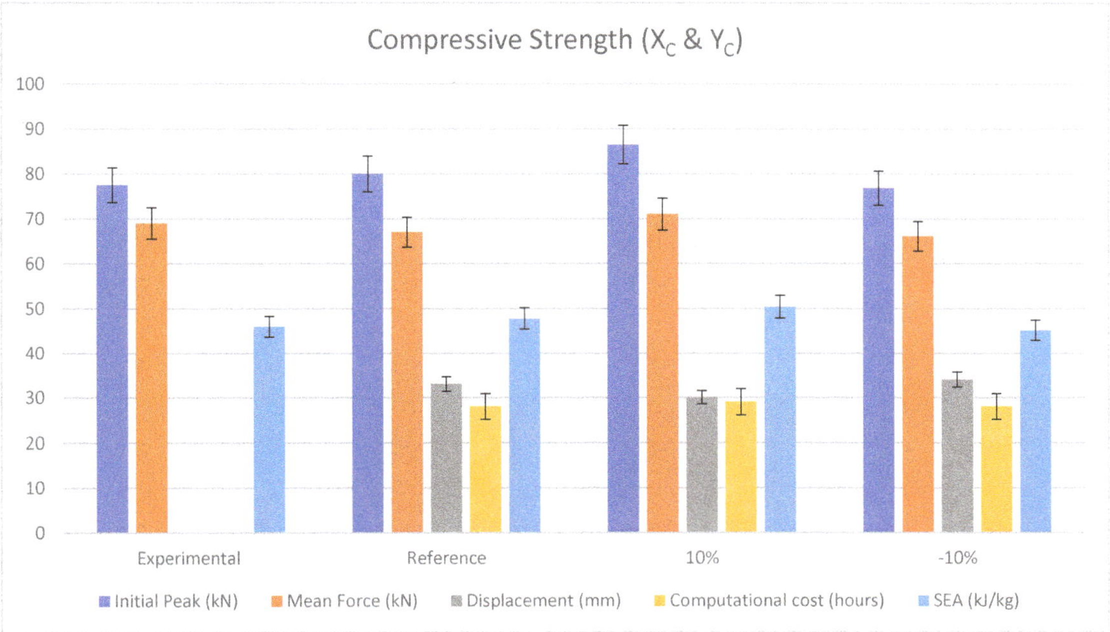

Figure 14. Compressive strength comparison [66].

The results showed that a reduction in compressive strength caused the mean crushing force value to drop along with the initial peak value.

5.1.3. Strain to Failure

The strain determines the element deletion, which is an important parameter for adjusting the model with experimental results. Mat_45-55 allows defining different failure strains for shear and tension/compression. The model response is governed by the deletion of elements, as previously mentioned. The failure of the elements is mainly affected by the strain to failure in compression (DFAILC) and failure in tension (DFAILT) that have been analysed by the increase and reduction of ±10% of these parameters.

Strain to Failure in Compression

In Figure 15, an important influence of the failure strain in compression with an increase and reduction of ±10% on the crushing efficiency is shown. The effect of a 10% increase in DFAILC results in an increase in the mean crush force of 6 kN. The initial peak was not affected, the displacement decreased by 4 mm, and the SEA value increased to 53.7 kJ/kg, with a 6 kJ/kg difference in comparison to the reference model. The effect of a 10% drop in DFAILC results in a drop in mean crushing force value by 5 kN, an increase in displacement by 7 mm, and a 10 kJ/kg drop in SEA value.

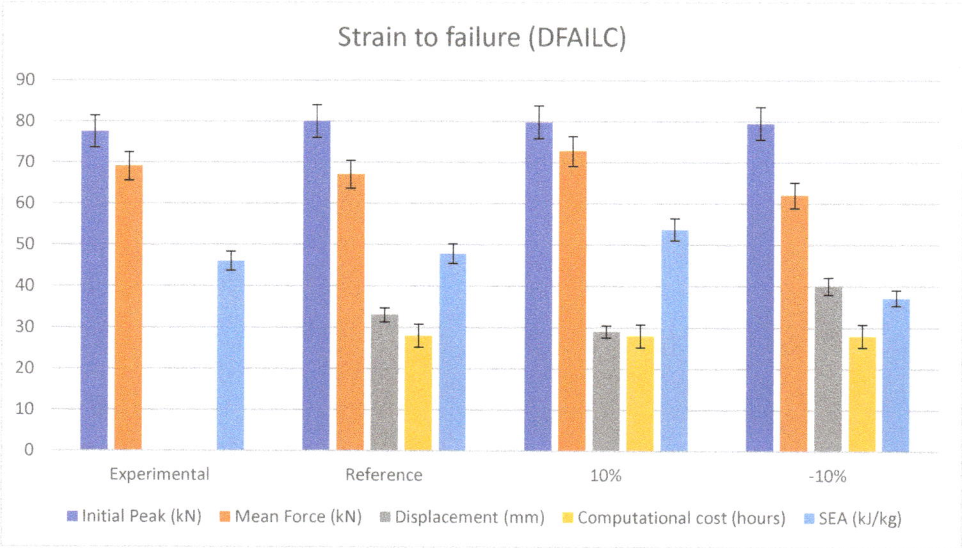

Figure 15. Strain to failure in compression (DFAILC) [66].

Strain to Failure in Tension

In Figure 16, an important influence of the failure strain in tension with increase and reduction of ±10% on the crushing efficiency is shown. The effect of a 10% increase in DFAILT results in an increase in mean crush force of 5 kN, an initial peak increase of 2 kN, a displacement decrease of 2 mm and a SEA value increase of 55 kJ/kg, with a 7 kJ/kg difference in comparison to the reference model. The effect of a 10% drop in DFAILT results in a drop in mean crushing force value of 2 kN, an increase in initial peak value by 5 kN, an increase in displacement by 2 mm, and a 3 kJ/kg drop in SEA value.

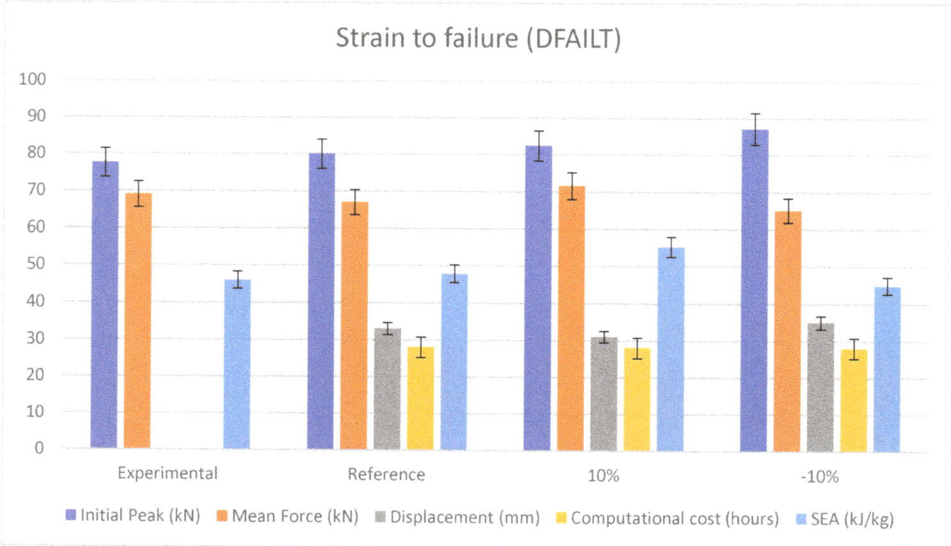

Figure 16. Strain to failure in tension (DFAILT) [66].

5.2. Delamination Model

The delamination between the shell was modelled with tiebreak option 8 described in Sections 2.2 and 4.1 in detail. The current study in this section is to determine the element size and the sensitivity of the delamination algorithms according to Table 2. To ensure the mesh is fine enough to avoid premature failure of the interface and instability, the model sensitivity towards the energy release rate will be analysed.

5.2.1. Tiebreak Contact Element Size Sensitivity

To capture the debonding of the plies, a simple model was created to simulate a DCB test for a Mode-I delamination test. Different mesh sizes were considered: 1.5 × 1.5 mm, 2.5 × 2.5 mm and 3.5 × 3.5 mm (see Figures 17 and 18).

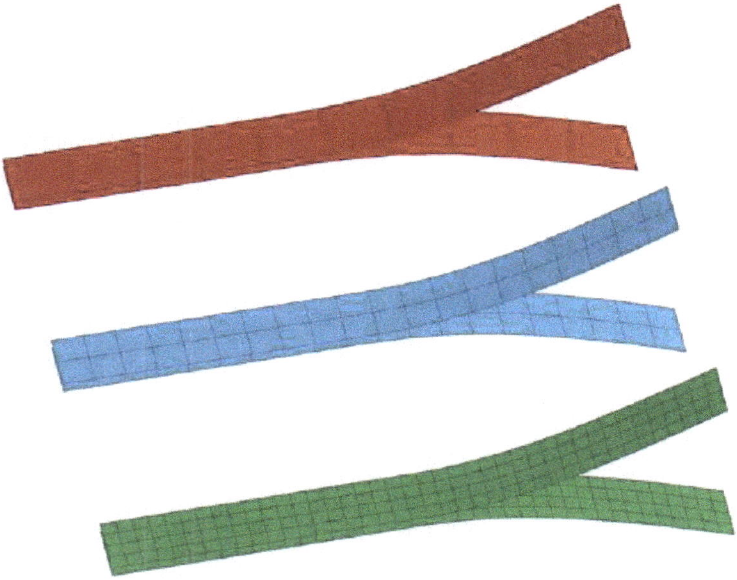

Figure 17. Tiebreak contact element size test [66].

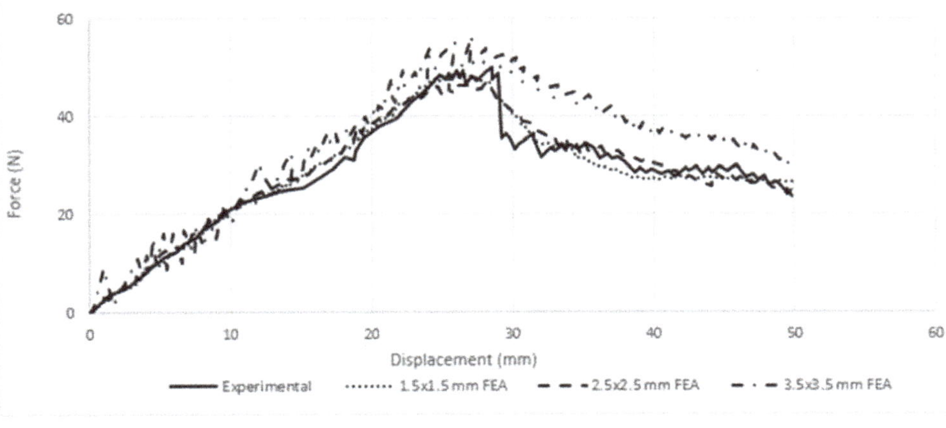

Figure 18. Force–distance curve of Mode-I delamination, experimental and FEA comparison [66].

To assure the obtained results would be relevant to experimental data, the FE results were compared with the experimental data. In Figure 18, all cases have captured the experimental curve; however, the 1.5 × 1.5 mm and 2.5 × 2.5 mm element sizes have closer values with minimal differences. The element size of 3.5 × 3.5 mm overestimated the results throughout, and 2.5 × 2.5 mm, which is the element size (see Section 4.5.), was concluded to be both accurate enough and computationally cost-efficient.

5.2.2. Delamination Resistance

The aim of this study is to examine the sensitivity and the importance of the effect of the tiebreak contact *PARAM* function on the energy absorption capability of the developed FEM. Once the progressive failure has been established, the debonding of the plies is ruled purely by normal stress, and according to Equation (13), the G_{CI} is governed by normal stress and *PARAM* from the contact card. An increase in *PARAM*, therefore, results in an increase in G_{CI} proportionally. This method has been used in this study to analyse its effect on Mode-I delamination resistance.

Figure 19 illustrates the effect of this phenomenon on crushing behaviour and energy absorption capability of the FEM by an increase and reduction of ±10%. The effect of a 10% increase in G_{CI} results in an increase in mean crush force by 6 kN, no change in initial peak, a displacement decrease of 5 mm and a SEA value increase of 54.5 kJ/kg, with a 6 kJ/kg difference in comparison to the reference model. The effect of a 10% decrease in G_{CI} results in a drop in mean crushing force value of 6 kN, no change in initial peak value, an increase in displacement of 7 mm, and a 5.5 kJ/kg drop in SEA value.

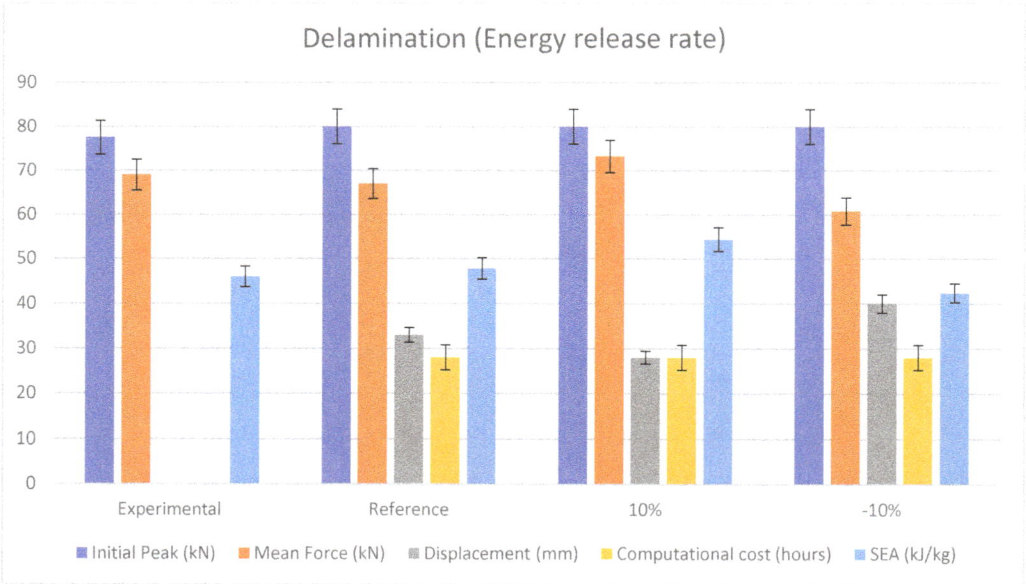

Figure 19. Delamination resistance comparison [66].

The results indicate that the Mode-I energy release rate has an important influence on the energy absorption capability, mean crushing force value, displacement and SEA value. To conclude this, it can be noted that modelling delamination as tiebreak option 8 has been established to be sensitive towards capturing and affecting Mode-I delamination. The input parameters have a major effect on delamination resistance. Therefore, it is essential to validate Tiebreak input parameters as it has been carried out in previous sections (see Sections 2.2, 4.1 and 5.2.1).

5.3. Friction

The coefficient of friction is one of the physical parameters that influences the progression of the simulation. In the literature, many values have been stated, varying from 0.1 to 0.3 for static and 0.1 to 0.2 for dynamic. The chosen values for static friction coefficient is 0.3 and 0.2 for the dynamic friction coefficient [3–7,43,53,59,60]. Both impactor to inner shell and inner shell to outer shell friction coefficients are set to 0.3 and 0.2 for static and dynamic, respectively. This is based on trial and error and based on previous researchers [3–7,43,53,59,60]. This combination enables a sensitive crushing performance. In this study, the sensitivity of the FEM to friction is studied. The two scenarios are the friction between the impactor and the inner shell and the friction between the shells.

5.3.1. Impactor to Shell

In Figure 20, the results from the friction between the impactor and the inner shell are compared based on an increase and decrease of ±10%. The influence of cthe oefficient of friction on mean crushing force is 2.5 and −3.5 kN. The only parameter that is not influenced by friction is computational cost. The initial peak force increased by 0.5 kN, which is minimal and small enough to neglect when increased by 10% and no change with 10% reduction. The main concentration was mean crushing force, displacement and SEA. With a 10% increase, the mean crushing force increased by 2.5 kN, the displacement dropped by 2 mm, and the SEA value increased by 3 kJ/kg. By decreasing the friction by 10%, the initial peak had no effect, the mean crushing force dropped by 3.5 kN, the displacement increased by 2 mm, and the specific energy absorption decreased to 45.3 kJ/kg, which is around 2.5 kJ/kg. This parameter can influence the energy absorption capability, and based on the literature, the optimum value is 0.3 and 0.2 for static and dynamic, respectively [59,60].

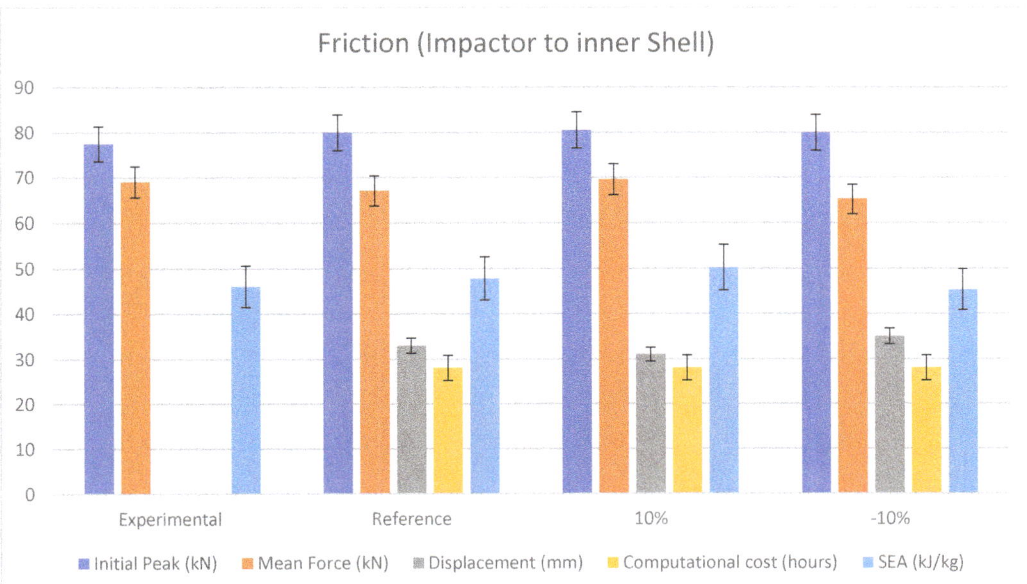

Figure 20. Impactor to inner-shell friction coefficient comparison [66].

5.3.2. Shell to Shell

Many researchers have used friction to simulate delamination, e.g., [59,60]. Friction influences the energy absorption capability. However, using friction influences the SEA value, increases friction between the shells, and causes a higher SEA value, and this compared with experimental data cannot be considered as a correct FEM, as the SEA comparison from the experimental and numerical data would be more than the error percentage. Friction influences the mean crush force and, consequently, affects SEA value. Due to this, a different approach was considered. Tiebreak option 8 was utilised instead of friction to model delamination as this contact card can define the Mode-I and Mode-II energy release rate, which simulates delamination. However, the correct friction between the shells must be implemented. This friction between the plies also exists in real scenarios, and its effect on the energy absorption capability must be addressed.

In Figure 21, the results from friction between the inner shell and the outer shell are compared based on an increase and decrease of ±10%. The influence of the coefficient of friction on mean crushing force is 3 and −3.5 kN. Throughout the study, the computational cost and initial peak force was unaffected. The main concentration was mean crushing force, displacement and SEA. With a 10% increase, the mean crushing force increased by 3 kN, the displacement dropped by 1 mm, and SEA value increased by 3.5 kJ/kg. By decreasing the friction by 10%, the initial peak had no effect, the mean crushing force dropped by 3.5 kN, the displacement increased by 3 mm, and the specific energy absorption decreased to 41.8 kJ/kg, which is around 6.5 kJ/kg. This parameter can influence the energy absorption capability, and based on the literature, the optimum value is 0.3 and 0.2 for static and dynamic, respectively [59,60].

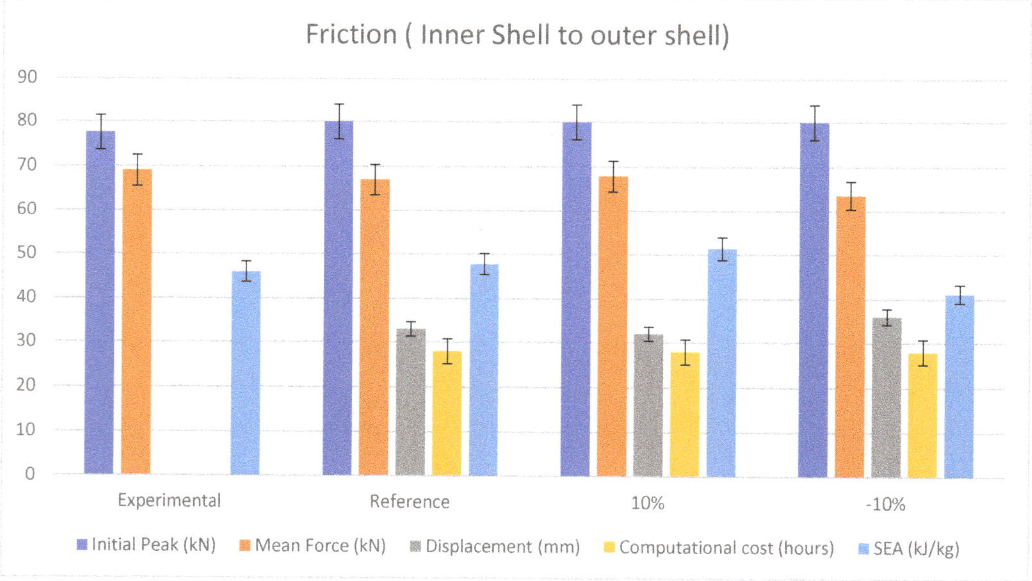

Figure 21. Inner-shell to outer-shell friction coefficient comparison [66].

5.4. Impact Velocity

In this section, various impact velocities are raised to study and analyse the developed model against its sensitivity towards capturing the initial peak, mean crush force, displacement (stroke) and specific energy absorption. In this study, the experimental data and reference model are compared with different impact velocities and applied kinetic energies. The cases are 4 m/s (0.84 kJ), 5 m/s (1.31 kJ), 6 m/s (1.89 kJ) and 8 m/s (3.4 kJ). Understanding how the model is robust with respect to the input kinetic energy would indicate the range of impact conditions predicted by the model.

Figure 22 shows the extracted results from the simulations. The simulation results illustrate a similar value or up to 0.4% difference in initial peak value, and the specific energy absorption, which indicates the energy absorption per crushed mass, is within 4% of the reference model. The mean crushing force is slightly affected, although the highest difference from the reference model is 3.5%.

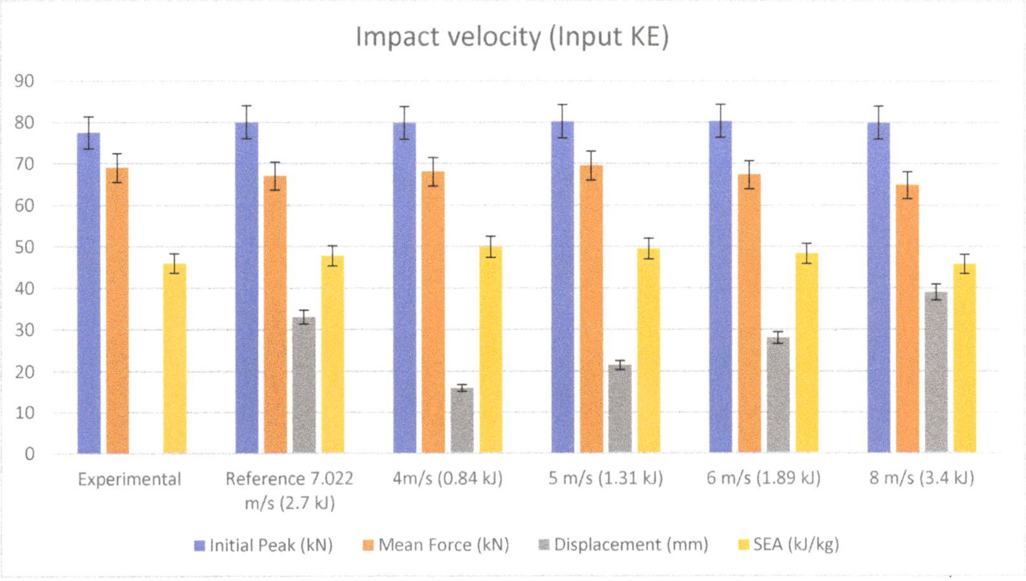

Figure 22. Impact velocity or kinetic energy input sensitivity data [66].

6. Modelling limitations

Based on the study carried out in Section 5, sensitivity towards physical parameters was studied based on the developed model. The FEA results are influenced by many input parameters. To develop a model that is capable of predicting energy absorption in an accurate enough format, a few extracted results must also be acceptable. These parameters are computational cost, initial peak force, mean crushing force, displacement and specific energy absorption. A relatively simple model approach leads to high efficiency, and this was the main concentration of this section. There is a balance between computational cost and final results that needs to be within an acceptable range of up to 10% [59].

In Mat_54, some parameters have no physical meaning, which means some limitations might cause the simulation model to be ruled by non-physical parameters. Therefore, the applicability of the model to another crush scenario must be carried out with care. A change in geometry and the mechanical properties mean that the model can be adapted to that specific crash scenario with care and some adjustments.

More advanced material cards are developed with solid elements to improve capturing delamination energy. Further studies are needed on the mechanical properties of the material and energy release rate in Mode-I and Mode-II. Most importantly, using solid elements rapidly increases computational costs compared with shell elements. This technique increases accuracy, although both testing and computational costs increase.

Increasing the material model complexity leads to further test data but is expensive and increases computational costs. Depending on case studies, from an engineering point of view, a finite element model with an accuracy of up to 10% is acceptable [59,60]. However, the developed model has its limitation in showing the fracture and crushing on a smaller scale, force–displacement characteristics, and debris wedge.

6.1. Fracture Morphology

Regarding crushed morphologies, to capture axial splits using Mat_54, the DFAILM, which is the failure strain in the matrix direction [47], can be utilised. Adjusting the value of DFAILM enables matrix splitting to occur (see Figure 23), as observed in the experiments. However, this parameter influenced the computational costs to be unreasonably high. Due to this fact and keeping a reasonable experimental efficiency, DFAILM values were retracted to original values. It is worth mentioning that, in this research, the main concentration was capturing energy absorption capability. DFAILM had a marginal effect on energy absorption capability, as mentioned above. In the material model Mat_54-55, the stacking sequence plays an important part in crushed morphologies. In Figure 24, the FEM crushed morphologies of $[0]_{12}$ are shown, which resembles the experimental crushed morphologies. However, using ±45 leads to inadequate failure modes if DFAILM values exceed 10% to obtain matrix splitting/separation (see Figure 23b).

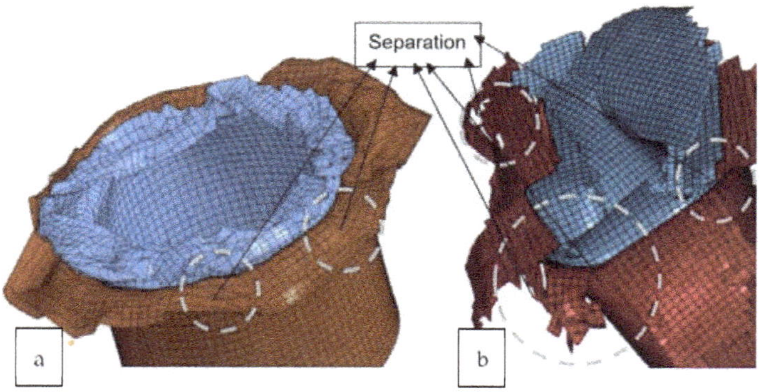

Figure 23. Effect of DFAILM on axial split, (**a**) reference model (**b**) 10% increase DFAILM [65].

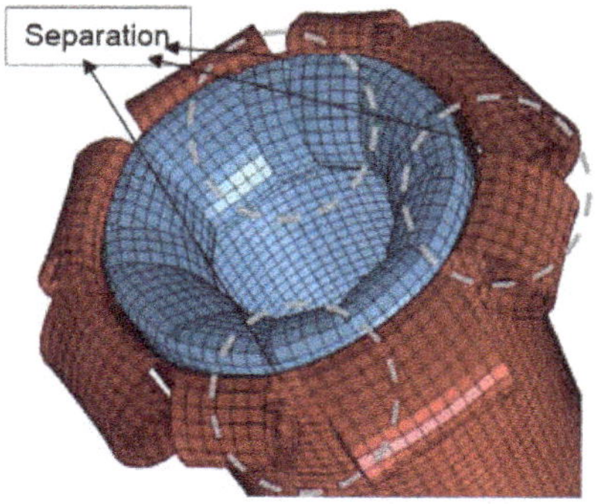

Figure 24. Effect of stacking sequence of $[0]_{12}$ configuration on petal formation of FEM crushed morphologies using DFAILM [59].

Figure 25 shows the comparison between the reference FEM and increase in DFAILM by 10% and 20%. The effect of DFAILM on the axial split is shown in Figure 23, which improves the crushed morphology representation. Although, the results regarding energy absorption capability, including SEA values, were unaffected by this parameter. At a 10% increase, the axial split occurs in this FEM, which causes a 138% increase in computational costs. Increasing DFAILM by 20% causes a 176% increase in computational costs.

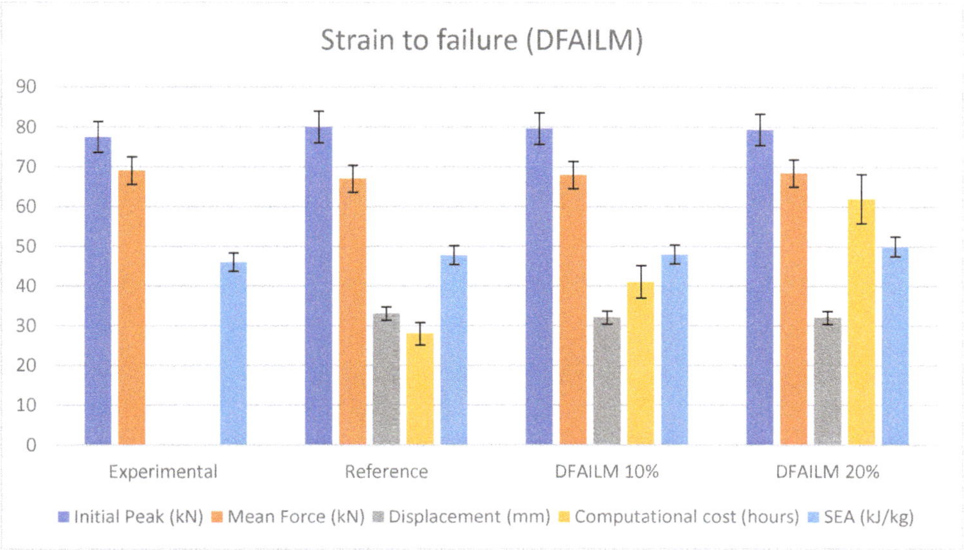

Figure 25. Strain to failure in matrix direction (DFAILM) [66].

In consideration of these results, capturing an axial split seems rather unreasonable due to its effect on computational costs. Increasing the DFAILM value by 10 to 20% had minimal or no effect on the results, and an axial split below 10% in this FEM does not occur.

6.2. Force–Displacement Characteristics

The visual differences in crushed morphologies and the crushing process of composite tubes and FEA are captured by the force–displacement curve diagram. Figure 26 shows a comparison between numerical and experimental data.

As described by Hall [17], the experimental data have a serrated shape where the positive slope segments, e.g., (ab) represents the increase in resistance load due to multiple crack propagations until point (b) is reached. Further crushing is initiated, causing a negative slope or drop in resistance force.

In the finite element (dotted black), the oscillations are governed by the element deletion between fronds, and the amplitudes are a function of the strain to failure of the element controlling the tearing and the mode-I delamination resistance.

At any stage of the crushing process, the experimental resistance force is governed by the weakest possible collapse mode(s). The curves from the FEA and experiment have different amplitudes and ranges (see Figure 26). The amplitudes of the curve and wavelength of FEA are governed by element deletion, and the element size affects the wavelengths. Nevertheless, this leads to collapse modes, which are not captured by the model properly.

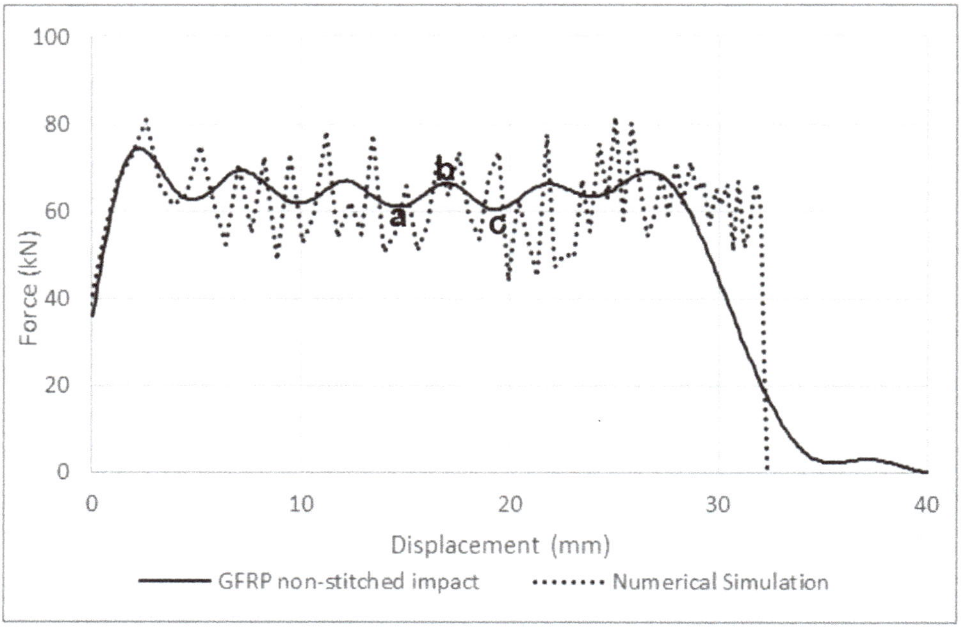

Figure 26. Force–displacement characteristic experimental and numerical comparison [66], (a–b) increase in load and(b–c) decrease in load.

6.3. Debris Wedge

During the progressive crushing process, the fractured material within the Mode-I opening of the tube becomes trapped in the Mode-I opening mode. This increases the friction and affects the crack growth and delamination resistance.

The FEM is not able to simulate this effect if element deletion takes place and does not interfere any longer with the remaining structure. This issue can be resolved by increasing delamination resistance or increasing friction between the shells if a multi-shell configuration is utilised.

Some researchers have tried modelling a debris wedge. McGregor [10] used a predefined debris wedge designed as rigid, and Mamalis [4] defined an intermediate layer trying to represent the pulverised material.

7. Results

Based on the finite element findings and calibrations in this paper and briefly explained in Section 3, the simulations were carried out under quasi-static and impact loading conditions.

7.1. Material and Experimental Setup

In this study, the composite sections were fabricated from glass/epoxy ($\rho = 2250$ kg/m^3) with a symmetric twelve-ply quasi-isotropic laminate design of $(-45/45/0/90/0/90)_S$ using hand lay-up techniques. The composite sections were 80 $\times$ 80 mm with a total thickness of 3 mm in size.

In quasi-static testing, a 500 kN load cell capacity hydraulic press was used with a 2 mm/second loading rate. All specimens were allocated with respect to the centre of the stroke for equal load distributions. The stroke displacement for all specimens was kept at 50 mm. The profile of load–displacement consists of load cell and stroke displacement.

In impact testing, a drop tower with an impactor mass of 108.4 kg was used at 2.0 m height and 7.022 m/s velocity, and a total energy of 2672 J was applied to each specimen. The mass is dropped from a pre-determined height of 2.0 m to initiate and record the load against time once it reaches the specimen. It will only stop when the applied energy of 2672 J is absorbed by the specimen. The hammer is then pulled back up by the machine. Once the impactor hits the tube leading edge, the load cell records the force history.

7.2. Crushed Morphologies

In this section, the crushed morphologies between experimental and numerical results are compared (see Figure 27). Part a and b represent the quasi-static loading condition, experimental and numerical, respectively. Part b and c represent the impact loading condition, experimental and numerical, respectively. The axial splitting, as mentioned before, is captured by FEA and shown with a large separation between the elements. The double-shell configuration can capture the inner and outer petals' formation in both cases.

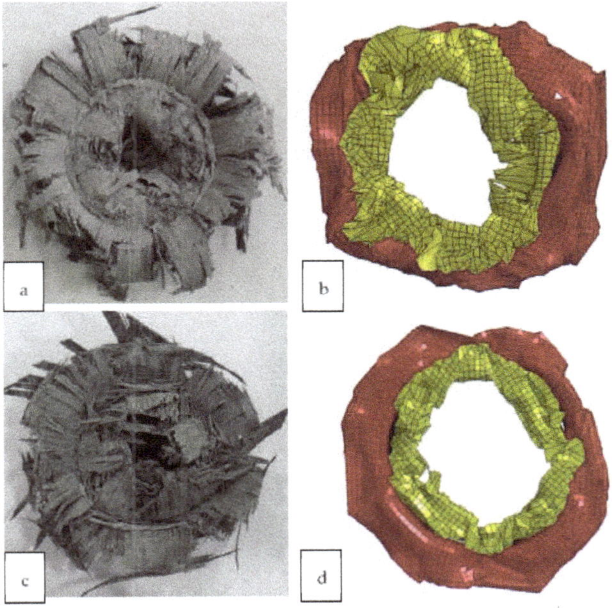

Figure 27. Plane view morphologies, experimental and numerical comparison, (**a**) crushed under quasi-static experimental, (**b**) crushed under quasi-static numerical, (**c**) crushed under impact loading experimental, (**d**) crushed under impact loading numerical [63,65].

7.3. Force–Displacement Graphs

In this section, the force–displacement graphs are shown in Figure 28. Part a represents the quasi-static loading condition, and part b shows the impact loading condition. In this case, the stiffness and the mean crushing force is the main concentration. The stiffness, which is the initial peak in the graph, is followed by the mean crushing force. In part a, the experimental data have a mean crush force of 100 kN, an FEA of 98 kN, and an initial peak value of 88 kN in both the FEA and experimental studies. In part b, the mean crush force is 69 kN for experimental and 68 kN for numerical, with an initial peak value of 78 kN in both cases. This shows that the FEM has captured and predicted the behaviour of the GFRP composite tubes under quasi-static and impact loading.

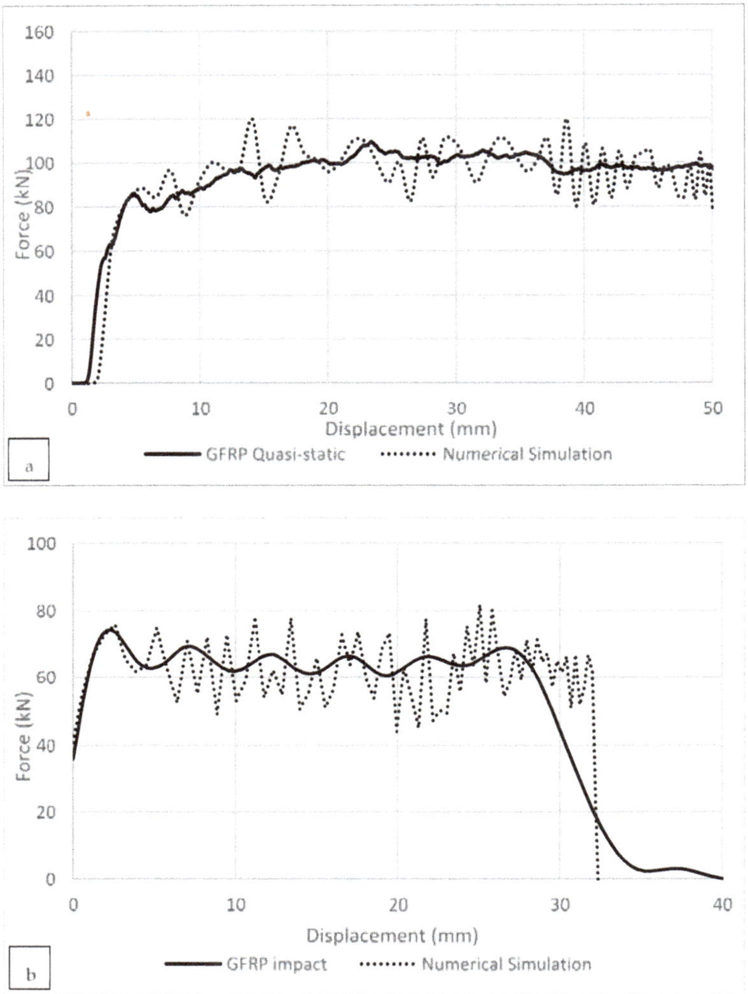

Figure 28. Force–displacement characteristic experimental and numerical comparison [66].

7.4. Specific Energy Absorption (SEA) Comparison

The SEA value comparison in this section is to determine the energy absorption capability of the finite element model compared with experimental studies. In this section, the differences between the numerical and experimental studies are compared. In Figure 29, the SEA value of the numerical model under quasi-static loading is 2.6% off compared with experimental data. The SEA value of numerical studies under impact loading is 3.8% off compared with experimental data. The aim of this paper is to achieve a 5% difference and have an efficient model in terms of computational costs and accuracy, which has been achieved.

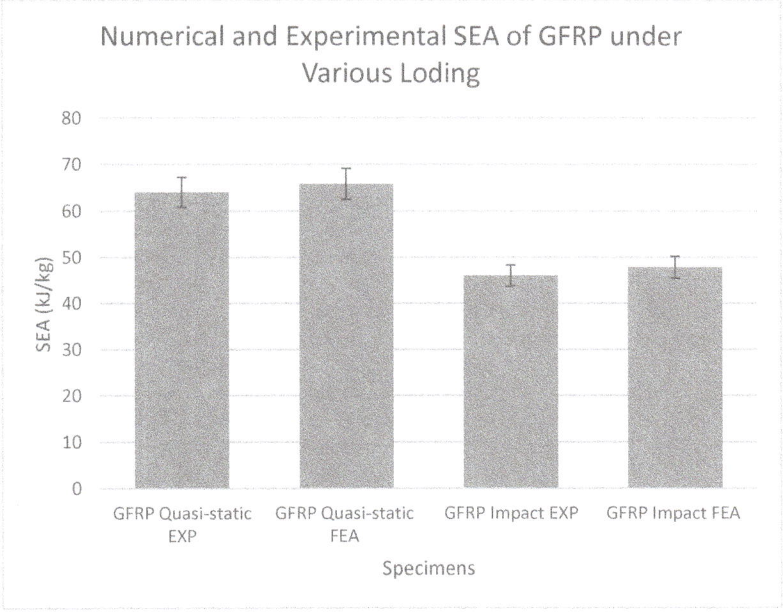

Figure 29. Numerical and experimental SEA of GFRP under Quasi-static and Impact loading conditions [66].

8. Conclusions

The aim of this paper was to develop a finite element model using LS-DYNA to simulate the crushing behaviour of composite tubes with a chamfer failure trigger mechanism under various loadings. Various shell configurations were studied, and a multi-layer shell element with double-shell configuration produced accurate enough results with a difference of less than 5% with minimal computational costs compared with other configurations. This configuration was used to predict energy absorption capability and specific energy abortion, and other considerations were deformation and damage progression of the composite tubes. Each shell element or layer can contain either a single ply or multiple plies. The layers were tied using tiebreak option 8 contact definitions. This contact card has the capability of modelling delamination between the layers through an energy-based approach. The material card of Mat_54 was used to represent each ply, and a few parameters in the material cards were studied to find the optimum configurations to match the experimental studies. SOFT and DFAILC (compression failure strain) were the main parameters that affected the energy absorption capability, and specific energy absorption was influenced by these parameters. The sensitivity of the model was studied against the material model, delamination model, friction and impact velocity. The results show that the model is sensitive towards minimal input change. The simulation results showed that the failure peak load, mean crushing force, and SEA all compared very well with the experimental results.

Numerical models are commonly used to predict the mechanical response of simple to complex geometries as this is a cost-effective approach compared to experimental results. FEM is utilised to predict various conditions and scenarios and relative optimisation takes place; however, numerical models, for instance, the one developed in this paper, have their limitations in showing the fracture and crushed zones on a smaller scale, force–displacement characteristics, and debris wedges, which are all contributing factors to energy absorption. Each developed numerical model has its range of validation, which means the model is capable of accurate prediction to a certain level, and the model is incapable of good prediction if the input values exceed the range or the input values are amended. It is critical to experimentally obtain this range if various scenarios are intended to be studied [66,67]. Therefore, Double Cantilever Beam (DCB), End-Notched Flexure (ENF), Three Point Bending, compression, tensile testing, and many more can be utilised to improve material behaviour and prediction. Maximising resolution of numerical models can be obtained for various application such as [68,69], by improving prediction of material behaviour i.e., the mechanical properties, via physical and non-physical parameters, the discrepancy between experimental and numerical data can be minimised.

The standard deviation between experimental and numerical studies is very low due to calibration of the model based on trial and error and the parametric study to improve resolution. However, different scenarios were conducted, impact and quasi-static, to further understand the discrepancy between a calibrated model and a different scenario without calibration; hence, in this paper, this was utilised as an indication of prediction. The calibrated model had a discrepancy of 5% in the experiment, and when the scenario changed, a similar discrepancy was observed (5%) in the relative experiment.

The calculation errors are all dependent on boundary conditions, mechanical properties inputs, mesh sensitivity and all other parameters discussed in this paper. The model is as accurate as the input values. The calculation error might occur in experimental studies, and measuring the crushed zones, for instance, can make a difference in SEA values.

In conclusion, the developed model in this study has shown to be capable of accurately capturing the crushing behaviour of the tubes with minimal differences whilst being computational cost-efficient. The numerical and experimental studies are in good agreement.

Author Contributions: Methodology, verification and validation, model development, analysis, and evaluation, writing and editing, A.R. writing, review, editing and supervising, H.G. All authors have read and agreed to the published version of the manuscript.

Funding: This research received no external funding.

Institutional Review Board Statement: Not applicable.

Informed Consent Statement: Not applicable.

Data Availability Statement: Not applicable.

Conflicts of Interest: The authors declare no conflict of interest.

Abbreviations

CFE	crush force efficiency
CFRP	carbon fibre reinforced plastic
GFRP	glass fibre reinforced plastic
DCB	double-cantilever beam
3ENF	three-point-end-notched flexure
KE	kinetic energy
ρ	density
A	cross-sectional area
E	Young's modulus
F	load
F_{max}	initial maximum load
F_m	mean load
P_L	peak load
G_{12}	shear modulus
G_{IC}	mode-i interlaminar fracture toughness
G_{IIC}	mode-ii interlaminar fracture toughness
SEA	specific energy absorption (kj/kg)
t	wall-thickness
D	diameter
C	circumference
L	axial length

V_R	volume reduction
V_f	fibre volume fraction
v	Poisson's ratio
β	weight factor
μ	coefficient of friction
Δ	displacement
δ	post crushing displacement
m	crushed specimen mass
V	volume of crushed specimen
s	cross-head distance
a, ALPH	nonlinear shear stress parameter
X_t	longitudinal tensile strength
X_c	longitudinal compressive strength
Y_t	transverse tensile strength
Y_c	transverse compressive strength
S_1	longitudinal tensile strength
DFAILC	max strain for fibre compression
DFAILT	max strain for fibre tension
DFAILM	max strain for matrix straining in tension and compression
DFAILS	max shear strain
EFS	effective failure strain
σ_n	normal stresses
σ_s	shear stresses
σ_1	longitudinal stress
σ_2	transverse stress
σ_{12}, S_{12}	transverse shear stress
σ_u	ultimate tensile stress
σ_b	flexural strength
τ_s, S_2	shear strength
NFLS	normal failure strength
SFLS	shear failure strength
SOFT	softening reduction factor for material strength in crashfront elements
TFAIL	time step size criterion for element deletion
PARAM	critical normal separation of the surface
FE	finite element
FEA	finite element analysis
FEM	finite element model
UD	unidirectional
θ	fibre orientation

References

1. Cook, R.D.; Malkus, D.S.; Plesha, M.E. *Concepts and Applications of Finite Element Analysis*; Wiley Publishing: Hoboken, NJ, USA, 1989.
2. Bisagni, C.; Pietro, G.P.; Fraschini, L.; Terletti, D. Progressive crushing of fiber-reinforced composite structural components of a Formula One racing car. *Compos. Struct.* **2005**, *68*, 491–503. [CrossRef]
3. Huang, J.; Wang, X. Numerical and experimental investigations on the axial crushing response of composite tubes. *Compos. Struct.* **2009**, *91*, 222–228. [CrossRef]
4. Fu, J.; Liu, Q.; Ma, Y.; Zhang, Z.A. A comparative study on energy absorption of flat sides and corner elements in CFRP square tube under axial compression. *Thin-Walled. Struct.* **2021**, *166*, 108080. [CrossRef]
5. Palanivelu, S.; Paepegem, W.; Degrieck, J.; Kakogiannis, D.; Ackeren, J.; Hemelrijck, D. Parametric study of crushing parameters and failure patterns of pultruded composite tubes using cohesive elements and seam. Part I: Central delamination and triggering modelling. *Polym. Test* **2010**, *29*, 729–741. [CrossRef]
6. Palanivelu, S.; Paepegem, W.; Degrieck, J.; Kakogiannis, D.; Ackeren, J.; Hemelrijck, D. Parametric study of crushing parameters and failure patterns of pultruded composite tubes using cohesive elements and seam. Part II: Multiple delaminations and initial geometric imperfections. *Polym. Test* **2010**, *29*, 803–814. [CrossRef]
7. Bussadori, B.P.; Schuffenhauer, K.; Scattina, A. Modelling of CFRP crushing structures in explicit crash analysis. *Compos. Part B-Eng.* **2014**, *60*, 725–735. [CrossRef]

8. Deleo, F.; Wade, B.; Feraboli, P.; Rassaian, M. Crashworthiness of Composite Structures: Experiment and Simulation. In Proceedings of the 50th AIAA Conference, Palm Springs, CA, USA, 4 May 2009.
9. Wang, Y.J.; Zhang, Z.J.; Xue, X.W.; Zhou, J.; Song, Z.X. Axial and lateral crushing performance of plate-lattice filled square sandwich tubes. *Compos. Struct.* **2021**, *274*, 114404.
10. McGregor, C.; Vaziri, R.; Xiao, X. Finite element modelling of the progressive crushing of braided composite tubes under axial impact. *Int. J. Impact. Eng.* **2010**, *37*, 662–672. [CrossRef]
11. McGregor, C.; Zobeiry, N.; Vaziri, R.; Poursartip, A. A constitutive model for progressive compressive failure of composites. *J. Compos. Mater.* **2008**, *42*, 2687–2716. [CrossRef]
12. Xiao, X. Modeling energy absorption with a damage mechanics based composite material model. *J. Compos. Mater.* **2009**, *43*, 427–444. [CrossRef]
13. Oshkovr, S.A.; Taher, S.T.; Oshkour, A.A.; Ariffin, A.K.; Azhari, C.H. Finite element modeling of axially crushed silk/epoxy composite square tubes. *J. Compos. Struct.* **2010**, *95*, 411–418. [CrossRef]
14. Kim, G.H.; Choi, J.H.; Kweon, J.H. Manufacture and performance evaluation of the composite hat-stiffened panel. *Compos. Struct.* **2010**, *92*, 2276–2286. [CrossRef]
15. Chatiri, M.; Gull, T.; Matzenmiller, A. An assessment of the new LS-DYNA layered solid element: Basics, patch simulation and its potential for thick composite structures analysis. In Proceedings of the 7th European LS-DYNA Conference, Salzburg, Austria, 14–15 May 2009.
16. Wagner, W. FE—Modeling of fiber reinforced polymer structures. In Proceedings of the 5th World Congress on Computational Mechanics, Vienna, Austria, 7–12 July 2002.
17. Feraboli, P. Development of a corrugated test specimen for composite material energy absorption. *J. Compos. Mater.* **2008**, *42*, 229–256. [CrossRef]
18. Greco, F.; Luciano, R. A theoretical and numerical stability analysis for composite micro-structures by using homogenization theory. *Compos. Part B Eng.* **2011**, *42*, 382–401. [CrossRef]
19. Sun, H.; Di, S.; Zhang, N.; Pan, N.; Wu, C. Micromechanics of braided composites via multivariable FEM. *Comput. Struct.* **2003**, *81*, 2021–2027. [CrossRef]
20. Tabiei, A.; Aminjikarai, B.S. A strain-rate dependent micro-mechanical model with progressive post-failure behavior for predicting impact response of unidirectional composite laminates. *Compos. Struct.* **2009**, *88*, 65–82. [CrossRef]
21. Crashworthiness Working Group of the CMH-17. Simulation of the quasi-static crushing of a fabric composite plate. *Abaqus Technol. Brief* **2011**. Available online: https://imechanica.org/files/Automotive-SIMULIA-Tech-Brief-11-Simulation-Quasi-static-Crushing-Full.pdf (accessed on 1 December 2021).
22. Feindler, N.; Drechsler, K.; Doll, J. Test method to analyse the energy absorption of composite material using flat coupon testing. In Proceedings of the 5th International Conference on Composites Testing and Model Identification, Lausanne, Switzerland, 14–16 February 2011.
23. Borovkov, A.; Palmov, V.; Banichuk, N.; Saurin, V.; Barthold, F.; Stein, E. Macro-failure criterion for the theory of laminated composite structures with free edge delaminations. *Comput. Struct.* **2000**, *76*, 195–204. [CrossRef]
24. Johnson, A.; Pickett, A. Impact and crash modelling of composite structures: A challenge for damage mechanics. In Proceedings of the 9th User Conference EURO-PAM, Darmstadt, Germany, 7–8 October 1999.
25. Greve, L.; Andrieux, F. Deformation and failure modelling of high strength adhesives for crash simulation. *Int. J. Fract.* **2007**, *143*, 143–160. [CrossRef]
26. Tang, C.Y.; Tsui, C.P.; Lin, W.; Uskokovic, P.S.; Wang, Z.W. Multi-level finite element analysis for progressive damage behavior of HA/PEEK composite porous structure. *Compos. Part B Eng.* **2013**, *55*, 22–30. [CrossRef]
27. Barut, A.; Madenci, E.; Tessler, A.; Starnes, J.H., Jr. A new stiffened shell element for geometrically nonlinear analysis of composite laminates. *Comput. Struct.* **2000**, *77*, 11–40. [CrossRef]
28. Mohite, P.M.; Upadhyay, C.S. Region-by-region modeling of laminated composite plates. *Comput. Struct.* **2007**, *85*, 1808–1827. [CrossRef]
29. Holzapfel, M.; Grundolf, K.; Adamski, P.; Elsenhans, H.; Doll, J.; Hambrecht, T. Untersuchungen zur Modellierung von Strukturen aus FKV unter Crashbelastung mit. Hilfe von Mehrschalen modellierungen. In Proceedings of the 7th German LS-DYNA Users Conference, Bamberg, Germany, 30 September 2008.
30. Feindler, N. Charakterisierungs- und Simulationsmethodik zum Versagensverhalten Energieabsorbierender Faserverbundstrukturen. Ph.D. Thesis, Technical University of Munich, Munich, Germany, 2012.
31. Joosten, M.W.; Dutton, S.; Kelly, D. Experimental and numerical investigation of the crushing response of an open section composite energy absorbing element. *Compos. Struct.* **2011**, *93*, 682–689. [CrossRef]
32. Hussain, N.N.; Regalla, S.P.; Rao, Y.V.D. Comparative Study of Trigger Configuration for Enhancement of Crashworthiness of Automobile Crash Box Subjected to Axial Impact Loading. *Procedia Eng.* **2017**, *173*, 1390–1398. [CrossRef]
33. Johnson, A.F. Modelling fabric reinforced composites under impact loads. *Compos. Appl. Sci. Manuf.* **2001**, *32*, 1197–1206. [CrossRef]
34. Abdallah, M.H.; Brainmah, A. Numerical design optimization of the fiber orientation of glass/phenolic composite tubes based on tensile and radial compression tests. *Compos. Struct.* **2021**, 114898, in press. [CrossRef]

35. Reiner, J.; Zobeiry, N.; Vaziri, R. A stacked sublaminate-based damage-plasticity model for simulating progressive damage in composite laminates under impact loading. *Thin-Walled Struct.* **2020**, *156*, 107009. [CrossRef]
36. Reiner, J. A practical approach for the non-local simulation of progressive damage in quasi-isotropic fibre-reinforced composite laminates. *Compos. Struct.* **2021**, *265*, 113761. [CrossRef]
37. Chambe, J.; Bouvet, C.; Dorival, O.; Ferrero, J.F. Energy absorption capacity of composite thin-wall circular tubes under axial crushing with different trigger initiations. *J. Compos. Mater.* **2019**, *54*, 1281–1304. [CrossRef]
38. Reiner, J.; Feser, T.; Waimer, M.; Poursartip, A.; Voggenreiter, H.; Vaziri, R. Axial crush simulation of composites using continuum damage mechanics: FE software and material model independent considerations. *Compos. Part B Eng.* **2021**, *225*, 109284. [CrossRef]
39. Xiao, X.; Botkin, M.E.; Jahnson, N.L. Axial crush simulation of braided carbon tubes using MAT58 in LS-DYNA. *Thin-Walled Struct.* **2009**, *47*, 740–749. [CrossRef]
40. Pinho, S.T.; Camanho, P.P.; De Moura, M.F. Numerical simulation of the crushing process of composite materials. *Int. J. Crashworthiness* **2004**, *9*, 263–276. [CrossRef]
41. Chiu, L.N.S.; Falzon, B.G.; Boman, R.; Chen, B.; Yan, W. Finite element modelling of composite structures under crushing load. *Compos. Struct.* **2015**, *131*, 215–228. [CrossRef]
42. Siromani, D.; Henderson, G.; Mikita, D.; Mirarchi, K.; Park, R.; Smolko, J.; Awerbuch, J.; Tan, T.-M. An Experimental Study on the Effect of Failure Trigger Mechanisms on the Energy Absorption Capability of CFRP Tubes under Axial Compression. *Compos. Part A* **2014**, *64*, 25–35. [CrossRef]
43. Siromani, D.; Awerbuch, J.; Tan, T.-M. Finite element modeling of the crushing behavior of thin-walled CFRP tubes under axial compression. *Compos. B Eng.* **2014**, *64*, 50–58. [CrossRef]
44. Seat Crashworthiness Improvement on Large Aeroplanes—Dynamic Testing 16g. 2018. Available online: https://www.easa.europa.eu/system/files/dfu/NPA%202013-20.pdf (accessed on 23 January 2018).
45. Tong, L.; Mouritz, A.P.; Bannister, M. *3D Fibre Reinforced Polymer Composites*; Elsevier: Amsterdam, The Netherlands, 2002; ISBN 9780080525822.
46. Boeing Demonstrates Use of Stitching Machine for Carbon Fiber Fabric. 1997. Available online: http://aviationweek.com/awin/boeing-demonstrates-use-stitching-machine-carbon-fiber-fabric (accessed on 18 September 2016).
47. Hallquist, J.O. *LS-DYNA Theory Manual*; Livermore Software Technology Corporation: Livermore, CA, USA, 2006.
48. Chang, F.K.; Chang, K.Y. Post-Failure Analysis of Bolted Composite Joints in Tension or Shear-Out Mode Failure. *J. Compos. Mater.* **1987**, *21*, 809. [CrossRef]
49. Hashin, Z. Failure Criteria for Unidirectional Fiber Composites. *J. Appl. Mech.* **1980**, *47*, 329–334. [CrossRef]
50. Tsai, S.W.; Wu, E.M. A General Theory of Strength for Anisotropic Materials. *J. Compos. Mater.* **1971**, *5*, 58. [CrossRef]
51. Zheng, X. Nonlinear Strain Rate Dependent Composite Model for Explicit Finite Element Analysis. Ph.D. Thesis, University of Akron, Akron, OH, USA, 2006.
52. Krishnamoorthy, S.K.; Hoptner, J.; Kopp, G.; Friedrich, H.E. Prediction of structural response of FRP composites for conceptual design of vehicles under impact loading. In Proceedings of the 8th European LS-DYNA Users Conference, Strasbourg, France, 23–24 May 2011.
53. Reuter, C.; Troster, T. Crashworthiness and numerical simulation of hybrid aluminium-CFRP tubes under axial impact. *Thin-Walled. Struct.* **2017**, *117*, 1–9. [CrossRef]
54. Mamalis, A.G. The static and dynamic axial collapse of CFRP square tubes: Finite element modelling. *Compos. Struct.* **2006**, *74*, 213–225. [CrossRef]
55. Yousefi, A.K. Buckling and Crack Propagation Analysis of Composite Laminated Structures. Master's Thesis, Kingston University, London, UK, 2016.
56. Matzenmiller, A.; Lubliner, J.; Taylor, R.L. A constitutive model for anisotropic damage in fiber composites. *Mech. Mater.* **1995**, *20*, 125–152. [CrossRef]
57. Bala, S.; Day, J. Tie-Break Contacts in Ls-Dyna, Livemore Software Technology Corporation. 2010. Available online: https://ftp.lstc.com/anonymous/outgoing/support/PRESENTATIONS/Crash_Guidelines.pdf (accessed on 1 December 2021).
58. Jayasinghe, P.D.N. New Development on Mode-I Delamination Failure of FRP Composites. Bachelor's Thesis, Kingston University, London, UK, 2016.
59. Ghasemnejad, H.; Blackman, B.R.K.; Hadavinia, H.; Sudall, B. Experimental studies on fracture characterisation and energy absorption of GFRP composite box structure. *Compos. Struct.* **2008**, *88*, 253–261. [CrossRef]
60. Ghasemnejad, H.; Hadavinia, H. Off-axis crashworthiness characteristic of woven glass/epoxy composite box structures. *J. Reinf. Plast. Comp.* **2010**, *29*, 2306–2330. [CrossRef]
61. Rabiee, A.; Ghasemnejad, H. Progressive Crushing of Polymer Matrix Composite Tubular Structures: Review. *Open J. Compos. Mater.* **2017**, *7*, 14–48. [CrossRef]
62. Rabiee, A.; Ghasemnejad, H. Effect of multi stitched locations on high speed crushing of composite tubular structures. *Compos. Part B Eng.* **2016**, *100*, 164–175. [CrossRef]
63. Rabiee, A.; Ghasemnejad, H. Improvement of Specific Energy Absorption of Composite Tubular Absorbers using Various Stitching Pattern Designs. *Appl. Compos. Mater.* **2020**, *27*, 717–738. [CrossRef]

64. Rabiee, A.; Ghasemnejad, H. Lightweight design to improve crushing behaviour of multi-stitched composite tubular structures under impact loading. *Thin-Walled Struct.* **2019**, *135*, 109–122. [CrossRef]
65. Rabiee, A.; Ghasemnejad, H. Laminate Tailoring of Composite Tubular Structures to Improve Crashworthiness Design at Off-Axis Loading. *Open J. Compos. Mater.* **2018**, *8*, 3. [CrossRef]
66. Rabiee, A. Lightweight Design of Multi-Stitched Composite Crash Absorbers to Improve Specific Energy Absorption Capability under Quasi-Static and Impact Loading. Ph.D. Thesis, Cranfield University, Cranfield, UK, 2018.
67. Omidali, M.; Khedmati, M.R. Numerical investigation on novel geometrical configuration for adhesively bonded T-joint between aluminium and sandwich panel. *Thin-Walled Struct.* **2018**, *131*, 122–134. [CrossRef]
68. Khosravani, M.R.; Weinberg, K. Characterization of sandwich composite T-joints under different ageing conditions. *Compos. Struct.* **2018**, *197*, 80–88. [CrossRef]
69. Ramaswamy, S.; Shariff, Z.A.; Munaf, A.A.; Leo, I.J.; Gnanaraj, S.J.P.; Jeshurun, S.B. Study on application of higher order lamination plate theory over various applications of natural fiber cross-ply composites. *Mater. Proc.* **2021**, in press. [CrossRef]

Journal of
Composites Science

Review

3D X-ray Microscopy as a Tool for in Depth Analysis of the Interfacial Interaction between a Single Carbon Fiber and an Epoxy Matrix after Mechanical Loading

Julia Vogtmann, Andreas Klingler, Thomas Rief and Martin Gurka *

Leibniz-Institut für Verbundwerkstoffe GmbH, Erwin-Schrödinger-Straße 58, 67663 Kaiserslautern, Germany; julia.vogtmann@ivw.uni-kl.de (J.V.); andreas.klingler@ivw.uni-kl.de (A.K.); thomas.rief@ivw.uni-kl.de (T.R.)
* Correspondence: martin.gurka@ivw.uni-kl.de

Abstract: The benefit of fiber-reinforced composites originates from the interaction between the fiber reinforcement and the matrix. This interplay controls many of its mechanical properties and is of utmost importance to enable its unique performance as a lightweight material. However, measuring the fiber–matrix interphase strength with micromechanical tests, like the Broutman test, is challenging, due to the many, often unknown boundary conditions. Therefore, this study uses state-of-the-art, high-resolution X-ray computed microtomography (XRM) as a tool to investigate post mortem the failure mechanisms of single carbon fibers within an epoxy matrix. This was conducted at the example of single carbon fiber Broutman test specimens. The capabilities of today's XRM analysis were shown in comparison to classically obtained light microscopy. A simple finite element model was used to enhance the understanding of the observed fracture patterns. In total, this research reveals the possibilities and limitations of XRM to visualize and assess compression-induced single fiber fracture patterns. Furthermore, comparing two different matrix systems with each other illustrates that the failure mechanisms originate from differences in the fiber–matrix interphases. The carbon fiber seems to fail due to brittleness under compression stress. Observation of the fiber slippage and deformed small fracture pieces between the fragments suggests a nonzero stress state at the fragment ends after fiber failure. Even more, these results demonstrate the usefulness of XRM as an additional tool for the characterization of the fiber–matrix interphase.

Keywords: interphase/interface; X-ray computed microtomography; damage mechanics; finite element analysis

Citation: Vogtmann, J.; Klingler, A.; Rief, T.; Gurka, M. 3D X-ray Microscopy as a Tool for in Depth Analysis of the Interfacial Interaction between a Single Carbon Fiber and an Epoxy Matrix after Mechanical Loading. *J. Compos. Sci.* **2021**, *5*, 121. https://doi.org/10.3390/jcs5050121

Academic Editor: Stelios K. Georgantzinos

Received: 1 April 2021
Accepted: 29 April 2021
Published: 4 May 2021

1. Introduction

Carbon fiber reinforced polymers (CFRP) possess a key role in lightweight structural design. Thereby, especially the mechanical properties of a CFRP strongly depend on the interaction between fiber and matrix. The force transfer from matrix to fiber occurs through the so-called interphase: a three-dimensional individual phase, in which the properties gradually change from matrix to fiber. Hence, it is influenced by both the properties of the matrix and the fiber [1,2] and formed by the interaction of the fiber surface with the matrix during processing [2,3]. However, its spatial dimensions are often not known or are often simply neglected. Therefore, in contrast to the interphase, an interface is understood as a geometrically well-defined 2D contact area between distinct phases such as the matrix and/or the fiber [4–6].

In the loaded case, additional to sufficient adhesive strength between fiber and matrix, also high cohesive strength in the matrix needs to prevail at every point along the fiber–matrix interphase to achieve a maximum stress transfer [4]. Therefore, the failure mechanisms of the fiber under mechanical loading depend strongly on the adhesive and cohesive interphase properties, which are geometrically, and in terms of mechanical properties, not well defined [1,7]. The discrepancy between actually measured and theoretically expected fiber–matrix interaction can be attributed to imperfections (pores/voids/difference in bulk elastic properties) either at the interface, in the interphase or of the constituents. Understanding the fiber–matrix interaction is of utmost importance [2,4,7] and many examples show that controlling the interphase leads to improved mechanical properties of the whole CFRP composites [8–10]. Describing the interphase in a composite material requires reliable and reproducible methods for characterization, but due to the small dimensions and the gradual transition to the volume properties of the matrix, its properties are difficult to measure [1]. The challenge for quantitative characterization of the interphase is not only specimen preparation itself but also the entire experimental procedure. There is no micromechanical test standard to reproducibly determine the interphase quality, e.g., by determining the force transfer between a single fiber and the surrounding matrix. Moreover, the various methods are not comparable with each other, due to differences in introduced stress states either caused by the nature of the test itself or the specific specimen dimensions, as the force can be applied either fiber-sided, as in the pull-out test and the microbond test, or matrix-sided, as in the Broutman test and the fragmentation test. The interpretation of their results is often difficult because boundary conditions such as the extension of the interphase itself or its adhesion to the fiber surface cannot be controlled reliably by the experiment or accurately modeled in such detail. For example, in the pull-out test, edge effects must be taken into account, which result in an increase in shear stress at the fiber's entry into the matrix resin. In the microbond test, the influence of the geometry of the matrix droplet and the position of the cutting edges on the resulting stress state has been demonstrated successfully [11]. All tests mentioned above are based on a description of the shear stress in the interface instead of the normal stress as the Broutman test [12,13].

Most importantly, the resulting stress distribution in the specimen is often complex and difficult to assess. Today, it is not possible to measure only adhesive properties without the influences of the specimen's geometry and boundary conditions, but since these quantities cannot be quantified, the real stress distribution in the interphase cannot be described. Instead, simplifications must be adopted to enable the description of the stress state to be approximated. Therefore, the properties of fiber and matrix have to be quantified. Quantifying the first failure at the interphase is a challenging task, after the first failure friction (slipping) affects the measured quantities, this complicates the interpretation of the results [14]. Optical investigation methods, such as stress-induced birefringence used during a pull-out or Broutman test are helpful for describing the micromechanical failure processes because they visualize the stress distribution along the interface at any time. However, optical evaluation of micromechanical failure processes are limited to transparent specimens. Other limitations include limited, region of interest (ROI) depth of field and resolution.

On the other hand, test methods like nondestructive imaging of the internal structure of a preloaded specimen via X-ray microscopy (XRM) are capable of supporting the identification of those stress distributions by the investigation of the fracture pattern post mortem in great detail. High-resolution 3D X-ray imaging of composite materials, especially for the analysis of structurally dependent mechanical properties has experienced significant growth over the last decade, due to enhancements in spatial resolution [15]. However, the application as an additional tool for characterizing CFRP interphases is a recently evolving topic [15,16]. The 3D visualization of failure, induced by mechanical loading can help to better understand the complex stress distribution in micromechanical tests, by combining them with a simulation-based analysis that reproduces the observed fracture pattern [4,17]. X-ray investigations of CFRP are especially challenging because of (I) the small diameter of the fibers and (II) the low contrast between carbon-based materials caused by similar atomic numbers. However, state-of-the-art microfocussed XRM enables in-line phase contrast and leads to very high-resolution at a distance by using optical and geometrical magnification.

The potential of the combination of micromechanical interphase testing with high-resolution XRM analysis for polymeric matrix composites was shown in [16], performing fragmentation tests on epoxy and vinylester matrix reinforced flax fiber composites. The authors observed fiber fragments inside failure zones, which were not detectable using classical, optical light microscopy. They showed that matrix failure can be observed at various locations, where fiber fracture occurred in all systems with a good adhesion between the constituents. In addition, many studies have shown the great potential in correlating nondestructive structural analyses with mechanical testing to achieve a deeper insight of the macro-mechanical fiber–matrix interaction [15,18,19].

Understanding the interface interactions between fiber and matrix, based on finite element (FE) simulations has still not been fully explored. Several authors use cohesive-zone-modeling to simulate the interface behavior via push-out/push-in tests in comparison to the experimentally assessed interphase behavior (friction and cohesive interaction of fiber and matrix). In [20] a cohesive/volumetric FE model, which includes an augmented Lagrangian treatment of the frictional contact between spontaneously created fracture surfaces was used. It was able to simulate the spontaneous initiation and propagation of the debonding crack front. In [21] a parametric FE simulation study of a push-in test illustrated load separation in two regimes. First, the cohesive fiber–matrix interaction was investigated and then, with increasing load, a frictional contact between the debonded part of the fiber and the matrix was simulated. Jäger et al. [22] accomplished an FE model to verify the data of cyclical loading push-out tests that allowed the determination of the contributions of elastic, plastic and frictional energy. Therefore, a model of several fibers to simulate the crack initiation and progress in a specific fiber–matrix interface was used. It was shown that the plastic deformation of the matrix had a huge impact on the crack initiation in the interface. However, the FE simulation of crack initiation in the fiber after debonding takes place, and the effects of matrix detachment are not yet sufficiently evaluated. The influence of fiber sizing on the fiber–matrix debonding via a unit-cell cohesive damage model for the pull-out test was shown by [23]. With the stress distribution model the debonding mechanism of the interface could be explained in a way that the interface debonded first from the matrix.

The present paper describes an in-depth 3D XRM analysis of single carbon fiber—epoxy matrix-based—delamination failure after a micromechanical compression test (Broutman test). The compressive stress applied to a tail-shaped Broutman specimen, with a single fiber embedded, generates a complex stress distribution within the interphase; shear stresses are superimposed by normal stresses [17,24]. This promotes an off-axis delamination between the fiber and the matrix. However, small variations in the geometry, e.g., caused by specimen preparation, can have large effects on the stress distribution within the specimen and cause unpredictable failure of the composite. The complexity of the stress state in the Broutman test specimen has been emphasized by various authors [17,24,25]. Furthermore, the toughness and stiffness of a matrix affect the deformation and failure behavior of FRPs [26], and thus the stress state under mechanical loading. Thereby, especially epoxy-based matrices suffer from high brittleness and low ductility, due to their highly cross-linked molecular network structure. To investigate the full potential of post mortem high-resolution XRM imaging for a better understanding of interphase failure, two different matrix systems were compared. The change from a more elastic to a more brittle matrix affected the failure mechanisms within the Broutman specimen. The toughness as well as the stiffness of the epoxy matrix was altered using a core-shell rubber nanoparticle modified epoxy resin.

In the present study the challenges and the benefits of XRM as a powerful tool to analyze single fiber failure on the microscale are demonstrated, using the example of a brittle and tough epoxy matrix. This should allow a better understanding of (i) the failure behavior of single carbon fibers within epoxy matrices under compressive loads and (ii) an assessment of the stress distribution within the Broutman test specimen. The gained experimental knowledge was then transferred to a simulation-based analysis.

2. Materials and Methods

2.1. Materials

The study focused on a neat DGEBA-based epoxy resin (Biresin CR144, Sika Deutschland GmbH, Stuttgart, Germany) that was cured by an anhydride curing agent (*Aradur 917* from Huntsman Advanced Materials GmbH, Basel, Switzerland [27]) and a toughened cycloaliphatic epoxy resin. The latter one being toughened with 16 wt.% (final system concentration) of core-shell rubber nanoparticles (*KaneAce MX553* from Kaneka Belgium N.V., Westerlo-Oevel, Belgium [28]). The reactivity was in both cases adjusted by using a 1-methylimidazol accelerator (*DY070* from Huntsman Advanced Materials GmbH, Basel, Switzerland [29]). Resin and curing agent were stoichiometrically mixed at 40 °C and cured in a three-step curing cycle: (1) 90 °C for 4 h, (2) 105 °C for 2 h, and (3) 140 °C for 4 h. The neat DGEBA-based epoxy matrix has a fracture toughness of 0.58 MPa$\sqrt{m}$, the toughened matrix of 1.4 MPa$\sqrt{m}$, which allows large scale energy dissipation via the activation of toughening mechanisms.

For the single carbon fiber (C-fiber) embedded in the specimen, a HTA40E13 (5131) [30] C-fiber from Toho Tenax (Tokyo, Japan) was used. The fiber was a high tenacity type one, having a diameter of 7 µm. The C-fiber was coated with an epoxy sizing. Material properties can be seen in Table 1. For better comparability with results obtained by light microscopy, the selected resin systems were transparent.

Table 1. Material data from the epoxy resin matrix and HTA40 E13 C-fiber, also used for the simplified debonding FE simulation.

		Epoxy Resin Matrix			
E [MPa]	**ν [−]**				
2150	0.343	Toughened			
2940	0.342	Data derived from tensile tests according to DEN EN ISO 527			
		HTA40 E13 C-fiber			
E_1 [MPa]	$E_2 = E_3$	$\nu_{12} = \nu_{13}$	ν_{23}	$G_{12} = G_{23}$	G_{13}
240,000 *	28,000 **	0.23 **	0.3 ***	50,000 **	28,000 ***

* Datasheet [27,30,31]. ** Literature [32]. *** Assumed.

2.2. Specimen Preparation

In this investigation, nonloaded (as prepared) as well as preloaded Broutman test specimens were post mortem analyzed via high-resolution 3D XRM. For X-ray imaging, the specimen size was reduced in several successive steps to achieve a specimen size suitable for acquisition of the internal structure. (Figure 1). The acquisition of the single fiber via XRM in the specimen having its original size was not possible (left, grey) due to the poor contrast of the constituents and the small fiber diameter relative to the specimen dimensions. Consequently, different specimen geometries and sizes were tested, until it was possible to visualize the embedded fiber. In a rotationally symmetric specimen (shown in the red box) the path length for the X-rays remained constant during rotation (between 0.95 and 1.2 mm), the latter resulting in an optimized imaging contrast. Two specimens were measured for each in Section 2.1 described system.

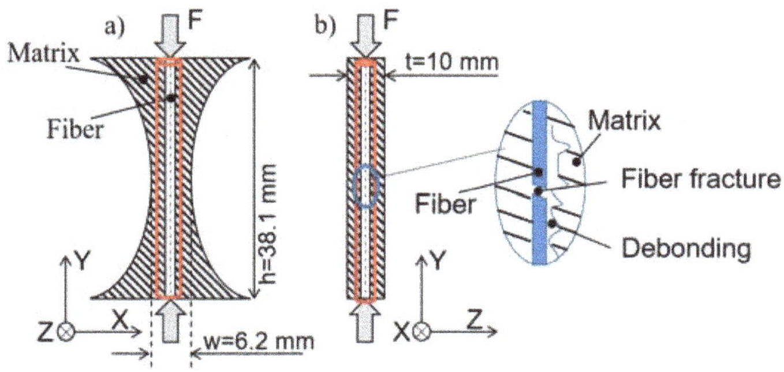

Figure 1. Schematic illustration of the Broutman test specimen geometry (grey), the prepared specimen geometry for X-ray scanning (red) (**a**) front and (**b**) side view. Visualization of the operating force, visualization of the expected resulting failure mechanism.

2.3. Broutman Test

The Broutman test [4,24] is based on a compressive load being applied along the direction of the embedded fiber (y-direction, cf. Figure 1). Because of the concave shape of the specimen, the highest compressive stress occurs in the smallest cross-section of the specimen. Due to the differences in the Poisson ratios of fiber and matrix, especially in the transverse direction, a debonding at the interphase is promoted (a brief model description of the test is given in Equation (1) in Supplementary Materials S1). The specimens were tested in a universal testing machine Zwick RetroLine with a 10 kN load cell. The crosshead speed was 1 mm/min. The performance of the Broutman test itself with the additional measurement set-up, were already described in [33]. Five specimens per configuration (C-Fiber with toughened polymeric matrix, C-Fiber with non-toughened polymeric matrix) were tested. During the test acoustic emission (AE) was applied to determine the fiber–matrix debonding force. Based on these results, the modelling parameters were selected. The fiber–matrix debonding force measured for the toughened matrix system was 5000 N and for the non-toughened matrix system 6500 N (more details are given in Figures S1 and S2 in S2 and the description in S3).

2.4. Optical Investigation of Specimen

An optical investigation of the Broutman test specimens preloaded and nonloaded was conducted. Transparent specimens were analyzed with an optical microscope (Leica DM 6000, Wetzlar, Germany), with and without polarized light and with different magnifications. For the analysis of the internal structure via computed tomography (CT), the contrast between the substituents must be high enough to identify all phases (matrix, fiber and interphase) in the specimen and the introduced fracture pattern. Standard CT relies on the absorption contrast for imaging of the different phases in a material. The attenuation depends on the density of the respective material, its atomic number and the energy of the incident photon [15]. Absorption contrast alone is not sufficient to distinguish between the C-fibers and carbon-based polymeric matrix [34]. With optical high-resolution 3D XRM, an additional contrast-mechanism, so called in-line phase contrast, can be utilized: depending on the material tested, the type of X-ray source and energy, the phase difference between X-ray waves traveling through different materials, a birefringence pattern in the Fresnel-regime manifests itself, leading to an increased feature detectability compared to simple absorption contrast, allowing for the detection of C-fibers in a polymeric matrix [15,19].

For nondestructive 3D imaging of the preloaded specimen, a high-resolution XRM (Zeiss Versa 520, Oberkochen, Germany) was used. Important details of this set-up are the small spot-size X-ray source, the specimen stage, enabling 360° rotation of the specimen, and the optical magnification unit, consisting of a scintillator screen, a microscope lens and the detector.

Due to the optical magnification in front of the detector, high-resolution at a distance can be achieved at a specified maximum resolution of 500 nm. The X-ray device has an acceleration voltage range of 40–160 kV within a maximum power of 10 W. A 1000 × 1000-pixel detector was used. The configuration of X-ray accelerating voltage, exposure time, geometrical and optical magnification, filtering and number of projections used is given in Table 2. As the diameter of the specimens varied slightly over the height in the specified range due to specimen preparation, the exposure time had to be adapted slightly to achieve the same number of counts to ensure a constant contrast. The exposure time used for the nonloaded specimen is marked with an (n), for the preloaded specimen with an (l). Hence, the settings for preloaded and nonloaded specimens can still be considered as identical. For the reconstruction, the software tool "Scout-and-Scan Control System Reconstructor" (Zeiss, Oberkochen, Germany) was used. The analysis of the volume was performed using VGSTUDIO Max 3.2.2.

Table 2. Different settings used for 3D X-ray microscopy analysis.

Settings	Config. 1	Config. 2	Config. 3	Config. 4	Config. 5
Accelerating Voltage [kV]	90	90	160	70	40
Exposure Time [s]	3	8 (n) 9 (l)	20	3 (n) 5 (l)	10
Objective	20x	40x	4x	20x	20x
Filter	Air	Air	HE6	Air	Air
Specimen-detector distance [mm]	22, 8	29	230	12, 3	12, 3
Resolution [nm]	560	281	570	650	650
Number of projections	3001	3001	3601	3001	3001

2.5. Modeling

To investigate the influence of a stochastically occurring debonding failure in the interface between fiber and matrix on the stress distribution within a single C-fiber, a simplified FE model was used. Boundary conditions in the fiber−matrix interface were changed to reflect different debonding scenarios, which led to different stress distributions in the C-fiber. Due to the brittle fracture behavior of the C-fiber these stress concentrations could imply fracture planes and explain the experimentally observed fracture pattern. As FEM software, Abaqus 2020 (SIMULIA) was used.

The simplified FE model is presented in Figure 2. Herein, only a section of a C-fiber embedded in epoxy resin was investigated. The HTA40E13 C-fiber with diameter of 7 µm was surrounded by epoxy resin matrix with a cylindrical shape, 100 µm in diameter. On the bottom surface boundary conditions constrained any rotations as well as an out-of-plane displacement. In-plane displacements were allowed to enable deformations due to the Poisson ratios of the materials, which is a key effect within the Broutman test setup. The top surface was constrained via a reference point on which the same boundary conditions applied, except that the top surface was pushed down by 3 µm. The interface as a 2D contact zone in between the C-fiber and the matrix was assumed as ideal bonding and realized with rough and hard contact conditions. Simulating the interface as a contact allowed the identification of contact stresses in normal and shear directions later on. The material data used for the simulation is presented in Table 1 (toughened), where the C-fiber data either were taken from the datasheet of the supplier, from literature or partially were assumed, as noted. To avoid misunderstandings, the effects of the interphase on material properties of the matrix as they would appear in between C-fiber and matrix were neglected in the FE model. The data for the matrix were measured, according to DIN EN ISO 527. The C-fiber was simulated anisotropic, the epoxy resin matrix isotropic, both with linear elastic material behavior.

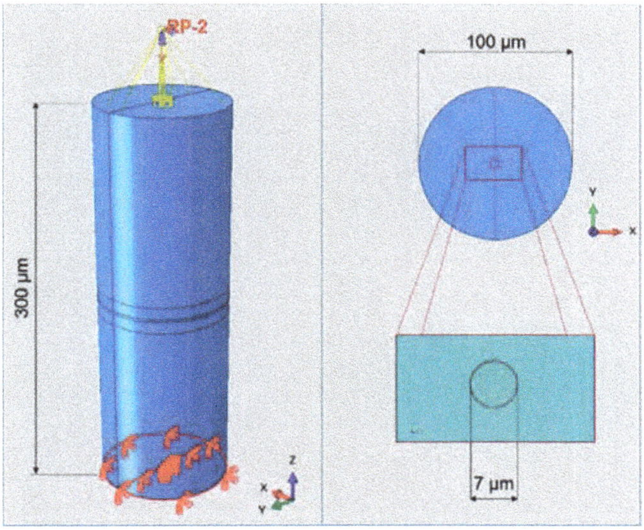

Figure 2. FE model for investigation of debonding on internal fiber stresses, setup with boundary conditions (**left**), top view with dimensions of the matrix (**top right**) and surrounded C-fiber with depicted mesh and fiber diameter (**bottom right**).

Debonding failure as it would appear during the Broutman test was realized by disabling certain elements in the contact surface from the C-fiber to the matrix. Thereby, a so-called cluster defect was assumed (see also Figure 3). Disabling these interface elements allowed no force transmission in between the C-fiber and matrix and led to stress concentrations around and within the thus debonded area. Some aspects have to be taken into account, such that the deactivation of elements leads to discontinuities at the edges, especially in corners. These discontinuities cannot be avoided due to the singularity effect caused by the change from not bonded to perfectly bonded. To minimize these discontinuities, which affect the overall results, certain measures can be used, such as the reduction of the element size or the element type itself. In our case, the element dimensions within the center of the defect were 0.25 µm in height and 0.34 µm in width, outside of the central section 0.50 µm in height with same width. The central section of the defect had a height of 2 µm (marked in Figure 3) and linear hexahedral elements were used (type C3D8) to allow for a reasonable short computing time of the FE simulation. This enabled the systematic investigation of different debonding scenarios and their influence on the stress distribution inside of the C-fiber. The results presented here have been verified by calculation with a quadratic element type, which exhibited a smaller zone being influenced by the singularities but showed higher stress peaks at these areas.

The geometry of the debonded area was chosen by a qualitative assumption of the stress distribution within the Broutman test specimen [4]. The concave shape of the Broutman test specimen leads to a concentration on opposite edges of the C-fiber. In consequence, circumferential stresses lead to two peaks at these edges, whereas in the axial direction the stresses are comparably high [17]. With the FE simulation setup though, several aspects, i.e., the different circumferential stresses, as well as a possible debonding on the opposite side, were in the first step neglected, due to the simplicity of the FE model and its interpretation.

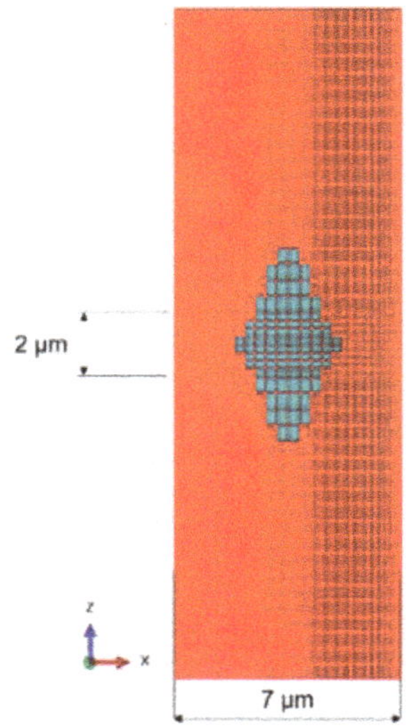

Figure 3. Depiction of the investigated debonding scenario of a cluster defect in the fiber−matrix interface with shown center section of 2 µm (only the C-fiber is shown).

3. Results

3.1. X-ray Imaging of as-Prepared Loaded and Nonloaded Broutman Test Specimen

In Figure 4, two X-ray projections of two Broutman test specimens are shown; a non-loaded specimen (a) and a preloaded specimen (b). The images were taken under identical imaging settings (Config. 4, Table 2), as described in Section 3.2 and schematically depicted in Figure 1b. The single C-fiber can be clearly distinguished with high contrast from the surrounding polymeric matrix. The as described machine settings led to an acquisition resolution of 650 nm. The mean diameter of the C-fiber can be estimated from the projections as approximately 7.5 µm, which is well in the range of the expected C-Fiber diameter, considering the resolution. Using high magnification and in-line phase contrast in the Fresnel-regime to enhance the overall contrast in the projection led to an excessive edge-contrast in the resulting projection, which appeared as a distinct black line between the C-fiber and surrounding matrix. Although the intensity of this artefact depended among other things on the acceleration voltage and magnification [15], it was not possible to completely eliminate it. Therefore, conclusions about the interface spatial dimensions between C-fiber and matrix are somehow speculative. However, the projection on the preloaded specimen in Figure 4b clearly showed four locations of fiber fracture (A, B, C and D) along the C-fiber with fracture sections of approximately 25.6 µm in length as a mean value. A characteristic fracture pattern could be seen with two fracture planes and a small fracture piece of approximately 2–8 µm in length in between. Figure 5 shows a high-resolution image of one of these C-fiber fracture sections of the projection shown in Figure 4. The volume view enabled a 3D representation of that region. This allowed the determination of the inclination angle of the fracture plane, which was in all observed cases about ±45° to the load direction.

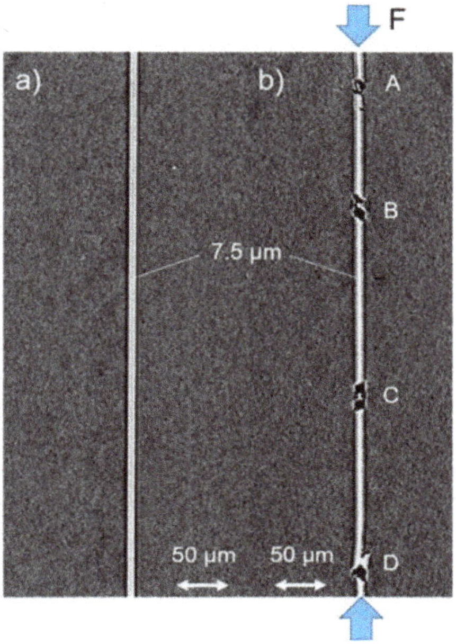

Figure 4. Post mortem X-ray projections of two Broutman test specimens taken under identical settings: (**a**) nonloaded and (**b**) preloaded. The C-fiber is clearly visible in the toughened polymeric matrix, additionally the in-line phase contrast artefacts are visible in both projections as a distinct black line at both sides of the fiber. A, B, C, D mark the locations of fiber fracture with two fracture planes each and a small fracture piece in between them. The blue arrows highlight the direction of force applied during tests (compression).

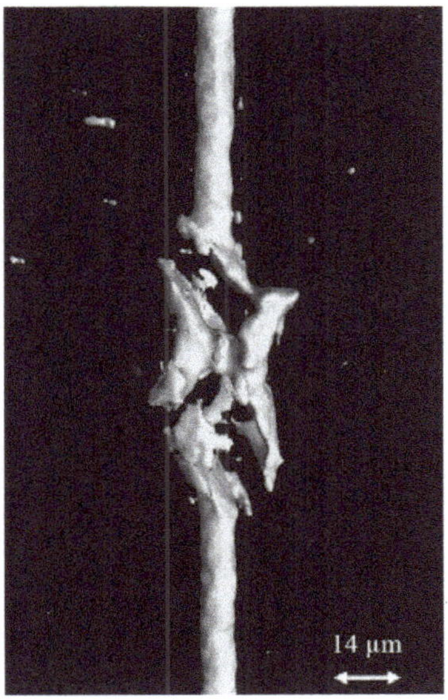

Figure 5. Example of one 3D location of fiber fracture image in high-resolution, taken by XRM, allowing the measurement of the inclination angle of the fracture planes and the size of the small fracture pieces between single fragments.

Figure 6 shows the comparison of differently acquired failure zones extracted from the middle of preloaded Broutman specimens (toughened matrix) one projection taken with XRM (a) and two images taken using optical light microscopy, with (b) and without (c) polarized light. The X-ray projection (Figure 6a) was taken with a magnification of 200 times, with the settings described in Config. 4 in Table 2. The image, taken with polarized light (Figure 6b), also shown at 200 times magnification and Figure 6c was taken with light microscopy with a magnification of 500 times. Due to prevailing residual stresses in the matrix, either induced by the manufacturing process, i.e., shrinkage phenomena, or the applied loads, the polarized light revealed the areas on the C-fiber surface, where the matrix was debonded during testing, but the fracture point itself was not resolved in sufficient detail, so the measurement of the exact fragment length or the inclination angle of the fracture plane was not possible. The images taken with light microscopy did not visualize the observed fracture pattern, due to limited magnification and depth resolution. The X-ray imaging however revealed the most details in high resolution, offered the possibility of a volumetric visualization and enabled the examination of nontransparent specimens. Therefore, a quantitative assessment of the fracture in the C-fiber in terms of fragment length and inclination angle of the fracture plane was easily possible. In our case, the mean fragment length (L) was 176 μm, measured with the light microscope without polarized light considering five fragments of two different images taken from the same specimen. The mean fragment length measured with XRM was 146 μm, considering five fragments of two different scans taken from the same specimen. Measuring the exact fragment length is important for correlation of the fragment length and the stress distribution as Park et al. did, by comparing the results of the Fragmentation test with those gained with the Broutman test [25]. Furthermore, only X-ray imaging can visualize those small fracture pieces which are pushed into the matrix. At some locations this induced a local cohesive matrix failure mechanism. (Figure 6, blue circle) The orange arrows mark the probably debonded areas.

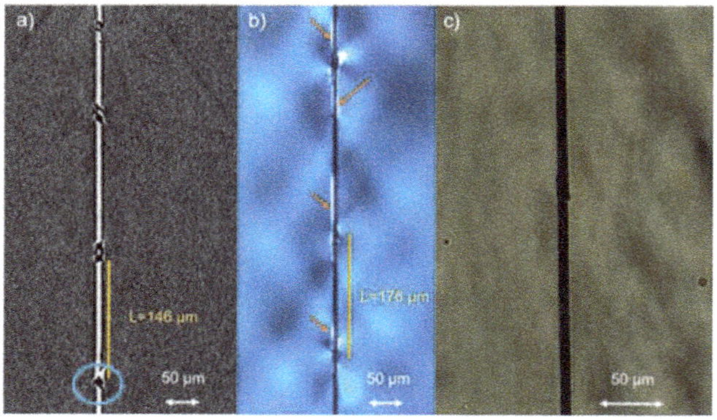

Figure 6. Comparison between visualization of the same preloaded toughened Broutman test specimen using (**a**) XRM, (**b**) polarized light microscopy and (**c**) light microscopy. The measured mean fragment length L for the respective ROI is shown for comparison. The probably debonded areas are marked by the orange arrows.

Figure 7 presents a projection of the toughened (a) and the non-toughened (b) specimen configuration for two specimens each. The fracture pattern of the non-toughened specimen was in most cases similar to the already described fracture pattern of the toughened specimen. The fracture section of the non-toughened specimens was approximately 22 μm in length as a mean value. The mean fragment length (L) was 183 μm, which was almost similar to the toughened system. The size of the small fracture pieces varied. The main visible difference between Figure 7a,b is the degree of fracture. In the toughened specimens, C-fiber slipping occurred. Additionally the small fracture pieces were pushed into the matrix, including cohesive matrix failure. The non-toughened specimens showed a smaller degree of fracture. After failure under compression, no further load seemed to be applied to the small fracture pieces. This can also be confirmed by the volume representation of individual fracture points of both systems, which is shown in Figure 8. A comprehensive interpretation of the results shown here will be given in Section 3.3. A comparison between the polarized light microscopy images of the toughened and non-toughened system is shown in the supplementary information (Figure S3 in S4).

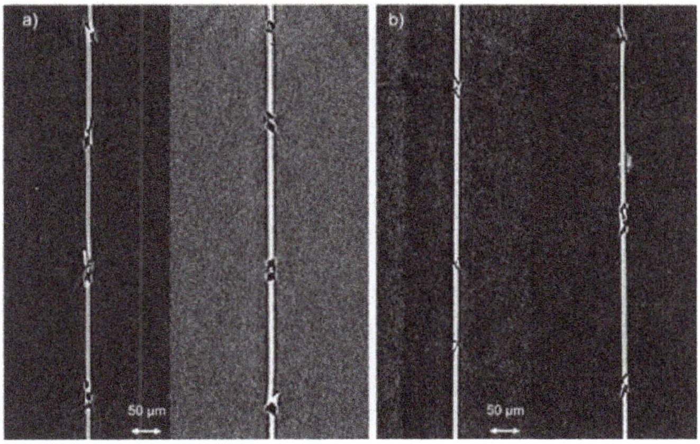

Figure 7. Post mortem X-ray projections of four Broutman test specimens taken under identical settings: with (**a**) toughened matrix and (**b**) non-toughened matrix. All show the already mentioned phase contrast effect and several locations of fiber fracture with two fracture planes each and a small fracture piece in between them.

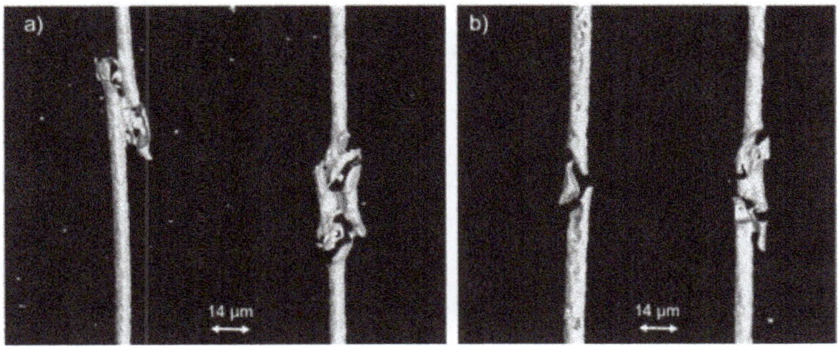

Figure 8. Example of one 3D image in high-resolution of a representative fiber fracture of each tested specimen, (**a**) with toughened matrix and (**b**) with non-toughened matrix, taken by XRM, allowing the measurement of the inclination angle of the fracture plane and the size of the small fracture pieces between single fragments.

3.2. FE Simulation of the Stress Distribution in the Partially Debonded Fiber—Matrix Interface and the C-Fiber

With the help of the FE model, described above in Section 2.5, characteristic changes of the stress distribution in the fiber can be observed as consequences of the localized debonding of the fiber—matrix interface. In the present case of a cluster defect in the fiber—matrix interface, this stress distribution is presented in Figure 9 by means of a cylindrical coordinate system in a frontal and a cut view from the edge of the defect to the center line. The cut view was chosen to better visualize the stress distribution in the C-fiber. The left part of Figure 9 presents the axial stress component S_{33}, where a stress distribution with tensional and compressive stress fields relative to the overall compressive load on the C-fiber can be observed. In the middle of the debonded area, a compressive stress concentration was present, whereas in the upper and lower boundary of the debonded area, tensional stress appeared. The difference between relative compression and tension however was within single MPa values.

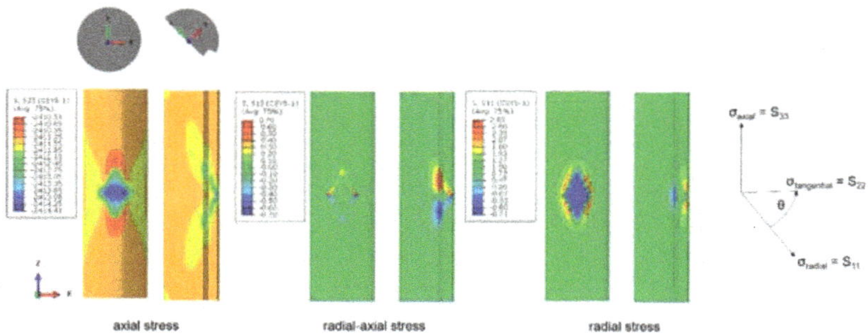

Figure 9. Effects of a fiber–matrix cluster debonding on fiber stresses, top view of the shown C-fiber segment; full fiber (**left**), cut view (**right**), axial stresses in MPa S_{33} (**top**), radial-axial shear stress in MPa S_{13} (**middle**) and radial stresses in MPa S_{11} (**bottom**).

The radial stress component S_{11} is depicted in the left part of Figure 9, showing a compression field in the middle of the debonded area, where no stress was transmitted to the matrix. Regarding the front view, singularities at the edge of the debonding can be observed, resulting in a peaked pattern of positive radial stresses. Nevertheless, the affected area of these singularities was small and only appearing at these edges, not influencing internal or superficial fiber stresses away from the singularities. In the cut view further positive stress concentrations can be observed leading to a radial expansion of the fiber.

In the middle section of Figure 9, the combined shear stress component S_{31} in radial-axial direction is shown. Distinctive maxima and minima can be seen, especially in the cut view. This distribution of S_{31} might explain the inclination angle of the fracture plane of $\pm 45^{\circ}$ relative to the surface of the C-fiber. By having two fracture planes, a separation of the C-fiber would lead to the fracture pattern as it can be observed in the X-ray projection in Figure 4. Accounting for the overall debonding stress, being present in the Broutman test specimen under compressive load would amplify the effects of the observed distributions of the stress components S_{33}, S_{11} and S_{31}, especially at the edges of the debonded area.

3.3. Discussion

The results shown in Section 3.1 revealed a regular fracture pattern, which raises some questions. For example, how the small fracture pieces between the fragment ends occur. The experimentally observed fracture pattern can be interpreted in several ways. The small fracture pieces between fragment ends could be an effect of secondary failure, initiated by the fragment ends remaining in contact during the test, due to the increasing compressive force and slip and crush against each other. This would imply that during fragmentation the normal stress at the fragment ends is nonzero. Several researchers have already assumed the slippage of fragment ends [24,25,35]. However, comparison of the different systems (toughened and non-toughened) shown in Figures 7 and 8 shows that the small fracture pieces in the fracture section are probably not a secondary failure effect. Rather, they appear to be part of the fracture pattern that occurs during compressive loading. The strength and the time period during which the fragment ends remain in contact and compress the small fracture pieces (or pushed them into the matrix) after C-fiber failure seems to have a correlation with the fiber–matrix interaction. In [24,25] a correlation between the inclination angle of the fracture plane and the resulting stress distribution was assumed, while [25] additionally observed that slippage of the broken fragment ends was caused by the high load bearing of the matrix during compression which resulted in yield phenomena which prevented matrix failure. Both assumed a nonzero normal stress at the fragment ends due to stress discontinuity caused by fiber fracture. The visible fracture pattern with the small fracture pieces in between assume a fracture plane not perpendicular to the fiber plane but influenced by the fiber fracture itself. The Broutman test itself can only be analyzed by determining the inclination angle of the fracture plane in the analytical modelling of interphase failure. The C-fiber failed in most visible fracture locations brittle, with an inclination angle of the fracture plane of $\pm 45^\circ$ but also fiber slipping could be observed. The visible fiber slippage could be attributed to transverse tensile stresses in the interphase according to [25]. Consequently, measuring the inclination angle of the fracture plane is of great importance for evaluation of the resulting stress distributions.

The inclination angle of the fracture plane was measured as $\pm 45^\circ$ at almost every fracture point in our investigation, this indicated a brittle C-fiber failure due to ductile shear failure under compression as proposed in [36].

By comparing the results of high-resolution XRM with polarized microscopy images a non-uniform delamination behavior between C-fiber and matrix can be supposed. This means that stress discontinuities along the C-fiber have to be considered, caused by a sequence of bonding and nonbonding regions, as stated in [32]. Observing C-fiber failure due to compression in several planes is an indication of a very good adhesion between the substituents [36,37]. Hence, for the considered specimens, a good adhesion can be assumed. Many authors conflict with each other with the expected stress distribution during the Broutman test and no consistent description could be found so far. In [24] it was shown that two different failure mechanisms compete during the Broutman test, fiber failure and fiber–matrix debonding. Fiber–matrix structures with a good interfacial adhesion always showed fiber failure due to the compressive stress before debonding occurred. They assumed that the shear properties of the interphase were determined during Broutman test.

Using high-resolution XRM can help as a post mortem tool to prevent misinterpretation of measured data. Using the Broutman equation (Equation (1) in S1), the calculated normal stress is 6.8 MPa (force at first detected failure 5000 N) for the toughened and 8.7 MPa (force at first failure detected 6500 N) for the non-toughened system. In [26] the macro-mechanical interphase behavior for the same fiber–matrix systems were already characterized in detail. From qualitative observation of SEM images it was deduced that toughening the polymeric matrix system resulted in better fiber–matrix adhesion. These qualitative findings are contradictory to the interfacial strength quantitatively determined by the Broutman test. This might indicate that the calculated debonding stress based on test data results misleads the understanding of the experimental results if no further analysis of the system is performed. Reasons for this ambiguity are manifold, such as geometric inaccuracies leading to different stress states. An alternative explanation would be that the toughened matrix system failure is more ductile, with a larger overall deformation than the non-toughened system. This causes an increase in the overall stress stored elastically in the system, which is then suddenly released leading to more damage to the C-fiber, which can be seen in the XRM images. However, since the force of the first failure, determined by AE, was higher in the non-toughened system, this could indicate that failure occurs gradually, whereas in the toughened system global failure (both cohesive and adhesive) occurs more or less simultaneously, starting from the center. These would correlate with the results shown in [25].

C-fiber buckling and C-fiber crushing are all time-dependent processes due to instabilities in the system. The failure mechanism analysis in our research does not depend on a dynamic process as our purely static FE simulation is well in line with the experimentally observed fracture pattern. The occurrence of a localized delamination alone leads to the observed fracture pattern.

4. Conclusions

Careful analysis of simple X-ray projections, together with high-resolution 3D XRM images of a Broutman specimen after mechanical loading allow the accurate description of important details of the observed fracture patterns of the single C-fiber. It was possible to assess the fragment length, the angle of inclination of the fracture plane and even the size of small fracture pieces between the broken fragments.

In comparison to classical light microscopy, the analysis via XRM is not limited by resolution depth and the optical transparency of the specimen. However, in the case of transparent specimens, polarized light microscopy provides additional information regarding the location and size of the debonded area of the interphase. To the authors' best knowledge, it is the first time a single carbon fiber within a matrix has been visualized via XRM in such a detailed manner.

With the combination of the presented techniques with regard to the single C-fiber failure mechanisms, a more profound interpretation of the failure process was possible, without assuming dynamic effects like a back-slippage of the C-fiber ends or other time-dependent failure mechanisms.

The C-fiber exhibited shear failure with an almost constant $\pm 45^{\circ}$ inclination angle of the fracture plane. The comparison between a toughened and a non-toughened matrix system indicated how the compression-introduced failure mechanism was influenced by the properties of the matrix system and the resulting fiber—matrix interaction. The degree of C-fiber fracture in the matrix gathered visually in the toughened system was higher than in the non-toughened. This could be an indication that the compression stress is stored in the matrix and suddenly released to the C-fiber, which leads to a higher degree of fracture in the toughened system.

Using a simple FE model, the stress distribution, prior to debonding, within the C-fiber was approximated. The FE simulation gave a first hint that a small debonded area in the fiber—matrix interface could be sufficient to introduce a stress distribution, which resulted in fiber failure with the observed fracture pattern. However, the limitation of the ROI during XRM analysis, the duration of the single exposure, and the phase contrast that provided further information about the interface are open issues that need further work. The combination of high-resolution 3D X-ray analysis of a micromechanical test, such as the Broutman test, together with a more refined simulation model shows a very promising approach to obtain a deeper understanding of the failure relevant processes in the interface and the fiber—matrix interphase itself.

5. Outlook

The results presented serve as a starting point for many further investigations. The FE model can be modified to investigate different debonding and failure mechanisms with more material-specific values and with additional improvements. The effect of a distributed debonding, i.e., two debonding locations on opposite sides on the C-fiber, were not considered here, as well as the effect of plasticity of the matrix or a cohesive zone modelling of the interphase instead of the simple representation of an infinitesimal interface.

In future activities, the variation of fiber or matrix properties or even the resulting interphase could be experimentally assessed, e.g., with high-resolution XRM. Therefore, different C-fiber types should be tested and the resulting failure patterns should be compared.

Most ambitious would be a redesign of the Broutman test specimen tail-shape to enable real 4D investigations of the interphase, by in situ testing in the XRM. With that, a better understanding of this micromechanical test could be achieved.

Supplementary Materials: The following are available online at https://www.mdpi.com/article/10.3390/jcs5050121/s1, S1: Micromechanical description of stress state according to Broutman, S2: Failure mechanism during Broutman test analyzed with additional methods: Figure S1: Post mortem polarized light microscopy images of the same Broutman test specimens (non-toughened) taken after different loading steps (4000 N and 6000 N). Figure S2: The cumulative number of Events of class 1, which describes a class of AE events which can be correlated (frequency dependent) with the fiber and the compression force applied in N for the toughened system, the non-toughened and two different reference systems. The further results of the tested systems and the test itself will be published elsewhere. S3: Supporting information about the Broutman test configurations and the test set-up; S4: Comparision between toughened and non-toughened matrix systems via polarized light microscopy: Figure S3: Post mortem polarized light microscopy image of the toughened and the non-toughened matrix system, indicating no visible difference in the degree of fracture.

Author Contributions: Conceptualization, J.V., A.K., T.R. and M.G.; methodology, J.V., A.K., T.R. and M.G.; formal analysis, J.V., T.R. and M.G.; investigation, J.V. and A.K.; data curation, T.R.; writing—original draft preparation, J.V., A.K., T.R. and M.G.; writing—review and editing, J.V. and M.G.; visualization, J.V. and T.R.; supervision, M.G.; project administration, M.G.; funding acquisition, M.G. All authors have read and agreed to the published version of the manuscript.

Funding: The authors gratefully acknowledge funding from the German Research Association (DFG), grant number: GU 1556/2-1.

Data Availability Statement: https://www.scientific.net/book/22nd-symposium-on-composites/978-3-0357-3453-9.

Acknowledgments: The support of Benjamin Kelkel who provided AE measurements during micromechanical testing and Christian Becker is highly appreciated.

Conflicts of Interest: The authors declare no conflict of interest. The funders had no role in the design of the study; in the collection, analyses, or interpretation of data; in the writing of the manuscript, or in the decision to publish the results.

References

1. Drzal, L.T.; Madhukar, M. Fibre-matrix adhesion and its relationship to composite mechanical properties. *J. Mater. Sci.* **1993**, *28*, 569–610. [CrossRef]
2. Jesson, D.A.; Watts, J.F. The interface and interphase in polymer matrix composites: Effect on mechanical properties and methods for identification. *Polym. Rev.* **2012**, *52*, 321–354. [CrossRef]
3. Pilato, L.A.; Michno, M.J. *Advanced Composite Materials*; Springer Science and Business Media LLC: Berlin/Heidelberg, Germany, 1994.
4. Broutman, L.J. Glass-resin joint strength and their effect on failure mechanisms in reinforced plastics. *Polym. Eng. Sci.* **1966**, *6*, 263–272. [CrossRef]
5. Remaoun, D.; Boutaous, A. Thermomechanical stress in the evolution of shear of fiber-matrix interface composite material. *Mater. Sci. Appl.* **2011**, *2*, 399–403. [CrossRef]
6. Riaño, L.; Chailan, J.-F.; Joliff, Y. Evolution of effective mechanical and interphase properties during natural ageing of glass-fibre/epoxy composites using micromechanical approach. *Compos. Struct.* **2021**, *258*, 113399. [CrossRef]
7. Zhang, X.; Fan, X.; Yan, C.; Li, H.; Zhu, Y.; Li, X.; Yu, L. Interfacial microstructure and properties of carbon fiber composites modified with graphene oxide. *ACS Appl. Mater. Interfaces* **2012**, *4*, 1543–1552. [CrossRef]
8. Mäder, E.; Gao, S.-L.; Plonka, R. Enhancing the properties of composites by controlling their interphase parameters. *Adv. Eng. Mater.* **2004**, *6*, 147–150. [CrossRef]
9. Sharma, M.; Gao, S.; Mäder, E.; Sharma, H.; Wei, L.Y.; Bijwe, J. Carbon fiber surfaces and composite interphases. *Compos. Sci. Technol.* **2014**, *102*, 35–50. [CrossRef]
10. Chen, L.; Jin, H.; Xu, Z.; Li, J.; Guo, Q.; Shan, M.; Yang, C.; Wang, Z.; Mai, W.; Cheng, B. Role of a gradient interface layer in interfacial enhancement of carbon fiber/epoxy hierarchical composites. *J. Mater. Sci.* **2015**, *50*, 112–121. [CrossRef]
11. Nishikawa, M.; Okabe, T.; Hemmi, K.; Takeda, N. Micromechanical modeling of the microbond test to quantify the interfacial properties of fiber-reinforced composites. *Int. J. Solids Struct.* **2008**, *45*, 4098–4113. [CrossRef]
12. Pisanova, E.; Zhandarov, S.; Mader, E. How can adhesion be determined from micromechanical tests? *Compos. Part A Appl. Sci. Manuf.* **2001**, *32*, 425–434. [CrossRef]
13. Lodeiro, M.J.; Maudgal, S.; McCartney, L.N.; Morrell, R.; Roebuck, B. *Critical Review of Interface Testing Methods for Composites*; NPL: Teddington, UK, 1998.
14. Pitkethly, M.; Favre, J.; Gaur, U.; Jakubowski, J.; Mudrich, S.; Caldwell, D.; Drzal, L.; Nardin, M.; Wagner, H.; Di Landro, L.; et al. A round-robin programme on interfacial test methods. *Compos. Sci. Technol.* **1993**, *48*, 205–214. [CrossRef]
15. Garcea, S.; Wang, Y.; Withers, P. X-ray computed tomography of polymer composites. *Compos. Sci. Technol.* **2018**, *156*, 305–319. [CrossRef]
16. Seghini, M.; Touchard, F.; Sarasini, F.; Chocinski-Arnault, L.; Mellier, D.; Tirillò, J. Interfacial adhesion assessment in flax/epoxy and in flax/vinylester composites by single yarn fragmentation test: Correlation with micro-CT analysis. *Compos. Part A Appl. Sci. Manuf.* **2018**, *113*, 66–75. [CrossRef]

17. Schuller, T.; Beckert, W.; Lauke, B.; Ageorges, C.; Friedrich, K. Single fibre transverse debonding: Stress analysis of the Broutman test. *Compos. Part A Appl. Sci. Manuf.* **2000**, *31*, 661–670. [CrossRef]
18. Yu, S.; Hwang, Y.H.; Hwang, J.Y.; Hong, S.H. Analytical study on the 3D-printed structure and mechanical properties of basalt fiber-reinforced PLA composites using X-ray microscopy. *Compos. Sci. Technol.* **2019**, *175*, 18–27. [CrossRef]
19. Wirjadi, O.; Godehardt, M.; Schladitz, K.; Wagner, B.; Rack, A.; Gurka, M.; Nissle, S.; Noll, A. Characterization of multilayer structures in fiber reinforced polymer employing synchrotron and laboratory X-ray CT. *Int. J. Mater. Res.* **2014**, *105*, 645–654. [CrossRef]
20. Lin, G.; Geubelle, P.; Sottos, N. Simulation of fiber debonding with friction in a model composite pushout test. *Int. J. Solids Struct.* **2001**, *38*, 8547–8562. [CrossRef]
21. Ojos, D.E.-D.L.; Ghisleni, R.; Battisti, A.; Mohanty, G.; Michler, J.; Sort, J.; Brunner, A. Understanding the mechanical behavior of fiber/matrix interfaces during push-in tests by means of finite element simulations and a cohesive zone model. *Comput. Mater. Sci.* **2016**, *117*, 330–337. [CrossRef]
22. Jager, J.; Sause, M.; Burkert, F.; Moosburger-Will, J.; Greisel, M.; Horn, S. Influence of plastic deformation on single-fiber push-out tests of carbon fiber reinforced epoxy resin. *Compos. Part A Appl. Sci. Manuf.* **2015**, *71*, 157–167. [CrossRef]
23. Singh, D.K.; Vaidya, A.; Thomas, V.; Theodore, M.; Kore, S.; Vaidya, U. Finite element modeling of the fiber-matrix interface in polymer composites. *J. Compos. Sci.* **2020**, *4*, 58. [CrossRef]
24. Ageorges, C.; Friedrich, K.; Schuller, T.; Lauke, B. Single-fibre Broutman test: Fibre–matrix interface transverse debonding. *Compos. Part A Appl. Sci. Manuf.* **1999**, *30*, 1423–1434. [CrossRef]
25. Park, J.-M.; Kim, J.-W.; Yoon, D.-J. Interfacial evaluation and microfailure mechanisms of single carbon fiber/bismaleimide (BMI) composites by tensile and compressive fragmentation tests and acoustic emission. *Compos. Sci. Technol.* **2002**, *62*, 743–756. [CrossRef]
26. Klingler, A.; Bajpai, A.; Wetzel, B. The effect of block copolymer and core-shell rubber hybrid toughening on morphology and fracture of epoxy-based fibre reinforced composites. *Eng. Fract. Mech.* **2018**, *203*, 81–101. [CrossRef]
27. HUNTSMAN. *Aradur 917 CH Safety Data Sheet*; HUNTSMAN: Woodlands, TX, USA, 2016.
28. Kaneka. *Kane Ace MX553 Safety Data Sheet*; Kaneka: Tokyo, Japan, 2017.
29. HUNTSMAN. *Accelerator DY 070 Safety Data Sheet*; HUNTSMAN: Woodlands, TX, USA, 2012.
30. HEXCEL. *HexForce G1157 D1300, Data Sheet*; HEXCEL: Stamford, CT, USA, 2017.
31. Sika. *Deutschland GmbH Biresin CR144, Data Sheet*; Sika: Baar, Switzerland, 2017.
32. Schürmann, H. *Konstruieren mit Faser-Kunststoff-Verbunden*; Metzler, J.B., Ed.; Springer: Berlin/Heidelberg, Germany, 2005.
33. 22nd Symposium on Composites. Available online: https://www.scientific.net/book/22nd-symposium-on-composites/978-3-0357-3453-9 (accessed on 29 March 2021).
34. Scott, A.; Mavrogordato, M.; Wright, P.; Sinclair, I.; Spearing, S. In situ fibre fracture measurement in carbon–epoxy laminates using high resolution computed tomography. *Compos. Sci. Technol.* **2011**, *71*, 1471–1477. [CrossRef]
35. Wood, J.R.; Wagner, H.D.; Marom, G. A Model for Compressive Fragmentation. *Adv. Compos. Lett.* **1994**, *3*, 096369359400300403. [CrossRef]
36. Peebles, L.H. *Carbon Fibers: Formation, Structure, and Properties*; CRC Press: Boca Raton, FL, USA, 2018; ISBN 978-1-351-08732-2.
37. Drzal, L.T. The Interphase in Epoxy Composites. In Proceedings of the Epoxy Resins and Composites II; Dušek, K., Ed.; Springer: Berlin/Heidelberg, Germany, 1986; pp. 1–32.

Journal of
Composites Science

MDPI

Article

Investigation of a Non-Equilibrium Energy Model for Resin Transfer Molding Simulations

Anthony Sherratt [1,*], Anthony G. Straatman [1], Christopher T. DeGroot [1] and Frank Henning [2]

1 Department of Mechanical and Materials Engineering, University of Western Ontario, London, ON N6A 5B9, Canada; astraatman@eng.uwo.ca (A.G.S.); cdegroo5@uwo.ca (C.T.D.)
2 Karlsruhe Institute of Technology (KIT), Institute of Vehicle System Technology (FAST), Department of Lightweight Technology (LBT), 76131 Karlsruhe, Germany; frank.henning@ict.fraunhofer.de
* Correspondence: asherra@uwo.ca

Abstract: Due to the high design freedom and weight specific properties carbon fiber reinforced plastics (CFRP) offer significant potential in light-weighting applications, specifically in the automotive sector. The demand for medium to high production quantities with consistent material properties has paved the way for the use of high-pressure resin transfer molding (HP-RTM). Due to high experimental cost and number of the operational parameters the development of numerical simulations to predict part quality is growing. Despite this, erroneous assumptions and simplifications limit the application of HP-RTM models, specifically with regards to the energy models used to model the heat transfer occurring during infiltration. The current work investigates the operating parameters at which the thermal non-equilibrium energy model's increased computational cost and complexity is worth added accuracy. It was found that in nearly all cases, using the thermal non-equilibrium is required to obtain an accurate prediction of the temperature development and resulting final properties within the mold after the infiltration process.

Keywords: non-equilibrium energy molding; mold filling; process simulation; high pressure resin transfer molding; openFOAM; porous media

Citation: Sherratt, A.; Straatman, A.G.; DeGroot, C.T.; Henning, F. Investigation of a Non-Equilibrium Energy Model for Resin Transfer Molding Simulations. *J. Compos. Sci.* **2022**, *6*, 180. https://doi.org/10.3390/jcs6060180

Academic Editor: Stelios K. Georgantzinos

Received: 1 June 2022
Accepted: 17 June 2022
Published: 20 June 2022

1. Introduction

Pressure on the automotive sector to reduce vehicle weight and thereby reduce energy usage and greenhouse gas emissions has paved the way for carbon fiber reinforced plastics (CFRP) due to their favorable weight-specific properties and design freedom [1–5]. However, significant upfront costs for tooling, increased manufacturing time and low recyclability has limited their application predominantly to the luxury sector [6–9]. Despite this, there is still a significant push to increase the use of CFRP components in automotive light-weighting.

Numerous methods exist for producing CFRP components, however, looking specifically at the requirements for automotive use (high design freedom, low cycle time and high automation potential), a subset of liquid composite molding, namely Resin Transfer Molding (RTM), is becoming increasingly popular [2,5,10]. The RTM manufacturing process starts with draping, wherein a sheet of dry fabric is cut to shape and laid in the bottom half of a rigid, heated mold. The mold is then closed and resin is injected. Once fully saturated, the resin is left to cure and once cured, the part is removed for post-processing. Given the large number of process parameters and high experimental cost, numerous numerical models have been developed to optimize the resin infiltration process, allowing the designer to vary parameters such as inlet pressure, injection location(s) and mold temperature to ensure the component is completely infiltrated [11–13]. However, with increased use of specialized fabrics, common assumptions such as thermal equilibrium between the solid (fiber) and liquid (resin) phases—limit the general application and accuracy of current RTM models.

During the infiltration stage of RTM, the temperature is constantly evolving due to the interaction between the preheated mold and preform, the cool injected resin and the exothermic nature of resin curing. This complex process has been simplified in RTM models using an equilibrium energy model, whereby the liquid (resin and/or air) and solid (fiber) phases are treated as though they are at the same temperature locally within the domain [14–18]. The main rationale for this assumption has been that, despite the different thermophysical properties of the fiber and resin, the resin has a slow velocity within the mold therefore permitting the different phases to locally be at effectively the same temperature. This assumption reduces the complexity of the formulation and the computational time for solving the energy equation, as it is a single equation which requires no phase-coupling. However, modern efforts [19,20] to reduce the overall cycle time for RTM manufacturing include derivatives such as high-pressure resin transfer molding (HP-RTM). In HP-RTM, resin is injected into the mold at significantly higher rates than standard RTM (i.e., 20–200 g/s [5]), which drastically reduces the infiltration time. In addition, to reduce the overall manufacturing time, highly reactive, fast curing resins are being used to reduce the curing stage [21]. This also has an effect on the infiltration stage, given that the resins cure more during infiltration, which can result in increased heat being released. The combination of a high injection rate and rapidly curing resin reduces the likelihood that the phases are in local thermal equilibrium. Given that the thermophysical properties of the resin and part warpage are also highly dependent on temperature and cure evolution, modelling the temperature evolution of the two phases separately can be crucial for obtaining accurate RTM simulations and to the authors' knowledge has not been investigated prior to this study [22,23].

The price of increased accuracy in thermal modelling is the requirement of a Local Thermal Non-Equilibrium (LTNE) formulation, whereby energy transport equations are solved for both the solid and fluid (resin) phases leading to added complexity and computational cost. Numerous studies have been conducted to determine the parameters that dictate when the assumption of thermal equilibrium is valid, however, multiple factors are at play. In the work done by Calmidi and Mahajan [24] and Straatman et al. [25] on highly-conductive porous foams, it was stated that the assumption of thermal equilibrium is not applicable when the conductivity ratio between the solid and fluid phases is substantial. This implies that for RTM processes that consider low-conductivity glass fibers, for example, the assumption of thermal equilibrium may be applicable, however, this assumption also depends on the infiltration velocity and its impact on the interstitial (convective) heat transfer coefficient, which impacts the thermal resistance between the phases. The relative temperatures of the phases are particularly important when the temperature is a threshold related to a phase change, which can be the case in modern HP-RTM. Quintard and Whitaker [26] proposed a method for determining when the assumption of local thermal equilibrium is applicable based on:

$$\frac{\langle T_\beta \rangle^\beta - \langle T_\alpha \rangle^\alpha}{\Delta \langle T \rangle} = \left(\frac{l_{\beta\alpha}}{L(t)} \right)^2 \{\mathbf{O}(1-10)\} \tag{1}$$

where $\langle T_\alpha \rangle^\alpha$ and $\langle T_\beta \rangle^\beta$ are the intrinsically averaged temperatures of the α and β phases, respectively, $\Delta\langle T \rangle$ is the temperature gradient across the domain, $l_{\beta\alpha}$ is the mix-node, small length scale and $L(t)$ is the length scale. Equation (1) suggests that an accurate prediction of $l_{\beta\alpha}$ is required for an accurate determination of when thermal equilibrium will occur, however, this property is strongly affected by the heat transfer coefficient between the phases. Due to the wide range of resins, fiber types and fiber weaves, and noting that this correlation was established for flow over a periodic unit cell subjected to stratified flow, this increases the potential error in determination of the heat transfer coefficient, thereby reducing confidence when using the right hand side of Equation (1) to predict thermal equilibrium for RTM. The left hand side of Equation (1), however, supports performing a non-dimensional analysis using a range of material properties and operating conditions commonly found in the manufacturing of RTM and RTM variant components. This analysis would provide a comprehensive set of guidelines that would allow for an accurate determination of when to use thermal equilibrium and thermal non-equilibrium energy models when modelling the RTM infiltration process.

The current study examines the use of a local thermal non-equilibrium (LTNE) energy model in RTM. The finite-volume based, multi-phase RTM model, originally developed by Magagnato et al. [3] in the open source tool box OpenFOAM®, is extended to include coupled transport equations for the resin (fluid) and fiber (solid) phases, wherein the fluid equation contains a heat release source term to account for resin curing, and the fluid and solid equations are linked through a bulk (interstitial) heat exchange term. Model verification is achieved using a 1D representative geometry to ensure model accuracy. A non-dimensional analysis is then presented to determine the appropriate material, geometric, and operating conditions where the LTNE energy model is worth the added computational cost and complexity. Finally, a sample case of a complex geometry simulated using both thermal equilibrium and LTNE models is presented to demonstrate the value of the LTNE approach.

2. Governing Equations and Models for Mold Filling Simulations

2.1. Conservation of Mass and Momentum

RTM mold filling simulations involve solutions of laminar, incompressible flow through a porous medium. The volume-averaged governing equations for the conservation of mass and momentum, following the approach given by Quintard et al. [27], are

$$\nabla \cdot \mathbf{v} = 0 \tag{2}$$

and

$$\rho_f \frac{\partial \mathbf{v}}{\partial t} + \frac{\rho_f}{\epsilon} \nabla \cdot (\mathbf{v}\mathbf{v}) = -\epsilon \nabla p_f + \mu_f \nabla^2 \mathbf{v} + \epsilon S \tag{3}$$

where $\mathbf{v}$ is the volume-averaged velocity, p_f is the fluid pressure and ϵ is the porosity. In OpenFOAM, the multi-phase model solves a single mass-weighted-averaged momentum equation for the two (or more) fluid phases, therefore, ρ_f and μ_f are the mass-weighted-averaged density and dynamic fluid viscosity, respectively, for the fluid phase(s). The source term, S, is included to account for the flow resistance from the fiber preform. This resistance is modelled by Darcy's law [2]:

$$\nabla p = \frac{\mu}{\mathbf{K}} \mathbf{v} \tag{4}$$

where $\mathbf{v}$ is the volume-averaged velocity, $\mathbf{K}$ is the permeability tensor, μ is the dynamic fluid viscosity and ∇p is the pressure gradient in the mold. This formulation of Darcy's law is applicable for flows where the Reynolds number based on the pore diameter is small ($Re_d < 1$). When considering flows at higher Reynolds numbers, the flow transfers into the Forchheimer flow regime [28], where the pressure drop is quadratically related to the fluid velocity. This is shown as

$$\nabla p = \frac{\mu \mathbf{v}}{\mathbf{K}} + C_f \rho \frac{\mathbf{v}^2}{\sqrt{K}} \tag{5}$$

where C_f is the inertia coefficient and ρ is the fluid density. To get Equation (5) to match the format required for the OpenFOAM momentum source, the volume averaged velocity can be factored out and the coefficients in the Forchheimer term can be combined to give

$$S = -\left(\frac{\mu}{\mathbf{K}} + \frac{1}{2}|\mathbf{v}|F\right)\mathbf{v} \tag{6}$$

where F is an inertia resistance term (i.e., $2C_f\rho/\sqrt{K}$) and the negative denotes the resistance to flow from the porous media.

The interface of the two phase flow (i.e., resin and air) is solved using the Volume-Of-Fluid (VOF) method, given as

$$\frac{\partial \alpha}{\partial t} = \nabla \cdot (\alpha \mathbf{v}) \tag{7}$$

where α is the fraction of which each volume is filled with the resin phase (α = 1) or air phase (α = 0).

2.2. Conservation of Energy

The goal for the proposed model is to accurately capture the transient temperature behaviour for RTM and RTM variants (i.e., HP-RTM). Operating parameter variations can lead to increased resin velocity, in the case of HP-RTM, as well as increased heat released when using highly reactive resins. To ensure generality, the model needs to accurately capture these variations while still maintaining computational efficiency. Given this, both a thermal equilibrium and non-equilibrium energy model are used and compared.

2.2.1. Thermal Equilibrium

In thermal equilibrium the local fluid and solid temperatures are assumed to be the same, allowing for a single energy equation to be solved to characterize both phases. The energy equation for both phases is given as

$$\tilde{\rho}\tilde{C_p}\frac{\partial T}{\partial t} + \rho_f C_{pf}(\mathbf{v} \cdot \nabla T) = \nabla \cdot \tilde{k} \nabla T + \epsilon S''' \tag{8}$$

where T is the bulk temperature of the fluid and solid, ρ_f is the fluid density, C_{pf} is the fluid specific heat, and S''' is the source term associated with the heat release from the resin curing. The effective density, $\tilde{\rho}$, specific heat, $\tilde{C_p}$, and thermal conductivity, $\tilde{k}$, are calculated using the rule of mixtures [29,30]

$$\tilde{\rho} = \frac{\rho_f \rho_s}{\rho_f w_f + \rho_s w_s}, \tag{9}$$

$$\tilde{C_p} = C_{pf} w_f + C_{ps} w_s, \tag{10}$$

and

$$\tilde{k} = \frac{k_f k_s}{k_f w_f + k_s w_s} \tag{11}$$

where the weight fractions of the fluid and solid phase, w_f and w_s, respectively, are defined as

$$w_f = \frac{\epsilon/\rho_s}{(\epsilon/\rho_s + (1-\epsilon)/\rho_f)}, \tag{12}$$

and

$$w_s = 1 - w_f. \tag{13}$$

2.2.2. Local Thermal Non-Equilibrium

In the local thermal non-equilibrium (LTNE) approach the fluid and solid phases are modelled separately. The governing energy equations for the fluid and solid phase, after volume averaging following the approach given by Quintard et al. [27], and are given as

$$\rho_f C_{pf}\left(\epsilon \frac{\partial T_f}{\partial t} + \mathbf{v} \cdot \nabla T_f\right) = \nabla \cdot (\epsilon k_f \cdot \nabla T_f) + a_{fs} h_{fs}(T_s - T_f) + \epsilon S''' \tag{14}$$

and

$$(1-\epsilon)\rho_s C_s \frac{\partial T_s}{\partial t} = \nabla \cdot ((1-\epsilon)\mathbf{k}_s \cdot T_s) + a_{fs} h_{fs}(T_f - T_s), \quad (15)$$

respectively. Here, ρ_f and ρ_s are the fluid and solid densities, respectively; C_{pf} and C_s are the fluid and solid specific heats, respectively; k_f is the fluid conductivity, $\mathbf{k}_s$ is the solid conductivity tensor; T_f and T_s are the fluid and solid temperatures, respectively; a_{fs} is the specific area of the porous media; and h_{fs} is the interstitial heat transfer coefficient. The source term, S''' is calculated based on the total reaction enthalpy (Δh) and the cure rate of the resin ($\dot{c}$), given as

$$S''' = \Delta h \dot{c} \quad (16)$$

2.3. Chemo-Rheological Models

During infiltration and curing, it is critical to accurately capture the reaction kinetics (modelling the progression of the cure degree) and the resulting variations in glass transition temperature and viscosity due to their use in the source terms of the momentum and energy transport equations. For a recent review of the experimental techniques used to characterize the parameters listed in the following models, the reader is to refer to an article by Bernath et al. [31] and Halley and Mackay [32]. Herein, we discuss the approaches currently used in RTM modelling.

2.3.1. Reaction Kinetics

There are many different reaction kinetics models that are used in industry, with the most common being the Kamal–Malkin kinetic model [31] due to its relative simplicity and small number of fitting parameters. The Kamal–Malkin kinetic model is given as

$$\dot{c} = \frac{dc}{dt} = (K_1 + K_2 \cdot c^m) \cdot (1-c)^n \quad (17)$$

where c is the cure degree, and m and n are fitting parameters. The reaction rate constants, K_1 and K_2, are given as

$$K_1 = A_1 \cdot exp\left(-\frac{E_1}{RT}\right) \quad (18)$$

$$K_2 = A_2 \cdot exp\left(-\frac{E_2}{RT}\right) \quad (19)$$

where A_1 and A_2 are pre-exponential factors, E_1 and E_2 are activation energies, R is the universal gas constant and T is the temperature. Due to its simplicity, the Kamal–Malkin model is unable to capture vitrification effects; i.e., premature solidification of the resin. Furthermore, due to the $(1-c)$ term, the model will predict that the resin will cure regardless of the temperature applied. Despite these drawbacks, the Kamal–Malkin model will be used in the present study due to its common use in commercial softwares. This will further highlight the effect of the non-equilibrium energy model.

2.3.2. Glass Transition Temperature

As mentioned previously, vitrification occurs when the resin solidifies prior to reaching is gel point. This is caused by the glass transition temperature approaching the current operating temperature in the mold. To capture this effect, a model based on the work by DiBeneditto [33,34] is used:

$$\frac{T_g - T_{g,0}}{T_{g,\infty} - T_{g,0}} = \frac{\lambda c}{1-(1-\lambda)c} \quad (20)$$

where T_g is the current glass transition temperature, $T_{g,0}$ and $T_{g,\infty}$ are the uncured and fully cured glass transition temperatures, respectively, c is the cure degree and λ is a fitting parameter.

2.3.3. Viscosity

Coupled with the significant dependency on the temperature and cure degree, the resin viscosity is a significant factor in the momentum source term caused by the porous media (see Equation (5)). To capture the evolution of the viscosity, the Castro–Macosko viscosity model is used [23], given as

$$\eta(T,c) = \eta_0 \left(\frac{c_g}{c_g - c} \right)^{C_1 + C_2 \cdot c} \tag{21}$$

where c_g is the cure degree at gelation, c is the cure degree, and C_1 and C_2 are fitting parameters. η_0, the viscosity of the uncured resin, can be found by

$$\eta_0 = B \cdot exp\left(\frac{T_b}{R \cdot T} \right) \tag{22}$$

where B and T_b are fitting parameters, R is the universal gas constant and T is the temperature.

3. Results

3.1. Grid Independence

Grid independence was also performed on the 2D domain shown in Figure 1 where the length, L, is 0.1 m and the height, h, is 0.005 m. The parameters of interest in the grid-independence study are the development of the thermal boundary layer and resultant temperature profile throughout the domain, therefore, grid independence was performed by increasing the number of control volumes in the thickness direction, h. The inlet and wall temperatures were fixed at 333.15 K and 393.15 K, respectively. The fiber volume fraction is 50% and the inlet velocity is 0.005 m/s. A total of five different mesh densities were used; three with uniform grid spacing and two with edge refinement towards the upper and lower walls.

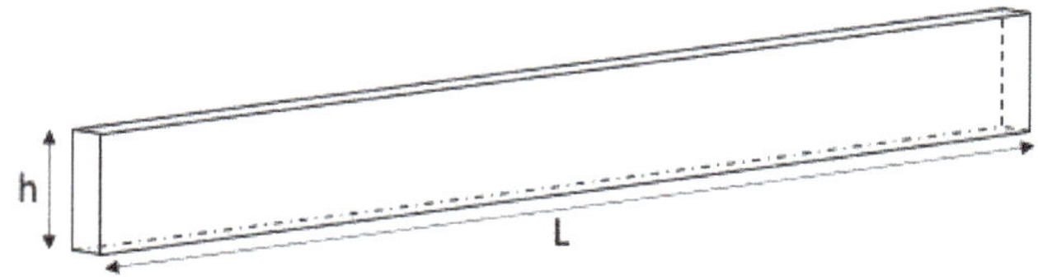

Figure 1. Simple 2D domain used for grid independence study.

Temperature profiles were extracted at two different locations along the length, 25 mm and 75 mm, and are shown in Figure 2a,b, respectively. At 25 mm, it is seen that the minimum temperature in the domain is consistent among the 5 grids, however, the temperature gradient at the wall varies dramatically with increased grid resolution. Correctly predicting this temperature gradient is critical as it dictates the heat flux into the domain. This is observed in the temperature profiles at 75 mm where there is a 30 K temperature discrepancy between the domains with 4 and 20 cells in the thickness direction. Grid independence was calculated based on the minimum temperature at 75 mm. Using the grid convergence index described by Celik et al. [35], it was found that with 20 cells in the thickness direction with edge refinement, the grid was converged to less than 1%.

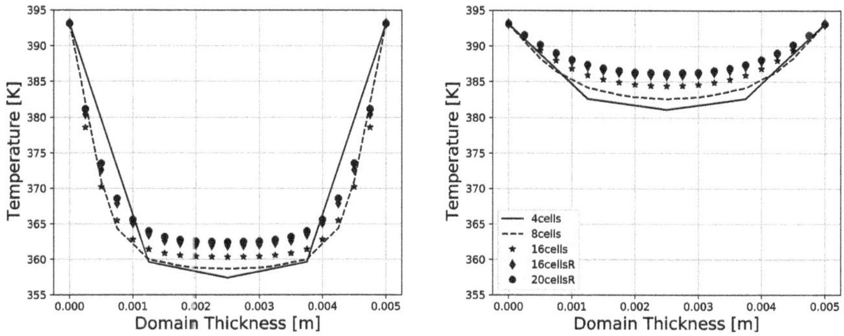

(a) (b)

Figure 2. Temperature profiles at (**a**) x = 25 mm and (**b**) x = 75 mm from the inlet after the domain was fully infiltrated. Note that the legend in (**b**) is applicable to both figures.

3.2. Temporal Resolution Study

The dependency of the source term, in both the thermal equilibrium and thermal LTNE energy models, on the simulation time step requires a time-step independence study be performed to ensure that the predicted temperature rise from the models is correct. This is achieved by modelling the curing stage and reducing the complexity of the simulation to 1D whereby *zeroGradient* boundary conditions are applied to all faces in the domain, shown in Figure 1. This allows the numerically predicted temperature rise to be directly compared to the analytical temperature rise using

$$\epsilon q = \tilde{C_p}(T_2 - T_1) \tag{23}$$

where ϵ is the porosity, q is the total specific heat released, $\tilde{C_p}$ is the effective specific heat and T_1 and T_2 are the initial and final mixture temperatures, respectively. For this verification test, the domain was assumed to be completely saturated with resin (i.e., $\alpha = 1$), with a uniform cure degree of 0%. The simulation then ran until a maximum cure degree of 60% was achieved, at which point the analytical and numerical temperature rise were compared.

The time-step range used in this study was 0.0025 s to 5 s to predict the resulting temperature rise error between the analytical determined and numerically predicted temperature. The results are given in Figure 3, which shows that the required time step to ensure numerical accuracy is 0.01 s. With this time step, it was found that there was a temperature rise error of 0.09%. This result verifies that the heat release model used in the thermal equilibrium and LTNE energy model are correctly predicting the heat release and resulting temperature rise.

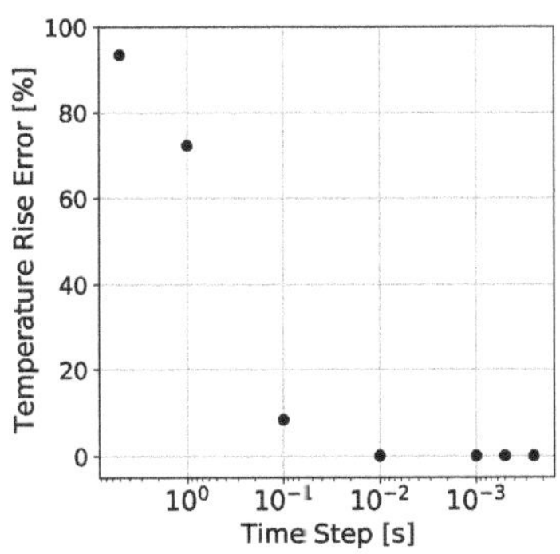

Figure 3. Reduction in temperature rise error between simulation and analytical results for decreasing time-step size.

3.3. Non-Dimensional Analysis

The increased accuracy of the LTNE energy model comes at the cost of longer computational time due to the additional energy equation being solved for the solid phase as well as the coupling required between fluid and solid energy equations. Given the number of variations on the RTM process, the balance between increased accuracy and computational requirements may not be inherently clear. To this end, a study was performed to determine when the LTNE energy model is necessary to provide accurate predictions.

To be able to accurately compare the predicted temperature development during infiltration, non-dimensional groups that contain the parameters that contribute most to the temperature evolution are determined. For RTM, these parameters are: (i) the permeability of the fabric, (ii) the thickness of the part, and (iii) the velocity of the resin being injected. These parameters can be used to form a Darcy number, Da, and a Peclet number, Pe. The Darcy number is given as

$$Da = \frac{K}{h^2} \tag{24}$$

where h is the thickness of the domain, and Pe is given as

$$Pe = \frac{vh}{\alpha} \tag{25}$$

where α is the thermal diffusivity, defined as

$$\alpha = \frac{k}{\rho C_p}. \tag{26}$$

The influence of Da and Pe are observed in variations of the fluid temperature, both when compared to the injection and mold temperature, as well as the solid temperature. These differences are captured using the non-dimensionalized fluid temperature, θ_f, defined as

$$\theta_f = \frac{T_f - T_{wall}}{T_{inj} - T_{wall}} \tag{27}$$

where T_f is the local fluid temperature, T_{wall} is the prescribed wall or mold temperature, and T_{inj} in the resin injection temperature. The fluid–solid temperature difference will be captured using the local thermal non-equilibrium parameter, $LTNE$, given as

$$LTNE = T_f - T_s \tag{28}$$

where T_s is the local solid temperature. A range of Darcy numbers based on the part thickness, h, $Da \in (4.210^{-5}, 9.510^{-5}, 3.810^{-4}, 10^{-3})$, and Peclet numbers numbers based on the inlet velocity, v, were varied from $Pe \in (400, 4000)$. These values were chosen to provide a comparison over a wide range of part thicknesses and injection velocities to understand the effect of using an LTNE model on RTM and HP-RTM manufacturing methods. Each Da and Pe number case is also run for two conductivity ratios (i.e., k_s/k_f of 0.2 and 125) which correspond to the use of either glass or carbon fibers within the mold. The boundary conditions used for all subsequent simulations are given in Table 1.

Table 1. Boundary conditions used for non-dimensional analysis cases.

Location	Field Type	Value
inlet	Velocity	*fixedValue*
	cure	0%
	T_f	363.15 K
	T_s	393.15 K
upperMold	Velocity	noSlip
	T_f	393.15 K
	T_s	393.15 K
lowerMold	Velocity	noSlip
	T_f	393.15 K
	T_s	393.15 K
wall	Velocity	noSlip
	T_f	393.15 K
	T_s	393.15 K
symmetry	Velocity T_f T_s	symmetry

3.4. Temperature Distribution and Development

The non-dimensional temperature (θ_f) and LTNE profiles are plotted as a function of the non-dimensional channel height in Figure 4. Since the profiles are all symmetric, each figure shows data contrasting two conductivity ratios; the left and right sides of each plot show data for conductivity ratios of 0.2 (black symbols) and 125 (grey symbols), respectively, for the same Da and channel location. Furthermore, the plots in Figure 4 are arranged such that the first and second columns show θ_f at locations of 25% and 75%, respectively, and the third and fourth columns show the LTNE at locations of 25% and 75%, respectively. Each row corresponds to a specific Da number, as denoted by the symbols in the legend.

Considering first the influence of Da number, the results show that as Da decreases, use of the LTNE model bears diminishing returns. The cases where $Da \in (9.5 \times 10^{-5}, 4.2 \times 10^{-5})$ for a $Pe \in (400, 2000)$ and a $k_s/k_f = 0.2$, show the local temperature difference for the fluid and solids phases is less than 5 K. In these cases, the low conductivities of the fluid and fibers reduce the influence of the *fixedValue* boundary condition at the mold wall. This, therefore, allows the resin to cool the fiber phase and bring them closer to thermal equilibrium. This is highlighted in Figure 4m–p, for a $Pe = 400$, where there is no discernible difference between the fluid and solid temperatures throughout the domain. This suggests that an equilibrium energy model could be used to model this case as the fluid and solid phases are in local equilibrium throughout infiltration. However, when considering a $k_s/k_f = 125$, the conductivity ratio has a significant effect on the LTNE, especially at the start of infiltration. For all Darcy numbers considered, it was found that the higher conductivity ratio increased the resin temperature significantly faster than the lower conductivity ratio cases. The end result of this rapid warming is seen in the θ_f profiles at $x/L = 0.75$, where for each Pe the resin is closer to the mold temperature. This suggests that for cases where a $Pe > 1000$ an LTNE energy model is needed. An additional consideration when performing a comparable case is the influence that this rapid increase in fluid temperature has on both the cure and viscosity development during infiltration. Given the increasingly large number of resins that can be used for RTM, this temperature rise could potentially result in premature curing or vitrification of the resin.

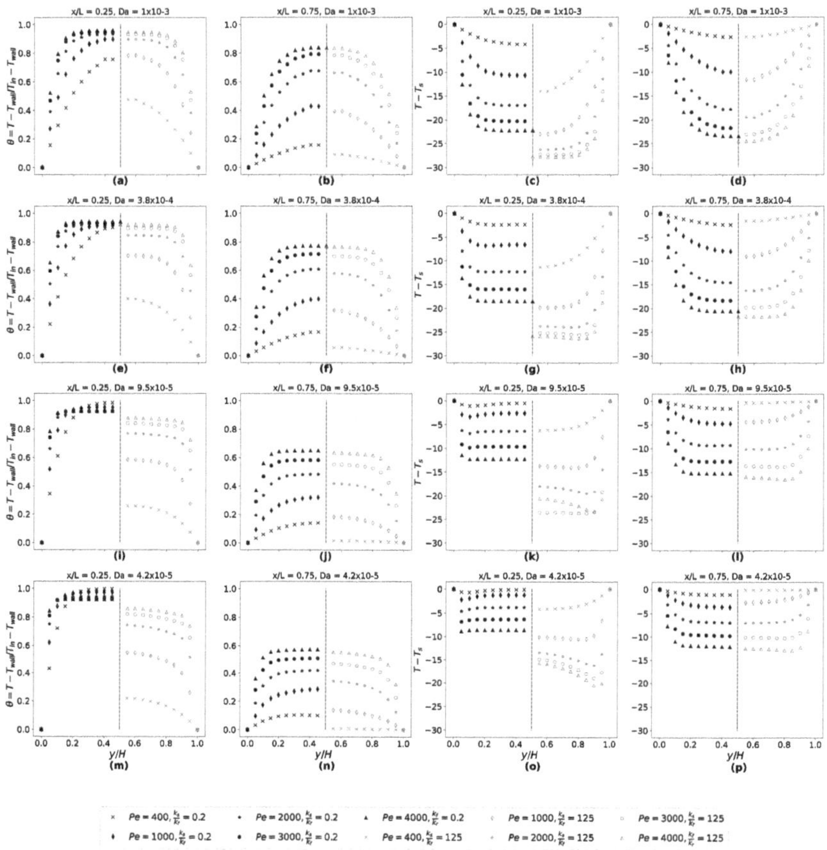

Figure 4. (**a**–**p**) Non-dimensional temperature and fluid–solid temperature difference for varying Pe and Da numbers. Values on the left hand side of the vertical dashed line (in black) are for a conductivity ratio of 0.2 and values on the right hand side of the dashed line (in grey) are for a conductivity ratio of 125.

As Da increases, an increase is observed in both the local θ_f and LTNE throughout the thickness of the part, even for the cases where $Pe = 400$. Although in these cases the Da was varied based on the height of the part, the increase in Da is also analogous to a reduction in fiber volume fraction (considering the high dependence the permeability of the fabric has on fiber volume fraction). This in turn reduces the influence of the fiber phase on the fluid temperature, which is observed in Figure 4a–d, where for each Pe the fluid temperature is closer to the mold temperature. The solid phase still contributes to the increase in fluid temperature, however, due to the transient nature of the flow, this influence is not as substantial as the mold temperature. The solid phase influence on the fluid temperature is also affected by the conductivity ratio. The higher the conductivity of the fibers, the less influence the fluid phase has on the temperature of the solid phase. This is due to the rapid conduction from the mold to the fibers throughout the domain, maintaining more of the fiber phase at the mold temperature. The result is an increase in the local fluid temperature due to the increased heat transfer from the solid phase to the fluid phase. This effect is highlighted in the LTNE plots in Figure 4, specifically at $x/L = 0.25$, which show increased LTNE between the fluid and solid phases for the higher fiber conductivity. Although this result is not intuitive given that the higher solid temperature would result in increased heat transfer from the fibers to the resin, the conductivity from the mold wall to the fiber phase maintains the fibers at a higher temperature when compared to the lower conductivity fibers.

The Pe number also has a significant effect on the local fluid temperature during mold filling due to its direct connection with filling time. Increasing the Pe number reduces the filling time and consistently results in larger differences between the fluid, solid and mold temperatures. Increases in Pe number also inherently increase the convective heat transfer between the solid and fluid phases, but this increase in heat transfer does not always lead to increases in fluid temperature because of the increased fluid velocity. The reduction in filling time from a Pe number of 400 to 4000 is 10 times (for a $Da = 1 \times 10^{-3}$). This in turn reduces the time that the increased convection heat transfer has to increase the resin temperature. The Da number also has an effect on the influence of the Pe number. As previously mentioned, reducing the Da number results in an increased effect of the mold temperature on the fluid temperature in the center of the domain. Coupling these two influences together, it is observed that for a $Pe > 1000$ number, when the Da number is reduced, the resulting fluid temperature profile both near the inlet and near end of the domain is closer to the mold temperature than the injection temperature.

These generic results highlight the utility of the LTNE energy model for RTM and HP-RTM processes. It is also important to note that additional attention needs to be directed at cases with large infiltration times since there exists the potential for premature curing of some resins. Furthermore, a longer infiltration time implies a longer cycle time to produce parts.

3.5. Fluid Property Development

Capturing the temperature development of the flow front also has implications on the predicted development of other fluid properties. Two critical properties, the fluid viscosity and cure degree, are heavily influenced by the history of the fluid temperature as it infiltrates the mold. This effect is captured by comparing the fluid property development for the $Da = 1 \times 10^{-3}$ and $Pe = 4000$ case with a $k_s/k_f = 125$, using the current LTNE and thermal equilibrium energy models. The coefficient values for the Kamal–Malkin reaction kinetic model, DiBeneditto glass transition temperature model, and Castro–Macosko viscosity model are given in Tables 2–4, respectively.

Table 2. Parameter values used for the Kamal–Malkin kinetic model.

Model Parameter	Value
A_1	3,862,141.7
A_2	105,920,589,010.0
E_1	62,877.7
E_2	321,915.1
m	1.571
n	1.63
R	8.314

Table 3. Parameter values used for the DiBeneditto glass transition temperature model.

Model Parameter	Value
$T_{g,0}$	243.0
$T_{g,\infty}$	406.09
λ	0.390

Table 4. Parameter values used for the Castro–Macosko viscosity model.

Model Parameter	Value
C_1	3.91
C_2	2.12×10^{-13}
B	1.414×10^{-12}
T_b	8.489×10^3
R	8.314
c_g	0.72

Figure 5 compares the fluid viscosity profile at two different locations within the domain, $x/L = 0.25$ and $x/L = 0.75$. It can be readily observed that the fluid viscosity near the mold walls is lower for the equilibrium model. This is due to the weighting factors (w_f and w_s) used to calculate the local fluid–solid mixture properties. Equations (12) and (13) show that the weighting factors are determined based on the density of the two phases. Considering that the density of fibers is nearly double that of the resin phase and coupled with the high conductivity ratio between the fluid and solid phases, it is expected that the model would predict a rapid increase in fluid temperature from the injection temperature to the mold temperature. This is captured in Figure 5a where it is shown that the viscosity of the fluid is lower near the wall when compared to the LTNE energy model. The effect of the rapid increase in temperature is further shown in Figure 5b, where the viscosity of fluid throughout the domain cross-section is substantially less for the equilibrium model than the LTNE model. In this respect, the LTNE model is also essential for properly predicting the pressures required for mold filling and the flow distribution since both are strongly affected by local viscosity.

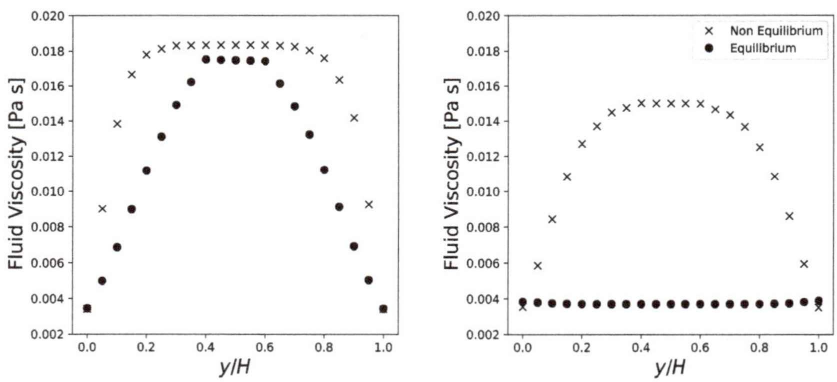

(a) (b)

Figure 5. Fluid viscosity development during infiltration for the thermal non-equilibrium and thermal equilibrium model at a location of (**a**) 25% and (**b**) 75% in the domain. Note that the legend in (**b**) is applicable to both figures.

The effect of the energy model on the cure degree development is shown in Figure 6. As with the fluid viscosity, there is a small discrepancy in the predicted cure degree at a $x/L = 0.25$, where the cure degree is higher at the walls in the equilibrium model due to the higher predicted temperature. At $x/L = 0.75$ this difference increases substantially, where the non-equilibrium model predicts the average cure degree to be 0.5% and the equilibrium model predicts 1.9%. The implications of this difference are significant. First, the source term added to the fluid temperature equation, Equation (16), will predict a larger heat release at the end of infiltration for the equilibrium energy model. This can lead to local rises in temperature in the domain, which can result in the prediction of premature curing. Furthermore, if the curing stage is to be simulated, this result would incorrectly predict a shorter curing time, which can have a significant impact on the material properties of the final parts. Secondly, in both Figures 5 and 6, it is observed that at $x/L = 0.75$ there is a transition in the viscosity and cure degree for the equilibrium case adjacent to the mold walls. The cause of this is the resin being in the domain at the mold temperature. There is an initial reduction in the resin viscosity near the mold walls as it warms, as seen in Figure 5a for both cases, but given the higher overall temperature in the domain for the equilibrium case, the resin reaches the mold temperature considerably faster. At $x/L = 0.75$ the resin viscosity is substantially lower for the equilibrium case but there are signs of premature curing near the walls given the local increase in fluid viscosity. Considering the cure profile in Figure 6, this increase in viscosity at the walls is due to a sharp increase in the local cure degree. Given the viscosity's dependency on the cure degree, this result is expected.

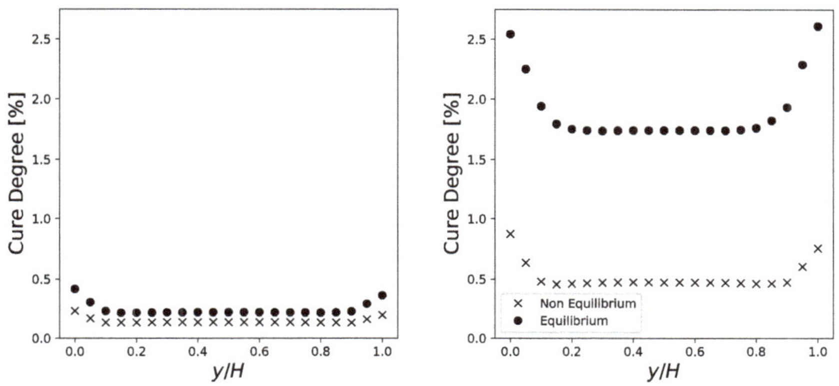

(**a**) (**b**)

Figure 6. Cure degree development during infiltration for the thermal non-equilibrium and thermal equilibrium model at a location of (**a**) 25% and (**b**) 75% in the domain. Note that the legend in (**b**) is applicable to both figures.

The implications of these results on predicting mold filling behaviour and cure degree development in RTM components are significant; the most significant being the resultant flow front predicted from the thermal equilibrium model. The local decrease in viscosity near the mold walls will result in a higher fluid velocity due its inclusion in Darcy's law. This advancing front near the walls can lead to predictions of void formation in the final part, especially for parts with a high *Da* or with multiple inlets resulting in multiple flow fronts meeting. There is also the potential for the prediction of premature curing of the resin at the mold walls. This can result in the prediction of local hot spots, depending on the resin and fiber properties, as well as inaccurate prediction of the required time for the curing stage. Finally, it was found that the computational cost of using an LTNE model was on the same order of magnitude as the thermal equilibrium. This is due to time scale of the heat transfer occurring when compared to the time scale of the fluid movement. The VOF method used to track the infiltration of the resin required a substantially lower time step than the energy model. This is a significant finding as it means the added computational cost of the LTNE model is reasonable given the increased accuracy when predicting the temperature development.

3.6. Complex Floor Geometry

Simulation of resin infiltration in a complex floor geometry is presented as a final case to further demonstrate the efficacy of the LTNE model compared to isothermal energy modelling and to thermal equilibrium modelling. The complex mold geometry is shown in Figure 7a. The computational mesh is shown in Figure 7b, where the domain has been discretized into 988,000 hexahedral elements. Meshing was performed using the results from the grid independence test in Section 3.1 where four cells per mm of part thickness, with edge refinement towards the upper and lower surfaces, was used. The boundary conditions, as well as the coefficients for the Kamal–Malkin reaction kinetics and Castro–Macosko viscosity models, are found in Tables 1, 2 and 4, respectively. The fiber volume fraction was set to a uniform value of 50% and the fiber orientation was assigned using a simple geometric approach that established in-plane and cross-plane components based on the local position of the upper and lower mold faces. While more sophisticated approaches are available for determining local fiber orientation, for the purpose of demonstrating the LTNE model, a simple approach was deemed suitable.

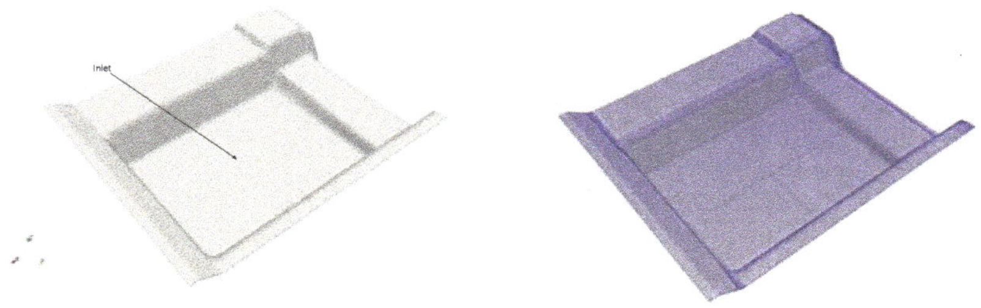

(a) Domain. (b) Discretized domain.

Figure 7. Complicated geometry used for energy modelling comparison. Image (**a**) shows an isometric view of the domain, while image (**b**) shows a surface mesh to illustrate the grid distribution with respect to geometric features.

3.6.1. Infiltration

The progress of the cure degree during infiltration for each of the four cases is shown in Figure 8. Columns one and two in Figure 8 consider the assumption of isothermal mold filling. These results also highlight the significant impact of operating temperature on the cure degree development during mold filling. As the operating temperature increases, the maximum cure degree in the domain is shown to also increase. This is further shown in Table 5, where a 3.6% difference in the maximum cure degree is shown between the two isothermal simulations. As expected, the areas of highest cure degree are at the edges of the component, as this corresponds to the resin that has been in the domain for the longest period of time. A major implication of this, in particular for the isothermal 363 K simulations, is an incorrect prediction of the required curing time needed to fully cure the resin. This will be discussed in further detail in Section 3.6.2.

Table 5. Minimum and maximum cure degree values in the domain after infiltration for each energy modelling approach.

Energy Modelling Approach	Maximum Cure Degree	Maximum Cure Rate
Isothermal 363 K	0.96%	0.3 %/s
Isothermal 393 K	4.6%	1.7 %/s
Thermal Equilibrium	4.5%	1.7 %/s
Thermal Non-Equilibrium	3.5%	1.6 %/s

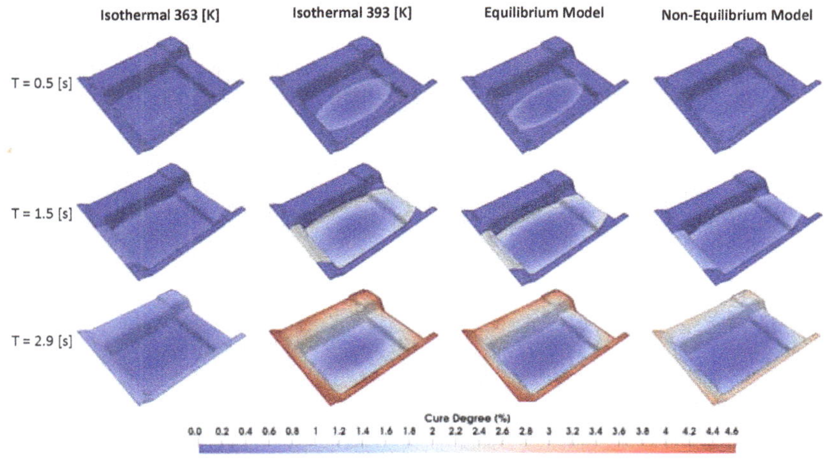

Figure 8. Predicted cure degree development during infiltration using different energy modelling approaches.

The most interesting results are evident when comparing the cure degree distribution for the equilibrium and LTNE energy models. Considering first the equilibrium energy model, it is observed that the maximum cure degree does not differ significantly compared to the isothermal 393 K case (4.5% and 4.6%, respectively, as shown in Table 5). This is also seen when comparing the cure degree distributions during infiltration. The cure degree contours shown in Figure 8 are nearly identical at all points during infiltration, with the only noticeable difference being at the bottom left hand edge of the part. In this region, the thermal equilibrium model is showing a lower cure degree compared to the isothermal 393 K case. The similarity in cure distribution is explained by the formulation used to determine the average material properties used in the equilibrium energy model (Equations (9)–(11)), in particular Equation (11), used to calculate the mixture conductivity. For these simulations the conductivity ratio between the solid and fluid phases (k_s/k_f) is 125. The high fiber conductivity forces a rapid increase from the injection temperature (363 K) to the mold temperature (393 K) immediately after the resin is injected. The infiltration then proceeds as though done isothermally at the prescribed mold temperature. As mentioned previously, this phenomenon is highly dependent on the thermophysical properties of the resin and fibers. For a lower conductivity ratio (i.e., using glass fibers) there would be a large discrepancy between the maximum cure degree with the domain when compared to the isothermal 393 K, as it would take longer for temperature to rise to the mold temperature in the domain.

This result is consistent with the predicted cure degree development for this complex geometry, where the Peclet number is 6780. Recall from Section 3.3 that for almost all cases where the Peclet number is greater than 1000 (with a conductivity ratio of 125), an LTNE model is needed to accurately capture the development of the fluid temperature during infiltration. The results predict a longer period of time where the cool injected resin warms to the mold temperature, when compared to the equilibrium model, as shown by the large region of less cured resin in center of the complex part. This suggests that the resin and fibers are not in local thermal equilibrium. The areas where the resin has warmed to the mold temperature and has been in the domain the longest are at the corners, which carry the highest resin cure degree of 3.5%. The lower predicted maximum cure degree means there is overall less variation in the cure degree distribution, when compared to the equilibrium energy model. The result is a more even development of the cure degree during the cure stage, which is discussed further in Section 3.6.2.

3.6.2. Curing

Results of the cure degree and temperature distribution at the end of the infiltration stage were used as the initial conditions for curing simulations. The resin was allowed to cure for 180 s, using the four energy modelling approaches, and the resulting cure degree and cure rate at 10 s are shown in Figures 9 and 10, respectively. Due to the large change in cure degree during the curing stage, the cure degree contour plot in Figure 9 only contains data at 10 s. Were a longer time range considered, the fine detail of the cure degree distribution would be lost. To address this, the cure degree range for each energy modelling approach (at times 10 s, 100 s, and 180 s) are shown in Table 6. In the case of Figure 10, the isothermal 363 K case had a significantly lower cure rate, which if included would have again resulted in the loss of the fine detail in the cure rate distribution.

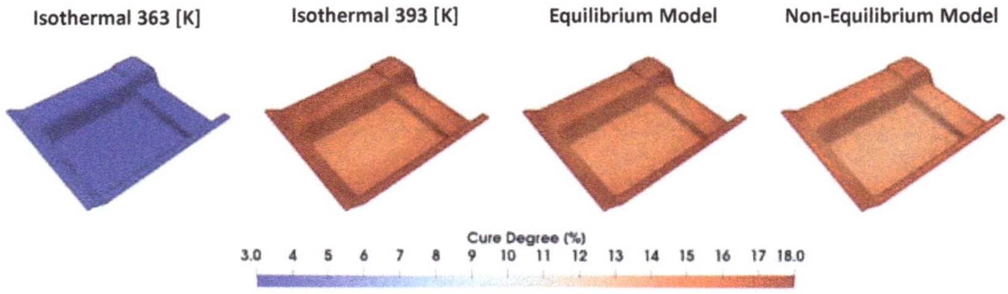

Figure 9. Predicted cure degree distribution at 10 s into the curing stage using different energy modelling approaches.

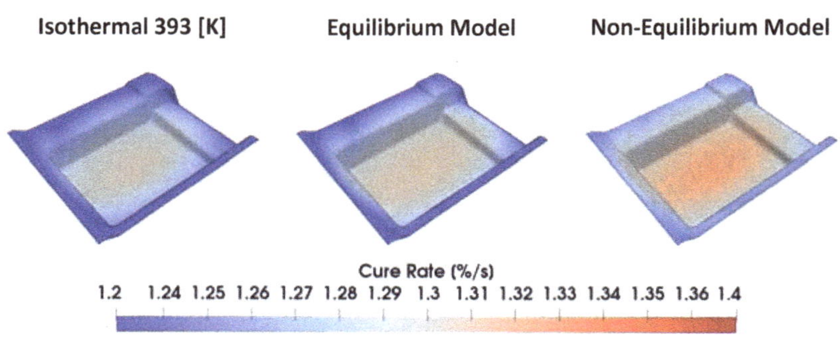

Figure 10. Predicted cure rate distribution at 10 s into the curing stage using different energy modelling approaches.

Columns one and two in Figure 9 show the cases of isothermal curing. The results further highlight the dependence of the cure degree development on the operating temperature. At the start of the curing stage there is a 12% difference in the average cure degree between the two isothermal cases. At the end, however, the discrepancy in average cure degree has grown to 41%, shown in Table 6. This is due to the lower operational temperature in the isothermal 363 K case which restricts the cure rate during the course of the curing stage. It should be noted that an operating temperature of 363 K is unlikely to be used in practice for this case given the additional time needed to cure the component, however, it reinforces the importance of optimizing the operating temperature through simulation and for manufacturing of CoFRP components.

Table 6. Cure degree distribution ranges for each case during the curing stage.

Energy Modelling Approach	Cure Degree Range (Min–Max)		
	10 s	100 s	180 s
Isothermal 363 K	3.4–4.3%	26.9–27.5%	40.9–41.3%
Isothermal 393 K	14.9–18.5%	68.5–69.2%	81.8–82.1%
Thermal Equilibrium	14.5–18.4%	68.4–69.2%	81.8–82.1%
Thermal Non-Equilibrium	12.5–17.6%	68.0–69.0%	81.6–82.0%

Consistent with the infiltration stage, the cure degree distribution for the isothermal 393 K and equilibrium model show nearly identical development during curing. This result is further highlighted in Table 6 where at each recorded time the minimum and maximum cure degree varies at most by 0.4% and in the case of 180 s it is shown that the cure degree ranges are identical. Again, this is due to the resin rapidly increasing from the injection temperature to the mold temperature when using the equilibrium model, resulting in the model behaving as though the process was done isothermally. An interesting outcome is the similarity in final cure degree distributions predicted by the isothermal 393 K, equilibrium, and LTNE models at 180 s. To fully understand this outcome, consideration first needs to be made with regard to the type of curing being performed. Bernath et al. [31] studied the influence of the type of cure modelling being used (i.e., isothermal versus non-isothermal) and the ability of the reaction kinetics model to capture the resulting cure degree progression as compared to experimental data. Their results showed the cure rate profile for isothermal curing begins with a peak in cure rate which quickly diminishes for the remainder of the cure cycle. For non-isothermal curing, however, there is a noticeable delay in the peak cure rate (the magnitude and time of which are highly dependent on the rate of temperature increase), due to the time needed for the resin to warm. The results of the present study are consistent with the findings of Bernath et al. [31]. For the isothermal 393 K and equilibrium cases (for which isothermal curing conditions are used) an initial peak in cure rate was observed (highlighted again by the higher cure degree at the beginning of the cure stage), which then quickly diminished during curing. In the case of the LTNE model, which simulates initially non-isothermal curing, the peak cure rate is observed post-infiltration, during the curing stage. This phenomenom is further highlighted in Figure 10, where the LTNE model is showing a cure rate nearly 8% higher throughout the component. The LTNE model then consistently predicts a higher cure rate for the entirety of the simulated cure cycle, resulting in a nearly identical cure degree distribution between the three cases.

A comment needs to be made on the limitation of the Kamal–Malkin kinetic model due to its reduction in accuracy as the glass transition temperature approaches the operating temperature. Were a more sophisticated kinetic model used, it is hypothesized that the LTNE case would predict a final cure degree distribution higher than that found in the isothermal 393 K and equilibrium models. This is due to the overall higher cure rate still present throughout the cure cycle in the LTNE energy model, as well as the results presented by Bernath et al. [31]. A consistent result seen in their work shows that in the case of isothermal curing, the final cure degree achieved is highly dependent on the operational temperature and is consistently less than 100%; a result not found in non-isothermal curing where the resin maintains the ability to nearly completely cure. A caveat to this, however, is the unknown effect of combining isothermal and non-isothermal curing. This combination is found in the LTNE case, where curing is initially non-isothermal, as the cool injected resin warms to the mold temperature. After 10 s of curing, the resin reaches the operating temperature of 393 K and continues the curing cycle isothermally. This likely slowed the overall peak cure rate, however, allowing the model to maintain an overall higher cure rate. An implication of this might be when predicting the final material properties as well as process-induced distortion (PID) [22]. The maximum achievable cure degree has a significant impact on the final material properties and dictates the amount of chemical shrinkage that occurs. Chemical shrinkage influences the build-up of residual stresses leading to PID. This further reinforces the importance of an accurate prediction of the temperature evolution given its influence on the cure development. It also highlights the importance of the reaction kinetic model utilized. As mentioned above, a limitation of the Kamal–Malkin kinetic model is that it always predicts full resin curing despite the operating temperature, which is known to not be the case. For a better understanding of the operating temperature on the maximum possible cure degree as well as the cure degree progress post-vitrification, the more sophisticated Grindling reaction kinetic model could be utilized.

4. Summary

The current study was aimed at studying the complexity viability of an LTNE energy model for obtaining accurate solutions of temperature development in RTM mold-filling simulations. A general LTNE model was implemented into an existing mold-filling code in OpenFOAM. A non-dimensional analysis was then conducted over a range of Da numbers, Pe numbers, and k_f/k_s ratios to show the non-dimensional fluid temperature and local thermal non-equilibrium development, for a wide range of part types and infiltration strategies. Results from the non-dimensional analysis show that for a $Da \in (9.5 \times 10^{-5}, 4.2 \times 10^{-5})$, $Pe \in (400, 1000)$ for a $k_s/k_f = 0.2$ and a $Pe = 400$ for a $k_s/k_f = 125$, the approximation of local thermal equilibrium is appropriate. However, for all other case runs it was shown that for an accurate prediction of the resin and solid temperature during mold filling, the use of an LTNE model was required.

To further understand the implications of using a thermal equilibrium model when a non-equilibrium model is required, a comparison case was run where $Da = 1 \times 10^{-3}$ and $Pe = 4000$ with a $kf/ks = 125$ using a equilibrium energy model from the literature and the proposed non-equilibrium model. The result further highlights the importance of using a thermal non-equilibrium model. The predicted resin cure degree for the case using the equilibrium model were substantially higher near the end of the domain due to the rapid increase in temperature throughout the domain. This resulted in a much lower resin viscosity, due to its dependency on the cure degree. The implications of this are an inaccurate prediction of the cure time needed in the curing stage, as well as the potential for a void to form if multiple flow fronts meet given the advancement in the resin front near the mold walls. It was also found that the additional computational cost was minimal, implying that the thermal non-equilibrium model is able to be used effectively in all RTM infiltration modelling.

A complex floor geometry was then used to evaluate three energy modelling approaches; isothermal mold filling (at operating temperatures of 363 K and 393 K), thermal equilibrium, and LTNE energy modelling. Both the infiltration and curing stages were modelled using the four different energy modelling approaches. The isothermal 363 K cases highlighted the effect of the operating temperature on the final cure degree, where a 3.6% difference in the maximum cure degree was found between the two isothermal mold-filling simulations. When considering the equilibrium energy model, it was found that the cure degree distribution was nearly identical to that predicted by the isothermal 393 K simulation. This was due to the weighting factors used to calculate the effective thermophysical properties employed in the energy transport equation. The most notable outcome was that of the LTNE energy model. The LTNE predicted a lower maximum cure degree as well as a larger area of less cured resin within the domain. This result highlights the increased accuracy in the predicted temperature development and resulting cure degree as the results show the fibers and resin are not in thermal equilibrium. The curing stage highlighted the importance of which energy model was used for the infiltration stage. In the case of the isothermal 363 K, isothermal 393 K, and equilibrium energy modelling, it was shown that curing occurred isothermally. The result of an initial peak in the cure rate during infiltration which quickly diminishes at the beginning of the curing stage. In the case of LTNE energy modelling, the cure degree development follows a more complex path, where it initially develops non-isothermally during infiltration and at the beginning of the curing stage. After 10 s of curing the resin has reached the prescribed mold temperature and then continues its curing under isothermal conditions. The effect of this has implications when considering the predicted material properties and PID from each case, given their dependency on the final cure degree. Moreover, it highlights the importance of the reaction kinetic model used. The selected Kamal–Malkin kinetic model predicts that resin would cure fully despite the operating temperature. Using a more complex kinetic model, such as the Grindling model, would provide better insights into the true final cure degree and cure degree development given the models ability to predict vitrification. Finally, it was also found that the use of an adaptive time step dependent on the current maximum cure rate and minimum cure degree, in conjunction with the LTNE model dramatically reduced the computational time of the curing stage. This further increases the applicability of an LTNE energy model for HP-RTM infiltration and curing simulation.

Author Contributions: Conceptualization, A.S., A.G.S. and C.T.D.; Formal analysis, A.S.; Funding acquisition, F.H.; Investigation, A.S.; Methodology, A.S., A.G.S. and C.T.D.; Resources, A.G.S.; Software, A.S.; Supervision, A.G.S., C.T.D. and F.H.; Visualization, A.S.; Writing—original draft, A.S.; Writing—review and editing, A.S., A.G.S., C.T.D. and F.H. All authors have read and agreed to the published version of the manuscript.

Funding: The authors thank the Natural Sciences and Engineering Research Council of Canada (NSERC) for supporting this study through funding grant CRDPJ-507776-16, as well as sponsors Honda R&D Americas, Hexion Inc., Zoltek Corp. and LAVAL International.

Institutional Review Board Statement: Not Applicable.

Informed Consent Statement: Not Applicable.

Data Availability Statement: Available upon request.

Conflicts of Interest: The authors declare no conflict of interest. The funders had no role in the design of the study; in the collection, analyses, or interpretation of data; in the writing of the manuscript, or in the decision to publish the results.

Abbreviations

The following abbreviations are used in this manuscript:

CFRP	Carbon fiber reinforced plastic
HP-RTM	High-pressure resin transfer molding
LTNE	Local thermal non-equilibrium
PID	Process-induced distortion
RTM	Resin transfer molding
VOF	Volume of fluid

References

1. Sullivan, R.A. Automotive carbon fiber: Opportunities and challenges. *JOM* **2006**, *58*, 77–79. [CrossRef]
2. Seuffert, J.; Kärger, L.; Henning, F. Simulating Mold Filling in Compression Resin Transfer Molding (CRTM) Using a Three-Dimensional Finite-Volume Formulation. *J. Compos. Sci.* **2018**, *2*, 23. [CrossRef]
3. Magagnato, D.; Seuffert, J.; Bernath, A.; Kärger, L.; Henning, F. Experimental and numerical study of the influence of integrated load transmission elements on filling behavior in resin transfer molding. *Compos. Struct.* **2018**, *198*, 135–143. [CrossRef]
4. Wittemann, F.; Maertens, R.; Bernath, A.; Hohberg, M.; Kärger, L.; Henning, F. Simulation of Reinforced Reactive Injection Molding with the Finite Volume Method. *J. Compos. Sci.* **2018**, *2*, 5. [CrossRef]
5. Henning, F.; Kärger, L.; Dörr, D.; Schirmaier, F.J.; Seuffert, J.; Bernath, A. Fast processing and continuous simulation of automotive structural composite components. *Compos. Sci. Technol.* **2019**, *171*, 261–279. [CrossRef]
6. Kasai, J. Life cycle assessment, evaluation method for sustainable development. *JSAE Rev.* **1999**, *20*, 387–394. [CrossRef]
7. Al-Lami, A.; Hilmer, P.; Sinapius, M. Eco-efficiency assessment of manufacturing carbon fiber reinforced polymers (CFRP) in aerospace industry. *Aerosp. Sci. Technol.* **2018**, *79*, 669–678. [CrossRef]
8. Verma, S.; Balasubramaniam, B.; Gupta, R.K. Recycling, reclamation and re-manufacturing of carbon fibres. *Curr. Opin. Green Sustain. Chem.* **2018**, *13*, 86–90. [CrossRef]
9. Shojaei, A.; Ghaffarian, S.R.; Karimian, S.M.H. Modeling and simulation approaches in the resin transfer molding process: A review. *Polym. Compos.* **2003**, *24*, 252–544. [CrossRef]

10. Hamdan, B.; Lafi, S.; Hassan, N.M. Optimizing the Manufacturing Processes of Carbon Fiber Epoxy Resin Composite Panels. *J. Manuf. Sci. Eng.* **2017**, *140*, 011003. [CrossRef]
11. Han, S.H.; Cho, E.J.; Lee, H.C.; Jeong, K.; Kim, S.S. Study on high-speed RTM to reduce the impregnation time of carbon/epoxy composites. *Compos. Struct.* **2015**, *119*, 50–58. [CrossRef]
12. Poodts, E.; Minak, G.; Mazzocchetti, L.; Giorgini, L. Fabrication, process simulation and testing of a thick CFRP component using the RTM process. *Compos. Part Eng.* **2014**, *56*, 673–680. [CrossRef]
13. Liu, B.; Bikerton, S.; Advani, S. Modeling and Simulation of RTM: Gate Control, Venting and Dry Spot Prediction. *Compos. Part A Appl. Sci. Manuf.* **1996**, *27*, 135–141. [CrossRef]
14. Bogetti, T.A.; Gillespie, J.W. Two-Dimensional Cure Simulation of Thick Thermosetting Composites. *J. Compos. Mater.* **1991**, *25*, 239–273. [CrossRef]
15. Bruschke, M.V.; Advani, S.G. A numerical approach to model non-isothermal viscous flow through fibrous media with free surfaces. *Int. J. Numer. Methods Fluids* **1994**, *19*, 575–603. [CrossRef]
16. Chang, W.; Kikuchi, N. Analysis of non-isothermal mold filling process in Resin Transfer Molding (RTM) and Structural Reaction Injection Molding (SRIM). *Comput. Mech.* **1995**, *16*, 22–35. [CrossRef]
17. Sorrentino, L.; Bellini, C. Validation of a Methodology for Cure Process Optimization of Thick Composite Laminates. *Polym. -Plast. Technol. Eng.* **2015**, *54*, 1803–1811. [CrossRef]
18. Sorrentino, L.; Esposito, L.; Bellini, C. A new methodology to evaluate the influence of curing overheating on the mechanical properties of thick FRP laminates. *Compos. Part Eng.* **2017**, *109*, 187–196. [CrossRef]
19. Bodaghi, M.; Cristóvão, C.; Gomes, R.; Correia, N.C. Experimental characterization of voids in high fibre volume fraction composites processed by high injection pressure RTM. *Compos. Part Appl. Sci. Manuf.* **2016**, *82*, 88–99. [CrossRef]
20. Bodaghi, M.; Simacek, P.; Correia, N.; Advani, S.G. Experimental parametric study of flow-induced fiber washout during high-injection-pressure resin transfer molding. *Polym. Compos.* **2020**, *41*, 1053–1065. [CrossRef]
21. Deléglise, M.; Grognec, P.L.; Binetruy, C.; Krawczak, P.; Claude, B. Modeling of high speed RTM injection with highly reactive resin with on-line mixing. *Compos. Part Appl. Sci. Manuf.* **2011**, *42*, 1390–1397. [CrossRef]
22. Bernath, A. Numerical Prediction of Curing and Process-Induced Distortion of Composite Ctructures. Ph.D. Thesis, Karlsruhe Institute of Technology, Karlsruhe, Germany, 2019.
23. Castro, J.M.; Macosko, C.W. Studies of Mold Filling and Curing in the Reaction Injection Molding Process. *AIChE J.* **1982**, *28*, 250–260. [CrossRef]
24. Calmidi, V.V.; Campmode, R.L.M. Forced Convection in High Porosity Metal Foams. *J. Heat Transfer.* **2000**, *122*, 557–565. [CrossRef]
25. Straatman, A.G.; Gallego, N.C.; Yu, Q.; Betchen, L.; Thompson, B.E. Forced convection heat transfer and hydraulic losses in graphitic foam. *J. Heat Transf.* **2007**, *129*, 1237–1245. [CrossRef]
26. Quintard, M.; Whitaker, S. Local thermal equilibrium for transient heat conlduction : theory and comparison with numerical experiments. *Inf. J. Heat Mass Transfer* **1995**, *38*, 2779–2796. [CrossRef]
27. Quintard, M.; Kaviany, M.; Whitaker, S. Two-medium treatment of heat transfer in porous media: Numerical results for effective properties. *Adv. Water Resour.* **1997**, *20*, 77–94. [CrossRef]
28. Vafai, K.; Tien, C. Boundary and inertia effects on flow and heat transfer in porous media. *Int. J. Heat Mass Transf.* **1981**, *24*, 195–203. [CrossRef]
29. Ruiz, E.; Trochu, F. Multi-criteria thermal optimization in liquid composite molding to reduce processing stresses and cycle time. *Compos. Part Appl. Sci. Manuf.* **2006**, *37*, 913–924. [CrossRef]
30. Shojaei, A.; Ghaffarian, S.R.; Karimian, S.M. Three-dimensional process cycle simulation of composite parts manufactured by resin transfer molding. *Compos. Struct.* **2004**, *65*, 381–390. [CrossRef]
31. Bernath, A.; Kärger, L.; Henning, F. Accurate cure modeling for isothermal processing of fast curing epoxy resins. *Polymers* **2016**, *8*, 390. [CrossRef]
32. Halley, P.J.; Mackay, M.E. Chemorheology of Thermosets-An Overview. *Polym. Eng. Sci.* **1996**, *36*, 593–609. [CrossRef]
33. DiBenedetto, A.T. Prediction of the glass transition temperature of polymers: A model based on the principle of corresponding states. *J. Polym. Sci. Part Polym. Phys.* **1987**, *25*, 1949–1969. [CrossRef]
34. Pascault, J.P.; Williams, R.J.; DiBenedetto, A.T. Glass transition temperature versus conversion relationships for thermosetting polymers. *J. Polym. Sci. Part Polym. Phys.* **1987**, *25*, 1949–1969. [CrossRef]
35. Celik, I.B.; Ghia, U.; Roache, P.J.; Freitas, C.J.; Coleman, H.; Raad, P.E. Procedure for estimation and reporting of uncertainty due to discretization in CFD applications. *J. Fluids Eng. Trans. ASME* **2008**, *130*, 0780011–0780014. [CrossRef]

Journal of
Composites Science

MDPI

Article

Study of a Bimodal α–β Ti Alloy Microstructure Using Multi-Resolution Spherical Indentation Stress-Strain Protocols

Natalia Millan-Espitia [1] and Surya R. Kalidindi [1,2,*]

[1] George W. Woodruff School of Mechanical Engineering, Georgia Institute of Technology, Atlanta, GA 30332, USA; natalia.millan@me.gatech.edu

[2] School of Computational Science and Engineering, Georgia Institute of Technology, Atlanta, GA 30332, USA

* Correspondence: surya.kalidindi@me.gatech.edu

Abstract: Recent investigations have highlighted the multi-resolution and high throughput characteristics of the spherical indentation experimental and analysis protocols. In the present work, we further demonstrate the capabilities of these protocols for reliably extracting indentation stress-strain (ISS) responses from the microscale constituents as well as the bulk scale of dual phase materials exhibiting bimodal microstructures. Specifically, we focus on bimodal microstructures produced in an α–β Ti6242 sample. Combining the multi-resolution indentation responses with microstructural statistics gathered from the segmentation of back-scattered electron images from the scanning electron microscope allowed for a critical experimental evaluation of the commonly utilized Rule of Mixtures based composite model for the elastic stiffness and plastic yield strength of the sample. The indentation and image analyses protocols described in this paper offer novel research avenues for the systematic development and critical experimental validation of composite material models.

Keywords: composite material; bimodal microstructure; titanium; Ti6242; spherical indentation; image segmentation; rule of mixtures; effective property

Citation: Millan-Espitia, N.; Kalidindi, S.R. Study of a Bimodal α–β Ti Alloy Microstructure Using Multi-Resolution Spherical Indentation Stress-Strain Protocols. *J. Compos. Sci.* **2022**, *6*, 162. https://doi.org/10.3390/jcs6060162

Academic Editor: Stelios K. Georgantzinos

Received: 17 May 2022
Accepted: 31 May 2022
Published: 2 June 2022

1. Introduction

Microstructures exhibiting two distinct morphologies in the arrangement of their phase constituents are generally referred to as bimodal microstructures [1,2]. Most commonly, these microstructures exhibit single-phase equiaxed grains (refers to regions of uniform crystal lattice orientation in the material microstructure) alongside grains displaying distinct dual-phase morphologies (e.g., lamellar, dendritic). Examples of such microstructures can be seen in two-phase steels, alpha-beta brasses, alpha-beta titanium alloys, and bulk metallic glass-matrix composites. This class of composite microstructures offers tremendous potential for several advanced technology applications since their effective properties can be tailored more easily to meet the designer-specified targets through the modulation of the relative volume fractions and morphological features of their phase constituents.

Our interest in this paper will be on the bimodal microstructures in α–β Ti alloys. It is well known that the microstructures in these alloys can be controlled through suitably designed heat treatments. Generally, one can take the material above the beta-transus temperature (i.e., temperature where the hcp α transforms completely into bcc β) and cool the material down at different cooling rates to produce a variety of distinct microstructures. One can also apply heat treatments at temperatures below the beta-transus temperature to allow for the formation and stabilization of fully α-phase regions known as primary α grains. Slow cooling generally leads to the formation of parallel secondary α-laths (i.e., colonies), whereas higher cooling rates favor the formation of the crisscrossed secondary α-laths (i.e., basket-weave morphology) [3,4]. It is therefore possible to produce a broad range of bimodal microstructures in α–β Ti alloys through tailored heat treatments. Bimodal microstructures in α–β Ti alloys have been reported to result in improved mechanical properties. As a specific example, a bimodal microstructure consisting of approximately 30 vol.% of equiaxed primary α grains and 70 vol.% of lamellar α–β has been found to provide an optimal combination of creep and fatigue properties for compressor disks operating at high temperatures [3]. Besides the influence of phase morphologies, the mechanical response of α–β Ti alloys is also strongly influenced by the grain-scale anisotropy of the hcp α, which is known to exhibit a pronounced crystal lattice orientation dependence [5–10].

Bimodal microstructures have been studied extensively in prior literature in efforts to understand and predict their effective properties [11–23]. Towards this goal, it is important to accurately measure the mechanical properties at different length scales of the material. Of primary interest are the effective properties at the scale of a representative volume element (RVE) of the material microstructure (i.e., length scales covering multiple grains and colonies) and properties at the scale of individual constituents (i.e., individual grains or colonies). At the smaller length scales, protocols involving micropillar compression [24–28], micro-cantilever beam bending [29–31], and micro-hardness tests [28,32–35] have been shown to provide valuable mechanical response information for the individual constituents in these materials. However, these protocols often incur high costs, require tedious sample preparation procedures, and produce limited amount of data [5,36]. Moreover, hardness tests produce hardness values that are not easily related to other standard mechanical properties of the constituents. This is due to the fact that the measured values often exhibit large variances in the results produced by different research groups due to the inherent variances in the test and analyses protocols employed [5,37–41]. At the larger length scales (i.e., RVE scale), the effective properties are typically measured using standard protocols such as uniaxial tension [42] and compression [43]. The lack of consistency in the experimental protocols employed at the different material length scales is likely to have contributed significantly to the large discrepancies in the multiresolution mechanical properties reported for several advanced material systems [13,14,34,35,38,39,44–51]. As a specific example, the micro-hardness measurements of Gupta et al. [48] on basket-weave grains in Ti64 specimens suggested uniaxial yield strength values in the range 950–1308 MPa [39], which are significantly higher than the macroscale tensile yield strengths measured in the range of 905–945 MPa by Sieniawski et al. [49] on similar samples (i.e., samples with similar processing histories). These discrepancies could arise from the differences in the loading histories (i.e., indentation versus tension) in the measurement protocols employed at the different material structure scales.

In efforts aimed at establishing consistent and reliable multi-resolution mechanical testing capabilities, Kalidindi and Pathak [52–54] have developed the spherical indentation protocols based on the Hertzian theory of elastic contact between two deformable isotropic solids with quadratic surfaces [55]. These protocols capture the mechanical responses in the sample at different length scales in the form of load-displacement curves, and subsequently convert them to indentation stress-strain curves (ISS). At the grain-scale, these protocols were successfully demonstrated to capture the dependence of the local mechanical response on the grain orientation in different material classes [5,7,56–60]. At the macroscale, a consistent set of measurement and analyses protocols were successfully demonstrated for the reliable evaluation of the effective mechanical properties of the material [37,61–66]. An attractive feature of these protocols is that they incur significantly less effort and cost compared to many of the other competing protocols mentioned above, while requiring relatively small material volumes.

Consistent multi-resolution mechanical test protocols are central to the systematic evaluation and refinement of composite theories capable of predicting the effective properties of a material based on the details of the material microstructure and the individual properties of the microscale constituents. As a specific example, the simple rule of mixtures (ROM) model is commonly employed [13,16–23], but has only been critically evaluated in only a few experimental studies [13,16,20–23]. As already discussed earlier, the main hurdle comes from the lack of consistent protocols that can be applied at the different material structure length scales. In recent work [20–22], the ROM model was confirmed to exhibit impressive accuracy in estimating the effective yield strength of ferrite-martensite steels [20–22], in spite of the highly simplified assumptions implied by this model.

In this work, we aimed to critically evaluate the performance of the ROM model for estimating the effective yield strength of a bimodal Ti6242 specimen. This was accomplished using multi-resolution spherical nanoindentation protocols on a bimodal sample exhibiting primary α and basket-weave α–β morphologies. Multi-resolution microstructure characterization of the samples was conducted using scanning electron microscopy and electron backscatter diffraction. It is shown that the consistent use of the multiresolution spherical indentation protocols on the bimodal Ti6262 microstructure produces grain-scale and macroscale measurements of the yield strengths in the sample that are fairly consistent with the ROM model.

2. Materials and Methods

2.1. Materials and Sample Preparation

The material chosen for this study was Ti-6Al-2Sn-4Zr-2Mo (Ti6242) due to its versatility in producing diverse microstructures through suitable heat treatments. The actual alloy composition was determined in a previous study by Pilchak et al. [67] to be Ti-5.93Al-2.01Sn-4.05Zr-1.88Mo-0.12Si, with the interstitial contents of oxygen, iron, and nitrogen being 0.107, 0.05, and 0.001 in wt.%, respectively. Atomic absorption and inductively coupled plasma mass spectrometry was employed to determine the metallic element composition, while inert gas fusion was employed for the measurement of the smaller elements. Specimens of dimensions 10.0 mm × 20.0 mm × 2.0 mm were cut using an electric discharge machining and placed into quartz tubes, which were subsequently backfilled with Argon to protect the samples from oxidation during the heat treatment process. Heat treatment was designed such that the microstructure was composed of primary-α and basket-weave morphologies with large enough grains that allowed the application of the spherical indentation protocols. Specifically, the specimen was heat-treated at 986 °C (10 °C below the beta transus for a bimodal microstructure) for 6 h, followed by a water-quenching step to achieve the desired bimodal microstructures comprising the basket-weave components. Subsequently, the sample was subjected to a stress-relief heat treatment (700 °C for 4 h) and air-cooled. The heat treatment was conducted in a Thermo Fisher Scientific, Lindberg/Blue 1100 °C Box furnace.

Specimen surfaces were prepared for microscopy and indentation following standard metallography protocols procedures [68]. These protocols removed any oxide or mechanically deformed layers. Specifically, for this work, it was important to minimize the height disparities on the sample surface due to the unavoidable differences in the polishing rates of the different microscale constituents present in the sample. Chemo-mechanical preparation steps utilized on these specimens included surface grinding using silicon carbide papers (starting with 800 grit and systematically going to 2400 grit), polishing using 9 μm, 3 μm and 1 μm diamond suspension on a Struers' (Copenhagen, Denmark) Tegramin Automatic Grinding Machine, and final polishing on the Buehler's (Lake Bluff, IL, USA) Vibremet 2 Vibratory polisher for 12 h with a medium consisting of one part of 0.06 μm colloidal silica, 4 parts of water, and 1 part of hydrogen peroxide.

2.2. Spherical Indentation

Mechanical characterization of the samples was conducted at room temperature on a Keysight Nano-Indenter G200 with the Continuous Stiffness Measurement (CSM) option. Two different tip sizes were used in this study: (i) a 15.2 μm radius tip was employed for the grain scale characterization of primary α and basket-weave (prior β) grains, and (ii) a 500 μm radius tip was used for the characterization of the bulk response of the sample over regions comprising a large number of grains. The grain-scale indentations were placed close to the centers of the grains, and only one indentation was conducted per grain to avoid proximity with the grain boundary and other previous indentations in the sample. For the indentations on the primary α grains, the average contact radius at yield (over grains of different orientations) was estimated to be about 550 nm (using equations presented in the next subsection). Since this is significantly smaller than the average size of the primary α grains of about 10 μm, these measurements are assumed to reflect the local grain-scale mechanical responses. For the indentations on the basket-weave grains, the average contact radius at yield was estimated to be about 640 nm. The deformation zone size in these indentations is significantly smaller than the prior β grain size of about 16 μm, but larger than the average α lath thickness of about 18 nm (lath spacing is significantly smaller in these microstructures). Therefore, the measurements in the basket-weave grains are assumed to reflect the effective properties of the basket-weave components. For the indentations with the larger tip, the average contact radius at yield was estimated to be about 25 μm. The deformation zone in these indentations are expected to include about 13 grains (mixtures of primary α and basket-weave components), and therefore an ensemble average of these measurements is assumed to reflect the bulk response of the sample. A total of 150 indentation tests were conducted for this work (about 50 tests on each grain scale morphology and about 50 tests for the bulk response).

2.3. Spherical Nanoindentation Analysis

The spherical indentation experimental and analysis protocols used in this work are largely based on Hertz theory, which is mainly focused on describing the deformation during frictionless contact between two approaching elastic bodies with quadratic surfaces [55]. The load-displacement relationship for such indentation is expressed as

$$P = \frac{4}{3} E_{eff} R_{eff}^{\frac{1}{2}} h_e^{\frac{3}{2}} \tag{1}$$

$$\frac{1}{E_{eff}} = \frac{1 - v_i^2}{E_i} + \frac{1 - v_s^2}{E_s} \tag{2}$$

$$\frac{1}{R_{eff}} = \frac{1}{R_i} + \frac{1}{R_s} \tag{3}$$

where P, E_{eff}, R_{eff}, and h_e denote the indentation load, effective elastic modulus, effective radius, and elastic indentation displacement, respectively (see Figure 1a).

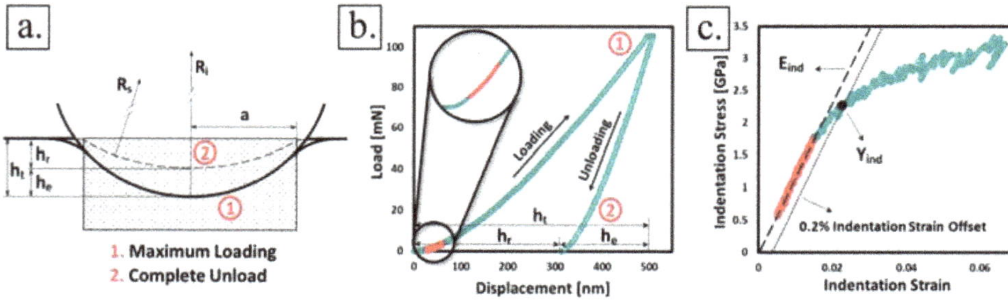

Figure 1. (**a**) Schematic description of the spherical indentation experiment depicting the loaded and unloaded configurations. (**b**) Measured load-displacement curve from a typical spherical indentation measurement, with a zoomed view of the early loading segment. (**c**) Indentation stress-strain (ISS) curve extracted from the load-displacement data with the elastic region of the deformation highlighted in red.

Furthermore, R, E, and v denote the radius, Young's modulus and the Poisson's ratio, respectively, while the subscripts i and s correspond to the indenter and sample, respectively. The value of E_{eff} is estimated from a segment selected in the initial elastic loading [53] (the bright colored segment in Figure 1b), and is assumed to be unchanged throughout the entire deformation of the material. In the analyses of the initial elastic segment for the determination of E_{eff}, R_s is assumed to represent infinity (i.e., $R_{eff} = R_i$) [53].

The analysis of the indentation load-displacement data starts with a zero-point correction, which identifies an effective point of initial contact between the indenter and the sample by finding a segment of the load-displacement curve that complies with Hertz's theory (i.e., Equation (1)). This correction accounts for the unavoidable surface artifacts such as roughness and oxide layers that affect the indentation measurements [53]. The zero-point correction is implemented using regression techniques that fit the measurements to the following equation derived from Hertz's theory [53].

$$S = \frac{3P}{2h_e} = \frac{3\left(\widetilde{P} - P^*\right)}{2\left(\widetilde{h} - h^*\right)} \tag{4}$$

where $(\widetilde{P}, \widetilde{h})$ denote the measured load-displacement values, $(P^*,\ h^*)$ denote the zero-point load and displacement corrections, respectively, and S denotes the CSM (obtained here using a superimposed load-unloading cycles of 2 nm amplitude and 45 Hz frequency) [69,70].

The use of CSM allows for the estimation of the contact radius, a, along the complete monotonic loading history, which can then be used to estimate the indentation stress and indentation strain values corresponding to every point on the measured load-displacement curve (see Figure 1b,c). These computations are performed using

$$a = \frac{S}{2E_{eff}} \tag{5}$$

$$\sigma_{ind} = \frac{P}{\pi a^2} \tag{6}$$

$$\varepsilon_{ind} = \frac{4}{3\pi}\frac{h_t}{a} \tag{7}$$

The indentation yield strength, Y_{ind}, is defined as the indentation stress at a 0.2% offset indentation plastic strain in the indentation stress-strain (ISS) curve (see Figure 1c).

2.4. Microstructure Characterization and Quantification

The back-scattered electron imaging signal from a Tescan Mira3 field emission scanning electron microscope (FE-SEM) was used for the characterization of the bimodal microstructure in the sample. Images with a constant view-field of 100 µm and size 2048 × 2048 pixels were captured using an accelerating voltage of 15 kV. The acquired grayscale images were analyzed through a series of image processing steps in order to label each pixel in the image as belonging to one of the two grain-scale morphologies (i.e., primary α or basket-weave). This process is generally referred to as image segmentation, and the sequence of image processing steps used in this process are referred as segmentation workflows [71–74]. The segmentation workflows utilized in this work were based on the framework proposed recently by Iskakov and Kalidindi [75]. They comprised the following steps: (i) gaussian global noise removal using the *imgaussfilt* function [76] with a smoothing kernel whose standard deviation was set to 0.9; (ii) global thresholding step with the *imquantize* function for which a single specified quantization value of 85 was selected [76]; and (iii) post-processing steps such as the *bwareaopen* function [76] to remove all connected objects that have fewer than 1000 pixels and the *imclose* function [76,77] which performs a dilation-erosion dual operation on the segmented areas using a disk of 3 pixel radius. Segmentation validation involved visual validation, as well as a more quantitative approach using precision and recall scores [78]. The quantitative validation of the segmented image was used to determine the parameter values mentioned above. An example of the application of our segmentation workflow is presented in Figure 2.

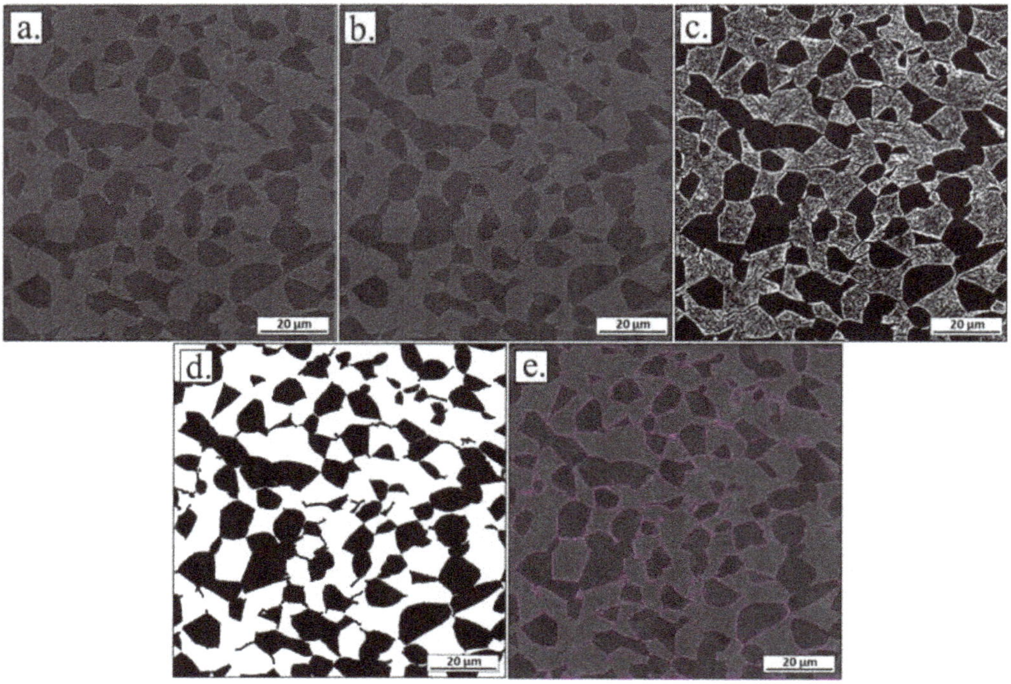

Figure 2. Segmentation workflow applied on a BSE micrograph with view-field of 100 µm. (**a**) Raw image from the scanning electron microscope device. (**b**) Image after a gaussian global noise removal correction. (**c**). Resulting micrograph after the thresholding step. (**d**) Final segmented microstructure representative. (**e**) Original image with the phase boundaries highlighted in magenta for the visual validation of the segmentation process.

The crystallographic orientations of the primary α phase grains were measured using electron back-scattered diffraction (EBSD) mapping in the SEM using an accelerating voltage of 20 kV. A total of 20 EBSD scans of adjacent and slightly overlapping areas of 150 µm × 150 µm were imaged using a 0.5 µm step-size. Using the TSL OIM Analysis 8 software, the images were suitably stitched to create a single large scan of approximately 680 µm × 545 µm (see Figure 3a). In order to extract the texture information for the primary α phase, the image quality parameter was used to filter out the pixels corresponding to the basket-weave grains as they exhibit low image quality values. The EBSD map showing the orientations of the primary α grains is shown in Figure 3b. It is seen that the primary α regions in the sample exhibit a pronounced texture. The area fractions corresponding to 10-degree bins in the declination angle (Φ) were computed and are summarized in Table 1.

Table 1. Fraction corresponding to each of the binned regions of the primary a phase, and the corresponding volume fractions of all constituents in the bulk microstructure.

Bimodal Microstructure Statistics		
Local State	**Fraction in α-Phase**	**Fraction in Bulk (*f*)**
α = 0°–10°	16%	6.8%
α = 10°–20°	6.1%	2.6%
α = 20°–30°	1.6%	0.7%
α = 30°–40°	2.0%	0.8%
α = 40°–50°	3.0%	1.3%
α = 50°–60°	9.7%	4.1%

Table 1. *Cont.*

Bimodal Microstructure Statistics		
Local State	**Fraction in α-Phase**	**Fraction in Bulk (f)**
α = 60°–70°	17.7%	7.5%
α = 70°–80°	14.7%	6.2%
α = 80°–90°	29.1%	12.3%
Basket-weave	0%	57.7%

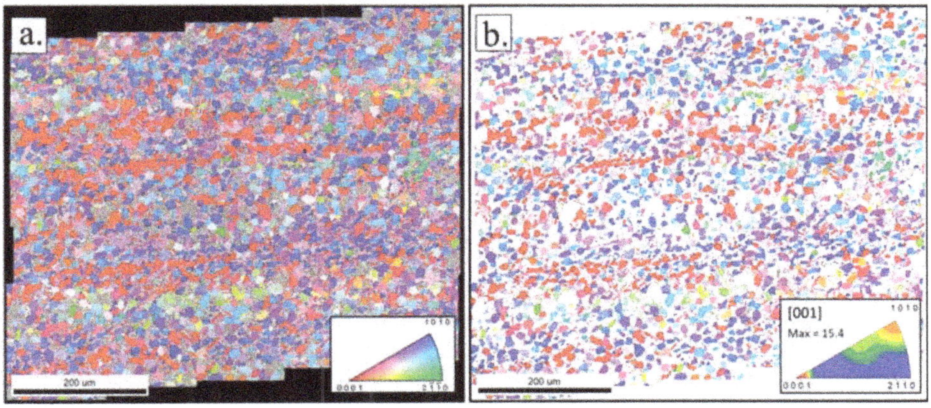

Figure 3. (**a**) Compilation of 20 EBSD scans from a Ti6242 bimodal microstructure with primary α and basket-weave grains. Areas with low image quality correspond to basket-weave grains. (**b**) Inverse Pole Figure map of the primary a grains only, with a harmonic tex.

2.5. Prediction of Effective Property

There exist several models in literature for the prediction of the effective elastic and plastic properties of composites based on microstructure statistics and the properties of their individual constituents. The classical Rule of Mixtures (ROM) is the simplest among these. For the bimodal titanium microstructure studied here, this model can be expressed as

$$P_{eff} = f_\alpha P_\alpha + f_{BW} P_{BW} \tag{8}$$

where P and f refer to the property and volume fractions, respectively, subscripts α and BW refer to the constituent morphologies (primary a and basket-weave, respectively), and P_{eff} denotes the effective property of the system.

This model has been evaluated extensively in literature [13,17–23] for predicting the effective yield strength of the composite. Due to the high plastic anisotropy associated with the primary α component of titanium alloys [5] (the yield strength of the basket-weave component does not exhibit a strong dependence on the lattice orientations of the α laths), in the present study we have re-written Equation (8) as

$$P_{eff} = f_{\alpha_{0^\circ-10^\circ}} P_{\alpha_{0^\circ-10^\circ}} + f_{\alpha_{10^\circ-20^\circ}} P_{\alpha_{10^\circ-20^\circ}} + \cdots + f_{\alpha_{80^\circ-90^\circ}} P_{\alpha_{80^\circ-90^\circ}} + f_{BW} P_{BW} \tag{9}$$

where the different f_α have already been tabulated in Table 1. The corresponding P_α and P_{BW} will be measured using the indentation protocols described above.

It is important to note that P_{eff} is also measured using the same indentation protocols in our study. Therefore, the consistent high-throughput multi-resolution spherical indentation stress-strain protocols combined with microscopy protocols offer a unique opportunity to critically evaluate composite models.

3. Results and Discussion

3.1. Microstructure Statistics

From an ensemble of 10 segmented BSE-SEM images, the average primary α volume fraction in the sample was determined to be 42.3 ± 4.1%. The remaining volume corresponding to the basket-weave morphology was computed as $f_{BW} = 1 - f_\alpha$. The inverse pole figure in Figure 3 and the values from Table 1 indicate that the primary α grain normals in our sample are predominantly aligned with the $[10\bar{1}0]$ and $[0001]$ directions in the hcp crystal. It is also clear that the primary α grains are arranged in a band-like structure, which is presumed to result from the prior thermo-mechanical deformation applied on the sample.

3.2. Spherical Nanoindentation Stress-Strain Measurements

The indentation stress-strain curves obtained from the application of the spherical indentation stress-strain protocols described above on the primary α grains, basket-weave grains, and the bulk measurements are presented in Figure 4. The values of the measured indentation moduli and indentation yield strengths in the primary α grains are correlated to their declination angles in Figure 5. In prior work, it has been shown that the influence of the other two Euler angles on these measurements is fairly low [5,7]. The results shown in Figure 5 are in good agreement with the values reported in a previous study, where similar protocols were applied on the primary α grains in a set of α-, near-α and α–β titanium alloys [5]. The highest measured indentation yield strength in the primary α grains in our sample was observed to be about 3 GPa, and corresponded to the grains with their c-axis oriented parallel to the indentation direction. On the other hand, the lowest indentation yield strength was measured to be about 1 GPa for grains with the c-axis perpendicular to the indentation direction. Following the strategy outlined earlier, these measurements were binned by the declination angle and added to Table 2. Indentation modulus and indentation yield strength for each bin were determined by applying a polynomial regression using generalized spherical harmonics (GSH) on the complete dataset [79], and then establishing the values at the center-point of each bin, as depicted in Figure 5.

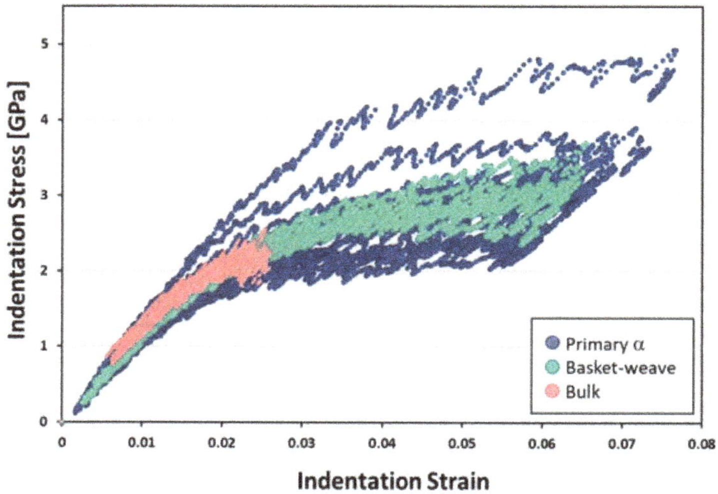

Figure 4. Examples of the indentation stress-strain curves measured in this study. The blue curves are from measurements on primary a grains, the green curves are from basket-weave grains, and the pink curves are bulk measurements.

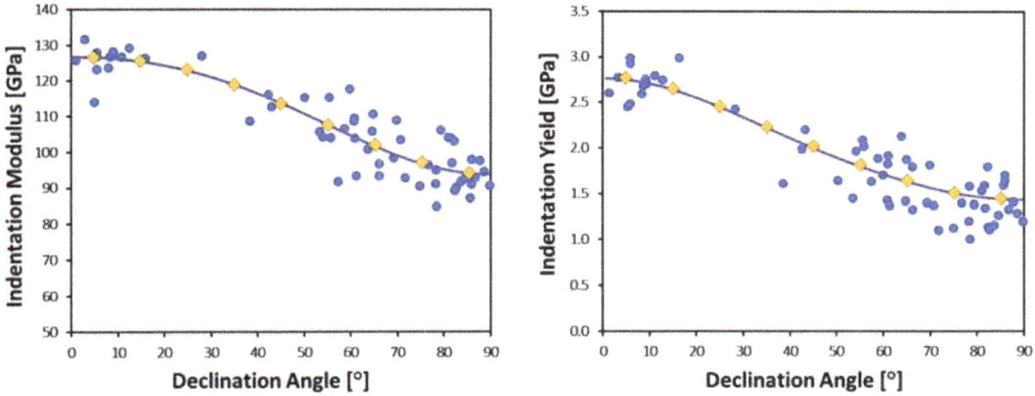

Figure 5. Results from the spherical indentations performed on the primary a grains, plotted as a function of the declination angle (Φ). Experimental results are shown as blue circles, and the values estimated from regression analysis at the mid points of each bin are depicted as yellow diamonds.

Table 2. Measured indentation moduli and indentation yield strengths for the different microscale constituents as well as the bulk responses of the material studied in this work. The last row provides the corresponding predictions from the application of the ROM model.

Spherical Indentation Measurements			
Length Scale	**Morphology**	E_{ind} [GPa]	Y_{ind} [GPa]
Constituents	**Basket-Weave**	121 ± 3.1	1.99 ± 0.12
	Φα = 0°–10°	125.2	2.73
	Φα = 10°–20°	124.1	2.65
	Φα = 20°–30°	121.7	2.47
	Φα = 30°–40°	118.4	2.25
	Φα = 40°–50°	111.9	2.03
	Φα = 50°–60°	106.6	1.79
	Φα = 60°–70°	100.9	1.60
	Φα = 70°–80°	96.3	1.44
	Φα = 80°–90°	93.7	1.37
Bulk	**Bimodal**	118 ± 2.6	1.96 ± 0.10
ROM Predictions			
Bulk	**Bimodal**	114	1.92

The average indentation modulus and the indentation yield strength for the basket-weave grains morphology were determined to be 121 GPa and 1.99 GPa, respectively (see Table 2). The corresponding standard deviations were computed as 3.12 GPa and 0.12 GPa, respectively. The measurements indicated a much-reduced variation in the mechanical properties exhibited by the basket-weave grains, when compared to the corresponding measurements on the primary α grains presented earlier. This observation suggests that the multiple α variants that coexist within a single basket-weave grain tend to homogenize the elastic and plastic properties to isotropic values at the grain-scale for this complex morphology.

The results from the application of the spherical indentation stress-strain protocols for the effective mechanical response are also summarized in Table 2. The average effective indentation modulus and the averaged effective indentation yield strength were measured to be 118 GPa and 1.96 GPa, respectively. The low standard deviations of these measurements (see Table 2) confirm that they reliably reflect the bulk response of the sample. Assuming an overall isotropic material response and using previously established conversion rules [37,52,63,65,80], the sample's Young's modulus and the uniaxial yield strength are estimated as 106 GPa and 996 MPa, respectively. These are in good agreement with prior reports in literature using conventional uniaxial test methods. For example, Bertrand et al. [81] reported an uniaxial yield strength of 948 MPa for a similar Ti6242 bimodal sample.

3.3. Evaluation of the Composite Model

The volume fraction information presented in Table 1 along with the mechanical properties measured by spherical indentation presented in Table 2 were used for the evaluation of the composite model described in Equation (9). The predictions from this simple model are also presented in Table 2. The predicted indentation modulus for this bimodal microstructure was calculated to be 114 GPa. This represents a 3.6% difference with respect to the measured modulus of 118 GPa. The predicted indentation yield strength was 1.92 GPa corresponding to a 2.1% difference from the experimentally measured bulk indentation yield strength of 1.96 GPa.

Prior applications of the ROM on multiphase materials have largely employed Equation (8) directly [13,15,16,20–23]. They implicitly assumed that the averaged values of the properties measured from multiple randomly placed indentations in each phase represented adequately the effective property for the respective phases. While this is likely to be true if one employs a very large number of indentations, in practice, this is not the case. In this study, we clearly noticed that Equation (9) produced a significantly better prediction for the bimodal microstructures, compared to the direct use of Equation (8). This is since the large area EBSD scan (see Figure 3) provides much more statistically reliable measure of the different texture components present in the sample. Since the different texture components (with the different declination angles) exhibit very different indentation properties (see Figure 5), Equation (9) represents a much more accurate application of ROM. Since the indentation properties of the basket-weave component did not exhibit a strong dependence on its texture components, there was no need to apply the same approach on the basket-weave component.

It is emphasized again that the consistent use of the spherical indentation stress-strain protocols at both the constituents' scale and the macroscale played an important role in providing a reliable set of measurements for our study. We believe that the protocols described in this work have opened several new research avenues for the critical evaluation and refinement of homogenization models for a broad range of heterogeneous (composite) materials. Furthermore, the high-throughput capabilities of the techniques described and the requirements of relatively small material volumes make these protocols extremely attractive for materials development efforts. This study opens new research avenues into high-throughput multi-resolution studies of the mechanical response of composite materials with complex microstructures.

4. Conclusions

This work demonstrates the systematic application of the multi-resolution spherical indentation and microstructure characterization and analyses protocols for generating the data required for an improved understanding of the mechanical response of a complex dual phase metallic alloy sample exhibiting a bimodal microstructure. The sample selected for this study not only comprised of two different crystal structures (i.e., thermodynamic stable phases), but also two different morphologies that exhibited distinct anisotropy in their local mechanical responses. Specifically, the indentation stress-strain curves extracted at the microscale quantified the degree of anisotropy in each constituent. For the primary α phase, it was found that the crystal lattice orientation played an important role in the anisotropic local mechanical response. At the macroscale, the larger micro-indentations were found to consistently and reliably capture the bulk or effective mechanical response of the sample. EBSD analysis allowed the correlation of the local mechanical responses in the individual primary α grains to their crystal lattice orientation. BSE-SEM images were able to provide reliable measures of the volume fractions of the two main constituents in the bimodal microstructures. All of the information from the microstructural analysis as well as the mechanical characterization was used to evaluate the commonly used Rule of Mixtures (ROM) models for the effective indentation modulus and indentation yield strength of the bimodal microstructures. It was found that the ROM predicted indentation modulus and indentation yield strength were within 4% of the experimentally measured properties.

Author Contributions: Conceptualization, N.M.-E. and S.R.K.; methodology, N.M.-E.; formal analysis, N.M.-E.; data curation, N.M.-E.; writing—original draft preparation, N.M.-E. and S.R.K.; supervision, S.R.K.; funding acquisition, S.R.K. All authors have read and agreed to the published version of the manuscript.

Funding: The authors acknowledge support from Air Force Office of Scientific Research (AFOSR) Grant FA9550-18-1-0330.

Conflicts of Interest: The authors declare no conflict of interest.

References

1. Lütjering, G. Influence of processing on microstructure and mechanical properties of ($\alpha + \beta$) titanium alloys. *Mater. Sci. Eng. A* **1998**, *243*, 32–45. [CrossRef]
2. Murayama, Y.; Hanada, S. High temperature strength, fracture toughness and oxidation resistance of Nb–Si–Al–Ti multiphase alloys. *Sci. Technol. Adv. Mater.* **2002**, *3*, 145–156. [CrossRef]
3. Polmear, I.; StJohn, D.; Nie, J.F.; Qian, M. *Light Alloys: Metallurgy of the Light Metals*, 5th ed.; Elsevier Science: Amsterdam, The Netherlands; Butterworth-Heinemann: Oxford, UK, 2017.
4. Banerjee, S.; Mukhopadhyay, P. *Phase Transformations: Examples from Titanium and Zirconium Alloys*; Elsevier: Amsterdam, The Netherlands, 2010.
5. Millan-Espitia, N.; Mohan, S.; Pilchak, A.L.; Kalidindi, S.R. Mechanical Responses of Primary-α Ti Grains in Polycrystalline Samples: Part I—Measurements of Spherical Indentation Stress–Strain Curves. *Integr. Mater. Manuf. Innov.* **2021**, *10*, 82–98. [CrossRef]
6. Britton, T.B.; Dunne, F.P.E.; Wilkinson, A.J. On the mechanistic basis of deformation at the microscale in hexagonal close-packed metals. *Proc. R. Soc. A Math. Phys. Eng. Sci.* **2015**, *471*, 20140881. [CrossRef]
7. Weaver, J.S.; Priddy, M.W.; McDowell, D.L.; Kalidindi, S.R. On capturing the grain-scale elastic and plastic anisotropy of alpha-Ti with spherical nanoindentation and electron back-scattered diffraction. *Acta Mater.* **2016**, *117*, 23–34. [CrossRef]
8. Perilla, J.M.; Gil Sevillano, J. Two-dimensional sections of the yield locus of a Ti-6%Al-4%V alloy with a strong transverse-type crystallographic a-texture. *Mater. Sci. Eng.* **1995**, *A201*, 103–110. [CrossRef]
9. Fundenberger, J.J.; Philippe, M.J.; Wagner, F.; Esling, C. Modeling and prediction of mechanical properties for materials with hexagonal symmetry (zinc, titanium and zirconium alloys). *Acta Mater.* **1997**, *45*, 4041–4055. [CrossRef]
10. Churchman, A.T. Preparation of single crystals of titanium and their mode of deformation. *Nature* **1953**, *171*, 706. [CrossRef]
11. Nemat-Nasser, S.; Hori, M. *Micromechanics: Overall Properties of Heterogeneous Materials*; Elsevier: Amsterdam, The Netherlands, 2013.
12. González, C.; Vilatela, J.; Molina-Aldareguía, J.; Lopes, C.; Llorca, J. Structural composites for multifunctional applications: Current challenges and future trends. *Prog. Mater. Sci.* **2017**, *89*, 194–251. [CrossRef]
13. Jinoch, J.; Ankem, S.; Margolin, H. Calculations of stress-strain curve and stress and strain distributions for an α-β Ti-8Mn alloy. *Mater. Sci. Eng.* **1978**, *34*, 203–211. [CrossRef]
14. Ankem, S.; Margolin, H. Finite element method (FEM) calculations of stress-strain behavior of alpha-beta Ti-Mn alloys: Part I. Stress-strain relations. *Metall. Trans. A* **1982**, *13*, 595–601. [CrossRef]
15. Ogden, H.; Holden, F.; Jaffee, R. Effect of Alpha Sohites on the Heat-Treatment Response of Ti-Mn Alloys. *JOM* **1955**, *7*, 105–112. [CrossRef]
16. Holden, F.; Ogden, H.; Jaffee, R. Heat treatment and mechanical properties of Ti-Mo alloys. *JOM* **1956**, *8*, 1388–1393. [CrossRef]
17. Hayden, H.; Floreen, S. The influence of martensite and ferrite on the properties of two-phase stainless steels having microduplex structures. *Metall. Trans.* **1970**, *1*, 1955–1959. [CrossRef]
18. Yegneswaran, A.; Tangri, K. Investigation of the early stages of deformation of two phase copper-aluminum alloys. *Metall. Trans. A* **1983**, *14*, 2407–2413. [CrossRef]
19. Yegneswaran, A.; Tangri, K. Deformation behaviour of two phase materials: Cu-Al system. *Met. Sci.* **1984**, *18*, 161–168. [CrossRef]
20. Davies, R. The deformation behavior of a vanadium-strengthened dual phase steel. *Metall. Trans. A* **1978**, *9*, 41–52. [CrossRef]
21. Davies, R. The mechanical properties of zero-carbon ferrite-plus-martensite structures. *Metall. Trans. A* **1978**, *9*, 451–455. [CrossRef]
22. Davies, R. Influence of martensite composition and content on the properties of dual phase steels. *Metall. Trans. A* **1978**, *9*, 671–679. [CrossRef]
23. Tseng, D.; Tangri, K. Deformation behavior of duplex Zircaloy-4-oxygen alloys. *Metall. Trans. A* **1982**, *13*, 1077–1082. [CrossRef]

24. Yu, Q.; Shan, Z.-W.; Li, J.; Huang, X.; Xiao, L.; Sun, J.; Ma, E. Strong crystal size effect on deformation twinning. *Nature* **2010**, *463*, 335–338. [CrossRef] [PubMed]
25. Xiao, L.I.N.; Yu, Q.; Sun, Q.; Sun, J.U.N. Size effect on deformation mode in micron-sized Ti-5Al single crystal loaded along $[2\bar{1}\bar{1}0]$ and [0001]. *Int. J. Mod. Phys.* **2012**, *24*, 2466–2471. [CrossRef]
26. Jun, T.-S.; Sernicola, G.; Dunne, F.P.; Britton, T.B. Local deformation mechanisms of two-phase Ti alloy. *Mater. Sci. Eng. A* **2016**, *649*, 39–47. [CrossRef]
27. Zhang, Z.; Jun, T.-S.; Britton, B.; Dunne, F.P. Intrinsic anisotropy of strain rate sensitivity in single crystal alpha titanium. *Acta Mater.* **2016**, *118*, 317–330. [CrossRef]
28. Ghassemi-Armaki, H.; Maaß, R.; Bhat, S.; Sriram, S.; Greer, J.; Kumar, K. Deformation response of ferrite and martensite in a dual-phase steel. *Acta Mater.* **2014**, *62*, 197–211. [CrossRef]
29. Gong, J.; Wilkinson, A.J. Micro-cantilever testing of ⟨a⟩ prismatic slip in commercially pure Ti. *Philos. Mag.* **2011**, *91*, 1137–1149. [CrossRef]
30. Gong, J.; Wilkinson, A.J. Anisotropy in the plastic flow properties of single-crystal α titanium determined from micro-cantilever beams. *Acta Mater.* **2009**, *57*, 5693–5705. [CrossRef]
31. Kumar, K.; Madhusoodanan, K.; Rupani, B. Miniature Specimen Technique as an NDT Tool for Estimation of Service Life of Operating Presurre Equipment. In Proceedings of the International Conference & Exhibition on Pressure Vessel and Piping, Chennai, India, 7–9 February 2006.
32. Mante, F.K.; Baran, G.R.; Lucas, B. Nanoindentation studies of titanium single crystals. *Biomaterials* **1999**, *20*, 1051–1055. [CrossRef]
33. Merson, E.; Brydson, R.; Brown, A. The effect of crystallographic orientation on the mechanical properties of titanium. *J. Phys. Conf. Ser.* **2008**, *126*, 012020. [CrossRef]
34. Liu, Y.; Ngan, A. Depth dependence of hardness in copper single crystals measured by nanoindentation. *Scr. Mater.* **2001**, *44*, 237–241. [CrossRef]
35. McElhaney, K.; Vlassak, J.J.; Nix, W.D. Determination of indenter tip geometry and indentation contact area for depth-sensing indentation experiments. *J. Mater. Res.* **1998**, *13*, 1300–1306. [CrossRef]
36. Kiener, D.; Motz, C.; Rester, M.; Jenko, M.; Dehm, G. FIB Damage of Cu and possible consequences for miniaturized mechanical tests. *Mater. Sci. Eng. A* **2007**, *459*, 262–272. [CrossRef]
37. Mohan, S.; Millan-Espitia, N.; Yao, M.; Steenberge, N.V.; Kalidindi, S.R. Critical Evaluation of Spherical Indentation Stress-Strain Protocols for the Estimation of the Yield Strengths of Steels. *Exp. Mech.* **2021**, *61*, 641–652. [CrossRef]
38. Rodrıguez, R.; Gutierrez, I. Correlation between nanoindentation and tensile properties: Influence of the indentation size effect. *Mater. Sci. Eng. A* **2003**, *361*, 377–384. [CrossRef]
39. Zhang, P.; Li, S.; Zhang, Z. General relationship between strength and hardness. *Mater. Sci. Eng. A* **2011**, *529*, 62–73. [CrossRef]
40. O'Neill, H. *Hardness Measurement of Metals and Alloys*; Chapman & Hall: London, UK, 1967.
41. Cahoon, J.; Broughton, W.; Kutzak, A. The determination of yield strength from hardness measurements. *Metall. Trans.* **1971**, *2*, 1979–1983. [CrossRef]
42. *E8/E8M−15a*; Standard Test Methods for Tension Testing of Metallic Materials. ASTM International: West Conshohocken, PA, USA, 2015.
43. ASTM International. Standard Test Methods of Compression Testing of Metallic Materials at Room Temperature. In *1990 Annual Book of ASTM Standards*; ASTM International: West Conshohocken, PA, USA, 2000; p. 8.
44. Pharr, G.M.; Herbert, E.G.; Gao, Y. The indentation size effect: A critical examination of experimental observations and mechanistic interpretations. *Annu. Rev. Mater. Res.* **2010**, *40*, 271–292. [CrossRef]
45. Qian, L.; Li, M.; Zhou, Z.; Yang, H.; Shi, X. Comparison of nano-indentation hardness to microhardness. *Surf. Coat. Technol.* **2005**, *195*, 264–271. [CrossRef]
46. Swadener, J.; George, E.; Pharr, G. The correlation of the indentation size effect measured with indenters of various shapes. *J. Mech. Phys. Solids* **2002**, *50*, 681–694. [CrossRef]
47. Johnson, K. The correlation of indentation experiments. *J. Mech. Phys. Solids* **1970**, *18*, 115–126. [CrossRef]
48. Gupta, A.; Khatirkar, R.K.; Kumar, A.; Parihar, M.S. Investigations on the effect of heating temperature and cooling rate on evolution of microstructure in an α + β titanium alloy. *J. Mater. Res.* **2018**, *33*, 946–957. [CrossRef]
49. Sieniawski, J.; Ziaja, W.; Kubiak, K.; Motyk, M. Microstructure and Mechanical Properties of High Strength Two-Phase Titanium Alloys. In *Titanium Alloys—Advances in Properties Control*; Sieniawski, J., Ed.; InTech: London, UK, 2013.
50. Jun, T.-S.; Armstrong, D.E.J.; Britton, T.B. A nanoindentation investigation of local strain rate sensitivity in dual-phase Ti alloys. *J. Alloys Compd.* **2016**, *672*, 282–291. [CrossRef]
51. Zhang, Z.; Jun, T.-S.; Britton, B.; Dunne, F.P. Determination of Ti-6242 α and β slip properties using micro-pillar test and computational crystal plasticity. *J. Mech. Phys. Solids* **2016**, *95*, 393–410. [CrossRef]
52. Pathak, S.; Kalidindi, S.R. Spherical nanoindentation stress–strain curves. *Mater. Sci. Eng. R Rep.* **2015**, *91*, 1–36. [CrossRef]
53. Kalidindi, S.R.; Pathak, S. Determination of the effective zero-point and the extraction of spherical nanoindentation stress–strain curves. *Acta Mater.* **2008**, *56*, 3523–3532. [CrossRef]
54. Pathak, S.; Shaffer, J.; Kalidindi, S. Determination of an effective zero-point and extraction of indentation stress–strain curves without the continuous stiffness measurement signal. *Scr. Mater.* **2009**, *60*, 439–442. [CrossRef]
55. Hertz, H.; Jones, D.E.; Schott, G.A. *Miscellaneous Papers*; Macmillan and Company: London, UK, 1896.

56. Weaver, J.S.; Kalidindi, S.R. Mechanical characterization of Ti-6Al-4V titanium alloy at multiple length scales using spherical indentation stress-strain measurements. *Mater. Des.* **2016**, *111*, 463–472. [CrossRef]
57. Kalidindi, S.R.; Vachhani, S.J. Mechanical characterization of grain boundaries using nanoindentation. *Curr. Opin. Solid State Mater. Sci.* **2014**, *18*, 196–204. [CrossRef]
58. Pathak, S.; Michler, J.; Wasmer, K.; Kalidindi, S.R. Studying grain boundary regions in polycrystalline materials using spherical nano-indentation and orientation imaging microscopy. *J. Mater. Sci.* **2011**, *47*, 815–823. [CrossRef]
59. Vachhani, S.J.; Kalidindi, S.R. Grain-scale measurement of slip resistances in aluminum polycrystals using spherical nanoindentation. *Acta Mater.* **2015**, *90*, 27–36. [CrossRef]
60. Pathak, S.; Kalidindi, S.R.; Mara, N.A. Investigations of orientation and length scale effects on micromechanical responses in polycrystalline zirconium using spherical nanoindentation. *Scr. Mater.* **2016**, *113*, 241–245. [CrossRef]
61. Khosravani, A.; Morsdorf, L.; Tasan, C.C.; Kalidindi, S.R. Multiresolution mechanical characterization of hierarchical materials: Spherical nanoindentation on martensitic Fe-Ni-C steels. *Acta Mater.* **2018**, *153*, 257–269. [CrossRef]
62. Khosravani, A.; Caliendo, C.M.; Kalidindi, S.R. New insights into the microstructural changes during the processing of dual-phase steels from multiresolution spherical indentation stress–strain protocols. *Metals* **2019**, *10*, 18. [CrossRef]
63. Weaver, J.S.; Khosravani, A.; Castillo, A.; Kalidindi, S.R. High throughput exploration of process-property linkages in Al-6061 using instrumented spherical microindentation and microstructurally graded samples. *Integr. Mater. Manuf. Innov.* **2016**, *5*, 192–211. [CrossRef]
64. Khosravani, A.; Cecen, A.; Kalidindi, S.R. Development of high throughput assays for establishing process-structure-property linkages in multiphase polycrystalline metals: Application to dual-phase steels. *Acta Mater.* **2017**, *123*, 55–69. [CrossRef]
65. Iskakov, A.; Yabansu, Y.C.; Rajagopalan, S.; Kapustina, A.; Kalidindi, S.R. Application of spherical indentation and the materials knowledge system framework to establishing microstructure-yield strength linkages from carbon steel scoops excised from high-temperature exposed components. *Acta Mater.* **2018**, *144*, 758–767. [CrossRef]
66. Bhat, A.; Neu, R. On the Constraint Factor Relating Uniaxial and Indentation Yield Strength of Polycrystalline Materials Using Spherical Microindentation. *Mater. Perform. Charact.* **2020**, *9*, 324–345. [CrossRef]
67. Pilchak, A.; Porter, W.; John, R. Room temperature fracture processes of a near-α titanium alloy following elevated temperature exposure. *J. Mater. Sci.* **2012**, *47*, 7235–7253. [CrossRef]
68. Pathak, S.; Stojakovic, D.; Doherty, R.; Kalidindi, S. Importance of surface preparation on the nano-indentation stress-strain curves measured in metals. *Mater. Res. Soc.* **2009**, *24*, 1142–1155. [CrossRef]
69. Li, X.; Bhushan, B. A review of nanoindentation continuous stiffness measurement technique and its applications. *Mater. Charact.* **2001**, *48*, 11–36. [CrossRef]
70. Vachhani, S.J.; Doherty, R.D.; Kalidindi, S.R. Effect of the continuous stiffness measurement on the mechanical properties extracted using spherical nanoindentation. *Acta Mater.* **2013**, *61*, 3744–3751. [CrossRef]
71. Collins, P.C.; Welk, B.; Searles, T.; Tiley, J.; Russ, J.C.; Fraser, H.L. Development of methods for the quantification of microstructural features in α+β-processed α/β titanium alloys. *Mater. Sci. Eng. A* **2009**, *508*, 174–182. [CrossRef]
72. Paredes-Orta, C.A.; Mendiola-Santibañez, J.D.; Manriquez-Guerrero, F.; Terol-Villalobos, I.R. Method for grain size determination in carbon steels based on the ultimate opening. *Measurement* **2019**, *133*, 193–207. [CrossRef]
73. Smith, T.; Bonacuse, P.; Sosa, J.; Kulis, M.; Evans, L. A quantifiable and automated volume fraction characterization technique for secondary and tertiary γ' precipitates in Ni-based superalloys. *Mater. Charact.* **2018**, *140*, 86–94. [CrossRef]
74. Payton, E.J.; Phillips, P.J.; Mills, M.J. Semi-automated characterization of the γ' phase in Ni-based superalloys via high-resolution backscatter imaging. *Mater. Sci. Eng. A* **2010**, *527*, 2684–2692. [CrossRef]
75. Iskakov, A.; Kalidindi, S.R. A Framework for the Systematic Design of Segmentation Workflows. *Integr. Mater. Manuf. Innov.* **2020**, *9*, 70–88. [CrossRef]
76. Higham, D.J.; Higham, N.J. *MATLAB Guide*; SIAM: Philadelphia, PA, USA, 2016; Volume 150.
77. Serra, J. *Image Analysis and Mathematical Morphology*; Academic Press: New York, NY, USA, 1982.
78. Diaz, A.; Patel, D.; Simmons, J.; Parthasarathy, T.A.; Kalidindi, S.R. Quantitative Assessment of Segmentation Workflows for 3D Serial Sectioned Microscopy Data Set. 2022; In preparation.
79. Lecomte, J. *Texture Analysis in Materials Science*; Bunge, H.-J., Ed.; Elsevier: Amsterdam, The Netherlands, 2015.
80. Patel, D.K.; Kalidindi, S.R. Correlation of spherical nanoindentation stress-strain curves to simple compression stress-strain curves for elastic-plastic isotropic materials using finite element models. *Acta Mater.* **2016**, *112*, 295–302. [CrossRef]
81. Bertrand, M.A.X.; Alexis, J.; Larignon, C.; Perusin, S. Titanium alloy Ti-6242 for high temperature structural application. Static and dynamic mechanical properties and impact of ageing. In *Proceedings of the MATEC Web of Conferences, 23 March 2020*; EDP Sciences: Paris, France, 2020; Volume 321, p. 11089.

Journal of
Composites Science

Article

Efficient Finite Element Modeling of Steel Cables in Reinforced Rubber

Martin Pletz *, Siegfried Martin Frankl and Clara Schuecker

Designing Plastics and Composite Materials, Department of Polymer Engineering and Science, Montanuniversitaet Leoben, 8700 Leoben, Austria; siegfried.frankl@unileoben.ac.at (S.M.F.); clara.schuecker@unileoben.ac.at (C.S.)
* Correspondence: martin.pletz@unileoben.ac.at

Abstract: Spiral steel cables feature complex deformation behavior due to their wound geometry. In applications where the cables are used to reinforce rubber components, modeling the cables is not trivial, because the cable's outer surface must be connected to the surrounding rubber material. There are several options for modeling steel cables using beam and/or solid elements for the cable. So far, no study that lists and evaluates the performance of such approaches can be found in the literature. This work investigates such modeling options for a simple seven-wire strand that is regarded as a cable. The setup, parameter calibration, and implementation of the approaches are described. The accuracy of the obtained deformation behavior is assessed for a three-cable specimen using a reference model that features the full geometry of the wires in the three cables. It is shown that a beam approach with anisotropic beam material gives the most accurate stiffness results. The results of the three-cable specimen model indicate that such a complex cable model is quite relevant for the specimen's deformation. However, there is no single approach that is well suited for all applications. The beam with anisotropic material behavior is well suited if the necessary simplifications in modeling the cable–rubber interface can be accepted. The present work thus provides a guide not only for calibrating but also for selecting the cable-modeling approach. It is shown how such modeling approaches can be used in commercial FE software for applications such as conveyor belts.

Keywords: finite element modeling; homogenization methods; steel cables; steel-cable-reinforced rubber

Citation: Pletz, M.; Frankl, S.M.; Schuecker, C. Efficient Finite Element Modeling of Steel Cables in Reinforced Rubber. *J. Compos. Sci.* **2022**, *6*, 152. https://doi.org/10.3390/jcs6060152

Academic Editor: Stelios K. Georgantzinos

Received: 3 March 2022
Accepted: 20 May 2022
Published: 24 May 2022

1. Introduction

Steel cables are an indispensable part of the infrastructure and many engineering applications because they reliably provide high strengths with low bending stiffness. They consist of individual steel wires that are wound into strands, which in turn are wound to form the cable. Since cables consist of many parallel thin wires, their tensile stiffness is very high, whereas the bending and compressive stiffness are low. Because of this helical topology of spiral cables, there is a coupling of tensile deformation and torsional deformation of the cable (see Figure 1). Steel cables have many design options in terms of steel grade and cable geometry. Much work has been done on computing the influence of those parameters on the cable stiffness, accounting for the trajectories of the individual wires and contact between them. Many analytical and semi-analytical solutions have been developed and are listed in the review papers by Utting and Jones [1], Cardou and Jolicoeur [2], and the works of Costello [3] and Feyrer [4]. For standard cable types, good agreement of existing cable models with experiments can be reached. Effects like wire–wire friction can be captured. Hysteresis effects, the nonlinearity of the cable stiffness, and cable failure have been studied as well. Most of the work of such cable models is setting up the geometry, particularly for a non-straight cable; see Wang et al. [5]. Recently, the finite element method (FEM) has become widely used for modeling the mechanical response of cables: Jiang et al. [6] modeled a seven-wire strand using cyclic symmetry, and Foti and de Luca di

Roseto [7] investigated the elastic–plastic effects of the wires. Furthermore, FEM provides a basis for newly developed simplified models; see Chen et al. [8] and Cao and Wu [9].

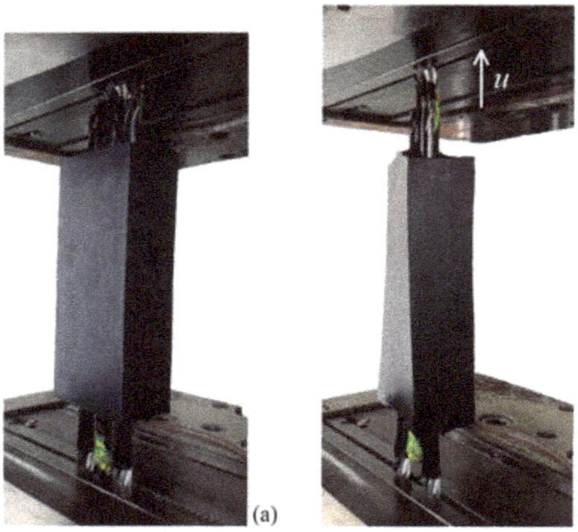

(a) (b)

Figure 1. A three-cable specimen: (**a**) unloaded and (**b**) vertically loaded with a twist due to a tension/torsion coupling of the cables.

Many of the mathematical cable models refer to tests of a seven-wire strand reported by Utting and Jones [10], who reported a distinct tension/torsion coupling. When testing cables, the constraints of the cable ends influence the test results. The cable ends can be free, clamped, or even welded together. This effect of the cable ends was studied by Chen et al. [11] for thick cables, and they showed that in cable tests and FEM simulations of cables, much care must be put into applying the loads.

Modeling steel cables in reinforced rubber on the one hand requires capturing the influence of the rubber penetrating the cable (see Bonneric et al. [12]). On the other hand, the outer surface of the cable needs to be connected to the rubber. This interface between cable and rubber is crucial for the failure of cable-reinforced rubber components, as modeled by Frankl et al. [13].

Cable-reinforced rubber components can be conveyor belts, for example; see Nordell [14], Fedorko et al. [15], and Frankl et al. [16]. Obtaining the stiffness of cables that are used in rubber components requires tests on cables that have been penetrated by rubber (rubberized cables). How to separately capture tensile, bending, and torsion stiffness and the tension/torsion coupling of the cable that is embedded in rubber is a big challenge. Nordell et al. [17] stated that they developed a special element in the commercial FEM code ANSYS based on principles described by Costello [3], but did not give any details about this element.

In the present work, a variety of such cable modeling approaches is evaluated for their use in rubber components using the commercial FEM code Abaqus [18]. Those efficient modeling approaches use solid elements, beam elements, or a combination of both. In some of those modeling approaches, an anisotropic material model is used to mimic the tension/torsion coupling of the cable. To not have to deal with uncertainties of tests, the results of a fully modeled rubberized cable are taken as the reference to evaluate the accuracy of these modeling approaches. In this reference model, all wires and the surrounding rubber are modeled with a linear elastic and a hyperelastic material model, respectively. This model is called a full-geometry model, in contrast to the efficient models that account for wires and rubber in a way such that the overall cable stiffness is captured.

To keep the computational cost low, a seven-wire rubberized strand is used as the cable. For this cable, the homogenized stiffness matrix is computed from a full-geometry simulation similar to what was reported by Cartraud and Messager [19]. The various cable-modeling approaches are then calibrated and their ability to capture the homogenized stiffness components of the full-geometry model is evaluated. Then, the modeling approaches are evaluated in a simple rubber shear specimen containing three cables. The loads applied in these specimens are similar to those in conveyor belts; see Nordell et al. [17]. The full-geometry version of the three-cable specimen is used as a reference, and the stiffness, deformation, and strains in the specimen are used to assess the performance of the efficient models.

2. Methods

This section introduces the homogenized cable stiffness S_{ij} and a range of efficient modeling approaches that attempt to mimic this stiffness matrix S_{ij}. The cables are regarded as linear elastic throughout this study. For typical cable loads, this is a good approximation despite the nonlinear elastic response of the rubber. This section further describes the setup of the single-cable FEM model that computes the S_{ij} matrix for a seven-wire rubberized strand and is used to calibrate and evaluate efficient cable-modeling approaches. Additionally, a FEM model of a three-cable specimen is introduced, which is used to evaluate the performance of the efficient cable models.

2.1. Cable Stiffness

Figure 2 illustrates the loads (normal load F_x, twist moment M_x, and flexural moment M_b) and corresponding deformations (longitudinal strain ε_x, twist per length φ_x, and curvature κ) of a cable, which is drawn as a cylinder. Longitudinal shear deformation is not considered and the elastic bending response is considered to be independent of the bending direction.

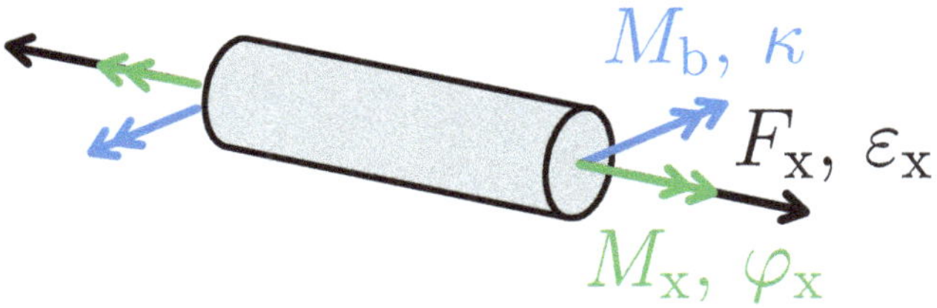

Figure 2. Definition of the loads and strains of a cable.

Since we regard small deformations, the stiffnesses in tension and compression and the stiffness for positive and negative twist are assumed to be the same. The elastic behavior of the cable can then be described by a stiffness matrix $\mathbf{S}$ (or S_{ij}) that couples the loads and deformations as

$$\begin{pmatrix} F_\mathrm{x} \\ M_\mathrm{x} \\ M_\mathrm{b} \end{pmatrix} = \begin{pmatrix} \frac{\partial F_\mathrm{x}}{\partial \varepsilon_\mathrm{x}} & \frac{\partial F_\mathrm{x}}{\partial \varphi_\mathrm{x}} & \frac{\partial F_\mathrm{x}}{\partial \kappa} \\ \frac{\partial M_\mathrm{x}}{\partial \varepsilon_\mathrm{x}} & \frac{\partial M_\mathrm{x}}{\partial \varphi_\mathrm{x}} & \frac{\partial M_\mathrm{x}}{\partial \kappa} \\ \frac{\partial M_\mathrm{b}}{\partial \varepsilon_\mathrm{x}} & \frac{\partial M_\mathrm{b}}{\partial \varphi_\mathrm{x}} & \frac{\partial M_\mathrm{b}}{\partial \kappa} \end{pmatrix} \begin{pmatrix} \varepsilon_x \\ \varphi_\mathrm{x} \\ \kappa \end{pmatrix} = \mathbf{S} \begin{pmatrix} \varepsilon_\mathrm{x} \\ \varphi_\mathrm{x} \\ \kappa \end{pmatrix} \tag{1}$$

with $\mathbf{S}$ defined as

$$\mathbf{S} = \begin{pmatrix} S_\mathrm{xx} & S_{\mathrm{x}\varphi} & S_\mathrm{xb} \\ S_{\varphi\mathrm{x}} & S_{\varphi\varphi} & S_{\varphi\mathrm{b}} \\ S_\mathrm{bx} & S_{\mathrm{b}\varphi} & S_\mathrm{bb} \end{pmatrix} \tag{2}$$

Inverting the stiffness matrix **S** yields the compliance matrix **C** (or C_{ij}). Similar to the definition of Young's modulus, we define stiffness parameters of the cable E_{xx}, $E_{x\varphi}$, ... , which correspond to the inverse of the components of **C**:

$$\begin{pmatrix} \varepsilon_x \\ \varphi_x \\ \kappa \end{pmatrix} = \mathbf{S}^{-1} \begin{pmatrix} F_x \\ M_x \\ M_b \end{pmatrix} = \begin{pmatrix} \frac{1}{E_{xx}} & \frac{1}{E_{x\varphi}} & \frac{1}{E_{xb}} \\ \frac{1}{E_{\varphi x}} & \frac{1}{E_{\varphi\varphi}} & \frac{1}{E_{\varphi b}} \\ \frac{1}{E_{bx}} & \frac{1}{E_{b\varphi}} & \frac{1}{E_{bb}} \end{pmatrix} \begin{pmatrix} F_x \\ M_x \\ M_b \end{pmatrix} \tag{3}$$

If all non-diagonal terms vanish (all terms except S_{xx}, $S_{\varphi\varphi}$, and S_{bb} are zero), the values E_{xx}, $E_{\varphi\varphi}$, and E_{bb} are the same as S_{xx}, $S_{\varphi\varphi}$, and S_{bb}, respectively. If this is not the case, it means that S_{xx} is the longitudinal stiffness that is observed when twist and bending strains are constrained during loading. On the other hand, E_{xx} corresponds to the longitudinal stiffness when twist and bending are free (M_x and M_b are zero).

Let us assume that we have a cable that has off-diagonal terms, which means that tension, twist, and bending are coupled. When we use an efficient cable model that cannot account for those coupling terms, we can fit either S_{xx}, $S_{\varphi\varphi}$, and S_{bb} or E_{xx}, $E_{\varphi\varphi}$, and E_{bb}. In the first case, the modeled cable has the same stiffness as the real cable when all other strains are set to zero during loading. For tension, this means that torsion φ_x and curvature κ are constrained during loading. When fitting E_{xx}, $E_{\varphi\varphi}$, and E_{bb}, the efficient cable model shows the same stiffness as the real cable during loading in one direction when strains in the other directions are unconstrained.

For an FEM model of a cable, the cable's stiffness matrix can be obtained by applying three orthogonal strain vectors similar to [19]. The stiffness matrix can be built from the resulting load vectors. If the applied strains $\varepsilon_{x,0}$, $\varphi_{x,0}$, and κ_0 are set to 1, the resulting load vectors constitute the columns of the stiffness matrix. Otherwise, the terms of the stiffness matrix must first be divided by the applied strain.

$$\begin{pmatrix} \varepsilon_x \\ \varphi_x \\ \kappa \end{pmatrix} = \begin{pmatrix} \varepsilon_{x,0} \\ 0 \\ 0 \end{pmatrix}, \begin{pmatrix} 0 \\ \varphi_{x,0} \\ 0 \end{pmatrix}, \begin{pmatrix} 0 \\ 0 \\ \kappa_0 \end{pmatrix} \tag{4}$$

2.2. Cable-Modeling Approaches

The cables can be modeled either by using solid elements that share nodes with the surrounding rubber, by using a beam that is coupled in some way to the corresponding rubber nodes, or with a combination of solid and beam elements (see Figure 3). The nodes of the beam in the solid/beam approach cannot transmit rotation since a direct nodal connection is used and the solid element nodes do not have rotational degrees of freedom. Each option requires a specific calibration of material parameters. These parameters are not physical but selected such that the whole modeled cable captures the target elastic response.

Our efficient cable-modeling approaches aim to reach S_{xx}, $S_{\varphi\varphi}$, $S_{x\varphi}$, and S_{bb} as closely as possible (Section 3.1 will show that S_{xb} and $S_{\varphi b}$ can generally be neglected). Some simplified approaches are also investigated that do not account for the coupling term $S_{x\varphi}$.

The first challenge is to independently capture the tensile and bending stiffness. This can be done in the following ways:

(a) Solid elements: Use a material that has different tensile and compressive stiffness.
(b) Beam elements: Set the radius of the beam such that S_{xx} and S_{bb} fit the target values.

Solid/beam approach: The whole bending stiffness is captured by the solid elements, whereas beam elements are used to capture the tensile stiffness that is not captured by the solid elements; see [16]. The beam elements have a very small cross-section such that the high tensile stiffness does not affect the overall bending stiffness.

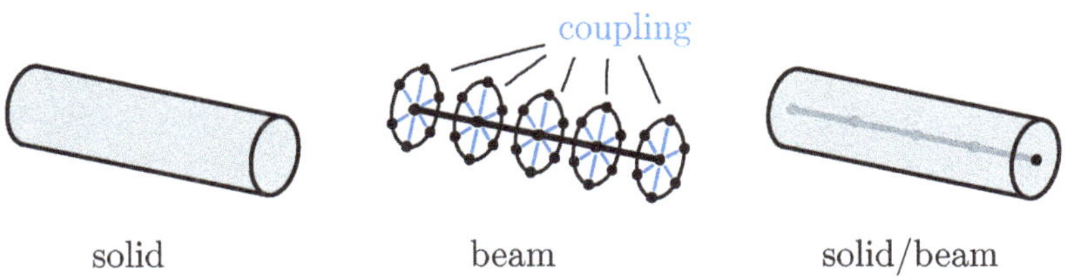

Figure 3. The three modeling approaches used for the cables: solid, beam, and a combination of both.

In the following, the cable-modeling approaches are presented and their material parameters are derived. For a linear elastic material model with cubic symmetry, the Young's modulus E, the shear modulus G, and the Poisson's ratio ν can be selected independently. For all models with linear elastic material, ν is set to zero in order to avoid unphysical effects in the cable deformation.

To account for the tension/torsion coupling of the cable, the linear elastic cubic material can be extended to a special kind of anisotropy that couples $\gamma_{\varphi z}$ and ε_{zz} in the cylindrical coordinate system of the cable:

$$\begin{pmatrix} \sigma_{xx} \\ \sigma_{yy} \\ \sigma_{zz} \\ \tau_{xy} \\ \tau_{xz} \\ \tau_{yz} \end{pmatrix} = \begin{pmatrix} E_{11} & 0 & 0 & E_K & 0 & 0 \\ 0 & E_{11} & 0 & 0 & 0 & 0 \\ 0 & 0 & E_{11} & 0 & 0 & 0 \\ E_K & 0 & 0 & G_{11} & 0 & 0 \\ 0 & 0 & 0 & 0 & G_{11} & 0 \\ 0 & 0 & 0 & 0 & 0 & G_{11} \end{pmatrix} \begin{pmatrix} \varepsilon_{xx} \\ \varepsilon_{yy} \\ \varepsilon_{zz} \\ \gamma_{xy} \\ \gamma_{xz} \\ \gamma_{yz} \end{pmatrix}, \quad \text{if } E_K = 0: \begin{cases} E_{11} = E \\ G_{11} = G \end{cases} \tag{5}$$

The parameter E_K accounts for the coupling of $\gamma_{\varphi z}$ and ε_{zz}. Whereas E_{11} and G_{11} are taken from the analytical calculations, E_K is calibrated to reach the target $S_{x\varphi}$ value of the efficient cable model. If E_K is set to zero, the coupling vanishes and the model corresponds to cubic material symmetry, with independent values of E and G, and the Poisson's ratio is set to zero, as in the cubic approach. This linear elastic material model is suited for both solid elements and beam elements. Alternatively, a hyperelastic material model with anisotropic stiffness is used in the solid models. This hyperelastic modeling approach can account for both the tension/torsion coupling and the tensile and bending stiffness and is introduced in the next section.

The material parameters of the homogenization approaches can be calculated analytically or calibrated using FEM models. The analytical calculations are based on the equations for a beam with circular cross-section and radius R. The tensile stiffness S_{xx}, torsional stiffness $S_{\varphi\varphi}$, and bending stiffness S_{bb} of the beam can be written as

$$S_{xx} = E\,A \quad \text{with } A = \pi\,R^2 \tag{6}$$

$$S_{\varphi\varphi} = G\,I_P \quad \text{with } I_P = 0.5\,\pi\,R^4 \tag{7}$$

$$S_{bend} = E\,I \quad \text{with } I = 0.25\,\pi\,R^4 \tag{8}$$

The elastic material parameters for the beam, solid, and solid/beam approaches are derived in the following. There, the S_{ij} components are written in the equations. To fit E_{ij} with the homogenized cable (see Section 2.1 for details), the S_{ij} terms in the formulas can be replaced by the corresponding E_{ij} terms, which then, of course, yields the corresponding stiffness for unconstrained loading.

2.2.1. Solid Approaches

There are two types of solid element approaches, where the material model is (a) linear elastic (with either cubic or anisotropic material symmetry) or (b) hyperelastic with anisotropic material response using the Holzapfel–Gasser–Ogden (HGO) formulation.

For linear elastic material, the Young's modulus and shear modulus of the material can be computed directly from the target values of tensile stiffness S_{xx} and torsional stiffness $S_{\varphi\varphi}$ as

$$E = \frac{S_{xx}}{A} \tag{9}$$

$$G = \frac{S_{\varphi\varphi}}{I_P} \tag{10}$$

This approach results in a bending stiffness S_{bend} that can be computed using the cable radius R by

$$S_{bend} = E\frac{\pi\, R^4}{4} = S_{bb} \tag{11}$$

This means that for the solid approach with cubic material, the axial stiffness S_{xx} and the torsional stiffness $S_{\varphi\varphi}$ can be made to fit while the bending stiffness is too high. The tension/torsion coupling can be captured when the anisotropic linear elastic material is chosen and the previously introduced coupling term E_K is calibrated.

The solid approach can account for S_{xx}, $S_{\varphi\varphi}$, S_{bb}, and $S_{x\varphi}$ when a material model with anisotropic behavior and a difference in its tensile and compressive stiffness is used. One such model is the Holzapfel–Gasser–Ogden (HGO) material model [20,21], which considers a hyperelastic matrix material model with fiber reinforcements. The HGO matrix uses the neo-Hookean model parameter C_{10}; the parameters k_1 and k_2 define the stiffness of the reinforcements. The parameter κ defines the level of dispersion of the fiber directions and lies between 0 for uniaxial orientation and 1/3 for evenly distributed fiber orientations. In the HGO model, the reinforcements only increase the stiffness of the material in the fiber direction under tension, but not under compression. It thus provides the possibility to reach a lower bending stiffness with solid elements, in contrast to the linear elastic anisotropic modeling approach. For this modeling approach, κ is set to zero to model uniaxial reinforcement. The parameter k_2 is an additional parameter to account for nonlinear effects and is set to $k_2 = 1$ in this work. The parameter D of the HGO model is set to zero, which is equivalent to incompressible material behavior. This deviates from the linear elastic cable models where ν is set to 0. Since, compared to the rubber, there are only small deformations in the cables, this inconsistency is expected to have a negligible effect on deformations and stresses. Within this HGO approach, the cable is modeled by solid elements and the orientation of the reinforcements is defined to be wound similar to the strands in a cable with a helix angle α_{HGO}. Note that this α_{HGO} of the HGO model approach can differ from the actual helix angle of the cable since it is calibrated to fit the stiffness components of the cable. The fitting parameters of the HGO approach, therefore, are this helix angle α_{HGO}, the material stiffness parameters of the matrix C_{10}, and the stiffness parameter of the reinforcements k_1.

2.2.2. Beam Approaches

When the cable is modeled using beam elements, the beam radius r can be used to also fit the bending stiffness of the cable. To that end, the equations for S_{xx} and S_{bb} can be formulated for the two unknowns, E and r:

$$S_{xx} = E\, r^2 \pi \tag{12}$$

$$S_{bb} = E\frac{\pi\, r^4}{4} \tag{13}$$

After eliminating r by inserting Equation (12) into Equation (13), the Young's modulus can be written as

$$E = \frac{S_{xx}^2}{4\, \pi\, S_{bb}} \tag{14}$$

This expression for E can be inserted into Equation (12) to yield the beam radius r:

$$r = \sqrt{\frac{S_{\mathrm{xx}}}{\pi E}} \tag{15}$$

Using Equation (10), the shear modulus of the beam can be calculated from the radius r and the desired $S_{\varphi\varphi}$ as

$$G = \frac{S_{\varphi\varphi}}{I_{\mathrm{P}}} = \frac{2\,S_{\varphi\varphi}}{\pi\, r^4} \tag{16}$$

This means that the beam approach can capture S_{xx}, $S_{\varphi\varphi}$, and S_{bb} by adjusting E, G, and r. Furthermore, $S_{\mathrm{x}\varphi}$ can be captured using a calibrated E_{K} of the linear elastic material model.

2.2.3. Solid/Beam Approaches

For a combination of solid and beam elements, the Young's modulus of the solid elements E_{s} is calculated from the bending stiffness:

$$E_{\mathrm{s}} = \frac{S_{\mathrm{bb}}}{I} \tag{17}$$

The shear modulus follows directly from the torsional stiffness; see Equation (10).

A beam radius r_{b} is applied to the beam elements, which is a factor of 1000 smaller than the actual cable radius. Thus, the contribution of the beam elements to the torsional and bending stiffness can be neglected. The Young's modulus for the beam elements is chosen such that the combination of solid stiffness and beam stiffness add up to the desired longitudinal stiffness S_{xx}:

$$E_{\mathrm{b}} = \frac{S_{\mathrm{xx}} - E_{\mathrm{s}}\, A}{\pi\, r_{\mathrm{b}}^2} \tag{18}$$

2.3. *Single Cable Models*

In this section, two kinds of single cable models are introduced. The first is a model with fully modeled steel wires and rubber. This model is used to obtain the cable stiffness matrix S_{ij}, which serves as a reference for the other models. The modeling with steel wires and rubber is referred to as full geometry in the following. The second kind of single cable models are the efficient cable models, which are set up to mimic the reference stiffness S_{ij} using solid elements, beam elements, or a combination of both.

The load definition and the computation of the stiffness are the same for the full-geometry cable model and the efficient single cable models. At the center of the two ends of the modeled cable, reference points are defined. All nodes of the two end surfaces (or end nodes in the case of the beam models) are rigidly coupled to the corresponding reference point. The load is applied at the right-side reference point while the left-side reference point is completely fixed. This constraint of the radial displacements introduces an additional stiffness to the model. It thus must be checked whether the modeled cable length L_{cable} is long enough for the influence of these end effects to vanish. The models are analyzed using the implicit nonlinear solver of Simulia Abaqus [18].

2.3.1. Full-Geometry Single-Cable Model

The full-geometry cable model uses a seven-wire rubberized strand as a very simple example of a cable. The strand geometry is defined in Figure 4 and Table 1. It is adapted from [7], but to provide a good-quality mesh [12], the rubber gap between the middle wire and the outer wires is increased. The steel of the wires is modeled linear elastically using a Young's modulus as given in [7], $E = 188$ GPa, and Poisson's ratio $\nu = 0.3$; the helix angle α of the strand is 11.8°. A perfect bond between the rubber and the steel wires is modeled (they share the same nodes in the interface) such that there can be no debonding or friction. The rubber is modeled as a Mooney–Rivlin hyperelastic material model with its parameters C_{10},

C_{01}, and d_1 taken from [22] as 0. MPa, 0. MPa, and 0. MPa, respectively. The full-geometry single-cable model uses two meshing options: (a) bilinear hexahedral elements with hybrid formulation (C3D8H) for the rubber elements and reduced integration (C3D8R) for the steel elements and (b) quadratic tetrahedral elements with hybrid formulation (C3D10H) both for rubber and steel elements.

The strand length L_{strand} of a cable is the axial distance at which one wire is completely wound around the cable axis. This length L_{strand} depends on the helix angle α and the distance between the cable axis and the axis of the wire. In the case of the seven-wire strand used, the strand length of the six outer wires is calculated as

$$L_{\text{strand}} = (r_{\text{i}} + r_{\text{o}} + d)\,\frac{2\,\pi}{\tan(\alpha)} = 118.8\text{ mm} \tag{19}$$

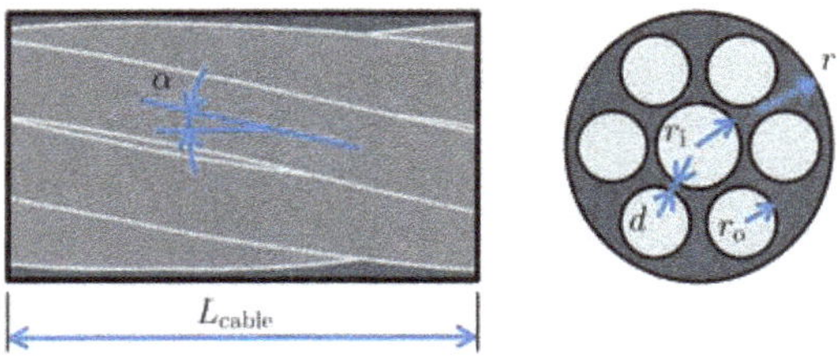

Figure 4. The geometry of the full-geometry single-cable model with its key parameters.

Table 1. Geometric parameters of the seven-wire strand.

Parameter Name	Value
Radius r (mm)	6
inner radius r_{i} (mm)	1.95
outer radius r_{o} (mm)	1.75
gap d (mm)	0.25
winding angle α (°)	11.8

The wires in the strand can be wound in two directions, referred to as the z-type and the s-type, where z is wound like a right-hand screw and s like a left-hand screw. All single cable models use the z-type (see Figure 4) and for the efficient models of s-cables needed in the tree-cable specimen, $S_{x\varphi}$ of the z-type cable is multiplied by -1.

2.3.2. Efficient Single-Cable Models

As stated in Section 2.2, the efficient cable models can be divided into approaches that use cubic material models and approaches that use some kind of anisotropic material response. The latter can be implemented using a linear elastic or hyperelastic material model to account for the tension/torsion coupling of the cables. For both the cubic and anisotropic approaches, cable models consisting of beam elements, solid elements, or both solid and beam elements can be set up. Table 2 defines the combinations of elements and material models used in this study.

Table 2. Options of the efficient modeling approaches in terms of elements and material models used.

	Linear Elastic		**Hyperelastic**
Approach	**Cubic (E, G, ν = 0)**	**Anisotropic (E, G, E_K)**	**HGO (C_{10}, k_1, α_{HGO})**
Solid	x	x	x
Beam	x	x	-
Solid/beam	x	-	-

The grey areas in Table 2 indicate the combinations that had not been implemented. The hyperelastic material model is used for solid elements only. The solid/beam does not use anisotropic solid material because there, the solid elements represent only a small part of the cable's tensile stiffness. This means that the tension/torsion coupling could not be achieved. The solid/beam approach does not use anisotropic material for the beam because the beam rotation cannot be transmitted through the nodes they share with the solid elements.

The parameters for the efficient cable models (E_K for the anisotropic solid and beam models and all parameters of the HGO model) are derived as described in Section 2.2. The calibration procedure of those parameters uses a Nelder–Mead algorithm [23]. For the length of the efficient single-cable models, 40% of the strand length (40% of 118.8 mm) have proven to be sufficient and the global mesh size is set to 1 mm. In the longitudinal direction, the length of the elements is set to 3 mm. For the anisotropic approaches that can feature tension/torsion coupling, S_{ij} components are used to obtain the model parameters. Note that for the solid and beam model, E_K is calibrated to best fit S_{couple}. With the solid HGO approach, all four S_{ij} components are used for the calibration. Although more increments are needed for convergence in the HGO approach, the resulting force and moment curves are approximately linear within the modeled load range. To ensure that the minimum found for the HGO approach in the calibration procedure is not a local one, three starting points of the C_{10}, k_1, and α_{HGO} parameters are evaluated: 600 MPa, 10,000 MPa, 5°; 600 MPa, 26,500 MPa, 11°; and 500 MPa, 10,000 MPa, 20°. They all yield approximately the same results as those stated in Section 3.2. The efficient single cable models are meshed by bilinear hexahedral elements with hybrid formulation (C3D8H) for the solid regions and linear beam elements (B31) for the beams.

2.4. Model of the Three-Cable Specimen

The cable modeling approaches are assessed using an FEM model of a three-cable shear specimen. The geometry and boundary conditions of the specimen are defined in Figure 5. The corresponding geometry parameters are defined in Table 3.

All nodes on the left faces of the two outer cables are fully constrained. Similar to the single-cable models, the right face nodes of the central cable are rigidly connected to a reference point that is used to apply the displacement load of 10 mm in the x-direction. During loading, all other displacements of the reference point except the rotation around the x-axis are constrained. When the center cable is pulled in the positive x-direction, the load is transferred through the rubber to the outer cables that are fixed on the left side. The cables are modeled as defined for the full-geometry cable model or the efficient cable models. The rubber properties defined in Section 2.3.1 are taken for the rubber region.

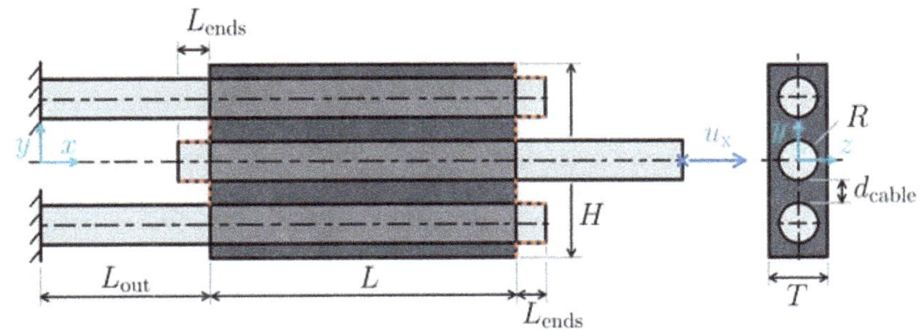

Figure 5. The geometry of the three-cable specimen. Contact is defined in the orange-dashed regions.

The orange-dashed lines in Figure 5 mark regions that can come into contact when the specimen is loaded. This contact is defined using a penalty algorithm and frictionless behavior. For the model with full geometry, quadratic hybrid tetrahedral elements (C3D10H) with a typical edge length of 1.25 mm are used. This model contains about

620,000 elements. The efficient cable models use bilinear hybrid hexahedral elements and reduced integration with a typical element edge length of about 1 mm and a swept mesh along the x-axis. In the sweeping direction, the element edge length is set to 3 mm. This swept mesh is important for the beam-type modeling approach, where the rubber nodes that lie on the outer cable surface are rigidly connected to the beam node that has the same x-coordinate, as illustrated in Figure 3. The models with solid and solid/beam approaches use about 77,000 elements and the beam approach models contain about 33,000 elements. The same element types as stated in Section 2.3.2 are used.

Table 3. Geometry parameters of the three-cable model.

Parameter Name	Value
length L (mm)	100
height H (mm)	62
depth T (mm)	20
outer cable length L_{out} (mm)	100
cable end length L_{ends} (mm)	10
cable distance d_{cable} (mm)	8

Several combinations of s-type and z-type cables are possible in the specimens. Here, the setup with bottom, central, and top cable s, z, and s is used, respectively. For this szs setup, the outer cables and the center cable want to rotate in opposite directions. Note that for a cable-reinforced component with a large number of parallel cables, the component will not be as free to twist as the three-cable specimen, and the stresses will be affected by the cable's tension/torsion coupling.

3. Results and Discussion

In this section, results are presented first for a single cable using the full-geometry and efficient modeling techniques. Afterward, results of three-cable specimen models are presented. The cable-modeling approaches are evaluated using the three-cable models in terms of the specimen's stiffness, deformation field, and strain fields in the rubber.

3.1. Full-Geometry Single-Cable Model

The full-geometry model of the seven-wire strand is used to obtain the components of the stiffness matrix that are used to evaluate the efficient cable models. Here, we look into the nonlinearity of the overall stiffness response of the full model and the influence of element type, mesh size, and cable length on the cable's stiffness, which is relevant for the full-geometry three-cable specimen of Section 3.3. The length of the modeled cable is quantified as a fraction of the strand length (the axial distance so that the outer wires are completely wound around the cable axis, as described in more detail in Section 2.3.1) and denoted as relative cable length.

To show the nonlinearity in the cable stiffness, two load cases are studied. A cable load ($\varepsilon_{x,0}$, $\varphi_{x,0}$, κ_0) of (5%, 0,0) and (0, 0, 10/m) is applied in load case A and load case B, respectively. Figure 6 shows the axial force F_x, the torsional moment M_x, and the bending moment M_b of the cable over the longitudinal strain ε_x (load case A, Figure 6a) and over the curvature κ (load case B, Figure 6b). For a mesh size of 0.5 mm and a relative cable length of 0.8, the F_x and M_x plots are approximately linear, whereas the M_b curve shows a slight nonlinearity towards higher curvatures.

The influence of cable length and mesh size are investigated for applied loads of $\varepsilon_x = 0.5\%$, $\varphi_x = 2$ rad/m, and $\kappa = 1$/m, which are applied individually. The longitudinal strain of 0.5% corresponds to a maximum Mises stress of $\sigma_{mises} = 960$ MPa in the central wire and a total force of $F_x = 60$ kN. We here assume that those loads cover the relevant range for the intended applications and that nonlinear effects that occur at higher loads can be neglected. The stiffness parameters are evaluated as secant stiffnesses of the loading curves and are plotted over the cable length and mesh size in Figure 7. The range and units of the individual stiffness parameters S_{xx}, $S_{\varphi\varphi}$, S_{bb}, $S_{x\varphi}$, and $S_{\varphi x}$ are quite different. To better visualize the dependency of those parameters on the cable length and mesh size, they are plotted relative to their respective most accurate values (such as those obtained for either highest cable length or smallest mesh size, as explained in the following).

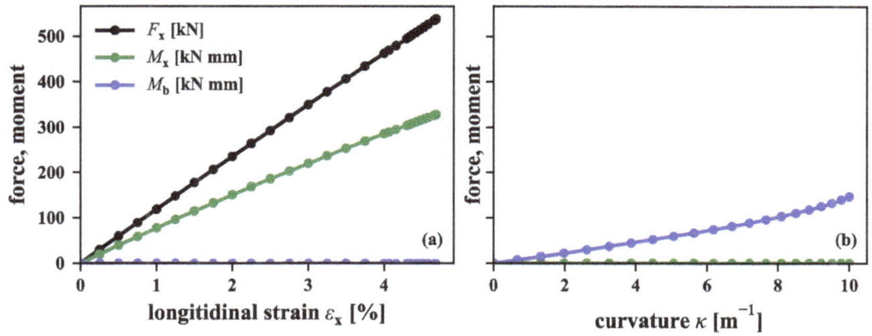

Figure 6. Tensile force, torsional moment, and bending moment plotted over (**a**) the longitudinal strain ε_x and (**b**) the curvature κ of the seven-wire strand model for a mesh size of 0.5 mm and a relative cable length of 0.8.

In Figure 7a, the cable length is varied for linear hexahedral elements with reduced integration and a fixed mesh size of 0.5 mm. The relative cable length of 1.4 is assumed to give the most accurate results, so those S_{ij} components are used to normalize the respective results of the other models. As expected, the rigid connection from the cable ends to their corresponding reference points introduces numerical artifacts that increase the evaluated S_{ij} components for smaller cable lengths. The bending stiffness S_{bb} is particularly sensitive to these cable end effects.

The element type and mesh size are varied in Figure 7b for a constant relative cable length of 1.2, since this is the length for which the stiffness parameters have already reached a plateau, as shown in Figure 7a. The results for the smallest mesh size of the quadratic tetrahedral elements (0.75 mm) are used to normalize S_{ij}. The curves for bilinear hexahedral and quadratic tetrahedral elements show that the finer the mesh size, the higher the computed stiffness components. For the hexahedral elements, no clear plateau of S_{ij} components is reached for the finest mesh size of 0.3 mm. This indicates that bilinear hexahedral elements would need to be much finer to accurately compute the cable's stiffness. The quadratic tetrahedral element results show a plateau at a mesh size of about 1 mm. Similar to the cable length study, the bending stiffness S_{bb} is most sensitive to the mesh size. The 1.25 mm mesh with quadratic tetrahedral elements (see the pictogram in Figure 7b) yields acceptable computation times and quite accurate results: The stiffness parameters are up to 4% lower than for a mesh size of 0.75 mm. Therefore, in the bigger three-cable specimen models with full geometry, quadratic tetrahedral elements with a mesh size of 1.25 mm are used.

Table 4 lists the model size, necessary RAM, and computation time for the full-geometry single-cable models of Figure 7. To keep the table short, only model parameters of the maximum and minimum cable length (length study) and mesh size (mesh study) are listed. For bilinear hexahedral elements, no mesh convergence is reached at a mesh size of 0.3 mm with computation times of 20 min. The finest quadratic tetrahedral element results took about 5 min to compute.

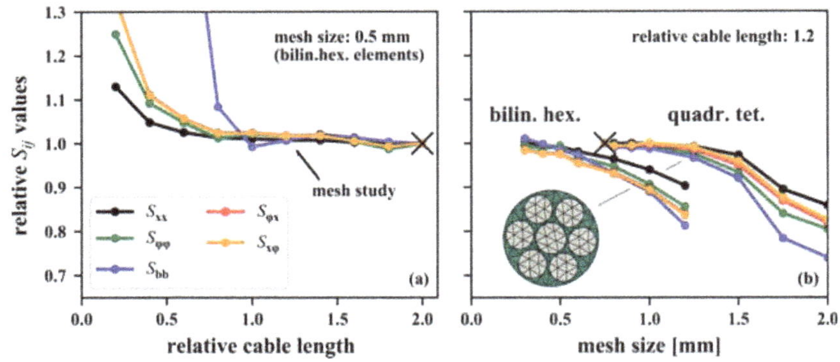

Figure 7. Relative stiffness components of the strand model with (**a**) varied cable length using bilinear hexahedral elements with a mesh size of 0.5 mm and (**b**) varied mesh size with both bilinear hexahedral and quadratic tetrahedral elements for a relative cable length of 1.2.

Table 4. Information on the model size, necessary RAM to load the full stiffness matrix, and computation time (four cores of a six-core Intel Xeon E5-1650 v3 @ 3.5 GHz workstation with 128 GB RAM) for the full-geometry single-cable model (tensile load case) with varied element type, relative cable length, and mesh size.

Mesh Type	Relative Length	Mesh Size (mm)	Nodes	Elements	Degrees of Freedom	Necessary RAM (MB)	Computation Time
Hex.	0.2/2	0.5	14,427/158,021	10,162/109,965	40,397/396,291	240/3985	0:07 min/4:54 min
Hex.	1.4	0.3/1.2	474,907/14,352	344,606/8744	1,203,365/35,068	19,867/205	20:19 min/0:08 min
Tet.	1.4	0.75/2	40,509/36,660	26,702/14,304	100,875/82,882	738/442	4:56 min/0:43 min

The finest mesh size of the tetrahedral elements is assumed overall to give sufficiently accurate S_{ij} components. Those S_{ij} components for the quadratic tetrahedral elements with a mesh size of 0.75 mm and a relative cable length of 1.2 are therefore extracted. The efficient cable models are set up to fit these components:

$$\begin{pmatrix} S_{\mathrm{xx}} & S_{\mathrm{x}\varphi} & S_{\mathrm{xb}} \\ S_{\varphi\mathrm{x}} & S_{\varphi\varphi} & S_{\varphi\mathrm{b}} \\ S_{\mathrm{bx}} & S_{\mathrm{b}\varphi} & S_{\mathrm{bb}} \end{pmatrix} = \begin{pmatrix} 11,793.2\ \mathrm{kN} & 7,746.3\ \mathrm{kN\,mm} & 0 \\ 7941.3\ \mathrm{kN\,mm} & 15,593.5\ \mathrm{kN\,mm^2} & 0 \\ 0.5\ \mathrm{kN\,mm} & -0.7\ \mathrm{kN\,mm^2} & 10,837.4\ \mathrm{kN\,mm^2} \end{pmatrix}. \tag{20}$$

Since the matrix is nearly symmetrical, we make it symmetrical by setting S_{bx} and $S_{\mathrm{b}\varphi}$, which are much smaller than the other components, to zero and introduce a parameter S_{couple} that we use in the following for both $S_{\mathrm{x}\varphi}$ and $S_{\varphi\mathrm{x}}$:

$$S_{\mathrm{xb}} = S_{\mathrm{bx}} = S_{\varphi\mathrm{b}} = S_{\mathrm{b}\varphi} = 0 \tag{21}$$

$$S_{\mathrm{couple}} = \frac{S_{\mathrm{x}\varphi} + S_{\varphi\mathrm{x}}}{2} \equiv S'_{\mathrm{x}\varphi} = S'_{\varphi\mathrm{x}} \tag{22}$$

This simplification results in only four S_{ij} parameters that should be reached in the efficient cable models; see Table 5. Inverting the simplified S_{ij} matrix yields E_{ij}. As mentioned in Section 2.1, the cubic modeling approaches can be fitted based on either S_{ij} or E_{ij}.

Table 5. Stiffness parameters S_{ij} and E_{ij} from the full cable model with quadratic tetrahedral elements (mesh size of 0.75 mm) and cable length per strand length of 1.2.

S_{xx}	11,793	kN	E_{xx}	7848	kN
$S_{\varphi\varphi}$	15,594	kN mm^2	$E_{\varphi\varphi}$	10,376	kN mm^2
S_{couple}	7843	kN mm	E_{couple}	−15,601	kN mm
S_{bb}	10,837	kN mm^2	E_{bb}	10,837	kN mm^2

3.2. *Efficient Single-Cable Models*

The cable-modeling approaches introduced in Section 2.2 are set up as described in Section 2.3. Table 6 lists the parameters that are either calculated analytically or calibrated using the cable FEM models. To obtain the model parameters, the stiffness parameter S_{ij} or E_{ij} are used. For the approaches that do not have a tension/torsion coupling, the parameters are calculated once with S_{ij} and once with E_{ij}.

Table 6. Material and geometry parameters for the efficient cable models from analytical calculation or calibration by FEM models.

	Fit Towards	***E* (GPa)**	***G* (GPa)**	**E_K * (GPa)**	
Solid	E_{ij}	69.39	5.097	-	
	S_{ij}	104.3	7.660	17.32	
	Fit towards	C_{10} * (GPa)	k_1 * (GPa)	k ** [1]	α_{HGO} * (°)
Solid, HGO	S_{ij}	0.7581	26.47	0.0	14.08
	Fit towards	*E* (GPa)	*G* (GPa)	*r* (mm)	E_K * (MPa)
Beam	E_{ij}	452.2	216.6	2.35	-
	S_{ij}	1021	735.1	1.917	531.5
Solid/beam	Fit towards	E_{solid} (GPa)	G_{solid} (GPa)	r_{beam} ** (mm)	E_{beam} (GPa)
	E_{ij}	10.65	5.097	0.006	58,740,000
	S_{ij}	10.65	7.660	0.006	93,628,000

* Calibrated to fit S_{ij}. ** Chosen values.

Figure 8 shows the components of S_{ij} and E_{ij} obtained for the cubic modeling approaches. The diagrams use a logarithmic scale with relative values normalized to the target S_{ij} or E_{ij} values stated in Table 5. Figure 8a,b show the stiffness values for cable model parameters calculated to fit S_{ij}. As expected, the stiffness parameters obtained for S_{ij} plotted in Figure 8a fit well to the target values, except for the bending stiffness in the solid approach, which is too high by a factor of 9. The fit of the beam and solid/beam approaches is equally good. The E_{ij} components for the same efficient cable models, however, are about 53% higher than the components of the full-geometry model. When the cable models are calibrated to E_{ij} (see Figure 8c,d), the E_{ij} components fit well, but components of S_{ij} are lower by about 36%. This shows that for a cable that has tension/torsion coupling ($S_{x\varphi} \neq 0$), an efficient cable model with cubic material can fit either S_{ij} or E_{ij} but not both at the same time. S_{xx} corresponds to the stiffness in tension with constrained torsion and E_{xx} to tension with free torsion. When using such a cable model with cubic material, the model's parameters should be calculated depending on the application of the model. If the application is unknown, an intermediate stiffness of S_{ij} and E_{ij} should be used for the models.

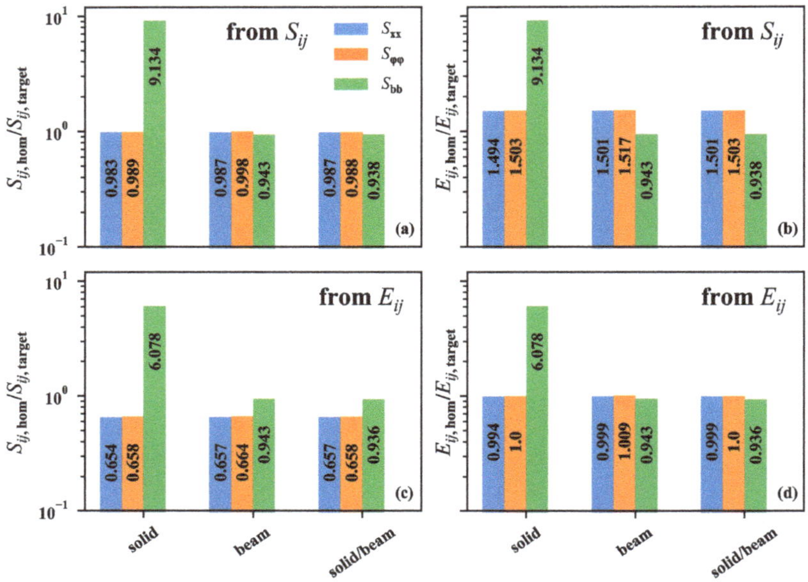

Figure 8. Components of S_{ij} and E_{ij} relative to their target values for varying cable-modeling approaches without anisotropic material behavior. The parameters of the models are calibrated to (**a**,**b**) fit S_{ij} or (**c**,**d**) fit E_{ij}.

Three efficient cable-modeling approaches that can account for the tension/torsion coupling are investigated, and their S_{ij} and E_{ij} components are plotted in Figure 9. Similar to the cubic approaches, the bending stiffness of the solid approach is too high by a factor of 7.8. The solid, solid–HGO, and beam approaches can capture S_{ij} well; see Figure 9a. The largest differences are observed for $S_{x\varphi}$ in the solid–HGO approach, which is 22% higher than the target value. Note that due to having only three calibration parameters in the HGO approach, the four independent stiffness parameters cannot all be fitted at the same time. Other HGO parameters such as D, k_2, and κ could be fitted as well but do not help to improve the accuracy of S_{ij}.

The components of the E_{ij} matrix of the solid and the beam approach in Figure 9b fit well to the target E_{ij} components, except for the E_{bb} of the linear elastic solid approach. For the solid–HGO approach, only the E_{bb} component fits well, whereas the other E_{ij} terms are lower by 46% to 60%. This is due to an amplification of the deviation of the S_{ij} components since the S_{ij} matrix is inverted to calculate E_{ij}. Furthermore, the convergence in the simulations with the solid HGO approach is bad, which requires many more iterations in the FEM simulation. The solid approach can therefore be used for applications where bending does not play a role, and the beam approach can be used in all cases where inaccuracies related to the coupling of the beam nodes to the rubber are acceptable—for example, because the rubber/cable interface is not of special interest in the model.

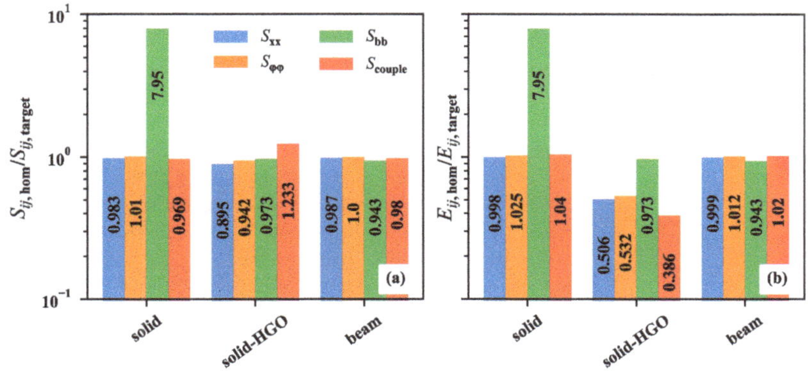

Figure 9. The three anisotropic efficient cable-modeling approaches with their relative stiffness values in terms of (**a**) S_{ij} and (**b**) E_{ij}. Note that the parameters of the efficient models were calculated or fit to reach S_{ij}.

3.3. Three-Cable Models

The three-cable shear specimens of the szs-type with efficiently modeled cables are now evaluated in terms of stiffness, deformation fields, and stress fields, and compared to the full-geometry results. The rubber between the outer and the central cable is sheared, and the forces in the cables causes them to rotate in opposite directions, which is only slightly hindered by the rubber.

3.3.1. Stiffness of the Three-Cable Specimens

Figure 10 shows the longitudinal force F_x and the end rotation of the central cable φ_x versus the end displacement u_x. The dashed black line and the solid lines illustrate the response of the full-geometry and efficient approaches, respectively. A linear relation of both F_x and φ_x with respect to u_x can be seen.

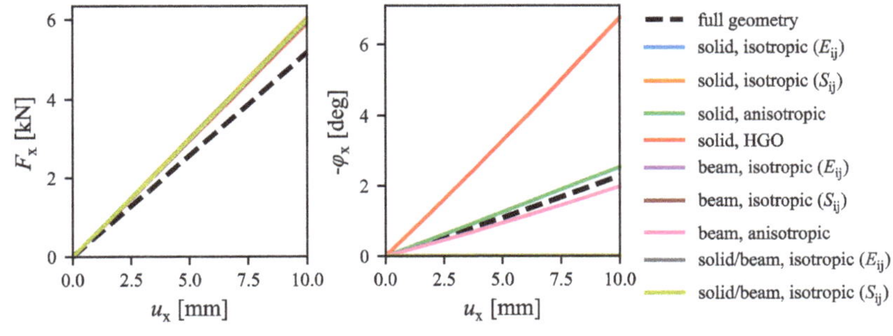

Figure 10. Force–displacement and end rotation–displacement plots of the three-cable szs-type specimen with free end rotation φ_x.

There is good agreement between the F_x curves of all the efficient cable-modeling approaches, but they all lie above the F_x of the model with full geometry by about 14%. This can partly be explained by the non-uniform strain fields in the full-geometry model. Furthermore, the results in Section 3.1 show that the mesh of the full-geometry three-cable model can underestimate the cable stiffness by up to 4%, which can also contribute to this difference. In addition, the full-geometry model has a layer of rubber around the wires that can be sheared (see Figure 4, where the gap from the six outer wires to the surface of the whole cable can be written as $r - r_i - 2\,r_o - d = 0.3$ mm). In the efficient models, this outer gap is assigned the same material properties as the rest of the cable, which is much stiffer than the rubber. When an efficient model is fitted to test data, the cable radius thus should be set to not include this layer of rubber to avoid this overestimation of stiffness in a rubber component. The end rotations of the central cable φ_x, plotted in Figure 10, are zero for all cubic modeling approaches since those models do not account for tension/torsion coupling. The anisotropic beam and anisotropic solid approaches show a good agreement with the full-geometry approach, whereas the solid–HGO approach overestimates the end rotation by about 150%.

The model size in terms of number of nodes, number of elements, degrees of freedom, necessary RAM to load the full stiffness matrix, and computation time are listed in Table 7 for all approaches in the three-cable model. There is a substantial difference in model size and computation time between the full-geometry approach and the efficient approaches, with the full-geometry model requiring about 50 GB of RAM to load the full stiffness matrix and a computation time of 4:35 h. The efficient three-cable models, on the other hand, need less than 2 GB of RAM and compute in about 2 min.

Table 7. Information on the model size, necessary RAM to load the full stiffness matrix, and computation time (four cores of a six-core Intel Xeon E5-1650 v3 @ 3.5 GHz workstation with 128 GB RAM) with varied model setup of the three-cable models.

	Nodes	Elements	Degrees of Freedom	Necessary RAM (MB)	Comp. Time
Full geometry	1,528,220	641,279	3,282,099	50,509	4:35 h
Solid	165,198	77,642	338,343	1953	2:16 min
Beam	74,053	33,990	165,081	843	0:46 min
Solid/beam	166,056	78,071	393,630	1961	1:45 min

3.3.2. Deformations of the Three-Cable Specimens

The tension/torsion coupling of the cables can cause a twisting of the specimen. One key result variable of this twist is the difference of the out-of-plane displacement u_z, which is plotted in Figure 11. If u_z is the same above and below the central cable, there is no twist—the displacement is merely a result of the Poisson effect in the rubber (especially the peaks at the right-hand side, which can be seen most clearly in Figure 11e). Otherwise, a twisting of the specimen occurs, which can be assessed by the u_z displacement at the top and bottom surface. Since such behavior can only be described by the anisotropic cable-modeling approaches, only one of the cubic approaches (solid/beam, fitted to S_{ij}) is shown for reference.

For the full-geometry model of Figure 11a, there is a distinct difference in the u_z fields above and below the central cable: On the top and the bottom of the specimen, a u_z of 0.7 mm and −0.7 mm is computed, respectively. Note that the highest values of u_z at the top face occur at about 60% along the length of the rubber block in the specimen. The results for the cubic cable-modeling approach (see Figure 11e) show a completely symmetric u_z field with respect to the xy-plane. The anisotropic solid approach in Figure 11b shows the same trend as the full-geometry model, but with a less pronounced twist of the specimen. The solid–HGO approach in Figure 11c, on the other hand, drastically overestimates the out-of-plane displacement of the specimen with u_z on the top and bottom of the specimen of 1.7 mm and −1.7 mm, respectively. The best agreement with the full-geometry model is obtained by the anisotropic beam approach of Figure 11d: The u_z fields are only slightly different above and below the center cable, with the top and bottom maximum values of u_z occurring on the right end of the rubber block.

The poor performance of the HGO model can be explained by the unwanted coupling factors inherent to this approach. In addition, the HGO approach requires the highest computational effort for a convergence of the simulation. The better fit of the anisotropic beam approach compared to the anisotropic solid approach can only be attributed to their difference in bending stiffness: Since the twisting of the specimen is associated with a bending deformation of the cables, the excessive bending stiffness of the solid approach affects these results.

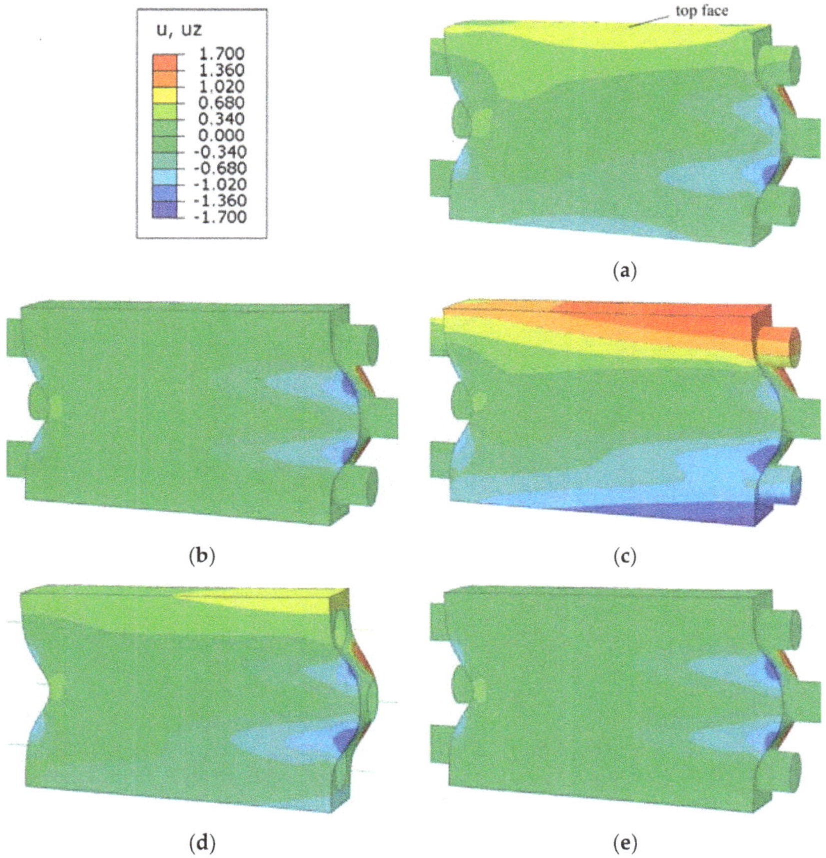

Figure 11. Contour plots of the out-of-plane displacements u_z in mm for (**a**) the full-geometry model and the four efficient cable-modeling approaches (**b**) solid, anisotropic, (**c**) solid, HGO, (**d**) beam, anisotropic, and (**e**) solid/beam, cubic (fit to S_{ij}) of the three-cable szs-type specimen with free end rotation.

3.3.3. Stresses in the Three-Cable Specimens

In many cases of reinforced rubber components, cable/rubber debonding and rubber failure is more relevant than deformations. Thus, the stresses in the rubber are evaluated in the following. It is assumed that the maximum principal stress σ_I is the best indicator for rubber failure. Figure 12 shows σ_I for the full-geometry model and various efficient modeling approaches. The specimen is cut in the plane of the cable axes. The main load of the rubber is shear between the central cable and the outer cables. These shear stresses, however, are not uniform and feature surface effects near the free surfaces at both ends of the rubber block (points A, A', B, and B'): The highest σ_I occurs at the junction of the center cable and the rubber on the right-hand side of the specimen (point A and point A'). Those maximum values of σ_I amount to 12.54 MPa, 9.74 MPa, 9.87 MPa, 10.17 MPa, and 9.11 MPa for the full-geometry, solid–anisotropic, solid–HGO, beam–anisotropic, and solid/beam (fitted to S_{ij}) approach, respectively. The junction of cable and rubber material imposes a singularity. This means that those stresses depend on the mesh size in the model, which must be accounted for in failure predictions. For relative comparisons like geometric studies with similar mesh, such results can be used nonetheless. There are also high stresses at the junction of the outer cables and the rubber on the left-hand side of the specimen (point B and point B′).

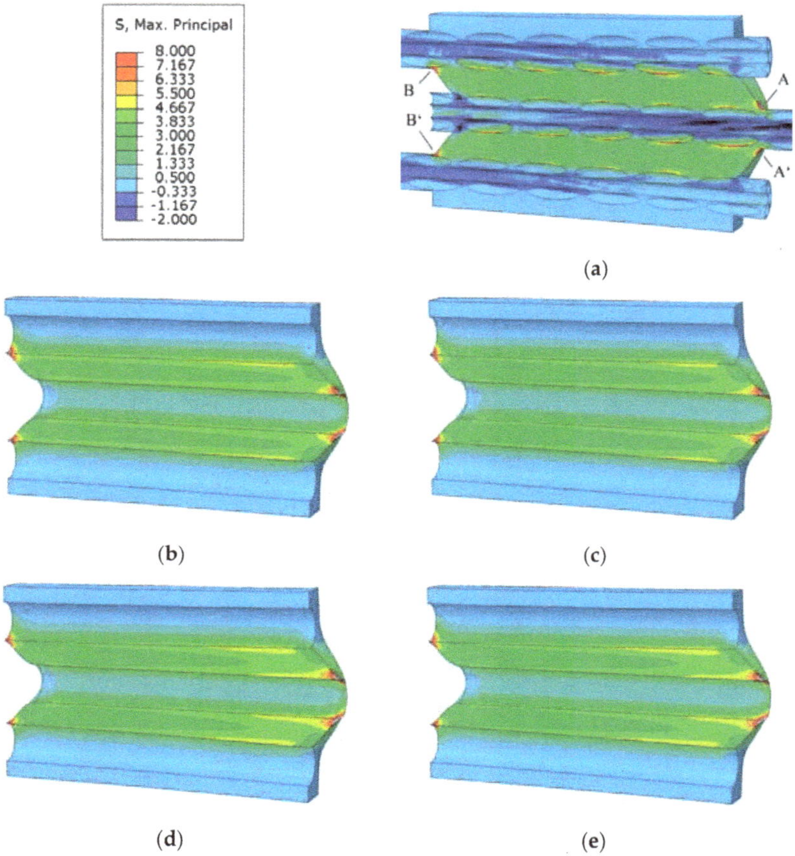

Figure 12. Contour plots of the maximum principal stress σ_I (MPa) for (**a**) the full-geometry model and the four efficient cable-modeling approaches (**b**) solid, anisotropic, (**c**) solid, HGO, (**d**) beam, anisotropic, and (**e**) solid/beam, cubic (fit to S_{ij}). The three-cable szs-type specimens are cut in the plane defined by z = 0.

The stress field in the full-geometry model shown in Figure 12a shows additional peaks where the cable wires reach farthest into the space between the center cable and the outer cables. This effect introduces another parameter to the model: If such a region coincides with the surface of the rubber block (is close to point A or A′), the stresses will be considerably higher. This effect is not studied here but should be considered when predicting the failure of cable–rubber specimens.

The stress fields in the models with efficiently modeled cables are more uniform. The stresses generally increase towards the right side of the specimen. Similar to the full-geometry approach, the highest maximum principal stresses occur at points A and A′. The highest maximum principal stress σ_I in the solid/beam approach (see Figure 12e) of 9.11 MPa is lower than that of the other modeling approaches (9.87 MPa to 10.17 MPa). This can be explained by the very low Young's modulus of the solid elements in the solid/beam approach, which leads to a shear deformation between the beam and the cable surface. The differences in the highest computed σ_I of the varying cable-modeling approaches are rather low, indicating that for such three-cable specimens, a stress-based failure assessment is not sensitive to the selection of the modeling approach. There is a slight increase in the σ_I peaks due to the coupling term. For example, the beam approach computes maximum σ_I = 10.17 MPa in the anisotropic approach and $\sigma_I = 10.02$ MPa in the cubic approach.

4. Conclusions

A variety of approaches for efficiently modeling the elastic response of a steel cable in a reinforced rubber component is introduced and evaluated both as a single cable and in a three-cable shear specimen. The aim is to reach an accurate representation of the high tensile stiffness, the high torsional stiffness, the low bending stiffness, and the tension/torsion coupling of steel cables. The modeling approaches considered consist of beam elements, solid elements, or a combination of both. In addition to an approach with linear elastic material behavior and cubic material symmetry, a special kind of anisotropic linear elastic material model is selected and fitted to capture the tension/torsion coupling. Furthermore, an approach using an anisotropic hyperelastic material model (HGO) is evaluated. The cable-modeling approaches are able to model the target stiffness of the cable to a varying extent:

- Solid linear elastic approaches: The bending stiffness is too high, but the other stiffness components are captured.
- Solid approach with anisotropic hyperelastic material: Only three parameters are available to fit four independent stiffness parameters. Not all four of them can be calibrated accurately at the same time. At least one of them differs up to 20% from the target value.
- Beam approach: All components of the target stiffness can be captured. However, the beam nodes are rigidly coupled to the rubber nodes at the cable surface, which is only valid if the cable is considerably stiffer than the rubber.
- Solid/beam approach: The tension/torsion coupling could be implemented in the beams, but the beam rotations would need to be coupled to the solid nodes. This coupling would induce numerical artifacts and is thus not implemented. Furthermore, the solid elements in the solid/beam approach have a very low Young's modulus that can lead to unphysical shear deformations inside the cable.

From those observations, the best modeling approach can be selected for a given application. The key questions for this selection is whether the cables experience a bending deformation and whether the tension/torsion coupling plays an important role in the model's application. Generally, the anisotropic beam approach is easy to calibrate and can capture the stiffness of the cable well. Inaccuracies introduced by the coupling of rubber nodes to the beam may not be acceptable, like in applications where evaluations at the rubber/cable interface require solid elements in the cable. In this case, the solid approach can be used if the cable bending is not relevant or the solid/beam approach if the tension/torsion coupling can be neglected. If both bending and tension/torsion coupling need to be captured, the HGO approach can be used, but it is associated with considerable discrepancies of all components of the stiffness matrix.

This work shows how to calculate and calibrate the geometric and material parameters of all cable modeling approaches and how to implement them. There is no approach that is suited for all possible applications, such as conveyor belts, hydraulic hoses, or tires. The modeling approach should be selected with care based on the type of loads the cables are exposed to in the application.

Author Contributions: Conceptualization, M.P., S.M.F. and C.S.; methodology, S.M.F. and M.P.; software, M.P.; validation, M.P.; formal analysis, M.P.; investigation, M.P. and S.M.F.; resources, M.P.; data curation, M.P.; writing—original draft preparation, M.P.; writing—review and editing, M.P., S.M.F. and C.S.; visualization, M.P.; supervision, C.S.; project administration, M.P.; funding acquisition, C.S. and M.P. All authors have read and agreed to the published version of the manuscript.

Funding: This research was supported by the Austrian Research Promotion Agency (FFG) within the BRIDGE framework as part of the project "Entwicklung einer Methodik zur Vorhersage des Versagens in elastomeren Gurten mittels Finite Elemente Simulation," grant agreement.

Institutional Review Board Statement: Not applicable.

Informed Consent Statement: Not applicable.

Data Availability Statement: The methods for modeling cables are derived and evaluated independently of the FEM software used so other researchers can implement them into the FEM software they are using. Thus, no additional data need to be provided.

Conflicts of Interest: The authors declare no conflict of interest.

References

1. Utting, W.S.; Jones, N. A Survey of Literature on the Behaviour of Wire Ropes. *Wire Ind.* **1984**, *51*, 623–629.
2. Cardou, A.; Jolicoeur, C. Mechanical Models of Helical Strands. *Appl. Mech. Rev.* **1997**, *50*, 1–14. [CrossRef]
3. Costello, G.A. *Theory of Wire Rope*; Mechanical Engineering Series; Springer: New York, NY, USA, 1997; ISBN 978-1-4612-7361-5.
4. Feyrer, K. Wire Ropes Under Tensile Load. In *Wire Ropes*; Springer: Berlin/Heidelberg, Germany, 2015; pp. 59–177. ISBN 978-3-642-54995-3.
5. Wang, X.-Y.; Meng, X.-B.; Wang, J.-X.; Sun, Y.-H.; Gao, K. Mathematical Modeling and Geometric Analysis for Wire Rope Strands. *Appl. Math. Model.* **2015**, *39*, 1019–1032. [CrossRef]
6. Jiang, W.G.; Yao, M.S.; Walton, J.M. A Concise Finite Element Model for Simple Straight Wire Rope Strand. *Int. J. Mech. Sci.* **1999**, *41*, 143–161. [CrossRef]
7. Foti, F.; de Luca di Roseto, A. Analytical and Finite Element Modelling of the Elastic–Plastic Behaviour of Metallic Strands under Axial–Torsional Loads. *Int. J. Mech. Sci.* **2016**, *115–116*, 202–214. [CrossRef]
8. Chen, Y.; Meng, F.; Gong, X. Full Contact Analysis of Wire Rope Strand Subjected to Varying Loads Based on Semi-Analytical Method. *Int. J. Solids Struct.* **2017**, *117*, 51–66. [CrossRef]
9. Cao, X.; Wu, W. The Establishment of a Mechanics Model of Multi-Strand Wire Rope Subjected to Bending Load with Finite Element Simulation and Experimental Verification. *Int. J. Mech. Sci.* **2018**, *142–143*, 289–303. [CrossRef]
10. Utting, W.S.; Jones, N. The Response of Wire Rope Strands to Axial Tensile Loads—Part II. Comparison of Experimental Results and Theoretical Predictions. *Int. J. Mech. Sci.* **1987**, *29*, 621–636. [CrossRef]
11. Chen, Z.; Yu, Y.; Wang, X.; Wu, X.; Liu, H. Experimental Research on Bending Performance of Structural Cable. *Constr. Build. Mater.* **2015**, *96*, 279–288. [CrossRef]
12. Bonneric, M.; Aubin, V.; Durville, D. Finite Element Simulation of a Steel Cable-Rubber Composite under Bending Loading: Influence of Rubber Penetration on the Stress Distribution in Wires. *Int. J. Solids Struct.* **2019**, *160*, 158–167. [CrossRef]
13. Frankl, S.; Pletz, M.; Schuecker, C. Incremental Finite Element Delamination Model for Fibre Pull-out Tests of Elastomer-Matrix Composites. *Procedia Struct. Integr.* **2019**, *17*, 51–57. [CrossRef]
14. Nordell, L.K. Steel Cord Belt and Splice Construction. *Bulk Solids Handl.* **1993**, *13*, 685–693.
15. Fedorko, G.; Molnar, V.; Dovica, M.; Toth, T.; Fabianova, J. Failure Analysis of Irreversible Changes in the Construction of the Damaged Rubber Hoses. *Eng. Fail. Anal.* **2015**, *58*, 31–43. [CrossRef]
16. Frankl, S.M.; Pletz, M.; Wondracek, A.; Schuecker, C. Assessing Failure in Steel Cable-Reinforced Rubber Belts Using Multi-Scale FEM Modelling. *J. Compos. Sci.* **2022**, *6*, 34. [CrossRef]
17. Nordell, L.; Qiu, X.; Sethi, V. Belt Conveyor Steel Cord Splice Analysis. *Bulk Solids Handl.* **1991**, *11*, 863–868.
18. Abaqus. *Dassault Systèmes*; Simulia: Johnston, RI, USA, 2020.
19. Cartraud, P.; Messager, T. Computational Homogenization of Periodic Beam-like Structures. *Int. J. Solids Struct.* **2006**, *43*, 686–696. [CrossRef]
20. Holzapfel, G.A.; Gasser, T.C.; Ogden, R.W. A New Constitutive Framework for Arterial Wall Mechanics and a Comparative Study of Material Models. *J. Elast. Phys. Sci. Solids* **2000**, *61*, 1–48. [CrossRef]
21. Gasser, T.C.; Ogden, R.W.; Holzapfel, G.A. Hyperelastic Modelling of Arterial Layers with Distributed Collagen Fibre Orientations. *J. R. Soc. Interface* **2006**, *3*, 15–35. [CrossRef] [PubMed]
22. Froböse, T.; Overmeyer, L.; Poll, G. *Verfahren zur Ermittlung der Materialparameter für die Auslegung von Stahlseil-Fördergurtverbindungen mit Hilfe der FEM*; Berichte aus dem ITA; PZH Verlag: Garbsen, Germany, 2017; ISBN 978-3-95900-153-3.
23. Gao, F.; Han, L. Implementing the Nelder-Mead Simplex Algorithm with Adaptive Parameters. *Comput. Optim. Appl.* **2012**, *51*, 259–277. [CrossRef]

Journal of
Composites Science

Article

Development of an Analytical Model to Predict Stress–Strain Curves of Short Fiber-Reinforced Polymers with Six Independent Parameters

Esha * and Joachim Hausmann

Leibniz-Institut für Verbundwerkstoffe (IVW), 67663 Kaiserslautern, Germany; joachim.hausmann@ivw.uni-kl.de
* Correspondence: esha@ivw.uni-kl.de

Abstract: Mechanical properties of fiber-reinforced polymers are sensitive to environmental influences due to the presence of the polymer matrix but inhomogeneous and anisotropic due to the presence of the fibers. Hence, structural analysis with mechanical properties as a function of loading, environment, design, and material condition produces more precise, reliable, and economic structures. In the present study, an analytical model is developed that can predict engineering values as well as non-linear stress–strain curves as a function of six independent parameters for short fiber-reinforced polymers manufactured by injection molding. These parameters are the strain, temperature, humidity, fiber content, fiber orientation, and thickness of the specimen. A three-point test matrix for each independent parameter is used to obtain experimental data. To insert the effect of in-homogenous and anisotropic distribution of fibers in the analytical model, microCT analysis is done. Similarly, dynamic mechanical thermal analysis (DMTA) is done to insert the viscoelastic effect of the material. The least mean square regression method is used to predict empirical formulas. The standard error of regression for the fitting of the model with experimental stress–strain curves is closely controlled below 2% of the stress range. This study provides user-specific material data for simulations with specific material, loading, and environmental conditions.

Keywords: analytical model; stress–strain curve; short fiber-reinforced thermoplastic

Citation: Esha; Hausmann, J. Development of an Analytical Model to Predict Stress–Strain Curves of Short Fiber-Reinforced Polymers with Six Independent Parameters. *J. Compos. Sci.* **2022**, *6*, 140. https://doi.org/10.3390/jcs6050140

Academic Editors: Francesco Tornabene and Thanasis Triantafillou

Received: 31 March 2022
Accepted: 9 May 2022
Published: 11 May 2022

1. Introduction

Short fiber-reinforced polymer (SFRP) parts are widely used in industries as they can be easily molded into complex shapes. However, the orientation of the fibers varies from one point to another in composite structures. In complex shapes such as the dome of a pressure vessel or plastic gears, plastic axle, bicycle seats, etc., the fiber angle varies locally due to the geometry of the structure, fabrication process, and type of fiber used in the composite material. This induces a strong heterogeneity throughout the structures, enhancing anisotropic mechanical behavior. For continuous fiber composites, finite element (FE) analysis is well developed to consider variations in fiber angle locally in stress analysis [1–5], whereas for SFRP parts, the micromechanical models in the FE analysis use the same empirical formulas that are used for continuous fiber composites with some modification. Due to the short fiber length, the randomness of the fiber arrangement significantly varies throughout the specimen. Hence, local fiber orientation distribution affects material characterization. Changes in the microstructure of short fibers should be considered in the calculation of mechanical properties in specimen level analysis and then in finite element analysis of the component.

For example, injection-molded short fiber-reinforced plates show fibers aligned towards the molding direction in outer peripheral layers and transversely deviated in the core layer. Therefore, heterogeneity and anisotropy in material properties should be considered in structural analysis. Mechanical properties of an injection-molded specimen are the combination of the mechanical properties of each layer. Such properties can be estimated by developing a model as a function of the degree of anisotropy of each layer. Engineering values created by such a model can insert the variation of local fiber orientation in the FE analysis of an injection-molded component [6]. In addition to fiber contribution, the polymer also contributes to sensitivity in mechanical properties due to environmental conditions such as temperature and humidity. Hence, in designing SFRP parts, the sensitivity of mechanical properties due to both the fiber and polymer matrix should be considered in material data. Such material data inserted in FE analysis will provide an economical, effective, and precise design.

One of the material data components is the stress–strain curve. The aim of this project is to develop a model that can provide the stress–strain curve considering inhomogeneity and anisotropy of the material due to fiber orientation distribution and sensitivity towards environmental conditions in addition to material selection. In stress analysis, the material is pulled with a certain pulling speed, which develops strain in it. Due to the strain, the resistance in the material increases, called stress. Hence, stress σ is dependent on strain ε. Therefore, the mathematical formula for this model should be as follows:

$$\sigma = f(\varepsilon) \quad (1)$$

Stress varies linearly until a certain point, which indicates elastic deformation. It follows Hook's law of elasticity. After the yield point, the material no longer follows Hook's law due to strain hardening or plastic deformation. Ramberg and Osgood used a three-parameter equation to predict stress–strain relationship beyond the yield point. The equation is as follows [7,8]:

$$\varepsilon = \frac{\sigma}{E_o} + k\left(\frac{\sigma}{\sigma_k}\right)^n \quad (2)$$

E_o is Young's modulus or the elastic modulus and σ_k is proof stress corresponding to the plastic strain k. Parameter n describes the bend of the stress–strain curve. The elastic part of the stress–strain curve, which is the first part of the equation, follows Hook's law of elasticity whereas the plastic part of the stress–strain curve, which is the second part of the equation, follows a power law of the non-dimensional stress ratio. This equation was designed initially for metals such as aluminum where k was generally accepted as 0.2% of the plastic strain and n is a material constant, which is calculated based on 0.01% to 0.2% proof stress [9,10]. This value gives the measure of work hardening or plastic deformation. It varies from 0–0.5 [11]. Equation (2) predicts the stress–strain curve until 0.2% of plastic strain for metals, but after that it cannot follow the curvature of the stress–strain curve accurately [10]. To overcome this limitation, the stress–strain curve is divided into elastic and plastic regions. Tayler series expansion of the Ramberg–Osgood (RO) equation is used for fitting the stress–strain curve in the plastic region [12]. However, this equation uses values of the plastic strain at ultimate and yield limit to calculate the value-modified material constant n [13]. Kamaya et al. [9] used the yield and ultimate strength value to develop a modified version of the RO equation with the help of the J integral. The accuracy of the curves varied from 2–10%. The RO equation requires values of k, n, and plastic strain.

To predict the stress–strain curve for a composite material, the fiber orientation distribution should be considered. Several numerical approaches were developed in LS-DYNA by overlapping fiber orientation distribution from an injection molding simulation model to the finite shell element of the anisotropic structural simulation. [14–17]. Dean et al. [18] tested a macro-mechanical model in which the average of the layerwise fiber orientation tensor in each direction (flow direction of injection molding and transverse direction) was inserted in a macro-mechanical invariant based on the anisotropic constitutive model mentioned in [18,19]. This FE simulation requires a huge computational time and cost.

An analytical model can be a more economical solution in the design of SFRP. Several analytical models were developed to predict the mechanical properties of SFRP using failure criteria usually used for unidirectional (UD) laminates such as the Tsai–Hill criterion [20], theory of linear elasticity for orthotropic material, Halpin—Tsai–Nielsen criterion [21], etc. [5,22,23] by considering the specimen as a pile of three UD laminates. The fiber orientation distribution within these layers varies according to the thickness of the injection-molded plate [18].

The thickness of the specimen has a significant influence on anisotropy as well as mechanical properties, especially in injection-molded plates. The fiber fraction ratio aligned to the flow direction is high in thin plates (1 mm) as compared to thick plates (>2 mm) [20,24]. Due to the difference in the fiber orientation distribution, there is a significant difference in the normalized modulus as the thickness of the specimen increases. The E modulus of a 0° fiber-oriented specimen decreases whereas at 90°, increases with an increase in the thickness of the specimen [20]. Moreover, temperature increases the difference in the tensile modulus with the thickness of the specimen for fiber angles greater than 30° [20]. This could be due to matrix-dominated behavior at a fiber angle higher than 30° since the matrix (polymer) is sensitive to temperature. Hence, a single model, which can insert synergetic effects of all influential parameters is required to predict the mechanical behavior of SFRP.

Therefore, in this project, an analytical model is developed, which provides material data (stress–strain curve) considering the influence of environmental, loading, material, and design condition.

$$\sigma = f(\varepsilon,\ temp,\ fo, fc, RH,\ th) \tag{3}$$

The aim of this project is to develop a formula as mentioned in Equation (3), which can predict the stress–strain curve at any arbitrary temperature (*temp*), fiber orientation (*fo*) and fiber content (*fc*), relative humidity (*RH*), and thickness of the specimen (*th*). Sensitivity in mechanical properties due to the fibers can be considered by inserting fiber orientation and fiber content in the formula. Temperature and humidity will add more pronounced viscoelastic behavior of the polymer in the formula.

If we compare a stress–strain curve of a metal and composite material, it is evident that in metals, the stress–strain curve transits quickly and steeply from the elastic to plastic region (Figure 1b). However, in SFRP, this transition is gradual and slow (Figure 1a). Therefore, the stress–strain curve for SFRP should be divided into three distinguished parts, which are defined as follows:

- Linear part: the part of stress–strain curve before the elastic limit (blue part in Figure 1a).
- Onset of the bend: This is the part of the stress–strain curve where the graph starts following a curve. The end-limit of this part will be the start of a line. This limit can be calculated through trial and error and can vary with material composition. This can also be referred to as the elasto–plastic region in the case of metals (yellow part in Figure 1a).
- Offset of the bend: This part of the stress–strain curve is approximately linear after the bend. This can also be referred to as the plastic region. This curve starts from the offset of the bend until fracture of the specimen (red part in Figure 1a).

These three parts can be predicted separately by using some mathematical functions.

The Ramberg–Osgood (RO) equation cannot be used here because it requires the value of the plastic strain. To calculate the plastic strain, values of stress and strain are required. However, in this project, both of these values for an arbitrary material condition are not available.

Therefore, a new empirical formula should be developed. The first step in designing the empirical formula is to predict the elastic modulus. Several models were developed to predict the elastic modulus, i.e., rule of mixture (ROM), inverse rule of mixture (IROM), and Halpin Tsai and Bowyer–Bader model [5,23,25]. All these models are designed for fiber laminates where fibers are continuous and compactly packed. All these models use the fiber volume fraction and other geometric parameters of the fiber. They require the elastic modulus of the fiber and matrix separately. This requirement of the measuring volume fraction ratio and fiber length makes the model user unfriendly and complicated. Moreover, equations involved in these models do not consider the change in temperature and moisture, which modifies not only the elastic modulus of the fiber and polymer but also the length of the fiber [20]. Hence, a model using the mass fraction ratio is more friendly and practical. There is no need for extra effort in converting the mass fraction to the volume fraction. A model described in [26] uses the mass fraction of the fiber. However, this model cannot accommodate variation in temperature.

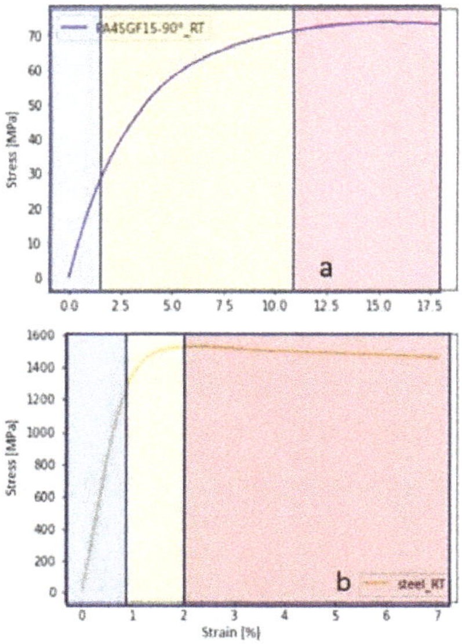

Figure 1. Exemplary stress–strain curves of SFRP (PA46 GF15) at (**a**) and metal (steel) at (**b**).

The modulus of elasticity varies with temperature if the fibers are in the transverse direction of the load. The longitudinal modulus of elasticity of the UD laminate GF PP is independent of temperature, whereas the transverse modulus of elasticity has a decaying tendency with increases in temperature [27]. An empirical formula mentioned in [27] to predict the elastic modulus has inserted a factor of the normalized temperature ratio with the melting point of the polymer. It predicts the E modulus quite accurately but cannot accommodate variations in the fiber angle. Zhai et al. tested the famous Mori–Tanka micromechanical model [28] to predict the elastic modulus with different temperatures and fiber angles [29]. The model requires the orientation tensor [30], elastic moduli of fiber, and matrix and fiber length ratio. This model is dependent on the shape factor of the fibers. Neither fiber angle nor temperature is an independent variable in this model, which is the requirement for this study. To add the variation of the elastic modulus of composite material due to temperature, viscous behavior of the polymer must be inserted in the form of mathematical functions.

In this project, a model is designed in which the relationship between the microstructural properties and mechanical properties of composite material is developed. This is done by studying the arrangement of fibers and matrix in the material at its micro-scale level and by studying temperature-dependent behavior of the polymer. Micro-computer tomography (μCT) analysis helps us to understand the fiber arrangement and its influence on mechanical properties. The stiffness of the polymer is dependent on the temperature. DMTA (Dynamic mechanical thermal analysis) describes the stiffness of the material over the whole temperature range. Parameters of these analyses should be included in the formula for the stress–strain curve in the form of some empirical equations, which will be described in Section 3. In Section 2, experimental methodology, observation of tensile tests, and the development of an analytical model will be discussed.

2. Methodology and Experiment

Short glass fiber-reinforced polyamide (PA46 GF) with different fiber contents (15, 30, and 60 percent by weight) was used to develop the stress–strain model. To add sensitivity of the material towards temperature and humidity in the model, DMTA was conducted. All specimens were pre-conditioned for 50% RH following the methodology described in standard DIN EN ISO 1110 for pre-conditioning of polyamide. Small dog bone specimens of type 1BA from standard DIN EN ISO 527-2 were milled from injection-molded plates with a dimension of 80 × 80 mm^2. The fiber angle of the specimen was varied by rotating the specimen with respect to the molding direction of the injection-molded plate. The number of specimens per plate was determined on the basis of prior fiber orientation distribution analysis.

2.1. Fiber Orientation Distribution Analysis

The fiber orientation in the injection-molded plates is not uniform throughout the thickness, width, and length of the plate [15,18,20]. Fibers at the outer periphery of the plates are aligned towards the molding direction, whereas at the core they are transversely deviated. Hence, fiber orientation distribution analysis is necessary to select the position for the specimens in the plate where maximum fibers are oriented towards the assigned direction (molding direction). Inhomogeneity of fiber distribution varies with the fiber content, thickness, and position of the specimen with respect to the plate.

Therefore, three specimens from each plate were milled as shown in Figure 2. A small section at the center of ex-centric and centric specimens was scanned by µCT. Fiber distribution and orientation tensor [30] were studied with the help of the software VG Studio from Volume Graphics®. Fiber angles with respect to the molding direction were compared in each layer of the specimen. Orientation tensor, index of anisotropy, and the variation of the core-shell layer thickness were calculated [30,31]. The index of anisotropy is a quantitative way to characterize a sample on a scale of 0 to 1 where 0 is isotropic and 1 is perfectly anisotropic [32]. The optimum number of specimens per plate for the tensile tests was determined by comparing all these factors in ex-centric and centric specimens for each test matrix point. Tensile tests for each ex-centric and centric specimen were also done to quantify the deviation in the mechanical properties within the plate. The specimens with the highest fraction of fibers orientated towards the assigned orientation (i.e., with respect to the molding direction) and highest index of anisotropy were chosen for further tensile tests.

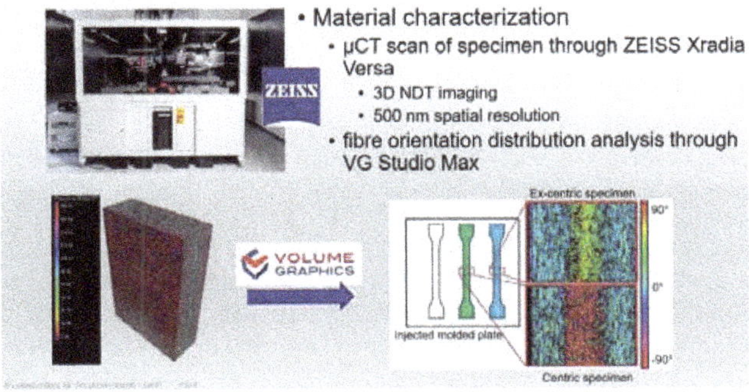

Figure 2. Material characterization to find out the optimized position of the specimens in the injection-molded plate.

2.2. Test Matrix

Three test points per independent variable were taken as shown in Table 1. With three experimental points, a linear function (or line) can be predicted. However, there are only two testing points for the thickness of the specimen. Hence, the thickness is a nominal variable rather than numerical.

Table 1. Test matrix to produce experimental data for the stress–strain model.

Fiber Content	Fiber Orientation	Thickness	Temperature	Humidity
15%, 30%, 60%	0°, 30°, 90°	2 mm, 3 mm	−20 °C, RT, 80 °C	50% RH

Fiber orientation in Table 1 is the assigned orientation of the specimen with respect to the molding direction. The number of specimens per plate was determined based on fiber orientation analysis mentioned in Section 2.1. In each specimen, most of the fibers are orientated towards the assigned orientation. The local fiber orientation distribution is not considered in this analytical model. Humidity is kept constant at 50% RH to reduce experimental efforts. However, the analytical model has a capability to insert humidity as numerical variable [33] Five tensile tests were conducted for each test matrix point to reduce data scattering. The same procedure was used for DMTA tests [33].

2.3. Experimental Observation

Tensile tests were performed on a uniaxial tensile testing machine with a load cell of 10 kN or 250 kN depending upon the fiber content of the material. Strain was calculated by digital image correlation from the video captured by a high-resolution 2D camera system. A climatic chamber was attached to perform tensile tests at high and low temperature. This chamber cannot control humidity. However, the effect of high temperature on the moisture content in the specimen was considered negligible if the specimens with a fiber content of 30% remained in the chamber for 8 min, those with a fiber content of 60% remained for 5 min, and those with a content of 15% remained for 17 min or less. None of the tests took more time than these. Hence, the relative humidity of the specimens was assumed to be constant throughout the tests. High and low temperature tests were closely controlled for the isothermal condition with a variation of ±2 °C, but the room-temperature test was not that closely controlled. It varied from 17–27 °C. To ensure uniform distribution of the heat, a thermocouple was attached in close proximity at the center of the specimen. All experimental stress–strain curves are plotted in Figure 3. From the graphs, it can be easily concluded that the fracture strain is lower at a lower temperature and vice versa due to the inverse relationship of stiffness and ductility.

As shown in Figure 3, the stiffness of the curves at −20 °C are quite high as compared to 80 °C of the same material. The water content in the specimens is frozen at low temperature such as −20 °C. This increases the stiffness of the material. Moreover, due to the sensitivity of the polymer to temperature, stiffness varies due to the glass transition temperature of the polymer. At low temperature, in this case, −20 °C, PA46 is in the glassy state. Therefore, the curve shows higher stiffness as compared to 80 °C where PA46 stays in a rubbery state.

Curves for the higher fiber content show higher stiffness. However, with lower orientation angles, the stress–strain curves show higher stiffness. Hence, it is concluded from this observation that the mathematical formula must have special parameters or procedures to accommodate the variation in stiffness with respect to the fiber content, fiber orientation as well as humidity and temperature. Since the same observations were noticed in the curves of storage modulus, a similar approach is used to develop an analytical model for the storage modulus [33]. Differences in Young's modulus between fiber orientations are due to the fiber content. The higher the fiber content of the specimen, the higher the difference in Young's modulus for the same fiber orientation. Similarly, the strength of the material increases linearly with the fiber content [34]. The variation of Young's modulus with respect to the fiber orientation within the same fiber content is not linear. This is well explained in the classical laminate theory. Young´s modulus of fiber-reinforced composites shows hyperbolic behavior if the fiber angle changes in regard to the loading direction [5,20]. Similar behavior is seen in the experimental curves of the storage modulus for the same material and specimens through DMTA [33]. Thus, Young´s modulus could be predicted by the storage modulus.

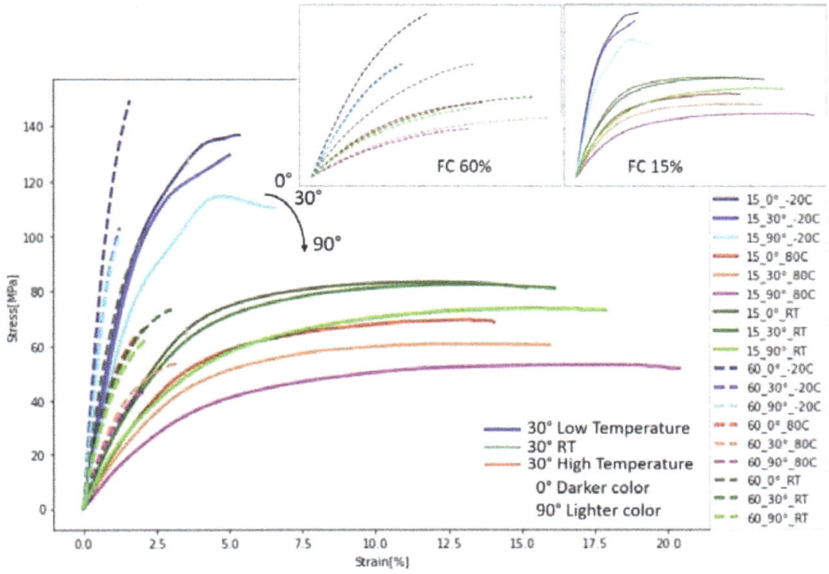

Figure 3. Experimental tensile test curves for PA46 GF60 and PA46 GF15 with all three fiber orientations (0°, 30°, and 90°) for 3 mm (in small boxes; curves are shown separately according to the fiber contents GF15 and GF60).

From the mathematical point of view, the stress–strain curves in Figure 3 can be divided into three parts as described earlier. The linear part can be described by the Young´s modulus Then, it follows a bend. In the case of SFRP, this onset of bending is longer as compared to metals. After this bend, the material shows purely plastic behavior that could be described as a straight line with lower/minimal slope. This approach will be used in designing empirical formulas.

2.4. Analytical Approach

An analytical approach for prediction of the stress–strain curve can be divided into two steps. First, the stiffness of the material should be predicted. This will help to predict the linear part of the curve. Later, the non-linear part of the curve can be assumed based on some mathematical formulas. Young´s modulus from tensile tests and the storage modulus from DMTA tests is similar based on the basic principle of testing [35,36]. In tensile tests, specimens are strained in a quasi-static condition. In DMTA tests, the specimen is subjected to oscillatory strain under varying temperature. Stiffness has been calculated at each temperature, which consists of the storage modulus and loss modulus. Therefore, the stiffness of the composite material should follow the same behavior as the storage modulus curve with respect to the temperature [37,38].

To analyze and validate the hypotheses, all values of storage modulus calculated from the DMTA tests were plotted against Young´s modulus determined by tensile tests for several fiber orientations, fiber contents, and temperatures as shown in Figure 4. A linear relationship can be observed between Young´s modulus and storage modulus by using the least mean square method of regression. Hence, Young´s modulus can be predicted with the help of the storage modulus model.

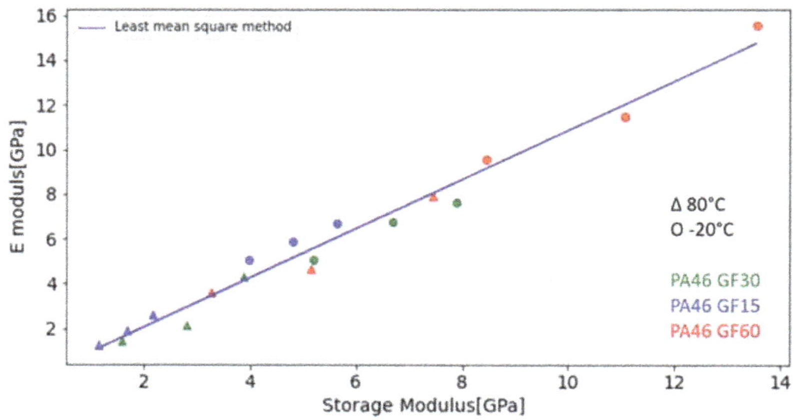

Figure 4. Experimental Young´s modulus plotted on the scale of the storage modulus of corresponding material data.

Experiments performed at room temperature were discarded because room temperature lies in the glass transition temperature range of PA46 (15–40 °C). In this region, the curve of the storage modulus is very sensitive to small temperature deviations. To predict the stiffness of the material, the storage modulus for arbitrary conditions is required. A separate analytical model for the storage modulus is designed that can predict the storage modulus with four independent parameters (i.e., fiber content, fiber orientation, temperature, and humidity). The development of this model called the storage modulus model is described in a conference paper [33]. The next step is to design the formula for the non-linear part of the stress–strain curve. The stiffness of the composite is the combination of fiber and matrix behavior. The fiber contributes towards stiffness of the composite whereas the polymer influences the toughness. As shown in Figure 3, the sensitivity of the stress–strain curve with respect to temperature and humidity is also due to the presence of the polymer. Therefore, the stress curve is a collection of the behavior contributed by both the fiber and polymer. Similar observations can be noticed in the storage modulus model [33]. In other words, the fiber content increases linearity, whereas the polymer influences the commencement of curvature in the stress–strain curve. The stress values in the elastic range are not exactly linear, which can be seen clearly from Figures 1 and 3. It is a combination of a line and a curve. This is due to the combination of the fiber and matrix. The plastic deformation of the matrix happens much faster as compared to the fibers. This has been investigated by plotting the rate of change of the E modulus throughout the test, which is expected to be constant or near constant at least at the elastic range of the stress–strain curve. This happens in metals, but in short fiber-reinforced polymer (especially PA46), this is not the case. Thus, a single mathematical function cannot predict the stress of the composite material. It can be considered as a combination of linear and non-linear functions in which the linear function influences the stiffness and the non-linear function influences the toughness of the curve. Linearity or slope of the stress–strain curve is governed by the fiber content. Each value of stress in Equation (3) can be assumed as weighted moving average of the fiber and polymer contribution biased by the fiber content by weight. Equation (3) can be described as:

$$\sigma = \frac{Fiber\ content[\%]}{100} f_l(\varepsilon) + \frac{100 - Fiber\ content[\%]}{100} f_{nl}(\varepsilon) \tag{4}$$

Here, $f_l(\varepsilon)$ and $f_{nl}(\varepsilon)$ are the mathematical functions for linearity and non-linearity of the curve, respectively. $f_l(\varepsilon)$ is a linear function with E modulus of the composite. This E modulus is predicted by the storage modulus model by simply inserting all arbitrary conditions. However, the non-linear function $f_{nl}(\varepsilon)$ can be a single or a combination of mathematical functions. The non-linear function will be estimated based on observations of experimental data of the material.

Since the experimental data were evaluated on the basis of standard ISO 527-2 [39], Young´s modulus is calculated using the range from 0.05% to 0.25% for the strain. Similarly, the elastic limit for the predicted curve should be 0.25% strain. As mentioned in Section 1, the predicted curve should also be divided into three parts. The offset of the bend in the stress–strain curve could start from a certain fraction of the fracture strain. For example, one-fourth of the fracture strain could be considered as offset of the bend. This value is calculated through trial and error or by observation. An initial estimate of the offset of the bend can be estimated by fitting the Holloman equation of strain hardening [40] to the available experimental data. Hence, the elaborated version of Equation (4) is as follows:

$$\left.\begin{array}{ll} \sigma_e = \frac{Fiber\ content}{100} f_l(\varepsilon) + \frac{100-Fiber\ content}{100} f_{nl1}(\varepsilon) & [\varepsilon < 0.25\%] \\ \sigma_{elpl} = \frac{Fiber\ content}{100} E_{a_{0.25\%\varepsilon}} \varepsilon + \frac{100-Fiber\ content}{100} f_{nl_1}(\varepsilon) & [0.25\% < \varepsilon < x\%\ FS] \\ \sigma_{pl} = f_{nl_2}(\varepsilon) + \sigma_{elpl_{x\%FS}} & [\varepsilon > x\%\ FS] \\ f_{nl1}(\varepsilon)\ \&\ f_{nl2}(\varepsilon) = any\ mathematical\ function, FS = fracture\ strain & \end{array}\right\} \quad (5)$$

Equation (5) shows the formulas used to predict three parts of the stress–strain curve, i.e., linear part, onset of bending and offset of bending, as illustrated by Figure 1. $E_{a_{0.25\%\varepsilon}}$ is the value of the apparent elastic modulus at 0.25% strain. The first part of the stress is referred to as σ_e, which indicates the elastic part of the stress–strain curve. The second part is σ_{elpl}, which refers to the onset of the bend of the stress–strain curve, which starts from 0.25%ε until x% of the fracture strain. This x value has been estimated by trial and error. The initial value has been estimated by plotting the Hollomon equation [40] on the graph of true stress and plastic strain. The strain values, from where the Hollomon equation fits linearly to the plastic strain is considered as the first initial estimate for the value of x. The last part of the stress–strain curve is the offset of the bend, which is denoted by σ_{pl}. The non-linear equation/s iterate from the strain values at x% of the fracture strain until fracture strain. $\sigma_{elpl_{x\%FS}}$ is the stress value at x% of fracture strain (FS), which is the initial value for the iteration. In this analytical model, the non-linear function f_{nl} is described by two separate mathematical functions for the onset and offset of the bend, that is f_{nl_1} and f_{nl_2}, respectively. Equation (5) was used to fit all experimental data, taking strain as an independent variable. For the non-linear function, a set of or single mathematical equations are used that can imitate the curvature nature of the onset and offset of the bend in the stress–strain curve. A curve-fitting method from python coding language was used to determine the best-fitting parameters for the mathematical functions mentioned in Equation (5).

To add temperature as an additional independent variable, the value of the fracture strain at the particular temperature is required. Similarly, the fitting parameters of all mathematical functions must be predicted. Hence, another code in python language was written to predict fracture strain and fitting parameters. The algorithm of this code is described in the next Section 2.5.

2.5. Analytical Model to Predict Fracture Strain and Fitting Parameters

To insert additional variables such as temperature, fiber content, fiber orientation, and thickness in Equation (5), fracture strain must be predicted first. Then, the fitting parameters for function f_{nl1} and f_{nl2} in Equation (5) can be predicted. The value of the fracture strain and the fitting parameters are predicted by using the algorithm shown in Figure 5. An analytical model for the storage modulus was developed based on the same test matrix as mentioned in Table 1 [33].

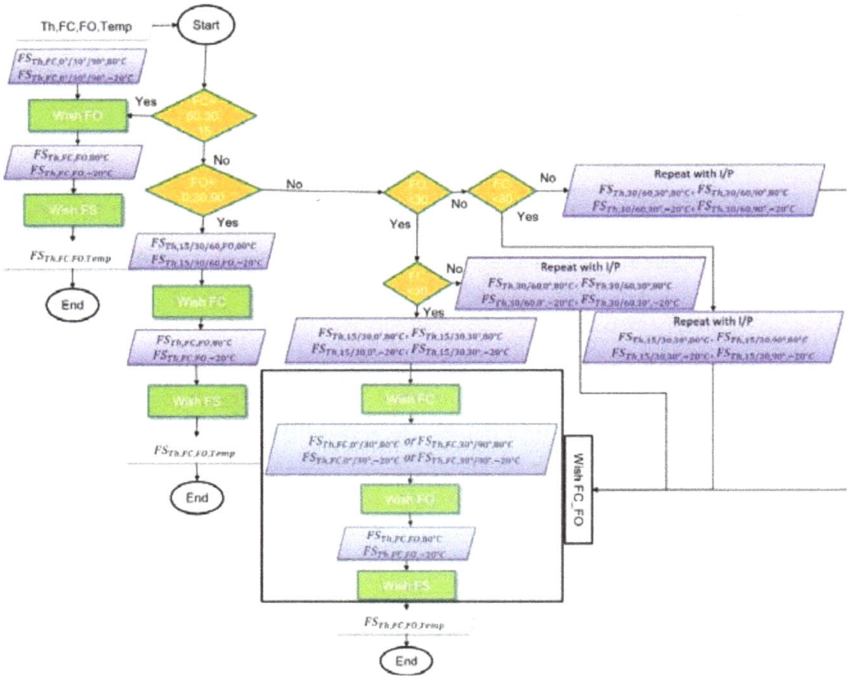

Figure 5. Algorithm used to predict the fracture strain (FS) for each material condition, Th: thickness of the specimen, FO: fiber orientation, FC: fiber content, Temp: temperature, Wish FO, Wish FC, and Wish FS are analytical models to predict values.

This algorithm as shown in Figure 5 is followed by the code to predict the fracture strain. There are three different models named Wish FO, Wish FC, and Wish FS. As the names says, these models provide the values of the fracture strain or any fitting parameter for the mathematical functions at an arbitrary temperature, fiber orientation or fiber content. From model Wish FS, the value of the fracture strain is predicted for different temperatures, whereas from Wish FO, the fracture strain is predicted for different fiber orientations. Similarly, Wish FC predicts the value of the fracture strain for different fiber contents. The workflow or procedure to predict the values from these models is described in the next paragraph.

Wish FS creates a normalized curve of the storage modulus with respect to temperature. Experimental fracture strain values for −20 °C and 80 °C are projected on this normalized curve. Hence, the normalized value for fracture strain at any temperature can be calculated by projecting on this normalized curve. The experimental fracture strain of room temperature has been excluded. Room temperature is in the range of 17 to 27 °C. The exact temperature was not recorded. The glass transition temperature for PA46 lies in this range. On the other hand, the storage modulus model is sensitive to temperature [33]. The more accurate the temperature values, the more accurate the prediction of the fracture strain. The experimental fracture strain of any two values of the fiber content or fiber orientation in the case of Wish FO will be projected on the normalized curve. Now the normalized value of fracture strain at any arbitrary fiber orientation or content can be calculated by mapping back the normalized value of the storage modulus at that particular fiber orientation or content.

The same algorithm will be run to calculate the fitting parameters for the mathematical functions. The algorithm in Figure 5 will again execute replacing the fracture strain with the required fitting parameter. For the 4P model with four independent variables, the algorithm will be executed separately for all fitting parameters of the mathematical functions for each iteration. Repetition and a combination of executing these three models (Wish FS, Wish FC, and Wish FO) can predict the fracture strain and fitting parameters at any arbitrary fiber orientation, fiber content, temperature, and thickness.

3. Comparison of Analytical Models with Experimental Data

Equation (5) is used to predict the stress–strain curve. First, strain is taken as an independent variable and the curve is predicted with the help of curve fitting by the least square method. The standard error of regression is calculated. This model is called the 1P model. Later, all other additional independent variables are inserted with the help of the algorithm mentioned in Figure 5. This model is called the 4P model. The value of the fracture strain will instruct the iteration code of the 4P model to stop iterating the values of stress at a certain strain value i.e., fracture strain. Apart from the fracture strain, the fitting parameters of all mathematical functions for any arbitrary material condition are also required. The same algorithm mentioned in Figure 5 is used to predict these values as well.

3.1. 1P Model with One Independent Variable

The 1P model predicts the stress–strain curve directly by fitting to experimental data. The fracture strain value of each model curve is taken from experimental data. Hence, the standard error of regression is quite low as shown in Figure 6. The standard error of regression of the model curves (red) is less than 2% of the stress range. To predict the stress–strain curve for any temperature between −20 °C and 80 °C, the 4P model is used as described in the next section.

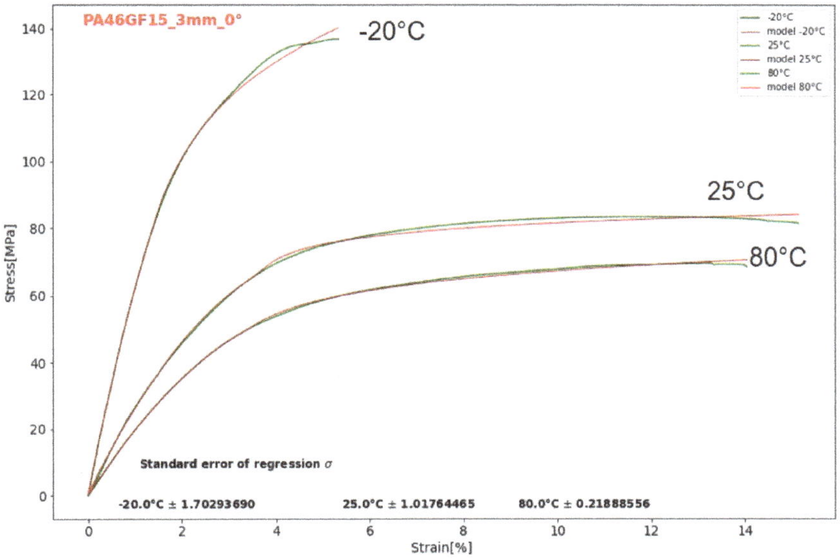

Figure 6. 1P stress–strain model with strain as an independent variable $\sigma = f(\epsilon)$. Green lines are experimental data and red lines are model data (1P model). The room temperature test is considered as 25 °C.

3.2. 4P Model with Four Independent Variables

The fracture strain and other fitting parameters of mathematical functions are predicted based on the procedure and concept described in the algorithm mentioned in Figure 5. Some of the parameters are predicted by the storage modulus model [33], for example, the stiffness of the material. The stiffness of PA46GF varies with temperature. Hence, the storage modulus model creates variation in the stiffness according to the viscous behavior.

Figure 7 shows verification of the 4P model with experimental data. Here, green curves are experimental stress strain curves whereas red curves are the 1P model, and colorful dotted lines are the 4P model. It shows that there is a cluster of curves near −20 °C and 80 °C. This verifies that the stress–strain model is working according to the viscous behavior of the material. For 50% RH conditioned PA46, the glass transition region lies between 15° and 40 °C. The variation in the stiffness below and above the glass transition temperature is less compared to the glass-transition temperature region. Therefore, the stress curves at temperatures ranging from 10 to 40 °C are uniformly spread out. Hence, incorporating the storage modulus model to predict the E modulus is a very important part of this analytical model.

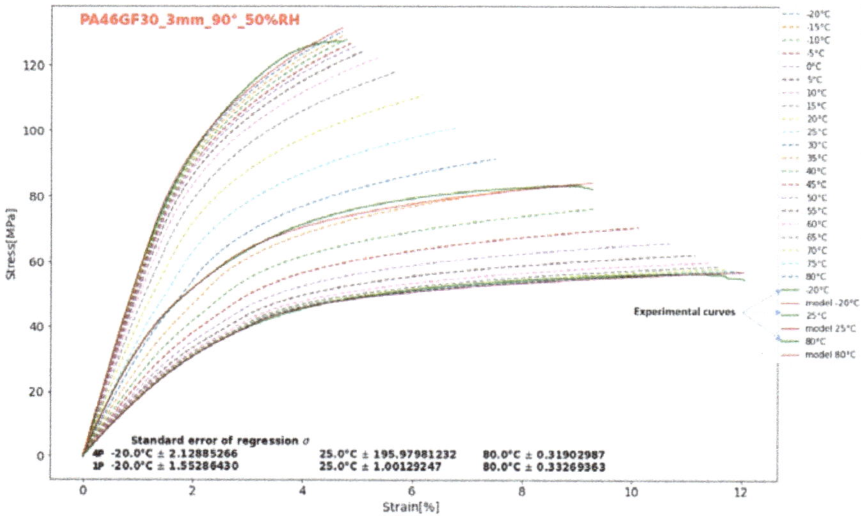

Figure 7. 4P stress–strain model, which provides the stress–strain graphs at any arbitrary fiber content, fiber orientation, and temperature $\sigma = f(\varepsilon, temp, fc, fo)$. Results for PA46GF30, 3-mm-thickness, and 90° orientation are displayed as an example. Green curves are experimental curves, red lines are model data with the 1P model, and dotted colorful data are 4P model with four independent parameters.

In the glass-transition region, the stress values are highly sensitive to the temperature as Young's modulus varies strongly with temperature. Room temperature lies in this region. However, the exact temperature of the laboratory was not measured. Hence, the standard error of regression of the 4P model at room temperature, which is assumed to be 25 °C, shows abnormally high values. The error has been accumulated from the 1P model to the 4P model due to four occurrences of the prediction of the fitting parameters and fracture strain. Some non-physical behavior of the curves is noticed at the higher temperature. For example, at the junction of two mathematical functions, there is a sudden drop of the curve, which is mathematically correct but physically incorrect with respect to the stress behavior of the material. Excessive bending of the stress–strain curve could be mathematically true based on the equations used but physically incorrect with respect to the material behavior. To reduce such non-physical behavior of the mathematical functions, some check functions were inserted in the 4P model. However, the other stress–strain curves for −20 °C and 80 °C show standard error of regression less than 2% of the stress ranges. The same model has been verified with experimental data for 2-mm-thick specimens and with the experimental data for dry specimens (room temperature condition).

4. Discussion and Conclusions

The paper proposed an analytical model that predicts the stress–strain curve for injection-molded SFRP considering five extra independent variables, namely, the fiber content, fiber orientation, temperature, humidity, and thickness of the specimen. The empirical formulas accommodate synergetic effects of each parameter by incorporating fiber properties with the help of μCT analysis and polymer properties with the help of DMTA analysis. This is the reason behind achieving an accuracy of 2% of the stress range in the prediction of mechanical properties. Mechanical characterization shows that the mechanical properties of the SFRP are strongly dependent on viscous behavior of the polymer. At temperatures higher than glass temperature, the polymer is in a rubbery phase. Hence, SFRP shows ductile behavior in the stress–strain curve whereas at temperatures lower than the glass transition temperature, it shows brittle behavior. In the glass transition region, the storage modulus varies almost linearly, and the stress–strain curves are equally spread out.

This model is developed for injection-molded short glass fiber-reinforced polyamide PA46. The following assumptions are considered:

- The stiffness of the material is predicted through the storage modulus. The frequency used for DMTA analysis is 10 Hz. It is assumed that the DMTA test at this frequency can be similar to the quasi-static tensile test.
- The variation of the stiffness of the material with respect to fiber orientation is assumed to be linear due to a lack of experimental points.
- It is assumed that the fracture strain and fitting parameters of all mathematical functions follow the same storage modulus curve from DMTA [33].

Further research can be done to determine the relationship of the frequency of the DMTA test to the tensile test. In this way, another model can also be designed to predict the fatigue behavior of SFRP. It would be interesting to develop a similar model for continuous fiber-reinforced composites and other matrices. The approach of the model will remain the same if the matrix material changes except for some modifications in the mathematical functions. Similarly, an analytical model for other loading conditions can also be developed, for example, compression or shear loading. Since this model has thickness as an independent variable, layer-wise mechanical properties for injection-molded specimens can also be extracted [6].

In engineering design, the stiffness, strength, yield strength, and ultimate strain are more important than the fracture strain. In most of the cases, a component is designed until the yield strain of the material with a factor of safety. Hence, this model provides these values with high accuracy, which is suitable for component development. This model can be inserted in FEA. In-situ material data provided from this analytical model according to the change in the testing environment can be mapped in each finite element. This will help the simulation engineer to design economical and reliable SFRP parts.

Author Contributions: E.: Conceptualization, experiment & methodology, programming, investigation and writing; J.H.: funding acquisition, conceptualization, supervision, review and editing. All authors have read and agreed to the published version of the manuscript.

Funding: The financial support for this study was provided by the Forschungsvereinigung Antriebstechnik e.V. (funding reference FVA 869 I).

Conflicts of Interest: The authors declare no conflict of interest. The funders had no role in the writing of the manuscript. The funders have accepted to publish the paper.

References

1. Gohari, S.; Sharifi, S.; Burvill, C.; Mouloodi, S.; Izadifar, M.; Thissen, P. Localized failure analysis of internally pressurized laminated ellipsoidal woven GFRP composite domes: Analytical, numerical, and experimental studies. *Arch. Civ. Mech. Eng.* **2019**, *19*, 1235–1250. [CrossRef]
2. Rief, T.; Schmidt, S. Microstructural Analysis of Fiber-Reinforced Injection Molding Material and Its Influence on Mechanical Properties. Available online: https://www.researchgate.net/publication/357281009_Microstructural_Analysis_of_Fiber_Reinforced_Injection_Molding_Material_and_its_Influence_on_Mechanical_Properties (accessed on 30 March 2022).
3. Wang, Y.; Wang, H.; Wei, J.; Lin, B.; Xu, J.; Fang, S. Finite element analysis of grinding process of long fiber reinforced ceramic matrix woven composites: Modeling, experimental verification and material removal mechanism. *Ceram. Int.* **2019**, *45*, 15920–15927. [CrossRef]
4. Gohari, S.; Mozafari, F.; Moslemi, N.; Mouloodi, S.; Sharifi, S.; Rahmanpanah, H.; Burvill, C. Analytical solution of the electro-mechanical flexural coupling between piezoelectric actuators and flexible-spring boundary structure in smart composite plates. *Arch. Civ. Mech. Eng.* **2021**, *21*, 33. [CrossRef]
5. Jones, R.M. *Mechanics of Composite Materials*, 2nd ed.; Taylor & Francis: Philadelphia, PA, USA, 1999; ISBN 978-1-56032-712-7.
6. Hausmann, J.E.; Schmidt, S.; Krummenacker, J. Transfer of mechanical properties from specimens to injection molded structures under consideration of local fiber orientation. In Proceedings of the European Congress and Exhibition on Advanced Materials and Processes-Euromat, Graz, Austria, 12–16 September 2021.
7. Ramberg, W.; Osgood, W.R. *Description of Stress-Strain Curve by Three Parameters*; National Advisory Committee for Aeronautics: Washington, DC, USA, 1943; p. 29.
8. Hill, H.N. *Determination of stress-Strain Relations from "OFFSET" Yield Strength Values*; National Advisory Committee for Aeronautics: Washington, DC, USA, 1944; p. 11.
9. Kamaya, M. Ramberg–Osgood type stress–strain curve estimation using yield and ultimate strengths for failure assessments. *Int. J. Press. Vessels Pip.* **2016**, *137*, 1–12. [CrossRef]
10. Rasmussen, K.J.R. Full-range stress–strain curves for stainless steel alloys. *J. Constr. Steel Res.* **2003**, *59*, 47–61. [CrossRef]
11. Stephens, R.I.; Fuchs, H.O. (Eds.) *Metal Fatigue in Engineering*, 2nd ed.; Wiley: New York, NY, USA, 2001; ISBN 978-0-471-51059-8.
12. Gadamchetty, G.; Pandey, A.; Gawture, M. On Practical Implementation of the Ramberg-Osgood Model for FE Simulation. *SAE Int. J. Mater. Manuf.* **2016**, *9*, 200–205. [CrossRef]
13. Probabilistic Model Code, Part-3: Resistance Model. 2006. Available online: https://www.jcss-lc.org/publications/jcsspmc/concrete.pdf (accessed on 30 March 2022).
14. Gruber, G.; Haimerl, A.; Wartzack, S. Consideration of orientation properties of Short Fiber Reinforced Polymers within Early Design Steps. *FEA Inf. Eng. J.* **2013**, *2*, 2167–2173.
15. Bauer, C. *Charakterisierung und Numerische Beschreibung des Nichtlinearen Werkstoff- und Lebensdauerverhaltens eines kurzglasfaserverstärkten Polymerwerkstoffes unter Berücksichtigung der im μCT Gemessenen Lokalen Faserorientierung*; IVW-Schriftenreihe; Als Manuskript gedruckt; Institut für Verbundwerkstoffe GmbH: Kaiserslautern, Germany, 2017; ISBN 978-3-944440-22-4.
16. Schöpfer, J.; Becker, F.; Maier, M.; Kolling, S. *Charakterisierung und Modellierung von Kurzfaserverstärkten Kunststoffen Teil 2: Simulationsmethoden mit LS-DYNA*; 9th LS-DYNA Forum; Material IV-Faserverstärkte Kunststoffe: Bamberg, Germany, 2010.
17. Nutini, M.; Vitali, M. Simulating Anisotropy with Ls-Dyna in Glass-Reinforced, Polypropylene-Based Components. 2010. Available online: https://www.dynamore.de/de/download/papers/forum10/papers/F-IV-01.pdf (accessed on 30 March 2022).
18. Dean, A.; Grbic, N.; Rolfes, R.; Behrens, B. Macro-mechanical Modeling and Experimental Validation of Anisotropic, Pressure- and Temperature-dependent Behavior of Short Fiber Composites. *Compos. Struct.* **2019**, *211*, 630–643. [CrossRef]
19. Dean, A.; Reinoso, J.; Sahraee, S.; Rolfes, R. An invariant-based anisotropic material model for short fiber-reinforced thermoplastics: Coupled thermo-plastic formulation. *Compos. Part A Appl. Sci. Manuf.* **2016**, *90*, 186–199. [CrossRef]
20. De Monte, M.; Moosbrugger, E.; Quaresimin, M. Influence of temperature and thickness on the off-axis behaviour of short glass fibre reinforced polyamide 6.6—Quasi-static loading. *Compos. Part Appl. Sci. Manuf.* **2010**, *41*, 859–871. [CrossRef]
21. Affdl, J.C.H.; Kardos, J.L. The Halpin-Tsai equations: A review. *Polym. Eng. Sci.* **1976**, *16*, 344–352. [CrossRef]
22. Price, C.D.; Hine, P.J.; Whiteside, B.; Cunha, A.M.; Ward, I.M. Modelling the elastic and thermoelastic properties of short fibre composites with anisotropic phases. *Compos. Sci. Technol.* **2006**, *66*, 69–79. [CrossRef]
23. Kalaprasad, G.; Joseph, K.; Thomas, S.; Pavithran, C. Theoretical modelling of tensile properties of short sisal fibre-reinforced low-density polyethylene composites. *J. Mater. Sci.* **1997**, *32*, 4261–4267. [CrossRef]
24. Friedrich, K. Microstructural efficiency and fracture toughness of short fiber/thermoplastic matrix composites. *Compos. Sci. Technol.* **1985**, *22*, 43–74. [CrossRef]
25. Facca, A.G.; Kortschot, M.T.; Yan, N. Predicting the elastic modulus of natural fibre reinforced thermoplastics. *Compos. Part Appl. Sci. Manuf.* **2006**, *37*, 1660–1671. [CrossRef]
26. Islam, M.A.; Begum, K. Prediction Models for the Elastic Modulus of Fiber-reinforced Polymer Composites: An Analysis. *J. Sci. Res.* **2011**, *3*, 225–238. [CrossRef]
27. Zhai, Z.; Gröschel, C.; Drummer, D. Tensile behavior of quasi-unidirectional glass fiber/polypropylene composites at room and elevated temperatures. *Polym. Test.* **2016**, *54*, 126–133. [CrossRef]
28. Mori, T.; Tanaka, K. Average stress in matrix and average elastic energy of materials with misfitting inclusions. *Acta Metall.* **1973**, *21*, 571–574. [CrossRef]

29. Lopez, D.; Thuillier, S.; Grohens, Y. Prediction of elastic anisotropic thermo-dependent properties of discontinuous fiber-reinforced composites. *J. Compos. Mater.* **2020**, *54*, 1913–1923. [CrossRef]
30. Advani, S.G.; Tucker, C.L. The use of tensors to describe and predict fiber orientation in short fiber composites. *J. Rheol.* **1987**, *31*, 751–784. [CrossRef]
31. Bernasconi, A.; Cosmi, F.; Dreossi, D. Local anisotropy analysis of injection moulded fibre reinforced polymer composites. *Compos. Sci. Technol.* **2008**, *68*, 2574–2581. [CrossRef]
32. Cosmi, F.; Bernasconi, A.; Sodini, N. Phase contrast micro-tomography and morphological analysis of a short carbon fibre reinforced polyamide. *Compos. Sci. Technol.* **2011**, *71*, 23–30. [CrossRef]
33. Esha; Hausmann, J. Development of an analytical model to predict storage modulus with four independent parameters. In Proceedings of the SAMPE Europe Conference 2021: Innovation towards Perfection, Baden, Switzerland, 29–30 September 2021; p. 8.
34. Saghir, F. Mechanical characterization of particulated FRP composite pipes: A comprehensive experimental study. *Polym. Test.* **2021**, *93*, 107001. [CrossRef]
35. Pielichowski, K.; Njuguna, J. *Thermal Degradation of Polymeric Materials*; Rapra Technology: Shawbury and Billingham, UK, 2005; ISBN 1-85957-498-X.
36. Hatakeyama, T.; Quinn, F.X. *Thermal Analysis: Fundamentals and Applications to Polymer Science*, 2nd ed.; Wiley: Chichester, NY, USA, 1999; ISBN 978-0-471-98362-0.
37. Franck, A. Viscoelasticity and Dynamic Mechanical Testing. Available online: http://www.tainstruments.com/pdf/literature/AAN004_Viscoelasticity_and_DMA.pdf (accessed on 30 March 2022).
38. Dynamic Mechanical Analysis (DMA) A Beginner's Guide. Available online: https://resources.perkinelmer.com/corporate/cmsresources/images/44-74546gde_introductiontodma.pdf (accessed on 30 March 2022).
39. ISO 527-1 2012. ISO Plastic-Determination of Tensile Properties, Part 1: General Principles. 2012. Available online: https://infostore.saiglobal.com/preview/98698754403.pdf?sku=858558_SAIG_NSAI_NSAI_2042422 (accessed on 30 March 2022).
40. Hollomon, J. Tensile deformation. *Trans. Metall. Soc. AIME* **1945**, *162*, 268–290.

Journal of
Composites Science

Article

Thermal Decomposition Characteristics of PEO/$LiBF_4$/LAGP Composite Electrolytes

Jacob Denney and Hong Huang *

Department of Mechanical and Materials Engineering, Wright State University, Dayton, OH 45435, USA; denney.3@wright.edu
* Correspondence: hong.huang@wright.edu

Abstract: Lithium-based batteries with improved safety performance are highly desired. At present, most safety hazard is the consequence of the ignition and flammability of organic liquid electrolytes. Dry ceramic-polymer composite electrolytes are attractive for their merits of non-flammability, reduced gas release, and thermal stability, in addition to their mechanical strength and flexibility. We recently fabricated free-standing solid composite electrolytes made up of polyethylene oxide (PEO), $LiBF_4$ salt, and $Li_{1+x}Al_xGe_{2-x}(PO_4)_3$ (LAGP). This study is focused on analyzing the impacts of LAGP on the thermal decomposition characteristics in the series of PEO/$LiBF_4$/LAGP composite membranes. It is found that the appropriate amount of LAGP can (1) significantly reduce the organic solvent trapped in the polymer network and (2) increase the peak temperature corresponding to the thermal degradation of the PEO/$LiBF_4$ complex. In the presence of LAGP, although the peak temperature related to the degradation of free PEO is reduced, the portion of free PEO, as well as its decomposition rate, is effectively reduced, resulting in slower gas release.

Keywords: lithium; LAGP; electrolyte; composite; thermal decomposition

Citation: Denney, J.; Huang, H. Thermal Decomposition Characteristics of PEO/$LiBF_4$/LAGP Composite Electrolytes. *J. Compos. Sci.* **2022**, *6*, 117. https://doi.org/10.3390/jcs6040117

Academic Editor: Stelios K. Georgantzinos

Received: 20 March 2022
Accepted: 11 April 2022
Published: 14 April 2022

1. Introduction

Lithium-ion batteries (LIBs), because of their high energy density and good discharge/charge cycle life, have evolved and dominated in markets from powering electronic devices and electric vehicles to energy storage in renewable grid systems. In these broad applications, occasional firing or explosive accidents resulting from LIBs have raised serious concerns. At present, battery researchers and manufacturers are dedicating to develop lithium-based batteries with improved safety performance, in addition to enhancing their power specs [1–7]. External protections, with the help of electronic accessories, like thermal fuse, pressure vents, and PTC element, have been integrated in LIB batteries to mitigate thermal runaway. However, internal protection based on the choice of safe materials is believed to be the "ultimate" solution to the safety issue. Recently, many key battery components have been developed towards minimizing thermal runaway, including surface-modified electrodes, non-flammable electrolytes, multifunctional separators, overcharging or flame-retardant additives, and thermally switchable current collectors.

Solid electrolytes, referring to ion-conducting ceramics or dry polymers, as well as polymer–ceramic composites, have many merits and are known for non-volatility, non-flammability, and high thermal decomposition temperatures, etc., to address the safety issues [8,9]. Composite polymer electrolytes (CPEs), which combine characteristics of polymer and ceramic electrolytes, with potential to overcome the drawbacks of each kind, are becoming more attractive [10–13]. For instance, the unique mechanical properties of CPEs provide enough strength to resist lithium dendrite penetration and meantime flexibility to ensure good interface contact. All-solid-state LIBs made up of CPEs are becoming more attractive and are recognized as a key technology for next-generation energy-storage systems.

Research on the thermal stability and safety of LIBs made up of dry CPEs is still in the infancy stage [14]. Extensive studies on CPEs had emphasized on increasing ionic conductivity, improving mechanical flexibility, and enhancing interfacial stability [15–23]. In the published reports, most thermal characteristics of CPEs presented are limited to temperatures below 100 °C, within which the glass transition temperature (Tg) and crystallization temperature (Tc) are known to be critical to ionic conductivities. In some papers, thermalgravimetric decomposition profiles up to 500 °C were presented, while in-depth analyses on the decomposition behaviors of CPEs are lack of discussion. Among sparse reports on thermal stabilities of PEO (polyethylene oxide)-based electrolytes, inconsistences or controversial conclusions are noted. Xia and Angulakshimi [24,25] observed that the addition of LiTFSI to PEO increased thermal decomposition temperature (T_{dec}) of PEO from 200 °C to 300 °C. However, Joost [26] found that T_{dec} of PEO was lowered in PEO/LiTFSI and that the decomposition has multiple stages. Similarly to CPE systems, Cheng [27] found that adding 30% LLZO increased the T_{dec} of PEO, while Piana [28] observed an opposite trend in the PEO/LAGP-based electrolytes.

Among various CPE formulations, active ceramic fillers with high ionic conductivities are favorable, especially when a high content of ceramic fillers is required. $Li_{1+x}Al_xGe_{2-x}(PO_4)_3$ (LAGP) is known for its high room-temperature ionic conductivity (10^{-3}–10^{-4} S/cm) and good stability with moisture. Furthermore, it is risk-free of thermal runaway and subsequent fires and allows high-temperature operation [29–32]. Researchers have studied the electrochemical characteristics of PEO/LiTFSI/LAGP systems and demonstrated their satisfactory cycling and rate performances in lithium-ion, as well as lithium batteries [33–36]. In our previous publications, we have reported the method of fabricating free-standing lithiated PEO with LAGP solid composite electrolytes and investigated the impacts of lithium salt and LAGP loading on electrical and mechanical properties [37,38]. To our knowledge, there is sparse in-depth research on thermal stability and safety matters in the PEO/Li-salt/LAGP electrolyte systems, especially with the composition of high LAGP loadings. This study is focused on understanding the impact of LAGP ceramic loadings on thermal decomposition characteristics of PEO and $LiBF_4$ in the PEO/$LiBF_4$/LAGP composite systems. $LiBF_4$ salt was selected in consideration of its favorable conducting merits in liquid electrolytes and its thermal properties fundamentally distinguished from LiTFSI. We performed in-depth quantitative analyses based on the thermogravimetric results, which will be presented in this paper.

2. Materials and Experimental

A series of free-standing membranes from pure PEO, PEO/$LiBF_4$, to lithiated PEO with different amounts of $Li_{1.4}Al_{0.4}Ge_{1.6}(PO_4)_3$ (LAGP) were fabricated and analyzed in this study. PEO at a molecular weight of 400,000, anhydrous $LiBF_4$, and anhydrous acetonitrile (AN) are all purchased from Sigma-Aldrich without further treatment. LAGP powders were synthesized via a two-step solid reaction method. The mixture of precursors was firstly heated at 600 °C for 6 h in air at a rate of 1 °C/min. The resultant powders were then subjected to milling in a high-energy shaker mill (SPEX Sample Prep 8000M Mill, Metuchen, NJ, USA) for 3 h before firing at 900 °C at a rate of 2 °C/min for 24 h in air. The as-synthesized LAGP powders were further milled for 24 h until the average particle size was in the sub-micron range.

Acetonitrile is the solvent used to dissolve PEO and $LiBF_4$. LAGP, $LiBF_4$, and PEO with predetermined compositions were mixed in AN at 50 °C in a dry room with a controlled moisture level. Afterwards, the viscous solution was transferred into a glove box (with moisture less than 5 ppm). Casting and drying were executed in the glove box at a pressure of 1 kPa followed by a constant argon flow at room temperature for up to five days. All of the membranes were sealed and stored in the glove-box prior to usage. Detailed synthesis process, structures, morphologies, and ionic conductivities of the electrolyte membranes were reported elsewhere [37].

In all of the electrolyte membranes, the molar ratio of EO to Li is fixed at 8.0 which was reported to have highest ionic conductivity [39]. This value corresponds to the mass ratio of 3.75 between PEO and LiBF4, i.e., $\frac{m_{PEO}}{m_{LiBF4}} = 3.75$. The composition of LAGP relative to (PEO + LAGP), i.e., $(\frac{m_{LAGP}}{m_{PEO}+m_{LAGP}})$, increases from 20 wt% to 60 wt%.

The nomenclature of the samples, the nominal weight percentage of each component and corresponding mass ratios are listed in Table 1.

Table 1. The nomenclature of the composite electrolyte membranes and the nominal composition of each component in the samples.

Sample Name	LAGP (wt%)	PEO (wt%)	$LiBF_4$ (wt%)	$\frac{m_{LAGP}}{m_{PEO}+m_{LAGP}}$	$\frac{m_{PEO}}{m_{LiBF4}}$
PEO	0	100	0		
PEO/$LiBF_4$	0	78.95	21.05		3.75
LAGP20	16.48	65.93	17.58	0.20	3.75
LAGP30	25.28	58.99	15.73	0.30	3.75
LAGP40	34.48	51.72	13.79	0.40	3.75
LAGP50	44.12	44.12	11.76	0.50	3.75
LAGP60	54.22	36.14	9.63	0.60	3.75

All of the samples were analyzed using the TA Instrument TGA Q5000. A few preliminary tests were performed to determine the appropriate temperature range. Excluding the LAGP portion, all other components completely decomposed below 450 °C. Hence the systematic TGA tests were set from room temperature to 450 °C. The heating rate was varied from 5 to 20 °C/min. At different heating rate, the characteristics of the TGA profile did not alter except the temperatures of weight loss uniformly shifted with the heating rate, which is common in thermal analyses. Hence, a high heating rate of 20 °C/min was selected in this study to align with the rapid thermal runaway process. The samples weighted between 15 mg and 20 mg were loaded into open platinum pans. The purge gas was nitrogen. The sample flow rate was 25 mL/min and the balance purge gas ran at a flow rate of 10 mL/min. TA Trios software was used to calculate the first derivative thermogravimetric data (DTG). The weight loss was determined from TGA profiles. The onset temperature, (T_{onset}), the peak temperature (T_p), and the max weight loss rate within each thermal event were determined from the DTG profiles.

3. Results and Discussion

The TGA and DTG profiles of the PEO/$LiBF_4$/LAGP series are shown in Figure 1a,b. For better comparison and visualization, the TGA plots superimpose all together while the DTG profiles are plotted with a constant offset on the y-axis. Three distinguishable thermal events are observed from the electrolyte membranes. An initial small amount of weight loss occurs in the temperature range of 60–160 °C. Major weight loss onsets in the vicinity of 230 °C for all the electrolyte membranes independent of LAGP content. The thermal decomposition completes at 430 °C or lower. Within 230 °C and 430 °C, the electrolyte membranes exhibit two distinct thermal events with a different peak temperature and max decomposition rate. When the temperature exceeds 430 °C, the weight remains constant, showing a long plateau. The amount of weight loss, onset and peak temperatures, and the weight loss rate appear to vary with membrane compositions, which will be analyzed and discussed in the following section.

Figure 2 plots the first event weight loss (below 160 °C) as a function of LAGP amount. With increasing LAGP amount in the CPE membranes, the weight loss within this temperature range shows a reducing trend. Commercial pure $LiBF_4$ salt with no special treatment is known to contain traces of HF acid, which is inevitable upon exposure to moisture in production and storage. According to Lu [40], about 2.1 wt% weight loss peaked at 73 °C was observed in TGA from pure anhydrous $LiBF_4$, resulting from the free HF acid removal. Based on the composition of $LiBF_4$ in our electrolyte membranes, the maximum HF acid amount is estimated to be less 0.4 wt%. However, seen in Figure 2, the PEO/$LiBF_4$ membrane has a total weight loss of 12.8 wt% below 160 °C. In the LAGP60 membrane, the weight loss decreases to 4.5 wt% but still more than ten times the estimated HF amount. Since all our membranes were fabricated in a dry room and stored in a glove box with moisture less than 5 ppm, this weight loss cannot be related to the removal of free HF or any uptake moisture.

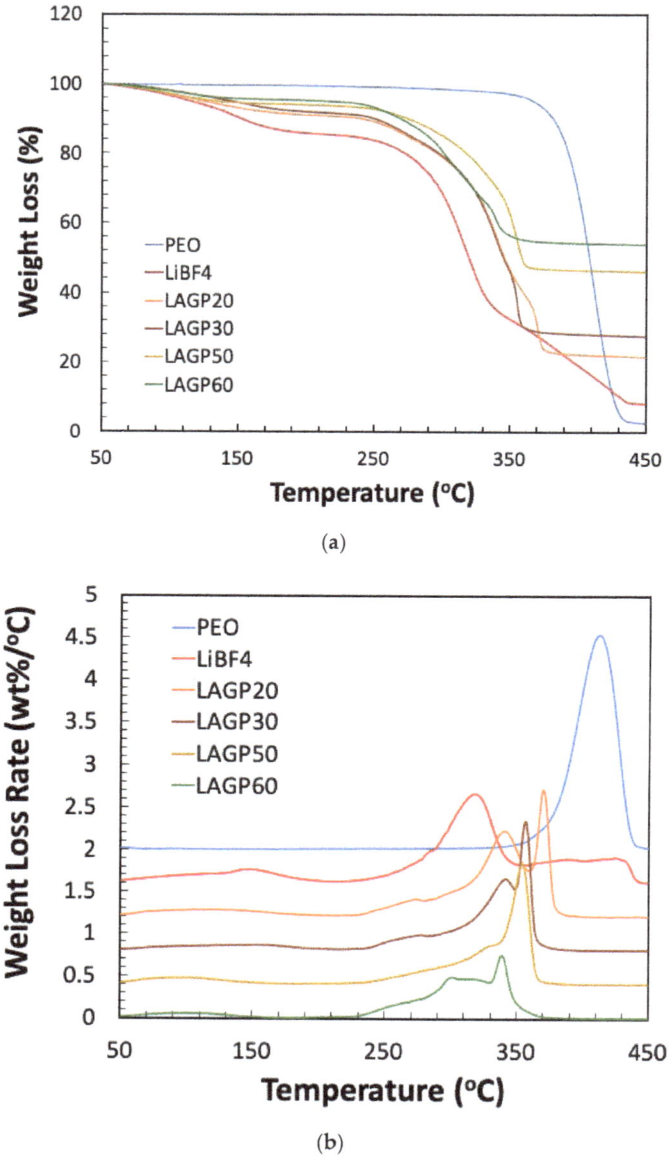

Figure 1. (**a**) TGA and (**b**) DTG profiles (offset by 0.4 interval for better visualization), obtained from the series of PEO/$LiBF_4$/LAGP composite electrolyte membranes. For comparison, pure PEO plots are included.

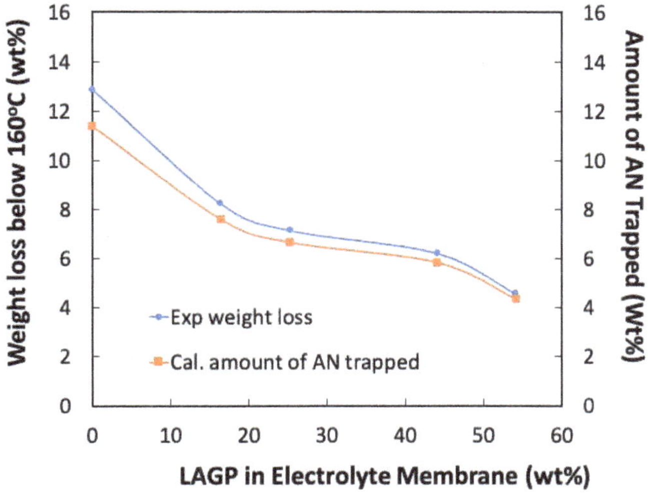

Figure 2. The experimental weight loss below 160 °C and the calculated amount of AN solvent trapped in the electrolyte membranes as a function of LAGP content.

In fabricating polymer electrolyte membranes, acetonitrile solvent was used to dissolve the PEO and $LiBF_4$ salt. Due to the strong interaction between $LiBF_4$ and acetonitrile, a certain amount of AN solvent could be trapped within the electrolyte membranes after the drying process. This phenomenon was reported previously [41,42]. The boiling point of acetonitrile is 82 °C. When it is associated with a lithium salt, the evaporation temperature will increase but can be completely evaporated under 160 °C [43]. Accordingly, we submitted that the weight loss between 60 °C and 160 °C observed in the electrolyte membranes corresponds to the amount of AN trapped. Based on the membrane composition and the weight loss, the amount of AN trapped is calculated and also plotted as a function of LAGP loading (see Figure 2). In the neat PEO membrane, there is only 0.35 wt% AN solvent trapped. By contrast, 11.4 wt% AN is trapped in the PEO/$LiBF_4$ electrolyte membrane. Interestingly, the addition of ceramic LAGP appears to weaken the $LiBF_4$-AN interaction, leading to the gradually reduced amount of AN in the membranes. The impact of LAGP is also reflected by the slightly decreased evaporation rate and inflection point as a function of LAGP content (see Figure 1b).

The major weight loss seen between 230 °C to 430 °C is the consequence of the thermal decomposition of $LiBF_4$ and PEO, as LAGP is known to be inactive to either $LiBF_4$ or PEO and remains thermally stable below 450 °C. Pure $LiBF_4$ powder or PEO membrane completes thermal decomposition in one event peaked at 277 °C and 417 °C, respectively, consistent with previous reports [40,44,45]. Seen in Figure 1, all the electrolyte membranes exhibit two distinct thermal events, with different weight-loss slopes (see Figure 1a) and two rate peaks (see Figure 1b), in this temperature range. Hereby, we refer to the two thermal events as stage-1 and stage-2 decompositions of the electrolyte membrane. The transition temperature ($T_{1\to2}$) from stage-1 to stage-2 is selected at the saddle point between the two weight-loss rate peaks. Stage-1 regime (230 °C to $T_{1\to2}$) is related with decomposition of $LiBF_4$, while stage-2 regime ($T_{1\to2}$ to 430 °C) is the characteristic of free PEO decomposition.

Figure 3a plots the weight loss at each stage, i.e., M_1 for stage-1 and M_2 for stage-2, as a function of the LAGP content. It is interesting to see that M_1 is more than the nominal amount of Li-salt, suggesting that a portion of PEO are complexed with $LiBF_4$ and degraded at the lower temperature. Considering the M_1 value is the sum of all the $LiBF_4$ (M_{LiBF4}) and complexed PEO portion ($M_{PEO,c}$) in the electrolyte, i.e., $M_1 = M_{LiBF4} + M_{PEO,c}$, the mass ratio of $\frac{M_1 - M_{LiBF4}}{M_{LiBF4}}$ will reflect the mass ratio between PEO and $LiBF_4$ in the complex. Therefore, $\frac{M_1 - M_{LiBF4}}{M_{LiBF4}} = \frac{M_{PEO,c}}{M_{LiBF4}}$. On the other hand, the M_2 value corresponds to the remanent PEO portion that are free or has much weak interaction with Li-salt. In we define $M_2 = M_{PEO,f}$, then $\frac{M_1 - M_{LiBF4}}{M_2} = \frac{M_{PEO,c}}{M_{PEO,f}}$.

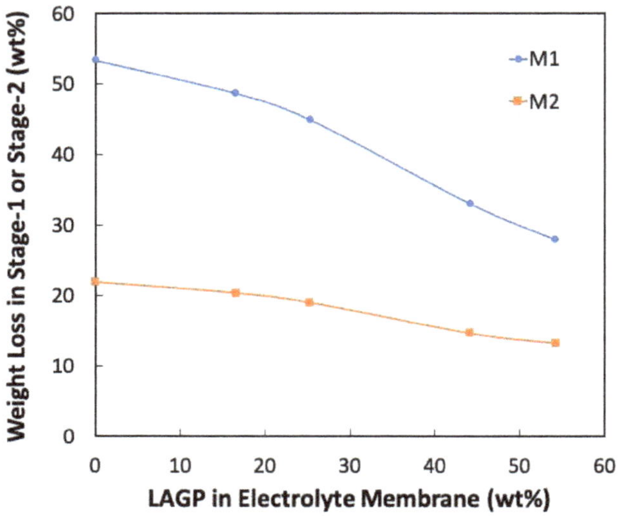

(**a**)

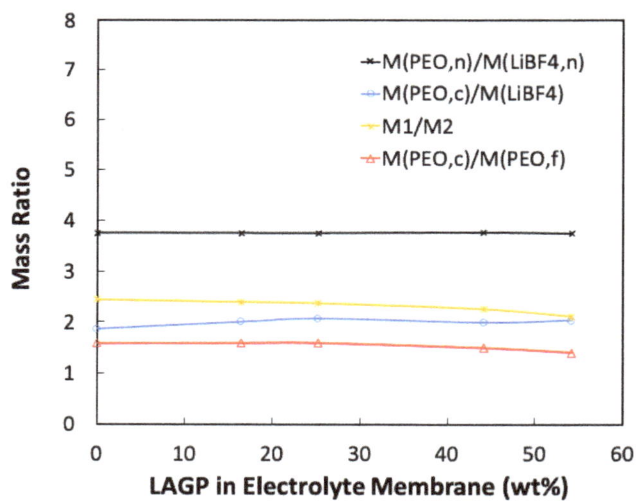

(**b**)

Figure 3. (**a**) Mass losses in stage-1 (M_1) and stage-2 (M_2) between 230 °C and 430 °C; (**b**) mass ratios, i.e., M_1/M_2, calculated $M_{PEO,c}/M_{LiBF4}$ in the complex, $M_{PEO,c}/M_{PEO,f}$, as well as nominal $M_{PEO,n}/M_{LiBF4,n}$ in the electrolytes membranes, in correlation with LAGP content.

Figure 3b show that the series of mass ratios, i.e., $\frac{M_1}{M_2}$, $\frac{M_{PEO,c}}{M_{LiBF4}}$, $\frac{M_{PEO,c}}{M_{PEO,f}}$, as a function of LAGP content in the membrane. For comparison, the nominal ratio between PEO and $LiBF_4$, $\frac{M_{PEO,n}}{M_{LiBF4,n}} = 3.75$, is also included in Figure 3b. Interestingly, all the mass ratios change insignificantly with LAGP content. From the values of $\frac{M_{PEO,c}}{M_{LiBF4}}$, it is determined that the EO/Li$^+$ molar ratio is in the range of 4.0–5.0 for all the electrolyte membranes. This observation corroborates well with the fact that PEO and lithium salt can readily form complexes with EO/Li ratio in the range of 4–6 [46,47]. The present results suggest that stable complexes of $(PEO)_nLiBF_4$ exist in all the PEO/$LiBF_4$/LAGP membranes and they are dominant in the thermal degradation within stage-1. Seen also in Figure 3b, $\frac{M_{PEO,c}}{M_{PEO,f}}$ changes insignificantly with LAGP content, indicating the portion between the complex PEO and free PEO is constant in all of the electrolyte membranes.

Figure 4 plots the peak decomposition temperatures related with the two thermal events. In the PEO/$LiBF_4$ membrane, the peak decomposition temperature of stage-1 ($T_{p,1}$) is at 318 °C, higher than pure $LiBF_4$ (277 °C), but much lower than free PEO (417 °C). The stage-2 decomposition at an almost constant rate is centered around 415 °C, characteristic of free PEO. Apparently, the "complexed" PEO appears less stable than "free" PEO, possibly catalyzed by $LiBF_4$. In the electrolyte membranes consisting of LAGP, $T_{p,1}$ increases to 345 °C in LAGP20, followed by a slightly decrease to 331 °C in LAGP50. $T_{p,1}$ recovers to 317 °C in LAGP60. This trend can be attributed to the impacts of excess LAGP on weakening interactions between PEO and $LiBF_4$. The presence of LAGP has a more significant impact on the free PEO decomposition at stage-2. The peak temperature of "free PEO" ($T_{p,2}$) is reduced to 370 °C in LAGP20 membrane and gradually recesses till to 339 °C in LAGP60 membrane. LAGP appears to accelerate the free PEO decomposition at reduced temperatures.

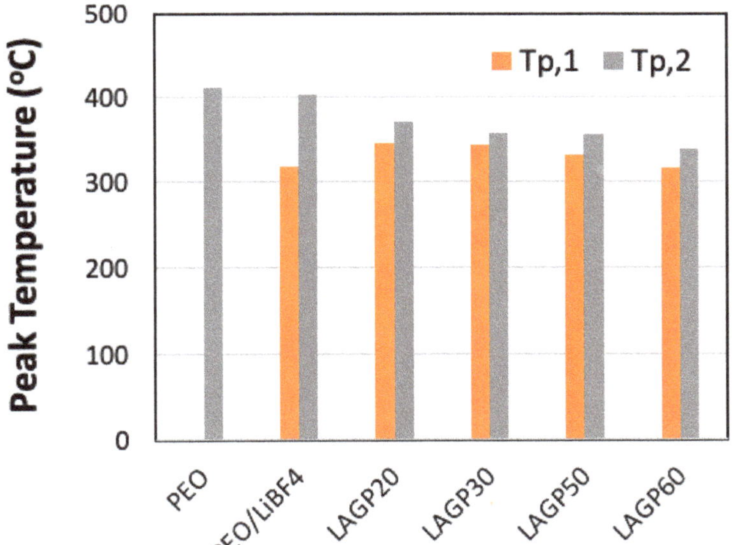

Figure 4. Two peak temperatures, $T_{p,1}$ and $T_{p,2}$, within the major decomposition occurred in all the membranes.

The 450 °C residue in all the electrolyte membranes should be made up of the nominal LAGP, LiF decomposed from $LiBF_4$, and PEO ashes. The actual amount of PEO ashes, calculated by subtracting the masses of LAGP and LiF from the residue, is plotted as a function of LAGP (see Figure 5). Neat PEO results in ash residue of 2.06 wt% at 450 °C. For comparison, the projected values, assuming the ash is proportional to the PEO amount, is also plotted in Figure 5. Results show that the actual PEO ash from the PEO/$LiBF_4$ electrolyte is only 1.02 wt%, about 60% of the projected value. The actual amount of PEO ashes continuously decreases upon increasing LAGP content and reaches almost zero in the LAGP60 membrane. This observation supports the findings discussed in the previous session that the presence of lithium-salt and LAGP ceramics catalytically accelerate PEO decomposition.

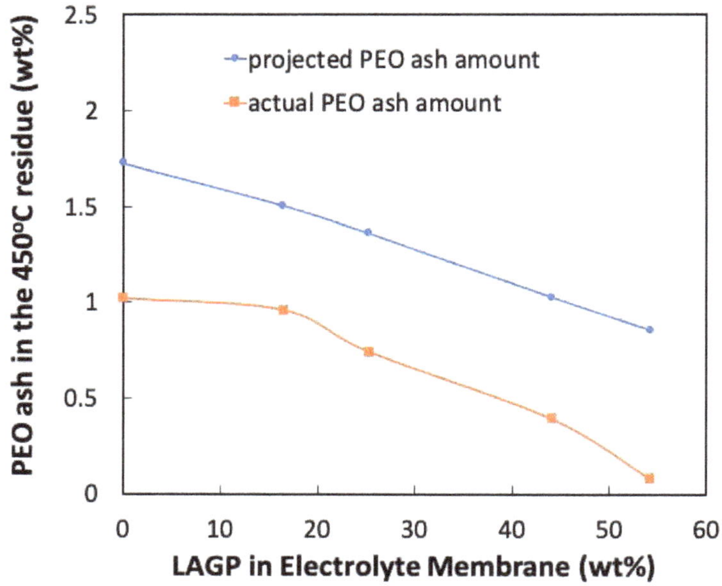

Figure 5. Projected and actual amount of PEO ashes in the 450 °C residue from all of the electrolyte membranes in relation with the nominal LAGP composition.

The impacts of LAGP on the thermal stability of the electrolyte membranes can be considered in the following aspects. Firstly, adding ceramics effectively reduces the active amount of polymer and lithium salt, and hence, the amount of gas release resulting from their decomposition [48]. This is readily observed from the TGA profiles. Seen in Figure 1a, the total released gas at 450 °C is 90 wt% for PEO/$LiBF_4$, 70 wt% for LAGP30, and only 30 wt% for LAGP60. Taking occurrence at 50% mass loss, the temperature increases from 320 °C for PEO/$LiBF_4$, to 345 °C for LAGP20 and 360 °C for LAGP50. Secondly, the major decomposition rate decreases and peak temperature increases in the presence of appropriate LAGP. Seen in Figure 1b, the maximum decomposition rate at the peak temperature is much lower in PEO/$LiBF_4$ compared with virgin PEO membrane. As LAGP composition increases, the maximum decomposition rate is continuously decreasing. For the stage-1 decomposition event, the peak rate is 1.0 wt% per °C occurring at 320 °C for PEO/$LiBF_4$, while the peak rate is reduced to 0.4 wt% per °C occurring at 336 °C for LAGP50. Thirdly, although the presence of LAGP reduces the peak decomposition temperature of the "free PEO" portion (about 40 wt% of total PEO composition), the presence of LAGP reduces the gas-release rate simultaneously. For instance, the stage-2 peak decomposition rate is 2.6 wt% per °C for the neat PEO membrane but this rate reduces to 1.3 wt% per °C for the LAGP50 electrolyte membrane. In general, a reduced decomposition rate can delay flame ignition and spread over the material [35]. We performed a flammability test on the PEO/$LiBF_4$/LAGP membranes. Neat PEO and PEO/$LiBF_4$ electrolyte membranes were easily ignited and quickly burned into ashes upon contact with flame. A significant reduction in flammability with self- extinguishing behavior was observed in the LAGP20 and LAGP30 membranes. When LAGP content is 50 wt% or greater, the composite electrolyte membranes could not be ignited even after repeated exposure and direct contact with the flame. This observation is consistent with Guo's results from the

PvDF-HFP/LiTFSI/LAGP system [35]. The thermally stable LAGP ceramic can act as a barrier capable of reducing/preventing ignition and flammability.

4. Conclusions

This study was intended to perform in-depth quantitative analyses on the series of free-standing PEO/$LiBF_4$/LAGP composite electrolyte membranes based on thermogravimetric results. Major thermal decomposition begins at 230 °C and completes at 430 °C, observed in all of these electrolyte membranes. It is found that the present membrane fabrication and drying process results in 11 wt% of AN solvent trapped in the PEO/$LiBF_4$ electrolyte membrane. The AN in the electrolyte membrane will not only lead to potential safety concerns but also affect electrochemical performance. The presence of LAGP can effectively reduce the amount of AN trapped in the membrane when drying at room temperature. The decomposition of PEO/$LiBF_4$ shows a maximum rate of 1.0 wt% per °C at 320 °C, much lower than neat PEO. The reduction can be attributed to the complex formation between PEO and Li-salt. Adding and increasing the amount of LAGP ceramic particles into the PEO/$LiBF_4$ electrolyte system, there are several benefits towards improving the thermal stability of the electrolyte membrane. Increasing the LAGP composition in the electrolyte membrane effectively reduces the net amount of flammable PEO and $LiBF_4$, as well as the solvent trapped in the electrolyte membrane. As a result, the net amount of gas release from the electrolyte membrane can be significantly reduced upon thermal decomposition. With an appropriate amount of LAGP (e.g., 20–30 wt%), the peak temperature reflecting decomposition of the PEO/$LiBF_4$ complex can be increased by several tens of degrees Celsius. Although the decomposition of free PEO portion in the electrolytes occurs at a lower temperature in the presence of LAGP, its peak decomposition rate is reduced, which is beneficial to delay flame ignition and spread and, therefore, to mediate the dramatic thermal runaway.

Author Contributions: Conceptualization, H.H.; methodology, J.D. and H.H.; formal analysis, H.H.; investigation, J.D. and H.H.; data curation, J.D. and H.H.; writing—original draft preparation, H.H.; writing—review and editing, H.H.; supervision, H.H. All authors have read and agreed to the published version of the manuscript.

Funding: This research received no external funding.

Conflicts of Interest: The authors declare no conflict of interest.

References

1. Balakrishnan, P.G.; Ramesh, R.; Kumar, T.P. Safety mechanisms in lithium-ion batteries. *J. Power Sources* **2006**, *155*, 401–414. [CrossRef]
2. Wen, J.; Yu, Y.; Chen, C. A Review on lithium-ion batteries safety issues: Existing problems and possible solutions. *Mater. Express* **2012**, *2*, 197–212. [CrossRef]
3. Wang, Q.; Ping, P.; Zhao, X.; Chu, G.; Sun, J.; Chen, C. Thermal runaway caused fire and explosion of lithium ion battery. *J. Power Sources* **2012**, *208*, 210–224. [CrossRef]
4. Feng, X.; Ouyang, M.; Liu, X.; Lu, L.; Xia, Y.; He, X. Thermal runaway mechanism of lithium ion battery for electric vehicles: A review. *Energy Storage Mater.* **2018**, *10*, 246–267. [CrossRef]
5. Liu, K.; Liu, Y.; Lin, D.; Pei, A.; Cui, Y. Materials for lithium-ion battery safety. *Sci. Adv.* **2018**, *4*, eaas9820. [CrossRef]
6. Feng, X.; Ren, D.; He, X.; Ouyang, M. Mitigating Thermal Runaway of Lithium-Ion Batteries. *Joule* **2020**, *4*, 743–770. [CrossRef]
7. Song, L.; Zheng, Y.; Xiao, Z.; Wang, C.; Long, T. Review on Thermal Runaway of Lithium-Ion Batteries for Electric Vehicles. *J. Electron. Mater.* **2022**, *51*, 30–46. [CrossRef]
8. Goodenough, J.B.; Singh, P. Review—Solid Electrolytes in Rechargeable Electrochemical Cells. *J. Electrochem. Soc.* **2015**, *162*, A2387–A2392. [CrossRef]
9. Manthiram, A.; Yu, X.; Wang, S. Lithium battery chemistries enabled by solid-state electrolytes. *Nat. Rev. Mater.* **2017**, *2*, 16103. [CrossRef]
10. Capuano, F.; Croce, F.; Scrosati, B. Composite polymer electrolytes. *J. Electrochem. Soc.* **1991**, *138*, 1918–1922. [CrossRef]
11. Agrawal, R.C.; Pandey, G.P. Solid polymer electrolytes: Materials designing and all-solid-state battery applications: An review. *J. Phys. D Appl. Phys.* **2008**, *41*, 3715–3725. [CrossRef]
12. Gao, Z.; Sun, H.; Fu, L. Promises, challenges, and recent progress of inorganic solid-state electrolytes for all-solid-state lithium batteries. *Adv. Mater.* **2018**, *30*, 1705702. [CrossRef]

13. Chen, L.; Li, Y.; Li, S.; Fan, L.; Nan, C.; Goodenough, J. PEO/garnet composite electrolytes for solid-state lithium batteries: From 'ceramic-in-polymer' to 'polymer-in-ceramic'. *Nano Energy* **2018**, *46*, 176–184. [CrossRef]
14. Wu, Y.; Wang, S.; Li, H.; Chen, L.; Wu, F. Progress in thermal stability of all-solid-state-Li-ion batteries. *InfoMat* **2021**, *3*, 827–853. [CrossRef]
15. Masoud, E.M.; El-Bellihi, A.-A.; Bayoumy, W.; Mousa, M. Organic–inorganic composite polymer electrolyte based on PEO–$LiClO_4$ and nano-Al_2O_3 filler for lithium polymer batteries: Dielectric and transport properties. *J. Alloy. Compd.* **2013**, *575*, 223–228. [CrossRef]
16. Choi, J.; Lee, C.-H.; Yu, J.-H.; Doh, C.-H.; Lee, S.-M. Enhancement of ionic conductivity of composite membranes for all-solid-state lithium rechargeable batteries incorporating tetragonal Li7La3Zr2O12 into a polyethylene oxide matrix. *J. Power Sources* **2015**, *274*, 458–463. [CrossRef]
17. Zhao, Y.; Huang, Z.; Chen, S.; Chen, B.; Yang, J.; Zhang, Q.; Ding, F.; Chen, Y.; Xu, X. A promising PEO/LAGP hybrid electrolyte prepared by a simple method for all-solid-state lithium batteries. *Solid State Ion.* **2016**, *295*, 65–71. [CrossRef]
18. Chen, B.; Huang, Z.; Chen, X.; Zhao, Y.; Xu, Q.; Long, P.; Chen, S.; Xu, X. A new composite solid electrolyte PEO/$Li_{10}GeP_2S_{12}$/SN for all-solid-state lithium battery. *Electrochim. Acta* **2016**, *210*, 905–914. [CrossRef]
19. Wang, W.; Yi, E.; Fici, A.J.; Laine, R.M.; Kieffer, J. Lithium ion conducting poly(ethylene oxide)-based solid electrolytes containing active or passive ceramic nanoparticles. *J. Phys. Chem. C* **2017**, *121*, 2563–2573. [CrossRef]
20. Blake, A.J.; Kohlmeyer, R.R.; Hardin, J.O.; Carmona, E.A.; Maruyama, B.; Berrigan, J.D.; Huang, H.; Durstock, M.F. 3D printable ceramic-polymer electrolytes for flexible high-performance Li-ion batteries with enhanced thermal stability. *Adv. Energy Mater.* **2017**, *7*, 1602920. [CrossRef]
21. Wang, X.; Zhang, Y.; Zhang, X.; Liu, T.; Lin, Y.; Li, L.; Shen, Y.; Nan, C. Lithium-salt-rich PEO/$Li_{0.3}La_{0.557}TiO_3$ interpenetrating composite electrolyte with three-dimensional ceramic nano-backbone for all-solid-state lithium-ion batteries. *ACS Appl. Mater. Interfaces* **2018**, *10*, 24791–24798. [CrossRef] [PubMed]
22. Zhu, L.; Zhu, P.; Fang, Q.; Jing, M.; Shen, X.; Yang, L. A novel solid PEO/LLTO-nanowires polymer composite electrolyte for solid-state lithium-ion battery. *Electrochim. Acta* **2018**, *292*, 718–726. [CrossRef]
23. Cha, J.H.; Didwal, P.N.; Kim, J.M.; Chang, D.R.; Park, C.-J. Poly(ethylene oxide)-based composite solid polymer electrolyte containing $Li_7La_3Zr_2O_{12}$ and poly(ethylene glycol) dimethyl ether. *J. Membr. Sci.* **2019**, *595*, 117538. [CrossRef]
24. Xia, Y.; Fujieda, T.; Tatsumi, K.; Prosini, P.P.; Sakai, T. Thermal and electrochemical stability of cathode materials in solid polymer electrolyte. *J. Power Sources* **2001**, *92*, 234–243. [CrossRef]
25. Angulakshmi, N.; Dhanalakshmi, R.B.; Kathiresan, M.; Zhou, Y.; Stephan, A.M. The suppression of lithium dendrites by a triazine-based porous organic polymer-laden PEO-based electrolyte and its application for all-solid-state lithium batteries. *Mater. Chem. Front.* **2020**, *4*, 933–940. [CrossRef]
26. Joost, M.; Kunz, M.; Jeong, S.; Schonhoff, M.; Winter, M.; Passerini, S. Ionic mobility in ternary polymer electrolytes for lithium-ion batteries. *Electrochim. Acta* **2012**, *86*, 330–338. [CrossRef]
27. Cheng, S.H.-S.; He, K.-Q.; Liu, Y.; Zha, J.-W.; Kamruzzaman, M.d.; Ma, R.L.-W.; Dang, Z.-M.; Li, R.K.; Chung, C. Electrochemical performance of all-solid-state lithium batteries using inorganic lithium garnets particulate reinforced PEO/$LiClO_4$ electrolyte. *Electrochim. Acta* **2017**, *253*, 430–438. [CrossRef]
28. Piana, G.; Bella, F.; Geobaldo, F.; Meligrana, G.; Gerbaldi, C. PEO/LAGP hybrid solid polymer electrolytes for ambient temperature lithium batteries by solvent-free, "one pot" preparation. *J. Energy Storage* **2019**, *26*, 100947. [CrossRef]
29. Thokchom, J.S.; Gupta, N.; Kumar, B. Superionic Conductivity in a Lithium Aluminum Germanium Phosphate Glass–Ceramic. *J. Electrochem. Soc.* **2008**, *155*, A915–A920. [CrossRef]
30. Kumar, B.; Kumar, J.; Leese, R.; Fellner, J.P.; Rodrigues, S.J.; Abraham, K.M. A solid state rechargeable long cycle life lithium-air battery. *J. Electrochem. Soc.* **2009**, *157*, A50. [CrossRef]
31. Chung, H.; Kang, B. Increase in grain boundary ionic conductivity of $Li_{1.5}Al_{0.5}Ge_{1.5}(PO_4)_3$ by adding excess lithium. *Solid State Ionics* **2014**, *263*, 125–130. [CrossRef]
32. Robinson, J.P.; Kichambare, P.D.; Deiner, J.L.; Miller, R.; Rottmayer, M.A.; Koenig, G.M., Jr. High temperature electrode-electrolyte interface formation between $LiMn_{1.5}Ni_{0.5}O_4$ and $Li_{1.4}Al_{0.4}Ge_{1.6}(PO_4)_3$. *J. Am. Ceram. Soc.* **2018**, *101*, 1087–1094. [CrossRef]
33. Jung, Y.-C.; Lee, S.-M.; Choi, J.; Jang, S.S.; Kim, D.-W. All Solid-State Lithium Batteries Assembled with Hybrid Solid Electrolytes. *J. Electrochem. Soc.* **2015**, *162*, A704–A710. [CrossRef]
34. Wang, C.; Yang, Y.; Liu, X.; Zhong, H.; Xu, H.; Xu, Z.; Shao, H.; Ding, F. Suppression of lithium dendrite formation by using LAGP-PEO (LiTFSI) composite solid electrolyte and lithium metal anode modified by PEO (LiTFSI) in all-solid-state lithium batteries. *ACS Appl. Mater. Interfaces* **2017**, *9*, 13694–13702. [CrossRef]
35. Guo, Q.; Han, Y.; Wang, H.; Xiong, S.; Li, Y.; Liu, S.; Xie, K. New class of LAGP-based solid polymer composite electrolyte for efficient and safe solid-state lithium batteries. *ACS Appl. Mater. Interfaces* **2017**, *9*, 41837–41844. [CrossRef]
36. Sung, B.-J.; Didwal, P.N.; Verma, R.; Nguyen, A.-G.; Chang, D.R.; Park, C.-J. Composite solid electrolyte comprising poly(propylene carbonate) and $Li_{1.5}Al_{0.5}Ge_{1.5}(PO_4)_3$for long-life all-solid-state Li-ion batteries. *Electrochim. Acta* **2021**, *392*, 139007. [CrossRef]
37. Lee, J.; Howell, T.; Rottmayer, M.; Boeckl, J.; Huang, H. Free-standing LAGP/PEO/LiTFSI composite electrolyte membranes for applications to flexible solid-state lithium-based batteries. *J. Electrochem. Soc.* **2019**, *166*, A416–A422. [CrossRef]

38. Lee, J.; Rottmayer, M.; Huang, H. Impacts of Lithium Salts on the Thermal and Mechanical Characteristics in the Lithiated PEO/LAGP Composite Electrolytes. *J. Compos. Sci.* **2021**, *6*, 12. [CrossRef]
39. Sircar, A.K.; Weissman, P.T.; Kumar, B.; Marsh, R. Evaluation of doped polyethylene oxide as solid electrolyte. *Thermochim. Acta* **1993**, *226*, 281–299. [CrossRef]
40. Lu, Z.; Yang, L.; Guo, Y. Thermal behavior and decomposition kinetics of six electrolyte salts by thermal analysis. *J. Power Sources* **2006**, *156*, 555–559. [CrossRef]
41. Xuan, X.; Zhang, H.; Wang, J.; Wang‡, H. Vibrational Spectroscopic and Density Functional Studies on Ion Solvation and Association of Lithium Tetrafluorobrate in Acetonitrile. *J. Phys. Chem. A* **2004**, *108*, 7513–7521. [CrossRef]
42. Foran, G.; Mankovsky, D.; Verdier, N.; Lepage, D.; Prébé, A.; Aymé-Perrot, D.; Dollé, M. The Impact of Absorbed Solvent on the Performance of Solid Polymer Electrolytes for Use in Solid-State Lithium Batteries. *iScience* **2020**, *23*, 101597. [CrossRef] [PubMed]
43. Commarieu, B.; Paolella, A.; Collin-Martin, S.; Gagnon, C.; Vijh, A.; Guerfi, A.; Zaghib, K. Solid-to-liquid transition of polycarbonate solid electrolytes in Li-metal batteries. *J. Power Sources* **2019**, *436*, 226852. [CrossRef]
44. Chen, P.; Liang, X.; Wang, J.; Zhang, D.; Yang, S.; Wu, W.; Zhang, W.; Fan, X.; Zhang, D. PEO/PVDF-based gel polymer electrolyte by incorporating nano-TiO_2 for electrochromic glass. *J. Sol-Gel Sci. Technol.* **2017**, *81*, 850–858. [CrossRef]
45. Barroso-Bujans, F.; Fernandez-Alonso, F.; Cerveny, S.; Parker, S.F.; Alegría, A.; Colmenero, J. Polymers under extreme two-dimensional confinement: Poly(ethylene oxide) in graphite oxide. *Soft Matter* **2011**, *7*, 7173–7176. [CrossRef]
46. Rodrigues, L.; Silva, M.; Veiga, H.; Esperança, M.; Costa, M.; Smith, M.J. Synthesis and electrochemical characterization of aPEO-based polymer electrolytes. *J. Solid State Electrochem.* **2012**, *16*, 1623–1629. [CrossRef]
47. Zheng, J.; Hu, Y. New insights into the compositional dependence of Li-ion transport in polymer–ceramic composite electrolytes. *ACS Appl. Mater. Interfaces* **2018**, *10*, 4113. [CrossRef]
48. Enotiadis, A.; Fernandes, N.J.; Becerra, N.A.; Zammarano, M.; Giannelis, E.P. Nanocomposite electrolytes for lithium batteries with reduced flammability. *Electrochim. Acta* **2018**, *269*, 76–82. [CrossRef]

Journal of
Composites Science

Article

FE Modelling and Simulation of the Size Effect of RC T-Beams Strengthened in Shear with Externally Bonded FRP Fabrics

Amirali Abbasi, Zine El Abidine Benzeguir, Omar Chaallal * and Georges El-Saikaly

Department of Construction Engineering, École de Technologie Supérieure, University of Quebec, Montreal, QC H3C 1K3, Canada; amirali.abbasi.1@ens.etsmtl.ca (A.A.); zine-el-abidine.benzeguir.1@ens.etsmtl.ca (Z.E.A.B.); georges.el-saikaly@etsmtl.ca (G.E.-S.)
* Correspondence: omar.chaallal@etsmtl.ca

Abstract: The objective of this study is to conduct a finite-element (FE) numerical study to assess the effect of size on the shear resistance of reinforced concrete (RC) beams strengthened in shear with externally bonded carbon fibre-reinforced polymer (EB-CFRP). Although a few experimental studies have been done, there is still a lack of FE studies that consider the size effect. Experimental tests are time-consuming and costly and cannot capture all the complex and interacting parameters. In recent years, advanced numerical models and constitutive laws have been developed to predict the response of laboratory tests, particularly for issues related to shear resistance of RC beams, namely, the brittle response of concrete in shear and the failure modes of the interface layer between concrete and EB-CFRP (debonding and delamination). Numerical models have progressed in recent years and can now capture the interfacial shear stress along the bond and the strain profile along the fibres and the normalized main diagonal shear cracks. This paper presents the results of a nonlinear FE numerical study on nine RC beams strengthened in shear using EB-CFRP composites that were tested in the laboratory under three series, each containing three sizes of geometrically similar RC beams (small, medium, and large). The results reveal that numerical studies can predict experimental results with good accuracy. They also confirm that the shear strength of concrete and the contribution of CFRP to shear resistance decrease as the size of beams increases.

Keywords: size effect; reinforced concrete beams; finite-element method; shear strengthening; externally bonded carbon fibre-reinforced polymer (EB-CFRP) composites

Citation: Abbasi, A.; Benzeguir, Z.E.A.; Chaallal, O.; El-Saikaly, G. FE Modelling and Simulation of the Size Effect of RC T-Beams Strengthened in Shear with Externally Bonded FRP Fabrics. *J. Compos. Sci.* **2022**, *6*, 116. https://doi.org/10.3390/jcs6040116

Academic Editor: Stelios K. Georgantzinos

Received: 23 March 2022
Accepted: 7 April 2022
Published: 12 April 2022

1. Introduction

In the last two decades, very few FE studies have been dedicated to RC beams strengthened in shear EB-FRP or any types of strengthening with composite materials [1] made of CFRP [2,3]. However, given the lack of accurate constitutive laws at that time, these early FE studies did not consider the bond between concrete and FRP, nor did they simulate the interaction between concrete and steel reinforcement [4–11]. Recently, some FE studies have concentrated on shear strengthening using embedded-through-section (ETS) and near-surface-mounted (NSM) techniques [12–14]. With recent advances in the development of high-performance FE programs and constitutive laws, numerical studies can better simulate and accurately predict the outcome of experimental tests in terms of load-deflection response, behavior of the interface between concrete and EB-FRP, and the strain distribution along fibres [13,15–35]. Nevertheless, among these studies, very few have considered either the size effect of EB-FRP-strengthened RC beams [33,34] or the crack band model along with the concrete smeared crack model. This was the main impetus to carry out this study to assess the size effect by means of a numerical approach, by implementing both crack models in modelling the concrete and by considering the interface behavior between EB-FRP and the concrete substrate.

Given their complex behavior under loading, as well as their brittle rupture without warning, shear failure in RC beams has long been a major concern in structural engineering.

Therefore, practicing engineers often privilege the sequence by which flexural failure occurs before shear failure. Lack of shear strength in RC beams can be due to various interacting factors. Neglecting the size effect in codes and guidelines and thereby overestimating the ultimate shear capacity in the design process is an example of the effect of such factors. In recent years, the trends towards using FRP composites for rehabilitation and strengthening of RC beams have become intensified due to the high strength-to-weight ratio and tensile strength of FRP composites, which can compensate for the shear-strength deficiency of existing RC beams. It has been established that when the beam size is increased, the shear strength decreases due to the so-called size effect [36–38]. Many parameters affect the size effect, either mitigating or amplifying it by controlling the width of the diagonal shear crack—for example, the rigidity of FRP sheets [39]. Even though comprehensive studies have been performed on the effect of size in RC beams, research studies related to the size effect on RC beams strengthened with EB-FRP are limited. FE analysis can be implemented instead of experimental testing to obtain an inside view of the shear-stress profile variation along the interface layer and the distribution of stress on the fibres during loading. Most analytical models proposed by codes and guidelines are based on experimental results and can be prone to errors (human error, defects in laboratory machines, restricted tools...). Therefore, the results obtained from these models are not as accurate as those from FE methods for capturing the shear resistance contribution of concrete and EB-FRP through appropriate evaluation of strain distributions on the fibres.

The FE method is a cost-effective and precise tool for replacing experimental tests as long as the models are simulated based on reliable and logical assumptions. A few FE studies have been performed on the size effect of RC beams strengthened in shear with EB-CFRP, but either their assumptions were very simplistic, such as perfect bonding between concrete and EB-CFRP, which does not reflect the response of such a beam (location of the shear crack), or they fail to mention the assumptions used in their simulations. As explained in the following sections, the developed 2D-FE model was preferred to 3D models because it is less time-consuming and simulates the propagation of the shear crack in concrete with higher precision. Note that the shear crack is a major parameter in predicting the size effect.

As illustrated in Figure 1, the shear contributions of EB-FRP predicted by ACI 440.2R 2017 for over 50 beams with different depths varying from 80 mm to 682 mm strengthened in shear with continuous U-wrap and strips were compared with their corresponding experimental tests (see Appendix A Table A1 for details). The beams were classified into three categories depending on their depth (Figure 1). As the depth of the specimens and their corresponding EB-FRP bond lengths increased, the ACI 440.2R (2017) guidelines clearly overestimated the shear contribution of EB-FRP, which may indicate the existence of an additional size effect due to the contribution of EB-FRP to shear resistance. In fact, the models of most guidelines overestimate the contribution of EB-CFRP to shear resistance in large specimens.

In the current study, nine RC-T beams without steel stirrups [39] were selected for simulation. The beams were grouped into three series (small, medium, large). In each series, one beam was considered a control (not strengthened with EB-FRP), and the others were strengthened with one and two layers of EB-FRP. The results from the simulated models were validated with experimental tests.

The objectives of the present study were to evaluate the size effect and the shear contributions of concrete and EB-FRP, as well as the effect of an increase in EB-FRP rigidity, on the three series of specimens (different sizes) through numerical investigation. Capturing the response of the interface layer between concrete and CFRP sheets, as well as the distribution of strain along the main fibre of CFRP fabrics during loading, is of paramount importance when using FEA, given their impact on the size effect. Therefore, the impact of the response of the interface layer, the strain distribution along the fibre, and the fibres intercepted by the main diagonal shear crack on the size effect will be studied carefully, along with the failure modes, the load-deflection response, and the pattern of shear cracks. The novelty of this study is to conduct FE research on the size effect and to show the development of the shear stress and strain in the interface layers and fibres during the loading process. Furthermore, by extracting the strain distribution curve on the fibres that intercepted the main shear crack, it would be possible to measure the distribution factor leading to the effective strain experienced, which is far lower than the effective strain introduced in codes and guidelines.

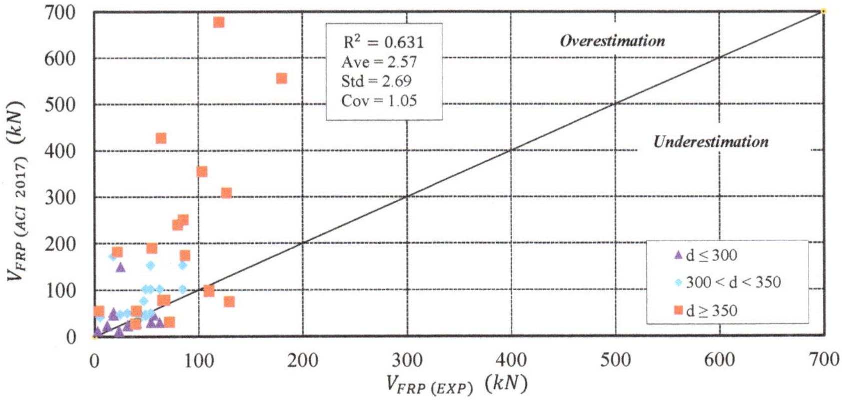

Figure 1. Comparison of the predicted ACI 440.2R 2017 code and experimental results.

2. Finite-Element Modelling

2.1. Suggested FE Modelling

The assumptions implemented for a simulation related to the types of crack models for concrete, steel bar, CFRP sheets, and the interface layer between concrete and CFRP are described in the following sections. Because the beam was not under stress in the normal direction to the plane of the beam, the plane stress model was used for concrete. Steel bars and CFRP sheets were modelled by 2D truss elements that contributed to transferring the stress in the direction of the truss element. Dynamic implicit analysis was implemented to overcome the convergence problem. Indeed, because the divergence occurred due to the brittle behavior of concrete and the nonlinearity of the interface layer between concrete and CFRP (delamination and debonding), general static solvers (static, general and static, Riks) cannot capture the nonlinearity of materials during imposed targeted displacement. Details of the implicit dynamic analysis implementation are described in [40].

2.2. Constitutive Models of Materials

2.2.1. Concrete Cracking Models

Various types of concrete cracking models can be used with FEA. The discrete crack model, the rotating smeared crack model, and the fixed smeared crack model are some examples. Considering the discrete crack model, a crack is introduced into the model geometry, where crack propagation occurs along the border of the element in FEA, proving its mesh objectivity. Furthermore, the location of the crack in the model must be defined in advance, which shows the dependency of this technique on how the precise initiation of the crack is predicted. Unlike the discrete crack model, there is no need to predefine the cracking initiation location in the smeared crack model because probable cracking zones and directions are recognized through the smeared crack technique. Elements lose their stiffness as the crack propagates in the smeared crack approach, whereas the stress-strain relation in concrete considers cracks a continuum and predicts the deletion of elements when a crack path is detected. The smeared crack approach can be classified into two categories: the rotating smeared crack approach and the fixed smeared crack approach. The differences between them are their theories for crack direction and their shear retention factor. The deficiency of the smeared crack model is that when element size decreases, it leads to zero energy dissipation in the softening part of the stress–strain curve in the tensile concrete material, resulting in strain localization [35]. To address strain localization, some limiters have been proposed, among which the crack band model implemented in the concrete damage plasticity framework has been proved to address mesh objectivity challenges resulting in convergence problems [41]. The function of the crack band model is to convert the width of the crack band to the cracking strain caused by the crack and softening behavior of the concrete in the tension, as shown in Figure 2.

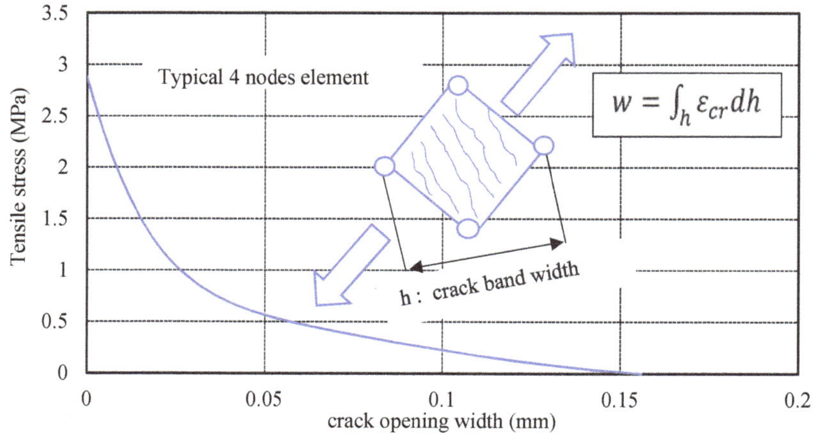

Figure 2. Relation between cracking width and tensile stress adopted from [42].

2.2.2. Concrete Response in Compression and Tension

Because the RC T-beams in the present study behave in their plane, a four-node plane stress element (CPS4) was implemented to simulate concrete. Various models have been proposed to represent the uniaxial behavior of concrete in compression, among which the model introduced by [43] (see Equation (1)) features a reasonable prediction of the ascending and softening parts of the concrete material curve, as shown in Figure 3.

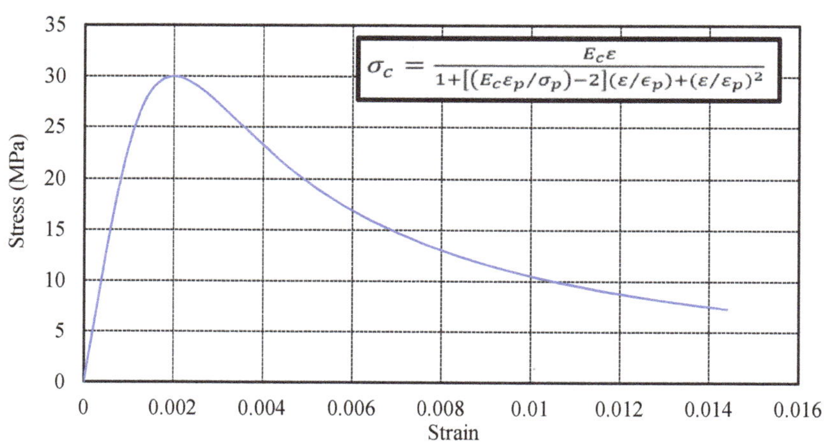

Figure 3. Stress–strain model for uniaxial compression in concrete introduced by [43].

$$\sigma_c = \frac{\alpha\varepsilon}{1 + \left[\left(\alpha\varepsilon_p/\sigma_p\right) - 2\right]\left(\varepsilon/\epsilon_p\right) + \left(\varepsilon/\varepsilon_p\right)^2} \tag{1}$$

where σ_p and ε_p are the maximum concrete compressive stress and strain obtained from experimental tests, equal to $f_c' = 30$ MPa and 0.002, respectively; E_c is the concrete modulus of elasticity $E_c = 4730\sqrt{f_c'}$ (MPa) according to [44]; and σ and ε are the applied compressive stress and corresponding strain during loading of the cylindrical specimen, respectively.

To define tensile concrete behavior in the descending and softening parts, the model introduced by [42] on the basis of numerous stress-crack displacement tests was implemented in this study as follows:

$$\frac{\sigma}{f_t} = \left[1 + \left(c_1 \frac{w}{w_{cr}}\right)^3\right] e^{\left(-c_2 \frac{w}{w_{cr}}\right)} - \frac{w}{w_{cr}}\left(1 + c_1{}^3\right) e^{-c_2} \tag{2}$$

$$w_{cr} = 5.14 \frac{G_f}{f_t} \tag{3}$$

$$f_t = 1.4\left(\frac{f_c' - 8}{10}\right)^{\frac{2}{3}} \tag{4}$$

$$G_f = \left(0.0469 d_a{}^2 - 0.5 d_a + 26\right)\left(\frac{f_c'}{10}\right)^{0.7} \tag{5}$$

where w is the crack width during loading; w_{cr} is the crack width at the moment when no stress can be transferred between the two sides of the crack; f_t is the maximum concrete tensile stress; σ is the tensile stress in the specimen during the stress-crack displacement test; G_f is the fracture energy, which in addition to Equation (5) can be obtained from the area of the stress-cracking displacement graph (Figure 2); $c_1 = 3$ and $c_2 = 6.93$ are constant parameters proposed by [42]); and d_a is the largest aggregate dimension.

2.2.3. Definition of Compressive and Tensile Damage to Concrete Damage Plasticity (CDP)

To define concrete damage in both compression and tension, represented by the softening part of the stress–strain curves, the proposed model introduced by [45] was considered as follows:

$$d_{t,c} = \begin{cases} \frac{(1-k)\varepsilon^P}{(1-k)\varepsilon^P + \sigma/E_0} \; if \; \dot{\bar{\varepsilon}}^P \geq 0 \\ \frac{\varepsilon^P - (\varepsilon - \bar{\varepsilon}^e_{cr})}{\varepsilon^P - (\varepsilon - \bar{\varepsilon}^e_{cr}) + \sigma/E_0} \; if \; \dot{\bar{\varepsilon}}^P < 0 \end{cases} \tag{6}$$

where $d_{t,c}$ is the damage parameter in both tension and compression, $\dot{\bar{\varepsilon}}^P$ is the plastic strain rate, k is the rate of inelastic strain when stiffness degrades ($\bar{\varepsilon}^P$) to inelastic strain when stiffness is constant (ε^P), and $\bar{\varepsilon}^e_{cr}$ is the cracking strain when the plastic strain rate is zero. The smeared crack model is implemented in the concrete damage plasticity (CDP) framework. Therefore, the stress–strain behavior of concrete in tension is transformed to stress-cracking displacement through the crack band model $\varepsilon_t^p = w_t/h$ [45]. Furthermore, the graphs in Figure 4a,b obtained from Equation (6) are applied for both tensile and compressive damage in concrete versus cracking displacement and plastic strain, respectively (for 10 mm element size). It has been proven that concrete damage plasticity (CDP) is able to show the response of concrete with high accuracy even for different types of concrete beams such as precast segmental concrete beams [46,47].

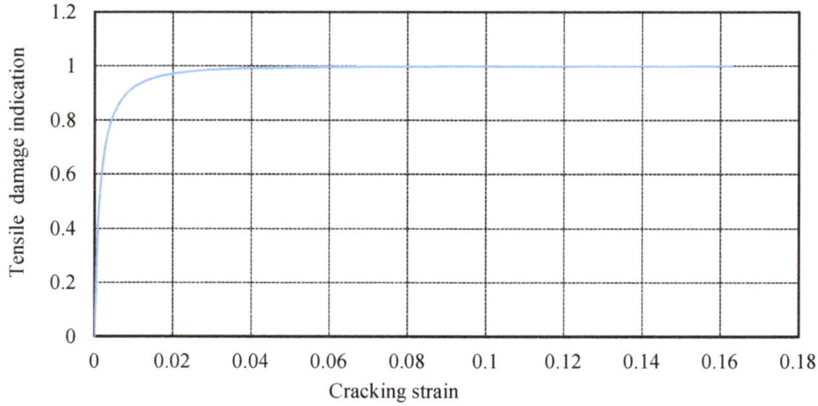

(a)

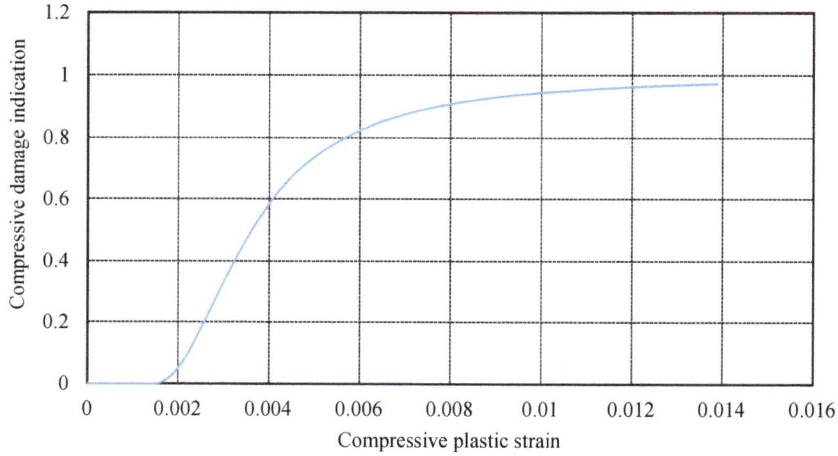

(b)

Figure 4. Tensile concrete damage model (for 10 mm element size): (**a**) compressive concrete damage models; (**b**) proposed by [45].

2.3. Bond-Slip Model for Concrete–Steel Reinforcement and Concrete–CFRP

To predict the ultimate shear capacity of an RC beam shear-strengthened with EB-FRP, the interaction between concrete and FRP composites should be defined precisely; otherwise, the software cannot identify the potential failure modes between concrete and CFRP, such as debonding and delamination. Early simulations assumed a perfect bond between the components of such beams, resulting in overestimation of the load-carrying capacity of the specimens. In addition, the perfect bond model between concrete and EB-CFRP has an effect on the distribution, direction, and position of shear cracks, leading to incorrect debonding and delamination. Because no slips were observed between concrete and longitudinal bars, a perfect bond model was assumed between the concrete and the longitudinal steel reinforcement. As for the bond between the concrete substrate and EB-CFRP, a two-dimensional, four-node cohesive element (COH2D4) that could capture both debonding and delamination failures in the model was implemented in ABAQUS. To define the properties of the cohesive elements, a simplified bond–slip law introduced by [48] was implemented in this study (Figure 5). The ascending and softening parts of the bond–slip curves were defined as follows:

$$\tau = \tau_{max}\sqrt{\frac{s}{s_0}}\ if\ s \le s_0 \quad (7)$$

$$\tau = \tau_{max} e^{-\alpha(\frac{s}{s_0}-1)}\ if\ s > s_0 \quad (8)$$

where $s_0 = 0.0195\beta_w f_t$, $G_f = 0.308\beta_w{}^2\sqrt{f_t}$, $\alpha = \frac{1}{\frac{G_f}{\tau_{max}s_0} - \frac{2}{3}}$, $\beta_w = \sqrt{\frac{2-(w_f/(s_f sin\beta))}{1+(w_f/(s_f sin\beta))}}$, and $\beta =$ fibre orientation. In the direction normal to the cohesive layer, which is representative of interface delamination, the following model was implemented for the cohesive layer to estimate the initial stiffness:

$$K_{nn} = \frac{1}{\frac{t_{concrete}}{E_{concrete}} + \frac{t_{epoxy}}{E_{epoxy}}} \quad (9)$$

where $t_{concrete}$ is the substrate thickness of concrete, t_{epoxy} is the thickness of epoxy, and $E_{concrete}$ and E_{epoxy} are the concrete and epoxy moduli of elasticity, respectively. The maximum tensile strength normal to the cohesive layer was also assumed equal to the maximum strength of concrete in tension.

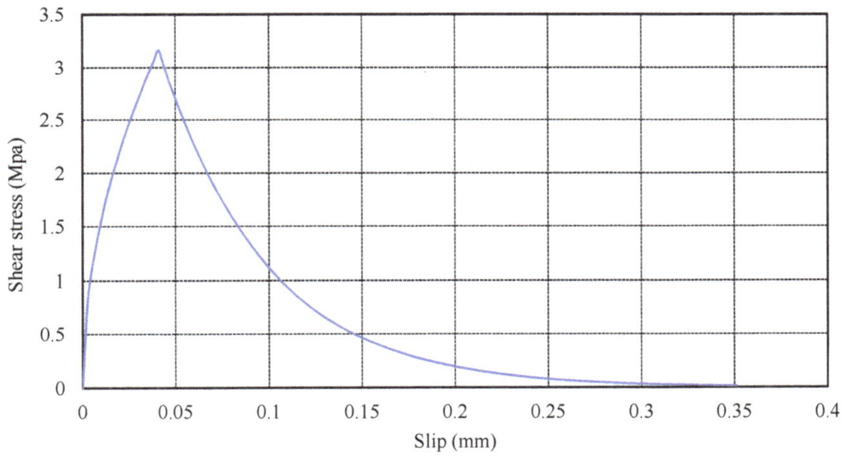

Figure 5. Bond–slip model between concrete and CFRP proposed by [48].

2.4. Modelling Internal Steel Reinforcement and EB-CFRP

To model the internal steel reinforcement and the external CFRP fabric, two-node 2D truss elements (T2D2) were implemented in the current study. Details of the simulation are illustrated in Figure 6. The elastic–plastic material was assigned to the steel reinforcement where bilinear response of the stress–strain behavior of steel bars in tension was assumed instead of nonlinear behavior after reaching the elastic limit to reduce calculation time (Figure 7a). As for EB-CFRP, the material was considered elastic until rupture in such a way that CFRP fibres could contribute to shear resistance through their tensile strength (Figure 7b) while their compression strength was zero. Based on [40], it was assumed that when FRP wrap is modelled by truss elements, the space between truss elements should be approximately $S_f = \frac{h_{f,e}}{20}$ to achieve reasonable agreement with continuous FRP fabrics. Therefore, the space between the truss elements modelling CFRP fibres was set to 10 mm, 10 mm, and 5 mm for large, medium, and small beams, respectively.

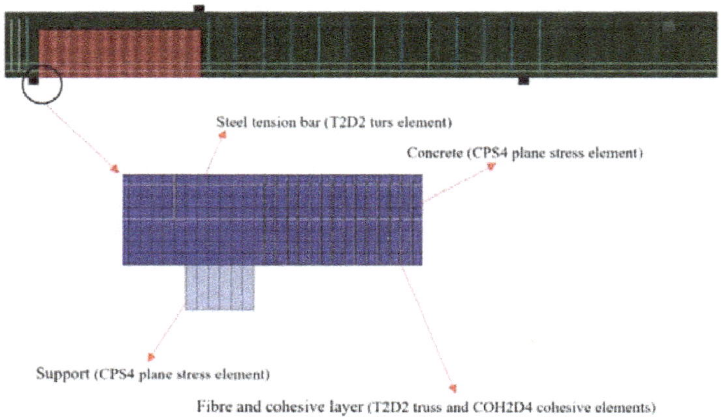

Figure 6. 2D simulation of the strengthened RC T-beams and their defined elements in ABAQUS.

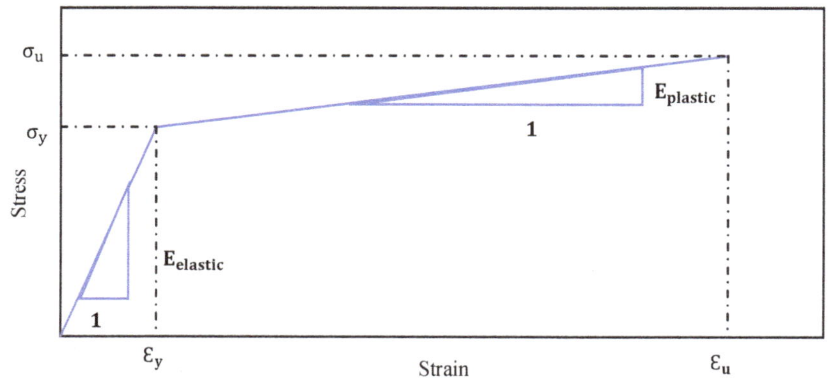

(**a**)

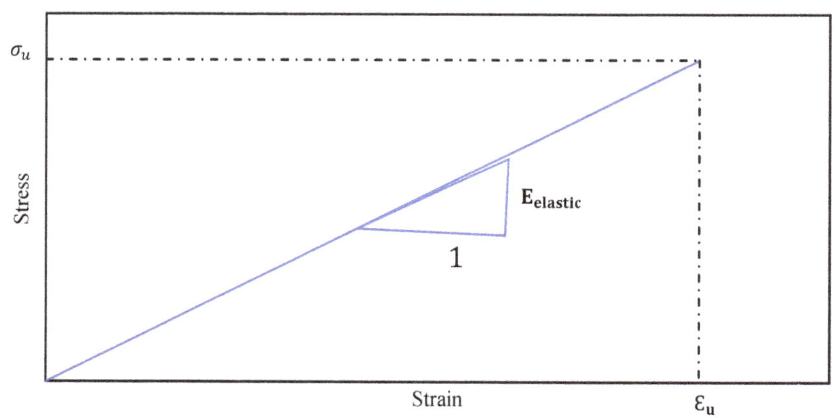

(**b**)

Figure 7. Stress–strain relation for (**a**) steel reinforcement and (**b**) CFRP fabrics.

3. Experimental Investigation

Nine RC T-beams without steel stirrups were selected from the experimental tests (control and strengthened using EB-CFRP) conducted by [39] to investigate the size effect by means of FEA. In addition, this study assessed the impact of increasing the rigidity of EB-CFRP on its contribution to the shear resistance of RC beams. The results are presented in terms of (1) load-deflection responses, (2) strain profiles along the normalized diagonal shear cracks, (3) strain profiles along the fibre direction, and (4) variation of interfacial shear stress profiles along the cohesive layer. Details of the geometry, steel reinforcement position, and configuration of EB-CFRP are illustrated in Figure 8. These beams were grouped into three series of RC T-beams that were geometrically similar, but of different sizes: large, medium, and small, abbreviated as L, M, and S, respectively. One beam in each series was not strengthened and served as a control beam. The specimens were subjected to a three-point loading scheme. The geometry and properties of the nine selected specimens are presented in Table 1.

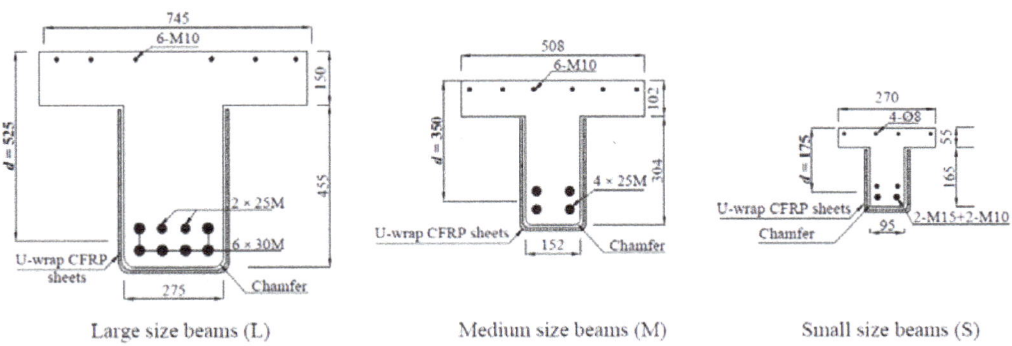

(a)

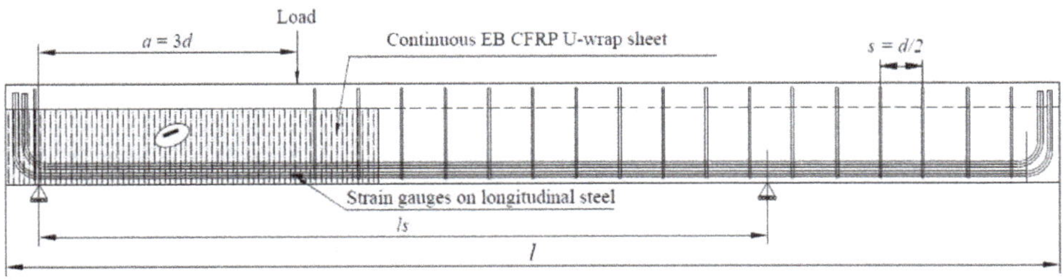

Beams	l	ls
Small	3000	1580
Medium	4520	3110
large	6400	4430

l : total length of beams (mm); ls : distance between supports (mm)

(b)

Figure 8. Details of beams: (**a**) cross-sections of large, medium, and small specimens (mm) and (**b**) elevation of beam and position of three-point loading ([39]).

Table 1. Geometry and property of material in studied beams.

		Series								
		S0-Con			S0–1L			S0–2L		
	Specimen	S.S0.Con	M.S0.Con	L.S0.Con	S.S0.1L	M.S0.1L	L.S0.1L	S.S0.2L	M.S0.2L	L.S0.2L
Concrete	f_c' (MPa)	30	30	30	30	30	30	30	30	30
	a/d	3	3	3	3	3	3	3	3	3
	Beam length, mm	3000	4520	6400	3000	4520	6400	3000	4520	6400
	Flange height, h_f, mm	55	102	150	55	102	150	55	102	150
	Flange width, b_f, mm	270	508	745	270	508	745	270	508	745
	Web height, h_w, mm	165	304	455	165	304	455	165	304	455
	Web width, b_w, mm	95	152	275	95	152	275	95	152	275
	Shear span	525	1050	1575	525	1050	1575	525	1050	1575
Steel Bars	Tensile bars	2 × M15 + 2 × M10	4 × M25	6 × M30 + 2 × M25	2 × M15 + 2 × M10	4 × M25	6 × M30 + 2 × M25	2 × M15 + 2 × M10	4 × M25	6 × M30 + 2 × M25
	Tensile yielding stress, MPa	420–440	470	420–470	420–440	470	420–470	420–440	470	420–470
	Modulus of elasticity E_s, GPa (T)	175–200	200	210–200	175–200	200	210–200	175–200	200	210–200
	Compressive bars	4 × ϕ8	6 × M10	6 × M10	4 × ϕ8	6 × M10	6 × M10	4 × ϕ8	6 × M10	6 × M10
	Compressive yielding stress, MPa	650	440	440	650	440	440	650	440	440
	Modulus of elasticity E_s, GPa (C)	215	200	200	215	200	200	215	200	200
CFRP Fabrics	Configuration	-	-	-	Ct-U	Ct-U	Ct-U	Ct-U	Ct-U	Ct-U
	Thickness of fabrics, t_{CFRP}, mm	-	-	-	0.066	0.107	0.167	0.132	0.214	0.334
	Modulus of elasticity E_f, GPa	-	-	-	231	231	234	231	231	234
	Tensile strength, MPa	-	-	-	3650	3650	3793	3650	3650	3793
	Number of layers	-	-	-	1	1	1	2	2	2

4. Validation with Experimental Tests

As mentioned earlier, the simulated model has been validated by the experimental tests carried out by [39]. The element size for discretization of the small beams (one- and two-layer strengthening) and the control beam was 5 mm, 10 mm, and 10 mm for small, medium, and large beams, respectively. These sizes have shown good agreement between numerical and experimental results.

4.1. Failure Modes, Crack Patterns, and Ultimate Load-Carrying Capacity

Negative strain shows as compression in concrete, which mainly occurs around the supports and the load plate. Figure 9 shows the numerical results, which illustrate the main shear crack distributions in all strengthened beams by means of the principal logarithmic plastic strain in the plane of the beams. As shown in Figure 9, regardless of size, the patterns of shear cracks for small, medium, and large beams strengthened with two CFRP layers were similar (Figure 9a,c–e), starting in the mid-depth of the web and extending to the support and the web/flange intersection to propagate horizontally towards the load plate. The results for the medium and large strengthened specimens with one CFRP layer (Figure 9b–d) followed the same trend.

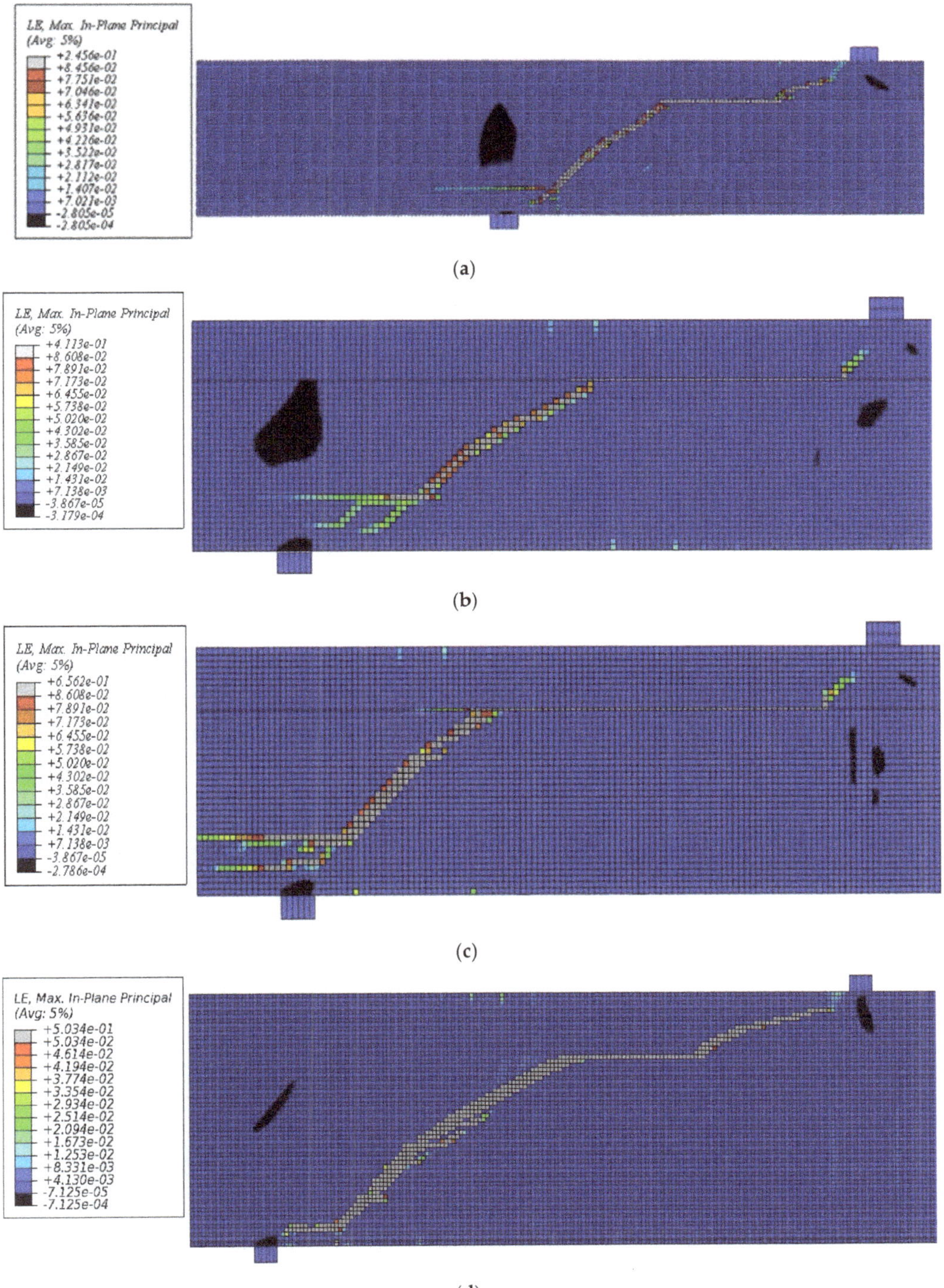

Figure 9. *Cont.*

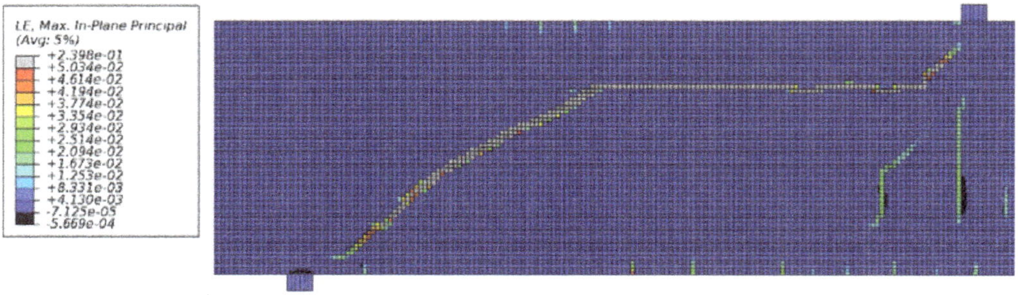

(e)

Figure 9. Crack pattern obtained from simulation in ABAQUS for specimen at complete failure: (**a**) specimen S.S0.2L; (**b**) specimen M.S0.1L; (**c**) specimen M.S0.2L; (**d**) specimen L.S0.1L; (**e**) specimen L.S0.2L.

Experimental tests of the control beams (small, medium, large) showed similar crack angle patterns with one single diagonal shear crack, appearing as a crack band at mid-height of the web and propagating toward the web soffit (support) and the flange (load application point) of the beam. As shown in Figure 10, the maximum crack angle occurred at mid-depth of the beams and then decreased as the crack extended towards the support and the load application point. The crack patterns predicted through numerical analysis were in good agreement with experimental results (Figure 10a–d). The failure thresholds of the control specimens with increasing load are presented in Table 2.

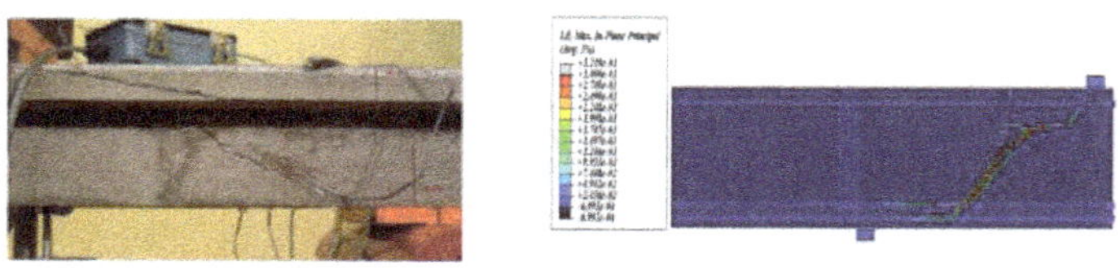

(a)

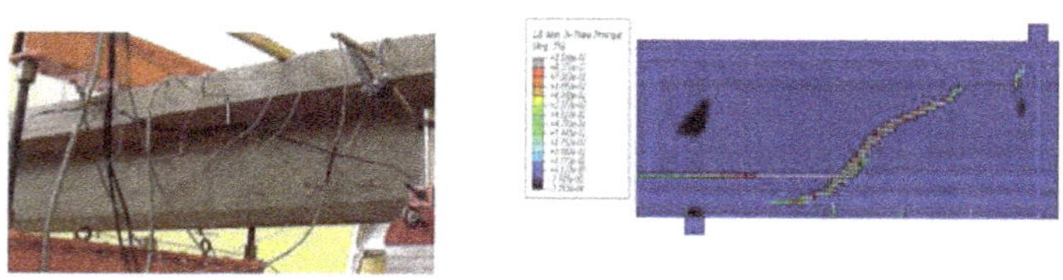

(b)

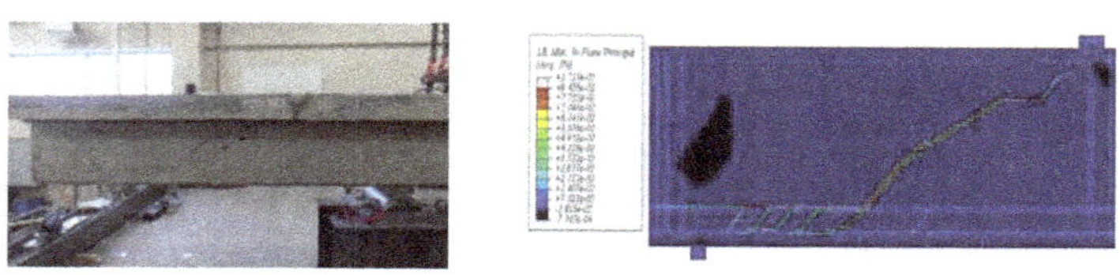

(c)

Figure 10. Crack pattern obtained from FEA and experimental testing for control beams at ultimate states: (**a**) S.0L.Con; (**b**) M.0L.Con; (**c**) L.0L.Con.

Table 2. Failing procedure of control specimens as increasing load for experimental and numerical tests.

	Imposed Load				
	Experimental Test		Numerical Prediction		
Specimen	Failure Load (kN)	Shear Cracks (kN)	Failure Load (kN)	Shear Cracks (kN)	$P_{num.}/P_{exp.}$
S.S0.Con	58	19 (31%)	62	25 (40%)	1.06
M.S0.Con	130	45 (35%)	128	60 (46%)	0.98
L.S0.Con	283	73 (25%)	293	89 (30%)	1.03

Note that the flexural and shear cracks in the small and medium specimens occurred at approximately the same ratio of the ultimate loads. However, this ratio decreased considerably for large beams, indicating the possible existence of a size effect in large specimens that reduced their shear strength capacities as depth increased. Numerical results showed that the ultimate load of the medium beam was 106% higher than that of the small beam, and that the ultimate load of the large beam was 372% and 128% higher than those of the small and medium ones, respectively. The failure loads occurred at 62, 128, and 293 kN, whereas shear cracks formed at 25 kN, 60 kN, and 89 kN applied loads for small, medium, and large beams, respectively. This was in good agreement with the experimental results (see Table 2). Single diagonal shear cracks formed in control beams (small, medium, and large), giving rise to shear failure in all specimens. This confirms the results obtained by [49]. Moreover, based on the experimental results, the shear crack angles in small, large, and medium beams were 42°, 37°, and 24° respectively, which are comparable with the numerical results (Figure 10a–d).

As shown in Figure 10, crack patterns in all control specimens were similar regardless of specimen size. However, the large beams featured more distributed minor cracks, probably due to wider cracking and the resulting loss of aggregate interlock (Figure 10c). Delamination of the interface layer occurred in all strengthened specimens when the stresses normal to the interface layer exceeded their maximum resistance (2.3 MPa). The delamination started from the top edge of the CFRP wrap located at the web/flange intersection and then extended horizontally and propagated vertically towards the top parts of the main diagonal shear crack. The stress normal to the interface layer at the web/flange intersection exceeded 2.3 MPa, which is the maximum strength in the normal direction of the interface layer.

As beam depths decreased from 525 mm to 350 mm and then to 175 mm, the behavior of the specimens changed from brittle to ductile, as indicated by the load-deflection response with a plateau (Figure 11a). Numerical results commonly overestimate the load-carrying capacity of a beam by approximately 6% because the bond between longitudinal bars and concrete is assumed perfect and an implicit dynamic is implemented to solve the model, thus amplifying deflection and load in the dynamic analysis [35,40]. However, as long as the parameters in the dynamic analysis are defined appropriately (time increment, loading time, and loading scheme), it can be an appropriate replacement for static analysis [35,40].

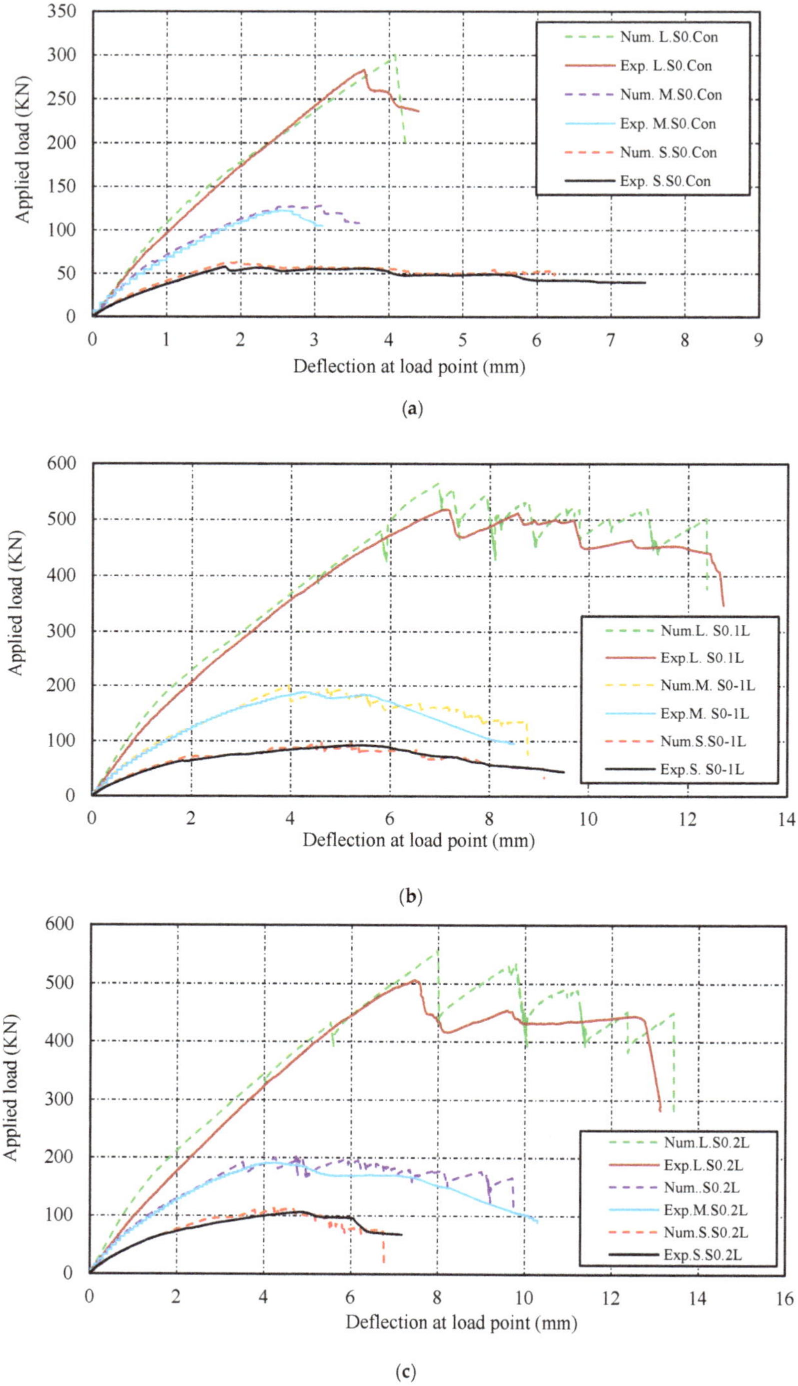

Figure 11. Numerical vs. experimental load-deflection response: (**a**) control beams; (**b**) beams strengthened with one CFRP layer; (**c**) beams strengthened with two CFRP layers.

The results of the numerical and experimental tests are summarized in Tables 3 and 4 in terms of ultimate load (P_{max}), deflection at P_{max}, shear contribution of CFRP (V_{CFRP}), shear gain due to CFRP (G_{CFRP}), maximum shear force (V_T), maximum shear stress in the section $\nu = V_T/(b_w \times d)$, loss of shear stress in medium and large specimens with respect to the small beam ν (%), and ultimate strain contributed by CFRP fabrics in each specimen.

Table 3. Comparison between experimental and numerical results in terms of load deflection and ultimate shear strength contributed by concrete and CFRP fabrics.

	P_{max} (kN)		Δ_{Pmax}(mm)		V_T(kN)		V_{CFRP} (kN)		P_{num}/P_{max}	Failure Mode	
Specimens	Exp.	Num.	Exp.	Num.	Exp.	Num.	Exp.	Num.		Exp.	Num.
S.S0.Con	58	62	1.8	1.9	38	41	-	-	1.06	Shear	Shear
M.S0.Con	130	128	2.6	3.1	86	85	-	-	0.98	Shear	Shear
L.S0.Con	283	293	3.7	4	182	189	-	-	1.03	Shear	Shear
S.S0.1L	93	98	5.4	5.2	62	65	23	24	1.05	Shear	Shear
M.S0.1L	189	195	4.2	3.9	125	130	39	45	1.03	Shear	Shear
L.S0.1L	518	550	7.1	7.2	334	354	151	165	1.06	Shear	Shear
S.S0.2L	106	112	4.9	4.23	71	75	32	34	1.05	Shear	Shear
M.S0.2L	191	202	4.1	4.8	127	134	40	49	1.05	Shear	Shear
L.S0.2L	506	543	7.5	8	326	349	144	160	1.07	Shear	Shear

Table 4. Comparison between numerical and experimental results in terms of shear gain and loss.

	Shear Gain Due to FRP (%)		Shear Stress in Concrete		Loss in Shear Stress with Respect to Control Beam (ν%)	
Specimens	Exp.	Num.	Exp.	Num.	Exp.	Num.
S.S0.Con	-	-	2.31	2.46	-	-
M.S0.Con	-	-	1.62	1.59	30	35
L.S0.Con	-	-	1.26	1.31	45	47
S.S0.1L	60	58	3.71	3.91	-	-
M.S0.1L	45	52	2.35	2.44	37	38
L.S0.1L	83	87	2.31	2.45	38	37
S.S0.2L	84	83	4.26	4.51	-	-
M.S0.2L	47	58	2.38	2.51	44	44
L.S0.2L	79	85	2.26	2.41	47	47

4.2. Load-Deflection Responses

Figure 11 compares the experimental and numerical results in terms of ultimate load-carrying capacities and displacements for the nine specimens. Note that shear strengthening with EB-CFRP fabrics showed higher levels of strength in specimens strengthened with one layer by about 58%, 52%, and 88% for small, medium, and large beams, respectively, with respect to control beams. Furthermore, the deflections corresponding to the maximum load (Δ_{pmax}) of specimens strengthened with one EB-CFRP layer increased by 173%, 25%, and 80% with respect to control beams, which could be attributed to the fact that CFRP fabrics control the deflection of specimens (Figure 11). Nevertheless, by adding two CFRP layers, no considerable additional deflections were observed in the specimens compared to those strengthened with one layer. The results of the load-deflection data obtained from numerical analysis are highly comparable with experimental observations, showing that the simulated model can predict laboratory tests with high accuracy.

5. FE Simulations and Results

This section is dedicated to FE simulations and analyses. The results are presented in terms of (a) shear strength for control and strengthened beams, (b) distribution of strain on the fibres along the diagonal shear crack, and (c) strain distributions along the CFRP fabric and interfacial shear stress at the cohesive layer.

5.1. Shear Strength and Loss in Control and Strengthened Beams

This section presents the FEA results for shear strength and shear loss due to the size effect. The size effect had an impact on strengthened beams in the way that the shear stress contributed by CFRP fabric decreased in specimens strengthened with one CFRP layer, from 1.45 MPa in S.S0.1L to 1.14 MPa in L.S0.1L. Table 4 compares the numerical and experimental results. The specimens of the third series, which were strengthened by two layers, resulting in higher CFRP rigidity, showed similar results, with shear stress decreasing from 2.05 MPa in S.S0.2L to 1.1 MPa in L.S0.2L. As illustrated in Figure 12, adding a second layer of EB-CFRP increased the shear stress in the small beam before delamination by 30%, that is, from 1.45 MPa to 2.05 MPa. This gain in shear stress decreased in the medium specimen by 7% and in the large specimen by 3%, indicating that the size effect has an impact on the shear stress contributed by both concrete and CFRP. Nevertheless, more investigations are required to clarify the relation between the size effect and the rigidity of CFRP as the specimen dimension increases.

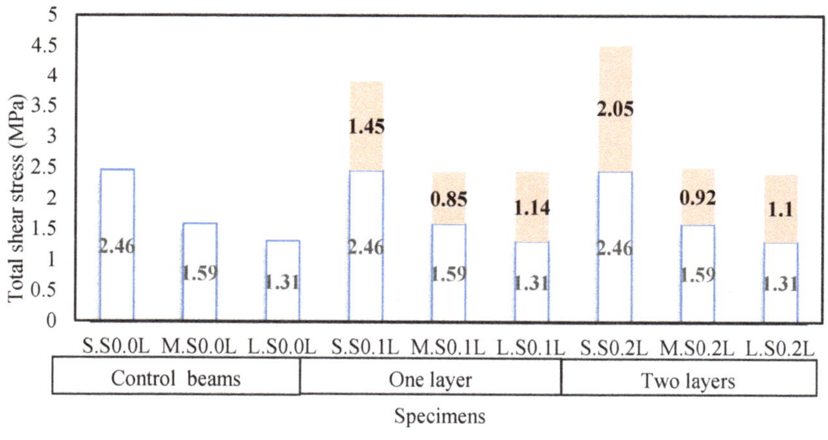

Figure 12. Shear stress contributed by concrete and CFRP fabric (FE results).

5.2. Distribution of Strain on the Fibres along the Diagonal Shear Crack by FEA

Assessing the strain profile along the main diagonal shear crack resulted in a better understanding of the behavior of fibres and their contribution to shear strengthening as the size of specimens and the rigidity of CFRP fabrics increased. By evaluating the strain distribution along the shear crack path, it is possible to locate the maximum crack width and to understand how debonding and delamination occur on both sides of the crack. The main diagonal shear cracks had an almost linear pattern for specimens strengthened with two layers of CFRP fabric and a semi-parabolic shape for those strengthened with one CFRP layer (Figures 9 and 10, respectively). Because the strain distribution on the fibres along the main diagonal shear crack constitutes the basis on which the distribution factors for the strain are established, the response of the strain in fibres intersected by the normalized shear crack is evaluated in this section. Note that the distribution factor D_{FRP}

introduced by [50] is a function of the average strain on the fibres intercepted by the shear crack and the maximum strain experienced by fibres at a specific load, as follows:

$$D_{FRP} = \frac{\sum_{i=1}^{n} \varepsilon_{FRP,i}}{n\varepsilon_{max}} \tag{10}$$

where $\varepsilon_{FRP,i}$ is the strain in the ith specific fibre intercepted by the shear crack, n is the number of fibres crossed by the diagonal shear crack, and ε_{max} is the maximum strain experienced by the fibre at ultimate state.

The formula introduced by [51] for the shear contribution of EB-FRP is:

$$V_f = 2f_{f,e}t_f w_f \frac{h_{f,e}(cot\theta + cot\beta)sin\beta}{S_f} \tag{11}$$

where $f_{f,e}$ is the effective stress in the *FRP* intercepted by the main shear diagonal shear crack that can be obtained through the distribution factor (D_{FRP}) given by [40].

$$f_{f,e} = E_f\varepsilon_{f,e} = E_f\varepsilon_{max}D_{FRP} \tag{12}$$

in which $\varepsilon_{f,e}$ is the effective strain in FRP wrap, E_f is the FRP modulus of elasticity, and D_{FRP} is the distribution factor obtained from Equation (10).

As shown in Figure 13a–f, the strain distributions along the shear crack are illustrated by four displacement levels corresponding to four phases: (1) the crack initiates at mid-depth of the web, (2) all fibres intercepted by shear cracks are in active phase (experiencing strain) just before initiation of delamination at the tip of the crack, (3) delamination at the cohesive layer is already formed and developed at mid-distance of shear crack paths (effective bond lengths start to decrease), and (4) one of the fibres intercepted by the shear crack is exposed to maximum strain during the loading process.

Regarding the series of small specimens, as displacements reached 2.23 mm and 2.79 mm in specimens strengthened with one and two layers, respectively, the shear crack initiated at mid-distance of the shear crack path from the end of the normalized distance. Then it propagated toward the bottom edge of the beams and the edge of the intersection between webs and flanges. When the displacements at the loading points reached 3.92 mm and 3.81 mm, the maximum strains on CFRP fabrics were 0.00637 and 0.00246 in S.S0.1L and S.S0.2L, respectively (Figure 13a,b). At this stage, shear cracks had completely formed, and all the fibres intersected by shear cracks were in the active phase (from the tips to the ends of the shear crack). The strain values on fibres intersected by the shear crack on the top part of the crack then dropped suddenly to zero due to delamination and to the short bond length compared to the bottom part of the crack. During that process, the cracks at the edge of the intersection between the flange and the web propagated horizontally.

The maximum strains experienced by fibres before entirely losing the CFRP shear contribution were 0.00866 and 0.00266 in S.S0.1L and S.S0.2L, respectively. These values are in good agreement with the corresponding experimental results—that is, 0.00714 and 0.00216, corresponding to 45% and 13% of CFRP ultimate strain, respectively. Note that the values of strains on fibres obtained from numerical analysis are larger than those obtained from experimental tests because dynamic implicit analysis was used to solve the models from which the amplified strains were recorded, whereas such an amplification did not exist in the static analysis [40]. Furthermore, strain gauges installed on EB-CFRP fabrics measure the average strains, which are lower than the maximum strain obtained from FEA [35]. For the medium beams, shear cracks appeared at mid-distance of the shear crack path when displacements at the loading points reached 3.36 mm and 3.15 mm in beams strengthened with one and two layers of CFRP fabric, respectively. When the displacements reached 4.73 mm and 5.44 mm in M.S0.1L and M.S0.2L, respectively (Figure 13c,d), the main diagonal shear cracks in both specimens became complete, and at this stage, all fibres crossed by shear cracks (from the tips to the ends of the cracks) experienced stress and strain.

The maximum strains experienced by the fibres just before delamination were 0.0032 and 0.0033 in M.S0.1L and M.S0.2L, respectively, as presented in Figure 13c,d, representing 20% and 21% of the ultimate strain of the fibres. After these maximum strains were reached, an inactive zone where more fibres lost their contribution to shear strengthening (zero strain) developed at the support. The maximum strains obtained from numerical analysis were in good agreement with experimental results (i.e., 0.00248 and 0.0027 in M.S0.1L and M.S0.2L, respectively). The same scenario was observed for the large specimens, from the initiation of shear cracks to the complete failure of EB-CFRP. Therefore, all fibres were in active modes as complete shear cracks formed, and at this stage, displacements reached 7.26 mm and 8.27 mm in L.S0.1L and L.S0.2L, respectively (Figure 13e,f)), just before delamination.

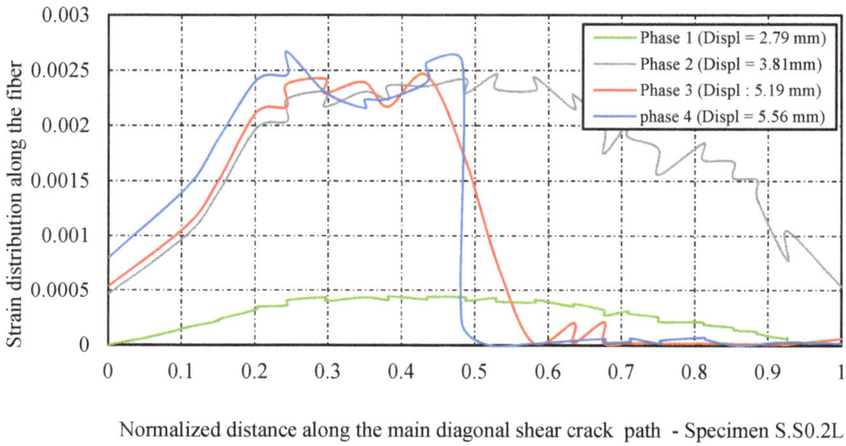

(**a**)

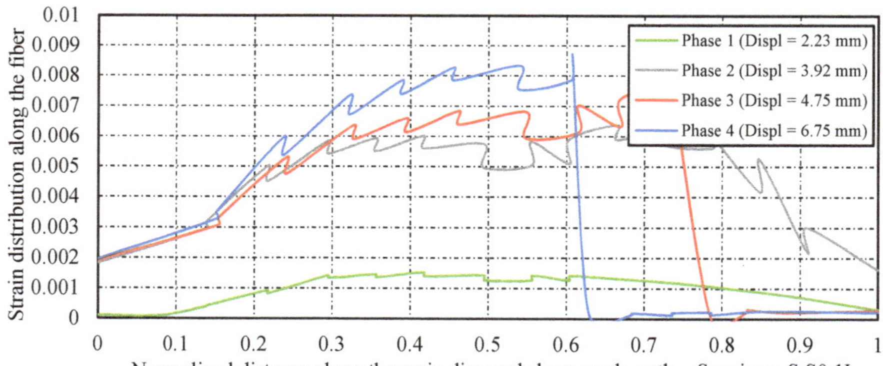

(**b**)

Figure 13. *Cont.*

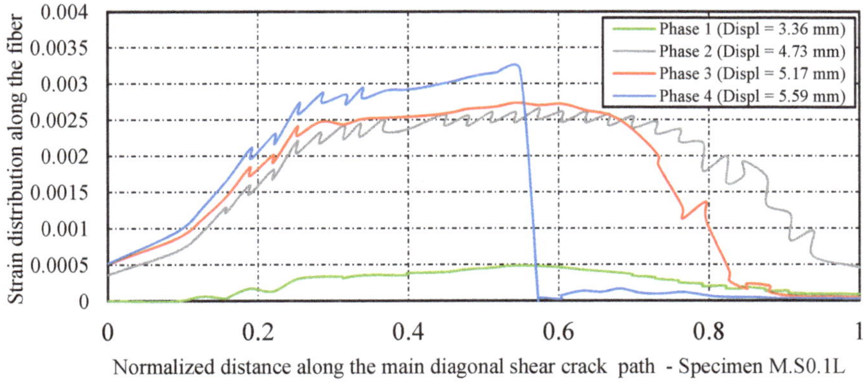

(c)

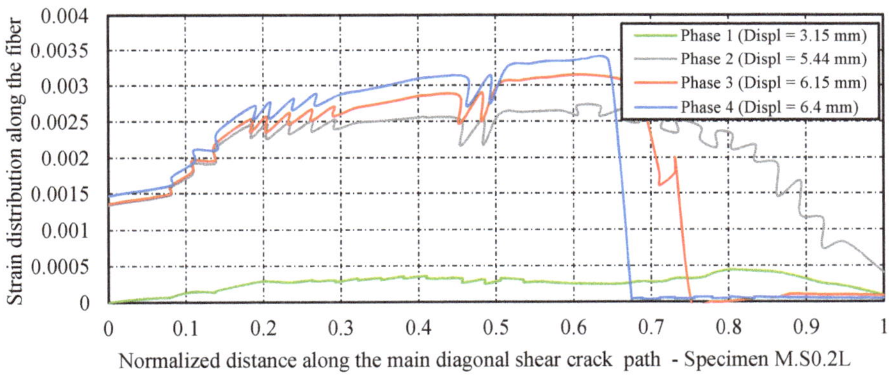

(d)

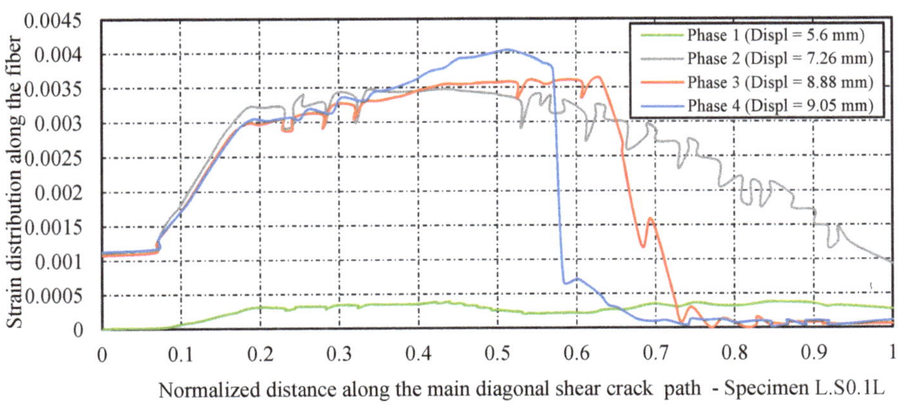

(e)

Figure 13. *Cont.*

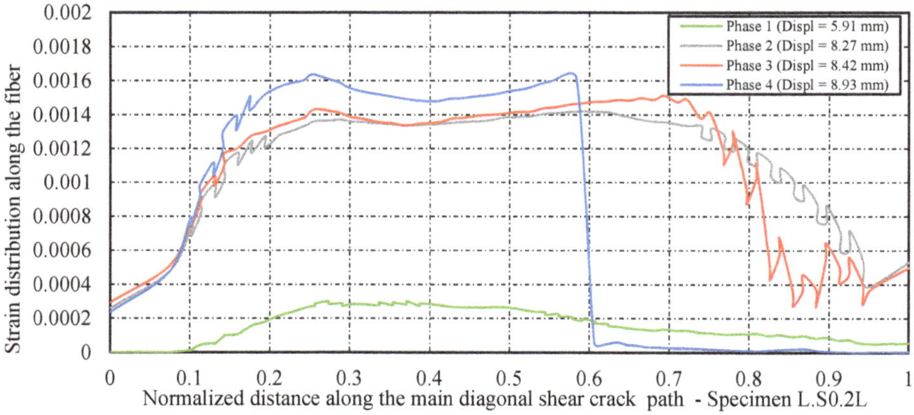

(f)

Figure 13. Distributions of strains on fibres crossed by normalized distance along the main diagonal shear path: (**a**) specimen S.S0.1L; (**b**) specimen S.S0.2L; (**c**) specimen M.S0.1L; (**d**) specimen M.S0.2L; (**e**) specimen L.S0.1L; (**f**) specimen L.S0.2L. Phase 1: initiation of main diagonal shear crack. Phase 2: all the fibres intersected by shear crack in an active phase. Phase 3: development of the loss of the shear contribution of the fibres at tips of the shear crack. Phase 4: the maximum strain recorded on fibres before the complete loss of the shear contribution of the fibres at the top part of the shear crack.

The maximum strains experienced by the fibres were 0.00415 and 0.00161 in L.S0.1L and L.S0.2L, respectively, which were very close to experimental values (0.00369 and 0.0016, respectively). Therefore, the maximum strains reached on EB-CFRP in large specimens (L.S0.1L, L.S0.2L) decreased in comparison to small beams by 53% and 40% in beams strengthened with one and two layers, respectively, resulting in a size effect on both concrete and CFRP shear contributions. Likewise, in the control beams (Figure 10a–c), the pattern of shear cracks at the final states in strengthened specimens as obtained from numerical analysis was in good agreement with experimental tests, confirming that the assumptions applied for simulation were accurate (Figure 9a–e).

5.3. Strain Distributions along the CFRP Fabric and Interfacial Shear Stress at the Cohesive Layer

By evaluating the strain distribution on fibres along with the normalized distance of the crack path, it is possible to locate the maximum vertical crack width. Those fibres that experience more strain before losing their shear contributions are located at the maximum crack width. Moreover, the vertical width of the crack can be calculated by summation of the interfacial slip along the two sides of the crack and the deformation of the CFRP fabric in the debonding area [35]. Therefore, after the maximum crack width was located and calculated, the strain distribution and the interfacial shear stress along the fibre intersected by the shear crack at its maximum width were evaluated to further investigate the size effect on the shear contribution of EB-CFRP.

The FEA strain profile along the fibre and the shear stress profile along the interface layer for L.S0.1L and L.S0.2L are presented in Figure 14a,b and Figure 15a,b, respectively. The results are presented in terms of strain development along the fibre direction and the interfacial shear stress along the interface layer in which debonding can be observed. Each graph shows the strain distributions and the interfacial shear stress at six displacement stages, in which the first three steps are related to initiation and development of the shear crack just before delamination, and the next three steps represent the initiation of delamination to complete loss of strain in the fibre. This yields six curves corresponding to six levels of displacement. As shown in Figure 14a,b and Figure 15a,b, there is a similarity between the strain distribution along the main fibre direction and the strain response obtained from the pullout tests.

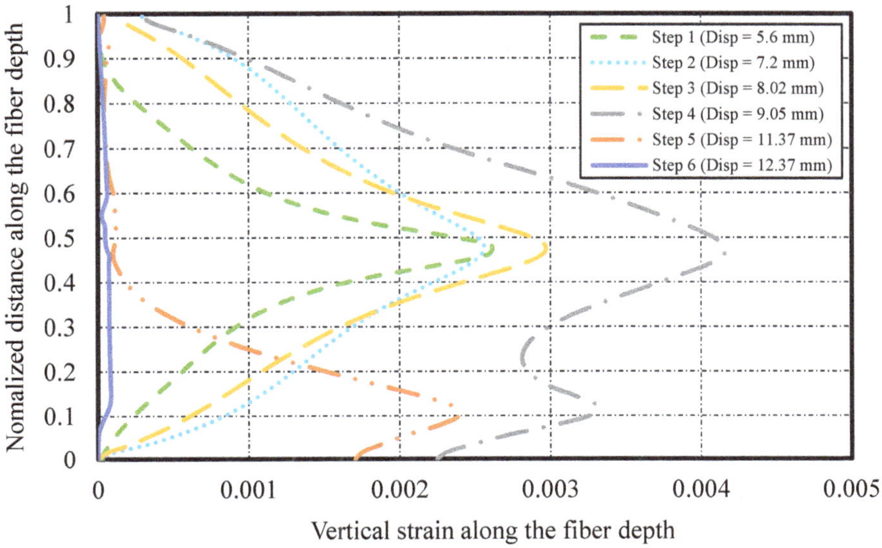

(a)

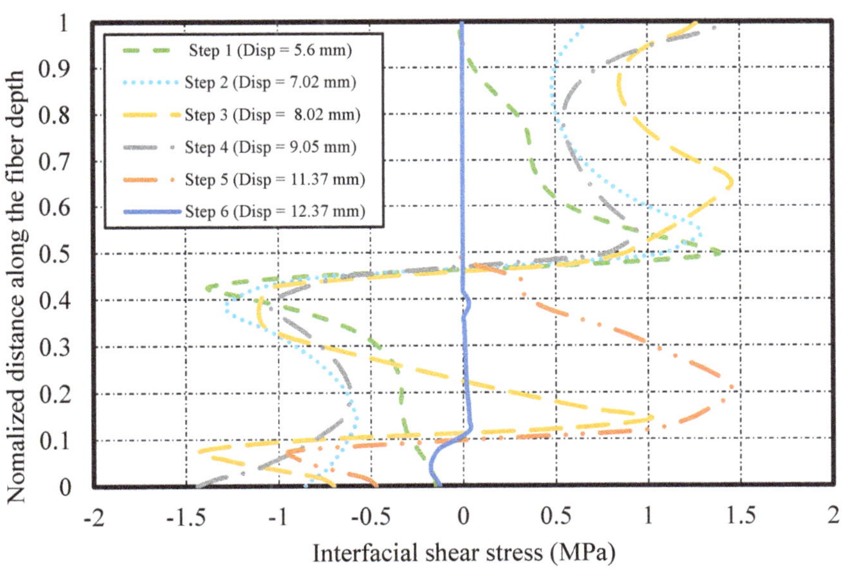

(b)

Figure 14. Strain profile and interfacial shear stress along the fibre and interface layer intercepted by maximum crack width (specimen L.S0.1L).

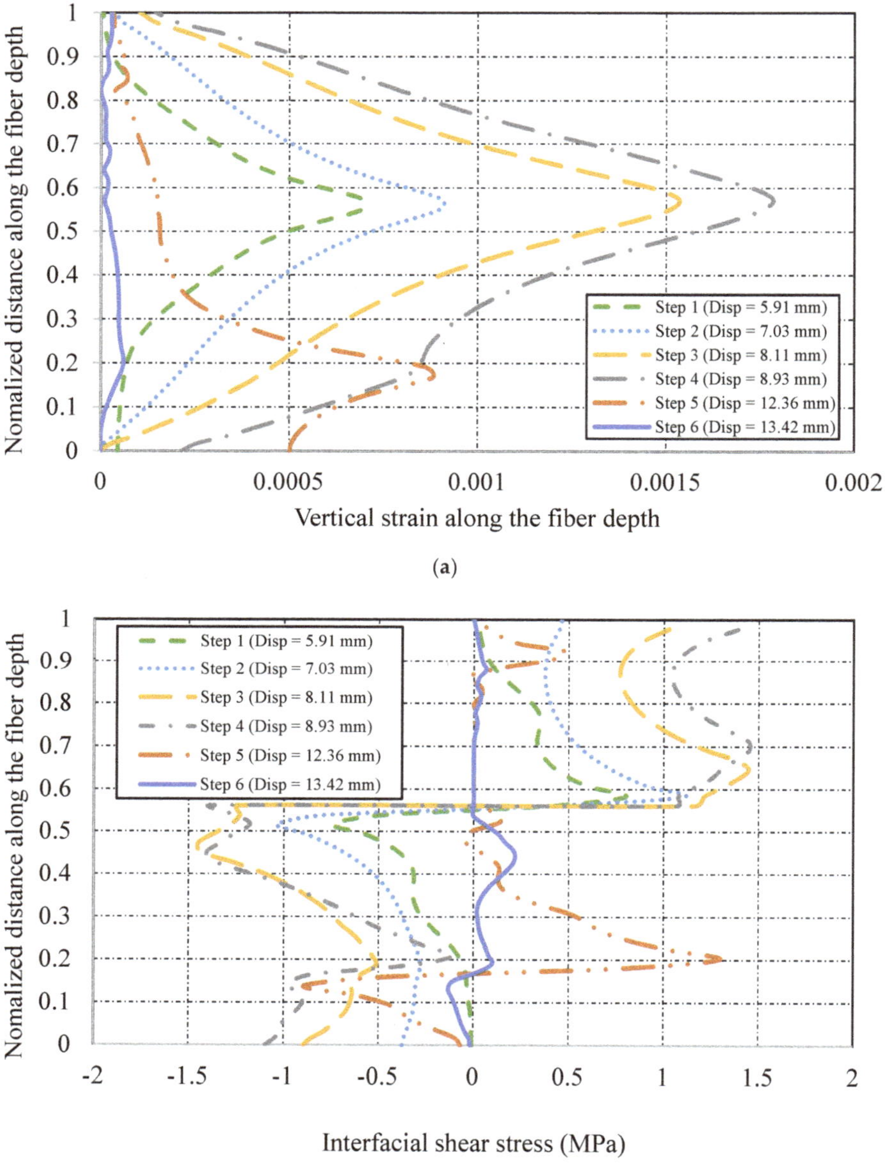

(a)

(b)

Figure 15. Strain profile and interfacial shear stress along the fibre and interface layer intercepted by maximum crack width (specimen L.S0.2L). Steps 1,2,3: the initiation and development of the shear crack just before the delamination procedures. Step 4,5,6: the procedure from the initiation of the delamination to the complete loss of strain in the fibre.

As soon as the shear crack appeared in L.S0.1L at 5.6 mm displacement, the fibre started to contribute to shear resistance. By increasing the load and the corresponding displacement, the amount of strain increased to a maximum of 0.00415. The maximum strain occurred at mid-depth of the specimen and then decreased gradually toward the top edge of the web/flange intersection. The amount of strain experienced by the lower part increased more than that on the top edge of the fibre because there was more effective bond length in the bottom crack part of the U-wrap configuration, in which the fibres below the shear crack were fixed.

Note that at the peak of the strain profile and when strain was constant, interfacial shear stress was zero, indicating the delaminated zone. At the third stage, during development of the delaminated area, this zone propagated in the top part of the crack, as evidenced by zero interfacial shear stress and zero strain in the strain profile (Figures 14 and 15). As the displacement reached 12.37 mm, complete delamination occurred in specimen L.S0.L1, and likewise in specimen L.S0.2L. In addition, the amount of strain on the fibres along the main diagonal shear crack decreased in all specimens strengthened with two CFRP layers compared to specimens strengthened with one CFRP layer. Unlike the medium specimens, where increasing rigidity did not significantly change the maximum strains on the fibres, the maximum strains in all the small and large specimens (strengthened with one and two CFRP layers) decreased with increasing size, indicating the existence of an additional size effect on CFRP shear contribution. However, there is a need for more investigations regarding the relation between the size effect and the increase in CFRP rigidity. To compare maximum strain results on the contribution of CFRP fabrics to the ultimate specimen shear strength, the following dimensionless value $V_T/(b_w \times d \times f_c')$ was introduced [34]. This formula defines a dimensionless unit of the ultimate shear capacity of the beam versus the maximum strain on fibres. Therefore, it can evaluate the impact of the size effect on the ultimate specimen shear capacity (Figure 16a,b).

The maximum vertical strains were measured at the widest parts of the shear cracks. As shown in Figure 16a,b, both small specimens (S.S0.1L, S.S0.2L) showed more shear contribution of EB-CFRP than the medium and large beams. This confirms the existence of a size effect because it was expected that by increasing beam size and consequently FRP bond length, more FRP shear contribution should be obtained. In addition, despite their longer effective bond lengths, large specimens experienced less vertical strain on the fibre than small specimens, which confirms the results of previous investigations [34,39] that as beam size increases, the shear strength contribution of CFRP decreases.

In conclusion, absorption of vertical strains through the fibres is greater in smaller than in larger specimens despite the fact that both beam sizes have the same ratio of CFRP fabric.

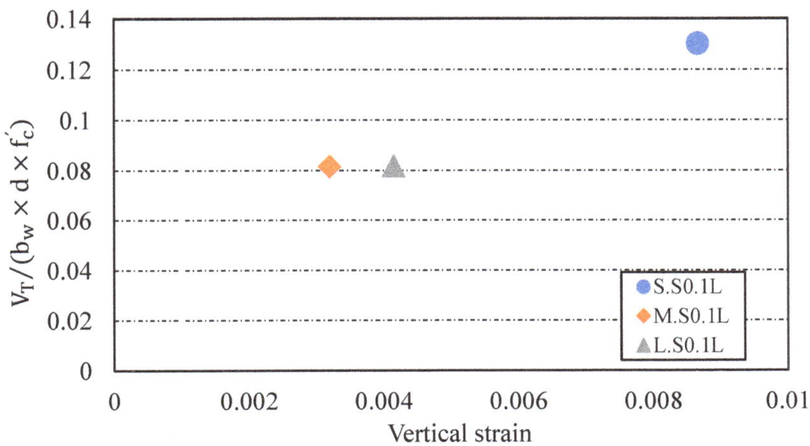

(a)

Figure 16. *Cont.*

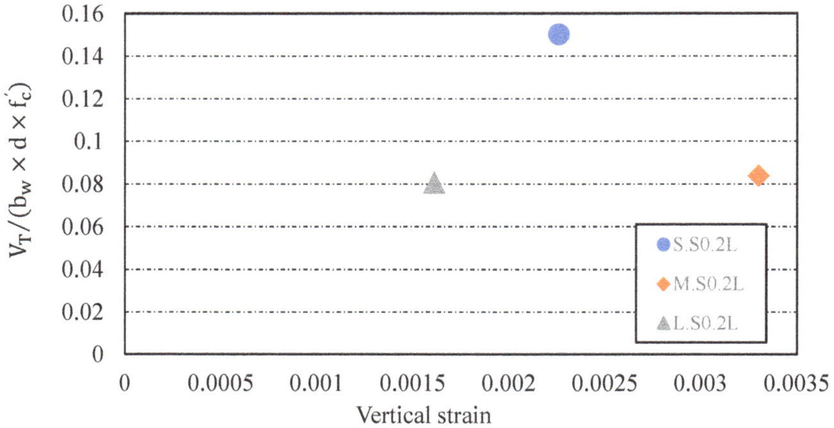

(b)

Figure 16. Correlation between maximum dimensionless shear capacity of specimens versus maximum strain along the fibre: (**a**) specimens strengthened by one layer of CFRP fabric; (**b**) specimens strengthened by two layers of CFRP fabric.

6. Conclusions

This research study has reported on advanced FE modelling of RC beams strengthened in shear with EB-FRP fabrics, with emphasis on the size effect on the shear contribution of RC beams shear-strengthened with EB-CFRP. The results obtained from FEA were in good agreement with the experimental results. Nine RC-T beams (three control beams and six strengthened beams in shear with EB-FRP) were simulated in ABAQUS. The results obtained from the numerical model were related to shear crack patterns, load-deflection curve, shear stress contributed by concrete and CFRP fabric, distributions of strains on fibres crossed by normalized distance along the main diagonal shear path during the loading process, and correlation between maximum dimensionless shear capacity of specimens versus maximum strain along the fibres, demonstrating that the proposed FEA is capable of capturing the response of the RC-T beams with high accuracy if the assumptions are defined properly. Furthermore, compared to experimental tests, FEA provided more precise observations and parameters during loading. The following conclusions can be drawn from this study:

- By increasing the rigidity of EB-CFRP in small specimens, the shear contribution of EB-CFRP showed a considerable increment, but in large specimens, EB-CFRP experienced a reduction in its absorption of shear stress due to the size effect.
- Delamination on the top parts of the diagonal shear cracks was the dominant failure mode compared to debonding, especially in medium and large specimens strengthened with EB-CFRP fabric.
- A reduction factor to account for size effect is of paramount importance. The reduction factor could be incorporated either into the effective strain or into the distribution factor, which are included in the model to express the shear contribution of EB-FRP.
- The delamination failure initiated suddenly around the tips of the shear crack, where the bond length was minimal.
- Considering strain profiles and interfacial shear stress along with the fibres and the interface layers, when the strain profile reached its peak value or became constant, the interfacial shear stress became zero.

Author Contributions: Conceptualization, A.A., Z.E.A.B., O.C. and G.E.-S.; methodology, A.A., Z.E.A.B., O.C. and G.E.-S.; software, A.A., Z.E.A.B., O.C. and G.E.-S.; validation, A.A., Z.E.A.B., O.C. and G.E.-S.; formal analysis, A.A.; investigation, A.A.; resources, O.C.; writing—original draft preparation, A.A.; writing—review and editing, O.C. and G.E.-S.; supervision, O.C. and G.E.-S.; project administration, O.C.; funding acquisition, O.C. All authors have read and agreed to the published version of the manuscript.

Funding: O.C. is funded by the National Science and Engineering Council (NERC) of Canada and by the Fonds de Recheche du Quebec-Nature & Technologie (FRQ-NT).

Institutional Review Board Statement: Not applicable.

Informed Consent Statement: Not applicable.

Data Availability Statement: The data supporting the finding of this study are available within the article.

Acknowledgments: The financial support of the Natural Sciences and Engineering Research Council of Canada (NSERC) and the Fonds de recherche du Québec–Nature et technologie (FRQNT) through operating grants is gratefully acknowledged.

Conflicts of Interest: The authors declare no conflict of interest.

Appendix A

Table A1. Experimental studies assessing the shear contribution of EB-FRP by ACI 440.2R 2017.

Specimen	Section	d_{fv}	h_f	h_w	b_w	d	nt_f	n	t_f	a/d	f'_c	ρ_{FRP}(%)	ρ_w(%)	ρ_s(%)	Configuration	β	Fibre	E_s	E_{FRP}	f_t (Mpa)	ϵ_{fu}	V_f(ACI)(N)	V_f(exp)(N)
												[52]											
T4S4-G90	T-	180	150	250	140	330	1.8	1	1.8	3	29	2.6	2.3	0.1	Ct-U	90	G	200	17.7	250	0.014124	45,878	49,000
T6S4-G90	T-	380	150	450	140	530	1.8	1	1.8	2.8	44	2.6	2.7	0.1	Ct-U	90	G	200	17.7	250	0.014124	96,854	110,000
												[53]											
SB-U1	Rect	155		180	75	155	0.11	1	0.11	2.5	27.4	0.1	5.4	0.28	St-U	90	C	200	235	4200	0.02	10,684	24,000
MB-U1	Rect	305		360	150	305	0.22	2	0.11	2.5	27.4	0.1	4.4	0.28	St-U	90	C	200	235	4200	0.02	42,049	5000
LB-U2	Rect	660		720	300	660	0.44	4	0.11	2.5	27.4	0.1	4.1	0.28	St-U	90	C	200	235	4200	0.02	181,984	22,000
												[54]											
S-Str	Rect	305		368	203	305	0.4191	1	0.4191	3	25.2	0.05	0.16	0	St-U	90	C	200	228	3450	0,015132	77,378	47,000
M-Str	Rect	457		546	305	457	0,4191	1	0.4191	3	32	0.05	0.16	0	St-U	90	C	200	228	3450	0.015132	174,101	87,000
L-Str	Rect	610		698.5	406	610	0.4191	1	0.4191	3	32	0.05	0.18	0	St-U	90	C	200	228	3450	0.015132	308,423	127,000
												[55]											
ED2 S0-1L	T-	100	55	165	95	175	0.066	1	0.066	3	25	0.14	3.76	0	Ct-U	90	C	200	231	3650	0.015801	12,196	23,000
ED1 S0-1L	T-	228	102	304	152	350	0.066	1	0.066	3	25	0.14	3.61	0	Ct-U	90	C	200	231	3650	0.015801	27,808	39,000
ED2 S1-1L	T-	100	55	165	95	175	0.066	1	0.066	3	25	0.14	3.76	0.38	Ct-U	90	C	200	231	3650	0.015801	12,196	3000
ED2 S0-2L	T-	100	55	165	95	175	0.132	2	0.066	3	25	0.28	3.76	0	Ct-U	90	C	200	231	3650	0.015801	24,393	32,000
ED1 S0-2L	T-	228	102	304	152	350	0.132	2	0.066	3	25	0.28	3.61	0	Ct-U	90	C	200	231	3650	0.015801	55,617	40,000
ED2 S1-2L	T-	100	55	165	95	175	0.132	2	0.066	3	25	0.28	3.76	0.38	Ct-U	90	C	200	231	3650	0.015801	24,393	12,000
ED1 S1-2L	T-	228	102	304	152	350	0.132	2	0.066	3	25	0.28	3.61	0.38	Ct-U	90	C	200	231	3650	0.015801	55,617	4000
												[56]											
G1-GFRP-1B	Rect	175		200	100	175	1.3	1	1.3	1.7	25	2.6	1.8	0.19	Ct-U	90	G	200	26.1	575	0.022031	47,502	18,000
G1-GFRP-2A	Rect	350		400	200	350	2.6	1	2.6	1.7	25	2.6	1.8	0.19	Ct-U	90	G	200	26.1	575	0.022031	190,008	55,000
G1-GFRP-3A	Rect	525		600	300	525	3.9	1	3.9	1.7	25	2.6	1.8	0.19	Ct-U	90	G	200	26.1	575	0.022031	427,518	64,000
G2-GFRP-2A	Rect	442		500	200	442	2.6	1	2.6	2	23.5	2.6	2.4	0.16	Ct-U	90	G	200	26.1	575	0.022031	239,953	80,000
G2-GFRP-3A	Rect	682		750	300	682	3.9	1	3.9	2	23.5	2.6	2.4	0.16	Ct-U	90	G	200	26.1	575	0.022031	555,366	180,000
G2-CFRP-1	Rect	196		250	100	196	1	1	1	2	23.5	2	2.4	0.16	Ct-U	90	C	200	95.8	986	0.010292	150,214	25,000
G2-CFRP-2	Rect	442		500	200	442	2	1	2	2	23.5	2	2.4	0.16	Ct-U	90	C	200	95.8	986	0.010292	677,497	120,000
												[57]											
4	Rect	89		102	114	85	1.539	1	1.539	3.53	42.9	2.7	2.61	0	Ct-U	45	G	200	16.7	342.35	0.0205	25,879	31,500
												[58]											
CO2	Rect	260		305	150	260	0.165	1	0.165	3.6	20.5	0.088	4	0	St-U	90	C	200	228	3648	0.016	31,299	40,000
CO3	Rect	260		305	150	260	0.165	1	0.165	3.6	20.5	0.22	4	0	Ct-U	90	C	200	228	3648	0.016	78,249	65,000
												[59]											
BT2	T-	260	100	305	150	360	0.165	1	0.165	3	35	0.22	2.28	0	Ct-U	90	C	200	228	3648	0.016	78,249	65,000
BT3	T-	260	100	305	150	360	0,165	1	0.165	3	35	0.44	2.28	0	Ct-U	90	C	200	228	3648	0.016	78,249	67,500
BT4	T-	260	100	305	150	360	0.165	1	0.165	3	35	0.088	2.28	0	St-U	90	C	200	228	3648	0.016	31,299	72,000

Table A1. *Cont.*

											[60]												
SO3-2	Rect	260		305	150	260	0.165	1	0.165	3	27.5	0.088	4.2	0	St-U	90	C	200	228	3648	0.016	31,299	54,000
SO3-3	Rect	260		305	150	260	0.165	1	0.165	3	27.5	0.132	4.2	0	St-U	90	C	200	228	3648	0.016	46,949	56,500
SO3-4	Rect	260		305	150	260	0.165	1	0.165	3	27.5	0.22	4.2	0	Ct-U	90	C	200	228	3648	0.016	78,249	67,500
SO4-2	Rect	260		305	150	260	0.165	1	0.165	4	27.5	0.088	4.2	0	St-U	90	C	200	228	3648	0.016	31,299	62,500
											[52]												
T6NS-C45	T-	390	150	450	140	540	0.7	1	0.7	2.9	44.1	2.6	2.81	0	St-U	45	C	200	230	3450	0.015	355,193	103,500
T6S4-C90	T-	390	150	450	140	540	0.7	1	0.7	2.9	44.1	0.08	2.81	0.1	St-U	90	C	200	230	3450	0.015	251,160	85,250
T6S4-G90	T-	390	150	450	140	540	1.8	1	1.8	2.9	44.1	2.6	2.81	0.1	Ct-U	90	G	200	17.7	265.5	0.015	99,403	109,900
T6S4-Tri	T-	390	150	450	140	540	2.1	1	2.1	2.9	44.1	3	2.81	0.1	Ct-U	45	G	200	8.1	121.5	0.015	75,054	129,200
T4NS-G90	T-	190	150	450	140	340	1.8	1	1.8	3.2	29	2.6	4.46	0	Ct-U	90	G	200	17.7	265.5	0.015	48,427	43,600
T4S4-G90	T-	190	150	450	140	340	1.8	1	1.8	3.2	29	2.6	4.46	0.1	Ct-U	90	G	200	17.7	265.5	0.015	48,427	48,650
T4S2-C45	T-	190	150	450	140	340	0.7	1	0.7	3.2	29	0.08	4.46	0.2	St-U	45	C	200	230	3450	0.015	173,043	17,800
T4S2-G90	T-	190	150	450	140	340	1.8	1	1.8	3.2	29	2.6	4.46	0.2	Ct-U	90	G	200	17.7	265.5	0.015	48,427	24,350
T4S2-Tri	T-	190	150	450	140	340	2.1	1	2.1	3.2	29	3	4.46	0.2	Ct-U	45	G	200	8.1	121.5	0.015	36,564	41,400
											[61]												
G5.5-1L	T-	254.1	88.9	355.6	92	343	0.10902	1	0.10902	2	37.9	0.237	3.6	1.1	Ct-U	90	C	203	231	3696	0.016	51,193	31,150
G5.5-2L	T-	254.1	88.9	355.6	92	343	0.21804	2	0.10902	2	37.9	0.475	3.6	1.1	Ct-U	90	C	203	231	3696	0.016	102,386	53,400
G8-1L	T-	254.1	88.9	355.6	92	343	0.10902	1	0.10902	2	37.9	0.237	3.6	0.76	Ct-U	90	C	203	231	3696	0.016	51,193	31,150
G8-2L	T-	254.1	88.9	355.6	92	343	0.21804	2	0.10902	2	37.9	0.475	3.6	0.76	Ct-U	90	C	203	231	3696	0.016	102,386	62,300
G8-3L	T-	254.1	88.9	355.6	92	343	0.32706	3	0.10902	2	37.9	0.712	3.6	0.76	Ct-U	90	C	203	231	3696	0.016	153,579	84,550
G16-1L	T-	254.1	88.9	355.6	92	343	0.10902	1	0.10902	2	37.9	0.237	3.6	0.38	Ct-U	90	C	203	231	3696	0.016	51,193	40,050
G16-2L	T-	254.1	88.9	355.6	92	343	0.21804	2	0.10902	2	37.9	0.475	3.6	0.38	Ct-U	90	C	203	231	3696	0.016	102,386	84,550
G24-1L	T-	254.1	88.9	355.6	92	343	0.10902	1	0.10902	2	37.9	0.237	3.6	0.25	Ct-U	90	C	203	231	3696	0.016	51,193	53,400
G24-2L	T-	254.1	88.9	355.6	92	343	0.21804	2	0.10902	2	37.9	0.475	3.6	0.25	Ct-U	90	C	203	231	3696	0.016	102,386	48,950
G24-3L	T-	254.1	88.9	355.6	92	343	0.32706	3	0.10902	2	37.9	0.712	3.6	0.25	Ct-U	90	C	203	231	3696	0.016	153,579	53,400

References

1. Tan, P. Ballistic protection performance of curved armor systems with or without debondings/delaminations. *Mater. Design* **2014**, *64*, 25–34. [CrossRef]
2. Kolanu, N.R.; Raju, G.; Ramji, M. A unified numerical approach for the simulation of intra and inter laminar damage evolution in stiffened CFRP panels under compression. *Compos. Part B Eng.* **2020**, *190*, 107931. [CrossRef]
3. Rozylo, P. Failure analysis of thin-walled composite structures using independent advanced damage models. *Compos. Struct.* **2021**, *262*, 113598. [CrossRef]
4. Kaliakin, V.N.; Chajes, M.J.; Januszka, T.F. Analysis of concrete beams reinforced with externally bonded woven composite fabrics. *Compos. Part B Eng.* **1996**, *27*, 235–244. [CrossRef]
5. Arduini, M.; Di Tommaso, A.; Nanni, A. Brittle failure in FRP plate and sheet bonded beams. *ACI Struct. J.* **1997**, *94*, 363–370.
6. Amir, M. Analytical study of reinforced concrete beams strengthened with web-bonded fiber reinforced plastic plates or fabrics. *ACI Struct. J.* **1998**, *95*, 12–16.
7. Vecchio, F.; Bucci, F. Analysis of repaired reinforced concrete structures. *J. Struct. Eng.* **1999**, *125*, 644–652. [CrossRef]
8. Lee, T.K.; Al-Mahaidi, R.S.H.; Taplin, G.R. Non-linear finite element modelling of shear-damaged concrete T-beams repaired with CFRP laminates. In Proceedings of the ACUN-2 International Composites Conference, Sydney, Australia, 14–18 February 2000; pp. 253–258.
9. Kachlakev, D.I.; Miller, T.H.; Potisuk, T.; Yim, S.C.; Chansawat, K. *Finite Element Modeling of Reinforced Concrete Structures Strengthened with FRP Laminates;* Oregon Department of Transportation Research Group: Klamath Falls, OR, USA, 2001.
10. Santhakumar, R.; Chandrasekaran, E.; Dhanaraj, R. Analysis of retrofitted reinforced concrete shear beams using carbon fiber composites. *Electron. J. Struct. Eng.* **2004**, *4*, 66–74.
11. Elyasian, I.; Abdoli, N.; Rounagh, H. Evaluation of parameters effective in FRP shear strengthening of RC beams using FE method. *Asian J. Civ. Eng.* **2006**, *7*, 249–257.
12. Godat, A.; Chaallal, O.; Neale, K.W. Nonlinear finite element models for the embedded through-section FRP shear-strengthening method. *Comput. Struct.* **2013**, *119*, 12–22. [CrossRef]
13. Qapo, M.; Dirar, S.; Jemaa, Y. Finite element parametric study of reinforced concrete beams shear-strengthened with embedded FRP bars. *Compos. Struct.* **2016**, *149*, 93–105. [CrossRef]
14. Shomali, A.; Mostofinejad, D.; Esfahani, M.R. Effective strain of CFRP in RC beams strengthened in shear with NSM reinforcements. In *Structures;* Elsevier: Amsterdam, The Netherlands, 2020; pp. 635–645.
15. Wong, R.S.; Vecchio, F.J. Towards modeling of reinforced concrete members with externally bonded fiber-reinforced polymer composites. *ACI Struct. J.* **2003**, *100*, 47–55.
16. Smith, S.T.; Otoom, O.; Foster, S. Finite element modelling of RC beams strengthened in shear with FRP composites. In Proceedings of the 2nd International Fib Congress, Naples, Italy, 5–8 June 2006.
17. Qu, Z.; Lu, X.-Z.; Ye, L.-P.; Chen, J.-F.; Rotter, J.M. Numerical modeling of FRP shear strengthened RC beams using compression field theory. In Proceedings of the Third International Conference on FRP Composites in Civil Engineering (CICE 2006), Miami, FL, USA, 13–15 December 2006; pp. 391–394.
18. Godat, A.; Neale, K.W.; Labossière, P. Numerical modeling of FRP shear-strengthened reinforced concrete beams. *J. Compos. Constr.* **2007**, *11*, 640–649. [CrossRef]
19. Godat, A.; Labossière, P.; Neale, K. Numerical prediction of shear crack angles for FRP shear-strengthened concrete beams. In Proceedings of the 4th International Conference on FRP Composites in Civil Engineering (CICE2008), Zurich, Switzerland, 22–24 July 2008; p. 6.
20. Lee, H.-K.; Ha, S.-K.; Afzal, M. Finite element analysis of shear-deficient RC beams strengthened with CFRP strips/sheets. *Struct. Eng. Mech.* **2008**, *30*, 247–261. [CrossRef]
21. Lu, X.; Chen, J.; Ye, L.; Teng, J.; Rotter, J. RC beams shear-strengthened with FRP: Stress distributions in the FRP reinforcement. *Constr. Build. Mater.* **2009**, *23*, 1544–1554. [CrossRef]
22. Chen, G.; Teng, J.; Chen, J.; Rosenboom, O. Interaction between steel stirrups and shear-strengthening FRP strips in RC beams. *J. Compos. Constr.* **2010**, *14*, 498–509. [CrossRef]
23. You, Y.-M.; Ayoub, A.; Belarbi, A. Three-dimensional nonlinear finite-element analysis of prestressed concrete beams strengthened in shear with FRP composites. *J. Compos. Constr.* **2011**, *15*, 896–907. [CrossRef]
24. Dirar, S.; Lees, J.M.; Morley, C. Phased nonlinear finite-element analysis of precracked RC T-beams repaired in shear with CFRP sheets. *J. Compos. Constr.* **2012**, *17*, 476–487. [CrossRef]
25. Godat, A.; Labossière, P.; Neale, K. Numerical investigation of the parameters influencing the behaviour of FRP shear-strengthened beams. *Constr. Build. Mater.* **2012**, *32*, 90–98. [CrossRef]
26. Godat, A.; Labossière, P.; Neale, K.W.; Chaallal, O. Behavior of RC members strengthened in shear with EB FRP: Assessment of models and FE simulation approaches. *Comput. Struct.* **2012**, *92*, 269–282. [CrossRef]
27. Imperatore, S.; Lavorato, D.; Nuti, C.; Santini, S.; Sguerri, L. Numerical modeling of existing RC beams strengthened in shear with FRP U-sheets. In Proceedings of the 6th International Conference on FRP Composites in Civil Engineering—CICE2012, Rome, Italy, 13–15 June 2012; pp. 13–15.
28. Sayed, A.M.; Wang, X.; Wu, Z. Modeling of shear capacity of RC beams strengthened with FRP sheets based on FE simulation. *J Compos. Constr.* **2013**, *17*, 687–701. [CrossRef]

29. Manos, G.; Theofanous, M.; Katakalos, K. Numerical simulation of the shear behaviour of reinforced concrete rectangular beam specimens with or without FRP-strip shear reinforcement. *Adv. Eng. Softw.* **2014**, *67*, 47–56. [CrossRef]
30. Qapo, M.; Dirar, S.; Yang, J.; Elshafie, M.Z. Nonlinear finite element modelling and parametric study of CFRP shear-strengthened prestressed concrete girders. *Constr. Build. Mater.* **2015**, *76*, 245–255. [CrossRef]
31. Ibars, E.O.; Ferreira, D.; Bernat, A.M.; García, J.M.B. Numerical analysis of reinforced concrete beams strengthened in shear by externally bonded (EB) fibre reinforced polymer (FRP) sheets. *Hormigón Acero* **2018**, *69*, 113–120. [CrossRef]
32. Al Jawahery, M.S.; Gulsan, M.E.; Albegmprli, H.M.; Mansoori, I.A.H.; Cevik, A. Experimental investigation of rehabilitated RC haunched beams via CFRP with 3D-FE modeling analysis. *Eng. Struct.* **2019**, *196*, 109301. [CrossRef]
33. Jin, L.; Xia, H.; Xuan-Ang, J.; Du, X. Size effect on shear failure of CFRP-strengthened concrete beams without web reinforcement: Meso-scale simulation and formulation. *Compos. Struct.* **2020**, *236*, 111895. [CrossRef]
34. Godat, A.; Qu, Z.; Lu, X.; Labossiere, P.; Ye, L.; Neale, K.W. Size effects for reinforced concrete beams strengthened in shear with CFRP strips. *J. Compos. Constr.* **2010**, *14*, 260–271. [CrossRef]
35. Chen, G.; Chen, J.; Teng, J. On the finite element modelling of RC beams shear-strengthened with FRP. *Constr. Build. Mater.* **2012**, *32*, 13–26. [CrossRef]
36. Kani, G.N.J. How safe are our large reinforced concrete beams? *J. Proc.* **1967**, *64*, 128–141.
37. Bazant, Z.P.; Kazemi, M.T. Size effect on diagonal shear failure of beams without stirrups. *ACI Struct. J.* **1991**, *88*, 268–276.
38. Collins, M.P.; Kuchma, D. How safe are our large, lightly reinforced concrete beams, slabs, and footings? *Struct. J.* **1999**, *96*, 482–490.
39. Benzeguir, Z.E.A.; El-Saikaly, G.; Chaallal, O. Size effect in RC T-beams strengthened in shear with externally bonded CFRP sheets: Experimental study. *J. Compos. Constr.* **2019**, *23*, 04019048. [CrossRef]
40. Chen, G. Behaviour and Strength of RC Beams Shear-Strengthened with Externally Bonded FRP Reinforcement. Ph.D. Thesis, The Hong Kong Polytechnic University, Hong Kong, China, 2010.
41. Bažant, Z.P.; Becq-Giraudon, E. Statistical prediction of fracture parameters of concrete and implications for choice of testing standard. *Cem. Concr. Res.* **2002**, *32*, 529–556. [CrossRef]
42. Hordijk, D. Local Approach to Fatigue of Concrete. Ph.D. Thesis, Delft University of Technology, Delft, The Netherlands, 1991.
43. Saenz, L.P. Discussion of "Equation for the Stress-Strain Curve of Concrete" by Desayi and Krishnan. *J. Am. Concr. Inst.* **1964**, *61*, 1229–1235.
44. ACI Committee. *Building Code Requirements for Reinforced Concrete (ACI 318–63)*; American Concrete Institute: Indianapolis, IN, USA, 1963.
45. Tao, Y.; Chen, J.-F. Concrete damage plasticity model for modeling FRP-to-concrete bond behavior. *J. Compos. Constr.* **2014**, *19*, 04014026. [CrossRef]
46. Tran, D.T.; Pham, T.M.; Hao, H.; Chen, W. Numerical investigation of flexural behaviours of precast segmental concrete beams internally post-tensioned with unbonded FRP tendons under monotonic loading. *Eng. Struct.* **2021**, *249*, 113341. [CrossRef]
47. Tran, D.T.; Pham, T.M.; Hao, H.; Chen, W. Numerical study on bending response of precast segmental concrete beams externally prestressed with FRP tendons. *Eng. Struct.* **2021**, *241*, 112423. [CrossRef]
48. Lu, X.; Teng, J.; Ye, L.; Jiang, J. Bond–slip models for FRP sheets/plates bonded to concrete. *Eng. Struct.* **2005**, *27*, 920–937. [CrossRef]
49. Pellegrino, C.; Modena, C. Fiber reinforced polymer shear strengthening of reinforced concrete beams with transverse steel reinforcement. *J. Compos. Constr.* **2002**, *6*, 104–111. [CrossRef]
50. Chen, J.-F.; Teng, J. Shear capacity of FRP-strengthened RC beams: FRP debonding. *Constr. Build. Mater.* **2003**, *17*, 27–41. [CrossRef]
51. Chen, J.; Teng, J. Shear capacity of fiber-reinforced polymer-strengthened reinforced concrete beams: Fiber reinforced polymer rupture. *J. Struct. Eng.* **2003**, *129*, 615–625. [CrossRef]
52. Deniaud, C.; Cheng, J.R. Shear behavior of reinforced concrete T-beams with externally bonded fiber-reinforced polymer sheets. *Struct. J.* **2001**, *98*, 386–394.
53. Leung, C.K.; Chen, Z.; Lee, S.; Ng, M.; Xu, M.; Tang, J. Effect of size on the failure of geometrically similar concrete beams strengthened in shear with FRP strips. *J. Compos. Constr.* **2007**, *11*, 487–496. [CrossRef]
54. Bae, S.; Tann, B.; Belarbi, A. Size effect of reinforced concrete beams strengthened in shear with externally bonded CFRP sheets. In Proceedings of the 6th International Conference on FRP Composites in Civil Engineering (CICE 2012), Rome, Italy, 13–15 June 2012; pp. 1–8.
55. Bousselham, A.; Chaallal, O. Experimental investigations on the influence of size on the performance of RC T-beams retrofitted in shear with CFRP fabrics. *Eng. Struct.* **2013**, *56*, 1070–1079. [CrossRef]
56. Nguyen-Minh, L.; Rovňák, M. Size effect in uncracked and pre-cracked reinforced concrete beams shear-strengthened with composite jackets. *Compos. Part B Eng.* **2015**, *78*, 361–376. [CrossRef]
57. Berset, J.-D. *Strengthening of Reinforced Concrete Beams for Shear Using FRP Composites*; Massachusetts Institute of Technology: Cambridge, MA, USA, 1992.
58. Khalifa, A.; Tumialan, G.; Nanni, A.; Belarbi, A. Shear strengthening of continuous RC beams using externally bonded CFRP sheets. In Proceedings of the 4th International Symposium on FRP for Reinforcement of Concrete Structures (FRPRCS4), Baltimore, MD, USA, 31 October–5 November 1999; pp. 995–1008.

59. Khalifa, A.; Nanni, A. Improving shear capacity of existing RC T-section beams using CFRP composites. *Cem. Concr. Compos.* **2000**, *22*, 165–174. [CrossRef]
60. Khalifa, A.; Nanni, A. Rehabilitation of rectangular simply supported RC beams with shear deficiencies using CFRP composites. *Constr. Build. Mater.* **2002**, *16*, 135–146. [CrossRef]
61. Chaallal, O.; Shahawy, M.; Hassan, M. Performance of reinforced concrete T-girders strengthened in shear with carbon fiber-reinforced polymer fabric. *Struct. J.* **2002**, *99*, 335–343.

Journal of
Composites Science

Article

The Impact of Fiber Orientation on Structural Dynamics of Short-Fiber Reinforced, Thermoplastic Components—A Comparison of Simulative and Experimental Investigations

Alexander Kriwet [1,*] and Markus Stommel [2]

[1] Structural Dynamics Powertrain, NVH CAE, NVH Powertrain, Mercedes-Benz AG, Mercedesstr. 122, 70372 Stuttgart, Germany
[2] Institute Polymer Materials, Leibniz-Institute for Polymer Research Dresden, Hohe Straße 6, 01069 Dresden, Germany; stommel@ipfdd.de
* Correspondence: alexander.kriwet@mercedes-benz.com

Abstract: The quality of the fiber orientation of injection molding simulations and the transferred fiber orientation content, due to the process–structure coupling, influence the material modeling and thus the prediction of subsequently performed structural dynamics simulations of short-fiber reinforced, thermoplastic components. Existing investigations assume a reliable prediction of the fiber orientation in the injection molding simulation. The influence of the fiber orientation models and used boundary conditions of the process–structure coupling is mainly not investigated. In this research, the influence of the fiber orientation from injection molding simulations on the resulting structural dynamics simulation of short-fiber reinforced thermoplastic components is investigated. The Advani–Tucker Equation with phenomenological coefficient tensor is used in a 3- and 2.5-dimensional modeling approach for calculating the fiber orientation. The prediction quality of the simulative fiber orientations is evaluated in comparison to experiments. Depending on the material modeling and validation level, the prediction of the simulated fiber orientation differs in the range between 7.3 and 347.2% averaged deviation significantly. Furthermore, depending on the process–structure coupling and the number of layers over the thickness of the model, the Kullback–Leibner divergence differs in a range between 0.1 and 4.9%. In this context, more layers lead to higher fiber orientation content in the model and improved prediction of the structural dynamics simulation. This is significant for local and slightly for global structural dynamics phenomena regarding the mode shapes and frequency response behavior of simulative and experimental investigations. The investigations prove that the influence of the fiber orientation on the structural dynamics simulation is lower than the influence of the material modeling. With a relative average deviation of 2.8% in the frequency and 38.0% in the amplitude of the frequency response function, it can be proven that high deviations between experimental and simulative fiber orientations can lead to a sufficient prediction of the structural dynamics simulation.

Keywords: short-fiber reinforced thermoplastic components; injection molding simulation; fiber orientation; structural dynamics; material modeling

Citation: Kriwet, A.; Stommel, M. The Impact of Fiber Orientation on Structural Dynamics of Short-Fiber Reinforced, Thermoplastic Components—A Comparison of Simulative and Experimental Investigations. *J. Compos. Sci.* **2022**, *6*, 106. https://doi.org/10.3390/jcs6040106

Academic Editor: Stelios K. Georgantzinos

Received: 9 March 2022
Accepted: 30 March 2022
Published: 1 April 2022

1. Introduction

Short-fiber reinforced thermoplastics are an essential group of engineering materials in modern vehicle powertrains due to their significant lightweight potential. Under Noise-Vibration-Harshness (NVH) aspects, short-fiber reinforced plastics offer good vibration isolation and thus noise isolation behavior. This is due to the favorable stiffness and damping behavior. For efficient prediction of the stiffness and damping of short-fiber reinforced plastics, established material models are based on multi-stage homogenization methods. The principle of all methods is that the microstructure and thus the properties of the composite are described with mathematical–physical models. Thereby, the consideration of the fiber orientation via the process–structure coupling is fundamental. By conditioning the material models with direction-dependent properties, discrete material parameters are assigned to each element of the simulation depending on the local fiber orientation tensor. Therefore, the quality of the prediction of the fiber orientation in the injection molding simulation influences the prediction quality of the structural dynamics simulation. Further information corresponding to the homogenization methods and the process–structure coupling is provided in [1–8]. Glaser shows in [9] that the prediction quality of structural dynamics simulations of a short glass fiber reinforced thermoplastic intake pipe can be significantly increased by considering the fiber distribution and thus the anisotropic stiffness and damping. The main focus of this study is on the procedure and advantages of the process–structure coupling. However, information and boundaries to calculate the fiber orientation are not provided. Studies by Schmachtenberg et al. [10] additionally use advanced experimental methods to calibrate the simulation models. Arping [11] and Kremer [12] use a similar method to homogenize the properties of the fiber-reinforced plastic component and recalibrate the models of the structural dynamics simulation through reverse engineering. Thereby, material modeling is the focus of the research. The simulative fiber orientation is considered a fixed boundary condition. The disadvantage is that errors caused by an insufficient simulative fiber orientation are compensated by the reverse engineering of the material model. Influences on the results caused by the quality of the simulative fiber orientation are largely not considered. State-of-the-art extended approaches additionally pursue the consideration of the material properties depending on the boundary conditions, such as frequency, temperature or humidity. There exists a large number of publications dealing with the structural dynamics design of fiber-reinforced plastic components under NVH aspects [13–18]. However, all existing publications have in common that the reliable prediction of the fiber orientation in the injection molding simulation is assumed. The boundary conditions and fiber orientation models used by the injection molding simulations are mainly not explained. However, there can be significant differences in the prediction quality of the fiber orientation, depending on which fiber model is used and how the models are calibrated. Furthermore, as a result of the process–structure coupling, there is always a loss of information about the fiber orientation between the injection molding model and the finite element (FE) model, which is not sufficiently considered in existing publications [1].

The aim of this research is to investigate the influence of the fiber orientation from the injection molding simulation on the resulting structural dynamics FE simulation of short-fiber reinforced thermoplastic components. In the first section of this paper, the state of art and the applied method for calculating the fiber orientation using injection molding simulations are explained. Furthermore, the method used to investigate the influence of the fiber orientation on the prediction quality of the structural dynamics simulation is presented. In the last section, the results of the simulative and experimentally investigated fiber orientation, as well as the prediction quality of the corresponding structural dynamics investigations, are discussed.

2. State of Art and Methods

In the first part of this section, the state of art for fiber orientation calculation with injection molding simulations is discussed. Following the proposed method for the simulative and experimental characterization of the fiber orientation is presented. In the last part, the focus is on the method for the experimental characterization by modal analysis and the structural dynamics simulation model of the short-fiber reinforced plastic components. In this research, the proposed methods are implemented for two types of short-fiber reinforced thermoplastics. On the one hand, polyamide 6.6 (PA66) is a common engineering plastic, and on the other hand, polyphthalamide (PPA) is a high-performance plastic, each with 50 wt.% short glass fiber reinforcement (GF50) [19,20]. As a representative composite component for investigating the influence of fiber orientation on structural dynamics, the so-called engine bracket is investigated. In modern combustion engines, the engine bracket is mounted on the crankcase and transmits the powertrain-induced operating vibrations into the vehicle structure. For further information about the functionality of the engine bracket, please refer to [15–18,21,22].

2.1. State of Art of Injection Molding Simulations for Fiber Orientation Calculation

The prediction of the fiber orientation in injection molding simulations is based on the so-called continuity equation according to Fokker-Planck [23,24]. This is derived from the velocity field of the fluid and thus the hydrodynamic forces acting on the fiber. Extended by a diffusion term to describe the fiber interaction in the fluid field, the Folgar–Tucker equation [25] is formed:

$$\frac{d\psi}{dt} = -\nabla\cdot(\dot{p}\psi) + D_i\nabla^2\psi, \tag{1}$$

ψ corresponds to the fiber probability density function (PDF) at time t, ∇ is the gradient operator, $\dot{p}$ is the change of the fiber orientation and D_i is the fiber interaction coefficient. The first part of the equation can be summarized as a hydrodynamic term and the other as a diffusion term. Applied to injection molding simulations of composite components, solving Equation (1) is numerically extremely cost-intensive. For this reason, the PDF is mainly substituted by an evolution equation of the fiber orientation tensor $\mathbf{A}$:

$$\frac{d\psi}{dt} \approx \frac{d\mathbf{A}}{dt} = \dot{\mathbf{A}} = \dot{\mathbf{A}}^h + \dot{\mathbf{A}}^d, \tag{2}$$

with

$$\mathbf{A} = \mathbf{A}_N = \int_0^{2\pi}\int_0^{\pi} p_N\psi(\Theta,\Phi)\sin\Theta d\Theta d\Phi. \tag{3}$$

Thus, $\dot{\mathbf{A}}^h$ corresponds to the hydrodynamic part and $\dot{\mathbf{A}}^d$ to the diffusion part of the Folgar–Tucker Equation. Equation (2) is numerically stable for calculating the fiber orientation of plastic components in injection molding simulations. On this basis, a large number of publications exist which follow different approaches for the formulation of the hydrodynamic and diffusive parts. Advani and Tucker formulate in [26] an extension of the Fokker–Planck Equation to consider the rate of change of the second-order fiber orientation tensor $\mathbf{A}$, which is calibrated by a phenomenological approach via the parameter C_I. Huynh shows in [27] that the prediction of the calculated fiber orientation can be further improved by a scalar reduction factor κ. Wang et al. [28] extend this reduction factor by a reformulation of the second-order fiber orientation tensor $\mathbf{A}$ through a decomposition of the eigenvalues λ and eigenvectors $\mathbf{e}$. This forms the reduced strain closure (RSC) model, which is an established method for calculating fiber orientation. A comparable established approach is shown by Phelps and Tucker in [29], whereby the phenomenological parameter C_I is replaced by a rotary diffusion tensor $\mathbf{C}$. This forms the so-called anisotropic rotary diffusion (ARD) model and thus the basis of advanced models [29–32]. An overview of macroscopic fiber orientation models is discussed in [24].

Furthermore, the decomposition of the second-order fiber orientation tensor $\frac{d\mathbf{A}_N}{dt}$ is dependent on the fiber orientation tensor of the next higher-order $\mathbb{A}_N$:

$$\frac{d\mathbf{A}_N}{dt} = f(\mathbb{A}_N). \tag{4}$$

Established methods use the formulation of a closure approximation. This approximation of the higher-order fiber orientation tensor $\mathbb{A}_N$ is based on mathematical assumptions [23,24,33]. In general, the increased information of the higher-order tensors is usually not fully captured by the closure formulations. Advani and Tucker show in [26] a linear summation of all products from the components of the second-order fiber orientation tensor $\mathbf{A}_{ij}$ and the Kronecker delta δ_{ij} to approximate the higher-order tensor. By neglecting the linear terms, the quadratic closure is formed and by combining the linear and quadratic approach, the hybrid closure is formed [26]. Furthermore, the so-called natural closure of Verleye and Dupret should be mentioned [34], where the fourth-order orientation tensor $\mathbb{A}_{ijkl}$ is defined as a function of the second-order tensor $f(\mathbf{A}_{ij})$. This forms the basis of various advanced approximations methods [24,35]. Advanced models pursue an exact formulation of the fourth-order fiber orientation tensor [36,37], the use of neural networks [38,39], distribution reconstruction methods [23,33,40,41] or the reconstruction of even higher-order fiber orientation tensors [33,42]. However, to ensure numerically cost-effective fiber orientation calculations in injection molding simulation, hybrid or natural closures are still widely used today [24]. An overview of existing closure formulations is discussed in [23,24,33].

2.2. Methodology for Simulative and Experimental Fiber Orientation Investigations

According to the state of art, the Advani–Tucker Equation is used as the basis for calculating the fiber orientation in this research. The hydrodynamic part is defined as:

$$\dot{\mathbf{A}}^{h} = (\mathbf{W}\cdot\mathbf{A} - \mathbf{A}\cdot\mathbf{W}) + \xi(\mathbf{D}\cdot\mathbf{A} + \mathbf{A}\cdot\mathbf{D} - 2\mathbb{A} : \mathbf{D}), \tag{5}$$

and the diffusion part as

$$\dot{\mathbf{A}}^{d} = 2\dot{\gamma}\cdot\boldsymbol{C}^{v}\cdot(\boldsymbol{I} - \boldsymbol{A}), \tag{6}$$

with $\mathbf{W}$ as vorticity tensor, ξ as particle shape function, $\mathbf{D}$ as strain rate tensor, $\dot{\gamma}$ as the magnitude of the strain rate tensor and $\boldsymbol{I}$ as a unit tensor. $\boldsymbol{C}^{v}$ corresponds to a phenomenological coefficient tensor for describing the fiber interaction and is calibrated through experimentally determined fiber orientations. Thus, the diffusion part is equivalent to a rotary diffusion approach by the appropriate definition of $\boldsymbol{C}^{v}$. To evaluate and calibrate the phenomenological parameters of the injection molding model, µCT investigations are carried out. For this purpose, material plates are created by varying the geometry or the process parameters, and µCT specimens are investigated at selected positions according to the fiber orientation. It is important to choose specimen positions that allow a representation of the microstructure and thus a reliable investigation of the fiber orientation. Under the usage of simple plate geometry, a small number of µCT specimens are taken in an evenly distributed way. When transferred to the injection molding simulation of short-fiber reinforced plastic components, significant differences in the prediction quality of the fiber orientation can occur [4,8,21]. As a result, extended methods include a recalibration of the default parameters of the simulation in comparison to µCT specimens taken from plastic components [8,43]. At the component level, µCT specimen positions are selected with significantly different geometric or material-specific characteristics in order to calibrate the simulation for robustness. The geometric aspects include, for example, different wall thicknesses, ribs or triple points. This allows a direct recalibration of the injection molding model. On the other hand, material-specific aspects are, for example, in- or ejection positions or impact points of the plastic melt. These aspects allow a recalibration of the injection molding model in case of process-related errors or damage analyses. Based on the experimentally determined fiber orientations, the simulation model parameters are calibrated with numerical or mechanistic methods. The injection molding material model is sufficient if the deviation between simulated and experimentally determined fiber orientation is minimized for several parameter variations [8,43]. Figure 1 schematically shows the procedure for a multi-stage calibration of the calculated fiber orientation from injection molding simulations.

In this research, injection molding simulations are performed using two commercial software programs. On the one hand, 2.5-dimensional (2.5D) injection molding simulations were conducted with the software CADMOULD®from SIMCON. The simulations were conducted in the scope of contract simulations by PART Engineering respectively. Thereby, the hydrodynamic part of the Advani–Tucker Equation equals a natural closure (NC) formulation to approximate the fourth-order tensor $\mathbb{A}_{ijkl}$. Furthermore, the fiber interaction tensor $\boldsymbol{C}^{v} = \boldsymbol{C}^{NC}$ in the diffusion part depends on the alignment factor in the middle layer

α_K, in the boundary layer α_R and on the rotational velocity factor β_R. This corresponds to the phenomenological coefficients in CADMOULD®, which are calibrated in comparison to µCT results. The initial coefficient tensors $\boldsymbol{C}_{0,CM}^{NC,S}$ for the respective material S originating from a database of the CADMOULD® software. These are defined as default values for both the PA66GF50 and the PPAGF50 material as $\alpha_K = 0.92$, $\alpha_R = 0.92$, $\beta_R = 0.15$. To maximize the calculated information of the fiber orientation from the injection molding simulation, 10 elements are defined uniformly over each wall thickness. In comparison, 3-dimensional (3D) injection molding simulations were conducted with the software MOLDFLOW® from AUTODESK. The simulations were conducted in the scope of contract simulations by Daimler Trucks AG respectively. Here, an orthotropic closure (OC) is defined in the hydrodynamic part and the coefficient tensor $\boldsymbol{C}^v = \boldsymbol{C}^{OC}$ depends on the scalar interaction coefficient C_I and the asymmetric coefficients of the rotational diffusion D_1, D_2, D_3. This corresponds to the phenomenological coefficients in MOLDFLOW®. The initial coefficient tensors $\boldsymbol{C}_{0,MF}^{OC,S}$ originated from a database of the company BASF SE. These are defined as default values $C_I = 1.0$, $D_1 = 1.0$, $D_2 = 0.8$, $D_3 = 0.15$. To maximize the fiber orientation information from the injection molding simulation with MOLDFLOW®, 14 elements are defined uniformly over each wall thickness.

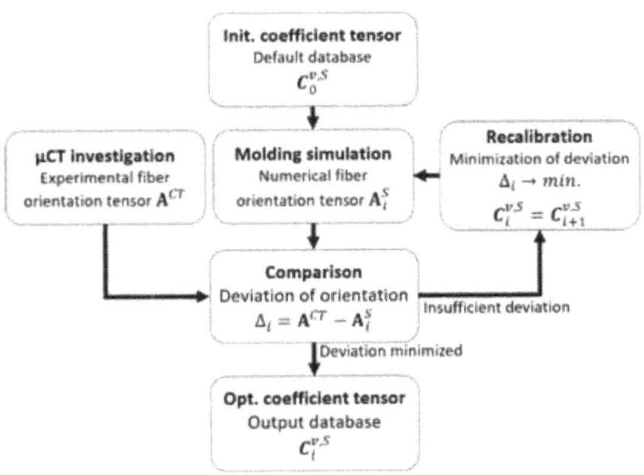

Figure 1. Method for calibrating the calculated fiber orientation of injection molding simulations in comparison to experimental µCT investigations.

To evaluate the prediction quality of the simulative fiber orientation, experimental µCT investigations are evaluated in comparison at the plate and component level, as seen in Figure 2. At the plate level, the µCT specimens are extracted equally distributed from the plate geometry. The dimensions (length × width × thickness) of the plate geometry are (180 mm × 180 mm × 2 mm). In comparison, µCT specimens at the component level are extracted with a focus on geometric aspects of the engine bracket. As a result, a µCT specimen is extracted from a side rib and the upper end of the bracket. The approximate dimensions of the engine bracket are (215 mm × 135 mm × 135 mm) with a thickness of the surfaces and ribs between 2–5 mm. In the current case, the dimensions of the µCT specimens both at the plate and component level are approximate (5 mm × 5 mm × 2 mm). Figure 2 provides an overview of the geometry on the plate and component level and shows the specimen positions. For evaluation, A_{11} corresponds to the component of the second-order orientation tensor $\mathbf{A}_{ij}$ in the main flow direction. A_{22} corresponds to the component transverse to the main flow direction and A_{33} to the component in the thickness direction. Figure 2 schematically shows the evaluation directions of the components of the second-order orientation tensor $\mathbf{A}_{ij}$. The µCT investigations are performed with the 3D-nanoCT Phoenix Nanotom from GE Sensing Technologies GmbH. The fiber orientation is evaluated using VGSTUDIO MAX software from Volume Graphics GmbH. To compare the simulative and experimentally determined fiber orientations, a process–structure coupling to FE models of the geometry of the μCT specimens is performed. Regarding the injection molding simulations, a discretization of 10 elements through the wall thickness of the FE model is defined to ensure comparability. The process–structure coupling is performed with the software CONVERSE® from PART Engineering GmbH. Figure 3 shows the proposed approach for evaluating and recalibrating the simulative fiber orientation from injection molding simulations in comparison to the experimental μCT results.

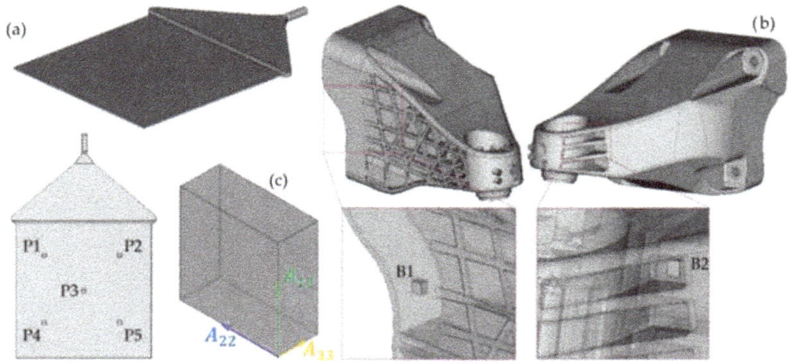

Figure 2. Geometry of the plate (**a**) and component level (**b**) and corresponding μCT specimen positions (P1–P5, B1–B2). Evaluation directions of the fiber orientation (**c**).

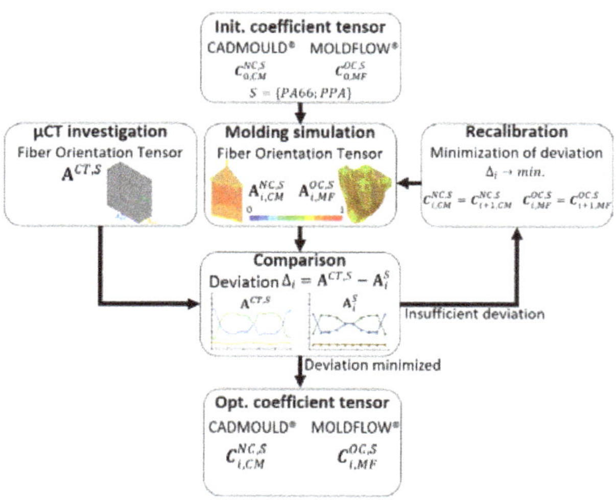

Figure 3. Proposed approach for calibrating and comparing the calculated fiber orientation of injection molding simulations to experimental μCT investigations.

2.3. Methodology for Experimental and Simulative Structural Dynamics Investigations

To investigate the structural dynamics of the short-fiber reinforced plastic components, modal analyses are carried out. Excitation of the system forces a structural response at the resonant frequency. This is characterized by transfer behavior. The response function represents the relationship between the out and input signals of the system. According to existing investigations, modal analyses are carried out with an elastic bearing of the plastic components [21]. The aim of the elastic bearing is that the impact effects of the surrounding system on the plastic component are minimized. Thus, the stiffness and damping of the bearing are defined, that there is no superposition on the resonance frequencies and amplitude of the response function of the plastic component. By applying the plastic component with bearing in a climate chamber, the temperature and humidity can be set reproducible.

The basis for the simulation of the structural dynamics of short-fiber reinforced, thermoplastic components is the material model of stiffness, damping and viscoelasticity. According to existing investigations, the Arbitrary-Reconsidered-Double-Inclusion (ARDI) material model is applied [4]. This corresponds to a two-stage homogenization method of the stiffness and damping from the properties of the composite. In the first step, the effective stiffness and damping are calculated from the properties of the fibers, the matrix and the interphase properties between them. In the second step, a material database is generated by transforming the effective stiffness and damping tensors over a discrete number of fiber distribution functions (ODF). An assignment of the properties is based on the minimum error deviation between the discrete orientation tensors of the material database and the orientation tensor of each element of the composite component. The orientation tensors of the component are taken from corresponding injection molding simulations, Section 2.2. Next, the geometry of the plastic component is discretized into an FE model. Here, the number of elements of the discretized model controls the provided fiber orientation tensor significantly from the injection molding simulation [1]. As a result, to investigate the influence of the fiber orientation on the prediction quality of the structural dynamics simulation, the surface mesh remains identical, and the number of elements over the wall thickness is iteratively increased. In the last step, the boundary conditions of excitation and bearing are modeled. To avoid additional simulative impacts, a two-dimensional node-force excitation is applied. Furthermore, an unbound numerical bearing is assumed. Figure 4 provides an overview of the method for the experimental and simulative characterization of the structural dynamics of the short-fiber reinforced plastic components.

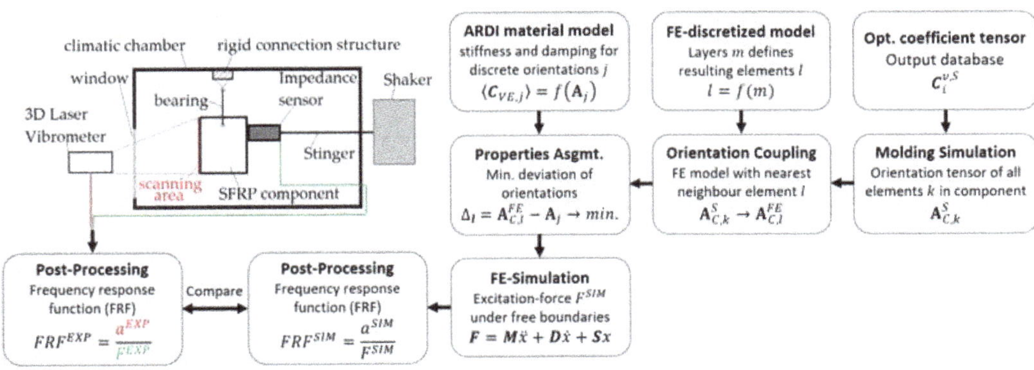

Figure 4. Method for the experimental and simulative characterization of the structural dynamics of short-fiber reinforced thermoplastic components.

Figure 5a shows the experimental setup for investigating the structural dynamics of the short-fiber reinforced, thermoplastic engine bracket by modal analyses in free-supported bearing. The engine bracket is suspended with 4 aramid ropes. Preliminary investigations confirmed that this leads to an optimum between stiffness and damping of the bearing method and thus minimizes a reaction on the engine bracket [44–46]. An electrodynamic shaker from BRÜEL & KJAER type 4809 [47] is used for excitation. This is equipped with a BRÜEL & KJAER type 8001 impedance measuring head and stinger to record the excitation force [48]. The structural dynamics response is recorded with a 3D laser vibrometer type PSV-500 from Polytec [49]. The measurement mesh consists of 145 scanning points with a focus on the qualitative evaluation of the mode shapes, as well as quantitative evaluation of the force reaction function (FRF), as shown in Figure 5b. In the present case, the experimental FRF is defined as the ratio between the spatially averaged acceleration spectrum of all scanning points on the front surface of the engine bracket and the excitation force. The experimental modal analysis is performed at constant conditioning of 23 °C and 0% relative humidity. To simulate the structural dynamics of the engine bracket, the surface geometry is discretized into 137,128 elements in the first step, as shown in Figure 5c. Next, the number of FE elements is varied via the component thickness to investigate the resulting influence. The simulative FRF is defined as the ratio between the spatially averaged acceleration spectrum of all nodes on the front surface of the engine bracket and the excitation force, as shown in Figure 5d. Figure 6 shows the proposed approach for evaluating and comparing the experimental and structural dynamics of the short-fiber reinforced thermoplastic engine bracket.

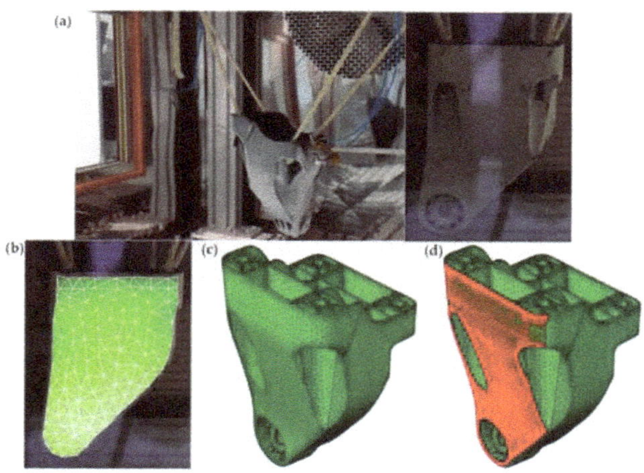

Figure 5. (**a**) Setup for the experimental investigation of the structural dynamics of the engine bracket. (**b**) Experimental scanning point mesh. (**c**) Simulative surface mesh of the engine bracket geometry. (**d**) Evaluation nodes for simulative FRF.

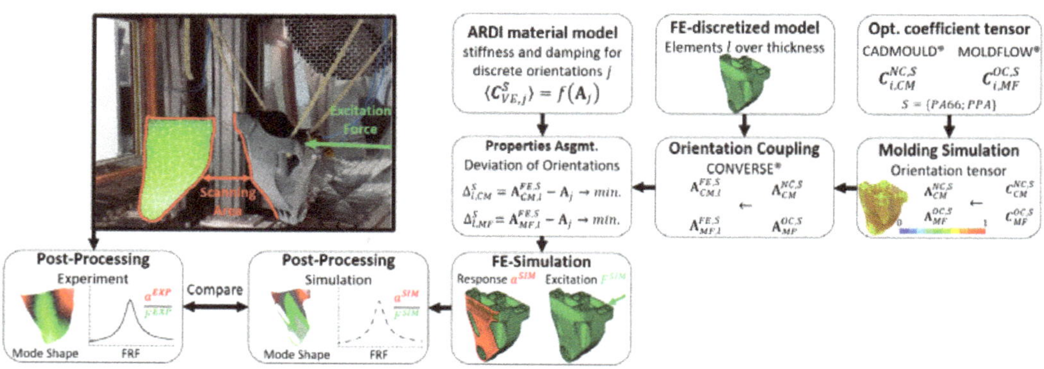

Figure 6. Proposed approach for the experimental and simulative characterization and comparison of the structural dynamics of the short-fiber reinforced thermoplastic engine bracket.

3. Results

3.1. Simulative and Experimental Fiber Orientation Investigations

Figure 7 shows the experimental fiber orientation tensors (CT) in comparison to the simulative fiber orientation tensors with CADMOULD® (CM) and MOLDFLOW® (MF) over the normalized position for the initial coefficient tensors $C_{0,CM}^{NC,1}$ and $C_{0,MF}^{OC,1}$ for the material PA66GF50 and for an evaluation on plate level. For quantitative comparison of the calculated fiber orientation with CM and MF to the CT results, the relative deviation of the averaged fiber orientation tensors is shown in Table 1.

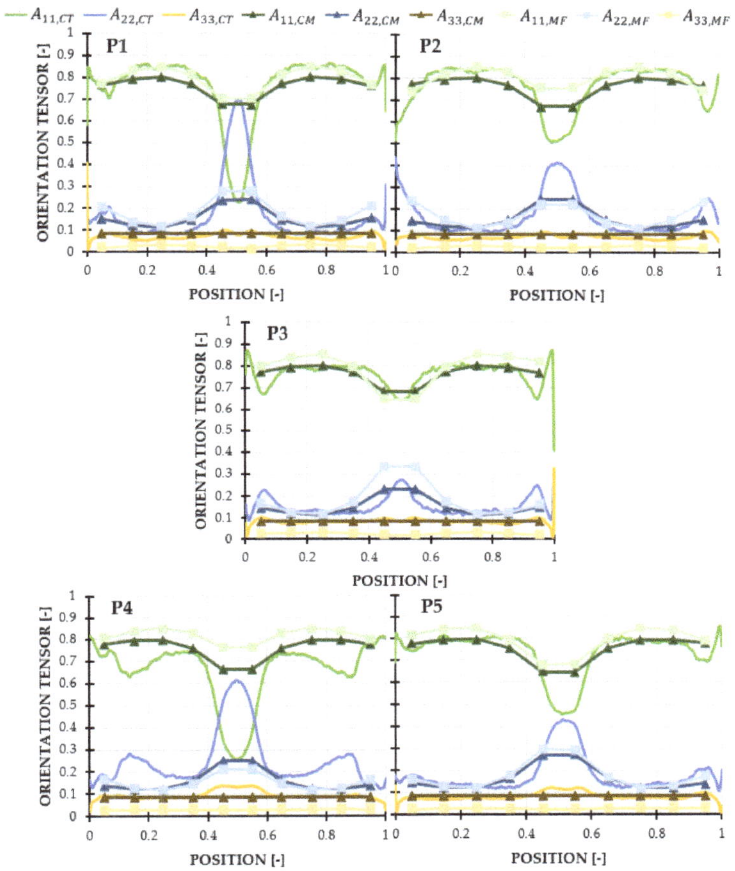

Figure 7. Experimental (CT) and simulative (CM, MF) components of the fiber orientation tensor $\mathbf{A}_{ij}$ plotted over normalized position for PA66GF50 at plate level (P1–P5).

Table 1. Relative deviation of the averaged orientation tensors from experiment $A_{ij,CT}$ and simulation $A_{ij,CM}$, $A_{ij,MF}$ of PA66GF50 on plate level (P1–P5).

Comp. \ Nr.	P1	P2	P3	P4	P5	Average
$A_{11,CT} \rightarrow A_{11,CM}$	8.9%	5.9%	2.9%	14.5%	4.0%	7.3%
$A_{22,CT} \rightarrow A_{22,CM}$	29.2%	21.9%	15.4%	60.4%	16.3%	28.7%
$A_{33,CT} \rightarrow A_{33,CM}$	16.9%	15.4%	4.5%	13.3%	15.5%	13.1%
$A_{11,CT} \rightarrow A_{11,MF}$	6.2%	5.7%	6.9%	20.4%	7.4%	9.3%
$A_{22,CT} \rightarrow A_{22,MF}$	21.9%	20.2%	18.9%	65.5%	11.3%	27.6%
$A_{33,CT} \rightarrow A_{33,MF}$	88.3%	91.9%	126.3%	117.4%	97.8%	104.4%

Figure 7 shows that the boundary layer equals the main part of the fiber orientation tensor of the μCT results. Furthermore, the majority of the μCT results have a pronounced middle layer. P1 and P4 even show a change in the main fiber orientation $A_{11,CT}$, whereby the fibers in the middle layer are oriented transverse to the main flow direction. The comparison between $A_{11,CT}$, $A_{22,CT}$ and $A_{11,CM}$, $A_{22,CM}$ shows, in general, a good prediction of the fiber orientation by the simulation with CM approach, as shown in Table 1. The comparison between $A_{11,CT}$, $A_{22,CT}$ and $A_{11,MF}$, $A_{22,MF}$ shows a good prediction, too. However, both the CM and MF results show, in general, that the development of the middle layer is not sufficiently represented. On the one hand, according to the evaluation method, elements are missing in the middle of the FE model. This results in a loss of information in the process–structure coupling since only a weighted average of the calculated fiber orientation tensor from the injection molding simulation is transferred [1]. On the other hand, the Advani—Tucker Equation with the rotary diffusion approach only allows a limited representation of the fiber orientation in the middle layer [8,34,38]. Furthermore, Figure 7 shows that the component $A_{33,MF}$ has a high deviation compared to $A_{33,CT}$. This difference can be explained by the method used to calculate the fiber orientation of the highly filled material coupled with the 3D modeling. As a result, the fiber orientation tensor is not only influenced by the phenomenological coefficient tensors, but also by the three-dimensional flow field and thus by the hydrodynamic part of the Advani–Tucker Equation, see [8]. Figure 8 shows the experimental (CT) and simulative (CM, MF) fiber orientation tensors for the coefficient tensors $C^{NC,1}_{0,CM}$, $C^{OC,1}_{0,MF}$ for the material PA66GF50 at the component level. Table 2 shows the corresponding deviations of the averaged fiber orientation tensors.

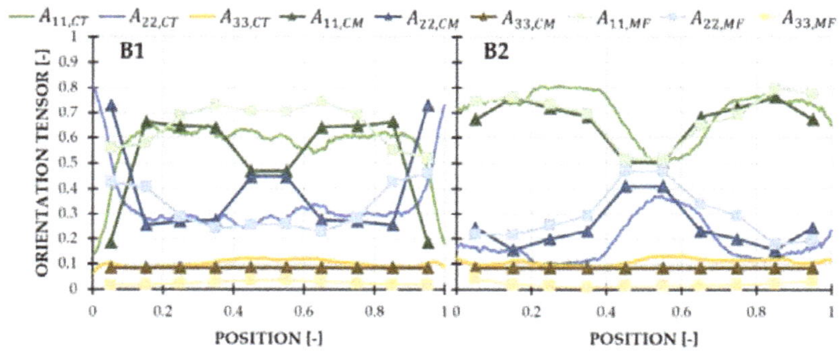

Figure 8. Experimental (CT) and simulative (CM, MF) components of the fiber orientation tensor $\mathbf{A}_{ij}$ plotted over normalized position for PA66GF50 at component level (B1–B2).

Table 2. Relative deviation of the averaged orientation tensors from experiment $A_{ij,CT}$ and simulation $A_{ij,CM}$, $A_{ij,MF}$ of PA66GF50 on component level (B1–B2).

Comp. \ Nr.	B1	B2	Average
$A_{11,CT} \rightarrow A_{11,CM}$	22.8%	7.9%	15.3%
$A_{22,CT} \rightarrow A_{22,CM}$	20.1%	29.0%	24.6%
$A_{33,CT} \rightarrow A_{33,CM}$	20.7%	30.9%	25.8%
$A_{11,CT} \rightarrow A_{11,MF}$	16.9%	9.4%	13.1%
$A_{22,CT} \rightarrow A_{22,MF}$	16.3%	36.1%	26.2%
$A_{33,CT} \rightarrow A_{33,MF}$	162.8%	281.2%	221.9%

Figure 8 shows the CT results of B1, where a pronounced boundary layer with a change in the main fiber orientation occurs. In comparison, no middle layer can be identified. Thus, a significantly different trend of the fiber orientation compared to the plate level can be determined at specimen position B1. Figure 8 also shows that the fiber orientation tensors of the CT result B2 are comparable to those in the plate level. Following this allows a robust check of the injection molding simulations according to the calculated fiber orientation. Figure 8 shows differences in the prediction quality of the CM and MF results compared to the CT results of specimen B1. Thereby, the CM results show a good prediction of the boundary layer, while the prediction of the MF results is not sufficient. An inverse prediction quality can be identified for the middle layer. On the other hand, the CM and MF results show a good prediction of the CT results at specimen B2. Thus, it can be assumed that good predictions of the fiber orientation can be achieved at the component level for similar flow conditions compared to the plate geometry. The quantitative comparison between $A_{11,CT}$, $A_{22,CT}$ and $A_{11,CM}$, $A_{22,CM}$ and $A_{11,MF}$, $A_{22,MF}$ shows a sufficient prediction of the fiber orientation, Table 2. Furthermore, it can be shown on the component level that $A_{33,MF}$ has a high deviation compared to $A_{33,CT}$. Finally, it should be mentioned that both at the plate and component level, the prediction quality of the CM and MF results could not be further improved by a parameter study of $\boldsymbol{C}_{i,CM}^{NC,1}$ and $\boldsymbol{C}_{i,MF}^{OC,1}$.

Next, the robustness of the injection molding simulations is checked in the presence of a material variation. Figure 9 shows the CT, CM and MF fiber orientation tensors for the coefficient tensors $\boldsymbol{C}_{0,CM}^{NC,2}$ and $\boldsymbol{C}_{0,MF}^{OC,2}$ for the material PPAGF50 at plate level. The CT results show a small boundary and significant middle layer. Thereby, the CM and MF results show a significant deviation compared to the CT results. It is assumed that this deviation is due to the loss of information of the evaluation method and the process–structure coupling. Further, the coefficient tensors $\boldsymbol{C}_{0,CM}^{NC,2}$ and $\boldsymbol{C}_{0,MF}^{OC,2}$ are insufficient. The evaluation of the relative deviation of $A_{11,CT}$, $A_{22,CT}$, $A_{33,CT}$ to $A_{11,CM}$, $A_{22,CM}$, $A_{33,CM}$ and $A_{11,MF}$, $A_{22,MF}$, $A_{33,MF}$ also show an insufficient prediction of the fiber orientation, as shown in Table 3.

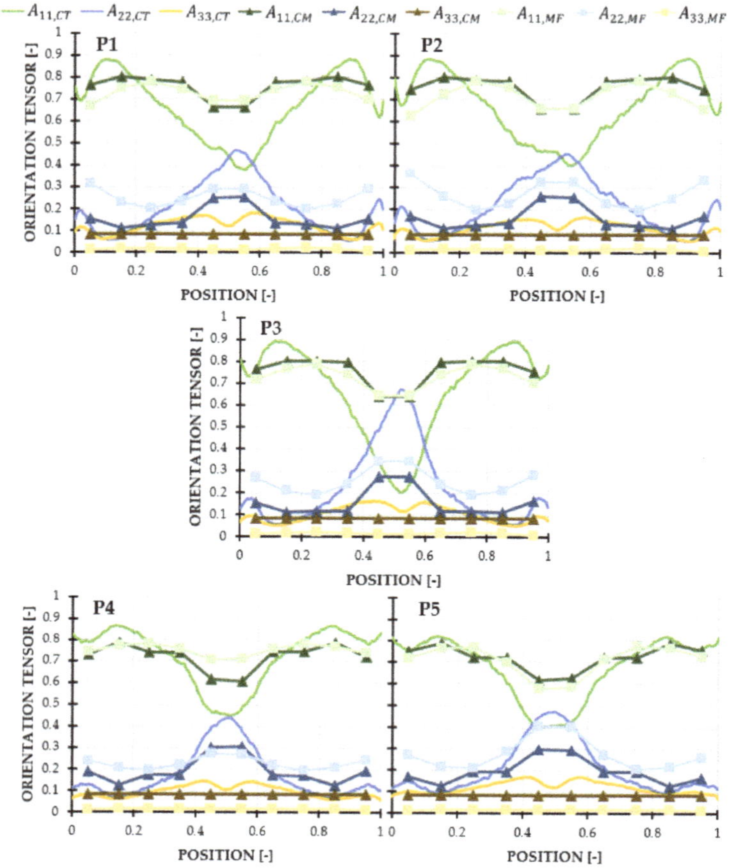

Figure 9. Experimental (CT) and simulative (CM, MF) components of the fiber orientation tensor $\mathbf{A}_{ij}$ plotted over normalized position for PPAGF50 at plate level (P1–P5).

Table 3. Relative deviation of the averaged orientation tensors from experiment $A_{ij,CT}$ and simulation $A_{ij,CM}$, $A_{ij,MF}$ of PPAGF50 on plate level (P1–P5).

Comp. \ Nr.	P1	P2	P3	P4	P5	Average
$A_{11,CT} \rightarrow A_{11,CM}$	14.9%	15.4%	19.0%	10.8%	10.5%	14.1%
$A_{22,CT} \rightarrow A_{22,CM}$	42.7%	49.6%	58.4%	29.4%	24.9%	41.0%
$A_{33,CT} \rightarrow A_{33,CM}$	48.9%	40.9%	40.5%	27.8%	43.0%	40.2%
$A_{11,CT} \rightarrow A_{11,MF}$	18.2%	20.4%	19.9%	12.3%	10.7%	16.3%
$A_{22,CT} \rightarrow A_{22,MF}$	37.9%	37.9%	44.4%	41.4%	30.6%	38.4%
$A_{33,CT} \rightarrow A_{33,MF}$	232.9%	299.2%	300.2%	219.4%	232.4%	256.8%

Figure 10 shows the CT, CM and MF fiber orientation tensors for the coefficient tensors $C_{0,CM}^{NC,2}$ and $C_{0,MF}^{OC,2}$ for the material PPAGF50 at the component level. The CT results of B1 show a significant boundary layer with a change of the main fiber orientation and a small developed middle layer. The CT results of B2 show a pronounced boundary and middle layer comparable to the PA66GF50. The comparison between CT and CM results shows with the difference in the relative deviation of $\Delta(A_{ij,CT} \rightarrow A_{ij,CM}) = 15.0\%$ an improved prediction of the fiber orientation in comparison to the investigations at the plate level, Tables 3 and 4. In this context, with a difference of $\Delta(A_{ij,CT} \rightarrow A_{ij,MF}) = -25.5\%$ the comparison between CT and MF results shows an increased deviation compared to the plate level. Thereby, it can be shown that the boundary and middle layer of specimen B1 are sufficiently predicted by the CM results but not by the MF results, as shown in Figure 10. However, the CM and MF results show a comparable prediction of the fiber orientation of specimen B2. The quantitative comparison between $A_{11,CT}$, $A_{22,CT}$, $A_{33,CT}$ and $A_{11,CM}$, $A_{22,CM}$, $A_{33,CM}$ shows a sufficient prediction of the fiber orientation, Table 4. In this context, the comparison between $A_{11,CT}$, $A_{22,CT}$, $A_{33,CT}$ and $A_{11,MF}$, $A_{22,MF}$, $A_{33,MF}$ shows an increased deviation due to the insufficient prediction quality of $A_{33,MF}$.

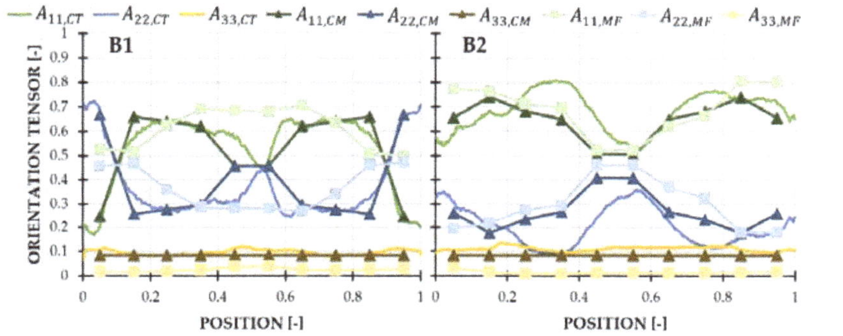

Figure 10. Experimental (CT) and simulative (CM, MF) components of the fiber orientation tensor $\mathbf{A}_{ij}$ plotted over normalized position for PPAGF50 at component level (B1–B2).

A parametrical study of the coefficient tensor $C_{i,CM}^{NC,2}$ shows, that the prediction quality of the fiber orientation can be improved at the plate level. The phenomenological coefficients of the optimized coefficient tensor $C_{i,CM}^{NC,2}$ are redefined as $\alpha_K = 0.11$, $\alpha_R = 0.9$, $\beta_R = 0.05$ Figure 11 shows the CT and CM fiber orientation tensors for the optimized coefficient tensor $C_{i,CM}^{NC,2}$ of the PPAGF50 material at plate level. Figure 11 shows that the prediction quality of the CM results of the boundary and middle layer is sufficiently improved. Furthermore, the comparison between $A_{11,CT}$, $A_{22,CT}$ to $A_{11,CM}$, $A_{22,CM}$ confirms that the prediction quality of the fiber orientation can be optimized, as shown in Table 5.

Table 4. Relative deviation provided in [23,50]. Figure averaged orientation tensors from experiment $A_{ij,CT}$ and simulation $A_{ij,CM}$, $A_{ij,MF}$ of PPAGF50 on component level (B1–B2).

Comp. \ Nr.	B1	B2	Average
$A_{11,CT} \rightarrow A_{11,CM}$	9.7%	10.9%	10.3%
$A_{22,CT} \rightarrow A_{22,CM}$	14.1%	15.3%	14.7%
$A_{33,CT} \rightarrow A_{33,CM}$	14.6%	33.1%	23.9%
$A_{11,CT} \rightarrow A_{11,MF}$	28.2%	12.6%	20.4%
$A_{22,CT} \rightarrow A_{22,MF}$	21.8%	18.6%	20.2%
$A_{33,CT} \rightarrow A_{33,MF}$	276.9%	417.6%	347.2%

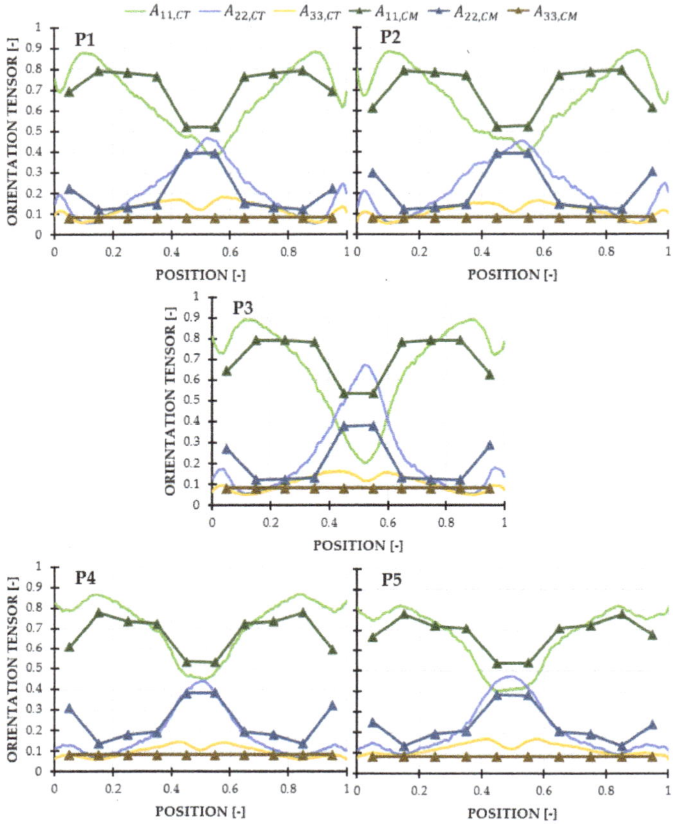

Figure 11. Experimental (CT) and simulative (CM) components of the fiber orientation tensor $\mathbf{A}_{ij}$ plotted over normalized position for PPAGF50 at plate level (P1–P5) in optimized condition $C_{i,CM}^{NC,2}$.

Table 5. Relative deviation of the averaged orientation tensors from experiment $A_{ij,CT}$ and simulation $A_{ij,CM}$ of PPAGF50 on plate level (P1–P5) in optimized condition $C_{i,CM}^{NC,2}$.

Comp. \ Nr.	P1	P2	P3	P4	P5	Average
$A_{11,CT} \rightarrow A_{11,CM}$	10.1%	11.2%	15.9%	9.0%	9.5%	11.2%
$A_{22,CT} \rightarrow A_{22,CM}$	29.4%	43.2%	49.1%	29.0%	23.4%	34.8%
$A_{33,CT} \rightarrow A_{33,CM}$	49.0%	40.9%	40.5%	27.8%	43.0%	40.2%

Next, the robustness of the CM simulation for PPAGF50 with optimized coefficient tensor $C_{i,CM}^{NC,2}$ is tested on the component level, as shown in Figure 12. Specimen B1 shows higher deviations between CT and CM results in the expression of the middle layer. Furthermore, the comparison between the CT and CM results at specimen B2 shows significantly higher deviations. This is due to the significant adjustment of the alignment factor α_K which leads to a higher deviation between CT and CM results, as shown in Table 6. In summary, it can be shown that an optimized coefficient tensor at the plate level does not necessarily constitute an improved prediction quality of the fiber orientation at the component level. Thus, limits of the calculation method of the fiber orientation and the modeling in injection molding simulation are reached. The default values of the initial coefficient tensors $C_{0,CM}^{NC,S}$ and $C_{0,MF}^{OC,S}$ are therefore considered as balanced optimum for the calculation of the fiber orientation at plate and component level with an averaged relative deviation of $(A_{ij,CT} \rightarrow A_{ij,CM})^{PA66} = 19.2\%$, $(A_{ij,CT} \rightarrow A_{ij,MF})^{PA66} = 67.1\%$ and $(A_{ij,CT} \rightarrow A_{ij,CM})^{PPA} = 24.1\%$, $(A_{ij,CT} \rightarrow A_{ij,MF})^{PPA} = 116.6\%$. This is an expected result since the suppliers of the material databases provide an extensive optimization of the phenomenological fiber interaction coefficient tensors, see [8,43]. If a significantly increased prediction quality of the simulative fiber orientation is necessary, case-specific optimizations of the component in injection molding simulations are required.

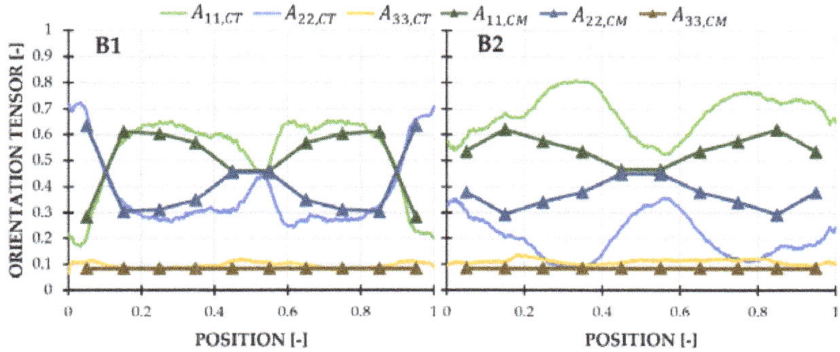

Figure 12. Experimental (CT) and simulative (CM) components of the fiber orientation tensor $\mathbf{A}_{ij}$ over normalized position for PPAGF50 at component level (B1–B2) in optimized condition $C_{i,CM}^{NC,2}$.

Table 6. Relative deviation of the averaged orientation tensors from experiment $A_{ij,CT}$ and simulation $A_{ij,CM}$ of PPAGF50 on component level (B1–B2) in optimized condition $C_{i,CM}^{NC,2}$.

Comp. \ Nr.	B1	B2	Average
$A_{11,CT} \rightarrow A_{11,CM}$	12.4%	31.7%	22.1%
$A_{22,CT} \rightarrow A_{22,CM}$	18.6%	45.1%	31.9%
$A_{33,CT} \rightarrow A_{33,CM}$	14.6%	33.1%	23.9%

Finally, the influence of the process–structure coupling is investigated. Thereby, the Kullback–Leibner divergence between the experimental and simulative orientation distribution is evaluated. The Kullback–Leiber divergence represents the similarity between probability distributions. Thus, low divergence equals a high similarity and vice versa. For simulative reconstruction of the orientation distribution, the maximum entropy method (MEM) is used. Further information on the Kullback–Leiber divergence and the MEM is provided in [23,50]. Figure 13 shows Kullback–Leibner divergences D_{KL} between CT, CM and MF fiber orientation distribution for PA66GF50 and PPAGF50 at the component level (B1–B2) plotted over the layers of the FE model. Figure 13 shows that the CM results at 10 layers have the smallest deviation compared to the CT results. In comparison, the deviation of the MF results at 10 layers is significantly increased due to a high deviation between $A_{33,CT}$ and $A_{33,CM}$. Furthermore, this influence is more pronounced for the PPAGF50. By reducing the number of layers in the process–structure coupling, the transferred fiber orientation content is reduced. This leads to an increased deviation between experimental and simulated fiber orientation distribution. In general, from 10 layers to 1 layer, there is an increased deviation due to a loss of fiber orientation content originating from the injection molding simulation. This is due to the process–structure coupling. By reducing the number of FE layers, a weighted average of the simulative fiber orientation tensors from the injection molding simulation is transferred. This averaging method significantly influences the deviation regardless of the material, the specimen position and the type of injection molding simulation.

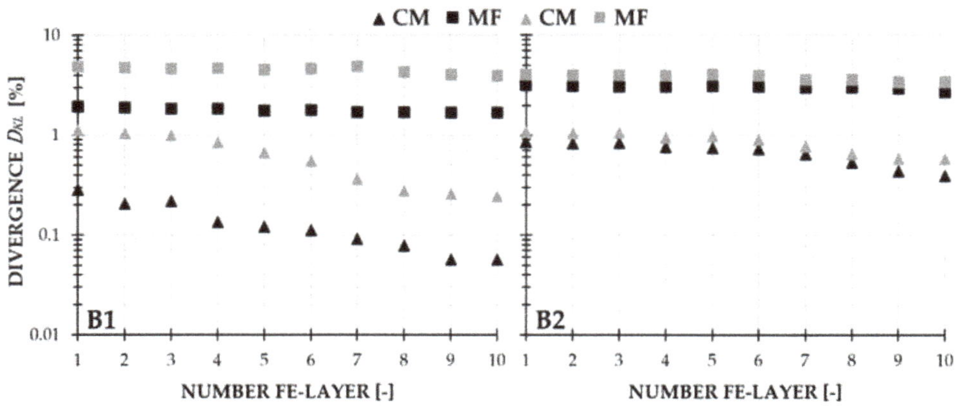

Figure 13. Kullback–Leibner divergences D_{KL} of the fiber orientation tensors of the simulation $A_{ij,CM}$, $A_{ij,MF}$ compared to the experiment $A_{ij,CT}$ of the PA66GF50 (black) and PPAGF50 (grey) on component level (B1–B2) and plotted over the number of layers of the FE-model.

3.2. Experimental and Simulative Structural Dynamics Investigations

Figure 14 shows the experimental FRF of the PA66GF50 and PPAGF50 engine bracket. The first relevant mode shape in the frequency range between 400–500 Hz corresponds to a global torsional mode, as shown in Table 7. Due to the higher stiffness of the PPAGF50, the resonance frequency of the torsional mode is shifted by 6.0% towards higher frequencies compared to the PA66GF50. Furthermore, the amplitude is increased by 5.7% due to the lower damping of the PPAGF50. A comparison via the Modal Assurance Criterion (MAC) shows comparability of 90.2% between the experiments. Above 1900 Hz, the second relevant mode shape is a global bending mode of the engine bracket, as shown in Table 7. Comparable to the first mode shape, the higher stiffness of the PPAGF50 leads to a shift of the resonance frequency by 5.5% towards higher frequencies. However, compared to the previous trend, the amplitude of the PPAGF50 is 61.3% lower than the PA66GF50. It can be assumed that the frequency-dependent structural damping is more pronounced than the material damping. This assumption is reliable by the fact that the MAC analysis shows comparability of 82.6% only. A similar comparison can be shown for the third relevant mode shape, a local surface vibration, as shown in Table 7.

Comparable to the second mode shape, the PPAGF50 shows a stiffer behavior and higher structural damping. Thus, the resonance frequency is shifted by 5.1% towards higher frequencies and the amplitude is 40.3% lower compared to the PA66GF50. Furthermore, the MAC analysis shows comparability of only 69.9%. This results in significantly different mode shapes. Pronounced mixed, local mode shapes characterize frequency ranges above 2300 Hz. A definite evaluation and interpretation of the resonance peaks are no longer ensured. Following, the three identified resonances according to Table 7 are used for further comparison with the simulation. Considering the deviations between the experimental results, it is assumed that not only a superposition of material and component properties occurs, but also the overall structure influences the results, see Section 2.3, Figure 5 and [21,46,51]. Thus, the experimental structural dynamics investigations represent a trend and not an exact behavior.

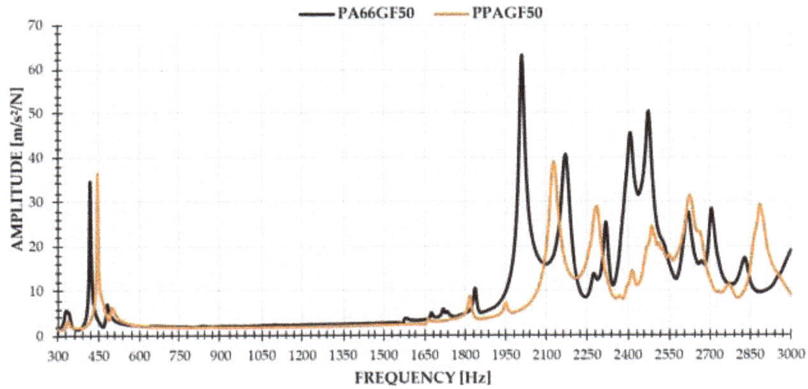

Figure 14. Experimental FRF of the modal analysis of the PA66GF50 and PPAGF50 engine bracket.

Table 7. Experimental mode shapes with resonance frequency, amplitude and MAC value of the modal analysis of the PA66GF50 and PPAGF50 engine bracket.

Mode-Shape with Relative Displacement [-] 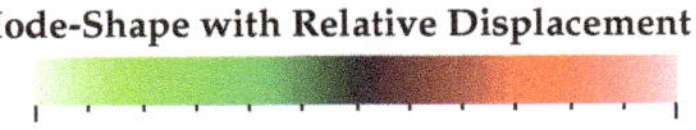 MIN. MAX.	Resonance Frequency [Hz]	Ampl. Res. Frequency [m/s²/N]	MAC Value [%]
Global torsion: +180° 0° −180° 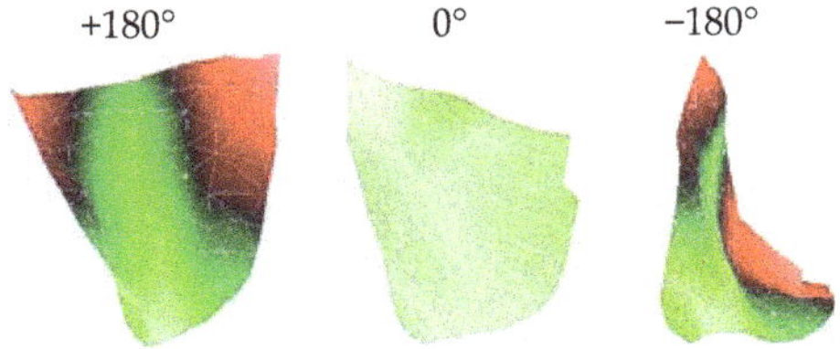	PA66GF50 419	PA66GF50 34.4	90.2
	PPAGF50 446	PPAGF50 36.5	
Global bending: +180° 0° −180° 	PA66GF50 2011	PA66GF50 63.1	82.6
	PPAGF50 2128	PPAGF50 39.1	
Local surface: +180° 0° −180° 	PA66GF50 2172	PA66GF50 40.6	69.9
	PPAGF50 2287	PPAGF50 29.0	

Concerning the proposed approach of Section 2.3 and the results of Section 3.1, more than six elements over the thickness do not lead to an increase in the fiber information content in the FE model, as shown in Figure 13. Furthermore, the engine bracket in the simulation is considered with a discretization between 1 to 6 element layers over the component thickness in this results section. The evaluation of the FRF of the simulations in the relevant frequency range allows the identification of the mode shapes, as shown in Table 8. Comparable to the experiment, the relevant global torsion and bending mode of the engine bracket can be identified in the simulation. Both in the simulation with fiber orientation from CM and MF results, a high MAC value can be verified for all layers. Furthermore, local surface mode shape can also be identified. However, with an average MAC value of 74.0%, the simulation shows noticeable differences compared to the experiment. Nevertheless, a comparison of all performed simulations with numerical fiber orientation from CM and MF results is sufficient. The evaluation of the simulations of the PPAGF50 engine bracket provides comparable mode shapes and MAC values for all layers. Corresponding to the experiments, this is an expected result, as the stiffness, damping and viscoelasticity of the PPAGF50 cause a shift in the resonance frequency and amplitude. Thus, the mode shape is only slightly affected. Furthermore, the mode shapes and MAC values of the simulation of the PPAGF50 engine bracket are not displayed.

Table 8. Mode shapes of the structural dynamics simulation and MAC values in comparison to the experiment of the PA66GF50 engine bracket.

<table>
<tr><th>Mode-Shape

MIN. MAX.</th><th>Sim.</th><th colspan="3">MAC Value [%]</th></tr>
<tr><td rowspan="4">Global torsion
+180° 0° −180°
</td><td rowspan="2">CM</td><td>1 Layer
89.4</td><td>2 Layer
89.6</td><td>3 Layer
90.1</td></tr>
<tr><td>4 Layer
90.5</td><td>5 Layer
90.4</td><td>6 Layer
91.6</td></tr>
<tr><td rowspan="2">MF</td><td>1 Layer
87.9</td><td>2 Layer
89.8</td><td>3 Layer
89.2</td></tr>
<tr><td>4 Layer
90.6</td><td>5 Layer
90.1</td><td>6 Layer
90.2</td></tr>
<tr><td rowspan="4">Global bending
+180° 0° −180°
</td><td rowspan="2">CM</td><td>1 Layer
82.4</td><td>2 Layer
82.7</td><td>3 Layer
83.8</td></tr>
<tr><td>4 Layer
83.8</td><td>5 Layer
84.1</td><td>6 Layer
84.7</td></tr>
<tr><td rowspan="2">MF</td><td>1 Layer
80.8</td><td>2 Layer
80.1</td><td>3 Layer
80.7</td></tr>
<tr><td>4 Layer
80.4</td><td>5 Layer
80.6</td><td>6 Layer
81.3</td></tr>
</table>

Table 8. *Cont.*

Mode-Shape  MIN. MAX.	Sim.	MAC Value [%]		
Local surface +180° 0° −180° 	CM	1 Layer 73.9	2 Layer 73.9	3 Layer 74.8
		4 Layer 73.9	5 Layer 74.2	6 Layer 75.1
	MF	1 Layer 73.7	2 Layer 73.6	3 Layer 73.0
		4 Layer 73.9	5 Layer 73.1	6 Layer 74.9

Figure 15 shows the experimental FRF of the PA66GF50 engine bracket compared to the structural dynamics simulation with fiber orientation tensor of the CM and MF results. Figure 15 shows for the global mode shapes (torsion, bending) the highest deviation of the FRF from the simulation with 1 layer compared to the experiment. This applies to the simulations with CM and MF fiber orientation, as shown in Table 9. This shows a good correlation to the deviations with a process–structure coupling of the fiber orientation, as shown in Figure 13. This trend is continued for a simulation with 2 and 3 layers. Thereby, increasing the number of layers generates an increased fiber orientation content and thus reduces the deviation between experimental and simulated FRF. Contrary, increasing the number of layers to 4 and 5 using the CM results shows a slight increase in the deviation between experimental and numerical FRF. It can be assumed that increases in the deviations are due to the weighted averaging of the orientation tensor in the process–structure coupling, see [1]. Further, this influences the material modeling and the structural dynamics simulation, as shown in Figure 6. However, by increasing the number of layers to 6, a reduction in the deviation between experimental and numerical FRF can be shown, using both the CM and MF results.

Table 9. Relative deviation of frequency and amplitude between simulations and experiment of the PA66GF50 engine bracket for identified mode shapes with fiber orientation from CM and MF results.

Mode Shape	Sim. \ Layer	1	2	3	4	5	6
		Relative deviation frequency [%] Relative deviation amplitude [%]					
torsion	CM	9.2 46.0	8.8 41.7	7.2 41.3	7.6 42.8	7.6 43.0	6.7 39.5
	MF	8.0 34.6	7.2 42.1	7.4 45.1	6.8 42.1	7.2 40.9	6.1 36.2
bending	CM	5.4 33.8	4.9 44.5	3.1 48.6	3.7 52.3	3.7 50.3	2.7 35.9
	MF	2.1 28.5	1.6 38.8	1.6 34.2	1.4 41.4	1.6 39.4	0.6 25.4
surface	CM	1.2 27.1	0.7 21.3	0.4 42.0	0.4 38.4	0.4 44.4	0.6 45.1
	MF	1.2 12.1	1.7 5.1	1.7 7.8	1.9 10.3	1.7 8.4	0.7 13.8

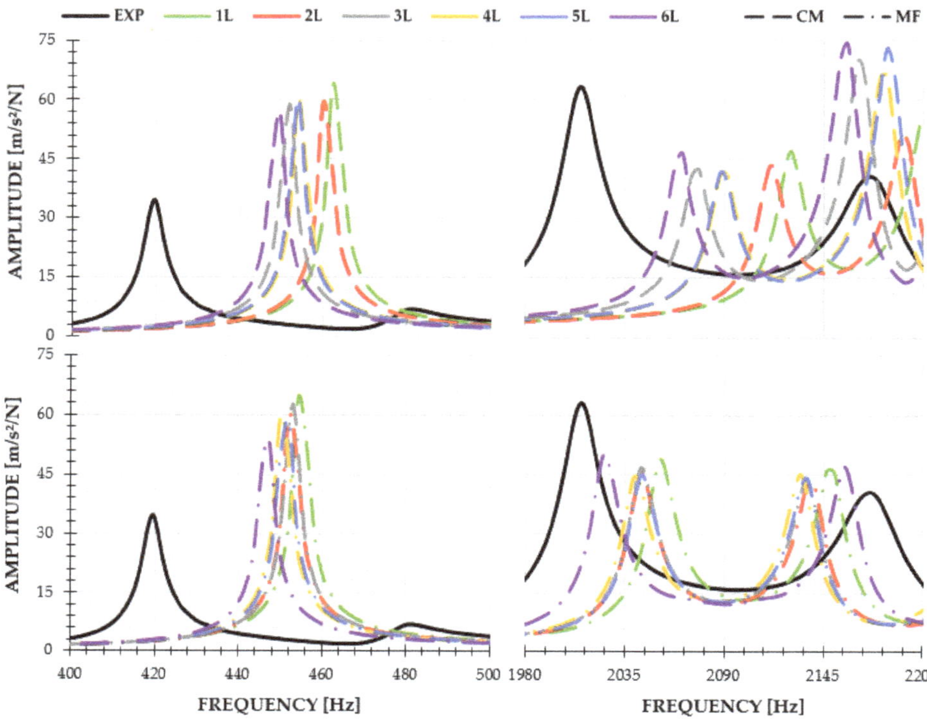

Figure 15. Experimental (EXP) and simulative (CM, MF) FRF of the structural dynamics investigations of the PA66GF50 engine bracket. Simulations are plotted as a function of used FE-layer (L).

A comparative evaluation of the local mode shape (surface) shows an inverse trend of the deviation when increasing the number of layers from 1 to 6, as shown in Figure 15 and Table 9. By increasing the number of layers, the boundary and middle layer of the fiber orientation are more pronounced in the process–structure coupling. Weighted over the component thickness, this leads to a reduction of the stiffness in the material model. As a result, the entire FRF shifts towards lower frequencies. The influence of the material modeling applies to both the local and the global structural dynamics phenomena and shows a good correlation to existing material modeling studies, see [4,7,15,21].

Figure 16 shows the experimental FRF of the PPAGF50 engine bracket compared to the simulation with fiber orientation tensor of CM and MF. Figure 16 shows that the simulation with CM results in the global torsion mode that increasing the layers from 1 to 6 leads to a shift in the resonance frequency. Thereby, increasing the layers improves the mapping of the fiber orientation, which positively affects the stiffness and damping of the material model. Furthermore, it can be shown that the simulations with 6 layers allow the best prediction of the structural dynamics, as shown in Table 10. However, this is not the case for the simulations with MF results. Thereby, the increase of layers leads to an increase in the deviation between experiment and simulation. Again, the fiber orientation of the MF results affects the stiffness and damping of the material model. The evaluation of the global bending mode and local surface mode shows a comparable trend for the CM and MF results. Increasing the layers reduces the deviation between simulated and experimental FRF, as shown in Figure 16 and Table 10. However, side resonances are formed in the simulation, which cannot be identified in the experiment. The influence of a high local deviation between experimental and simulated fiber orientation becomes significant and leads to different local stiffness and damping in the material model.

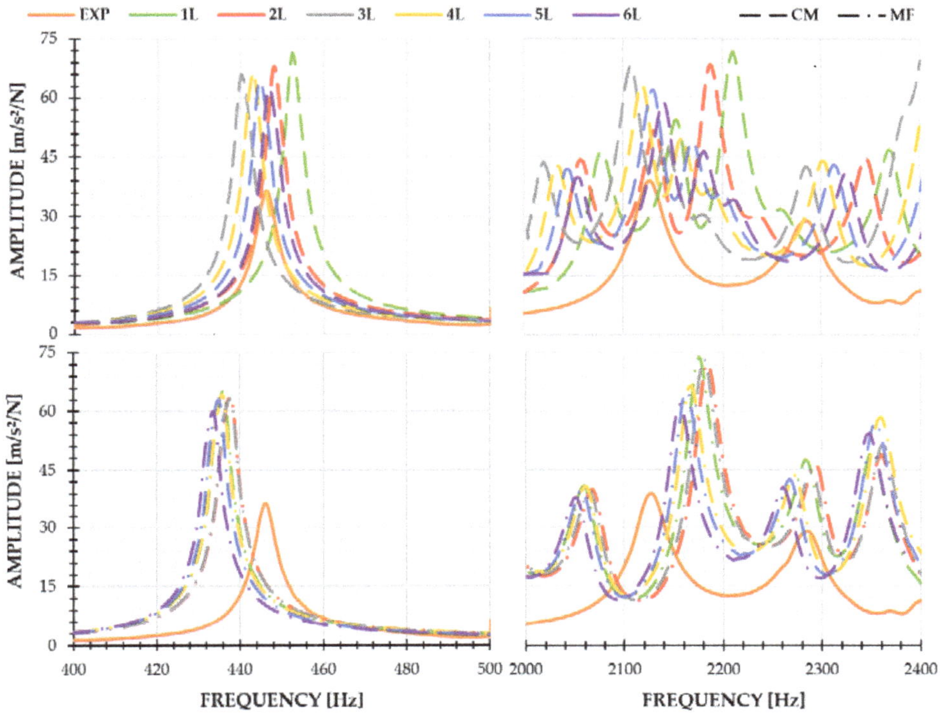

Figure 16. Experimental (EXP) and simulative (CM, MF) FRF of the structural dynamics investigations of the PPAGF50 engine bracket. Simulations are plotted as a function of used FE-layer (L).

Table 10. Relative deviation of frequency and amplitude between simulations and experiment of the PPAGF50 engine bracket for identified mode shapes with fiber orientation from CM and MF results.

Mode Shape	Sim. \ Layer	1	2	3	4	5	6
		Relative deviation frequency [%] Relative deviation amplitude [%]					
torsion	CM	1.4 8.8	0.4 46.2	1.5 44.1	0.8 45.0	0.4 43.0	0.1 41.2
	MF	2.3 43.7	1.9 42.5	1.9 41.8	2.4 42.9	2.6 41.7	2.9 38.6
bending	CM	3.7 45.5	2.7 42.8	1.0 42.4	0.4 38.2	0.1 36.9	0.6 33.6
	MF	2.2 47.0	2.5 46.0	2.4 46.7	1.8 41.1	1.6 39.9	1.3 36.8
surface	CM	3.4 38.5	2.4 35.4	0.2 31.9	0.6 34.1	1.0 32.8	1.5 29.2
	MF	2.7 42.0	3.1 40.9	3.1 43.6	3.1 50.3	2.9 49.3	2.6 46.7

Concerning existing works of literature, the results of this research significantly show the importance of reproducibly describing the boundary conditions of the injection molding simulation, the process–structure coupling and the material modeling. With the used framework of the integrative simulation, relative averaged deviations of 2.8% in the frequency and 38.0% in the amplitude of the FRF can be proven in the structural dynamics simulation, as shown in Tables 9 and 10. Furthermore, it can be shown that high deviations between experimental and simulative fiber orientation tensors can lead to a sufficient prediction of the structural dynamics simulation. This can be verified for global and local structural dynamics phenomena.

4. Conclusions

This contribution investigates the influence of fiber orientation from injection molding simulation on the structural dynamics simulation of the short-fiber reinforced thermoplastic components, e.g., engine brackets made of PA66GF50 and PPAGF50. The Advani–Tucker Equation with rotary diffusion approach and phenomenological fiber interaction coefficient tensor was used to calculate the numerical fiber orientation. The comparison of the experimental and simulative fiber orientations shows a sufficient prediction of the PA66GF50. Optimizations of the interaction coefficients on the plate and component level constitute no further improvements. On the other hand, optimizations of the interaction coefficients of the PPAGF50 on the plate level lead to an improvement and, on the component level, to an insufficient prediction of the simulative fiber orientation compared to the experiments. Simulations of the structural dynamics of the PA66GF50 and PPAGF50 engine bracket show that increasing the layers over the component thickness of the model leads to an improved prediction quality compared to the experiment. This can be shown for global and local structural dynamics phenomena. However, from a simulative point of view, the difference in the FRF with different layers is significantly smaller than the loss of information of the fiber orientation due to a process–structure coupling. In summary, the following conclusions can be derived:

- Optimizing the fiber interaction coefficients at the plate level does not necessarily impose an improved prediction at the component level.
- The process–structure coupling significantly influences the transferred fiber orientation content with the corresponding number of layers.
- Simulative fiber orientations with high deviation compared to the experiments can provide a sufficient prediction in structural dynamics simulation.
- The prediction quality of the structural dynamics simulation is slightly affected by the fiber orientation and significantly by the corresponding material model of stiffness, damping and viscoelasticity.

Author Contributions: Methodology, A.K. and M.S.; software, A.K.; investigation, A.K.; validation, A.K. and M.S.; writing—original draft preparation, A.K.; writing—review and editing, M.S.; visualization, A.K.; supervision, M.S. All authors have read and agreed to the published version of the manuscript.

Funding: This research received no external funding.

Institutional Review Board Statement: Not applicable.

Informed Consent Statement: Not applicable.

Data Availability Statement: The data is not publicly available.

Conflicts of Interest: The authors declare no conflict of interest.

References

1. Stommel, M.; Stojek, M. *Korte, FEM zur Berechnung von Kunststoff- und Elastomerbauteilen*, 2nd ed.; Carl Hanser Verlag: München, Germany, 2018.
2. Christensen, R.M. A critical evaluation for a class of micro-mechanics models. *J. Mech. Phys. Solids* **1990**, *38*, 379–404. [CrossRef]
3. Kaiser, J.-M. Beitrag zur Mikromechanischen Berechnung Kurzfaserverstärkter Kunststoffe–Deformation und Versagen. Ph.D. Thesis, Saarland University, Saarbrücken, Germany, 2013.
4. Kriwet, A.; Stommel, M. Arbitrary-Reconsidered-Double-Inclusion (ARDI) Model to Describe the Anisotropic, Viscoelastic Stiffness and Damping of Short Fiber-Reinforced Thermoplastics. *J. Compos. Sci.* **2020**, *4*, 37. [CrossRef]
5. PART Engineering GmbH. *CONVERSE Documentation V4.0.6*; PART Engineering GmbH: Bergisch Gladbach, Germany, 2019.

6. Calmels, S. *Introduction to e-Xstream Solutions for the Automotive Industry*; E-Xstream Engineering: Luxembourg, 2018.
7. Korte, W. Workflow for Applying the Abaqus Multiscale Modeling Approach. In Proceedings of the Simulia 2020 3DEXPERIENCE German, Virtual, 26 November 2020.
8. Wonisch, A.; Raesch, C.; Schreiber, W.; Jin, J.; Fertig, D.; Wüst, A.; Glaser, S. Injection molding simulations with optimized fiber orientation model parameters. In Proceedings of the CONNECT! European Moldflow User Meeting, Frakfurt, Germany, 3–4 June 2014.
9. Glaser, S. GFK-Bauteile—Berechnung des dynamischen Verhaltens mittels FEM. *KU Kunststoffe* **2001**, *7*, 86–90.
10. Schmachtenberg, E.; Krumpholz, T.; Franck, A.; Vorländer, M. Akustik von Kunststoffbauteilen. *Kunststoffe* **2004**, *9*, 230–235.
11. Arping, T.-W. Werkstoffgerechte Charakterisierung und Modellierung des Akustischen Verhaltens Thermoplastischer Kunststoffe für Körperschallsimulationen. Ph.D. Thesis, Rheinisch-Westfälische Technische Hochschule Aachen (RWTH), Aachen, Germany, 2010.
12. Kremer, H. Materialdatenermittlung Thermoplastischer Kunststoffe für Körperschallsimulationen auf Basis von Reverse Engineering. Ph.D. Thesis, Rheinisch-Westfälische Technische Hochschule Aachen (RWTH), Aachen, Germany, 2010.
13. Calmels, S.; Mathieu, S.; Lesueur, M. *Anisotropic Damping Behaviour of Reinforced Plastic Parts for NVH Simulations*; E-Xstream Engineering: Luxembourg, 2016.
14. Gieß, M. Untersuchungen zur Akustischen Formteilauslegung. Ph.D. Thesis, University Siegen, Siegen, Germany, 2018.
15. Raschke, K.; Korte, W. Faserverstärkte Motorbauteile besser berechnen. *Kunststoffe* **2019**, *109*, 184–189.
16. Kriwet, A.; Stommel, M. Investigation of operational vibrations on the structural dynamics of glass fiber reinforced thermoplastic components. In Proceedings of the MSE Conference, Darmstadt, Germany, 22–25 September 2020.
17. Kriwet, A.; Stommel, M. A Contribution to Predict the Structural Dynamics of Mounted Short Glass Fiber Reinforced Thermoplastic Components. In Proceedings of the 14th WCCM-ECCOMAS Congress, Virtual, 11–15 January 2020.
18. Kriwet, A.; Raschke, K.; Stommel, M. Charakterisierung der Wechselwirkung der Lagerstelle und Aggregatgehäuse auf die Strukturdynamik von Kunststoffbauteilen. *Schwingungen* **2021**, *2021*, 129–142.
19. Ultramid®. *A3WG10—PA66-GF50. CAMPUS®—A Material Information System for the Plastics Industry*; CWFG GmbH: Frankfurt, Germany, 2021. Available online: https://www.campusplastics.com/material/pdf/156435/UltramidA3WG10?sLg=en (accessed on 16 February 2022).
20. Ultramid®. *Advanced T1000HG10 UN—PA6T/6I-GF50. CAMPS®—A Material Information System for Plastics Industry*; CWFG GmbH: Frakfurt, Germany, 2021. Available online: https://www.campusplastics.com/material/pdf/156477/UltramidAdvancedT1000HG10UN?sLg=en (accessed on 16 February 2022).
21. Urban, F.; Middendorf, P. Macroscopic Modeling of the Linear-Viscoelastic Vibration Behavior of Short-Fiber Reinforced Plastics. In Proceedings of the SAMPE 2020, Virtual, 3 August 2020.
22. BASF SE. Weltweit Erster Motorträger aus Kunststoff in Neuer Mercedes GL-Klasse aus Ultramid®. Available online: https://documents.basf.com/36a902759d1a3b974a33b8a6e56df1af61b3d115/Ultramid_first_plastic_engine_support_Mercedes_GL_de.pdf (accessed on 13 February 2022).
23. Breuer, K.; Stommel, M.; Korte, W. Analysis and Evaluation of Fiber Orientation Reconstruction Methods. *J. Compos. Sci.* **2019**, *3*, 67. [CrossRef]
24. Kugler, S.K.; Kech, A.; Cruz, C.; Osswald, T. Fiber Orientation Predictions—A Review of Existing Models. *J. Compos. Sci.* **2020**, *4*, 69. [CrossRef]
25. Folgar, F.; Tucker, C.L. Orientation Behavior of Fibers in Concentrated Suspensions. *J. Reinf. Plast. Compos.* **1984**, *3*, 98–119. [CrossRef]
26. Advani, S.G.; Tucker, C.L. The Use of Tensors to Describe and Predict Fiber Orientation in Short Fiber Composites. *J. Rheol.* **1987**, *31*, 751–784. [CrossRef]
27. Huynh, H.M. Improved Fiber Orientation Prediction for Injection-Molded Composites. Master's Thesis, University of Illinois Urbana-Champaign, Champaign County, IL, USA, 2001.
28. Wang, J.; O'Gara, J.F.; Tucker, C.L. An objective model for slow orientation kinetics in concentrated fiber suspensions. Theory and rheological evidence. *J. Rheol.* **2008**, *52*, 1179–1200. [CrossRef]
29. Phelps, J.H.; Tucker, C.L., III. An anisotropic rotary diffusion model for fiber orientation in short- and long-fiber thermoplastics. *J. Non-Nowton. Fluid Mech.* **2009**, *156*, 165–176. [CrossRef]
30. Tseng, H.C.; Chang, R.Y.; Hsu, C.H. Phenomenological improvements to predictive models of fiber orientation in concentrated suspensions. *J. Rheol.* **2013**, *57*, 1597–1631. [CrossRef]
31. Tseng, H.C.; Chang, R.Y.; Hsu, C.H. The use of principal spatial tensor to predict anisotropic fiber orientation in concentrated fiber suspensions. *J. Rheol.* **2017**, *62*, 313–320. [CrossRef]
32. Bakharev, A.; Yu, H.; Ray, S.; Speight, R.; Wang, J. *Using New Anisotropic Rotational Diffusion Model to Improve Prediction of Short Fibers in Thermoplastic Injection Molding*; ANTEC: Orlando, FL, USA, 2018.
33. Jack, D.A. Advanced Analysis of Short-Fiber Polymer Composite Material Behavior. Ph.D. Thesis, University of Missouri, Columbia, MO, USA, 2006.
34. Verley, V.; Dupret, F. Numerical prediction of the fiber orientation in complex injection molded parts. *Trans. Eng. Sci.* **1994**, *4*, 303–312.

35. Chung, D.H.; Kwon, T.H. Invariant-Based Optimal Fitting Closure Approximation for the Numerical Prediction of Flow-Induced Fiber Orientation. *J. Rheol.* **2002**, *46*, 169. [CrossRef]
36. Montgomery-Smith, S.; He, W.; Jack, D.A.; Smith, D.E. Exact tensor closures for the three-dimensional Jeffery's equation. *J. Fluid Mech.* **2011**, *680*, 321–335. [CrossRef]
37. Montgomery-Smith, S.; Jack, D.; Smith, D.E. The Fast Exact Closure for Jeffery's equation with diffusion. *J. Non-Newton. Fluid Mech.* **2011**, *166*, 343–353. [CrossRef]
38. Qadir, N.U.; Jack, D.A. Modeling fibre orientation in short fibre suspensions using the neural network-based orthotropic closure. *Compos. Part A Appl. Sci. Manuf.* **2009**, *40*, 1524–1533. [CrossRef]
39. Jack, D.A.; Schache, B.; Smith, D.E. Neural network-based closure for modeling short-fiber suspensions. *Polym. Compos.* **2010**, *31*, 1125–1141. [CrossRef]
40. Jack, D.A. Investigating the Use of Tensors in Numerical Predictions for Short-Fiber Reinforced Polymer Composites. Master's Thesis, University of Missouri, Columbia, MO, USA, 2003.
41. Jack, D.A.; Smith, D.E. Assessing the Use of Tensor Closure Methods with Orientation Distribution Reconstruction Functions. *J. Compos. Mater.* **2004**, *38*, 1851–1872. [CrossRef]
42. Jack, D.A.; Smith, D.E. An invariant based fitted closure of the sixth-order orientation tensor for modeling short-fiber suspensions. *J. Rheol.* **2005**, *49*, 1091–1115. [CrossRef]
43. Webelhaus, K.; Becker, F. *Automatische Optimierung von Naturfaserverstärkten Bauteilen mit Varimos® 2*; Fachsymposium zur Verarbeitung von Biokunststoffen: Würzburg, Germany, 2014.
44. Kanirope®GmbH. Aramidseil ARABRAID ⌀6mm Meterware (lfm) 8-Fach Geflochten von Kanirope®. Available online: https://www.kanirope.de/aramidseil-arabraid-6mm-meterware-lfm-8-fach-geflochten-kanirope#product_info (accessed on 13 February 2022).
45. Kanirope®GmbH. Aramidseil, Aramidschnur (Geflochten)—Werkstoffeigenschaften der Aramidfasern. Available online: https://www.kanirope.de/shop/aramidseile-geflochten (accessed on 13 February 2022).
46. Klenk, M. Entwicklung Eines Versuchsaufbaus zur Präzisen Charakterisierung des Schwingverhaltens von Bauteilen aus Kurzfaserverstärktem Kunststoff. Master's Thesis, Universität Stuttgart, Stuttgart, Germany, 2019.
47. Hottinger Brüel & Kjaer®. Modal and Measurement Exciters—Vibration Exciter Type 4809. Available online: https://www.bksv.com/-/media/literature/Product-Data/bp0231.ashx (accessed on 13 February 2022).
48. Hottinger Brüel & Kjaer®. Product Data Impedance Heads—Types 8000 and 8001. Available online: https://www.bksv.com/-/media/literature/Product-Data/bp0244.ashx (accessed on 13 February 2022).
49. Polytec GmbH. PSV-500-3D Scanning Vibrometer. Available online: https://www.polytec.com/de/vibrometrie/produkte/full-field-vibrometer/psv-500-3d-scanning-vibrometer (accessed on 13 February 2022).
50. Kullback, S.; Leibler, R.A. On Information and Sufficiency. *Ann. Math. Stat.* **1951**, *1951*, 79–86. [CrossRef]
51. Ertürk, E. Entwicklung und Validierung Einer Methodik zur Struktursimulation Ineinander Spritzgegossener Kunststoffbauteile. Master's Thesis, Universität Stuttgart, Stuttgart, Germany, 2020.

Journal of
Composites Science

MDPI

Article

Derivation and Validation of Linear Elastic Orthotropic Material Properties for Short Fibre Reinforced FLM Parts

Christian Witzgall, Harald Völkl and Sandro Wartzack *

Engineering Design, Friedrich-Alexander-Universität Erlangen-Nürnberg, 91058 Erlangen, Germany; witzgall@mfk.fau.de (C.W.); voelkl@mfk.fau.de (H.V.)
* Correspondence: wartzack@mfk.fau.de

Abstract: Additively manufactured parts play an increasingly important role in structural applications. Fused Layer Modeling (FLM) has gained popularity due to its cost-efficiency and broad choice of materials, among them, short fibre reinforced filaments with high specific stiffness and strength. To design functional FLM parts, adequate material models for simulations are crucial, as these allow for reliable simulation within virtual product development. In this contribution, a new approach to derive FLM material models for short fibre reinforced parts is presented; it is based on simultaneous fitting of the nine orthotropic constants of a linear elastic material model using six specifically conceived tensile specimen geometries with varying build direction and different extrusion path patterns. The approach is applied to a 15 wt.% short carbon-fibre reinforced PETG filament with own experiments, conducted on a Zwick HTM 5020 servo-hydraulic high-speed testing machine. For validation, the displacement behavior of a geometrically more intricate demonstrator part, printed upright, under bending is predicted using simulation and compared to experimental data. The workflow proves stable and functional in calibration and validation. Open research questions are outlined.

Keywords: short fibre reinforced plastics; SFRP; fused layer modeling; FLM; fused deposition modeling; FDM; material models; simulation

Citation: Witzgall, C.; Völkl, H.; Wartzack, S. Derivation and Validation of Linear Elastic Orthotropic Material Properties for Short Fibre Reinforced FLM Parts. *J. Compos. Sci.* **2022**, *6*, 101. https://doi.org/10.3390/jcs6040101

Academic Editor: Stelios K. Georgantzinos

Received: 10 February 2022
Accepted: 4 March 2022
Published: 22 March 2022

Publisher's Note: MDPI stays neutral with regard to jurisdictional claims in published maps and institutional affiliations.

1. Introduction

1.1. Motivation

Additively manufactured parts for structural applications have gained importance in recent years in both research and industrial applications, shifting from models and prototypes towards end use parts and products [1,2]. This development is accompanied by the need for suitable processes and materials as well as reliable simulation models, as the Design for Additive Manufacturing (DfAM) pre-process is largely based on virtual product development [2,3]. Through virtual product development, fulfilment of requirements and an efficient development process can be ensured preceding the physical product [4].

In terms of material choice, especially fibre reinforced polymers (FRP) have gained importance due to their favourable specific stiffness and strength and wide range of applications [5–12] while their material costs in AM can be lower than for metal AM materials, although overall cost including machining and labour has to be considered [13]. Concerning the choice of FRP, Such et al. [14] cite a patent [15] which considers short FRP a "sweet spot" between mechanical performance (best: endless fibre) and manufacturability (best: no fibre at all). Both short fibre and endless fibre FRP can be manufactured using the Fused Layer Modeling (FLM) process, which allows for higher design freedom when compared to other FRP-capable processes such as Laminated Object Manufacturing (LOM; limited fibre orientation capabilities) or stereolithography (SLA; very limited fibre orientation capabilities and low fibre weight fraction, only with short fibre FRP) [16]; moreover, correlating with low material costs, FLM is commonly more readily available when compared to other

AM processes such as Laser Sintering (LS). For these cost-efficiency, mechanical property and design freedom reasons, this contribution focuses on the derivation and application of material models for short fibre FRP FLM-*printed* structural parts. To introduce the principle of FLM, the hot-end of a FLM printer and its schematic illustration is given in Figure 1a. Extrusion beads of heated thermoplastic filament are laid side-by-side on a printing platform, initially; and on top of preceding layers, subsequently. In-plane orientation of beads is divided into contours (outer border walls) and infill (inner structure between border walls). Overhangs are upheld by support structures which are removed after printing.

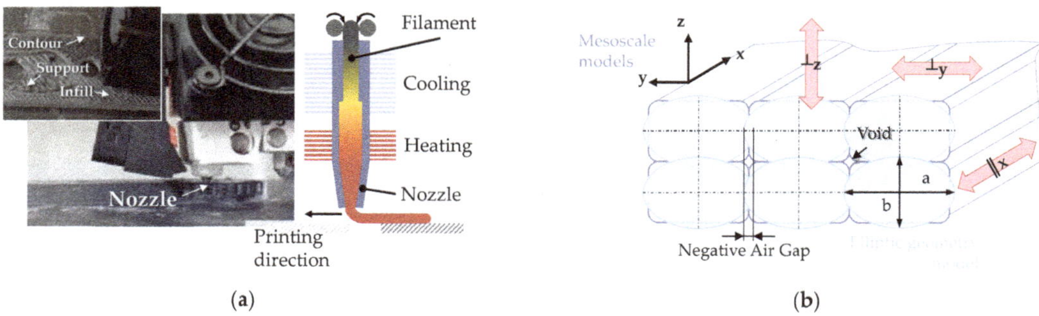

Figure 1. FLM printing: (**a**) Photo of short 15 wt.% carbon-fibre reinforced PETG-CF15 getting printed on a Raise3D Pro2 Plus (**left**) and corresponding schematic of FLM hot-end (**right**); (**b**) Illustration of various current FLM material modeling approaches as described in the text.

1.2. State of the Art and Objectives

For the effective design of structural parts, several attempts have been made to enable simulation-based design, which predominantly address the mechanical characterization of standard test specimens, mostly tensile testing, to derive a certain set of material parameters. The following section provides an overview of existing approaches.

There has been broad research in the mechanical characterization of FLM test specimen which to a large extent focused on unreinforced polymers, mostly acrylonitrile butadiene styrene (ABS). Montero, Ahn et al. [17] in 2001 scrutinized ABS tensile specimen with various geometries for material characterization and compared these FLM results with injection-molded samples. FLM samples yielded 65–72% of the strength of injection-molded parts. Furthermore, from testing experience, build rules for FLM parts were derived: (1) tensile loads should be carried axially along the "fibres"; (2) radii shall be built using contours (instead of lines/grid rasters); (3) negative air gap (bead overlap), Figure 1b, increases strength and stiffness (air gap larger than −0.002 inches, i.e., 0.0508 mm); (4) "Shear strength between layers is greater than shear strength between roads" [17]; and (5) bead width and temperature do not affect strength, but have to be considered concerning build time, surface quality and wall thickness. In the following year 2002, Ahn, Montero et al. [18] published a contribution concerning the anisotropic material properties of FLM ABS. The air gap, road width, model temperature, ABS color and raster orientation were varied and their influence on mechanical properties (tensile and compressive strength) studied; this led to a complement to their building rules: (6) Tensile loaded area tends to fail easier than compression loaded area (later tests indeed showed a tendency towards that, however, there is variation in results [19–21]). A large impact of FLM process parameters on mechanical properties was also found by Pei et al. [22] as well as Ning et al. [10], largely confirming Ahn and Montero's build rules, though Ning et al. additionally point out that there is a sweet spot in printing temperature: a temperature too low provokes weak layer interbonding, while a temperature that is too high implicates more pores, decreasing tensile properties. Li et al. [23] proposed a geometrical ellipse-based theoretical material model for FLM parts, which was successfully validated using tensile tests (3.1 to 7.1% of deviation between material model and experiment). This model accounts for voids produced by the FLM process, Figure 1b, and is, in turn, based on a Stratasys® ABS P400 material model by Matas et al. [24]. Bellini et al. [25] proposed a method to derive orthotropic material models for FLM parts with pre-defined infill rasters of [0° 90° +45° −45°] and Delaunay-Triangulation based "domain decomposition" infill. This method (which will be adapted and altered later in this contribution) is based on different build-up and infill orientations in six specimen geometries to derive the nine orthotropic material constants. Domingo-Espin et al. [26] also used Bellini's orientation for a [+45° −45°] infill and validated the method successfully with a hook-like geometry made of polycarbonate (PC). The authors pointed out, however, that the quality of fit of the orthotropic model depends on build orientation and even proposed an isotropic material model for certain cases. Rodríguez et al. [27] proposed a unit-cell based mesoscale method to model FLM-printed ABS and validated their model with less than 10% deviation from experiment. Lee et al. [28] also used ABS to compare FLM, 3D printing (ink jet based) and "nano composite deposition system" (mechanical micro machining included) concerning raster orientation, air gap, bead width, color, and model temperature with focus on compression strength under different build directions (axial and transverse). Compressive strength for FLM was found to be 11.6% higher for axially than for transversely printed specimen. Other, more recent research on unreinforced FLM specimen encompasses Cantrell et al. [29], who included digital image correlation (DIC) in their tensile testing of different infill lay-ups ([+45 −45], and [0 90]) of FLM ABS specimen. It was found that there is little effect of these stacking sequences on Young's modulus and Poisson's ratio; however, shear modulus and shear yield strength were largely affected with variation between results ranging up to 33%. Specimen made of PC behaved similarly; specimen printed upright (tension direction orthogonal to layers), on-edge (flat, but flipped 90° along the longest axis) and flat (tension direction parallel to layers) produced variations of similar magnitude [29]. A Finite Element (FE)-mesostructural model for ABS P400 was conceived and applied by Somireddy et al. [30]. The positive impact of a negative air gap was confirmed therein ("tightly packed") and a Classical Laminate Theory (CLT)-model was derived to be applied on a 2D FE model of a component. Sheth et al. used a representative volume cell simulation approach with 4 by 4 roads (extrusion beads) and tested this model under varying angles from 0° in steps of 15° to 90°, which showed very good accordance.

Addressing short carbon-fibre reinforced ABS with varying fibre weight percentage (10, 20, 30 and 40 wt.%), Tekinalp et al. [31] found that there is up to 91.5% alignment of fibres in printing direction which leads to significant tensile strength and stiffness increases against unreinforced specimen (+115% and +700%, respectively). The authors also reported nozzle clogging above 30 wt.% fibre content and intensely scrutinized the effect of short fibre reinforcement on voids in and between beads; within beads, voids increased, between beads, voids decreased. These benefits of using FRP within the FLM process sparked interest and further research: Duty et al. [32] scrutinized Big Area Additive Manufacturing (BAAM, nozzle diameter 2.5–7.6 mm instead of the more commonly used 0.2–0.8 mm in FLM) with 13 wt.% short carbon fibre reinforced ABS; anisotropy, stiffness and strength were again increased significantly when compared to unreinforced polymers. A comprehensive overview of various fibre-matrix combinations with different fibre weight fractions was given by Brenken et al. [7]. Mostly, axial to transverse (plane, i.e., within printer platform) stiffness and strength are listed, except for interlayer stiffness and strength by Love et al. [33] For completeness, it is stated that further reviews on 3D printing of FRP have been written by Kabir et al. [11] and Parandoush et al. [16].

To conclude, it can be stated that current specimen-based research largely focuses on comparison between axial and transverse tensile properties (largely within the printer platform plane) or the transfer of specifically built specimen properties (e.g., individual lay-ups in infill) to rather geometrically limited FLM parts. Other approaches include mesoscale models, whose applicability to more complicated geometries and infill patterns, which exploit the large DfAM design freedom, is yet to be shown; and/or 2D models, which may be questionable in case of large through-thickness stresses. A more general approach for deriving material models and transferring these to arbitrary geometries seems necessary to enhance practical applicability and flexibility.

1.3. Objectives and Novelty of This Contribution

This contribution therefore intends to meet the following objectives: (a) Provide a systematic approach to FLM FRP material characterization, which is also transferable to more complicated part geometries, including calibration and validation; (b) Simultaneous fitting of an orthotropic material model (nine constants with moduli and Poisson's ratios) for arbitrary lay-ups in infill instead of pre-defined infills and (c) scrutiny of the fitting by comparison of orthotropic material model results, which are obtained from calibration within a certain range of error to the experiments, during validation to analyse the effectivity of the method.

The novelty of the work is, firstly, that the z-direction of the printed layer structure is also to be investigated in detail. Secondly, the consideration of component experiments for the validation of the material models is a particular feature of this research. Multiaxial stress states and orthotropic properties are considered during validation. Such investigations have rarely been published before, but promise to offer a great value to enhance the practical applicability of the derived material models.

2. Materials and Methods

Figure 2 presents the overall methodology of this contribution, which consists of four steps: specimen concept and printing (1), experiments (2), fitting of the orthotropic material model (3) and validation simulation/experiment (4).

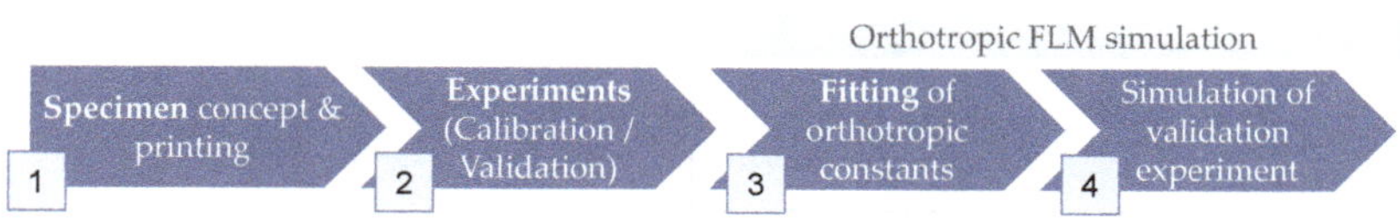

Figure 2. Overarching methodology of this contribution.

In the first step, specimens are conceived, which are intended to reflect the orthotropic constants, such as a longitudinally printed 0° tension rod for Young's modulus E_x parallel to the extrusion beads in axial direction. With these, calibration experiments are conducted (2) which yield force, displacement, stress and strain results. These results, in turn, are used to fit the orthotropic material model parameters of the specimens' FE simulation models simultaneously for all geometries/infill patterns (3). The orthotropic material model is then applied for validation of the results using a different geometry ("XX-rib"): an orthotropic simulation model fully depicting internal extrusion paths is compared to the force-displacement curve derived from the validation experiment.

2.1. Specimen Geometry, Slicing and Printing

Figure 3a presents the specimen geometry used, which is a Becker tensile bar and can be used for both high-speed and quasi-static testing [34,35] and thus allows for flexibility for further extensions towards high-speed testing. In this contribution, only quasi-static testing is performed. Experiments at elevated strain rates and in the fatigue range have already been successfully carried out in [36,37] on injection-molded SFRP with this specimen geometry. The tensile bars have a thickness of 4 mm so these can be tested on testing equipment with a higher force range.

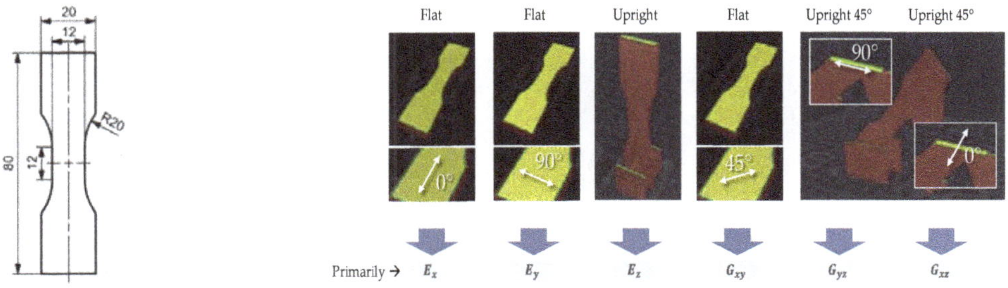

(a) (b)

Figure 3. Specimen geometries: (**a**) Becker specimen, technical drawing; (**b**) Specimen alignment and infill and their allocation to orthotropic constants, sliced by Raise3D ideaMaker software [38].

The orientation on the build platform and infill of the specimen are an extension of the work by Bellini et al. [25]. Six specimens are used overall, Figure 3b:

- The tensile bar is printed flat on the printing platform with longitudinal infill (Flat 0°, F0° in the following), intended to primarily yield Young's modulus in longitudinal direction E_x;
- Same orientation, but with perpendicular infill (F90°) for Young's modulus in perpendicular direction in-plane E_y;
- Upright position with infill printed in the same direction as F90°, thus called Upright 90° (U90°), intended to explain interlayer modulus E_z;
- Flat printed position with 45° infill within plane (F45°) for shear modulus G_{xy};
- Diagonally printed position (45° to plane) with parallel infill to walls, thus called U45°-90° for shear modulus G_{yz};
- And diagonally printed position, yet with perpendicular infill to walls (like F0°), thus called U45°-0° for shear modulus G_{xz}.

Thus, for all specimens, 0°-direction in infill is used when a Young's modulus or shear modulus concerning the x-direction of the material model is to be calibrated; 90° in infill for y-direction; and 45° in in-plane infill or build orientation for shear. For each tpecimen, a FE simulation depicting both outer geometry and infill is set up using the approach presented by the authors [39]. This process is described in Section 2.2.

The specimens were printed on a Raise3D Pro2 Plus FLM printer using a 0.6 mm nozzle (instead of smaller nozzles, to avoid clogging by short fibres) at 254 °C nozzle and 60 °C build plate temperature. Infill extrusion width percentage was 105% (overlap/negative air gap). As printer filament, 1.75 mm FormFutura CarbonFil (15 wt.% short carbon fibre reinforced PETG [40]) filament was used. Support and wall/infill structure were both printed using CarbonFil and separated manually. Printing speed was 40 mm/s for most parts and reduced to 20 mm/s for upright samples. Both calibration (tensile) and validation (XX-rib) specimens were printed with this same slicing template.

2.2. Simulation and Fitting of Orthotropic Material Parameters

The setup of the specimen simulations is based on two inputs: First, the real, printed Becker samples are measured in each dimension (length, width, height) and the mean over each specimen geometry's samples dimensions is calculated (one length, width, and height for each specimen geometry F0°, F90°, U90°, F45°, U45°-90° and U45°-0°, respectively). CAD models with this real geometry are created–not of the nominal geometry; thus, later FE models fit the real geometries appropriately. Second, for each geometry, the building source (G-Code) from the slicing software is used as an input for the mapping of inner material trajectories. A FE model is set up with in-plane element size aligned with printed beads width (for the 0.6 mm nozzle utilized this is approximately 0.6 mm). Element layer heights in z-direction, in turn, correspond to printed layer heights of 0.2 mm. Thus, infill orientation can be depicted properly. Boundary conditions and loads are configured as in the physical experiment (fixed at the bottom, given displacement at the top area on both sides of the tensile bar). The final, mapped FE model for F0° is presented in Figure 4 alongside the extrusion paths from building source (outer walls in red, inner walls in green, infill in yellow).

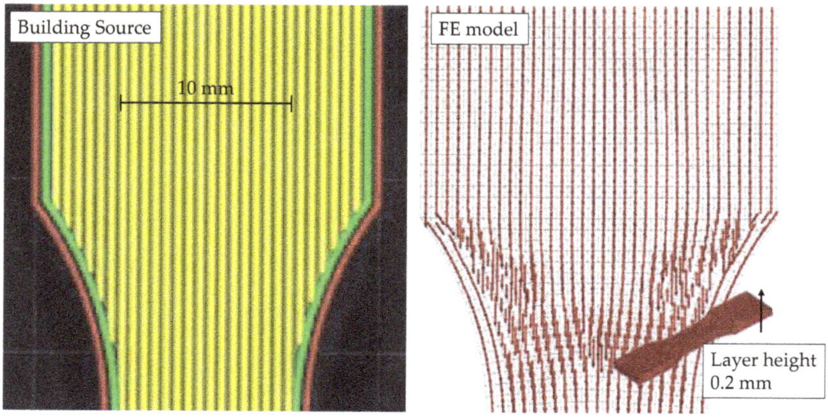

Figure 4. Building source (G-Code, **left**) mapped to FE model (**right**) for the first specimen (F0°) using the approach presented by the authors [39].

The mapping of the extrusion paths from building source to the FE mesh show accordance in the areas of both wall (red/green) and infill (yellow). The same procedure is repeated for all other specimen orientations and infill patterns, as presented again exemplarily for F45° in Figure 5.

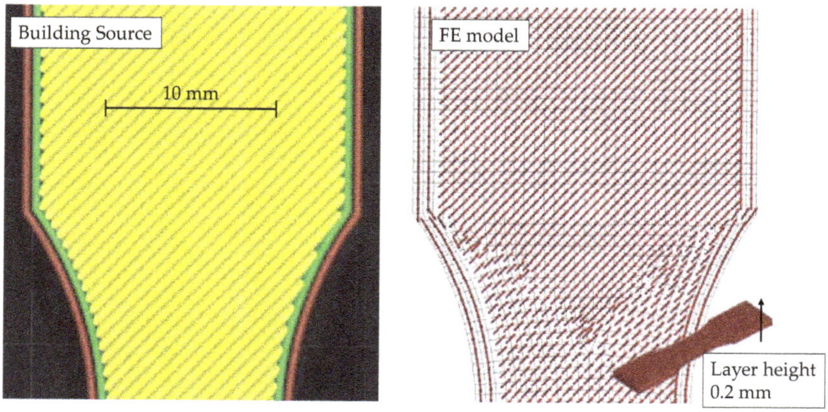

Figure 5. Building source (G-Code, **left**) mapped to FE model (**right**) for the fourth specimen (F45°) using the approach presented by the authors [39].

Based on the six simulations, a fitting workflow is established, Figure 6. All specimen simulations are fed by the same orthotropic material model with its nine constants, which at the same time constitute the design variables for fitting.

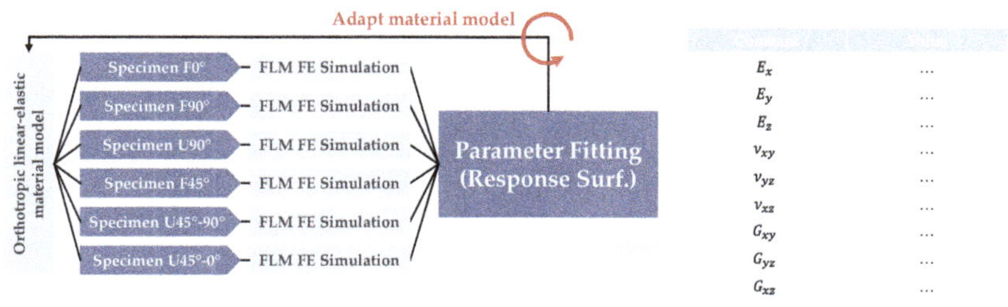

Figure 6. Fitting method in detail: F = flat, U = upright specimen as in Figure 3.

For each of the specimen simulations, ten equidistant force-displacement-pairs, longitudinal strain-displacement-pairs, and transverse strain-displacement pairs (overall $3{\cdot}10 = 30$ pairs) are extracted as outputs and compared to the averaged corresponding pairs of the respective experiments. The method to select these ten force-displacement pairs follows the steps below:

- Average the experiment data to obtain one force-displacement, one longitudinal strain-displacement, and one transverse strain-displacement curve for each of the six samples (18 curves total).
- Find the linear parts of the curves (from displacement = 0 to the end of the linear part), for reference of the experimental data see the results section (Section 3). This is done to allow the linear elastic model to fit to the actual linear section of the curve. To obtain this linear limit, the following steps were taken and are proposed as a solution: (1) Smooth the average curve by a moving average of 10 measurement pairs; (2) Calculate the slope between each curve point and its predecessor dy/dx; (3) Calculate the curvature by calculating the "slope's slope", in turn; and (4) find the first occurrence where the percentage difference in curvature is smaller than a certain threshold (here, 0.75% were used arbitrarily). The threshold depends on the desired "strictness" of linearity; the smaller, the stricter. To avoid considering the initial, rather noisy data within the first part of the experiment, the linearity detection starts after 10% of the experiment curve data. (4) Finally, 10 equidistant displacement points are selected from the linear span of deformation $[s_{lin,min}; s_{lin,max}]$. Using a mapping function, the closest (minimum-difference) data pairs from the averaged, but still discrete experiment data are selected.

Results of this method for stress-strain curves are depicted in Figure 7 for the six specimen and in detail for U90° (right). Detected linear curve segments are shown as red-dashed lines, the equidistant value pairs scattered as red "x" markers.

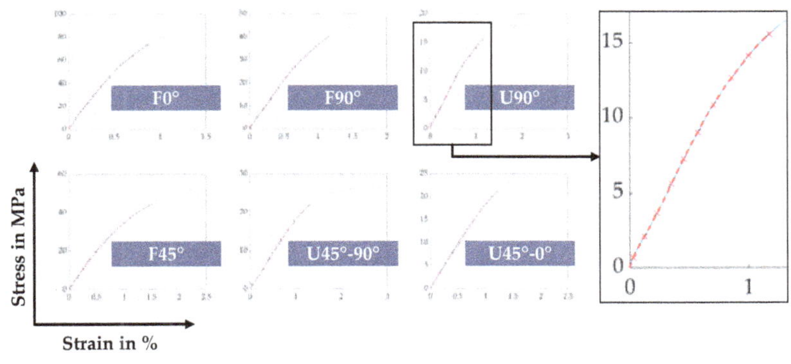

Figure 7. Results of the detection of linear area within experimental averaged results.

Concerning the fitting itself, for each specimen an error measure between simulation and experiment curves is calculated for later comparison of fitted material models. Mean Absolute Percentage Error (MAPE) is selected as error measure, as it allows condensation of individual errors into one aggregate number per geometry and pair category (that is, three error measures per specimen, for reaction force, longitudinal, and transverse strain, respectively; $6 \cdot 3 = 18$ error measures in total) and does not "average out" negative and positive deviations between simulation and experiment, as simple Mean Percentage Error (MPE) would. At the same time, the percentage error is independent of scale, which is important as strains are much smaller in absolute terms than forces. Overall fitting target is then to minimize the unweighted average of all error measures. That omission of weighting implies that equal importance is paid to all specimen geometries and their quantitative results (force, longitudinal/transverse strains). Fitting of the strains alongside force-displacement further allows to calibrate the Poisson's ratios of the orthotropic material model in addition to the Young's and shear moduli. An overview of the aggregation approach is finally presented in Figure 8.

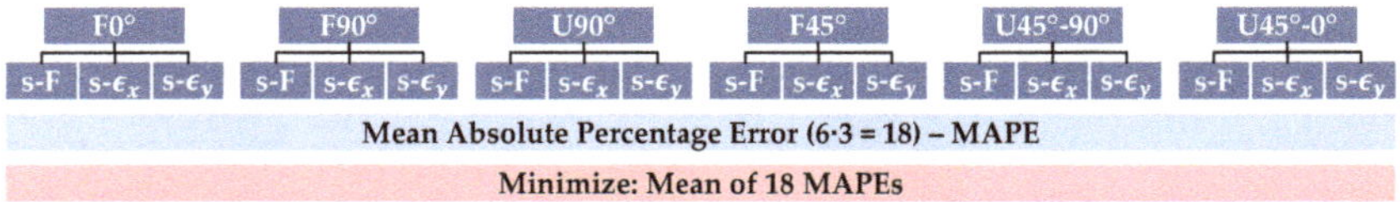

Figure 8. Fitting: Aggregation of error measures in detail.

For fitting of the material parameters, response surface based optimization from ANSYS Software [41] is used with the built-in multi-objective genetic algorithm (MOGA) for optimization. Force value "tolerance" for target value search in the algorithm was set to 10 N (<1% of expected force values), strain tolerance to 0.01% (<1% of expected strain values).

2.3. Validation

To evaluate the prediction quality of the calibrated material model, a three-point bending test is compared in experiment and simulation for validation. The specimen geometry used is a cross-ribbed beam called XX-rib, dimensions specified in Figure 9a and CAD model in Figure 9b, manufactured upright. Because the FLM-printed specimens show quite large deviations from the nominal shape, these are measured at several points. For the simulation, wall thicknesses are used which correspond to the average geometric dimensions of all measured specimens. This average specimen geometry has a length of 60.3 mm, a height of 8.5 mm and a width of 20.6 mm. The wall thickness of the cover plate is 2.6 mm, that of the side walls 2.1 mm and that of the ribbing 2.2 mm. The setup of the simulation, with the specimen resting unrestrained on the supports and a load applied by the flexure fin, is shown in Figure 9c. The contact interfaces between the steel tool and the specimen are modeled with a coefficient of friction of $\mu = 0.3$.

2.4. Conduct and Evaluation of Experiments

The experiments to characterize the tensile specimens are carried out on a servo-hydraulic high-speed testing machine, Zwick HTM 5020. It allows tests to be conducted with loads of up to 50 kN and test speeds from the quasi-static range to 20 m/s. For the characterization presented here, a load cell suitable for forces up to 10 kN is used, which provides a sufficiently fine resolution of the low test forces to be expected.

The test setup is shown in Figure 10 and consists essentially of the lower, fixed clamping and the upper clamping, which is set in motion by the hydraulic piston from the crosshead. The optical measuring system GOM ARAMIS 3D HHS is used for contactless measurement of the movement of the upper clamping and the deformation of the specimen itself. The measuring system works according to the principle of digital image correlation

(DIC), whereby the strain on the sample surface is measured by distorting a stochastic grey value pattern, the displacement of the upper fixture as a translation of discrete points.

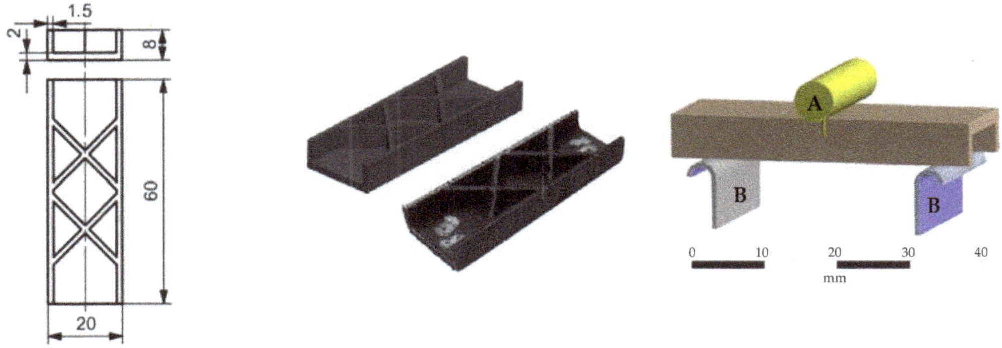

(**a**) (**b**) (**c**)

Figure 9. XX-rib specimen geometry: (**a**) Specimen, technical drawing with main dimensions; (**b**) 3D model of and FLM-printed XX-rib specimen; (**c**) Bending load case for XX-rib specimen as applied in FE simulation. All contacts modeled using a coefficient of friction of $\mu = 0.3$.

Figure 10. Experimental setup with lower and upper clamping and speckled specimen.

The tensile tests are performed at a speed of 1 mm/s and recorded at a sampling rate of 1 kHz. The experiments of the targeted 6 specimen geometries are repeated 5 times each. Force values are recorded image-synchronously so that a load value is available for each recorded deformation state. This allows the creation of force-displacement as well as stress-strain curves. From the knowledge of the stress-strain relationship, Young's modulus and Poisson's ratio are calculated for the different specimen types following DIN EN ISO 527 [42,43]. The stiffness values obtained in this way are to be used as starting values for the subsequent optimization of the material model.

Since several specimen types have a border of the parallel area pointing in the longitudinal dsirection even if the orientation is different (e.g., visible in Figure 5), a correction of the measured Young's modulus must be carried out for these specimens. This treats the parallel perimeter area and the differently oriented inner area like a parallel connection of springs. The respective area fractions of the total sample cross section are obtained from an analysis of the fracture surfaces of the destroyed samples. The following applies here:

$$E_{F90^\circ} = \alpha \cdot E_x + \beta \cdot E_y \tag{1}$$

With E_{F90° being the overall stiffness of the specimen F90° including longitudinal borders and a transversal infill, E_x and E_y being the stiffness in each direction. α and β are the respective are fractions of the border layers and the filling.

In this way, the Young's moduli in x, y and z directions are calculated from the experiments on samples F0°, F90° and U90°. To determine the values of the shear modulus, the samples F45°, U45°-90° as well as U45°-0° are used for the off-axis tensile test as well as the Poisson's ratio [44]. In accordance to Bellini et al. [25], the shear modulus is calculated from the measured stiffness as follows:

$$G_{xy} = \frac{E_{45}}{2 \cdot (1 + \nu_{xy})} \tag{2}$$

Although a linear elastic material model will be calibrated and used in the simulation, it is expected that the material behavior in the real experiments be noticeably non-linear. Therefore, the linear range found according to Figure 7 is used for the calibration of the material model.

To validate the material model, 3-point bending tests are carried out on cross-ribbed beams, called XX-rib, cf. Figure 9. The steel supports with a spacing of 46 mm are filleted with a radius of 2 mm, the fin with a radius of 5 mm. The resulting force is measured by using a load cell in the upper piston. The indentation of the fin, like the movement of the upper restraint of the tensile tests, is measured by means of digital image correlation, resulting in force-displacement curves. The validation experiment is repeated for 7 specimens each.

3. Results

This section presents the results of the investigations carried out, namely those of the experiments necessary for the calibration and the parameter fitting in the simulation. Furthermore, the results of experiments and simulations for validation are presented.

3.1. Calibration

3.1.1. Experiments

The results of the tensile tests carried out are summarized in Figure 11 in the form of force-displacement curves (a) and stress-strain curves (b) for the different orientations.All curves displayed are averaged over the number of repeated experiments. The highest results by far are achieved by the sample F0°, reinforced in the tensile direction, with an average maximum stress of 85.8 MPa. As expected, the upright-printed sample U90° performs weakest, showing a mean maximum stress of 18.3 MPa. When looking at the curves, an overall non-linear behavior is noticeable, with a linear range being identified for all samples at the beginning of the loading. This linear range is used to calibrate the Young's moduli of the material model.

According to the calculation procedures described in Section 2.4, the Young's moduli and shear moduli as well as the Poisson's ratios were determined as shown in Table 1. The values shown are averaged over the number of experiments.

The parameters determined in this way are used as starting values for the subsequent simulative parameter fitting.

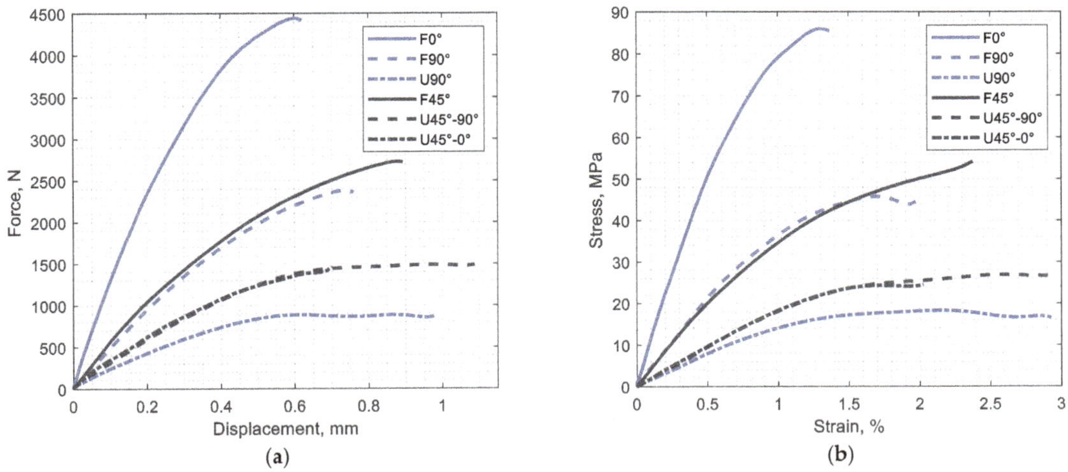

Figure 11. Results of experimental testing, averaged: (**a**) Force-displacement curves; (**b**) stress-strain curves.

Table 1. Material parameters determined from experiments, rounded.

E_x	E_y	z	ν_{xy}	ν_{yz}	ν_{xz}	G_{xy}	G_{yz}	G_{xz}
MPa	**MPa**	**MPa**	-	-	-	**MPa**	**MPa**	**MPa**
8212	2615	1437	0.28	0.11	0.10	883	637	634

3.1.2. Simulation and Parameter Fitting

Following the description in Section 2.2, for each of the six specimens, displacement-force, displacement-longitudinal strain, and displacement-transverse strain curves (totalling $6{\cdot}3 = 18$ curves) were fitted at 10 displacement points (i.e., $18{\cdot}10 = 180$ output parameters overall) using the 9 orthotropic constants of the material model as input parameters for all the six FE specimen setups. The displacement-strain curves are intended to derive the Poisson's ratios primarily, while the displacement-force curves should explain Young's and shear moduli. Figure 12 presents the results for each of the 18 curves with the specimen geometries listed horizontally and the three curve types vertically.

Overall, the fitting quality is adequate (MAPE < 20%) in most cases, as also presented quantitatively in Table 2. The material parameters of the best response surface method optimized fit are given in Table 3. In comparison to the start values (in italics in Table 3), larger deviations occur in E_y. This seems logical as in the experimental specimen, there is a certain fraction of 0° walls in the cross section which overstates longitudinal stiffness in comparison to E_y which depicts purely 90° infill (no walls). The analytical estimation displayed in Equation (1) seems to still overestimate the value. Poisson's ratio ν_{xz} and shear modulus G_{xz} are greatly increased during the fitting, which can be explained again by a large proportion of walls within the cross section of specimen U45°-0° (refer Figure 3b, right). U45°-0° is thus not sufficient to explain the XZ-constants. In comparison, expectedly, the completely-0° specimen F0° yields a relative fitted-measured-difference in E_x of just −0.7% (8153 MPa fitted vs. 8212 MPa measured). For the lying-down and upright specimens F0°, F90°, U90° and F45° fitting is visually accurate in the graphs above overall. For F90° the MAPE in force and transverse strain are higher in comparison (25% and 35%, respectively). In terms of displacement-force, this is explainable as the detected "linear" curve segment is not really linear and shows a disadvantage of the "linear-detection" method: small changes in local curvature might, nonetheless, yield large overall curvature. In terms of transverse strain, the poorer fitting might be due to a compromise of the optimization of other parameters (e. g., longitudinal stiffness of walls in F0°). For the upright-diagonal specimen U45°-90°, displacement-force yields poorer fitting quality, partly due to the linearity issue (first three data points are fitted quite well with MAPE < 25%). For U45°-0°, transverse strain results are less precisely fitted, in accordance with U90° (large influence of 0° infill).

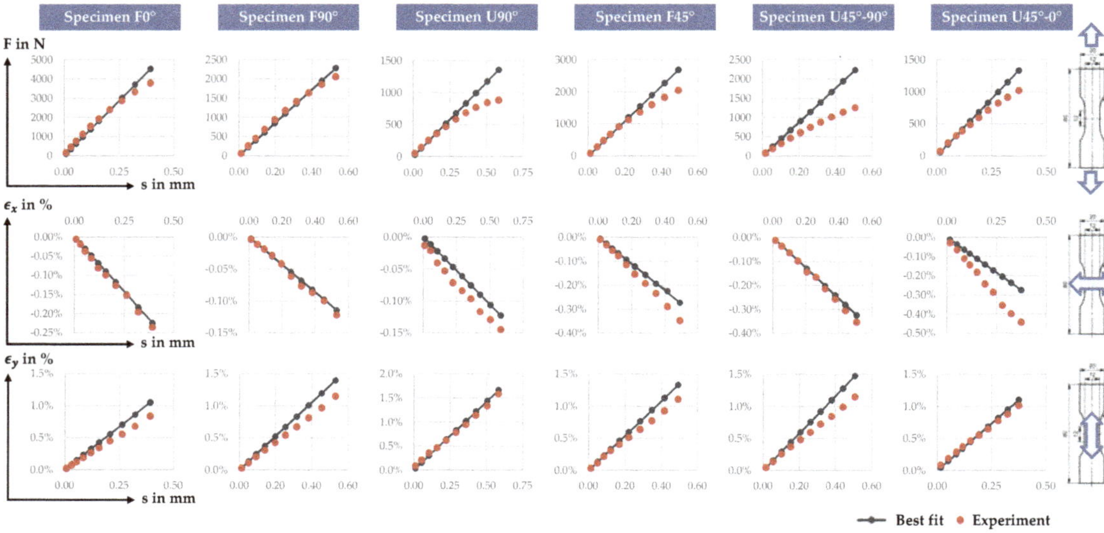

Figure 12. Fitting of force-displacement, transverse strain-displacement, and longitudinal strain simulation results (grey lines) to experimental results (ten red points each, identified above).

Table 2. Mean absolute percentage error (MAPE) for each fitting, rounded to full percent (larger values than 30% are highlighted in grey).

	F0°	F90°	U90°	F45°	U45°-90°	U45°-0°
MAPE $(F-s)$	14%	9%	25%	13%	49%	17%
MAPE $(\varepsilon_x - s)$	10%	8%	35%	19%	6%	42%
MAPE $(\varepsilon_y - s)$	20%	21%	14%	14%	21%	12%

Table 3. Material parameters determined from simulation-based fitting, rounded. Experimental start values from Table 1 are repeated in italics.

E_x	E_y	E_z	ν_{xy}	ν_{yz}	ν_{xz}	G_{xy}	G_{yz}	G_{xz}
MPa	**MPa**	**MPa**	-	-	-	**MPa**	**MPa**	**MPa**
8153	1949	1549	0.31	0.17	0.36	1096	642	1120
8212	*2615*	*1437*	*0.28*	*0.11*	*0.10*	*883*	*637*	*634*

To scrutinize whether specimen geometries influence the desired parameters (Figure 3b) and whether the measured strains at least partly determine Poisson's ratios, a correlation coefficient analysis is conducted using sample points from the response surface. Correlation coefficients between the input parameters and orthotropic constants are given in Table 4.

Table 4. Linear correlation coefficients between input and output parameters. Blue: Correlation coefficient > 0.8 or < −0.8; Orange: Correlation coefficient in [0.6; 0.8] or in [−0.8; −0.6].

		E_x	E_y	E_z	ν_{xy}	ν_{yz}	ν_{xz}	G_{xy}	G_{yz}	G_{xz}
F (reaction forces at boundary)	F0°	0.99	0.40	0.32	0.23	0.23	0.38	0.43	−0.08	0.15
	F90°	0.52	0.98	0.40	0.66	0.50	0.39	0.44	0.27	0.31
	U90°	0.24	0.35	1.00	0.53	0.61	0.65	0.63	0.32	0.44
	F45°	0.64	0.73	0.58	0.61	0.57	0.61	0.84	0.26	0.36
	U45°-90°	0.10	0.46	0.77	0.58	0.66	0.67	0.62	0.81	0.69
	U45°-0°	0.14	0.44	0.76	0.59	0.71	0.65	0.59	0.73	0.81
ϵ_x (transverse mean strain)	F0°	0.05	−0.63	−0.45	−0.92	−0.63	−0.46	−0.41	−0.38	−0.35
	F90°	0.06	−0.81	−0.36	−0.87	−0.61	−0.29	−0.33	−0.38	−0.36
	U90°	0.05	−0.09	−0.82	−0.47	−0.64	−0.80	−0.56	−0.42	−0.50
	F45°	−0.09	−0.75	0.02	−0.45	−0.20	0.11	0.28	−0.10	−0.06
	U45°-90°	−0.43	−0.01	−0.59	−0.10	−0.12	−0.27	−0.23	0.56	0.19
	U45°-0°	−0.39	−0.08	−0.60	−0.15	−0.15	−0.27	−0.23	0.52	0.30
ϵ_y (longitudinal mean strain)	F0°	−0.56	0.25	0.30	0.50	0.49	0.35	0.36	0.79	0.51
	F90°	−0.39	0.23	0.19	0.45	0.43	0.27	0.23	0.44	0.36
	U90°	−0.29	−0.05	−0.48	0.07	0.20	0.01	−0.10	0.45	0.41
	F45°	−0.32	0.61	0.27	0.59	0.48	0.25	0.02	0.64	0.52
	U45°-90°	0.55	−0.08	0.05	−0.11	−0.12	0.01	0.08	−0.85	−0.40
	U45°-0°	0.59	0.19	0.25	0.09	0.00	0.20	0.27	−0.62	−0.51

Young's moduli E_x, E_y and E_z correlate with F0°, F90° and U90° reaction forces nearly perfectly. Similarly, G_{xy}, G_{yz} and G_{xz} are closely linearly correlated with the reaction forces of F45°, U45°-90° and U45°-0°. Poisson's ratios correlate with the transverse strains especially of F0°, F90° and U90° (transverse strain is measured with negative sign, thus smaller – larger absolute – transverse strain correctly correlates with larger ratios). The former also correlate considerably with reaction forces. In contrast to Young's moduli, no clear assignment of a single, individual specimen to a certain ratio can be made. Shear modulus G_{yz} positively correlates with longitudinal strain for F0° and F45° and negatively with longitudinal strain for U45°-90°. Young's modulus E_y also correlates with the longitudinal strain of F45°. Longitudinal strain correlation is very much dependent on the overall geometry and infill (displacement is pre-defined in the FE model) and overall variations are small (10% variation in yields about 1% variation in ϵ_y of F90°, for example).

3.2. Validation

3.2.1. Experiments

The results of the bending tests are shown as force-displacement curves in Figure 13. It shows the averaged curve as well as the results of the individual tests, which have a certain scatter. The values of the maximum force are between 253 N and 289 N, the indentation reached between 0.73 mm and 0.86 mm. For the averaged curve, the maximum force is 276 N at an indentation of 0.79 mm.

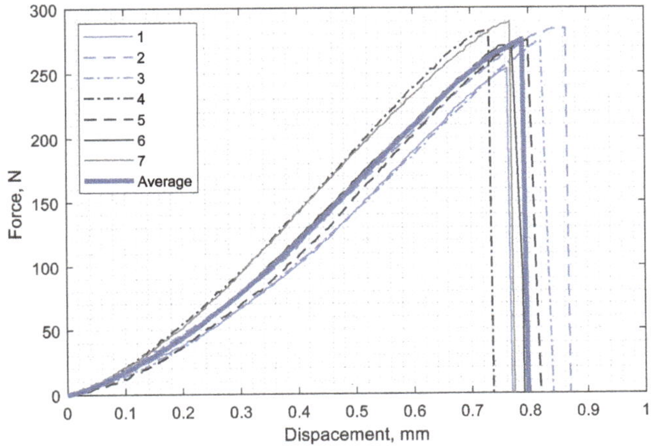

Figure 13. Force-displacement curves for single specimen 1 to 7 as well as averaged.

The scatter of the individual experiments may be due to the geometric inequality and the deviating wall thicknesses of the additively manufactured samples. The averaged force-displacement curve is therefore used for the subsequent comparison with the simulation.

3.2.2. Simulation and Material Model Validation

Application of the fitted material model (Table 3) in the simulation of the XX-rib demonstrator yields the following force-displacement-curve, Figure 14.

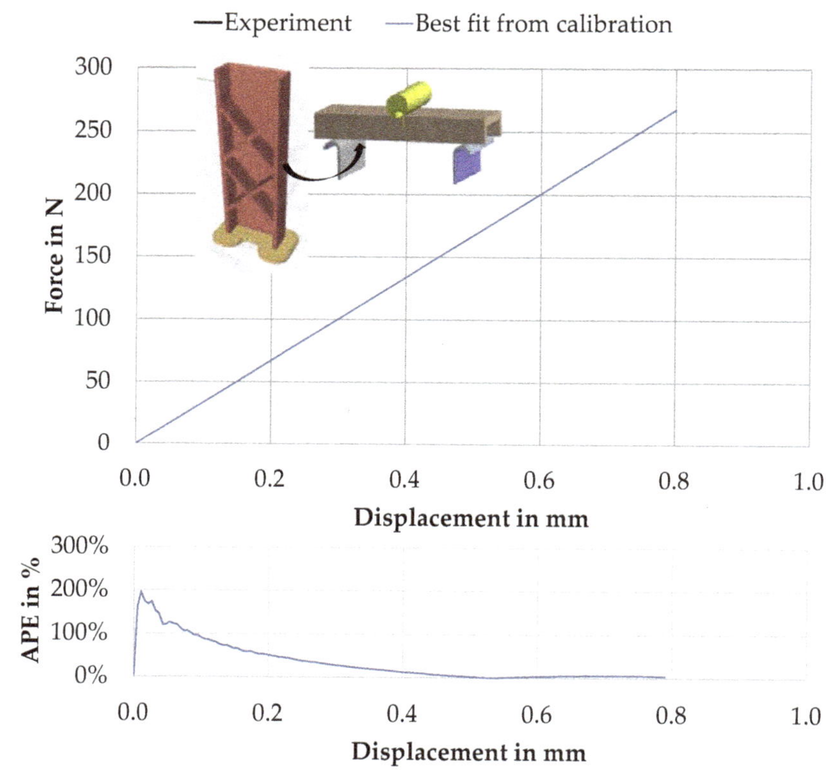

Figure 14. Calibrated material model applied to upright-printed XX-rib specimen bending simulation compared to experimental result (force-displacement-curve above, average percentage error (APE) below; MAPE = 35%).

Despite the XX-rib being printed in an upright manner and tested under bending–thus, transferring the load rather inhomogeneously both between and within layers under different infill and contour directions–the overall match between simulation and experiment is considered satisfactory (overall MAPE is 35%). Within 0 mm to 0.5 mm displacement (maximum longitudinal strain approximately 1.8%) the FE model stiffness is higher than the physical demonstrator's stiffness. In this interval, the majority of the overall MAPE is caused (≈52.6%). From there on to 0.8 mm displacement, this is inversed. The MAPE in this interval is small (≈4.0%).

4. Discussion: Advantages, Disadvantages, and Open Research Questions of Simultaneous Parameter Fitting

Simultaneous fitting of the orthotropic material constants offers advantages in comparison to the frequently pursued simple testing of longitudinal and transverse Young's modulus, which merely yields these two constants (plus Poisson's ratio ν_{xy} in-plane, if measured): (1) A comprehensive, orthotropic, linear elastic material model with its 9 constants can be derived; (2) The obtained material model is capable of predicting the structural behavior under more complicated build direction, load case and wall/infill preconditions, as was demonstrated for the XX-rib demonstrator; (3) The approach is systematic, based on FLM-printable tensile test specimen. It is fully simulation-based and, therefore, integrable into digital workflows within product development.

The printing effort for the elaborate tensile specimens, especially U45°-0° and U45°-90° with their large proportion of support structures, constitutes a downside of the approach; moreover, for stability, both specimens require printed walls which in the case of U45°-0° are largely orthogonal to the desired infill direction.

Overall simulation effort is satisfactory and can be done on a regular workstation within a few hours, as the RSM allows for a limited number of simulations to obtain the response surface (instead of, for example, direct optimization using a simulation solver run at a time for each parameter set update). Naturally, the linear elastic material model only allows for linear simulation outcomes and, therefore, only a limited precision at approximating real, non-linear XX-rib specimen behavior; however, prediction quality appears satisfactory.

Open research questions are therefore: (i) Other, more easily printable specimen for deriving G_{yz} and G_{xz} should be conceived to simplify and accelerate printing (less support) as well as interpretation of fitting results (less orthogonal-wall proportion in U45°-0°, thus more direct derivation of G_{xz}). This measure would improve fast applicability of the approach. (ii) The linearity measure could be improved to account for slow, but steady variations in curvature change. (iii) A more sophisticated material model than linear elastic could be fitted to better account for the actual, non-linear FLM material behavior. (iv) Other demonstrator parts should be tested in the future to have more reliable data on prediction quality. (v) Process parameter influence on the structural behavior and material model should be scrutinized more deeply, as these have a significant influence in this regard [10] and are frequently adapted, for example to minimize warping [45]. (vi) For future work, it should furthermore be checked whether an experimental model validation is also possible with further tensile tests, for example with other specimen geometries, in order to enable the validation in laboratories where there are no possibilities to carry out bending tests. (vii) Finally, it will be essential to carry out investigations in the area of fatigue in order to ensure the long-term, operationally safe use of components made of fibre reinforced FLM materials.

5. Conclusions

FLM material models play a significant role in designing and ensuring the functional properties of structural parts. In this contribution, a systematic approach to derive orthotropic FLM material models was motivated, described, and applied using 15 wt.% short carbon fibre reinforced PETG filament (FormFutura CarbonFil). For calibration, six specimen geometries intended to explain the six moduli of the orthotropic material model were conceived and printed as further development of Bellini et al. [25]. All orthotropic constants were fitted simultaneously using the RSM method. Quantitatively, correlation analysis showed the intended correlation between specimen geometry reaction forces and the respective material parameters. Poisson's ratios were fitted using longitudinal and transversal, averaged experimental strain data. The correlation analysis indicated that multiple specimen reaction forces were influenced by these ratios. Overall calibration quality was satisfactory with the majority of MAPEs below 20%. Validation was conducted using an upright-printed, rib-stiffened bending specimen ("XX-rib") to ensure that orthotropic constants apart from "simple" longitudinal and transversal Young's moduli are necessarily used to explain material behavior. Validation quality was satisfactory observing the fitting graph. MAPE was higher for lower displacements and small for larger displacements, showing the original non-linear behavior of the real specimen. In the described way, material parameters of an orthotropic model were identified, which are summarized in

Table 5. The model has been successfully validated for longitudinal strains of up to 1.8% and can be used for simulation.

Table 5. Overview of the identified material parameters for 15 wt.% short carbon-fibre reinforced PETG filament (Formfuture CarbonFil).

E_x	E_y	E_z	ν_{xy}	ν_{yz}	ν_{xz}	G_{xy}	G_{yz}	G_{xz}
MPa	**MPa**	**MPa**	-	-	-	**MPa**	**MPa**	**MPa**
8153	1949	1549	0.31	0.17	0.36	1096	642	1120

The overall approach looks promising to be used as a systematic approach for deriving FLM material models in virtual product development. Open research questions were outlined.

Author Contributions: Conceptualization, C.W., H.V. and S.W.; Data curation, C.W. and H.V.; Formal analysis, C.W. and H.V.; Funding acquisition, S.W.; Investigation, C.W. and H.V.; Methodology, C.W. and H.V.; Project administration, S.W.; Resources, C.W. and H.V.; Software, H.V.; Supervision, S.W.; Validation, C.W. and H.V.; Visualization, C.W. and H.V.; Writing—original draft, C.W. and H.V.; Writing—review and editing, C.W., H.V. and S.W. All authors have read and agreed to the published version of the manuscript.

Funding: We acknowledge financial support by Deutsche Forschungsgemeinschaft and Friedrich-Alexander-Universität Erlangen-Nürnberg within the funding programme "Open Access Publication Funding".

Acknowledgments: The authors would like to thank the students Robert Schade, Markus Lindner and Dominik Klose for their valuable contributions in the printing, testing and simulation of the material specimens used in this work.

Conflicts of Interest: The authors declare no conflict of interest.

References

1. Lachmayer, R.; Lippert, R.B.; Fahlbusch, T. *3D-Druck Beleuchtet*; Springer: Berlin/Heidelberg, Germany, 2016.
2. Thompson, M.K.; Moroni, G.; Vaneker, T.; Fadel, G.; Campbell, R.I.; Gibson, I.; Bernard, A.; Schulz, J.; Graf, P.; Ahuja, B.; et al. Design for Additive Manufacturing: Trends, opportunities, considerations, and constraints. *CIRP Ann.* **2016**, *65*, 737–760. [CrossRef]
3. Verein Deutscher Ingenieure. *Additive Fertigungsverfahren: Grundlagen, Begriffe, Verfahrensbeschreibungen*; 25.020(VDI 3405); Beuth Verlag: Berlin, Germany, 2014.
4. Vajna, S.; Weber, C.; Zeman, K.; Hehenberger, P.; Gerhard, D.; Wartzack, S. *CAx für Ingenieure: Eine Praxisbezogene Einführung*, 3rd ed.; Springer Vieweg: Wiesbaden, Germany, 2018.
5. Tan, P. Numerical simulation of the ballistic protection performance of a laminated armor system with pre-existing debonding/delamination. *Compos. Part B Eng.* **2014**, *59*, 50–59. [CrossRef]
6. Rozylo, P. Failure analysis of thin-walled composite structures using independent advanced damage models. *Compos. Struct.* **2021**, *262*, 113598. [CrossRef]
7. Brenken, B.; Barocio, E.; Favaloro, A.; Kunc, V.; Pipes, R.B. Fused filament fabrication of fiber-reinforced polymers: A review. *Addit. Manuf.* **2018**, *21*, 1–16. [CrossRef]
8. Shofner, M.L.; Lozano, K.; Rodríguez-Macías, F.J.; Barrera, E.V. Nanofiber-reinforced polymers prepared by fused deposition modeling. *J. Appl. Polym. Sci.* **2003**, *89*, 3081–3090. [CrossRef]
9. DeNardo, N.M. Additive Manufacturing of Carbon Fiber-Reinforced Thermoplastic Composites. Master's Thesis, West Purdue University, Lafayette, IN, USA, 2016.
10. Ning, F.; Cong, W.; Hu, Y.; Wang, H. Additive manufacturing of carbon fiber-reinforced plastic composites using fused deposition modeling: Effects of process parameters on tensile properties. *J. Compos. Mater.* **2017**, *51*, 451–462. [CrossRef]
11. Kabir, S.M.F.; Mathur, K.; Seyam, A.-F.M. A critical review on 3D printed continuous fiber-reinforced composites: History, mechanism, materials and properties. *Compos. Struct.* **2020**, *232*, 111476. [CrossRef]
12. Kolanu, N.R.; Raju, G.; Ramji, M. A unified numerical approach for the simulation of intra and inter laminar damage evolution in stiffened CFRP panels under compression. *Compo. Part B Eng.* **2020**, *190*, 107931. [CrossRef]
13. Thomas, D.S.; Gilbert, S.W. *Costs and Cost Effectiveness of Additive Manufacturing, NIST Special Publication 1176*; NIST National Institute of Standards and Techology: Gaithersburg, MD, USA, 2014.
14. Such, M.; Ward, C.; Potter, K. Aligned Discontinuous Fibre Composites: A Short History. *JMC* **2014**, *2*, 155–168. [CrossRef]

15. Armiger, T.E.; Edison, D.H.; Lauterbach, H.G.; Layton, J.R.; Okine, R.K. Composites of Stretch Broken Aligned Fibers of Carbon and Glass Reinforced Resin. European Patent EP0272088A1, 15 December 1987.
16. Parandoush, P.; Lin, D. A review on additive manufacturing of polymer-fiber composites. *Compos. Struct.* **2017**, *182*, 36–53. [CrossRef]
17. Montero, M.; Roundy, S.; Odell, D.; Ahn, S.; Wright, P. Material Characterization of Fused Deposition Modeling (FDM) ABS by Designed Experiments. In Proceedings of the Rapid Prototyping and Manufacturing Conference, Cincinnati, OH, USA, 15–17 May 2001; pp. 1–21.
18. Ahn, S.; Montero, M.; Odell, D.; Roundy, S.; Wright, P.K. Anisotropic material properties of fused deposition modeling ABS. *Rapid Prototyp. J.* **2002**, *8*, 248–257. [CrossRef]
19. Wu, W.; Geng, P.; Li, G.; Di, Z.; Zhang, H.; Zhao, J. Influence of Layer Thickness and Raster Angle on the Mechanical Properties of 3D-Printed PEEK and a Comparative Mechanical Study between PEEK and ABS. *Materials* **2015**, *8*, 5834–5846. [CrossRef]
20. Sood, A.K.; Ohdar, R.K.; Mahapatra, S.S. Experimental investigation and empirical modelling of FDM process for compressive strength improvement. *J. Adv. Res.* **2012**, *3*, 81–90. [CrossRef]
21. Hernandez, R.; Slaughter, D.; Whaley, D.; Tate, J.; Asiabanpour, B. Analyzing the Tensile, Compressive, and Flexural Properties of 3D Printed ABS P430 Plastic Based on Printing Orientation Using Fused Deposition Modeling. In *Solid Freeform Fabrication 2016, Proceedings of the 27th Annual International Solid Freeform Fabrication Symposium, Austin, TX, USA, 8–10 August 2016*; Bourell, D.L., Crawford, R., Seepersad, C., Beaman, J.J., Fish, S., Harris, M., Eds.; University of Texas at Austin: Austin, TX, USA; pp. 939–950.
22. Pei, E.; Lanzotti, A.; Grasso, M.; Staiano, G.; Martorelli, M. The impact of process parameters on mechanical properties of parts fabricated in PLA with an open-source 3-D printer. *Rapid Prototyp. J.* **2015**, *21*, 604–617. [CrossRef]
23. Li, L.; Sun, Q.; Bellehumeur, C.; Gu, P. Composite Modeling and Analysis for Fabrication of FDM Prototypes with Locally Controlled Properties. *J. Manuf. Process.* **2002**, *4*, 129–141. [CrossRef]
24. Matas, J.F.R. Modeling the Mechanical Behavior of Fused Deposition Acrylonitrile-Butadiene-Styrene Polymer Components. Master's Thesis, University of Notre Dame, Notre Dame, IN, USA, 1999.
25. Bellini, A.; Güçeri, S. Mechanical characterization of parts fabricated using fused deposition modeling. *Rapid Prototyp. J.* **2003**, *9*, 252–264. [CrossRef]
26. Domingo-Espin, M.; Puigoriol-Forcada, J.M.; Granada, A.A.G.; Llumà, J.; Borros, S.; Reyes, G. Mechanical property characterization and simulation of fused deposition modeling Polycarbonate parts. *Mater. Des.* **2015**, *83*, 670–677. [CrossRef]
27. Rodríguez, J.F.; Thomas, J.P.; Renaud, J.E. Mechanical behavior of acrylonitrile butadiene styrene fused deposition materials modeling. *Rapid Prototyp. J.* **2003**, *9*, 219–230. [CrossRef]
28. Lee, C.S.; Kim, S.G.; Kim, H.J.; Ahn, S.H. Measurement of anisotropic compressive strength of rapid prototyping parts. *J. Mater. Process. Technol.* **2007**, *187–188*, 627–630. [CrossRef]
29. Cantrell, J.T.; Rohde, S.; Damiani, D.; Gurnani, R.; DiSandro, L.; Anton, J.; Young, A.; Jerez, A.; Steinbach, D.; Kroese, C.; et al. Experimental characterization of the mechanical properties of 3D-printed ABS and polycarbonate parts. *Rapid Prototyp. J.* **2017**, *23*, 811–824. [CrossRef]
30. Somireddy, M.; Czekanski, A. Mechanical Characterization of Additively Manufactured Parts by FE Modeling of Mesostructure. *J. Manuf. Mater. Process.* **2017**, *1*, 18. [CrossRef]
31. Tekinalp, H.L.; Kunc, V.; Velez-Garcia, G.M.; Duty, C.E.; Love, L.J.; Naskar, A.K.; Blue, C.A.; Ozcan, S. Highly oriented carbon fiber–polymer composites via additive manufacturing. *Compos. Sci. Technol.* **2014**, *105*, 144–150. [CrossRef]
32. Duty, C.E.; Kunc, V.; Compton, B.; Post, B.; Erdman, D.; Smith, R.; Lind, R.; Lloyd, P.; Love, L. Structure and mechanical behavior of Big Area Additive Manufacturing (BAAM) materials. *Rapid Prototyp. J.* **2017**, *23*, 181–189. [CrossRef]
33. Love, L.J.; Kunc, V.; Rios, O.; Duty, C.E.; Elliott, A.M.; Post, B.K.; Smith, R.J.; Blue, C.A. The importance of carbon fiber to polymer additive manufacturing. *J. Mater. Res.* **2014**, *29*, 1893–1898. [CrossRef]
34. Becker, F. Entwicklung Einer Beschreibungsmethodik für das Mechanische Verhalten Unverstärkter Thermoplaste bei Hohen Deformationsgeschwindigkeiten. Ph.D. Thesis, Martin-Luther-Universität Halle-Wittenberg, Halle, Germany, 2009.
35. Kunkel, F. Zum Deformationsverhalten von Spritzgegossenen Bauteilen aus Talkumgefüllten Thermoplasten unter Dynamischer Beanspruchung. Master's Thesis, Otto-von-Guericke-Universität Magdeburg, Magdeburg, Germany, 2017.
36. Witzgall, C.; Wartzack, S. An Investigation of Mechanically Aged Short-Fiber Reinforced Thermoplastics under Highly Dynamic Loads. In Proceedings of the DFX 2016 27th Symposium Design for X, Jesteburg, Germany, 5–6 October 2016; pp. 135–146.
37. Witzgall, C.; Giolda, J.; Wartzack, S. A novel approach to incorporating previous fatigue damage into a failure model for short-fibre reinforced plastics. *Int. J. Impact Eng.* **2022**, *18*, 104155. [CrossRef]
38. *IdeaMaker: 3D Slicer Software*; Raise 3D Technologies I; Raise 3D Technologies, Inc.: Irvine, CA, USA, 2020.
39. Völkl, H.; Mayer, J.; Wartzack, S. Strukturmechanische Simulation additiv im FFF-Verfahren gefertigter Bauteile. In *KfAF 2019 Konstruktion für die Additive Fertigung 2019*, 1st ed.; Lachmayer, R., Rettschlag, K., Kaierle, S., Eds.; Springer: Berlin/Heidelberg, Germany, 2020; pp. 143–157.
40. FormFutura. CarbonFil. Available online: https://www.formfutura.com/shop/product/carbonfil-2788?category=458 (accessed on 26 January 2022).
41. *ANSYS*; ANSYS Help; SAS IP, Inc.: Cheyenne, WY, USA, 2017.
42. *ISO 527-1:2012-06*; Kunststoffe—Bestimmung der Zugeigenschaften—Teil 1: Allgemeine Grundsätze. ISO; Beuth: Berlin, Germany, 2012.

43. *DIN EN ISO 527-4 1997*; Kunststoffe—Bestimmung der Zugeigenschaften—Teil 4: Prüfbedingungen für isotrop und anisotrop faserverstärkte Kunststoffverbundwerkstoffe. ISO; Beuth Verlag: Berlin, Germany, 1997.
44. Pierron, F.; Vautrin, A. Accurate comparative determination of the in-plane shear modulus of T300/914 by the iosipescu and 45° off-axis tests. *Compos. Sci. Technol.* **1994**, *52*, 61–72. [CrossRef]
45. Alsoufi, M.; El-Sayed, A. Warping Deformation of Desktop 3D Printed Parts Manufactured by Open Source Fused Deposition Modeling (FDM) System. *Int. J. Mech. Mechatron. Eng.* **2017**, *17*, 7–16.

Journal of
Composites Science

MDPI

Article

On Numerical Models for Cube Drop Test of Bladder Fuel Tank for Aeronautical Applications

Domenico Cristillo [1], Francesco Di Caprio [1,*], Claudio Pezzella [2], Carmen Paciello [2], Simone Magistro [3], Luigi Di Palma [4] and Marika Belardo [1]

[1] CIRA—Italian Aerospace Research Centre, Via Maiorise, 81043 Capua, CE, Italy; d.cristillo@cira.it (D.C.); m.belardo@cira.it (M.B.)
[2] Step Sud Mare SRL, Via Ex Aeroporto c/o Consorzio Il Sole, LOTTO X1, 80038 Pomigliano d'Arco, NA, Italy; claudio.pezzella@stepsudmare.com (C.P.); carmen.paciello@stepsudmare.com (C.P.)
[3] Aero Secur S.p.a., Via Delle Valli 46, 04011 Aprilia, LT, Italy; simone.magistro@sekur.aero
[4] Maregroup, Via Ex Aeroporto c/o Consorzio Il Sole, LOTTO X1, 80038 Pomigliano d'Arco, NA, Italy; luigi.dipalma@maregroup.it
* Correspondence: f.dicaprio@cira.it; Tel.: +39-082-362-3538

Abstract: For some categories of aircraft, such as helicopters and tiltrotors, fuel storage systems must satisfy challenging crash resistance requirements in order to reduce or eliminate the possibility of fuel fires and thus increase the chances of passenger survival. Therefore, for such applications, fuel tanks with high flexibility (bladder) are increasingly used, which are able to withstand catastrophic events and avoid fuel leakages. The verification of these capabilities must be demonstrated by means of experimental tests, such as the cube drop test (MIL-DTL-27422). In order to reduce development costs, it is necessary to execute experimental tests with a high confidence of success, and, therefore, it is essential to have reliable and robust numerical analysis methodologies. The present work aims to provide a comparison between two explicit FE codes (i.e., Abaqus and Ls-Dyna), which are the most frequently used for such applications according to experimental data in the literature. Both codes offer different material models suitable for simulating the tank structure, and therefore, the most suitable one must be selected by means of a specific trade-off and calibration activity. Both are able to accurately simulate the complex fluid–structure interaction thanks to the use of the SPH approach, even if the resulting sloshing capabilities are quite different from each other. Additionally, the evolution of the tank's deformed shape highlights some differences, and, in particular, Abaqus seems to return a more natural and less artificial behavior. For both codes, the error in terms of maximum impact force is less than 5%, but, even in this case, Abaqus is able to return slightly more accurate results.

Keywords: SPH; hyperelastic; drop test; fuel tank; FEM; explicit analysis; bladder; tiltrotor

Citation: Cristillo, D.; Di Caprio, F.; Pezzella, C.; Paciello, C.; Magistro, S.; Di Palma, L.; Belardo, M. On Numerical Models for Cube Drop Test of Bladder Fuel Tank for Aeronautical Applications. *J. Compos. Sci.* **2022**, *6*, 99. https://doi.org/10.3390/jcs6030099

Academic Editor: Stelios K. Georgantzinos

Received: 14 February 2022
Accepted: 18 March 2022
Published: 21 March 2022

1. Introduction

The runway-independent aircraft concept is progressively becoming more prominent in civil aviation thanks to its ability to improve public mobility in areas where airport infrastructures are not highly developed. Horizon 2020 Clean Sky 2 FRC IADP Next Generation of Civil Tiltrotor Technology Demonstrator (NGCTR-TD) is an example of European-funded research in the sector of VTOL aircraft. NGCTR-TD is the innovative Civil Tiltrotor Technology Demonstrator [1,2] with a configuration that will go beyond current architectures for this type of aircraft. NGCTR-TD is currently under development by Leonardo Helicopter (helicopter division of Leonardo Company, leader in Italian aerospace and defense), and it is shown in Figure 1.

Figure 1. NGCTR tiltrotor, courtesy of Leonardo Helicopters (https://svppbellum.blogspot.com/2020/01/il-next-generation-civil-tiltrotor.html, accessed on 28 January 2022).

Because the tiltrotor is able to operate as both a helicopter and an aircraft, the airworthiness specifications are a combination of CS-25 and CS-29 Airworthiness Requirements, and in some cases, specific tailoring is necessary. For their VTOL capability, tiltrotors have to comply with crashworthiness requirements according to CS-29, similarly to helicopters. The fuel storage system of the NGCTR-TD consists of bladder tanks installed inside the wing structures. Crashworthiness is one of the most demanding aspects to consider in the design of VTOL fuel tanks in order to satisfy passenger survivability requirements. Among the various requirements, tanks have to successfully pass a drop test from a height equal to 15.2 m according to CS 29.952 [3] by showing no leakage after the drop. Moreover, the tanks must possess sufficient flexibility and foldability in order to be installed through suitable access doors placed on the lower skin of the wing and to be used for inspection and maintenance activities. NGCTR-TD has 14 fuel bladder tanks, all located in the wing structure [4]. As an example, the V22 has 16 fuel tanks, 10 integrated into the wings and 6 in the fuselage, holding from 5489 to 13,779 L of fuel [5]. The NCGTR-TD fuel storage system is based on bladder tanks manufactured from very flexible and resistant rubberized materials. Each tank provides a fuel barrier against fuel and fuel vapor leakages and mechanical resistance in case of a crash, puncture or penetration. All tanks are completed with co-cured metallic flanges and valves to connect the fuel storage system with fuel and venting lines.

Not only must the tanks be designed according to crashworthiness requirements, but the wing itself must have crashworthy features, such as the ability to separate itself from the fuselage to ensure that a crash is survivable. This is necessary to prevent the fuselage from being crushed due to the inertial load related to the wings, fuel and nacelles and to let passengers escape from the aircraft. This feature is also present in military tiltrotors, such as Bell V-22 tiltrotor [6]. The civil application should require an experimental test to prove the effectiveness of any solutions devoted to breaking the wing in specific sections. Experimental and numerical studies of a full-scale regional fuselage drop test without any wings installed [7–9] have demonstrated that the fuselage damage is located mainly on the lower section of the fuselage, while the upper part is not critically damaged [10,11]. No evidence on the structural behavior of the wing during a crash has been obtained under crash conditions. In the 1960s, the US Army introduced the first military regulation (MIL-DTL-27422) that defined the certification requirements for fuel storage systems to be installed in a helicopter. The crashworthiness of a tiltrotor tank is based on CS-29 and MIL-DTL-27422 specifications, similarly to a helicopter [12]. As described in MIL-DTL-27422, the drop impact test must be conducted to verify the structural behavior in a dynamic regime of the fuel tank, both in the standalone configuration (cube drop test) and in the partially integrated configuration with the surrounding structure [13]. The Federal Aviation Administration (FAA) funded research on crash-resistant fuel systems for commercial aircraft with the aim of providing data that are useful in reducing the possibility of fire due to crash events [14,15]. Anghileri [16] investigated the fuel tank crashworthiness from experimental and numerical points of view. Fuel sloshing during crash events was studied by means of numerical models, which were validated by comparing their outputs to experimental results [17]. Using parallel computing, Li et al. [18] investigated the dynamic behavior of a fuel tank in a dynamic impact event with the ground, demonstrating that such an algorithm is able to run with a high speed-up ratio and parallel efficiency. Luo et al. [19] studied the crashworthiness of a fuel tank for helicopter applications by means of the finite element method (FEM) in order to verify the energy absorption capabilities of the textile layer and protection frame. Kim et al. [20] used the commercial FE code Ls-Dyna to simulate the dynamic response of fuel tanks installed in a rotorcraft. An accurate study based on analytical, numerical and experimental results was performed to investigate the structural performance of tanks with energy-absorbing fixtures under dynamic pressure and blast load conditions [21,22].

The activities aimed at developing the next-generation fuel storage system of NGCTR-TD are part of the DEFENDER project, and within the design and analysis work package, an innovative aspect is represented by the setup of high-fidelity methodologies and models to support the tank design, especially its performance against crash and impact loads. The final aim is to provide models that are validated by means of a full-scale crash test on the most critical tank plus the wing surrounding its composite structure. The activity will be performed with an incremental approach, which passes through numerical–experimental validation, the simulation of a cube-like tank test and, finally, a full-scale test. The study presented herein is one of the basic steps foreseen for the final objective of the high-fidelity crashworthiness simulation. In particular, in this work, two numerical models for the prediction of the structural response of a cube-like tank under impact load conditions were developed and verified by comparing their results with experimental data available in the literature.

2. Experimental Test Case Description

In order to validate the numerical model, experimental data available in the literature [23] were used. Yang et al. performed two drop tests on a fuel tank filled with water (to replace real fuel). The fuel tank was realized with a flexible structure made of an external fabric nylon layer and an inner sealing layer. The tank was 760 mm long and wide and its height is about 600 mm. In order to execute the drop test, according to the specification, the tank was filled with about 350 kg of water. In the drop impact test, the tank was lifted to a specific height by means of a crane and dropped on a rigid impact surface instrumented with four load cells with a capacity equal to 2000 kN. The impact surface was a steel plate and was 80 mm thick; therefore, it can be reasonably considered rigid. They performed two drop tests using two different drop heights: 15.2 and 19.8 m. The impact velocities related to these drop heights are 17.3 and 19.7 m/s, respectively. For the following numerical comparison, only the impact velocity equal to 17.3 m/s was taken into account, because it is related to a standard drop height, as reported in MIL-DTL-27422.

3. Numerical Model Description

The validation of the experimental test, as previously described, was performed by means of a numerical model, realized with two different software packages: Abaqus and LS Dyna. The experimental test was simulated by means of a 3D FE model composed of three parts: the fuel tank, the ground and the fuel. In particular, the fuel tank was modeled by using shell elements with a thickness of 2 mm and fuel with SPH elements, and the ground was discretized by means of rigid elements in Abaqus and a rigid wall in Ls-Dyna.

For both ABAQUS and Ls-Dyna models, the same mesh density was adopted in order to remove any dependency by the discretization technique. The tank structure was discretized by means of 31728 shell elements, while the number of smooth particle hydrodynamics (SPH) elements used for modeling the fuel was 537936. The platform used for the numerical simulations is an HP Z840 workstation, equipped with an Intel Xeon E5-2620 v3 CPU @ 2.40 GHz and 128 GB RAM. Considering that the simulated event time is

200 ms (even if the most relevant part is up to 50 ms), the computational time is about 4.5 h for Abaqus and about 4 h for Ls-Dyna (both codes run adopting the parallel solution with 4 cores). Obviously, this small difference is also related to the requested output; therefore, the difference in computational time is not too relevant. Figure 2 reports the adopted numerical model with schematic detail of SPH elements and the fluid–structure interaction.

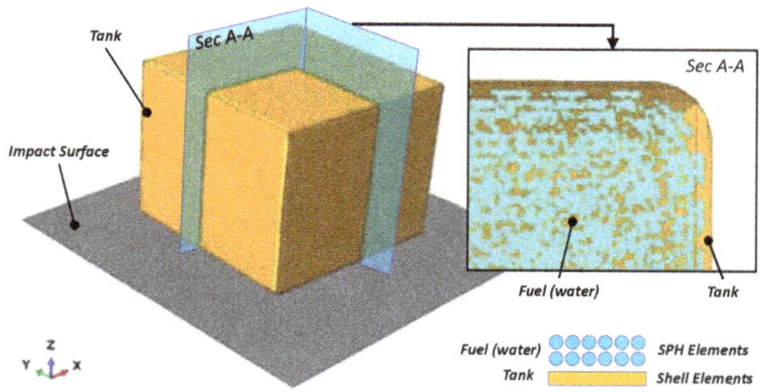

Figure 2. FE model.

3.1. Boundary Conditions

The tank is subject to a free-fall form and height equal to 15.2 m; therefore, the impact velocity is about 17.3 m/s. The ground was simulated by means of rigid elements.

Further, from the reference work, it can be assumed that the impact angle is not equal to zero, but it is worth noting that such a condition is quite hard to obtain. The impact angle is about 3° around both the X and Y axes. Figure 3 schematically reports the adopted boundary conditions.

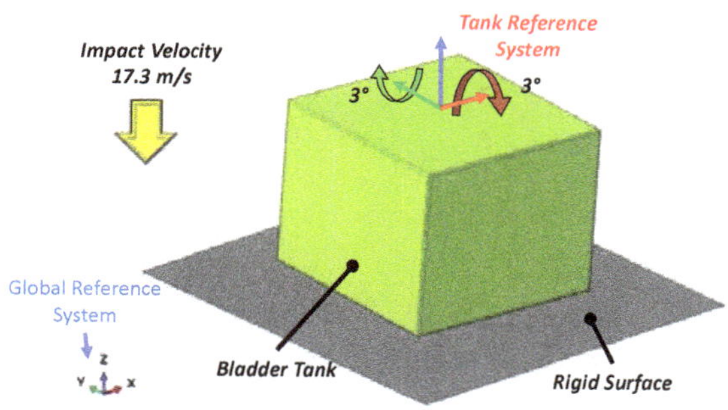

Figure 3. Boundary conditions.

3.2. Smooth Particle Hydrodynamics Method (SPH) and Water Material Model

Smoothed particle hydrodynamics (SPH) are represented by three-dimensional elements with three degrees of freedom and defined by their center of mass. These elements have their own shape functions that depend on the connectivity of the particles. The interpolation distance between the particles, called smoothing length, furnishes the location and provides information about transmission among the different particles.

In the SPH approach, the water was simulated as particles with the same dimensions and distances without mass. The velocity and energy of the particles at any time can be solved by means of a function $f(x)$. The value of this function can be approximated for particle data by Equation (1):

$$\langle f(x_i) \rangle = \sum_{j=1}^{N} \frac{m_j}{\rho_j} f(x_j) W(x_i - x_j, h) \tag{1}$$

where h is the smoothing length, m_i and ρ_i are the mass and density of the particle, x_i and x_j are the positions of particles, and W is the kernel interpolation.

In particular, we consider two particles (i and j) to be nearest neighbors. If the distance between particles i and j is lower than the radius of the sphere of influence of j, then i is connected to j. The dimension of the sphere of influence is a multiple of the particle's smoothing length, and the multiplication factor depends on the kernel used to create the smooth particle's shape function.

Therefore, the value of a variable for particle j can be obtained by adding the contributions of particle i contained within the influence radius. The interpolation kernel used in this analysis was proposed by Monaghan [24].

The behavior of the fluid material inside the fuel tank is determined by the equation of state (EOS), which relates the pressure and the material volume change rate in a physical state.

The equation for the conservation of energy equates the increase in internal energy per unit mass, E_m, to the rate at which work is being performed by stresses and the rate at which heat is being added. In the absence of heat conduction, the energy equation can be written as:

$$\rho\frac{\partial E_m}{\partial t} = (p - p_{bv})\frac{1}{\rho}\frac{\partial \rho}{\partial t} + S : \dot{e} + \rho\dot{Q} \tag{2}$$

where p is the pressure stress, defined as positive in compression; p_{bv} is the pressure stress due to the bulk viscosity; S is the deviatoric stress tensor; $\dot{e}$ is the deviatoric part of the strain rate; and $\dot{Q}$ is the heat rate per unit mass.

The equation of state is assumed for the pressure as a function of the current density, ρ, and the internal energy per unit mass, E_m.

Since the internal energy can be eliminated, it is possible to write the EOS as a p versus V relationship (where V is the current volume) or, equivalently, a p versus $1/\rho$ relationship that is unique to the material described by the equation-of-state model. This unique relationship is called the Hugoniot curve and is the locus of p-V states achievable behind a shock (see Figure 4).

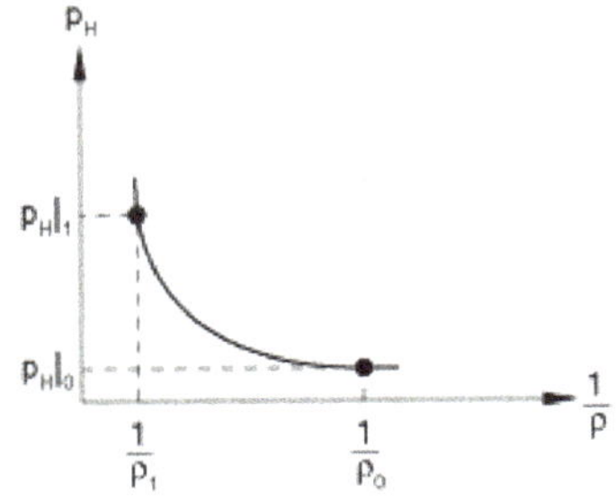

Figure 4. A schematic representation of a Hugoniot curve.

The Hugoniot pressure, p_H, is a function of density only and can be determined, in general, by fitting experimental data.

There are different formulations of EOS, and the most common form for the water description is the linear *Us–Up* equation, which is given by:

$$pH = \frac{\rho_0 c_0^2 \eta}{(1 - s\eta)^2} \tag{3}$$

where c_0 and s define the linear relationship between the shock velocity, *Us*, and the particle velocity, *Up*, as follows:

$$U_s = c_0 + sU_p \tag{4}$$

With the above assumptions, the linear $Us-Up$ Hugoniot form is written as

$$p = \frac{\rho_0 c_0^2 \eta}{(1-s\eta)^2}\left(1 - \frac{\Gamma_0 \eta}{2}\right) + \Gamma_0 \rho_0 E_m \tag{5}$$

where $\rho_0 {c_0}^2$ is equivalent to the elastic bulk modulus at small nominal strains.

3.3. Tank Material Model

The fuel tank structure is a crash-resistant composite fabric composed of a rubber layer and a nylon layer. A different formulation is used for each of these materials.

Rubber materials have very small compressibility compared to their shear flexibility. This behavior can be modeled by means of hyperplastic material. In fact, the hyperplastic material model is isotropic and nonlinear, is valid for materials that exhibit an instantaneous elastic response to large strains and requires that geometric nonlinearity be accounted for during the analysis step since it is intended for finite-strain applications.

Hyperelastic materials are described in terms of a "strain energy potential", $U(\varepsilon)$, which defines the strain energy stored in the material per unit of reference volume (volume in the initial configuration) as a function of the strain at that point in the material. There are several forms of strain energy potentials available in Abaqus to model approximately incompressible isotropic elastomers: the Arruda–Boyce form, the Marlow form, the Mooney–Rivlin form, the neo-Hookean form, the Ogden form, the polynomial form, the reduced polynomial form, the Yeoh form and the van der Waals form. For this material description, the Marlow form is used. A strain energy potential is constructed that will reproduce the test data exactly and that will have reasonable behavior in other deformation modes.

The form of the Marlow strain energy potential is:

$$U = U_{dev}(\bar{I}_1) + U_{vol}\left(J^{el}\right) \tag{6}$$

where U is the strain energy per unit of reference volume, with U_{dev} as its deviatoric part and U_{vol} as its volumetric part; $\bar{I}_1$ is the first deviatoric strain invariant and is defined as:

$$\bar{I}_1 = \bar{\lambda}_1^2 + \bar{\lambda}_2^2 + \bar{\lambda}_3^2 \tag{7}$$

where the deviatoric stretch $\bar{\lambda}_i = J^{-\frac{1}{3}}\lambda_i$ is the total volume ratio, J^{el} is the elastic volume and λ_i is the principal stretch. The deviatoric part of the potential is determined by providing either uniaxial, equibiaxial or planar test data, while the volumetric part is determined by providing the volumetric test data, defining Poisson's ratio or specifying lateral strains together with uniaxial, equibiaxial or planar test data.

Additionally, the fabric layer has very low compressibility compared to its shear flexibility, and for this reason, the hyperelastic material model was also adopted for the fabric layer.

The most suitable mathematical model is the Ogden form. A strain energy potential is constructed that will reproduce the test data exactly and that will have reasonable behavior in other deformation modes.

The form of the Ogden strain energy potential is:

$$U = \sum_{i=1}^{N} \frac{2\mu_i}{\alpha_i^2}\left(\bar{\lambda}_1^{\alpha_i} + \bar{\lambda}_2^{\alpha_i} + \bar{\lambda}_3^{\alpha_i} - 3\right) + \sum_{i=1}^{N} \frac{1}{D_i}\left(J^{el} - 1\right)^{2i} \tag{8}$$

where $\overline{\lambda}_i$ is the deviatoric principal stretch $\overline{\lambda}_i = J^{-\frac{1}{3}}\lambda_i$; λ_i is the principal stretch; N is a material parameter; and μ_i , α_i and D_i are temperature-dependent material parameters. The initial shear modulus and bulk modulus for the Ogden form are given by:

$$\mu_0 = \sum_{i=1}^{N} \mu_i,\ K_0 = \frac{2}{D_1} \tag{9}$$

Since the fabric layer is stiffer and stronger than the rubber layer, its behavior is predominant, and for this reason, the Ogden material model was used to reproduce the structural behavior of the tank skin.

3.4. Tank Material Properties

The tank structure is made in a composite material that has a fabric nylon layer and an inner rubber layer aimed at ensuring impermeability to the fuel. Generally, the rubber layer is co-cured on the fabric layer, and the connection is very strong; possible failures involve both layers. The global response of the tank material can be related to a hyperplastic material, and therefore, a suitable material model was adopted in both Abaqus and Ls-Dyna. The best approach to defining the structural behavior is by means of obtaining the experimental stress–strain curve in a uniaxial tensile test. The density of the woven material is 1150 kg/m^3, and the average thickness of the woven material is 2 mm. The stress–strain curve of the woven material, used as input, is reported in Figure 5.

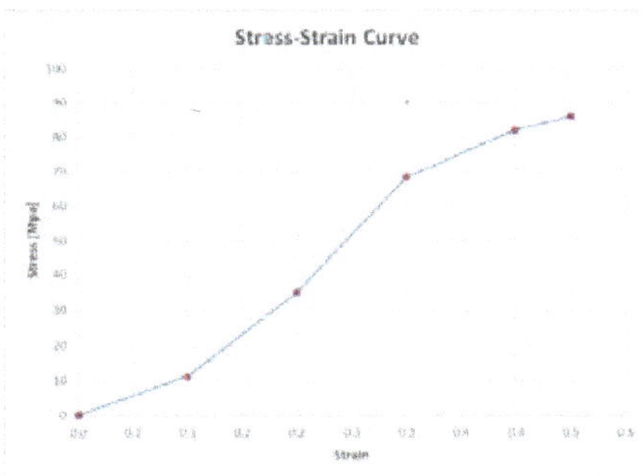

Figure 5. Stress–strain curve of the woven material.

4. Numerical Results

In the following, the comparison between Ls-Dyna and Abaqus results is shown. Figure 6 shows the deformed shape of the fuel tank in terms of global displacement (vectorial sum) for several time steps in order to provide a sort of deformation time-history.

The next figure shows that, generally, the two models behave quite similarly to each other. In particular, since the structure is very flexible, it begins to deform and is crushed in the lower area (the one that first comes into contact with the rigid ground). The internal fluid, under the action of inertial forces, pushes the tank structure downwards, introducing a moderate deformation state to the upper part. The magnitude of the global displacement is comparable between the two models, and therefore, the overall deformation of the structure does not present striking differences. However, it is equally evident that the structural response starts to exhibit some differences at around 40 ms. In fact, the Abaqus model seems to be softer, and therefore, the upper central part is subject to greater displacements. This is explained by the fact that the liquid begins to press towards the lateral walls in the bottom region (see also the following figures that report a sectional view). The fluid then generates an empty region that is unable to support the tank structure, and therefore, it collapses. On the other hand, in the LS-Dyna model, this peculiarity is not evident, and this is associated with the greater rigidity of the structure, which is therefore unable to accommodate fluid dislocations.

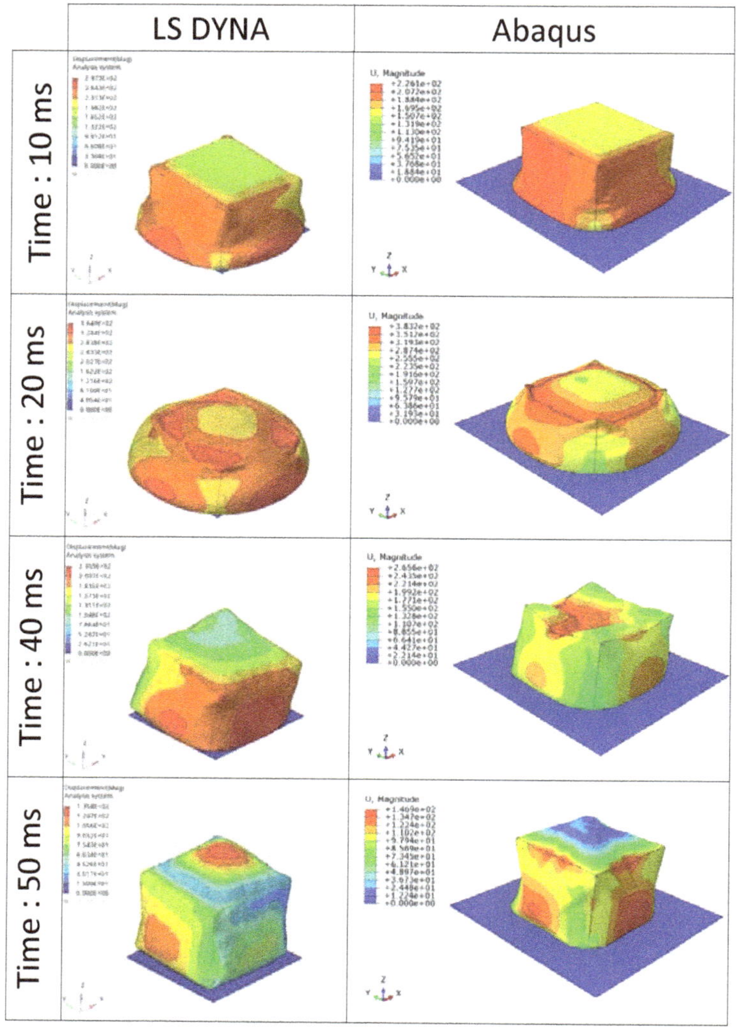

Figure 6. Global displacement contour plot. Left: Ls-Dyna model; right: Abaqus model.

Figure 7 shows the time histories of the strain of the bladder fuel tank for both the Ls-Dyna and Abaqus models. The distribution of the strain is quite different between the two models, even if, in terms of absolute values, they are still comparable. The biggest differences occur after the rebound stage. At 20 ms, the LS-Dyna model has a wide band with fairly high strain values (about 38%), so the area subject to high stretching is quite extensive. In the Abaqus model, on the other hand, only the areas close to the vertical edges show high strain values (around 32%). It is reasonable to assume that the largest strain values should be concentrated just in these areas, because the structure will try to eliminate the existing bends in order to accommodate fluid sloshing. The parts far from the edges, on the other hand, will be subject almost exclusively to a tensile stress state.

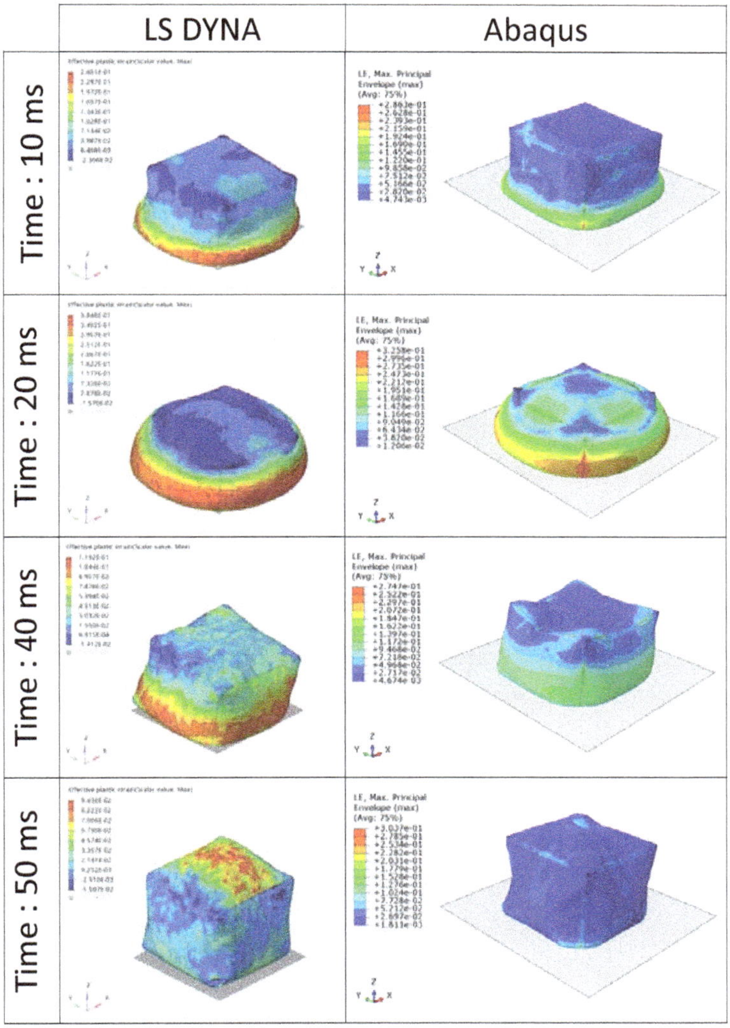

Figure 7. Strain distribution on tank structure. Left: Ls-Dyna model; right: Abaqus model.

After the rebound, at 40 ms, in the LS-Dyna model, a wide region with high strain values remains (obviously, the values are smaller than those related to the maximum crushing step). On the other hand, in the Abaqus model, the regions with high strain values are always very reduced and localized in the bending areas. However, it should be noted that at 40 ms, the deformation state of the bottom regions presents comparable values between the two models (about 12% for Ls-Dyna and 14% for Abaqus). The top of the Ls-Dyna structure has an average strain state of about 4%, while in Abaqus, this area is not notably stressed since the average strain is about 0.4%. All of this, again, highlights the greater flexibility of the Abaqus model compared to the Ls-Dyna model.

Finally, at 50 ms, although the colors could be misleading, the maximum values on the top are comparable and are equal to about 10%. This is due to the fluid, which, after the rebound, is concentrated towards the area with minimum resistance, i.e., the central area, and pushes the tank from the inside upwards.

Figure 8 shows a section of the tank in which it is possible to see the global displacement of the SPH elements. The sectional view provides a clear overview of the behavior of the fluid inside the tank. Prior to 20 ms, there are no significant differences between the two models, as observed in the previous results. Even in this case, the greatest differences occur after the rebound phase. The Abaqus model allows for greater mobility of fluid particles, and thus, the sloshing is much greater. In fact, it is possible to note a clear mixing of the particles. Further, the 50 ms frame can explain the strain distribution on the top region. In Abaqus, the particles are more able to concentrate towards the central region, and therefore, in this area, the pressure is much greater. It is equally important to note that both models are able to accurately simulate the fluid–structure interaction. The particles remain contained within the tank and are able to slide easily on the internal walls. For example, at 10 ms, it is evident that there are large tank portions in which the fluid is not in direct contact with the structure.

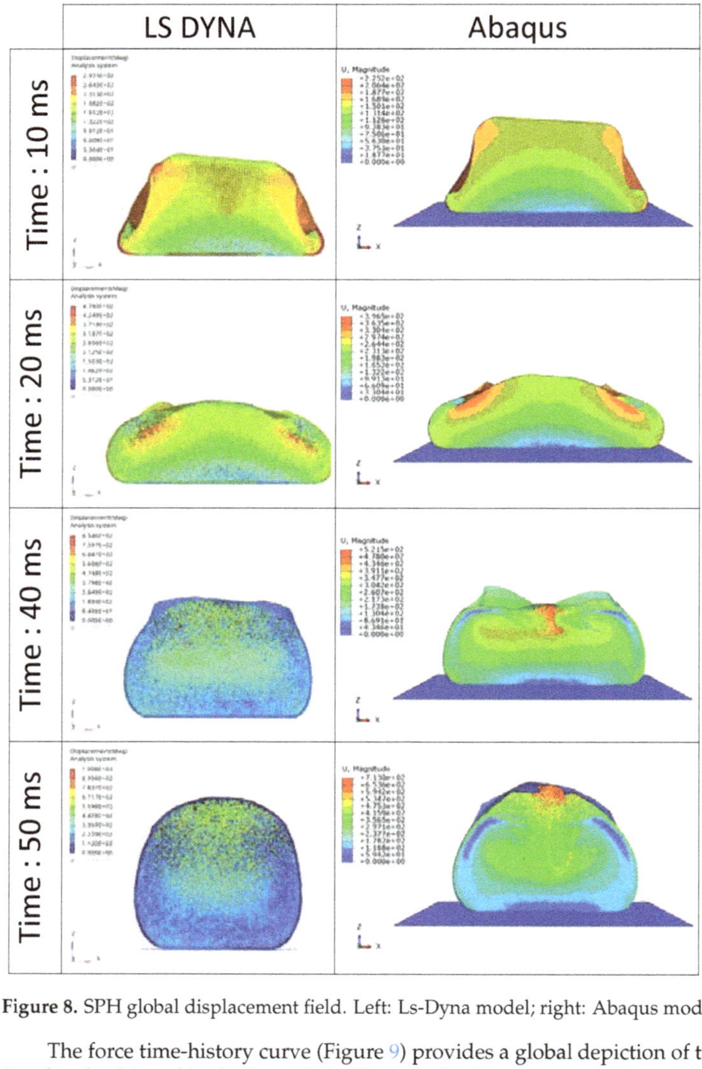

Figure 8. SPH global displacement field. Left: Ls-Dyna model; right: Abaqus model.

The force time-history curve (Figure 9) provides a global depiction of the good correlation level achieved by both models. The whole phenomenon is dominated by two force peaks, which are related to the first contact time and to the maximum crushing time. In reference to this, the Abaqus model is more accurate since the Ls-Dyna model estimates the presence of a third peak (although with a reduced value). This could be related to the lower liquid fluidity (less sloshing), which leads to the concentration of inertial forces in more than two events. In addition, the overestimation of the second force peak by the LS-Dyna model could be related to the fact that the greater force is due to greater mass (therefore, more fluid is concentrated in the contact region due to a lower mobility capability and therefore a decreased ability to spread itself over a larger surface).

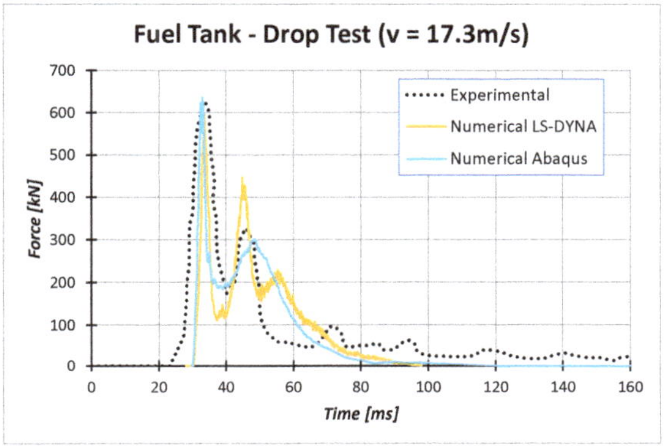

Figure 9. Numerical–experimental comparison in terms of impact force time-history.

In particular, the first experimental impact load peak is about 622 kN, whereas the value determined from the Abaqus model is 637.15 kN (+2.40%), and the value related to the Ls-Dyna model is 595.29 kN (−4.33%). As said before, the second force peak is well estimated by the Abaqus model since the error is about −7.45%, and it is quite overestimated by the Ls-Dyna model, which has an error of about +36%. Finally, both models overestimate the contact time by about 10 ms.

Another aspect to underline is that the experimental results do not show zero force after the rebound (an event that should occur at around 60 ms). This could be due to a poor damping capacity of the impact surface, since such structures are subject to significant rebounds if no catastrophic failures occur. Therefore, the comparison after that time step is not realistic.

5. Conclusions

This paper provides a comparison between two numerical models developed by means of two different explicit FE codes: Abaqus and LS-Dyna. The quality of the numerical results was verified by numerical–experimental comparison with respect to experimental data available in the literature related to the standard cube drop test for fuel tank certification.

- Having reliable and robust numerical models is a key factor in both realizing a good design and reducing the costs related to product certification. In fact, this research activity is motivated by the need to define numerical models applicable to the design and verification phases of the fuel storage system of the NGCRT-TD. In order to perform a fair comparison, the developed models have the same peculiarities; therefore, both simulate the tank structure by means of shell elements and a hyperelastic material model, and both models use the SPH approach for modeling the fluid. In particular, the most suitable mathematical formulation for the tank structure is the Ogden model. The material calibration curve was established using experimental tensile test data on samples consisting of a nylon layer and a rubber layer (the latter is essentially introduced to ensure the sealing capability).
- The results show that both codes are able to achieve an excellent correlation level with comparable computation costs. For both models, the error related to the maximum

impact force is smaller than 5% (a slightly better result was obtained with the Abaqus model). Both codes overestimate the duration of the main crash event (i.e., up to the rebound) to a similar extent. More significant differences can be found regarding the second peak of the impact force (relative to the instant of maximum crushing). In this case, Abaqus is clearly more accurate since the second peak is estimated with an error of about 7%.

- Regarding the structural response of the tank, the Abaqus model is able to simulate greater flexibility. This allows both better accommodation of the movements of the internal fluid and therefore a more realistic deformation mode, and it provides a strain distribution with reduced average values localized in a few regions, such as the edges.
- Finally, as regards the dynamics of the fluid particles, minimal differences are found up to the maximum crushing time, but after that, the differences are more significant. The Abaqus model is able to simulate a high state of particle mixing, and therefore, the resulting sloshing is more extensive and more realistic. It would be interesting to experimentally study the real degree of sloshing in such crash events. This is, obviously, a key aspect since, due to the high material flexibility, the fluid is able to stretch the tank structure with its pressure.

Author Contributions: Conceptualization, D.C., F.D.C., C.P. (Claudio Pezzella), C.P. (Carmen Paciello), S.M. and M.B.; methodology, D.C., F.D.C., C.P. (Claudio Pezzella) and C.P. (Carmen Paciello); validation, D.C., C.P. (Carmen Paciello) and S.M.; formal analysis, D.C. and C.P. (Carmen Paciello); data curation, D.C., F.D.C. and S.M.; writing—original draft preparation, D.C., F.D.C., C.P. (Claudio Pezzella), C.P. (Carmen Paciello), S.M. and M.B; writing—review and editing, F.D.C.; funding acquisition L.D.P.; supervision, L.D.P. All authors have read and agreed to the published version of the manuscript.

Funding: This research was funded by the Clean Sky 2 Joint Undertaking under the European Union's Horizon 2020 research and innovation program, Grant Agreement number: 738078—DEsign, development, manufacture, testing and Flight qualification of nExt geNeration fuel storage system with aDvanced intEgRated gauging and self-sealing capabilities (DEFENDER).

Institutional Review Board Statement: Not applicable.

Informed Consent Statement: Not applicable.

Data Availability Statement: Not applicable.

Conflicts of Interest: The authors declare no conflict of interest.

References

1. Sillers, P. T-WING: In Conversation with CIRA. Available online: https://www.cleansky.eu/clean-skys-t-wing-in-conversation-with-cira (accessed on 28 January 2022).
2. Perry, D. ANALYSIS: Leonardo Helicopters Advances on Next-Gen Tiltrotor. Available online: https://www.flightglobal.com/analysis/analysis-leonardo-helicopters-advances-on-next-gen-tiltrotor/131350.article (accessed on 28 January 2022).
3. European Aviation Safety Agency. "Certification Specifications for Large Rotorcraft–CS-29", Amendment 10. December 2021. Available online: https://www.easa.europa.eu/document-library/certification-specifications/cs-29-amendment-10 (accessed on 10 March 2022).
4. Belardo, M.; Marano, A.D.; Beretta, J.; Diodati, G.; Graziano, M.; Capasso, M.; Di Palma, L. Wing Structure of the Next-Generation Civil Tiltrotor: From Concept to Preliminary Design. *Aerospace* **2021**, *8*, 102. [CrossRef]
5. Freudenruch, C. How the V-22 Osprey Works. HowStuffWorks. Available online: https://science.howstuffworks.com/osprey.html (accessed on 28 January 2022).
6. Cronkhite, J.D.; Tanner, A.E. Tilt Rotor Crashworthiness. In Proceedings of the 41st American Helicopter Society (AHS) Forum, Ft. Worth, TX, USA, 15–17 May 1985. Available online: https://vtol.org/store/product/tilt-rotor-crashworthiness-1489.cfm (accessed on 28 January 2022).
7. Di Palma, L.; Di Caprio, F.; Chiariello, A.; Ignarra, M.; Russo, S.; Riccio, A.; De Luca, A.; Caputo, F. Vertical Drop Test of Composite Fuselage Section of a Regional Aircraft. *AIAA J.* **2020**, *58*, 474–487. [CrossRef]
8. Perfetto, D.; Lamanna, G.; Manzo, M.; Chiariello, A.; di Caprio, F.; di Palma, L. Numerical and Experimental Investigation on the Energy Absorption Capability of a Full-Scale Composite Fuselage Section. *Key Eng. Mater.* **2019**, *827*, 19–24. [CrossRef]
9. Riccio, A.; Saputo, S.; Sellitto, A.; Russo, A.; Di Caprio, F.; Di Palma, L. An Insight on the Crashworthiness Behavior of a Full-Scale Composite Fuselage Section at Different Impact Angles. *Aerospace* **2019**, *6*, 72. [CrossRef]

10. Riccio, A.; Saputo, S.; Sellitto, A.; Di Caprio, F.; di Palma, L. A numerical-experimental assessment on a composite fuselage barrel vertical drop test: Induced damage onset and evolution. *Compos. Struct.* **2020**, *248*, 112519. [CrossRef]
11. Caputo, F.; Lamanna, G.; Perfetto, D.; Chiariello, A.; Di Caprio, F.; Di Palma, L. An experimental and numerical crash-worthiness study of a full-scale composite fuselage section. *Aeronaut. Astronaut.* **2020**, *59*, 685–703. [CrossRef]
12. Giavotto, V.; Caprile, C.; Sala, G. The design of helicopter crashworthiness. Energy absorption of aircraft structures as an aspect of crashworthiness. In Proceedings of the AGARD 66th Meeting of the Structures and Material Panel, Luxembourg, 1–6 May 1988; pp. 6.1–6.9.
13. Paciello, C.S.; Pezzella, C.; Belardo, M.; Magistro, S.; Di Caprio, F.; Musella, V.; Lamanna, G.; Di Palma, L. Crashworthiness of a Composite Bladder Fuel Tank for a Tiltrotor Aircraft. *J. Compos. Sci.* **2021**, *5*, 285. [CrossRef]
14. Harris, F.D.; Kasper, E.F.; Iseler, L.E. US Civil Rotorcraft Accidents, 1963 through 1997. NASA STI Program, NASA/TM-2000-209597, USAAMCOM-TR-00-A-006. 2000. Available online: https://citeseerx.ist.psu.edu/viewdoc/download?doi=10.1.1.976.8762&rep=rep1&type=pdf (accessed on 28 January 2022).
15. Robertson, S.H.; Johnson, N.B.; Hall, D.S. *A Study of Transport Airplane Crash-Resistant Fuel Systems*; Robertson Aviation LLC: Tempe, AZ, USA, 2002.
16. Anghileri, M. Crash behavior of helicopter fuel tank structures. *US Army* **1998**, *64*, 95–159.
17. Anghileri, M.; Castelletti, L.-M.; Tirelli, M. Fluid–structure interaction of water filled tanks during the impact with the ground. *Int. J. Impact Eng.* **2005**, *31*, 235–254. [CrossRef]
18. Li, Z.; Jin, X.L.; Chen, X.D. The simulation of dual layer fuel tank during the impact with the ground based on parallel computing. *J. Comput. Nonlinear Dyn.* **2007**, *2*, 366–373.
19. Luo, C.; Liu, H.; Yang, J.-L.; Liu, K.-X. Simulation and Analysis of Crashworthiness of Fuel Tank for Helicopters. *Chin. J. Aeronaut.* **2007**, *20*, 230–235. [CrossRef]
20. Kim, H.-G.; Kim, S. Numerical simulation of crash impact test for fuel cell group of rotorcraft. *Int. J. Crashworthiness* **2014**, *19*, 639–652. [CrossRef]
21. Wang, Y.; Liew, J.R. Blast performance of water tank with energy absorbing support. *Thin-Walled Struct.* **2015**, *96*, 1–10. [CrossRef]
22. Wang, Y.; Liew, J.R.; Lee, S.C. Structural performance of water tank under static and dynamic pressure loading. *Int. J. Impact Eng.* **2015**, *85*, 110–123. [CrossRef]
23. Yang, X.; Zhang, Z.; Yang, J.; Sun, Y. Fluid–structure interaction analysis of the drop impact test for helicopter fuel tank. *SpringerPlus* **2016**, *5*, 1573. [CrossRef] [PubMed]
24. Monaghan, J.J.; Lattanzio, J.C. A refined particle method for astrophysical problems. *Astron. Astrophys.* **1985**, *149*, 135143.

Journal of
Composites Science

MDPI

Article

Systematic Approach for Finite Element Analysis of Thermoplastic Impregnated 3D Filament Winding Structures—Advancements and Validation

Jonathan Haas [1,*], Daniel Aberle [1], Anna Krüger [1], Björn Beck [1], Peter Eyerer [1], Luise Kärger [2] and Frank Henning [1,2]

[1] Fraunhofer Institute for Chemical Technology, Department of Polymer Engineering, Joseph-von-Fraunhofer-Str. 7, 76327 Pfinztal, Germany; d-aberle@web.de (D.A.); anna.krueger.ei@gmail.com (A.K.); bjoern.beck@ict.fraunhofer.de (B.B.); peter.eyerer@ict.fraunhofer.de (P.E.); frank.henning@ict.fraunhofer.de (F.H.)

[2] Karlsruhe Institute of Technology, Institute of Vehicle System Technology, Rintheimer Querallee 2, 76131 Karlsruhe, Germany; luise.kaerger@kit.edu

* Correspondence: jonathan.haas@ict.fraunhofer.de

Abstract: This work aims to enhance and validate a systematic approach for the structural finite element (FE) analysis of thermoplastic impregnated 3D filament winding structures (fiber skeletons). The idealized modeling of geometrically complex fiber skeletons used in previous publications is refined by considering additional characteristic dimensions and investigating their mechanical influence. Moreover, the modeling approach is transferred from the meso- to the macro-level in order to reduce modeling and computational effort. The properties of meso- and macro-level FE models are compared using the example of simple loop specimens. Based on the results, respective application fields are defined. In the next step, the same modeling approach is applied to a more complex, three-dimensional specimen—the inclined loop. For its macro-level FE model, additional material characterization and modeling, as well as enhancements in the modeling of the geometry, are proposed. Together with previously determined effective composite properties of fiber skeletons, these results are validated in experimental tensile tests on inclined loop specimens.

Keywords: 3D filament winding; commingled yarn; fiber skeleton; geometry modeling; finite element analysis; structural simulation

Citation: Haas, J.; Aberle, D.; Krüger, A.; Beck, B.; Eyerer, P.; Kärger, L.; Henning, F. Systematic Approach for Finite Element Analysis of Thermoplastic Impregnated 3D Filament Winding Structures—Advancements and Validation. *J. Compos. Sci.* **2022**, *6*, 98. https://doi.org/10.3390/jcs6030098

Academic Editor: Stelios K. Georgantzinos

Received: 26 February 2022
Accepted: 15 March 2022
Published: 19 March 2022

1. Introduction

The 3D filament winding technology, also referred to as 3D Skeleton Winding technology (3DSW), is a design and production approach that allows lightweight potentials to be exploited to a particularly high degree. Its concept consists of winding thermoset or thermoplastic impregnated reinforcement fibers (e.g., glass or carbon fibers) onto a winding form, whereby (after curing or solidification of the polymer matrix) a so-called fiber skeleton is created. In comparison to the conventional filament winding process for the production of rotationally symmetric composite structures, the robot-based coreless 3D filament winding process offers greater geometric flexibility, allowing more complex, topology-optimized fiber skeletons to be realized. The impregnated fibers are usually wound around load introduction locations and support points (mostly implemented as metallic inserts) so that the fiber orientation is ideally aligned with the load paths. This allows occurring loads to be transferred directly into the continuous fibers in a form-fit manner, leading to high utilization of the fibers' mechanical properties. Depending on the application, the fiber skeletons are used either as pure skeletal structures [1–3] or as local reinforcements within molded parts or laminates [4–6]. In both cases they generally serve to carry significant mechanical loads while minimizing the component's mass [4,7,8].

Quantitative methods are needed that allow precise estimates of the fiber skeletons' load bearing capacity and load-dependent deformation. While the mechanical behavior of wound ring reinforcements under internal pressure (with a constant, rectangular cross-sectional area) can be estimated on the basis of analytical equations [9], finite element (FE) analysis is generally used for the structural assessment of more complex winding structures. Depending on the size of the fiber skeleton considered and the failure modes investigated, different FE modeling approaches are suitable. In the simulation of large fiber skeletons (e.g., lightweight installation profiles, ship cabins or building structures) highly simplified geometry models are used. An overview of this modeling approach is given in [10]. As can be seen in [1,2,11–13] the fiber strings are usually modeled as one-dimensional rod or beam elements, while the fiber deflection points are represented by rigid connections of these elements. This approach enables precise estimates of elastic deformations of large wound trusses at low modeling and computational effort. However, it is not suited for precise simulations of the failure mechanisms typically occurring at the fiber deflection points that usually determine the fiber skeletons' bearing capacity under tensile loading [8,9,14]. Since those are of great importance in the dimensioning of small fiber skeletons often serving as local reinforcement in structural components, in this case, more realistic geometry models are created. As can be seen in [15,16], the (idealized) orientation of the fibers is mapped and represented in the geometry models—notably at the fiber deflection points. In this context shell and/or solid elements are used, enabling the consideration of orthotropic material properties and the precise analysis of stress distributions. Thus, failure criteria and damage evolution models for unidirectionally reinforced composites can be implemented and, consequently, besides elastic deformations, failure-critical locations as well as critical loads and failure sequences are also evaluated. In this context, most widely used are the failure criteria of Hashin [17], Puck [18] and the Maximum Stress Criterion as well as the degradation models according to Lapczyk et al. [19] and Puck [18]. In order to also simulate relative movement and/or delamination between individual winding layers, mesoscopic geometry models are created in [5,9,20], that is, all windings are modeled individually and their interactions are defined either by isotropic separation layers or contact definitions (bonded/frictional/frictionless). Validation studies for such detailed approaches to the FE simulation of fiber skeletons show that the agreement between calculated and measured mechanical behavior greatly depends on a realistic geometry model. Thus, it is found in [5] that the consideration of geometry modifications caused by subsequent pressing of the wound fiber skeletons is crucial to obtain precise simulation results. In [15], it is shown that the tensile loading capacity of a wound loop is strongly influenced by the loop's thickness (in loop plane), but degressively increases with it. A theoretical rationale for this is given in [9] where the relevance of geometric non-linearities is pointed out in this context.

In [21], the authors of the present paper introduce a systematic approach to the structural FE simulation of fiber skeletons. It is summarized in the following two paragraphs. The advancements to this approach generated in the present paper are listed in bullet point form at the end of this section.

First, as can be seen in Figure 1a, it is shown that fiber skeletons generally have four different structural constituents (structural constituents: phases within a structure showing clearly distinguishable mechanical properties): the impregnated roving (A), the roving-roving interface (B), the insert-roving interface (C) and the insert (D). Consequently, all modeling and investigation steps are carried out on a mesoscopic level. Using the example of the chosen materials (windings consisting of a polypropylene-glass fiber commingled yarn, inserts made of aluminum), the respective mechanical behavior of the different structural constituents is characterized experimentally. On the basis of these measurements, material models for the structural constituents are selected and parametrized. The elastic behavior and strength of (A) are defined by transversely isotropic constants. Local initial failure is determined by the Maximum Stress Criterion and the subsequent local degradation is defined by a reduction of the elastic constants following the Material Property Degradation Method (MPDG) [22–24]. The mechanical behavior of (B) is defined by a bilinear cohesive zone model (CZM) according to [25]. It covers elastic and plastic deformation as well as detachments. (C) is not experimentally examined in [21]. It is simplified as a frictionless contact, as it is assumed not to be a significant influence in the experiments performed. The elastic behavior of (D) is assumed to be isotropic. No failure model is defined for it, as aluminum inserts within fiber skeletons typically do not show signs of failure.

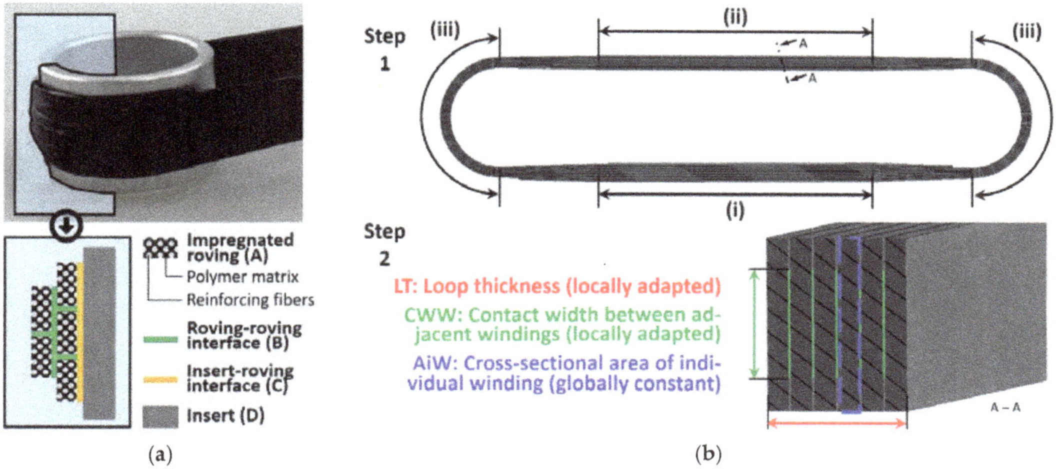

Figure 1. (**a**) Structural constituents within fiber skeletons, adapted from [21]; (**b**) Distinction of characteristic areas in a simple loop (shafts (i, ii) and wraps (iii)) and definition of characteristic dimensions, adapted from [21].

Besides the isolated consideration and characterization of the structural constituents described above, the geometric modeling of the fiber skeletons is also an essential part of the systematic approach presented in [21]. Geometry modeling is of particular importance in the structural simulation of fiber skeletons, since certain dimensions have a considerable influence on the stress distribution [14] and consequently also on the calculated load bearing capacity [5,15]. Detailed geometry modeling of fiber skeletons is challenging as—due to the manufacturing process—they usually do not show clearly identifiable contours and shapes. In the 3D filament winding process, a limb yarn is wound around inserts or stretched freely into space, depending on the location in the fiber skeleton. As a result, the windings assume cross-sectional shapes that are geometrically complex to describe and may vary over the course of the winding path. Consequently, geometry models of fiber skeletons must be an idealization of reality—this is especially the case with respect to skeleton models that are not (yet) physically present and thus cannot be measured. Figure 1b graphically summarizes the geometry modeling approach developed in [21]. Similar to geometry modeling based on repeating unit cells (RUC), as used for example in the structural simulation of conventionally filament wound (rotationally symmetric) composite structures [26,27], the fiber skeleton considered (simple loop) is first divided into characteristic areas in which the windings assume clearly distinguishable arrangements. In this regard the two shafts (i, ii)—which contain different numbers of windings due to the final overlap applied in the winding process—as well as the two identical bolt wrappings (iii) are distinguished. In the second step, characteristic dimensions are defined which are assumed to have an important influence on the mechanical behavior of the fiber skeleton and are therefore modeled as realistically as possible in each of the characteristic areas. Here, the local loop thickness (LT), the local contact width between adjacent windings (CWW) and the globally constant cross-sectional area of the individual winding (AiW) are considered. To ensure simple modeling of these characteristic dimensions, the windings are modeled with a rectangular cross-section. The described modeling approach strives to be as precise as necessary to accurately represent the skeleton's mechanical behavior while being as simple as possible to implement.

While performing the work described in [21] and while evaluating the first validation results, some potential improvements and extensions to the developed approach were identified which are addressed in this paper and briefly summarized below.

- During modeling of the geometry, **more potentially relevant dimensions** were found (Section 2.2.2). They are included in this paper's models and their mechanical effects are examined (Section 3.2.2).
- It has been shown in [8,21] that delamination has an influence on the mechanical behavior (i.e., the load-displacement curve) of tension-loaded simple loops. However, this influence decreases at higher winding numbers. This indicates that at higher winding numbers, mesoscopic modeling may be abandoned in favor of a time-efficient **macroscopic modeling approach** (Section 2.2.2), which neglects the roving–roving interface (B). Both modeling approaches are compared using the example of simple loops in this paper (Section 3.2.5).
- The loops' stiffness measured in N/mm, hereafter referred to as spring constant K, was overestimated in the simulations in [21]. This deviation as well as possible reasons are examined experimentally (Section 3.2.3). Based on the findings, the fiber-parallel Young's modulus ($E_{||}$) of the impregnated roving (A) is adapted to the loops' **effective elastic behavior**.
- After intensive consideration of simple loop structures, the knowledge is transferred to the modeling and simulation of **inclined loops** and thus to three-dimensional fiber skeletons (Section 3.3). Since relative movements between windings and inserts may occur in this case, the insert-roving interface (C) is characterized in this context (Section 3.1).

2. Materials and Methods

2.1. Specimens and Mechanical Testing

2.1.1. Materials

The materials correspond to the ones used in [21]. All windings as well as the plates for the friction tests are made from the polypropylene-glass fiber (PP-GF) commingled yarn Twintex ® RPP60 1870 B provided by the manufacturer Owens Corning. It has a fiber volume content of 35 vol.-% and a linear density of 1870 tex (g/km). The polypropylene filaments contained in it are colored black. All inserts and the slider for the friction tests are made from aluminum. Table 1 lists the elastic constants and strength values used in the FE simulations.

Table 1. Elastic constants and strength values used in the FE simulations, adapted from [21].

Elastic Constants [a]							
PP-GF roving	$E'_{\|\|}$ [MPa] [b] 23,953	$E_{\perp}$ [MPa] 3750	$G_{\|\|\perp}$ [MPa] 1225	$G_{\perp\perp}$ [MPa] 1125	$\nu_{\|\|\perp}$ [-] 0.32	$\nu_{\perp\|\|}$ [-] 0.32	$\nu_{\perp\perp}$ [-] 0.59
Aluminum	E [MPa] 71,000	ν [-] 0.33					
Strength Values [a]							
PP-GF roving	$R'_{\|\|t}$ [MPa] [b] 987.9	$R_{\|\|c}$ [MPa] 274.0	$R_{\perp t}$ [MPa] 6.7	$R_{\perp c}$ [MPa] 44.6	$R_{\|\|\perp}$ [MPa] 17.0	$R_{\perp\perp}$ [MPa] 17.8	

[a] | |: parallel to fiber orientation, ⊥: transverse to fiber orientation, t: tension, c: compression; [b] Values are redetermined in this paper, see Sections 3.2.3 and 3.2.4.

With two exceptions, the material properties given in Table 1 correspond to the measurements and assumptions from [21]. One exception is the fiber-parallel Young's modulus

($E_{||}$) of the PP-GF roving (A) which is adapted in Section 3.2.3 based on the observed deviations between calculated and measured spring constants of simple loops under tensile loading. The other exception concerns the fiber-parallel tensile strength ($R_{||t}$) which is redetermined in Section 3.2.4. Both adaptations are carried out using FE models of the simple loop and are later validated using an FE model of the inclined loop.

To investigate the influence of fiber twists on the spring constant K of simple loops, a comparable PP-GF commingled yarn with an artificial twist of 40 turns per meter is produced in a cooperation of the companies Comfil ApS (Gjern, Denmark) and Culimeta Textilglas-Technologie GmbH and Co. KG (Bersenbrück, Germany).

2.1.2. Test Specimens

The simple loop specimen, shown in Figure 2, is used in this paper to investigate the influence of different geometric dimensions, to compare meso- and macroscopic geometry modeling and to investigate the spring constant K. The tensile tests on simple loops with two and six windings described in [21] are repeated in this work with optimized production parameters and supplemented by loops with ten windings. In order to investigate a possible cause for the deviation between the calculated and the measured spring constant K, all specimens are additionally produced in a variation with artificially created fiber twist. The production of the simple loop specimens corresponds to the description in [21].

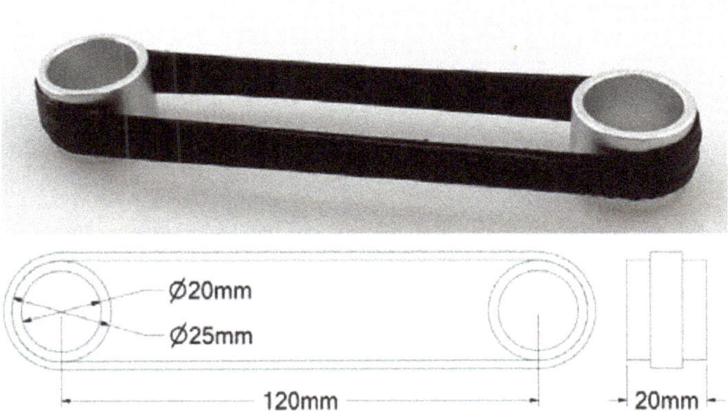

Figure 2. Simple loop specimen with dimensions, adapted from [21].

The inclined loop, shown in Figure 3a, is used to investigate more challenging skeleton geometries as well as inclined load transfers typically occurring in three-dimensional fiber skeletons. It is a part of the fiber skeleton developed and investigated in [4]. As the name suggests, the inclined loop differs from the simple loop in that the two inserts are not positioned within the same plane. The specimen considered here shows further differences: the two inserts are of different size, are concave in shape and both completely wrapped twice (double-eye winding on each insert). This configuration originates from the investigations on fiber skeletons with inclined load transfers performed in [4] and is not modified in this paper. The production of the inclined loop is also carried out according to the descriptions in [21], but the winding form shown in Figure 3b and an adapted robot program are used. The inclined loop specimen is dimensioned with four windings. As in the case of the simple loop, a final overlap is applied during the winding process, so that the shaft in the front contains four windings while the one in the back contains five windings. A detailed overview of the winding numbers at every location of the specimen is given in Section 2.2.3.

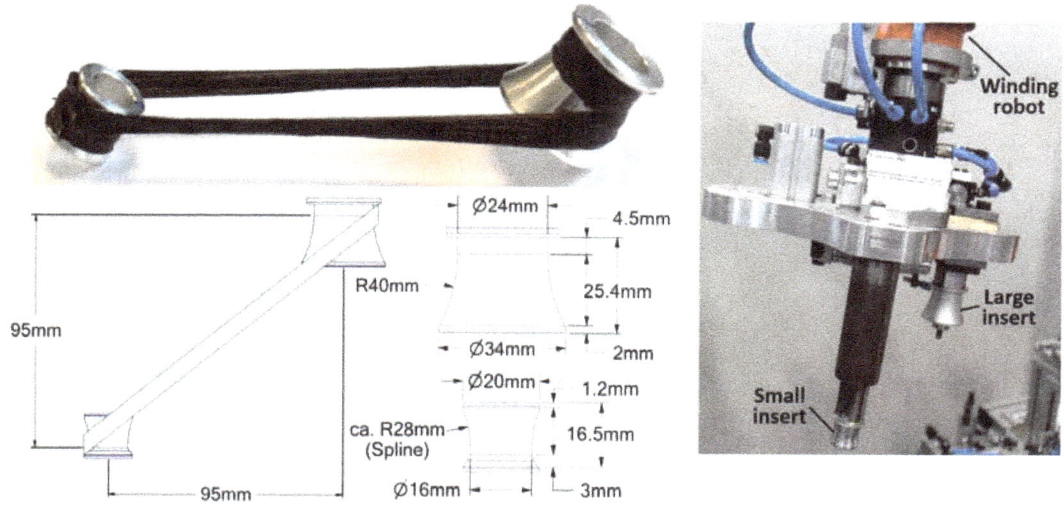

(a) (b)

Figure 3. (**a**) Inclined loop specimen after [4] with dimensions; (**b**) Corresponding winding form.

Since significant movement between the windings and the inserts cannot be ruled out in the case of the inclined loop, the insert-roving interface (C) is characterized experimentally in this work. It is known from previous studies that no significant adhesion occurs between thermoplastic impregnated windings and metallic inserts (especially when using polypropylene matrices), so this contact is considered to be frictional. Consequently, the characterization is based on the DIN EN ISO 8295 standard [28] for the determination of friction coefficients. As can be seen in Figure 4c, plate specimens are used for this purpose which are press-consolidated from the material of the windings and afterwards cut according to standard dimensions. The standard slider, which slides over the plates in the friction test, is made of the material of the inserts (aluminum).

(a) (b) (c)

Figure 4. Test setups. (**a**) Simple loop, test setup according to [21]; (**b**) Inclined loop; (**c**) Friction test.

2.1.3. Test Equipment and Procedures

As indicated in Section 2.1.2, tensile tests are performed on simple loops with two, six and ten windings as well as on inclined loops with four windings. As in [21], these are carried out at a testing speed of 5 mm/min on a Hegewald and Peschke inspect table 50 testing machine equipped with a 50 kN load cell and a video extensometer. The corresponding test setups are shown in Figure 4a,b. In both cases, the bottom insert is fixed while the top insert is pulled vertically upwards. The friction tests are performed on a Hegewald and Peschke inspect table 5 testing machine equipped with a 10 N load cell following the DIN EN ISO 8295 standard [28]. The illustration of the test setup in Figure 4c shows that the aluminum slider (200.05 g) is connected to the load cell by a rope. The rope is deflected by about 90° by a pulley so that the slider can slide over the horizontally oriented composite plate. For an approximate estimation of the friction influence of the pulley, additional idle tests are carried out in which the rope is connected only to the pulley. The test speed is set to 100 mm/min in all friction tests. The friction is measured in parallel and transverse to the fiber orientation in the plate specimens.

2.2. *Finite Element Modeling*

All FE models in this paper are prepared, computed and analyzed with the commercial FE software Ansys Mechanical 2020 R1.

2.2.1. Material Modeling

The material models described and used for the structural constituents (A), (B) and (D) in [21] are adopted in this paper. To represent potential friction effects in the insert-roving interface (C), the basic Coulomb friction model implemented in ANSYS Mechanical [29] is added. Its central relation is given in Equation (1):

$$\tau_{\text{lim}} = \mu P + b. \tag{1}$$

τ_{lim} represents the limit frictional stress, while μ and P are the coefficient of friction and the normal pressure acting between the contact partners, respectively. b represents a contact cohesion providing sliding resistance even without normal pressure. It is set to zero in this work, as no adhesion between inserts and windings is observed.

If the frictional stress of a contact is below the calculated τ_{lim}, the contact partners adhere to each other (sticking state). If, however, τ_{lim} is exceeded, larger tangential displacements between the contact partners are tolerated while the frictional stress remains constant (sliding state). In this simplified model, no distinction is made between static and dynamic coefficients of friction. The resulting relation between frictional stress (τ) and tangential displacement (δ) is shown in Figure 5. τ is determined by a penalty stiffness, so that very small displacements δ are assumed in the sticking state of a contact [30].

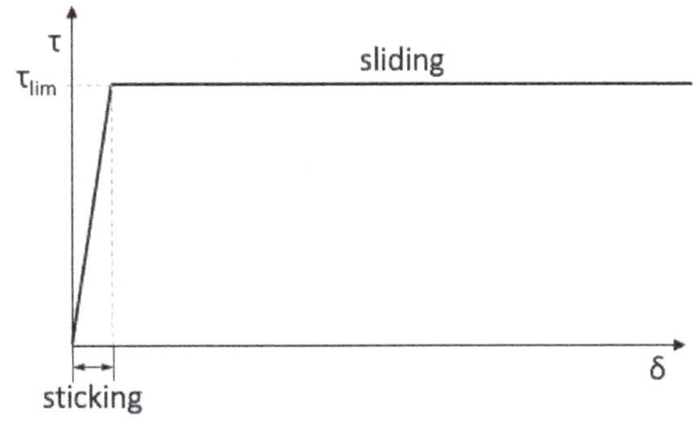

Figure 5. Relation of frictional stress τ and tangential displacement δ in the Coulomb friction model, adapted from [30].

2.2.2. FE Modeling of the Tensile Tests on Simple Loops

As described in Section 2.1.2, much of the research in this paper is done using simple loop specimens as examples. For this type of specimen, two different FE model types are generated whose main difference lies in the geometry modeling: meso-level models in which each winding is modeled individually and macro-level models in which all windings are combined, so that the roving–roving interface (B) is neglected.

Meso-Modeling

The generation of the meso-models is carried out as described in [21] with minor modifications. A detailed description of the modeling is therefore not repeated here. Instead, the changes made compared to [21] are listed below.

- In [21], discrepancies between simulation results and measurements, as well as uncertainties associated with material characterization, are found. In Sections 3.2.3 and 3.2.4 of the present work, potential causes are investigated and the material parameters $E_{||}$ and $R_{||t}$ of the PP-GF roving (A) are adapted by means of reduction factors. The resulting values are given in Table 1.
- As shown in Figure 7, further characteristic dimensions are considered and modeled according to the respective measurements given in Table 2. Their mechanical influence—and thus their relevance for precise FE simulations of fiber skeletons—is evaluated in Section 3.2.2.

Table 2. Characteristic dimensions of simple loops with two, six and ten windings.

		OWS [mm] [a] (Std. Dev.)	TL [mm] [b] (Std. Dev.)	AiW [mm²] [c]	LT [mm] (Std. Dev.)	CWI [mm²] [c]	CWW [mm] (Std. Dev.)
2 windings	(i)	0.85 (0.20)	15.80 (7.34)	2.47	1.49 (0.24)	-	0.33 (0.24)
	(ii)	0.85 (0.20)	15.80 (7.34)	2.47	1.14 (0.16)	-	0.00 (0.00)
	(iii)	-	-	2.47	0.90 (0.11)	215.56	1.96 (1.22)
6 windings	(i)	1.11 (0.22)	23.64 (6.39)	2.47	3.34 (0.16)	-	0.92 (0.32)
	(ii)	1.11 (0.22)	23.64 (6.39)	2.47	2.55 (0.12)	-	0.49 (0.19)
	(iii)	-	-	2.47	1.91 (0.20)	304.22	3.50 (0.41)
10 windings	(i)	1.18 (0.15)	20.78 (6.67)	2.47	2.82 (0.18)	-	1.60 (0.40)
	(ii)	1.18 (0.15)	20.78 (6.67)	2.47	2.48 (0.11)	-	1.37 (0.41)
	(iii)	-	-	2.47	1.94 (0.08)	499.84	4.64 (0.65)

[a] Without distinction between the characteristic areas (i) and (ii); [b] Without distinction between the transitions (i)–(iii) and (ii)–(iii); [c] Calculated values.

- To reduce calculation time, the symmetric geometry model is halved.
- Since meshing complex skeleton models with hexahedral SOLID186 elements often leads to poor element quality, all models studied are consistently meshed with tetrahedral SOLID187 elements. SOLID187 is a 10-node 3D solid element with a quadratic shape function and three degrees of freedom per node (x, y and z direction).
- The insert–roving interface (C) is no longer represented by a frictionless contact, but by the friction model described in Section 2.2.1 in combination with the coefficient of friction $\mu_{S\perp}$ determined in Section 3.1.
- Geometric non-linearity is considered by activating the corresponding analysis setting in ANSYS Mechanical ("large deflection"). This ensures that deformation-induced changes of the stiffness matrix are determined and adjusted iteratively [31]. Even though the simple loop specimens deformed by only a few millimeters before total failure, the assumption of geometric linearity can lead to unrealistic deformations and stress peaks in the FE simulations. This applies in particular in the context of delamination.

The resulting meso-models are shown in Figure 6. Table 2 lists the characteristic dimensions of the simple loop specimens. The loop thickness (LT) and the contact width between adjacent windings (CWW) are determined based on micrographs as described in [21] with the difference that metal foil is used instead of polyimide foil to enable a more precise measurement of the roving–roving-interface (B). The calculated rectangle height is larger than the measured CWW, which is why special surfaces are defined to model the contact between the windings (marked in green). In addition, 2D images of the loops are made with an optical scanner, from which the offset between wrap and shaft (OWS) and the transition length between wrap and shaft (TL) are measured using the public domain image processing software *ImageJ*. The area of an individual winding (AiW) is adopted from [21] and the contact area between wrap and insert (CWI) results from the rectangle height in area (iii).

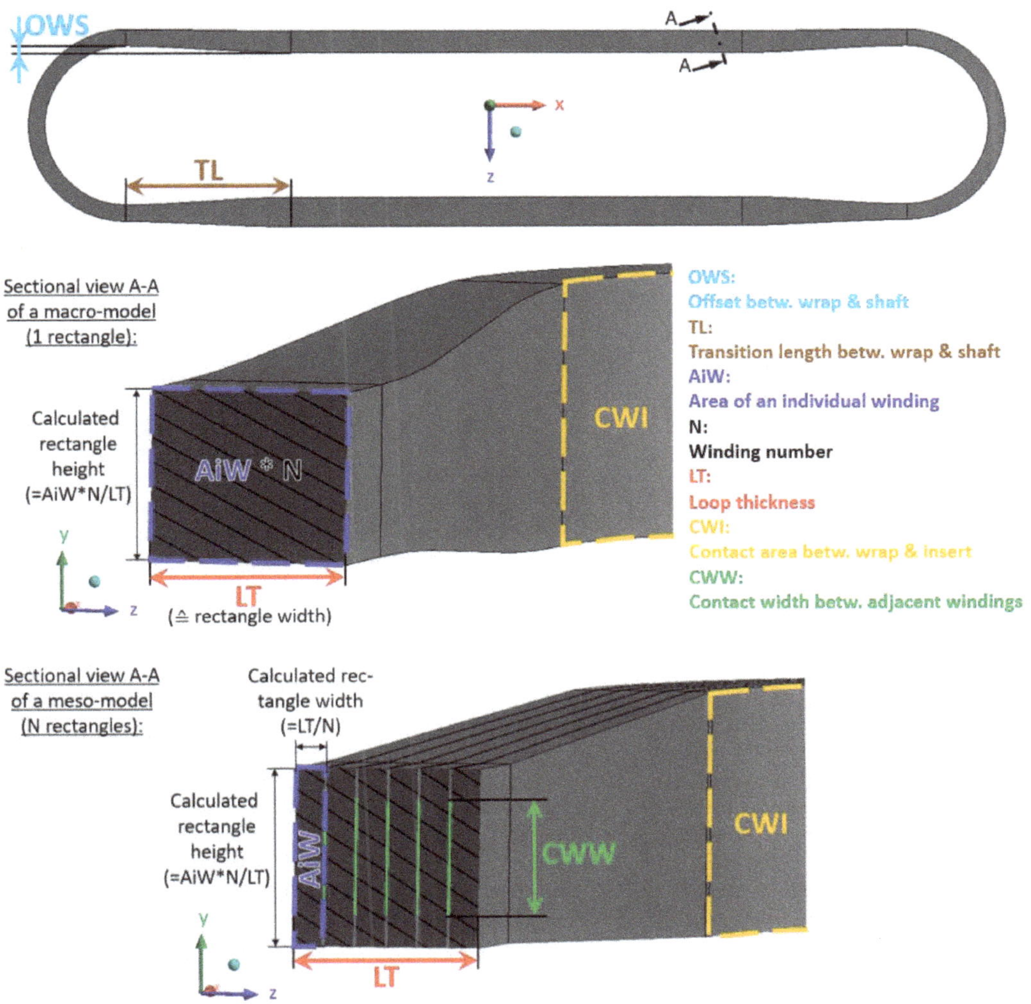

Figure 6. Meso-models of the simple loops: fiber orientation and characteristic areas (**top**), close-up view of insert area (**bottom**).

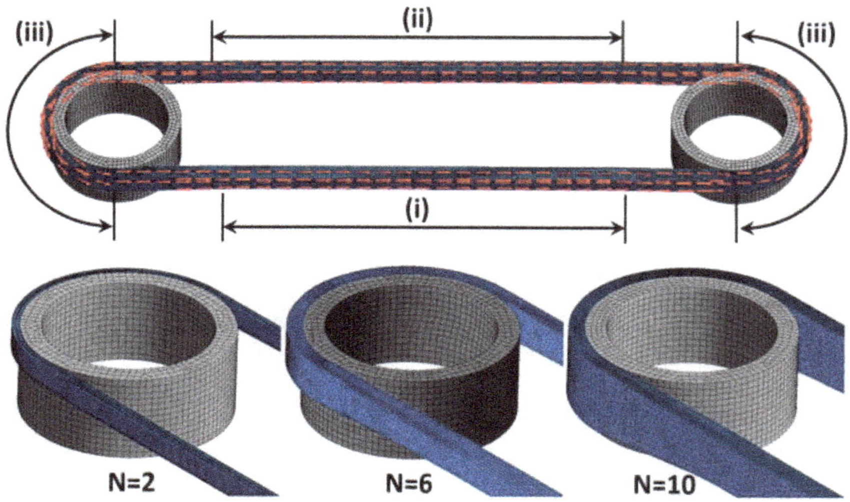

Figure 7. Overview of the characteristic dimensions considered in this work and illustration of the windings' rectangular modeling.

Macro-Modeling

To ensure comparability, macro-modeling of the simple loops is based as closely as possible on the meso-models described above. The characteristic areas shown in Figure 6 and the characteristic dimensions specified in Table 2 apply in the same way. The windings are modeled with a rectangular cross-section as well. In contrast to the meso-models, however, there is only one rectangular cross-section per characteristic area, since the interfaces between the windings are neglected. The local cross-sectional area corresponds to the cross-sectional area of an individual winding (2.47 mm^2) multiplied by the number of locally present windings. With this value and the local loop thickness given in Table 2 (corresponds to the width of the rectangle), the local height of the loop (height of the rectangle) results are unambiguous. This is graphically explained with the characteristic dimensions in Figure 7. Another difference between the meso- and macro-models concerns the transitions between the shafts (i, ii) and the wraps (iii). In the meso-models, these must be modeled as straight-line connections to maintain close contact between the adjacent windings. Due to the neglection of the roving-roving interface (B), this restriction does not apply to the macro-models, so that the notches associated with straight-line modeling can be avoided here. Therefore, the transitions are modeled based on curves, in this case, which tangentially fade into the shafts and wraps (see Figure 7). This also leads to a more realistic representation of the fiber skeletons' actual shape. As with the meso-models, the material properties specified in Table 1 are used and aligned with the (idealized) fiber orientation by adjusting the elements' local coordinate system. The insert–roving interface (C) and all analysis settings are also defined identically. Furthermore, the macro-level geometry models are also meshed with tetrahedral SOLID187 elements, and the load introduction is also carried out by displacing one insert while the other one is fixed.

2.2.3. FE Modeling of the Tensile Tests on Inclined Loops

From past studies it is known that delamination is unlikely to occur in wound loop structures with eye windings. For this reason, geometry modeling on the meso-level is skipped and a macro-model of the inclined loop specimen is generated instead. As can be seen in Figure 8, in this case as well, the geometry modeling starts with a segmentation of the fiber skeleton into characteristic areas. As with the simple loops, the two shafts (i, ii) are distinguished here, since they contain different winding numbers. The two outer wraps of the inserts (iii, iv) are modeled differently, since the two inserts are not identical. The inner wraps (v, vi) are added as additional characteristic areas to model the eye windings. The shafts and the outer wraps are connected by tangential transitions in this case as well. Again, the modeling of the windings is based on rectangular cross-sections as indicated by the equations shown in Figure 7. This time the measurements of the inclined loop, given in Table 3, are used. The measurement of the inclined loop is carried out using the optical 3D measuring system ATOS 5 provided by GOM GmbH. Thus, as shown in Figure 8, the outer surfaces of the skeleton are comprehensively measured, but no section views are available. Thus, the loop height (height of the rectangle) is measured in this case and the loop thickness is derived from it. This again requires the cross-sectional area of a single winding (2.47 mm^2) and the number of locally present windings given in Table 3.

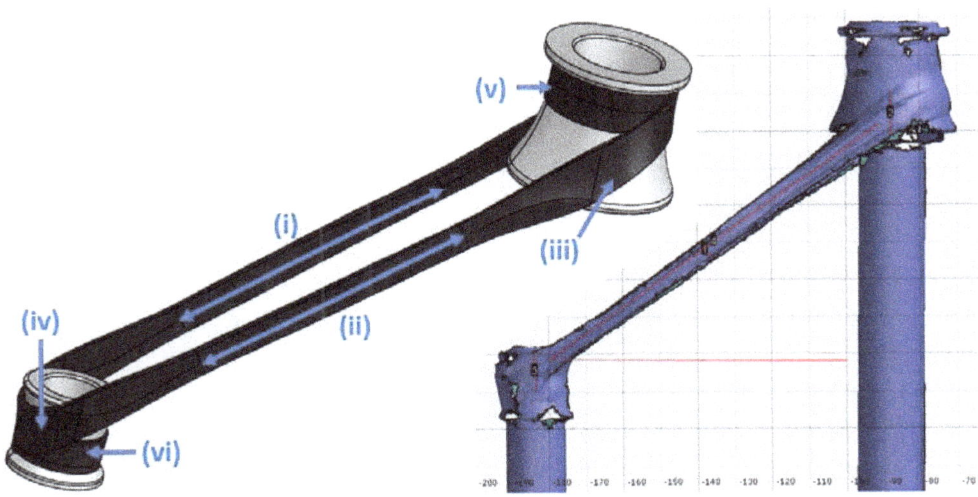

Figure 8. Macroscopic geometry model of the inclined loop with characteristic areas (**left**), screenshot of the 3D measuring (**right**).

Table 3. Characteristic dimensions and locally present winding numbers of the inclined loop.

	OWS [mm] [a]	TL [mm] [a]	AiW [mm^2] [b]	LH [mm] [c] (Std. Dev.)	CWI [mm^2] [b]	CWW [mm]	Inclination [°] [d] (Std. Dev.)	N [-]
(i)	0.00	29.00	2.47	5.64 (0.31)	-		36.58 (0.71)	5
(ii)	0.00	29.00	2.47	5.21 (0.06)	-		36.58 (0.71)	4
(iii)	-	-	2.47	10.84 (1.16)	477.35	Neglected in	36.58 (0.71)	6
(iv)	-	-	2.47	10.39 (0.31)	245.40	macro-model	36.58 (0.71)	7
(v)	-	-	2.47	9.20 (0.16)	710.78		-	2
(vi)	-	-	2.47	6.88 (0.29)	272.84		-	2

[a] Values specified based on the recommendations defined in Section 3.2.2; [b] Calculated values; [c] The optical 3D measuring system measures the local loop height, from which the local loop thickness is calculated; [d] Without distinction between (i)–(iv).

The inclined orientation of the loop and the concave insert designs require more complex geometry modeling than with the simple loop. The key adaptations are listed below:

- As shown in Figure 9a, the cross-sections of the outer wraps follow the insert shapes and are therefore not rectangular but rectangle-like: they have a constant wall thickness and parallel edges; however, two of the four edges are not straight. This modeling approach enables a close contact between insert and windings while keeping implementation simple. The inner wraps are modeled in the same way and are connected to the outer wraps by bonded contacts.

- The inserts have different diameters. The two outer wraps therefore do not form a semicircle around the inserts (180° deflection), as is the case with the simple loop. Instead, a larger deflection angle is modeled at the large insert and a smaller one at the small insert. This can also be seen in Figure 9a.
- Analogously to the simple loop, the shafts are modeled with a rectangular cross-section. However, in this case, the shafts are twisted so that their ends match with the angled ends of the outer wraps. The twist of the shafts is shown in Figure 3a (real specimen) and in Figure 9b (geometry model).
- Due to the twist in the shafts, the OWS and the TL are difficult to measure. The OWS is therefore not considered in the geometry model. Instead, the free shafts are simply centered between the two wrap endings they connect. Based on the recommendations in Section 3.2.2, the TL is set to 25% of the distance between the insert centers.
- The windings are not aligned perpendicularly to the insert axes, but at an incline to them. As can be seen in Figure 9c it is ensured that the outer wraps and the shafts, as well as the transitions connecting them, are all oriented at a uniform angle. Thus, unrealistic bending moments in the windings are avoided.

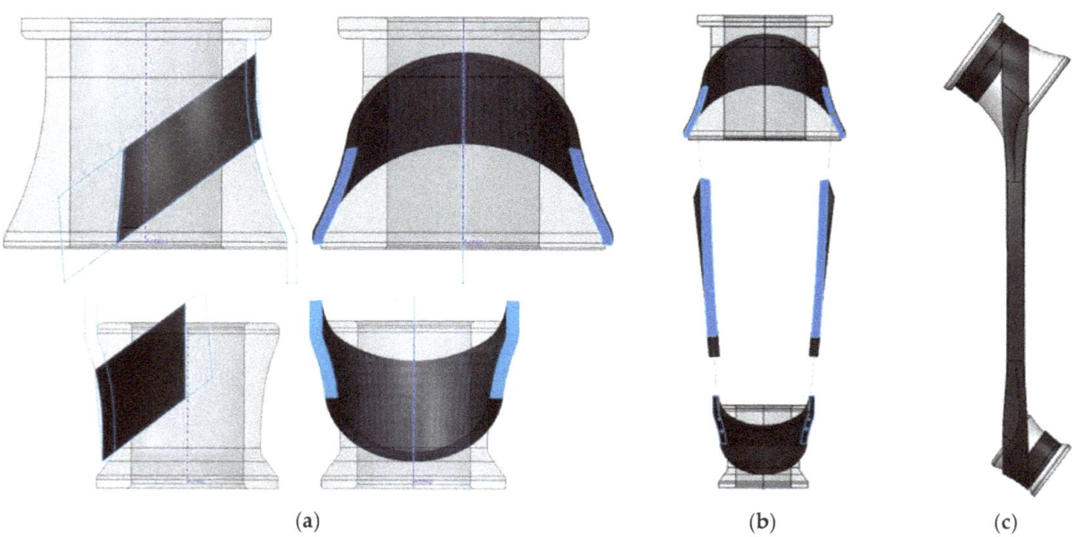

Figure 9. Detailed illustration of the inclined loop model. (**a**) Outer wraps; (**b**) Twisted shafts; (**c**) Inclination angle.

The rest of the modeling (material modeling, load introduction, analysis settings) corresponds to the descriptions given in Section 2.2.2.

2.2.4. Geometry, Mesh and Load Step Size Dependencies

To investigate the influence of the different characteristic dimensions shown in Figure 7 on the calculated mechanical behavior of the simple loops, a sensitivity study was performed using a generic macro-model of a simple loop specimen. First, the reference model was generated and subjected to a simulation of a tensile test, as described in Section 2.2.2. Then the characteristic dimensions were in-/decreased individually by +25% and −25%, while ensuring that the cross-sectional area of the windings is identical in all models. The models created in this way were then simulated analogously to the reference model so that the effects of the geometry modification can be seen by direct comparison of the results.

To avoid mesh and load step size dependent results, respective convergence studies are performed prior to the simulations described in Section 2.2. For this purpose, the mentioned reference model is simulated with different mesh and load step resolutions.

3. Results and Discussion

3.1. Characterization of the Insert-Roving Interface (C)

Figure 10 shows some load-displacement measurements of the friction tests used to characterize the insert-roving interface (C). A typical course of friction tests can be seen: it starts with a load peak representing the exceedance of the limit friction stress, then the load settles at a nearly constant level which reflects dynamic sliding.

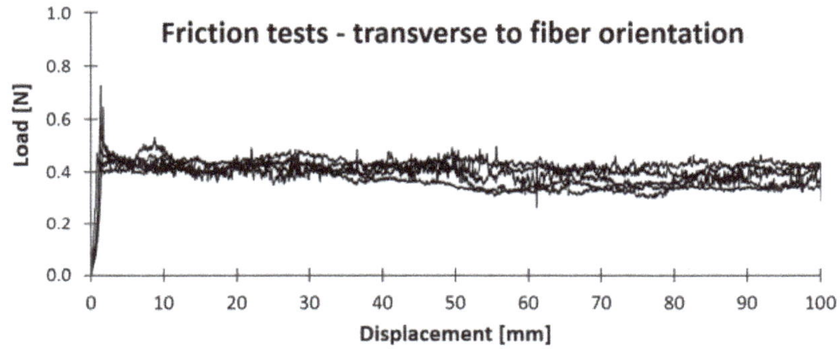

Figure 10. Experimental results of the friction tests on PP-GF plate specimens with transverse fiber orientation.

The static and dynamic coefficients of friction given in Table 4 are determined after Equations (2) and (3), respectively:

$$\mu_S = \frac{F_S}{F_N}, \quad (2)$$

$$\mu_D = \frac{F_D}{F_N}. \quad (3)$$

Table 4. Coefficients of friction determined between pressed PP-GF plate specimens and the standard slider made from aluminum.

Friction Coefficients Parallel to Fiber Orientation		Friction Coefficients Transverse to Fiber Orientation	
$\mu_{S\|\|}$ [-] (Std. Dev.)	$\mu_{D\|\|}$ [-] (Std. Dev.)	$\mu_{S\perp}$ [-] (Std. Dev.)	$\mu_{D\perp}$ [-] (Std. Dev.)
0.21 (0.03)	0.16 (0.02)	0.25 (0.05)	0.19 (0.02)

μ_S and μ_D represent the static and dynamic friction coefficient while F_S and F_D are the static and dynamic frictional force. F_N represents the weight of the slider.

For F_S, the maximum load at the start of the measurement curves is used. F_D is defined as the average load measured in the displacement range between 10 and 60 mm. The idle friction load of the pulley is subtracted from both values.

As significant relative movements between windings and inserts are primarily expected transverse to fiber orientation and as the friction model described in Section 2.2.1 only considers static friction, the $\mu_{S\perp}$-value of 0.25 is relevant for this work.

3.2. Investigations on Simple Loop Specimens

3.2.1. Mesh and Load Step Convergence

As described in Section 2.2.4, a generic macro-model of the simple loop is used to investigate discretization influences on the calculated mechanical behavior. Figure 11a shows the calculated normalized failure load F_{max} as a function of the elements' edge length (red curve). Besides, the percentage deviation compared to the convergent F_{max} is given (green curve). It can be seen that using an edge length of 0.45 mm or smaller, F_{max} varies less than 0.1%. The third curve indicates the number of element layers over the loop thickness (LT) in area (iii) (blue curve). It seems that a minimum of six layers is required to achieve convergence. This rule is confirmed using other macro-models, while the element edge length required to attain convergence varies between the models. Therefore, all macro-models in this work are meshed with six element layers in the wrap areas.

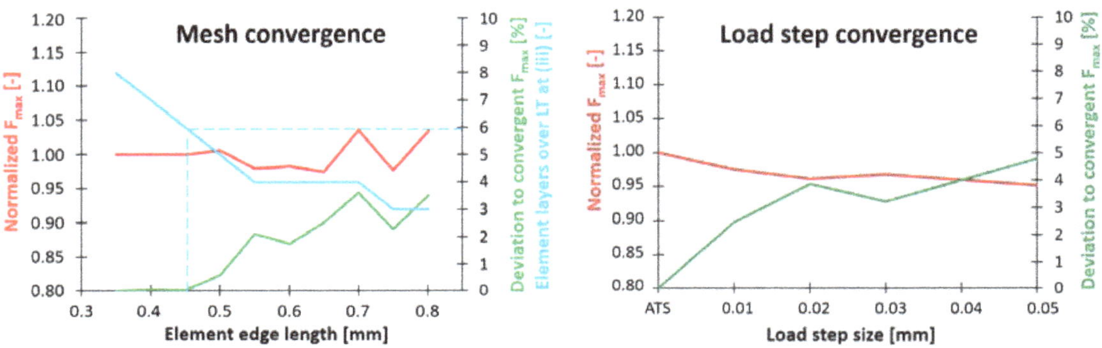

(a) (b)

Figure 11. Convergence studies. (**a**) Influence of element edge length; (**b**) Influence of load step size.

In a similar way, the influence of the load step size on the calculated mechanical behavior is investigated. Again, the generic macro-model is used as a reference. Non-linear behavior is expected due to fiber fracture (represented by the Maximum Stress Criterion and the MPDG). For its calculation, an implicit iterative solver (Newton-Raphson) and stepwise load application are used. As can be seen in Figure 11b, the precision of the non-linear calculation depends on the resolution of the load steps, which are defined by the displacement of an insert (see Section 2.2.2). As the step size decreases, the calculated F_{max} (red curve) approaches the convergent solution determined using the Automated Time-Stepping feature (ATS) (automatic load adjustment is implemented as automatic time adjustment in ANSYS Mechanical, as loads are generally defined in dependency of time). ATS automatically adjusts the load step size to the current situation of an analysis (e.g., non-linearities) [29]. If necessary, extremely fine load step resolutions are realized. As the deviation from the convergent solution (green curve) indicates, this enables a notable increase in precision here, even compared to fine constant load step sizes. Consequently, all simulations in this work are performed using the ATS feature. It is assumed that it is also suitable for the calculation of non-linearities based on delamination (represented by the CZM in the meso-models).

The mesh convergence study is performed using the ATS feature. The load step convergence is performed using a mesh with six element layers over LT in area (iii).

3.2.2. Mechanical Influence of the Characteristic Dimensions

The results of the sensitivity study described in Section 2.2.4 are summarized in Figure 12. Since the macro-model on which the study is based generally predicts a linear elastic deformation followed by abrupt total failure, the mechanical behavior can be summarized well by specifying the calculated spring constant K and the failure load F_{max}.

The sensitivity of the varied characteristic dimensions is evaluated based on the resulting K- and F_{max}-deviations compared to the reference model. The variation of LT affects both K (moderately) and F_{max} (more significantly). The deviations of F_{max} are consistent with the stress distribution analyses on tensile-loaded wound loops in [14]: in order to achieve high tensile-load capacity, it is recommended to keep LT small. LT's influence on K can be explained as follows: the larger the LT, the stronger the wrap area (iii) gets squeezed in the direction of the tensile load (x-direction in Figure 7); since the fibers are transverse to the loading direction at (iii) and $E_{\perp}$ is much smaller than $E_{||}$, K decreases with increasing LT. CWI can only be decreased, as it corresponds to the total contact area between the windings and the inserts in the reference model. The decrease of CWI does not cause any significant deviations of K or F_{max}. This is in accordance with the Coulomb friction model described in Equation (1) in which the contact area can be cancelled out of τ_{lim} as well as P [29]. The variation of TL has a significant effect on F_{max}, but not on K. The effect on F_{max} is attributed to the different notch shapes that result from the variation of TL, as shown in Figure 13b. Although the transitions are modeled tangentially in the macro-models, the notch shape still affects the stress distribution in this failure-critical area. It is found that F_{max} decreases, if TL is modeled smaller than measured (−10% and −25%). However, no significant deviation arises when TL is modeled larger than measured (+25%). Finally, the variation of OWS also affects F_{max}, but not K. The reason for the dependence between OWS and F_{max} is that high OWS values induce curvatures within the windings, as can be seen at the top of Figure 7. As the loop is subjected to tensile loading, these curved regions align with the load direction. This is accompanied by a local superposition of tensile and bending loads resulting in stress peaks.

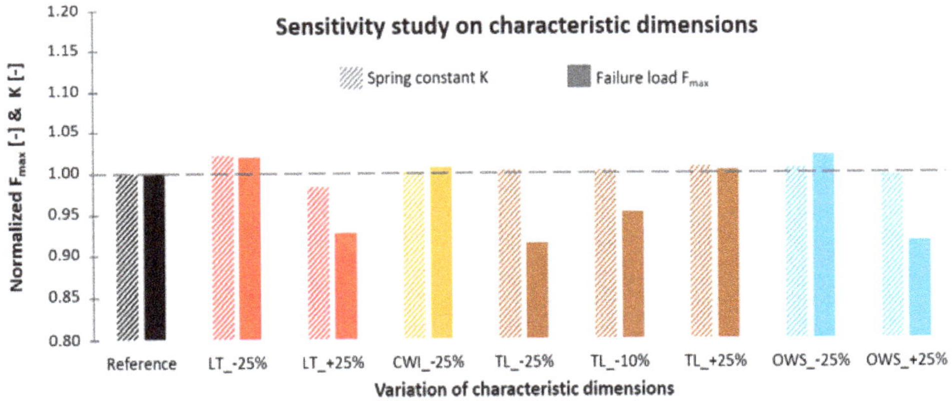

Figure 12. Influence of varied characteristic dimensions on the mechanical behavior of a macro-model.

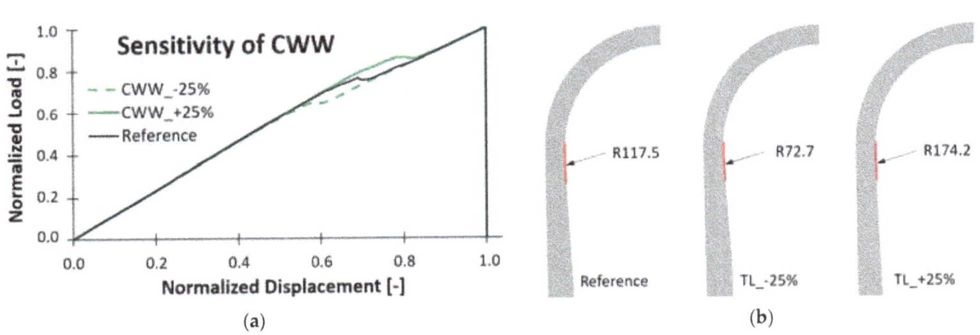

Figure 13. (**a**) Influence of CWW variations on the load-displacement-curve of a meso-model; (**b**) Notch shapes resulting from TL variations.

A separate consideration is required for the dimension CWW, which only occurs in meso-models and is therefore investigated using a generic meso-model of the simple loop. Due to delamination, a non-linear deformation is expected even before F_{max} is reached, so that the evaluation of the mechanical influence is based on load-displacement curves in this case. As shown in Figure 13a, the variation of CWW has a direct effect on the onset of delamination, recognizable by the drops in the curves.

The recommendations for geometry modeling derived from the above-described sensitivity study are summarized in Table 5.

Table 5. Recommendations for the determination and modeling of characteristic dimensions.

Char. Dimension	Mechanical Relevance	Recommendations
LT	High	Should be measured in micrographs from each characteristic area and modeled accordingly.
CWI	Low	A measurement is not required. CWI must be greater than zero. It is recommended to use the contact area resulting from the height of the windings.
TL	High	Should be measured by means of 2D/3D scans and modeled accordingly. If this is not possible, it must be ensured that TL is modeled rather too large than too small: 25 % of the distance between the insert centers can be used as a rule of thumb.
OWS	High	Should be measured by means of 2D/3D scans and modeled accordingly. If this is not possible, the shaft should be centered between the wrap endings it connects, knowing that curvatures of the windings and associated stress peaks might be neglected.
CWW	High	Should be measured in micrographs from each characteristic area and modeled accordingly.

3.2.3. Investigation of the Spring Constant K and Adaptation of the Fiber-Parallel Young's Modulus $E_{||}$

To quantify the discrepancy between calculated and measured spring constants observed in [21], both variables are plotted against the winding number in Figure 14a. The black curve represents the mean values of the spring constants measured in the tensile tests on simple loops. The red curve shows the calculated spring constants obtained by computing the respective macro-models while applying the fiber-parallel Young's modulus ($E_{||}$) of 26,518 MPa which was originally measured on press-consolidated PP-GF plates in [21]. It can be seen that both variables are linearly related to the winding number, while the curve of the measured spring constants has an inferior slope and is constantly approximately 10% below the curve of the calculated spring constants. These ratios are multiplied by the originally characterized $E_{||}$–value of 26,518 MPa to obtain a plot indicating the adapted $E_{||}$-values required to recreate the simple loops' effective elastic behavior. The result is the orange solid curve in Figure 14b. Since the curve is virtually constant, the mean value of 23,953 MPa is henceforth adopted as the *adapted fiber-parallel Young's modulus* $E'_{||}$ of the impregnated PP-GF roving (A), which is assumed to be valid for loop specimens with winding numbers between 2 and 10. The described adaptation of $E_{||}$ is a temporary solution serving to promptly consider the observed spring constant deviations in the context of FE analyses of fiber skeletons. The authors are aware that it is a simplification which does not replace an in-depth investigation of the underlying physical phenomena. The general applicability of this simplification is tested in Section 3.3 by applying it to the inclined loop.

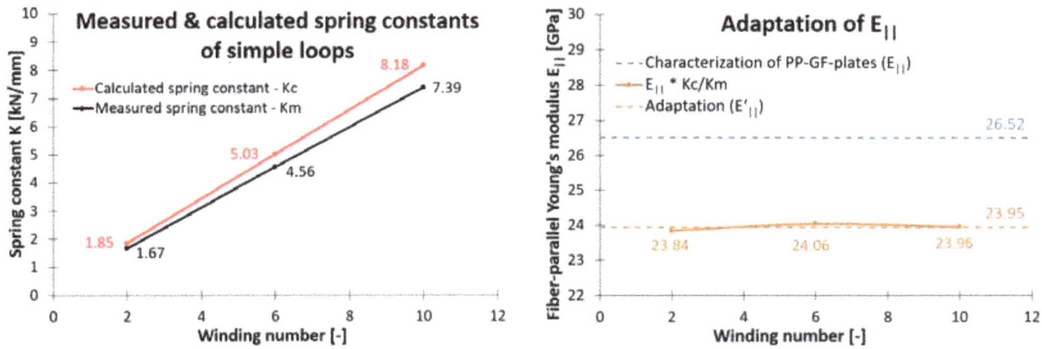

(a) (b)

Figure 14. (**a**) Comparison of measured and calculated spring constants; (**b**) Corresponding adaptation of $E_{||}$.

One potential cause for the above-described spring constant deviation—namely the influence of twisted fibers—is investigated in this work. For this purpose, comparison specimens are made from an artificially twisted commingled yarn, as described in Section 2.1. The twist of 40 turns per meter (Z40) is chosen to be sufficiently high to ensure that it significantly exceeds the process-induced fiber twist which is estimated to be well below five turns per meter and cannot be fully prevented even in the reference samples. The normalized spring constants of both kinds of specimen are given in Figure 15. It is shown that the increased degree of fiber twist generally has a negative effect on the loop's stiffness. The spring constants of the specimens with increased fiber twist are 3–10% below those of the reference specimens. Besides the non-ideal fiber orientation, the increased proportion of broken reinforcement fibers is another side effect of the increased fiber twist that certainly contributes to the reduced spring constants. It should be noted, though, that the standard deviations are at a similar order of magnitude as the spring constant deviations. Thus, it is confirmed that fiber twist can be a possible cause of the observed deviations between calculated and measured spring constants. However, the relatively small reduction of the spring constants compared to the high degree of fiber twist suggests that it is not the only and probably not the most significant influence. Another possible cause worthwhile investigating in follow-up work is the potential occurrence of inter-fiber-fractures in the wrap area (iii) during the tensile tests.

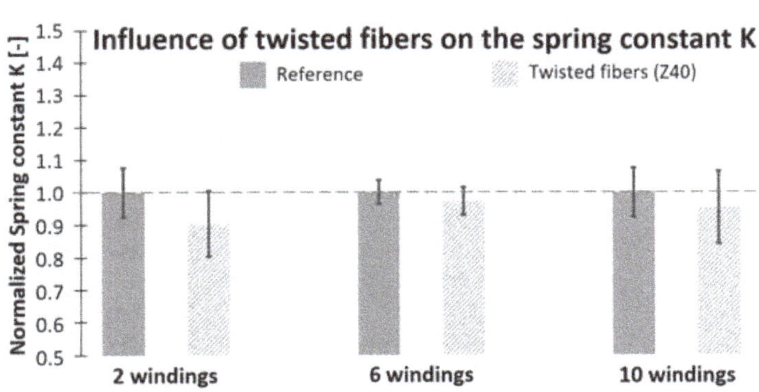

Figure 15. Spring constants measured on simple loops with (40 turns/m) and without (<5 turns/m) twisted fibers.

3.2.4. Adaptation of the Fiber-Parallel Tensile Strength $R_{||t}$

Next to the fiber-parallel Young's modulus of the impregnated roving (A) described in Section 3.2.3, its fiber-parallel tensile strength $R_{||t}$ is also adapted in this paper. The reason for this adaptation is the uncertainty associated with the wide range of results obtained when determining $R_{||t}$ by different methods [21]. Thus, the value found in tensile tests on press-consolidated plates (affected by stress peaks in the clamping area) is 609.0 MPa, whereas the value calculated according to the rule of mixture given in [14] (assuming ideal homogenization) is 1232.2 MPa.

A simple approach to this issue, which is widely used in the design of filament wound structures, is the empirical determination of a reduction factor [9]. This factor serves to reduce the homogenized $R_{||t}$-value calculated according to the rule of mixture, so that—based on experimental experience—precise FE simulations of the load-bearing capacity of filament wound structures are enabled.

Following this approach, the macro-models of the simple loops with six and ten windings (simple loops with two windings are not included, as their failure behavior is strongly influenced by delaminations) are simulated using the homogenized $R_{||t}$-value of 1232.2 MPa and the resulting failure loads are compared to the ones measured in the tensile tests. The average deviation is taken as the adaptation factor which is then multiplied by 1232.2 MPa. The result of this calculation is 987.9 MPa and is henceforth considered as the *adapted fiber-parallel tensile strength* $R'_{||t}$, which is used in all subsequent simulations. The reduction factor is approx. 0.8 and thus agrees very well with the value given in [9]. Like $E'_{||}$, $R'_{||t}$ is also validated by application to the inclined loop in Section 3.3.

3.2.5. Comparison of Meso- and Macroscopic Models

A quantitative comparison between the meso- and macro-models of the simple loops is made on the basis of the respectively calculated load-displacement curves. Additionally, two qualitative results—the failure sequence and the failure location—are compared. In order to evaluate the precision of the models, the corresponding measurements and observations from the tensile tests are included in the comparisons.

Figure 16 provides an overview of the deformation and failure behavior of the simple loops with two windings—in the tensile tests as well as in the simulations. Figure 17 summarizes the associated load-displacement curves. As already found in [21], the failure behavior observed in the tensile tests is dominated by delamination. With only two windings, the area of the roving–roving interface is too small to withstand the tensile load until pure fiber fracture is reached. Consequently, the specimens either fail by abrupt/stepwise total delamination or by local delamination followed by premature fiber fracture. In the measured curves (black), both delamination and fiber fracture are manifested by abrupt drops in the measured load. Curves with only one load drop can be classified as abrupt total delamination, while curves with several load drops can either represent a local delamination with subsequent fiber fracture or a stepwise total delamination. The macro-model cannot reproduce delamination processes. It therefore significantly overestimates both the spring constant and the failure load of the simple loop with two windings. It locates the fiber fracture at one end of the wrapping area (iii), as expected in theory [14] and confirmed in several studies [5,8,21], including the present one. This failure-critical point is henceforth referred to as the wrapping flank. The meso-model can reproduce delamination processes and thus, as already shown in [21], achieves a good qualitative agreement with the experimentally observed failure behavior. Quantitative agreement is also good: both the spring constant and the failure load lie within the scatter of the measured curves. The onset of delamination occurs slightly too early. It should be noted that symmetrical modeling of the roving–roving interface results in symmetrical delamination, that is, both ends of the loop delaminate simultaneously. To reproduce the asymmetrical delamination observed in the experiments (either the first or the last deposited winding delaminates first), it is advisable to slightly increase CWW at one end. As shown in Figure 16, this leads to a local delamination of one loop end, followed by premature fiber fracture at the wrapping flank.

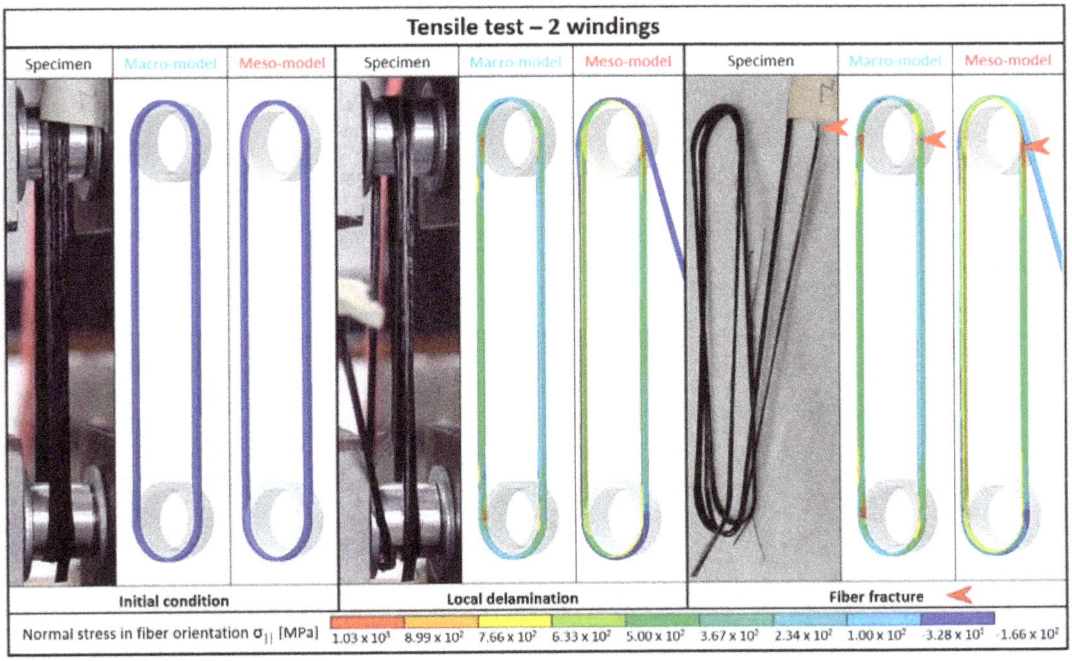

Figure 16. Deformation and failure behavior of simple loops with two windings (specimens and models).

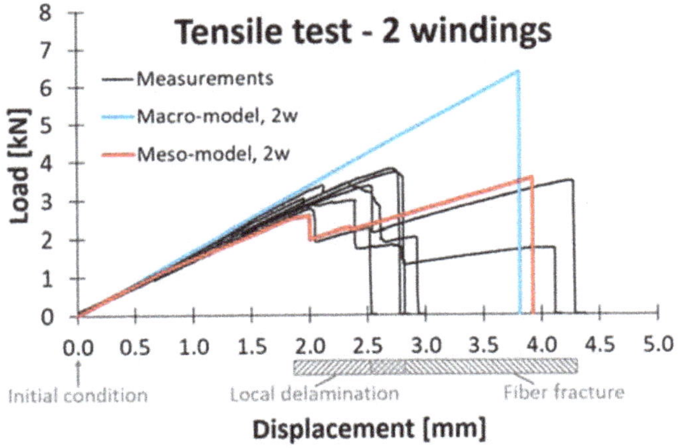

Figure 17. Load-displacement curves from tensile tests on simple loops with two windings (measured and calculated).

The mechanical behavior and the associated load-displacement curves of the (real and modeled) simple loops with six windings are shown in Figures 18 and 19. Although it is hardly recognizable in the figures, local delamination of the loop's ends also occurs in this case. However, due to the increased area of the roving–roving interface, total delamination is not observed. The local delamination causes a relatively small reduction of the interface area here, manifested by minor drops in the measurement curves. Total failure follows in the form of a fiber fracture at the wrapping flank. This less delamination-dominated behavior is better reproduced by the macro-model. Both the spring constant and the failure load are in the right order of magnitude and the fiber fracture is predicted at the correct location. The ignorance of delamination processes leads to a rather conservative prognosis of the displacement at total failure. The meso-model shows similar results. The only difference is that—as delamination is considered—the displacement lies further within the scatter of the measurement curves. In this case as well, CWW is slightly increased on one end of the loop, to properly reproduce the asymmetric delamination.

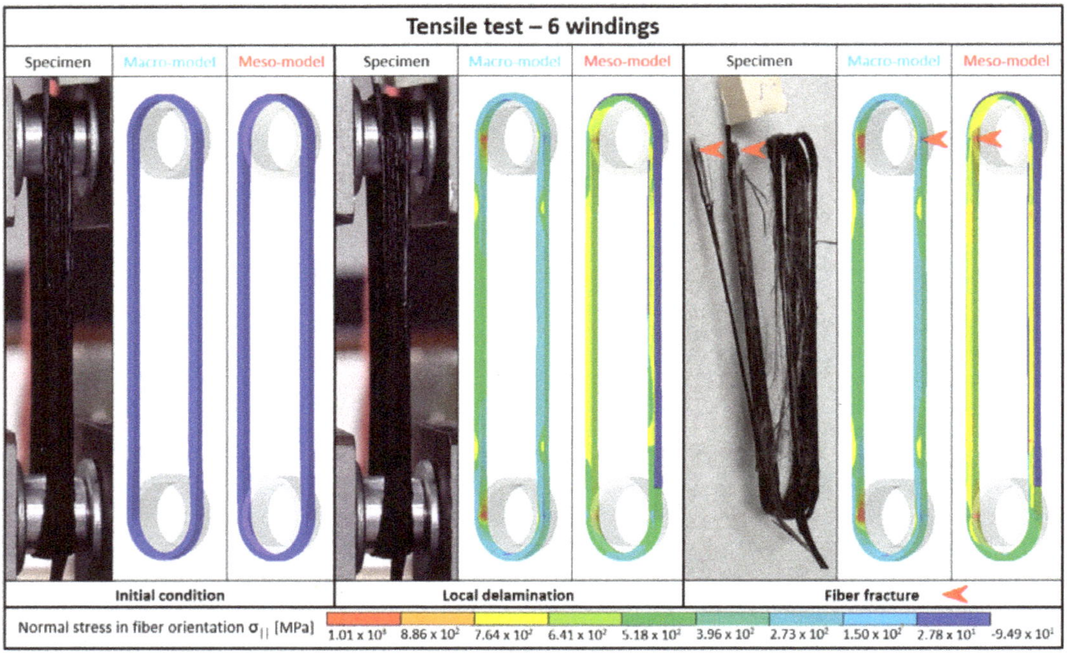

Figure 18. Deformation and failure behavior of simple loops with six windings (specimens and models).

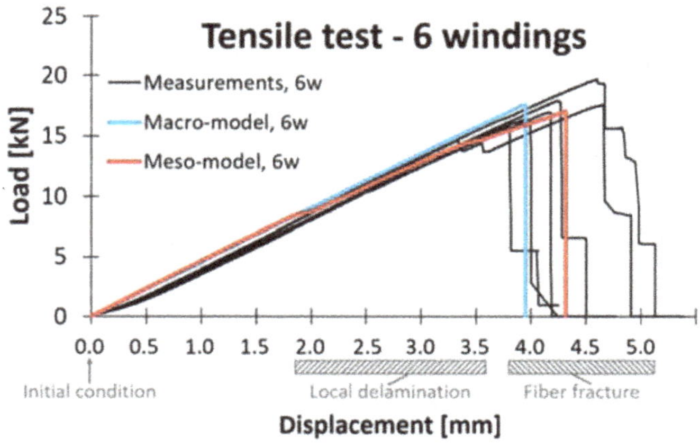

Figure 19. Load-displacement curves from tensile tests on simple loops with six windings (measured and calculated).

The numerical and experimental results obtained for simple loops with ten windings are summarized in Figures 20 and 21. The findings and observations that can be drawn here essentially correspond to the descriptions in the last paragraph (results of the simple loops with six windings). The results of the macro- and meso-models are even more similar in this case, and both agree well with the measurements from the tensile tests.

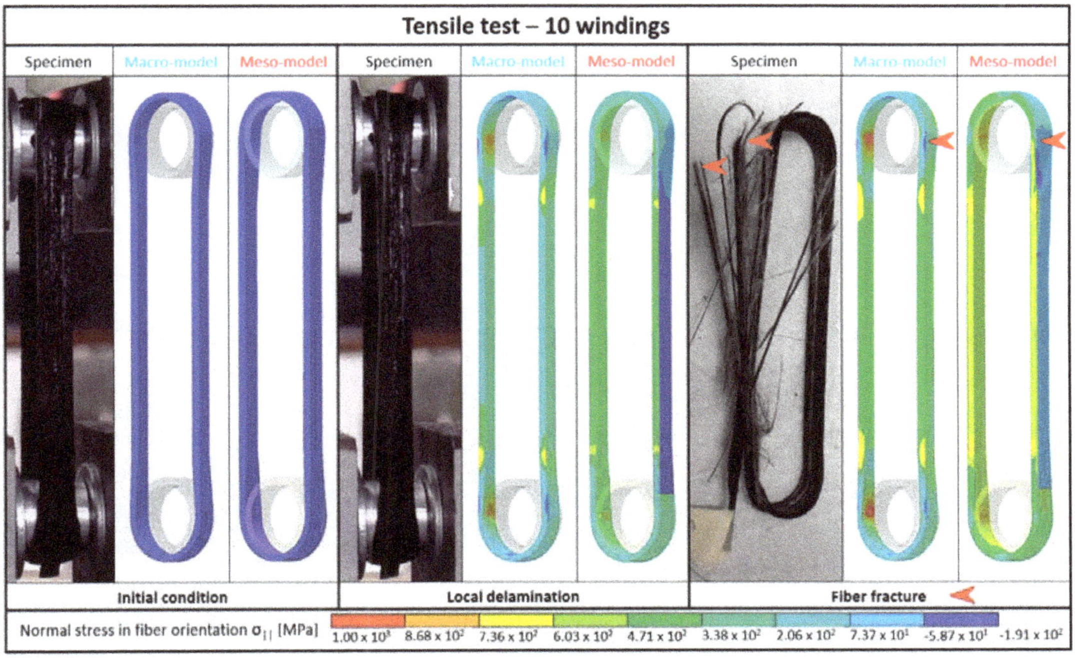

Figure 20. Deformation and failure behavior of simple loops with ten windings (specimens and models).

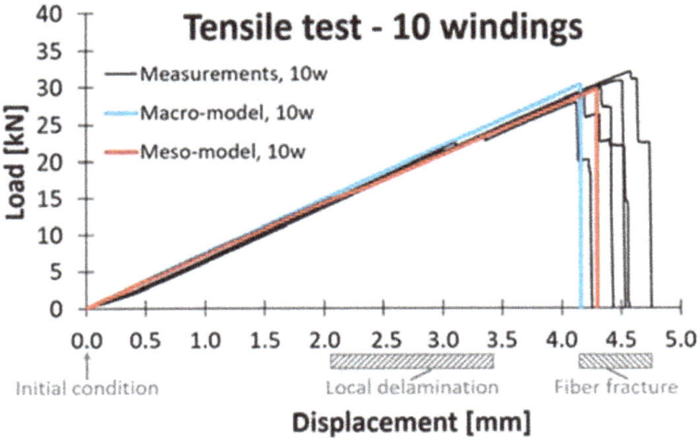

Figure 21. Load-displacement curves from tensile tests on simple loops with ten windings (measured and calculated).

3.3. Validation of the Presented Approach Using the Example of the Inclined Loop

To validate the presented approach using the example of the inclined loop, the relative tangential displacement between the windings and the inserts (sliding distance) is evaluated as a further reference, next to the sequence and location of failure as well as the load-displacement curves. The results are shown in Figures 22–24.

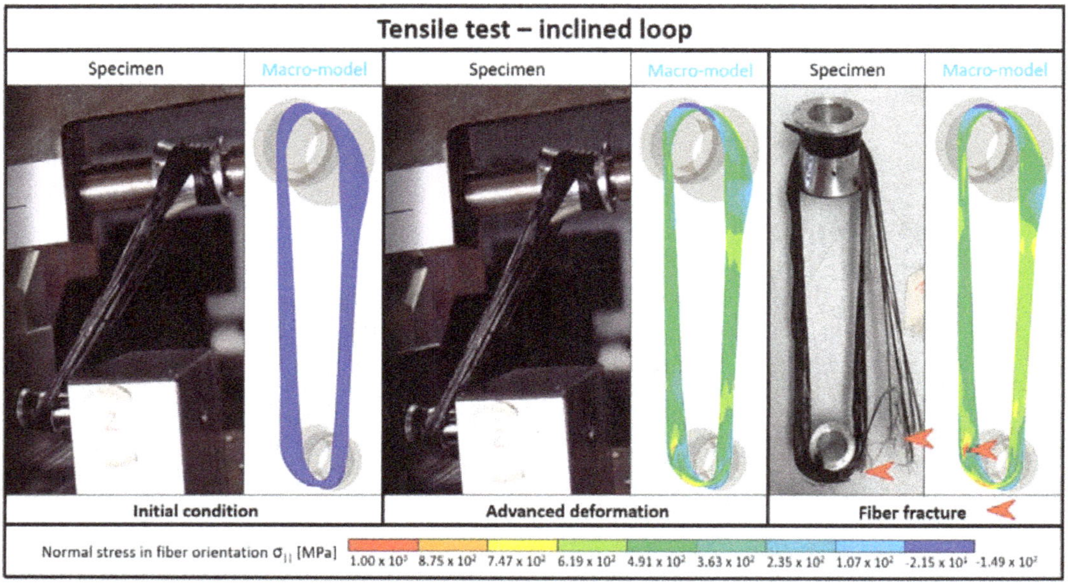

Figure 22. Deformation and failure behavior of inclined loops (specimens and model).

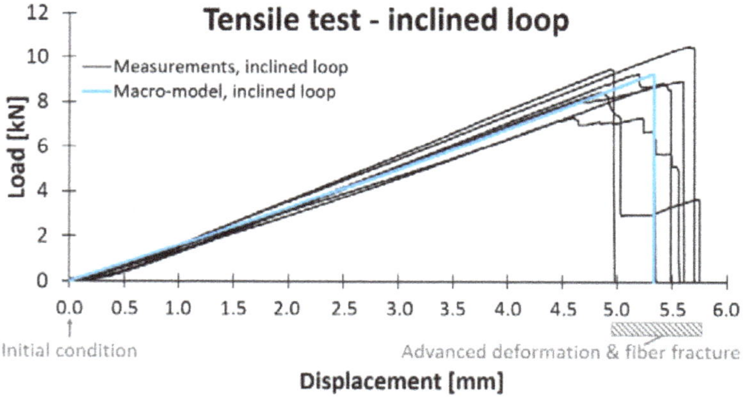

Figure 23. Load-displacement curves from tensile tests on inclined loops (measured and calculated).

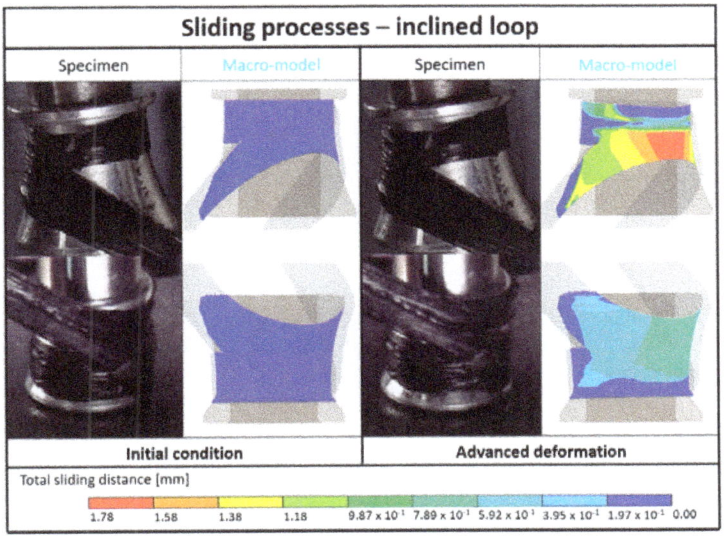

Figure 24. Sliding processes at the inserts of the inclined loop (specimens and model).

As expected, no recognizable delamination occurs in the tensile tests. Instead, elastic deformation is observed until fiber fracture occurs at one of the wrapping flanks of the small insert. In some cases, not all fibers fail at the same time, which can be seen from the different-sized drops at the end of the measurement curves and is possibly related to an uneven load distribution provoked by the concave insert geometry in combination with the sliding processes occurring at the small insert. The total sliding distance (large and small insert combined) directly before fiber fracture varies between 0.6 and 3.0 mm in the tensile tests, while the sliding processes at the small insert generally account for the larger share of it (74% on average). From video recordings of the tensile tests, it can be seen that stepwise fiber fracture is less likely to occur on specimens with a small sliding distance. Since no delamination occurs, the macro-model is well suited to reproducing the mechanical behavior of the inclined loop. The spring constant as well as the failure load and the maximum displacement lie centrally in the scatter range of the measurements. The location of the fiber fracture is also predicted correctly at one of the wrapping flanks of the small insert. These results indicate that the adaptations $E'_{||}$ and $R'_{||t}$, as well as the presented geometry modeling approach, generally enable a good reproduction of the inclined loop's mechanical behavior. To validate the modeling of the insert–roving interface (C), the calculated total sliding distance is considered. At 2.44 mm it is in the upper fourth of the measured range. Since the friction model applied only considers the static friction coefficient (which is higher than the dynamic one), the simulation should, in theory, underestimate the sliding distance. Thus, it can be assumed that the surface of the press-consolidated PP-GF plates used in the friction tests might be smoother than the contact area of the PP-GF windings. Besides, it should be noted that the model locates the greater share (73%) of the total sliding distance at the large insert. This deviation from the measured results is assumed to be related to the idealized geometry modeling approach pursued.

4. Conclusions

In the present work, a systematic approach for FE analysis of thermoplastic impregnated fiber skeletons, initially presented in [21], was advanced and validated. The meso-level geometry modeling procedure was supplemented by further characteristic dimensions whose mechanical influence was subsequently investigated using the example of simple loops. From the results obtained, recommendations for more complex modeling tasks could be derived. With regard to more time-efficient modeling and simulation, the geometry modeling was transferred to the macroscopic level. In a comparison of meso- and macro-modeled simple loops, the application areas of the two approaches were distinguished. Being able to reproduce fiber fracture as well as delamination, meso-models have the potential to provide accurate predictions of deformation and failure for all specimens studied in this paper. Macro-models have the advantage of reduced modeling and computational effort but cannot reproduce delamination and may therefore provide inaccurate results for fiber skeletons that are prone to local or total detachments. These simulation-based findings were confirmed by experimental tensile tests on simple loop specimens. Consequently, it is recommended to rely on macro-modeling only, as long as no major delamination is expected. If this cannot be assured, modeling at the meso-level is required. Based on this finding, the procedures and models developed up to this point were used to generate a macro-model of an inclined loop. In the context of a three-dimensional load transfer and fiber skeleton, a mechanical characterization and modeling of the insert-roving interface (C), as well as some enhancements in the geometry modeling procedure, had to be added. The entirety of the models and procedures developed was validated by experimental tensile tests on inclined loop specimens. Both the qualitative failure sequence observed in the experiments, as well as the measured failure loads and displacements, were reproduced well in the simulations. In order to be able to apply the developed FE analysis approach in the sense of a design engineer, that is, to design and dimension fiber skeletons that are not yet physically existent and cannot be measured, a better understanding and empirical data are required regarding their characteristic dimensions, which are the basis of the proposed geometry modeling procedure.

Consequently, further work will be dedicated to investigating the influence of different process and design parameters on the characteristic dimensions of fiber skeletons. Moreover, the proposed approach will be applied in the context of insert design optimization.

Author Contributions: Conceptualization, J.H.; methodology, J.H. and D.A.; validation, J.H.; investigation, J.H., D.A. and A.K.; writing—original draft preparation, J.H.; writing—review and editing, D.A., B.B., P.E., L.K. and F.H.; visualization, J.H. and D.A.; supervision, F.H.; project administration, J.H. and B.B.; funding acquisition, J.H., B.B. and P.E. All authors have read and agreed to the published version of the manuscript.

Funding: This research was funded by the German Federal Ministry of Education and Research (Bundesministerium für Bildung und Forschung—BMBF), grant number 03VP06670. The APC was funded by the Fraunhofer Publication Fund.

Data Availability Statement: The data presented in this study is available on request from the corresponding author.

Acknowledgments: The authors would like to thank the companies Comfil ApS (Denmark) and Culimeta Textilglas-Technologie GmbH & Co. KG (Germany) for generously providing the artificially twisted PP-GF commingled yarn used in this work.

Conflicts of Interest: The authors declare no conflict of interest. The funders had no role in the design of the study; in the collection, analyses, or interpretation of data; in the writing of the manuscript, or in the decision to publish the results.

References

1. Minsch, N.; Müller, M.; Gereke, T.; Nocke, A.; Cherif, C. 3D truss structures with coreless 3D filament winding technology. *J. Compos. Mater.* **2018**, *53*, 2077–2089. [CrossRef]
2. Büchler, D.; Elsken, T.; Glück, N.; Sier, M.; Bludszuweit, S. Ultraleichte Raumzelle—Entwicklung von ultraleichten Großstrukturen in Faserverbund- und Hybridbauweise für den Schiffbau. In *Statustagung Schifffahrt und Meerestechnik—Tagungsband der Statustagung 2011*; GmbH, F.J., Ed.; Schriftenreihe Projektträger Jülich: Jülich, Germany, 2011; pp. 25–45.
3. Reichert, S.; Schwinn, T.; La Magna, R.; Waimer, F.; Knippers, J.; Menges, A. Fibrous structures: An integrative approach to design computation, simulation and fabrication for lightweight, glass and carbon fibre composite structures in architecture based on biomimetic design principles. *Comput. Des.* **2014**, *52*, 27–39. [CrossRef]

4. Beck, B.; Tawfik, H.; Haas, J.; Park, Y.-B.; Henning, F. Automated 3D Skeleton Winding Process for Continu-ous-Fiber-Reinforcements in Structural Thermoplastic Components. In *Advances in Polymer Processing 2020*; Hopmann, C., Dahlmann, R., Eds.; Springer Vieweg: Berlin, Germany, 2020; pp. 150–161.
5. Botzkowski, T.; Galkin, S.; Wagner, S.; Sikora, S.P.; Kärger, L. Experimental and numerical analysis of bolt-loaded open-hole laminates reinforced by winded carbon rovings. *Compos. Struct.* **2016**, *141*, 194–202. [CrossRef]
6. Holzinger, M.; Loy, C.; Gruber, M.; Kugler, K.; Bühler, V. 3D winding of tailor-made thermoplastic rods for locally reinforced injection molded components. In Proceedings of the 5th International Conference & Exhibition on Thermoplastic Composites ITHEC 2020, Bremen, Germany, 13–15 October 2020.
7. Minsch, N.; Müller, M.; Gereke, T.; Nocke, A.; Cherif, C. Novel fully automated 3D coreless filament winding technology. *J. Compos. Mater.* **2018**, *52*, 3001–3013. [CrossRef]
8. Huber, T. Einfluss Lokaler Endlosfaserverstärkungen auf das Eigenschaftsprofil Struktureller Spritzgießbauteile. Ph.D. Thesis, Karlsruher Institut für Technologie (KIT), Karlsruhe, Germany, 2014.
9. Schwarz, M. Gezielte Steifigkeits- und Festigkeitssteigerung von Maschinenbauteilen Durch Vorgespannte Ringarmierungen aus Faser-Kunststoff-Verbunden. Ph.D. Thesis, Technische Universität Darmstadt, Darmstadt, Germany, 2007.
10. Morozov, E.; Lopatin, A.; Nesterov, V. Finite-element modelling and buckling analysis of anisogrid composite lattice cylindrical shells. *Compos. Struct.* **2011**, *93*, 308–323. [CrossRef]
11. Woods, B.K.; Hill, I.; Friswell, M. Ultra-efficient wound composite truss structures. *Compos. Part A Appl. Sci. Manuf.* **2016**, *90*, 111–124. [CrossRef]
12. Solly, J.; Früh, N.; Saffarian, S.; Aldinger, L.; Margariti, G.; Knippers, J. Structural design of a lattice composite cantilever. *Structures* **2018**, *18*, 28–40. [CrossRef]
13. Bellini, C.; Di Cocco, V.; Iacoviello, F.; Sorrentino, L. Performance index of isogrid structures: Robotic filament winding carbon fiber reinforced polymer vs. titanium alloy. *Mater. Manuf. Process.* **2021**, 1–9. [CrossRef]
14. Schürmann, H. *Konstruieren mit Faser-Kunststoff-Verbunden*, 2nd ed.; Springer: Berlin/Heidelberg, Germany, 2007.
15. Krystek, J.; Kottner, R.; Bek, L.; Las, V. Validation of the adjusted strength criterion LaRC04 for uni-directional composite under combination of tension and pressure. *Appl. Comput. Mech.* **2010**, *4*, 171–178.
16. Havar, T. Beitrag zur Gestaltung und Auslegung von 3D-Verstärkten Faserverbundschlaufen. Ph.D. Thesis, Universität Stuttgart, Stuttgart, Germany, 2007.
17. Hashin, Z. Failure Criteria for Unidirectional Fiber Composites. *J. Appl. Mech.* **1980**, *47*, 329–334. [CrossRef]
18. Puck, A. *Festigkeitsanalyse von Faser-Matrix-Laminaten: Modelle für die Praxis*; Hanser: Munich, Germany, 1996.
19. Lapczyk, I.; Hurtado, J.A. Progressive damage modeling in fiber-reinforced materials. *Compos. Part A Appl. Sci. Manuf.* **2007**, *38*, 2333–2341. [CrossRef]
20. Kärger, L.; Botzkowski, T.; Galkin, S.; Wagner, S.; Sikora, S.P. Stress analysis and design suggestions for multi-loop carbon roving rosettes to reinforce bolt-loaded open-hole laminates. In Proceedings of the 17th European Conference on Composite Materials ECCM17, Munich, Germany, 26–30 June 2016.
21. Haas, J.; Hassan, O.N.; Beck, B.; Kärger, L.; Henning, F. Systematic approach for finite element analysis of thermoplastic impregnated 3D filament winding structures—General concept and first validation results. *Compos. Struct.* **2021**, *268*, 113964. [CrossRef]
22. Camanho, P.P.; Matthews, F.L. A Progressive Damage Model for Mechanically Fastened Joints in Composite Laminates. *J. Compos. Mater.* **1999**, *33*, 2248–2280. [CrossRef]
23. Tan, S.C.; Perez, J. Progressive Failure of Laminated Composites with a Hole under Compressive Loading. *J. Reinf. Plast. Compos.* **1993**, *12*, 1043–1057. [CrossRef]
24. Tan, S.C. A Progressive Failure Model for Composite Laminates Containing Openings. *J. Compos. Mater.* **1991**, *25*, 556–577. [CrossRef]
25. Alfano, G.; Crisfield, M.A. Finite element interface models for the delamination analysis of laminated composites: Mechanical and computational issues. *Int. J. Numer. Methods Eng.* **2001**, *50*, 1701–1736. [CrossRef]
26. Ye, J.; Yan, Y.; Li, Y.; Luo, H. Parametric mesoscopic and multi-scale models for predicting the axial tensile response of filament-wound structures. *Compos. Struct.* **2020**, *242*, 112141. [CrossRef]
27. Zhang, Y.; Xia, Z.; Ellyin, F. Two-scale analysis of a filament-wound cylindrical structure and application of periodic boundary conditions. *Int. J. Solids Struct.* **2008**, *45*, 5322–5336. [CrossRef]
28. *DIN EN ISO 8295:2004-10*; Plastics—Film and Sheeting—Determination of the Coefficients of Friction (ISO 8295:1995). Beuth: Berlin, Germany, 2004.
29. ANSYS, Inc. *Theory Reference—Release 2020 R2*; ANSYS Inc.: Canonsburg, PA, USA, 2020.
30. Rust, W. *Nichtlineare Finite-Elemente-Berechnungen mit ANSYS Workbench: Strukturmechanik: Kontakt, Material, große Verformungen*; Springer Vieweg: Wiesbaden, Germany, 2020.
31. ANSYS Inc. *Structural Analysis Guide—Release 2020 R2*; ANSYS Inc.: Canonsburg, PA, USA, 2020.

Journal of
Composites Science

MDPI

Article

Influence of Moisture Diffusion on the Dynamic Compressive Behavior of Glass/Polyester Composite Joints for Marine Engineering Applications

Oumnia Lagdani [1], Mostapha Tarfaoui [1,2], Marwane Rouway [3,4], Houda Laaouidi [1], Sara Jamoudi Sbai [4,5], Mohamed Amine Dabachi [6], Abdelwahed Aamir [7] and Mourad Nachtane [8,*]

1 IRDL Laboratory, ENSTA Bretagne, UMR-CNRS 6027, 29200 Brest, France; oumnialagdani@gmail.com (O.L.); mostapha.tarfaoui@ensta-bretagne.fr (M.T.); laaouidi.h@gmail.com (H.L.)
2 Green Energy Park (IRESEN/UM6P), km 2 R206, Benguerir 43150, Morocco
3 LPMAT Laboratory, FSAC, Hassan II University, Casablanca 20100, Morocco; marwanerouway@gmail.com
4 REMTEX Laboratory, ESITH, Casablanca 20000, Morocco; jamoudi.sa@gmail.com
5 LIMAT Laboratory, FSBM, Hassan II University, Casablanca 20000, Morocco
6 High School of Technology (ESTC), Hassan II University of Casablanca, BP 8012, Casablanca 20000, Morocco; ma.dabachi@ensem.ac.ma
7 MEET Laboratory, FST, Hassan I University, BP 577, Settat 26002, Morocco; profaamir2012@gmail.com
8 Arts et Métiers ParisTech Metz, CNRS, University of Lorraine, LEM3-UMR 7239 CNRS, 57070 Metz, France
* Correspondence: mourad.nachtane@ensta-bretagne.org

Abstract: Thermoset polymers offer great opportunities for mass production of fiber-reinforced composites and are being adopted across a large range of applications within the automotive, aerospace, construction and renewable energy sectors. They are usually chosen for marine engineering applications for their excellent mechanical behavior, including low density and low-cost compared to conventional materials. In the marine environment, these materials are confronted by severe conditions, thus there is the necessity to understand their mechanical behavior under critical loads. The high strain rate performance of bonded joints composite under hygrothermal aging has been studied in this paper. Initially, the bonded composite specimens were hygrothermal aged with the conditions of 50 °C and 80% in temperature and relative humidity, respectively. After that, gravimetric testing is used to describe the moisture diffusion properties for the adhesively bonded composite samples and exhibit lower weight gain for this material. Then, the in-plane dynamic compression experiments were carried out at different impact pressures ranging from 445 to 1240 s^{-1} using the SHPB (Split Hopkinson Pressure Bar) technique. The experimental results demonstrated that the dynamic behavior varies with the variation of strain rate. Buckling and delamination of fiber are the dominant damage criteria observed in the sample during in-plane compression tests.

Keywords: moisture diffusion; compressive behavior; SHPB technique; glass/polyester composite joints

Citation: Lagdani, O.; Tarfaoui, M.; Rouway, M.; Laaouidi, H.; Sbai, S.J.; Dabachi, M.A.; Aamir, A.; Nachtane, M. Influence of Moisture Diffusion on the Dynamic Compressive Behavior of Glass/Polyester Composite Joints for Marine Engineering Applications. *J. Compos. Sci.* **2022**, *6*, 94. https://doi.org/10.3390/jcs6030094

Academic Editor: Stelios K. Georgantzinos

Received: 13 February 2022
Accepted: 11 March 2022
Published: 16 March 2022

Publisher's Note: MDPI stays neutral with regard to jurisdictional claims in published maps and institutional affiliations.

1. Introduction

Glass fiber composites are usually chosen for marine engineering applications due to their excellent mechanical behavior and particularly, their low densities and cost compared to traditional materials [1–5]. However, the most structural requirement of composites is the ability to maintain a high proportion of its load-carrying performance over prolonged time under extreme environmental circumstances, such as temperature changes, humidity, oxidation, microbial attack, etc. [6–8]. Over the past few decades, the strengthening and repair of marine engineering structures using glass fiber reinforced polymer (GFRP) laminates have gained much attention [9,10]. As the science of polymers has advanced, composite bonding has developed into a major joining method for many applications, because it affords a lot of benefits compared to metallic joints, such as high fatigue resistance, galvanic corrosion elimination and uniform stress distribution across the joint [11–14]. The exploring of the hygrothermal aging effect on the mechanical performance of bonded joints has increasingly become recognized in the last years. The water aging influence on interface debonding has been studied in the literature. Komai et al. [15] demonstrated that water absorption leads to strength loss of the fiber/matrix interfaces of aligned reinforced carbon/epoxy composites. Ashik [16] conducted an experimental study of moisture uptake and mechanical characteristics of hybrid natural and glass fiber reinforced composites. The result shows that the addition of coconut fiber can improve strength and serve as an alternative material to glass fiber. Selzer and Friedrich [17] observed that the bonding of fiber/resin deteriorates with increasing moisture absorption. Due to the individual set of properties that synthetic polymers offer, their applications in the outdoor environment are steadily growing.

Polymers have replaced conventional materials in the marine industry. Many research papers have shown the impact of seawater on polymeric composites which leads to a degradation of performance [18]. Buehler and Seferis [19] noted interface delamination in carbon-glass reinforced epoxy due to water absorption. However, there are limited works on damage mechanisms that can change fatigue resistance, even in the absence of important seawater deterioration. So, it is substantial to better comprehend the extent and mechanisms of seawater deterioration. Schutte [20] demonstrated that hygrothermal decomposition of glass fiber composites is primarily due to the decomposition of the glass fibers, the polymer matrix and the interface. However, Mijović et al. [21] showed that vinylester reinforced composites display excellent durability over epoxy-based composites. Khalilullah [22] characterized the effect of moisture and hygroscopic bulking of silicone/phosphorus composite film, also the moisture uptake increases with temperature for the film. Therefore, vinylester and epoxy resins are usually chosen due to their low cost, thermal aging resistance and high sustainability [23]. Jiang et al. [24,25] experimentally and numerically investigated the impact of humidity and hygrothermal degradation in GFRP adhesives. The results revealed that samples lost mass under 40 °C conditions. Grace et al. [26] presented a new approach to characterize anisotropic moisture absorption in polymer composites using gravimetric absorption. Benyahia et al. [27] investigated the mechanical behavior of composite pipes at different temperatures between −40 °C and 80 °C where the pipes degrade when temperature increases. Sassi and his colleagues [28–32] investigated the dynamic behavior of bonded composite joints at elevated strain rates using an SHPB machine, the experimental results have shown a strong material sensitivity to strain rates. Moreover, damage investigations have revealed that the failure mainly occurred in the adhesive/adherent interface because of the brittle nature of the polymeric adhesive. Results have shown good agreement with the dependency of the dynamic parameters on strain rates. Despite these efforts, the influence of moisture diffusion on the dynamic compressive behavior of glass/polyester composite joints remains immature. Hence, the importance of understanding the mechanical behavior of these materials that are subjected to critical loads for an extended period. In this context, this paper highlights the impact of hygrothermal aging on the compressive responses of bonded composite joints at an elevated strain rate using the Split Hopkinson Pressure Bar (SHPB) machine.

In summary, the outline of this article is as follows: Section 2 provides the material description and methods, followed by the experimental methodology utilized for the mechanical characterization in Section 3. Finally, Section 4 draws pertinent conclusions as well as some prospects for future developments.

2. Materials and Methods

2.1. Materials

The composite used in this work was made of a 45° biaxial glass fiber mat with 0.29 mm in thickness reinforced polyester polymer. The composite part was joined with an adhesive of polyvinylester with 1 mm in thickness purchased from NORPOL. The mechanical properties of materials are listed in Table 1 [28]. The polyester polymer is used in the maritime industry as an equivalent to other types of polymers, due to their cost being slightly less than epoxy resin and also it has an excellent performance in corrosive environments and at extreme temperatures; therefore, it is perfectly adapted to naval applications.

Table 1. Composite and adhesive mechanical properties.

Properties	Composite	Adhesive
Density (kg/m^3)	1960	1960
Young's modulus (MPa)	E_1 = 48,110, E_2 = E_3 = 11,210	3100
Poisson's ratio	ν_{12} = ν_{13} = 0.28, ν_{23} = 0.34	0.3
Shear modulus (MPa)	G_{12} = G_{13} = 4420, G_{23} = 5000	-

2.2. Methods

2.2.1. Hygrothermal Aging Test

The samples were tested and performed by hygrothermal aging to investigate the moisture absorption properties and to measure the gain in weight of the samples in the function of time. The dimension of samples is 13 mm × 13 mm × 9 mm as shown in Figure 1. To perform the SHPB test, the samples must be prepared where both sides of the specimen are recommended to be planned and oriented with a high level of accuracy. The mass of each sample was calculated every 96 h in a regular period using a Precisa XT220A analytical balance with a tolerance of 0.0001 g. To record the weight variation after a specified interval of time, each specimen is taken out of the conditioned chamber, weighed quickly and returned into the chamber. The moisture uptake M absorbed by each sample is measured as a function of its initial weight w_i and final weight w_f [25,29] as described below:

$$M = \frac{w_f - w_i}{w_i} \times 100 \quad (1)$$

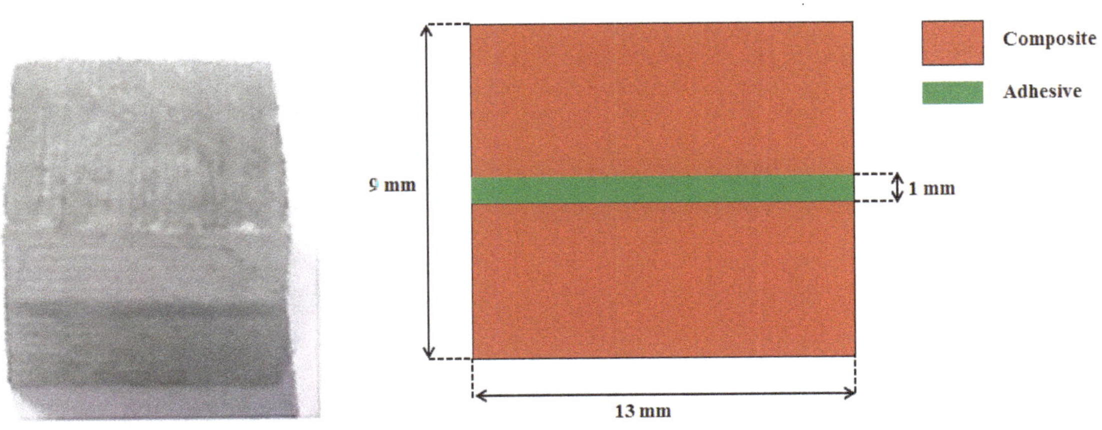

Figure 1. Adhesively bonded composite joints sample.

Before studying the aging of an adhesive, it is necessary to ensure that the polymerization is complete to avoid crosslinking phenomena during aging [2]. The DSC (Differential Scanning Calorimetry) test can determine an adequate curing cycle (polymerization of the glue). Likewise, these tests carried out on polymerized samples make it possible to determine a glass transition temperature T_g. From these data, it is then possible to set aging conditions. In general, it is necessary, for the aging temperature, to be lower than the T_g of an aged material saturated with water to overcome complex aging phenomena in a rubbery state. In addition, it is recommended that the aging temperature be much lower than the post-cure temperature as shown in Figure 2. The bonded composite specimen was tested under the moisture diffusion effect with 50 °C and 80% of temperature and relative humidity, respectively, for different periods.

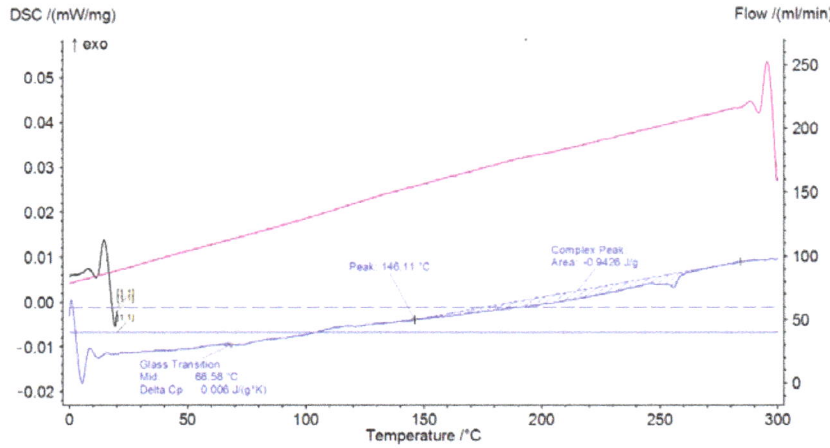

Figure 2. Glass transition temperature of the specimen.

2.2.2. Moisture Absorption

Many researchers investigated the performance degradation of marine water on polymeric composites [30,31]. However, the marine environment could lead to chemical bond hydrolysis and mechanical degradation and also a loss in interfacial stress transmission due to matrix plastification, in addition, the interface changes between the fiber and the matrix [18,32].

Various diffusion models have been developed to comprehend the behavior of homogeneous and heterogeneous materials, the best known and most commonly used of which is the one-dimensional isotropic Fick model [33] as seen in Figure 3, but divergence from this model are frequently remarked. On the other hand, the Langmuir-type model can accurately describe the water absorption of fiber-reinforced epoxy composite [34].

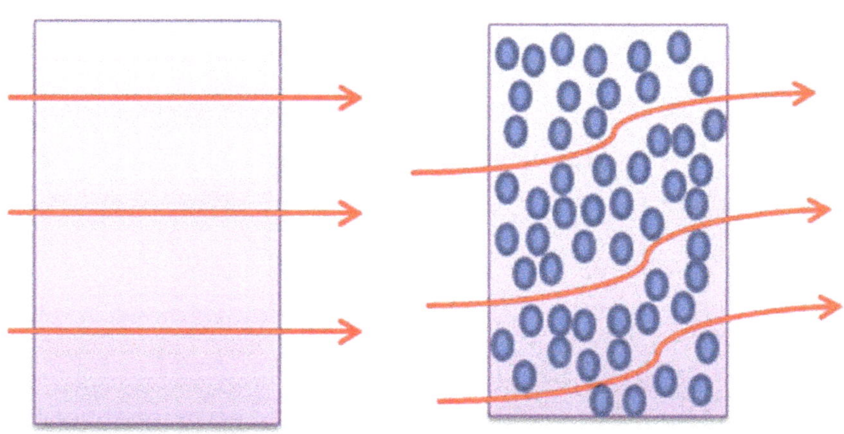

(a) (b)

Figure 3. Comparative diagram of diffusion phenomena. (**a**) Homogeneous material, (**b**) Heterogeneous material.

Experimental data were expressed as the moisture uptake $M(t)$. However, the study was conducted using average concentration $C(t)$ as the main parameter. To compare the experimental and numerical results, the $M(t)$ was converted to $C(t)$. The mean concentration was evaluated using Equation (2), where the densities of the material and water at 60 °C are $\rho_{mat} = 1.9\ g/cm^3$ and $\rho_{water} = 0.9\ g/cm^3$, respectively.

$$C(t) = \frac{\rho_{mat}}{\rho_{eau}} M(t) \quad (2)$$

The average concentration C_i is computed by using the finite element model at each integration point. So, to compare the results, C_i is evaluated at each increment point for the entire test tube using Equation (3), where C_i and V_i represent the concentration and volume at the integration point, respectively. The equation was executed using a Python script that extracts the appropriate information from the Abaqus results database.

$$C_i(t) = \frac{\sum_{i=1}^{n} C_i V_i}{\sum_{i=1}^{n} V_i} \quad (3)$$

Figure 4 presents the variation of the moisture uptake curve in the function of the square root of time, which usually concurs with the Fick law. The numerical results remain below the experimental and analytical values which confirm the model correlates satisfactorily with the theoretical and experimental results. The experimental data presented in Figure 4 represents the average of 10 tested samples. The moisture process of bonded joints may be defined into three period phases. In the first phase, the moisture uptake rate grows rapidly with time, achieving to moisture absorption of 0.239 wt% in 49 h. The rate of water absorption in bonded joints was changed linearly with aging time. The moisture uptake in this phase is principal because of composite material defects, such as voids and micro-cracks in the matrix. Into the second phase, the moisture decelerates until the saturation point at 0.2875 wt% in 400 h. The moisture absorption mechanisms include delamination of the fiber/resin interface and hydrolysis of the resin. At the third phase, the moisture diffusion of the composite has achieved equilibrium after approximately 420 h of aging.

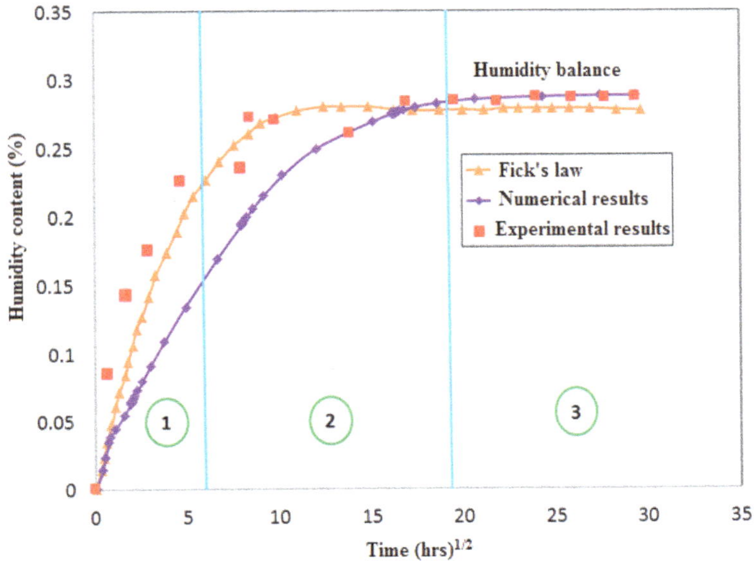

Figure 4. Bonded joint moisture absorption, a comparison of analytical (Fick) and numerical models with experimental data.

Numerous authors have described the volume variation of the specimens under hygrothermal loading. Gazit [35] showed that the increase in sample weight is linearly proportional to the dimensions of all reinforced samples. According to the polarity of water, it can create hydrogen bonds with the hydroxyl groups. Consequently, the hydrogen bonds between chains may be interrupted to raise the length of the hydrogen bond between segments. These mechanisms reduce the glass transition temperature T_g of the wetted specimen and is known as plasticization and swelling of the polymer matrix, which leads to microstructural damage, such as fibers delamination and matrix cracking [36,37]. In contrast, matrix swelling is important in the region with high resin uptake. This induces decoherence of the fiber/resin interface. In another work, Lee [38] demonstrates that the absorption of water produces plasticization and swelling of the resin and a reduction of the glass transition temperature. These generally affect the modulus of the composite material and can be accelerated by increasing the temperature. As observed in Figure 5, the numerical results of moisture diffusion at different aging times (0 h, 216 h, 648 h and 864 h).

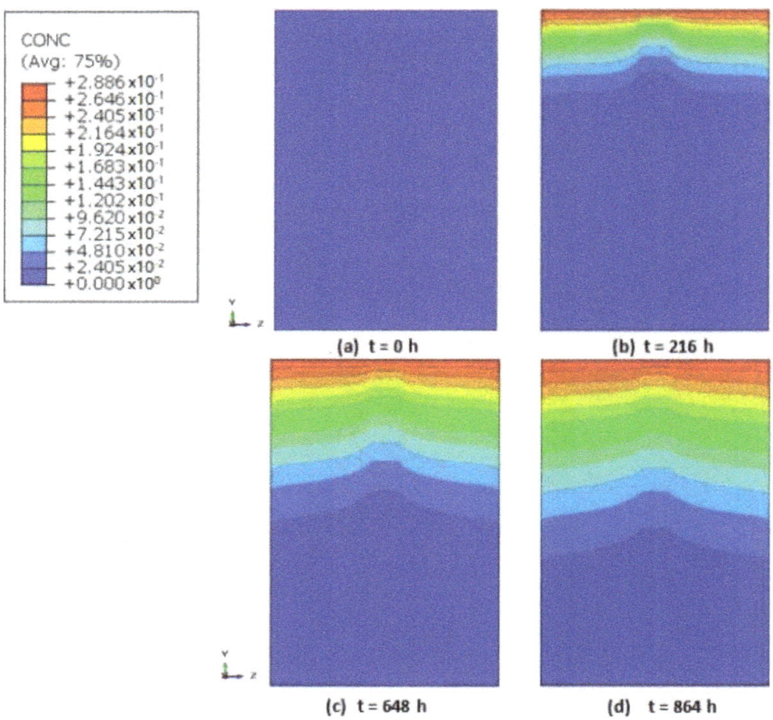

Figure 5. Illustration of numerical results at different aging times.

3. Dynamic Response in Moisture Absorption

3.1. Dynamic Compression Test

Many techniques were employed to evaluate the mechanical characteristics of the composites at high strain rates. The most utilized machines were the Split Hopkinson Pressure Bar (SHPB) for strain rates between 500 and 104 s^{-1} [39–41]. Figure 6 illustrates a photograph of the used SHPB machine with different parts of the system including striker bar, input, and output bars. To register the experiment signals, incident and transmitted bars attached with two strain gauges were installed. The samples were embedded between two bars with a diameter of 20 mm in each bar. The striker, input and output bars have a length of 0.8 m, 3 m and 2 m, respectively. The bars are correctly aligned and can move easily on the base. The striker is projected onto the input bar at a specific speed.

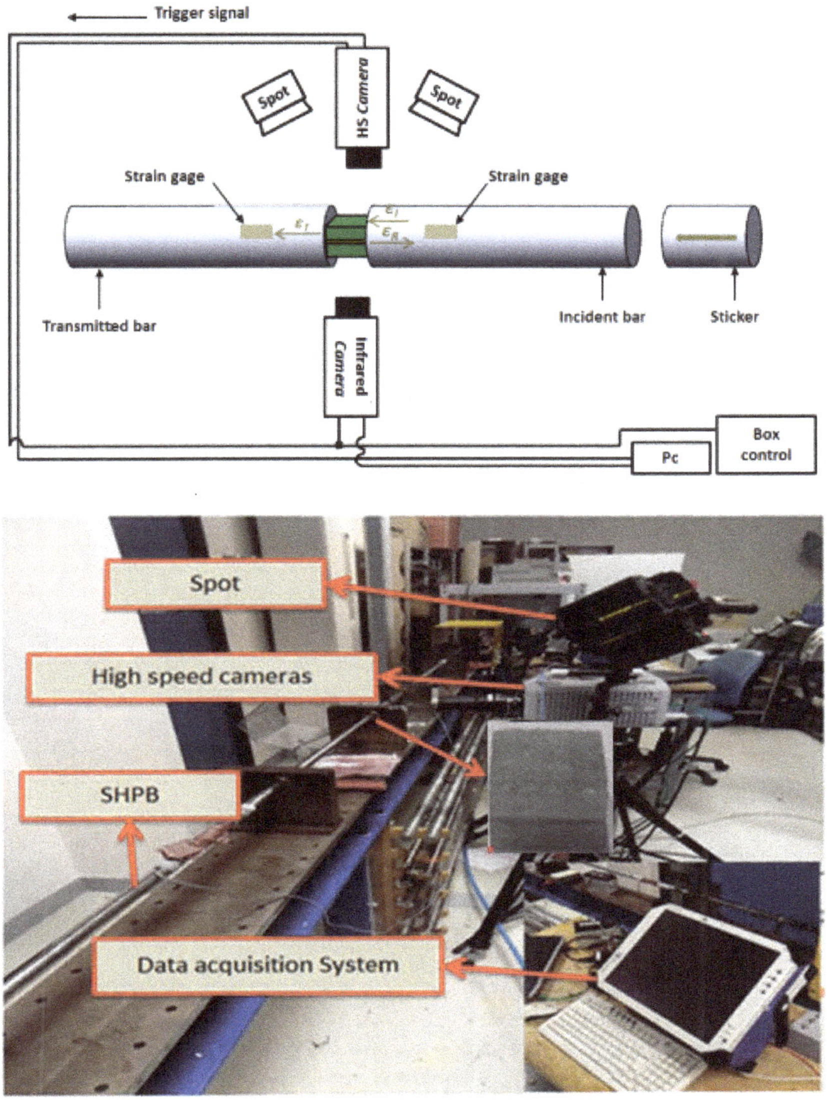

Figure 6. SHPB machine.

3.2. Mechanical Behavior

The thermomechanical properties of a composite are affected by moisture and temperature. However, an increase in moisture and temperature can augment the molecular displacement of the material and can significantly alter the shape and volume of composites. Such deformation is further sophisticated, especially with the dynamic compression of the material. The impact pressure effect on the stress/strain behavior of bonded composite under in-plane dynamic compression is shown in Figure 7.

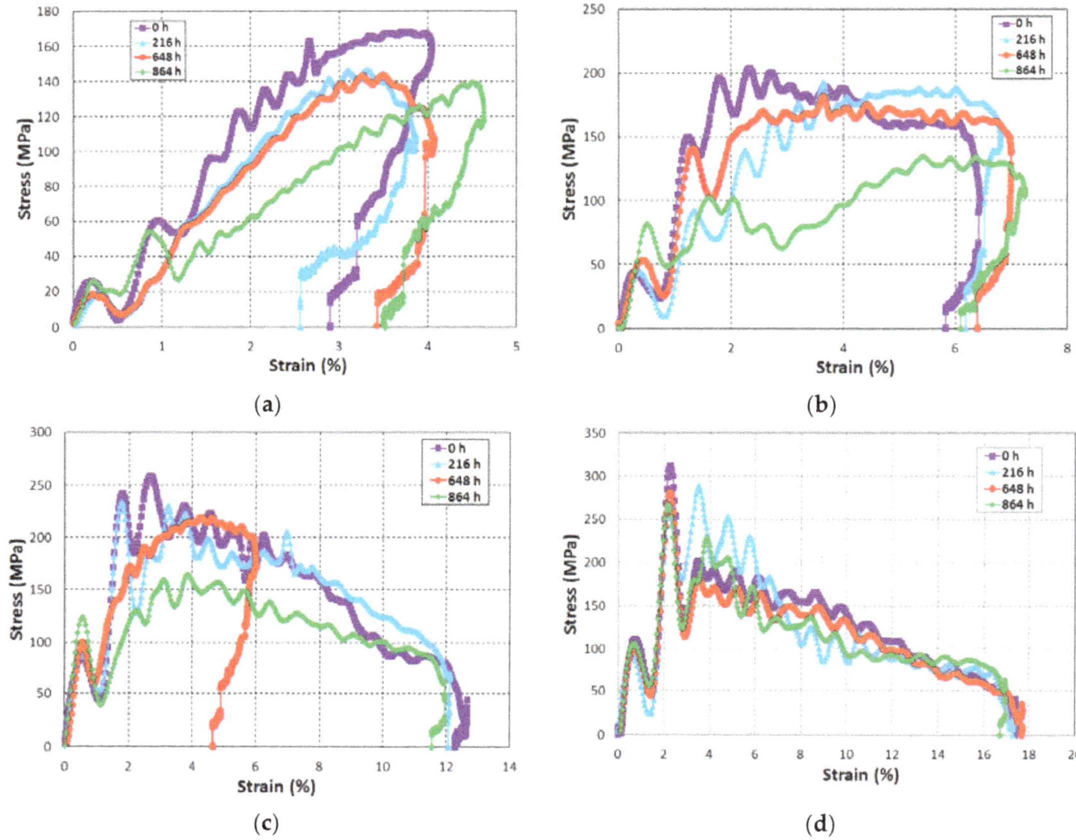

Figure 7. Stress/strain curves of the sample under in-plane loading for different pressure and aging times. (**a**) 1 bar. (**b**) 2 bar. (**c**) 3 bar. (**d**) 4 bar.

Figure 7 presents the stress–strain responses for various impact pressures from 1 bar to 4 bar with various aging times of 0 h, 216 h, 648 h, and 864 h. It can be noted that the dynamic compressive behavior of the composite is strongly affected by strain rate and aging time. Moreover, the effect of moisture on the stress–strain curves is more significant at a low strain rate than at a high strain rate. This clearly shows that the strain rate response is sensitive to the entry pressure P in the chamber of compressed air. At the ultimate stress, the sample damage appears and loses its loadbearing capacity. It is apparent that for all impact tests, the stress–strain tendency was approximately the same during the linear elastic phase, with no damage at small strains. In addition, it can be shown that in the undamaged case, the samples regain their original state with negligible plastic deformation. On the other hand, in the damaged samples, the first peak is observed in the elastic response which shows the beginning of microscopic damage modes, such as matrix cracking. The matrix cracking of samples with in-plan direction encourages micro-buckling and fiber twisting, and leads to debonding, fiber separation, delamination and fiber breaking towards the final failure observed with the appearance of the second peak for the strain rate curves. It can be concluded that the first region of nonlinearity in the stress–strain curves are mainly due to the viscoelastic nature of the polyester polymer, while the nonlinearity noticed before the final failure is caused by the cracking of the matrix, and the elastic modulus reduced with aging time due to the rising of moisture uptake in the composite.

Figure 8 shows the evolution of strain rate as a function of time for different aging times from 0 h to 864 h and with various impact pressures on the bonded joints. It can be observed that the damage becomes significant only for high-pressure impact due to the second peak presented in the signal and it becomes increasingly important as the impact pressure increases, indicating the accumulated failure modes in the material subjected to dynamic loading. In contrast, for low impact pressure, only the residual plastic deformation occurs due to matrix cracks. Thus, the noticeable fact of increasing the strain rate is manifested by changes in the damage modes. The samples tested were damaged by fiber twisting at low strain rates, and by interfacial debonding and delamination at high strain rates.

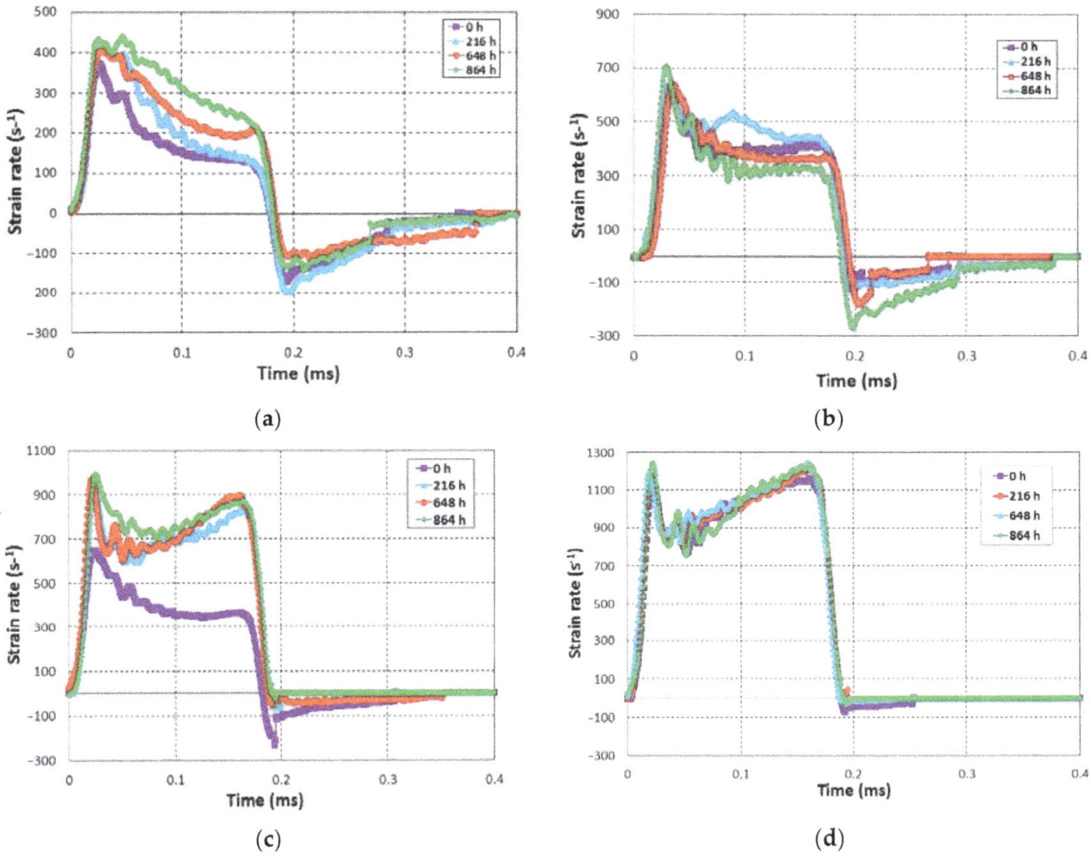

Figure 8. Strain rate versus time curves of the sample at different aging times and pressure under in-plane loading. (**a**) 1 bar. (**b**) 2 bar. (**c**) 3 bar. (**d**) 4 bar.

According to our knowledge, there is no empirical study proposing a constitutive model that reflects the impact of strain rate on bonded joints under dynamic conditions with the effect of aging due to the limited availability of reliable experimental data of dynamic tests. However, the effect of strain rate on the damage of metals and composites has been investigated in many works [42,43]. Many research works have recently been proposed to model the dynamic behavior of the material subjected to impact testing [44,45].

In this paper, empirical laws have been proposed relating to the dependence of the dynamic properties of bonded joints to the impact pressure and strain rate under dynamic in-plane compression loading for different aging times. To optimize the graphical representations, the deformation rate and the maximum stress are given in relation to the impact pressure. Figure 9 shows the trends obtained from the strain rate dependencies of impact pressure and aging time. The evolution of the strain rate is approximated by a quadratic equation and presents two phases. In the first phase (1 bar $\leq$ P $\leq$ 2 bar) the increase is less pronounced and in the second phase the increase is accelerated. On the other hand, we can see that the stiffness of the material is sensitive to the deformation rate and the aging time. It is to be noticed that these nonlinear equations were derived from empirical plots of the dynamic characterization of specimens for different aging times under dynamic compression. The obtained equations are similar to the quadratic equations of glass/epoxy composite in dynamic compression [42,43]. In addition, Tarfaoui et al. [46] employed the non-linear evaluation of damage versus strain rate for dynamic modeling under pressure impact.

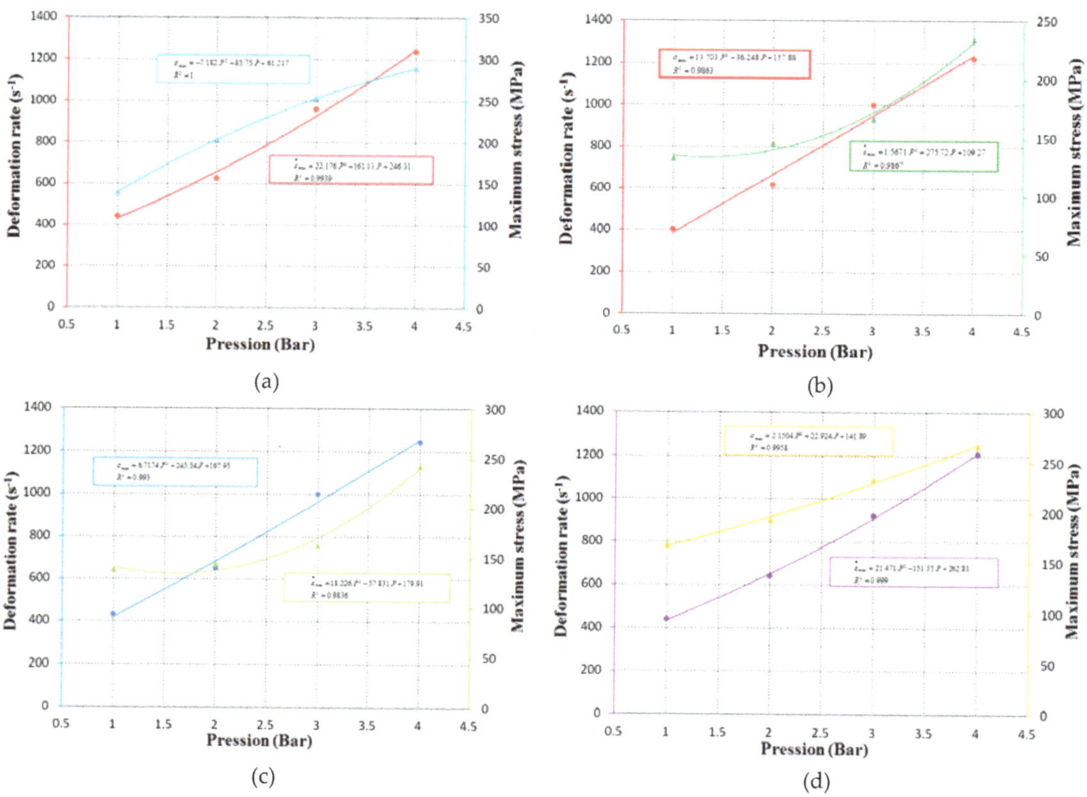

Figure 9. Variation of maximum stress/strain rate in function of aging time. (**a**) t = 0 h. (**b**) t = 216 h. (**c**) t = 648 h. (**d**) t = 864 h.

3.3. Theoretical Characterization of Absorbed Energy

The incident impact energy is the complete energy available at the start and represents the kinetic energy contributed by the impactor. At the interface between the bar and sample, a portion of this energy is absorbed by the specimen and can produce damage or plastic deformation in various forms and also can generate heat, which is associated with the occurrence of microscopic/macroscopic damage. The rest of the energy consists of reflected and transmitted energy and can be measured from the deformation profile. The absorbed energy W_{abs} is expressed as follows:

$$W_{abs} = W_{inc} - (W_{trans} - W_{ref}) \quad (4)$$

With W_{inc}, W_{ref} and W_{trans} representing the incident, reflected and transmitted energy, respectively, and their expressions are given by:

$$W_{inc} = \frac{A}{\rho c} \int_0^t \sigma_i^2(t)dt \quad (5)$$

$$W_{ref} = \frac{A}{\rho c} \int_0^t \sigma_r^2(t)dt \quad (6)$$

$$W_{trans} = \frac{A}{\rho c} \int_0^t \sigma_t^2(t)dt \quad (7)$$

Figure 10 presents typical absorbed energy by the sample at an impact pressure of 4 bar for an aging time of t = 648 h. The energy versus time curves representative of the aged bonded joint sample indicates that the incident and reflected energy increases quickly compared to the transmitted and absorbed energy during the propagation of the deformation wave and their values are maintained stable after achieving a required value, and the apparition of the maximum absorption energy indicating the presence of damage to the joint. Much research indicates that there is a relationship between damage modes and the characteristics of the absorbed energy [47]. The results demonstrated that the impact energy can influence the absorbed energy and can be described as follows:

- As the impact energy increases, the number of damage cracks is large and its distribution becomes more uniform, while the energy absorbed also increases as shown in Figure 10.
- Most of the incident energy is absorbed due to the effect of aging and the low impedance of composites.

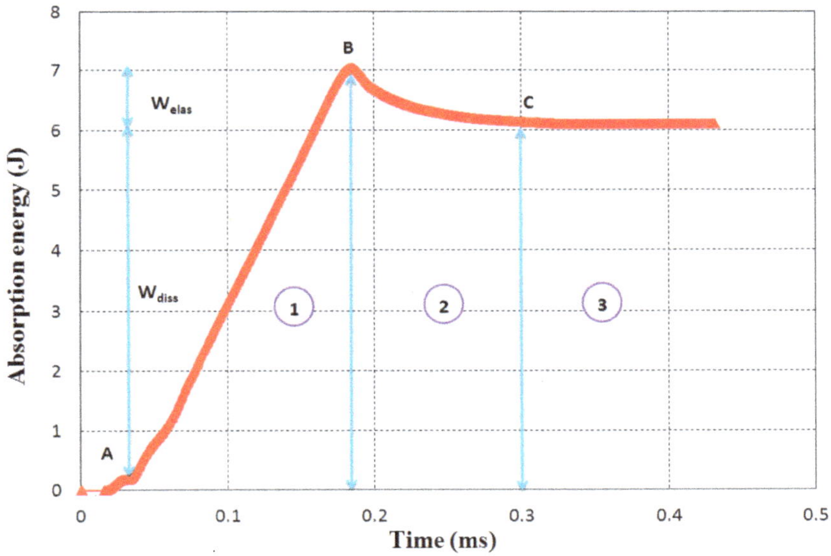

Figure 10. Typical profile of absorbed energy W_{abs}.

Moreover, the incident energy in the initial part increases quickly with the stress wave propagation, whereas the absorbed energy of the bonded joint has a small increase with the propagation of the stress wave. This is because the majority of the incident energy is absorbed by the specimens and just a minor fraction of the incident energy is transferred into the transmitted bar. In Figure 10, the presence of different zones can be identified:

- Zone 1 represents the loading phase (AB): The sample is absorbing energy at point B ($W_{abs} = y_B - y_A$), and this energy is composed of an elastic component and an unrecovered dissipative component.
- Zone 2 represents the progressive discharge part: The elastic energy is realized up to point C ($W_{elas} = y_B - y_C$).
- Zone 3 represents the final stage of the charge/discharge cycle: W_{abs} corresponds to the energy permanently dissipated in the material.

In this way, the absorbed energy is decomposed into an unrecovered elastic part and an inelastic part, and can be expressed by the equation below:

$$W_{abs} = W_{elas} + W_{diss} \tag{8}$$

W_{abs} is the energy absorbed by the sample other than the stored deformation energy that is exhausted to create damage. The inelastic part is related to the damage generation in different forms, such as matrix fracture, fiber rupture, and delamination or decohesion between the fiber and matrix. The curve of the absorbed energy tends towards a fixed value at the end of the first cycle where the transmitted signal is reflected by the incident, and permanently corresponds to the energy dissipated by the damage in the sample. Figure 11 shows the absorption energy for a compression test for an aging time of 216 h at different impact pressures. The fluctuating profiles represent the release and storage of deformation energy throughout the experimental process. It can be easily observed that at low impact energy, a considerable amount of the input energy of W_{elas} is stored in the loading phase and released in the discharge phase. On the other hand, W_{abs} has an increase as the impact pressure is raised, which in turn influences the release of W_{elas} through the discharge phase. The last one (W_{elas}) is the largest fraction of W_{abs} at low impact energy, while W_{diss} takes a larger fraction of W_{abs} as the impact energy increases. It should be noted that W_{elas} is zero when the sample has macroscopic damage. One can also note that the loading portions of the curves for the several impact pressures are coherent and overlapped. Additionally, an increase is observed in slope with the increase of impact pressure, with the suggestion that the reaction is conditioned by the deformation rate.

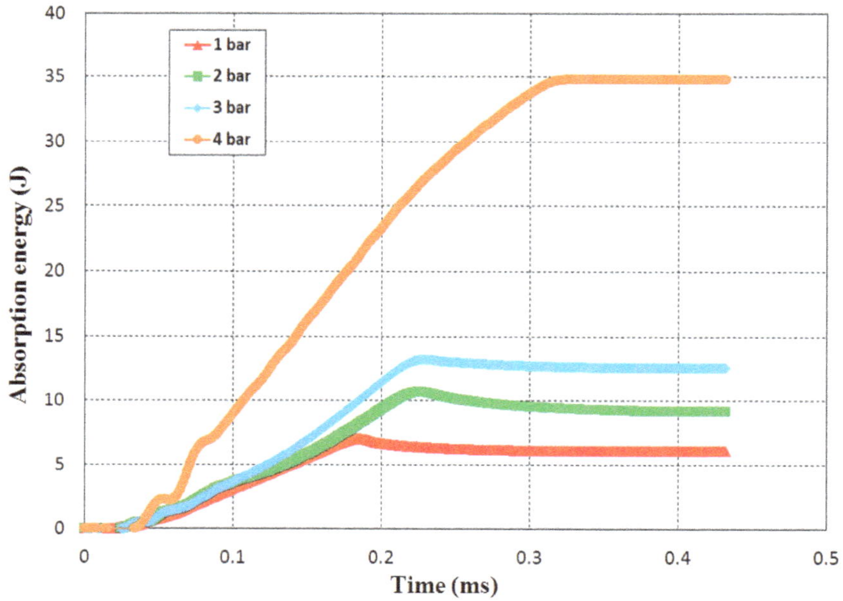

Figure 11. Absorbed energy for an aging time t = 216 h.

3.4. Damage Mechanisms and Failure Modes

A camera with a high-speed system is employed to visualize the damage mode of the samples. Figure 12 shows the damage pattern of the bonded composite loaded in an in-plane direction. It can be noted that the damage is observed as cracks across the diagonal of the cube. In addition, matrix fractures and intense delamination are produced in the preferred inter-laminar planes. To understand the damage behavior of the bonded joined under dynamic loading, a camera of high-speed is used to capture the loading of the sample at different impact pressures. The damage is observed only at high impact pressures. For low impact pressure, only residual plastic deformations due to matrix fractures were present. So as the damage mechanism is developed, it is expressed in the form of matrix/fiber failure, fiber pullout, and delamination of the ply pack.

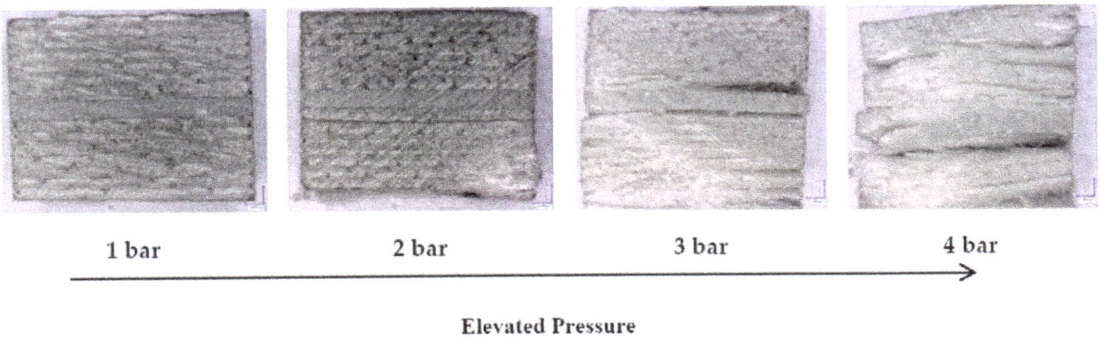

Figure 12. Optical micrographs of bonded composite under dynamic compression for $t = 216$ h.

The fiber and matrix can be observed to be highly bonded before hygrothermal aging. The beginning of the damage includes a "V" shaped shear band and the development of delamination at its tip. Microcracks were observed at the interface of matrix/fiber after 216 h, as shown in Figure 13. By increasing the hygrothermal aging time, the interface disbonding grows and propagates as long cracks at the 864 h aging time. The obvious interfacial degradation effects of moisture absorbed in the bonded composites on the interfacial area between the adhesive and the composite are detected. In these results, the hygrothermal influence causes chemical changes in the resin matrix, plasticization and dimensional changes, such as swelling, and deterioration of the strength of the adhesive/composite interface. The bond between the fiber and matrix weakens and the interface weakens as the moisture absorption increases. All these factors reduce the impact resistance of the composite.

Figure 13. Optical micrographs of adhesively bonded composite loaded through at pressure of 3 bar.

3.5. Microstructural Analysis of Adhesively Bonded Composite

Figure 14 illustrates the microstructure of adhesive bonded composite as remarked using SEM. There were no significant morphological differences. The materials contain the elements of C, O, Mg, Al, Si, K and Ca. It is observed that the crack propagation rate of bonded composite joints becomes more prominent with the increase in strain rate. The results of reaching such raised strength and initial modulus can be justified by the fact that the composite materials become comparatively brittle and harder, which leads to high strength and modulus [36]. The results show also that the failure stresses degrade for various aging times for the samples subjected to in-plane compression because temperature and humidity considerably modify and weaken the composite. The strain rate is seen to be increased for various aging times. These results can be explained by the plasticization of the matrix being the principal factor in high strain rate properties when materials are not totally immersed in corrosion enclosure. Therefore, the matrix is more responsive to the strain rate and thus, plays a more important role than the fiber in identifying the material performance response to the high strain rate of polymer matrix composites. From the above discussion, it is further revealed that the crack propagation model changes as the deformation rate increases, and the higher the deformation rate, the more complex the crack propagation model appears.

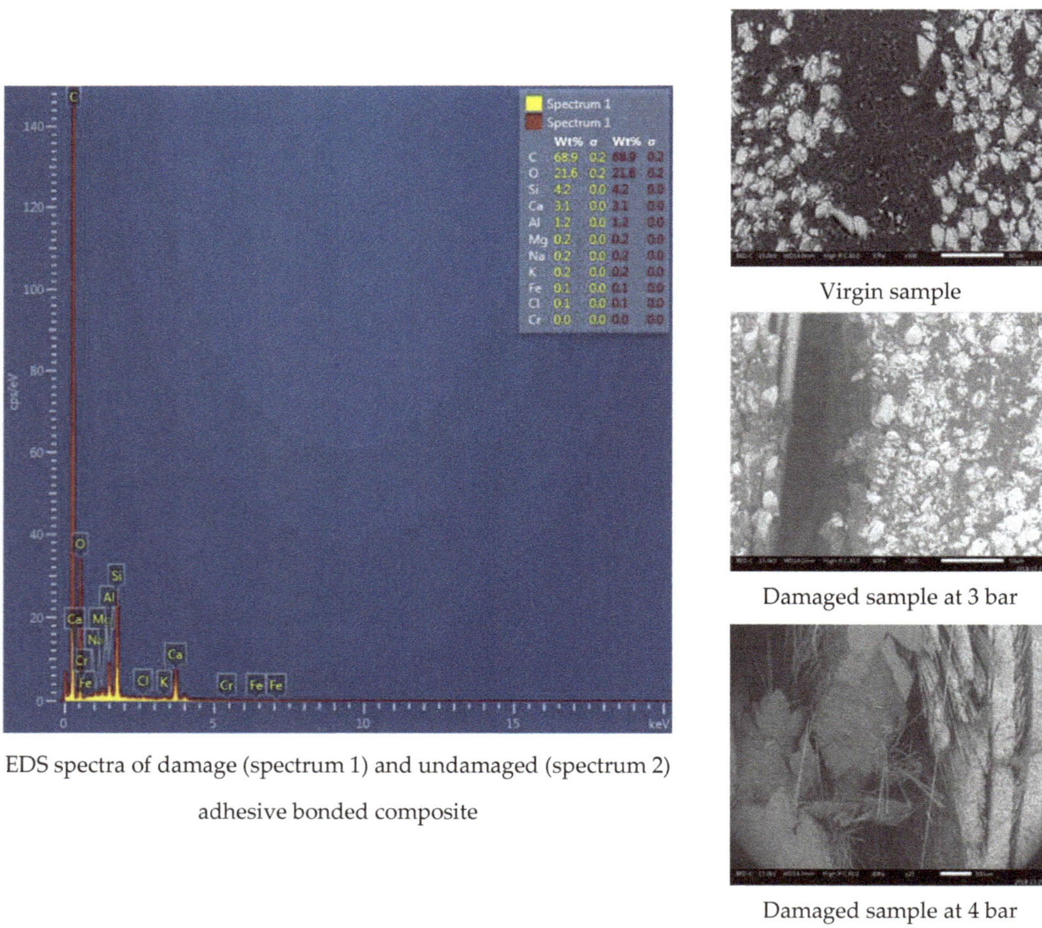

EDS spectra of damage (spectrum 1) and undamaged (spectrum 2)

adhesive bonded composite

Virgin sample

Damaged sample at 3 bar

Damaged sample at 4 bar

Figure 14. SEM micrographs of the composite of bonded adhesive at 864 h.

4. Conclusions

In the present paper, the influence of moisture diffusion on the dynamic compressive behavior of glass/polyester composite was investigated. The failure modes were viewed by SEM and scanning electron microscopy for different aged specimens. The results obtained show that the hygrothermal influence induces chemical changes in the resin matrix, plasticization and dimensional changes, such as swelling, and also the weakened adhesive/composite interfaces. In addition, the bond between the fiber and the matrix is weakened as the humidity uptake increases. All of these elements would affect the dynamic behavior and the impact strength of an assembly of composite materials. The conclusions from this investigation can be summarized as follows:

1. Matrix plasticization is the major factor in the high strain rate properties when materials are partially or fully wet-immersed in a room temperature bath.
2. The matrix is more sensitive to deformation rate, and therefore, plays a more important role than the fiber in identifying the high deformation rate material behavior of polymer matrix composites.
3. The deterioration of the interface fiber/matrix leads to a decrease in the favorable impact of the plasticization of the matrix.
4. An overall increase in material properties occurs due to absorption of moisture, except in high-temperature baths, due to the extreme degradation of the interface.
5. There is an amount of absorbed moisture that gives optimum material properties that correspond to the highest matrix plasticization with minimal degradation of the interface between fiber and matrix.

Author Contributions: O.L.: Investigation, Methodology, Writing—original draft; M.T.: Methodology, Investigation, Project administration, Conceptualization, Writing—original draft, Writing—review & editing, Funding acquisition; M.R. Methodology, Investigation, Writing—original draft, Writing—review & editing; H.L., S.J.S., M.A.D., A.A.: Validation, Visualization; M.N.: Methodology, Investigation, Writing—original draft, Writing—review & editing, Conceptualization, Data curation, Supervision, Validation, Visualization. All authors have read and agreed to the published version of the manuscript.

Funding: This research received no external funding.

Institutional Review Board Statement: Not applicable.

Informed Consent Statement: Not applicable.

Conflicts of Interest: The authors declare no conflict of interest.

References

1. Nachtane, M.; Tarfaoui, M.; Saifaoui, D.; El Moumen, A.; Hassoon, O.H.; Benyahia, H. Evaluation of Durability of Composite Materials Applied to Renewable Marine Energy: Case of Ducted Tidal Turbine. *Energy Rep.* **2018**, *4*, 31–40. [CrossRef]
2. Mangalgiri, P.D. Composite Materials for Aerospace Applications. *Bull. Mater. Sci.* **1999**, *22*, 657–664. [CrossRef]
3. Nachtane, M.; Tarfaoui, M.; Saifaoui, D.; El Moumen, A.; Boudounit, H. Caractérisation Mécanique d'une Hydrolienne En Matériau Composite Dans Un Environnement Marin. In Proceedings of the 2017 23e Congrès français de Mécanique-CFM, Lille, France, 28 August–1 September 2017; Association Française de Mécanique (AFM): Courbevoie, France, 2017.
4. Nachtane, M.; Tarfaoui, M.; El Moumen, A.; Saifaoui, D. Numerical Investigation of Damage Progressive in Composite Tidal Turbine for Renewable Marine Energy. In Proceedings of the 2016 International Renewable and Sustainable Energy Conference (IRSEC), Marrakech, Morocco, 14–17 November 2016; pp. 559–563.
5. Nachtane, M.; Tarfaoui, M.; Goda, I.; Rouway, M. A Review on the Technologies, Design Considerations and Numerical Models of Tidal Current Turbines. *Renew. Energy* **2020**, *157*, 1274–1288. [CrossRef]
6. Maziz, A.; Tarfaoui, M.; Gemi, L.; Rechak, S.; Nachtane, M. A Progressive Damage Model for Pressurized Filament-Wound Hybrid Composite Pipe under Low-Velocity Impact. *Compos. Struct.* **2021**, *276*, 114520. [CrossRef]
7. Maziz, A.; Tarfaoui, M.; Rechak, S.; Nachtane, M.; Gemi, L. Finite Element Analysis of Impact-Induced Damage in Pressurized Hybrid Composites Pipes. *Int. J. Appl. Mech.* **2021**, *13*, 2150074. [CrossRef]
8. Nachtane, M.; Tarfaoui, M.; Ledoux, Y.; Khammassi, S.; Leneveu, E.; Pelleter, J. Experimental Investigation on the Dynamic Behavior of 3D Printed CF-PEKK Composite under Cyclic Uniaxial Compression. *Compos. Struct.* **2020**, *247*, 112474. [CrossRef]

9. Shahawy, M.A.; Beitelman, T.; Arockiasamy, M.; Sowrirajan, R. Experimental Investigation on Structural Repair and Strengthening of Damaged Prestressed Concrete Slabs Utilizing Externally Bonded Carbon Laminates. *Compos. Part B Eng.* **1996**, *27*, 217–224. [CrossRef]
10. Nachtane, M.; Tarfaoui, M.; El Moumen, A.; Saifaoui, D. Damage Prediction of Horizontal Axis Marine Current Turbines under Hydrodynamic, Hydrostatic and Impacts Loads. *Compos. Struct.* **2017**, *170*, 146–157. [CrossRef]
11. Sassi, S.; Tarfaoui, M.; Yahia, H.B. In-Situ Heat Dissipation Monitoring in Adhesively Bonded Composite Joints under Dynamic Compression Loading Using SHPB. *Compos. Part B Eng.* **2018**, *154*, 64–76. [CrossRef]
12. Sassi, S.; Tarfaoui, M.; Yahia, H.B. Thermomechanical Behavior of Adhesively Bonded Joints under Out-of-Plane Dynamic Compression Loading at High Strain Rate. *J. Compos. Mater.* **2018**, *52*, 4171–4184. [CrossRef]
13. Laaouidi, H.; Tarfaoui, M.; Nachtane, M.; Lagdani, O. Modal Analysis of Composite Nozzle for an Optimal Design of a Tidal Current Turbine. *J. Nav. Archit. Mar. Eng.* **2021**, *18*, 39–54.
14. Lagdani, O.; Tarfaoui, M.; Nachtane, M.; Trihi, M.; Laaouidi, H. Modal Analysis of an Iced Offshore Composite Wind Turbine Blade. *Wind Eng.* **2021**, *46*, 134–149. [CrossRef]
15. Komai, K.; Minoshima, K.; Shibutani, T.; Nomura, T. The Influence of Water on the Mechanical Properties and Fatigue Strength of Angle-Ply Carbon/Epoxy Composites. *JSME Int. J. Ser. 1 Solid Mech. Strength Mater.* **1989**, *32*, 588–595. [CrossRef]
16. Ashik, K.P.; Sharma, R.S.; Guptha, V.J. Investigation of Moisture Absorption and Mechanical Properties of Natural/Glass Fiber Reinforced Polymer Hybrid Composites. *Mater. Today Proc.* **2018**, *5*, 3000–3007. [CrossRef]
17. Selzer, R.; Friedrich, K. Inluence of Water Up-Take on Interlaminar Fracture Properties of Carbon Fibre-Reinforced Polymer Composites. *J. Mater. Sci.* **1995**, *30*, 334–338. [CrossRef]
18. Wu, L.; Hoa, S.V.; Ton-That, M.-T. Effects of Water on the Curing and Properties of Epoxy Adhesive Used for Bonding FRP Composite Sheet to Concrete. *J. Appl. Polym. Sci.* **2004**, *92*, 2261–2268. [CrossRef]
19. Buehler, F.U.; Seferis, J.C. Effect of Reinforcement and Solvent Content on Moisture Absorption in Epoxy Composite Materials. *Compos. Part A Appl. Sci. Manuf.* **2000**, *31*, 741–748. [CrossRef]
20. Schutte, C.L. Environmental Durability of Glass-Fiber Composites. *Mater. Sci. Eng. R Rep.* **1994**, *13*, 265–323. [CrossRef]
21. Mijović, J.; Zhang, H. Molecular Dynamics Simulation Study of Motions and Interactions of Water in a Polymer Network. *J. Phys. Chem. B* **2004**, *108*, 2557–2563. [CrossRef]
22. Khalilullah, I.; Reza, T.; Chen, L.; Mazumder, A.M.H.; Fan, J.; Qian, C.; Zhang, G.; Fan, X. In-Situ Characterization of Moisture Absorption and Hygroscopic Swelling of Silicone/Phosphor Composite Film and Epoxy Mold Compound in LED Packaging. In Proceedings of the 2017 18th International Conference on Thermal, Mechanical and Multi-Physics Simulation and Experiments in Microelectronics and Microsystems (EuroSimE), Dresden, Germany, 3–5 April 2017; Institute of Electrical and Electronics Engineers IEEE: Piscataway Township, NJ, USA, 2017; pp. 1–9.
23. Megel, M.; Kumosa, L.; Ely, T.; Armentrout, D.; Kumosa, M. Initiation of Stress-Corrosion Cracking in Unidirectional Glass/Polymer Composite Materials. *Compos. Sci. Technol.* **2001**, *61*, 231–246. [CrossRef]
24. Jiang, X.; Kolstein, H.; Bijlaard, F.; Qiang, X. Effects of Hygrothermal Aging on Glass-Fibre Reinforced Polymer Laminates and Adhesive of FRP Composite Bridge: Moisture Diffusion Characteristics. *Compos. Part A Appl. Sci. Manuf.* **2014**, *57*, 49–58. [CrossRef]
25. Nachtane, M.; Tarfaoui, M.; Saifaoui, D.; Hilmi, K. Hygrothermal and Mechanical Performance Evaluation of Glass-Polyester Composite for Renewable Marine Energies. In Proceedings of the 13ème Congrès de Mécanique (cmm2017), Meknès, Morocco, 4–11 April 2017.
26. Grace, L.R.; Altan, M.C. Characterization of Anisotropic Moisture Absorption in Polymeric Composites Using Hindered Diffusion Model. *Compos. Part A Appl. Sci. Manuf.* **2012**, *43*, 1187–1196. [CrossRef]
27. Benyahia, H.; Tarfaoui, M.; El Moumen, A.; Ouinas, D.; Hassoon, O.H. Mechanical Properties of Offshoring Polymer Composite Pipes at Various Temperatures. *Compos. Part B Eng.* **2018**, *152*, 231–240. [CrossRef]
28. Sassi, S.; Tarfaoui, M.; Yahia, H.B. An Investigation of In-Plane Dynamic Behavior of Adhesively-Bonded Composite Joints under Dynamic Compression at High Strain Rate. *Compos. Struct.* **2018**, *191*, 168–179. [CrossRef]
29. Firdosh, S.; Murthy, H.N.; Angadi, G.; Raghavendra, N. Investigation of Water Absorption Characteristics of Nano-Gelcoat for Marine Application. *Prog. Org. Coat.* **2018**, *114*, 173–187. [CrossRef]
30. Shah, O.R.; Tarfaoui, M. Effect of Damage Progression on the Heat Generation and Final Failure of a Polyester–Glass Fiber Composite under Tension–Tension Cyclic Loading. *Compos. Part B Eng.* **2014**, *62*, 121–125. [CrossRef]
31. Hassoon, O.H.; Tarfaoui, M.; Alaoui, A.E.M.; El Moumen, A. Experimental and Numerical Investigation on the Dynamic Response of Sandwich Composite Panels under Hydrodynamic Slamming Loads. *Compos. Struct.* **2017**, *178*, 297–307. [CrossRef]
32. Gargano, A.; Pingkarawat, K.; Pickerd, V.; Delaney, T.; Das, R.; Mouritz, A.P. Effect of Seawater Immersion on the Explosive Blast Response of a Carbon Fibre-Polymer Laminate. *Compos. Part A Appl. Sci. Manuf.* **2018**, *109*, 382–391. [CrossRef]
33. Pang, S.; Tao, W.; Liang, Y.; Liu, Y.; Huan, S. A Modified Method of Pulse-Shaper Technique Applied in SHPB. *Compos. Part B Eng.* **2019**, *165*, 215–221. [CrossRef]
34. Zhao, H.; Gary, G. On the Use of SHPB Techniques to Determine the Dynamic Behavior of Materials in the Range of Small Strains. *Int. J. Solids Struct.* **1996**, *33*, 3363–3375. [CrossRef]
35. Gazit, S. Dimensional Changes in Glass-Filled Epoxy Resin as a Result of Absorption of Atmospheric Moisture. *J. Appl. Polym. Sci.* **1978**, *22*, 3547–3558. [CrossRef]

36. Haque, A.; Hossain, M.K. Effects of Moisture and Temperature on High Strain Rate Behavior of S2-Glass–Vinyl Ester Woven Composites. *J. Compos. Mater.* **2003**, *37*, 627–647. [CrossRef]
37. Sassi, S.; Tarfaoui, M.; Nachtane, M.; Ben Yahia, H. Strain Rate Effects on the Dynamic Compressive Response and the Failure Behavior of Polyester Matrix. *Compos. Part B Eng.* **2019**, *174*, 107040. [CrossRef]
38. Lee, M.C.; Peppas, N.A. Water Transport in Graphite/Epoxy Composites. *J. Appl. Polym. Sci.* **1993**, *47*, 1349–1359. [CrossRef]
39. Khosravani, M.R.; Weinberg, K. A Review on Split Hopkinson Bar Experiments on the Dynamic Characterisation of Concrete. *Constr. Build. Mater.* **2018**, *190*, 1264–1283. [CrossRef]
40. Tarfaoui, M.; Nachtane, M. Can a Three-Dimensional Composite Really Provide Better Mechanical Performance Compared to Two-Dimensional Composite under Compressive Loading? *J. Reinf. Plast. Compos.* **2019**, *38*, 49–61. [CrossRef]
41. Nachtane, M. Staking Lay-up Effect on Dynamic Compression Behaviour of e-Glass/Epoxy Composite Materials: Experimental and Numerical Investigation. *Adv. Mater. Lett.* **2018**, *9*, 816–822.
42. Randles, P.W.; Nemes, J.A. A Continuum Damage Model for Thick Composite Materials Subjected to High-Rate Dynamic Loading. *Mech. Mater.* **1992**, *13*, 1–13. [CrossRef]
43. Park, S.W.; Schapery, R.A. A Viscoelastic Constitutive Model for Particulate Composites with Growing Damage. *Int. J. Solids Struct.* **1997**, *34*, 931–947. [CrossRef]
44. Li, Y.; Ramesh, K.T.; Chin, E.S.C. Viscoplastic Deformations and Compressive Damage in an A359/SiCp Metal–Matrix Composite. *Acta Mater.* **2000**, *48*, 1563–1573. [CrossRef]
45. Yang, L.M.; Shim, V.P.W.; Lim, C.T. A Visco-Hyperelastic Approach to Modelling the Constitutive Behaviour of Rubber. *Int. J. Impact Eng.* **2000**, *24*, 545–560. [CrossRef]
46. Tarfaoui, M.; Choukri, S.; Neme, A. Effect of Fibre Orientation on Mechanical Properties of the Laminated Polymer Composites Subjected to Out-of-Plane High Strain Rate Compressive Loadings. *Compos. Sci. Technol.* **2008**, *68*, 477–485. [CrossRef]
47. Liu, D.; Schulz, P.; Templeton, D.; Raju, B. Dynamic Failure and Energy Absorption of Composite Materials with Geometrical Control. In Proceedings of the 16th International Conference on Composite Materials, Kyoto, Japan, 8–13 July 2007.

Journal of
Composites Science

MDPI

Article

The Formulation of the Quadratic Failure Criterion for Transversely Isotropic Materials: Mathematical and Logical Considerations

Shuguang Li *, Mingming Xu and Elena Sitnikova

Faculty of Engineering, University of Nottingham, Nottingham NG8 1BB, UK; mingming.xu@nottingham.ac.uk (M.X.); elena.sitnikova@nottingham.ac.uk (E.S.)
* Correspondence: shuguang.li@nottingham.ac.uk

Abstract: The quadratic function of the original Tsai–Wu failure criterion for transversely isotropic materials is re-examined in this paper. According to analytic geometry, two of the troublesome coefficients associated with the interactive terms—one between in-plane direct stresses and one between transverse direct stresses—can be determined based on mathematical and logical considerations. The analysis of the nature of the quadratic failure function in the context of analytic geometry enhances the consistency of the failure criterion based on it. It also reveals useful physical relationships as intrinsic properties of the quadratic failure function. Two clear statements can be drawn as the outcomes of the present investigation. Firstly, to maintain its basic consistency, a failure criterion based on a single quadratic failure function can only accommodate five independent strength properties, viz. the tensile and compressive strengths in the directions along fibres and transverse to fibres, and the in-plane shear strength. Secondly, amongst the three transverse strengths—tensile, compressive and shear—only two are independent.

Keywords: quadratic failure function; Tsai–Wu criterion; failure envelope; strength; transverse shear strengths; transversely isotropic materials

Citation: Li, S.; Xu, M.; Sitnikova, E. The Formulation of the Quadratic Failure Criterion for Transversely Isotropic Materials: Mathematical and Logical Considerations. *J. Compos. Sci.* **2022**, *6*, 82. https://doi.org/10.3390/jcs6030082

Academic Editor: Stelios K. Georgantzinos

Received: 8 February 2022
Accepted: 1 March 2022
Published: 7 March 2022

1. Background

Ever since Tsai and Wu proposed their failure criterion [1] based on a quadratic failure function, the polynomials employed to construct failure criteria have been mostly kept to the second order. Whilst one could legitimately argue for higher orders of polynomials along the line, as was suggested in [2], or to partition the failure function according to the failure modes, as was attempted in [3], many are on the verge of abandoning all such theories developed on a phenomenological basis, having been discouraged by their unsatisfactory performance. However, a sensible question does not seem to have ever been asked: 'Has the quadratic failure function been understood well enough before increasing or abandoning any further efforts along this line of development?' It is true that quadratic functions are well understood in analytic geometry as a branch of mathematics, and yet it will be revealed in the present paper that the established mathematics has not been properly utilised in the important subject of composite failure. The well-established conclusions of analytic geometry have simply not been appropriately recognised in the formulations of failure criteria for composites based on quadratic functions. Therefore, before this aspect has been appropriately examined and evaluated, it would surely be a premature decision to propose more complicated arrangements for the failure function or to ditch every account of phenomenological criteria based on macroscopic stresses or strains completely.

A thorough re-examination of popular existing failure criteria for modern composites [4] by two of the authors of this paper suggested that whilst the predictions of these theories fell short of the expectation of the users, none of them had been formulated consistently enough at a basic level. Phenomenological approaches based on macroscopic stresses or strains are bound to have their limitations. However, blaming the phenomenological nature alone for the deficiencies in the existing criteria is a far too quick and easy escape. Once one moves into the micromechanics of composites, there will be many more complications involved due to the consideration of both mechanics and materials, let alone the necessary mathematics and logic. As an example, representative volume elements and unit cells as basic and also necessary means of micromechanical investigations were subjected to a serious and systematic examination in [5]. It was revealed that their formulation and implementation are full of pitfalls and any oversight leads to an erroneous interpretation of genuine material behaviours. If the mathematics and logic cannot be sorted out at a macroscopic level, where there are only six macroscopic stresses to deal with, it is hard to imagine that the study of composites failure can be accomplished at a microscopic scale, where one can easily be swamped with new parameters, sometimes of little physical meaning. Whilst researchers should not be discouraged from undertaking explorations into micromechanics, it is worthwhile to re-examine the aspect of the rationality of conventional and phenomenological failure criteria, not aiming to generate any new criterion, but to tidy up the basic mathematics and logic underlying existing theories and to eliminate any irrational elements before presenting them on a footing of a basic level of consistency. By then, one could make an objective decision on how applicable such rationally formulated failure criteria are and how far such a phenomenological theory can be used reliably. This may even lay a firm basis for more sophisticated developments, as mentioned above.

2. A Critical Review of the Tsai–Wu Criterion

A reasonably comprehensive literature review on the subject of the Tsai–Wu failure criterion is given in a recent publication [6] and will be waived in this paper. The criterion was originally proposed as [1]:

$$\begin{aligned} &F_1\sigma_1 + F_2\sigma_2 + F_3\sigma_3 + F_{11}\sigma_1^2 + F_{22}\sigma_2^2 + F_{33}\sigma_3^2 + 2F_{23}\sigma_2\sigma_3 + 2F_{13}\sigma_1\sigma_3 + 2F_{12}\sigma_1\sigma_2 \\ &+F_{44}\tau_{23}^2 + F_{55}\tau_{13}^2 + F_{66}\tau_{12}^2 - 1 = 0 \end{aligned} \tag{1}$$

in the context of generally orthotropic materials. The left-hand side is a quadratic failure function. A careful examination of (1) suggests that many terms, such as linear and bilinear expressions of shear stresses, are missing from the failure function for it to be a complete quadratic polynomial of six stress components. These terms have been dropped because the failure function must be even for each shear stress as a basic requirement of objectivity. Therefore, the failure function in the form of a quadratic function as given in (1) is complete as far as homogeneous orthotropic materials are concerned.

The coefficients F_1, F_{11}, F_{44}, etc., in (1) should be determined by the strengths of the material obtained through standard or special experiments. In fact, most of them have been explicitly expressed in terms of conventional strengths. For instance, those coefficients that are of particular relevance to the present discussion are listed below as:

$$\begin{gathered} F_1 = \frac{1}{\sigma_{1t}^*} - \frac{1}{\sigma_{1c}^*},\ F_2 = \frac{1}{\sigma_{2t}^*} - \frac{1}{\sigma_{2c}^*},\ F_{11} = \frac{1}{\sigma_{1t}^*\sigma_{1c}^*},\ F_{22} = \frac{1}{\sigma_{2t}^*\sigma_{2c}^*}, \\ F_{44} = \frac{1}{\left(\tau_{23}^*\right)^2} \text{ and } F_{66} = \frac{1}{\left(\tau_{12}^*\right)^2} \end{gathered} \tag{2}$$

where σ_{1t}^* and σ_{1c}^* are the tensile and compressive strengths of the material in the direction along the fibres, respectively; σ_{2t}^* and σ_{2c}^* are the tensile and compressive strengths of the material in the direction transverse to the fibres, respectively; and τ_{23}^* and τ_{12}^* are the shear strength transverse and parallel to the fibres, respectively. These conventional strength properties of the composite should be obtained when appropriate specimens are loaded under uniaxial stress states or pure shear stress states in their material's principal axis according to available standards [7,8]. However, the coefficients of the interactive terms

F_{23}, F_{13} and F_{12} have been left somewhat loose. Restrictions were introduced in [1] on the ranges of their values as:

$$\begin{aligned} F_{23}^2 &< F_{22}F_{33} \\ F_{13}^2 &< F_{11}F_{33} \\ F_{12}^2 &< F_{11}F_{22}. \end{aligned} \tag{3}$$

Given the inequality form of (3), these restrictions are insufficient to fully quantify F_{23}, F_{13} and F_{12}. However, at this point, certain mathematically inaccurate statements were made in [1], and they need to be corrected, since the clarification of this matter will eventually lead to the development that will be presented in this paper.

In general, the failure envelope is defined in a six-dimensional stress space. If (1) is re-written as:

$$\begin{aligned} &F_1\sigma_1 + F_2\sigma_2 + F_3\sigma_3 + F_{11}\sigma_1^2 + F_{22}\sigma_2^2 + F_{33}\sigma_3^2 + 2F_{23}\sigma_2\sigma_3 + 2F_{13}\sigma_1\sigma_3 + 2F_{12}\sigma_1\sigma_2 \\ &= 1 - \left(F_{44}\tau_{23}^2 + F_{55}\tau_{13}^2 + F_{66}\tau_{12}^2\right) = K \end{aligned} \tag{4}$$

it can be considered that the presence of any non-vanishing shear stress can be equivalently considered as a reduction in the value of the constant term in (1) from 1 to a smaller value, denoted as K, which could be negative under some stress states ($K = 0$ implies that failure is due to shear stresses alone). The nature of the failure function can be sufficiently comprehensively discussed within the three-dimensional subspace of direct stresses.

It was claimed in [1] that the conditions of (3) were to keep the failure envelope closed on the basis that materials should exhibit finite strengths. This was misleading on two accounts. Firstly, the conditions of (3) do not ensure the closeness of the failure envelope. According to analytic geometry [9], each of the three conditions ensures that the intersection of the failure envelope with the respective coordinate plane as a locus is an ellipse. Take coordinate plane 1–2 for example. In this case, the locus is obtained when $\sigma_3 = 0$ as:

$$F_1\sigma_1 + F_2\sigma_2 + F_{11}\sigma_1^2 + F_{22}\sigma_2^2 + 2F_{12}\sigma_1\sigma_2 = 1 - F_{66}\tau_{12}^2. \tag{5}$$

The condition $F_{12}^2 < F_{11}F_{22}$ ensures that (5) is an ellipse in the 1–2 plane [9]. The same argument can be made for the remaining two conditions in (3). The precise physical interpretation of (3) is that strength is finite under any plane stress condition in each of the principal planes of the material, which is a reasonable observation amongst all available experimental data, as was also argued in [6]. However, the conditions of (3) do not exclude the possibility of the failure envelope being open in the 3D space. Consider a real composite, for example, T300/PR-319, which was one of the composites involved in WWFE-II [10] and was also employed in [6] as one of the cases studied. With F_{12} given as:

$$F_{12} = -\frac{1}{2}\sqrt{F_{11}F_{22}} \tag{6}$$

which is the same as that introduced in [11], and the measured transverse shear strength as $\tau_{23}^* = 45$ MPa, the failure envelope (4) can be proven to be an ellipsoid in the subspace of direct stresses. It is well known that no strength properties can be practically obtained without experimental error, although clear indications of such errors are not usually provided in the published data. A rare example of published raw data for strength properties can be found in [12], where the variability was as high as 30%, although the authors used a different type of composite (AS4/8552). Suppose that there was a 10% experimental error in τ_{23}^* and its value dropped to 40.5 MPa whilst satisfying (3) perfectly well. Then, the failure envelope would turn into an hourglass-like hyperboloid. As will become clear later in this paper, logically, the conditions in (3) serve as a set of necessary conditions for the failure envelope to be closed, whilst the sufficient conditions provided later in this paper are more restrictive [9].

The second issue is that the claim made in [1] prohibited the openness of the failure envelope or any infinite strength. On one hand, a claim of an infinite strength under some specific conditions can be equally as scientifically unfounded as a claim of a finite strength under any condition; on the other hand, either claim could be a reasonable assumption under the appropriate considerations. If a material under a specific stress state could sustain a stress level much higher than its strength under any uniaxial stress state, it would not be unreasonable to assume the strength under that special condition was practically infinite. Realistically, the available means for testing materials under multiaxial stress states only allow for stress within a limited range. Without loading the material to failure, one cannot be certain whether the strength is infinite. In view of this, a more meaningful question to ask is not whether a claim of finite or infinite strength is correct, but whether a claim has been made consistently within the theoretical framework.

In [11], Tsai and Hahn introduced (6), and their justification was that the Tsai–Wu criterion should reproduce the von Mises criterion for isotropic materials of equal tensile and compressive strengths. If so, the claim of a finite strength under all conditions would lead to a logical contradiction, because the failure envelope for the von Mises criterion is in fact open [13], since it is a cylinder in the space of three principal stresses intersecting a coordinate plane in closed ellipses. Its axis corresponds to the hydrostatic stress state under which the material is assumed to have infinite strength.

An attempt to reconcile the contradictive issues described above was made in [6]. Without violating the conditions in (3), it was concluded that an elliptic paraboloidal envelope offers a reasonable compromise for logical consistency. Mathematically, there exists a unique elliptic paraboloid sitting on the borderline between the ellipsoids and the elliptic hyperboloids. Employing an elliptic paraboloid as the failure envelope, given its open appearance, allows infinite strength, but only under a unique stress ratio [6] in triaxial compression. There appears to be two ways to lock the failure envelope in an elliptic paraboloid. One method is to abandon (6) and to employ the expression of F_{12} as was obtained in [6]:

$$F_{12} = -\frac{1}{2}\sqrt{\delta}\sqrt{F_{11}F_{22}} \tag{7}$$

where

$$\delta = 4 - \frac{F_{44}}{F_{22}} = 4 - \frac{\sigma_{2t}^{*}\sigma_{2c}^{*}}{\left(\tau_{23}^{*}\right)^{2}}. \tag{8}$$

This leads to a method of determining F_{12} in terms of known strength properties, eliminating the need for an additional strength measured under combined stress states. However, a critical drawback of this approach is that δ as defined in (8) can be negative for some materials, as shown in [6]; based on this, it was concluded that the Tsai–Wu criterion was inapplicable to these materials, which remained an unsatisfactory aspect of this development.

Alternatively, an elliptic paraboloidal failure envelope can also be achieved by the slight modification of transverse strengths, given the wide variability in experimental data. This consideration results in the development that will be presented in this paper, leading to the determination of the transverse shear strength whilst delivering the necessary logical consistency. There have been a number of other approaches proposed to determine the transverse shear strength from the transverse tensile and/or compressive strengths, e.g., in [14–16], which will be discussed later in this paper, as well as those summarised in [6].

It is now clear that the conditions in (3) should be satisfied; however, the satisfaction of (3) is not enough. It should also be noted that the three conditions in (3) are equally critical in term of necessity, because violating any of them will allow an open locus in the respective coordinate plane, implying an infinite strength under a plane stress condition. The conditions in (3) are also equally restrictive for F_{23}, F_{13}, and F_{12} in specifying their ranges.

3. The Quadratic Failure Function for Transversely Isotropic Materials

Similar to [6], as well as in [1,3,14–16], the present paper will be confined only to transversely isotropic materials, which cover a large class of unidirectionally fibre-reinforced composites. In this case, the transverse isotropy offers the following relationships:

$$\begin{gathered} F_2 = F_3,\ F_{22} = F_{33},\ F_{13} = F_{12}, \\ F_{44} = 2(F_{22} - F_{23}) \text{ or } F_{23} = F_{22} - \tfrac{1}{2}F_{44} \text{ and } F_{55} = F_{66}. \end{gathered} \tag{9}$$

The Tsai–Wu criterion reduces to [1,6,11]:

$$\begin{gathered} F_1\sigma_1 + F_2(\sigma_2+\sigma_3) + F_{11}\sigma_1^2 + F_{22}\left(\sigma_2^2+\sigma_3^2\right) + 2F_{23}\sigma_2\sigma_3 + 2F_{12}\sigma_1(\sigma_3+\sigma_2) \\ +2(F_{22}-F_{23})\tau_{23}^2 + F_{66}\left(\tau_{13}^2+\tau_{12}^2\right) - 1 = 0 \end{gathered} \tag{10}$$

and there are seven seemingly independent coefficients $F_1,\ F_2,\ F_{11},\ F_{22}, F_{23}$ (or F_{44}), F_{12} and F_{66} to be determined from experimental data.

The relationships in (9) are a natural consequence of the transverse isotropy. However, a follow-up question would have a far-reaching effect. The third relationship in (9) suggests that F_{13} and F_{12} make equal contributions to the failure function as material properties, regardless of the stress state. The justification for this is, of course, the identical strength characteristics of the material in directions 2 and 3. In general, given the anisotropy of the material, F_{12} and F_{23} are not expected to be the same. However, should the difference between F_{12} and F_{23} be quantitatively associated with the degree of anisotropy? This is a question that has never been asked properly in the literature to the best of our knowledge, let alone has it been addressed. An attempt will be made and justified in this paper.

Having achieved what was presented in [6], the objective of the present paper is to address some intrinsic relationships in the quadratic failure function that will have profound implications for the understanding of the strength of transversely isotropic materials in general. The consequence is that a quadratic failure function can only accommodate five independent strength properties, and two out of the seven coefficients as involved in (10) can be expressed in terms of the remaining five in general, provided that the material is homogeneous and transversely isotropic, and the failure function is a single quadratic function.

4. Characteristics of the Quadratic Failure Function for Transversely Isotropic Materials

4.1. Necessary and Sufficient Conditions for an Elliptic Paraboloid

As was argued in Section 2, as well as in [6], the failure envelope should be allowed to be an elliptic paraboloid to maintain a basic level of consistency in the logic behind the Tsai–Wu criterion. The failure envelope for transversely isotropic materials in the three-dimensional subspace of direct stresses is given as:

$$\begin{gathered} F_1\sigma_1 + F_2(\sigma_2+\sigma_3) + F_{11}\sigma_1^2 + F_{22}\left(\sigma_2^2+\sigma_3^2\right) + 2F_{23}\sigma_2\sigma_3 \\ +2F_{12}\sigma_1(\sigma_3+\sigma_2) = K. \end{gathered} \tag{11}$$

This defines a quadric surface in the $\sigma_1 - \sigma_2 - \sigma_3$ space in general, of which the following invariants can be introduced [9].

$$A = \begin{vmatrix} F_{11} & F_{12} & F_{12} & \frac{1}{2}F_1 \\ F_{12} & F_{22} & F_{23} & \frac{1}{2}F_2 \\ F_{12} & F_{23} & F_{22} & \frac{1}{2}F_2 \\ \frac{1}{2}F_1 & \frac{1}{2}F_2 & \frac{1}{2}F_2 & -K \end{vmatrix} \tag{12}$$

$$J = \begin{vmatrix} F_{11} & F_{12} \\ F_{12} & F_{22} \end{vmatrix} + \begin{vmatrix} F_{22} & F_{23} \\ F_{23} & F_{22} \end{vmatrix} + \begin{vmatrix} F_{22} & F_{12} \\ F_{12} & F_{11} \end{vmatrix} = 2\left(F_{11}F_{22} - F_{12}^2\right) + F_{22}^2 - F_{23}^2 \tag{13}$$

$$D = \begin{vmatrix} F_{11} & F_{12} & F_{12} \\ F_{12} & F_{22} & F_{23} \\ F_{12} & F_{23} & F_{22} \end{vmatrix} = F_{11}{F_{22}}^2 + 2F_{23}{F_{12}}^2 - 2F_{22}{F_{12}}^2 - F_{11}{F_{23}}^2 = (F_{22} - F_{23})\Delta \quad (14)$$

where

$$\Delta = F_{11}(F_{22} + F_{23}) - 2F_{12}^2. \quad (15)$$

In general, quadric surfaces can be a range of different shapes, and for complete categorisation a few more invariants [9] are required. The above conditions are sufficient to determine if the surface is an elliptic paraboloid. Most quadric surfaces, such as paraboloidal cylinders and various hyperboloids, lead to physically impermissible behaviours, e.g., possible infinite strength under a plane stress condition, or infinite strengths at an infinite number of different stress conditions, some involving tensile stresses, etc., as elaborated in [6]. For instance, the elliptic cylinder defines the failure envelope for a class of materials that show infinite strength under a triaxial stress state at a certain stress ratio both in tension and compression—a typical ductile behaviour. It is inapplicable to modern composites, which are mostly brittle, i.e., the material failure characteristics show significant differences under tension and compression. The permissible quadric surfaces are limited to an ellipsoid and an elliptic paraboloid. As argued earlier and will be further elaborated later, an ellipsoid as a failure envelope for (11) cannot be the right choice.

The elliptic paraboloid, which is open at only one end and allows infinite strength only at one specific triaxial compressive stress ratio, remains the only viable choice. The conditions that are both necessary and sufficient for a quadratic function to be an elliptic paraboloid are [9]:

$$A < 0,\ J > 0 \text{ and } D = 0. \quad (16)$$

Whilst the inequalities are discriminants, the satisfaction of which will be shown later, the equation, i.e., $D = 0$, offers an additional relationship purely from a mathematical point of view, which will be exploited below. Given (14), the first condition in (3) rules out the possibility of $F_{22} - F_{23} = 0$; therefore, $D = 0$ is equivalent to $\Delta = 0$, i.e.,

$$F_{12}^2 = \frac{1}{2}F_{11}(F_{22} + F_{23}) \quad (17)$$

which agrees with what was obtained in [6]. This means that F_{23} and F_{12} are not both independent.

Given the relationship between F_{23} and F_{12} shown in (17), one could be tempted to shed the remaining responsibility of the full determination of F_{23} and F_{12} to experiments, i.e., to have one of these values measured experimentally. Unfortunately, there are no reliable experimental means to directly determine either of them. The best one can do is to determine the transverse shear strength, from which F_{44} can be evaluated, and then obtain F_{23} using the fourth relationship in (9). Then, F_{12} can be determined using (17), which is in line with [6]; however, as discussed in Section 2, this approach is inefficient.

On the other hand, the available resources in terms of mathematics and logic have not been exhausted. The consideration of mathematical and logical consistency will offer another much-desired condition, and will be pursued in the next subsection.

4.2. Determination of F_{12} and F_{23}

Equation (17) shows F_{12} as a function of F_{23}, which is a parabola on the F_{23}-F_{12} plane, as sketched in Figure 1. The domain of the function can be obtained from the first condition in (3). For transversely isotropic materials, it becomes:

$$F_{23}^2 < F_{22}^2. \quad (18)$$

In regard to the domain, the range of the function happens to coincide with the third condition in (3), i.e., $F_{12}^2 < F_{11}F_{22}$, given that F_{11} and F_{22} are both positive and, hence, $F_{23} < F_{22}$. The permissible values of F_{23} and F_{12} can only be within the green rectangle shown in Figure 1, with the parabola passing its two corners on the right and touching its side on the left.

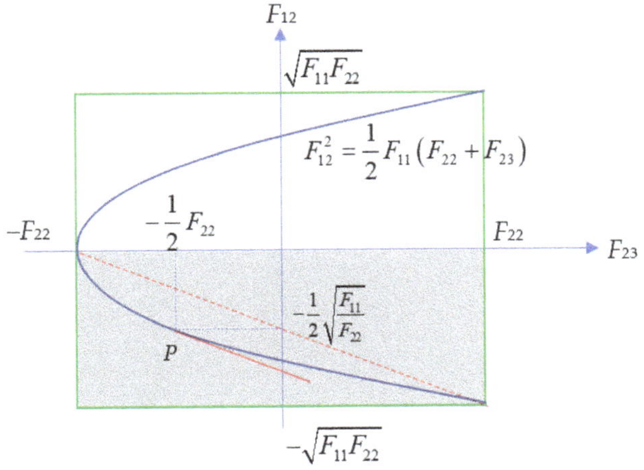

Figure 1. The relationship between F_{12} and F_{23} and their permissible ranges.

Any point on the parabola in Figure 1 satisfies (17). To determine F_{12} and F_{23} completely means to fix a point on the parabola, referred to as point P. This requires one additional consideration that can be obtained by examining (17) in the context of its physical implications. The coefficients F_{12} and F_{23} are expressed one way or another in terms of strength properties, which are obtained experimentally; hence, errors are inevitable. Any perturbation due to an experimental error in one of the coefficients will have to be met by a variation in the other in order to keep (17) satisfied. If P is selected close to the tip of the parabola, where dF_{12}/dF_{23} approaches infinity, it would require a large variation in F_{12} to correct a small perturbation in F_{23}, making the failure criterion extremely sensitive to small errors in the evaluation of F_{23}. Similarly, if P is placed close to the open end of the parabola, although the derivative is neither infinite nor zero, the failure criterion is bound to be more sensitive to F_{12} than to F_{23}. Between these two extremes, according to the mean-value theorem [9], there exists an unbiased point where the failure criterion will be equally sensitive to the perturbations in F_{12} and F_{23}. The question now is how to determine it logically.

The parabola shown in Figure 1 will be subject to further examination, as its intricacy has yet to be decoded. F_{12} as a function of F_{23} is apparently not single valued. The top and bottom halves of the parabola both provide a single valued branch. Following the relevant discussion in [6] on the sense of F_{12}, for the quadratic form (11) to be an elliptic paraboloid with an opening on the triaxial compression side, the value of F_{12} must be negative; therefore, it is the lower half of the parabola in Figure 1 that should be considered as the permissible branch for p to fall upon. For this branch, the legitimate range of F_{12} is $(0, -\sqrt{F_{11}F_{22}})$ as F_{23} varies within its range of $(-F_{22}, F_{22})$, covering the zone shaded in Figure 1. Both F_{12} and F_{23} can now be normalised with respect to the breadths of their single valued ranges as follows:

$$\widetilde{F}_{12} = \frac{F_{12}}{\sqrt{F_{11}F_{22}}} \text{ and } \widetilde{F}_{23} = \frac{F_{23}}{2F_{22}}. \tag{19}$$

$\widetilde{F}_{12}$ varies from 0 to −1 over a non-dimensional unit distance, while $\widetilde{F}_{23}$ takes its value from $-1/2$ to $1/2$, also covering a non-dimensional unit distance. Thus, both F_{12} and F_{23} are normalised in a consistent and unbiased manner. Accordingly, the expression of Δ as given in (15) can also be normalised as:

$$\widetilde{\Delta} = \frac{\Delta}{2F_{11}F_{22}} = \frac{1}{2} + \widetilde{F}_{23} - \widetilde{F}_{12}^2. \tag{20}$$

For $\widetilde{\Delta}$ to be equally sensitive to $\widetilde{F}_{12}$ and $\widetilde{F}_{23}$ in their unnormalized form, it is reasonable to expect that:

$$\frac{\partial\widetilde{\Delta}}{\partial\widetilde{F}_{12}} = \frac{\partial\widetilde{\Delta}}{\partial\widetilde{F}_{23}}. \tag{21}$$

From (20), one obtains:

$$\frac{\partial\widetilde{\Delta}}{\partial\widetilde{F}_{12}} = -2\widetilde{F}_{12} \text{ and } \frac{\partial\widetilde{\Delta}}{\partial\widetilde{F}_{23}} = 1 \tag{22}$$

Thus, (21) leads to:

$$\widetilde{F}_{12} = -\frac{1}{2} \text{ or } F_{12} = -\frac{1}{2}\sqrt{F_{11}F_{22}} \text{ as given in (6).} \tag{23}$$

Since $\Delta = 0$, its nondimensionalised form $\widetilde{\Delta}$ as defined in (20) should also vanish. Given (23), the relationship in (17) leads to:

$$\widetilde{F}_{23} = -\frac{1}{4} \text{ or } F_{23} = -\frac{1}{2}F_{22}. \tag{24}$$

Thus, both F_{12} and F_{23} are determined as shown above. Whilst the relationship in (17) represents a rigorous consideration from analytic geometry, the relationship in (21) results from the logical consideration of the unbiased sensitivity of the failure criterion to F_{12} and F_{23}. The normalisations made to F_{12} and F_{23} in (19) ensure that they can be compared as like with like.

From (24), given the fourth condition of (9) and the expressions of F_{44} and F_{22} in (2), a natural consequence is:

$$F_{44} = 2(F_{22} - F_{23}) = 3F_{22} \text{ or } \tau_{23}^* = \sqrt{\frac{\sigma_{2t}^*\sigma_{2c}^*}{3}}. \tag{25}$$

This suggests that amongst the three transverse strengths—tensile, compressive and shear—of a transversely isotropic material, only two are independent as long as the failure criterion is based on a single quadratic function. Therefore, based on the elaboration presented above, the fully rationalised Tsai–Wu criterion can be delivered as follows:

$$\begin{aligned} &F_1\sigma_1 + F_2(\sigma_2+\sigma_3) + F_{11}\sigma_1^2 + F_{22}(\sigma_2^2+\sigma_3^2-\sigma_2\sigma_3+3\tau_{23}^2) \\ &-\sqrt{F_{11}F_{22}}\sigma_1(\sigma_3+\sigma_2) + F_{66}(\tau_{13}^2+\tau_{12}^2) = 1 \end{aligned} \tag{26}$$

where the coefficients F_1, F_2, F_{11}, F_{22} and F_{66} are determined by conventional strength properties, as given in (2). It is clear that under triaxial compression at the following stress ratio:

$$\sigma_1 : \sigma_2 : \sigma_3 = -\sqrt{\frac{F_{22}}{F_{11}}} : -1 : -1 \tag{27}$$

whilst all shear stresses vanish, criterion (26) will never be satisfied, i.e., failure will never take place, implying an infinite strength under this special and unique stress state. In fact, the position holds even in presence of shear stresses. The triaxial compression at the ratio given in (27) has a reinforcing effect on the shear strengths. This can be seen clearly if (26) is re-written under this particular stress state as:

$$3F_{22}\tau_{23}^2 + F_{66}\left(\tau_{13}^2 + \tau_{12}^2\right) = 1 + \left(F_1\sqrt{\frac{F_{22}}{F_{11}}} + 2F_2\right)\sigma \tag{28}$$

where σ is the magnitude of the transverse stresses (i.e., non-negative) of the triaxial compressive stress state.

For 2D stress states in the 1–2 plane, (26) is simplified to:

$$F_1\sigma_1 + F_2\sigma_2 + F_{11}\sigma_1^2 + F_{22}\sigma_2^2 - \sqrt{F_{11}F_{22}}\sigma_1\sigma_2 + F_{66}\tau_{12}^2 = 1. \tag{29}$$

An important observation to make is that to predict the failure of a material under a 3D stress state from (26), the same set of strength properties is required as that for a 2D stress state from (29). It should be emphasised here that the rationalisation above has to be carried out on the criterion in its 3D form before it is reduced to 2D. Although (29) is a well-known form [11], its rational justification can be obtained only through the route of manipulating 3D stress states, as presented above.

4.3. Verification and Discussion

The point on the parabola given by (23) and (24) is special in multiple ways. It was argued above that it renders failure criterion (10) to be equally sensitive to F_{12} and F_{23} under the condition that the failure envelope is an elliptic paraboloid. The slope to the parabola at this point happens to be equal to the slope of the secant between the two extremes of the single valued branch concerned, i.e., the tip of the parabola and the bottom right end. The tangent and secant are marked in Figure 1 by solid and dashed red lines, respectively. Therefore, this is the point whose existence has been asserted by the mean-value theorem [9].

In order to interpret the results above and to verify the normalisations introduced in (19), all the terms in the failure criterion (10) are normalised as follows, which is similar to procedure used in [11].

$$\widetilde{F}_1\widetilde{\sigma}_1 + \widetilde{F}_2(\widetilde{\sigma}_2 + \widetilde{\sigma}_3) + \widetilde{\sigma}_1^2 + \widetilde{\sigma}_2^2 + \widetilde{\sigma}_3^2 - \widetilde{\sigma}_2\widetilde{\sigma}_3 - \widetilde{\sigma}_1(\widetilde{\sigma}_3 + \widetilde{\sigma}_2) + \widetilde{\tau}_{23}^2 + \widetilde{\tau}_{13}^2 + \widetilde{\tau}_{12}^2 = 1 \tag{30}$$

where

$$\begin{array}{c} \widetilde{\sigma}_1 = \sqrt{F_{11}}\sigma_1,\ \widetilde{\sigma}_2 = \sqrt{F_{22}}\sigma_2,\ \widetilde{\sigma}_3 = \sqrt{F_{22}}\sigma_3, \\ \widetilde{\tau}_{23} = \sqrt{F_{44}}\tau_{23} = \sqrt{2(F_{22} - F_{23})}\tau_{23}, \\ \widetilde{\tau}_{13} = \sqrt{F_{66}}\tau_{13},\ \widetilde{\tau}_{12} = \sqrt{F_{66}}\tau_{12} \\ \widetilde{F}_1 = \frac{F_1}{\sqrt{F_{11}}},\ \widetilde{F}_2 = \frac{F_2}{\sqrt{F_{22}}} \end{array} \tag{31}$$

The values of $\widetilde{F}_{12}$ and $\widetilde{F}_{23}$ as obtained in (23) and (24) are also incorporated above. All terms of the similar nature tend to contribute to F in the same way, whilst the differences due to the anisotropy of the material in all second order terms are absorbed in the normalisation scheme. The contributions of F_{12} and F_{23} obtained in (23) and (24) have indeed been equalised in their normalised form.

The outcomes presented in (23) and (24) rest heavily on the normalisations in (19). If one is prepared to accept (19) as the correct normalisations, which equalise the contributions of $\widetilde{F}_{12}$ and $\widetilde{F}_{23}$ to the failure function as the major premise for the deductive reasoning as presented above, conclusions (23) and (24) will be the logical consequence.

The existence of a relationship amongst transverse strengths, as shown in (24), can also be argued as follows. There are only two independent strength properties for isotropic materials of different tensile and compressive strengths for 3D stresses in general, as well as for any plane stress conditions as special cases, according to the established Raghava–Caddell–Yeh criterion [17], which has been more rationally formulated in [18]. When a transversely isotropic material is under a plane stress condition in its transverse plane, i.e., the 2–3 plane, the material is effectively isotropic; therefore, the Raghava–Caddell–Yeh criterion in its 2D form should be applicable. If so, it will only require two strength properties. For the Tsai–Wu criterion to reproduce the Raghava–Caddell–Yeh criterion under this condition, the third strength property cannot be independent.

With F_{12} and F_{23} as determined in (23) and (24), one can verify conditions in (16) as follows:

$$A = -\frac{3}{16}F_{22}\left(\sqrt{F_{22}}F_1 + 2\sqrt{F_{11}}F_2\right)^2 < 0 \text{ provided } \frac{-F_1}{\sqrt{F_{11}}} \neq 2\frac{F_2}{\sqrt{F_{22}}} \tag{32}$$

$$J = F_{22}(2F_{11} + F_{22}) - \frac{1}{2}F_{11}F_{22} - \frac{1}{4}F_{22}^2 = \frac{3}{4}F_{22}(2F_{11} + F_{22}) > 0 \tag{33}$$

$$D = (F_{22} - F_{23})\left(F_{11}(F_{22} + F_{23}) - 2F_{12}^2\right) = \frac{3}{2}F_{22}\left(\frac{1}{2}F_{11}F_{22} - \frac{1}{2}F_{11}F_{22}\right) = 0 \tag{34}$$

Hence, (11) defines an elliptic paraboloid. The inequality of (32) is met by most composites of practical importance and the only alternative is $A = 0$, which arises only under special conditions, as will be discussed later.

As a further verification, reducing the materials to isotropic materials of unequal tensile and compressive strengths, whilst (33) and (34) still hold:

$$A = -\frac{27}{16}F_{11}^2F_1^2 < 0 \text{ since } F_1 \neq 0. \tag{35}$$

In this case, Equation (10) leads to:

$$\begin{aligned}&\left(\frac{1}{\sigma_t^*} - \frac{1}{\sigma_c^*}\right)(\sigma_1 + \sigma_2 + \sigma_3)\\&+\frac{1}{\sigma_t^*\sigma_c^*}\left(\sigma_1^2 + \sigma_2^2 + \sigma_3^2 - \sigma_2\sigma_3 - \sigma_1\sigma_3 - \sigma_1\sigma_2 + 3\tau_{23}^2 + 3\tau_{13}^2 + 3\tau_{12}^2\right) = 1\end{aligned} \tag{36}$$

where σ_t^* and σ_c^* are the tensile and compressive strengths of the isotropic material, respectively. This reproduces the Raghava–Caddell–Yeh criterion [17], for which the failure envelope is a circular paraboloid demonstrating infinite strength against hydrostatic compression, but not hydrostatic tension.

As a case of further specialisation for isotropic materials of equal tensile and compressive strengths, i.e., $\sigma_t^* = \sigma_c^* = \sigma^*$, (36) further reduces to the von Mises criterion [13]:

$$\sigma_1^2 + \sigma_2^2 + \sigma_3^2 - \sigma_2\sigma_3 - \sigma_1\sigma_3 - \sigma_1\sigma_2 + 3\tau_{23}^2 + 3\tau_{13}^2 + 3\tau_{12}^2 = (\sigma^*)^2 \tag{37}$$

where σ^* is the strength of the material, which the same for both tension and compression. This corresponds to the case when the condition for $A > 0$ does not hold. Given the expression of A in (32), the only alternative is $A = 0$. In this case, (10) turns into a cylindrical surface according to [9]. It can be seen that a cylindrical surface can be mathematically considered as an extreme case of an elliptic paraboloidal surface, allowing the von Mises criterion to be seen as a special case of the Tsai–Wu criterion.

As a further point of discussion, (24) shows that $\delta = 1$ according to (8). Equation (7), as obtained in [6], naturally leads to (23), reproducing (6) as Tsai and Hahn suggested in [11]. However, the only justification given in [11] was that it allowed the Tsai–Wu criterion to reproduce the von Mises criterion for isotropic materials of equal tensile and compressive strengths, whereas in this paper, (23) and (24) were obtained simultaneously from systematic and logical deduction. In [6], it was suggested that δ was a material constant that could vary from between materials. Relationship (24) asserts further that δ is a universal constant, i.e., $\delta = 4 - \frac{\sigma_{2t}^*\sigma_{2c}^*}{\left(\tau_{23}^*\right)^2} = 1$, for all transversely isotropic materials, brittle or ductile, including completely isotropic materials as a special case. This is apparently observed in isotropic materials obeying the Raghava–Caddell–Yeh criterion [17] and the von Mises criterion [13], as special cases. Any variation in δ, similar to those shown in [6], should be attributed to the variability of the experimentally measured transverse strength properties. This resembles a universal relationship amongst Young's moduli and Poisson's ratios, $\frac{\nu_{12}E_2}{\nu_{21}E_1} = 1$, although the underlying considerations are rather different. Raw experimental data would show significant variability as well if they were all independently measured.

4.4. Comparisons of Predicted Transverse Shear Strengths with Measured Values and Those from Other Theories

In the Puck failure criterion [14,15], the transverse shear strength is a derived quantity that is equal to the transverse tensile strength. In comparison to (25), it corresponds to the special case of $\sigma_{2c}^* = 3\sigma_{2t}^*$. The transverse shear strengths obtained according to the Puck criterion and the present derivation are compared in Table 1 over a range of composites from [10,19,20] where experimental data are available. Relative to the measured values of τ_{23}^* in the third row of Table 1, those obtained from (25) outperform those from the Puck criterion, except for S-2 glass/Epoxy 2 and G40-800/5026, as highlighted in Table 1. For these two cases, one is only marginally worse than that predicted by the Puck criterion, whilst the other is identical to that predicted from the Puck criterion, because $\sigma_{2c}^* = 3\sigma_{2t}^*$ holds, although the experimental shear strength of 57 MPa is nearly 20% below $\sqrt{\frac{\sigma_{2t}^*\sigma_{2c}^*}{3}}$ for this particular material.

Given that $\sqrt{\frac{\sigma_{2t}^*\sigma_{2c}^*}{3}}$ is the strength of the material under pure transverse shear, as predicted from failure criterion (26), the degree of its agreement with the experimentally obtained transverse shear strength—i.e., the comparison between the measured values in the third row of Table 1—and that predicted from (25), as shown in the fourth rows of Table 1, can serve as a basic level of validation for the criterion (26). When applying (26) to problems involving general stress states, one should be prepared for discrepancies of a magnitude comparable with those between the figures in the third and fourth rows in Table 1, as an indication of the level of accuracy that criterion (26) is capable of offering.

Another relationship amongst transverse strengths can be found in [16], and is given as follows:

$$(\tau_{23}^*)^2 = \left(\frac{1+\frac{\sigma_{2t}^*}{\sigma_{2c}^*}}{3+5\frac{\sigma_{2t}^*}{\sigma_{2c}^*}}\right)\sigma_{2t}^*\sigma_{2c}^* \text{ or } \frac{(\tau_{23}^*)^2}{\sigma_{2t}^*\sigma_{2c}^*} = \frac{1+\frac{\sigma_{2t}^*}{\sigma_{2c}^*}}{3+5\frac{\sigma_{2t}^*}{\sigma_{2c}^*}}. \quad (38)$$

It is apparently a more complicated expression than that given in (25). The derivation in [16] was based on the following failure criterion stemming from Hashin's matrix failure criterion [3] under a plane stress state in the transverse plane before it was split into tensile and compressive modes.

$$F_2(\sigma_2+\sigma_3) + F_{22}(\sigma_2+\sigma_3)^2 + F_{44}\left(\tau_{23}^2 - \sigma_2\sigma_3\right) = 1. \quad (39)$$

This is an incomplete quadratic form. The contributions from fibre direction as well as the interactive term associated with F_{12} have been excluded a priori.

The interaction between the stresses represented by F_{12} resembles the role of Poisson's ratios in the generalised Hooke's law in the theory of elasticity. It might be true that in the solutions to many elastic problems, the contributions from Poisson's ratios are limited in terms of magnitude. Whilst ignoring such contributions would not lead to excessive error, the absence of Poisson's ratios would prevent the appropriate understanding of many important mechanical behaviours of materials, such as the anticlastic behaviour, free edge effects, etc.

Table 1. Experimentally measured strengths and predicted transverse strengths according to different theories, all in MPa.

Category	Row No.	Strength	Type of Composite Considered								
			AS4 3501-6	**T300 BSL914C**	**E-Glass LY556**	**E-Glass MY750**	**IM7 8551-7**	**T300 PR-319**	**A-SE poxy 1**	**S-2 Glass Epoxy 2**	**G40-800 5026**
Experimentally measured strengths	1	σ_{2t}^*	48	27	35	40	68	40	38	56.5	70
	2	σ_{2c}^*	200	200	114	145	185	125	150	180	210
	3	τ_{23}^*	55	N/A	50	50	57	45	50	40	57
Predicted τ_{23}^* from σ_{2t}^* and σ_{2c}^*	4	Present from Equation (25).	56.57	42.43	36.47	43.97	64.76	40.82	43.59	58.22	70.00
	5	Puck [14,15], which is equal to σ_{2t}^*.	48	27	35	40	68	40	38	56.5	70
	6	Christenson, as shown in Equation (39) [16].	50.93	43.82	33.50	42.56	58.59	36.91	41.90	54.48	56.12

Similarly, to reveal the special relationships elaborated in this paper, it is important to incorporate the interactive terms in the failure criterion. The approach to treating the interactive terms can also be viewed as a meaningful indicator of the consistency of the formulation. For instance, in the well-known Hashin criterion [3], the interactive term F_{12} was ignored based on the assumption that failure is determined by stresses on the failure plane, which was inherited from the Mohr criterion [21] for isotropic materials; however, this particular assumption is inapplicable to anisotropic composites, as was argued in [22]. Whilst the numerical errors could be insignificant in many cases, as Hashin also stated in [3], the exclusion of such interactions would not allow the intricate relationships obtained in this paper to be revealed. From another perspective, including one interactive coefficient (F_{23}) in the failure function, whilst excluding another (F_{12}), as in [3], can at least be viewed as a degree of inconsistency. The numerical closeness of the results from a theory to experimental data in one aspect or another is important; however, we believe that the mathematical and logical consistency of the theory is even more important.

The numerical values obtained from (38) [16] over the same range of materials as above are also shown in the fifth row of Table 1. Relative to the experimental data in the third row of Table 1, the predictions from (25) listed in the fourth row of Table 1 outperform those from (38) for five out of eight materials. The trend is further shown graphically in Figure 2, where the ratio $\frac{(\tau_{23}^*)^2}{\sigma_{2t}^*\sigma_{2c}^*}$ is plotted as a function of the ratio $\frac{\sigma_{2t}^*}{\sigma_{2c}^*}$. The curve corresponding to (38) does not seem to be more representative than the straight line obtained according to (25), understanding the variability in measured transverse strengths as argued in [16]. The predictions are not more relevant to carbon composites (black symbols) than to glass composites (green symbols) either, as was claimed in [16].

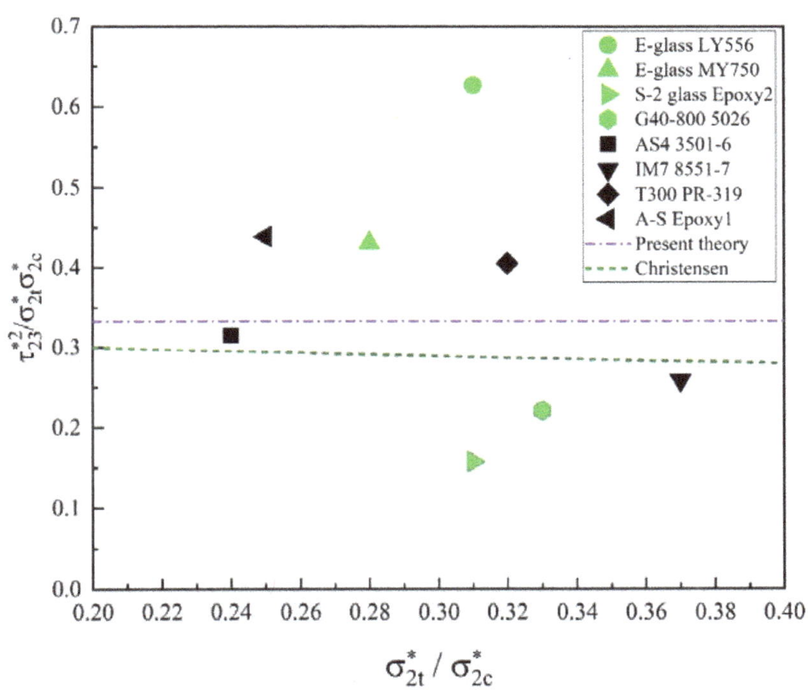

Figure 2. Graph of $\frac{(\tau_{23}^*)^2}{\sigma_{2t}^*\sigma_{2c}^*}$ versus $\frac{\sigma_{2t}^*}{\sigma_{2c}^*}$.

Logically, the major premise underlying (38) is failure criterion (39), which is at its best a special form of the criterion based on the full quadratic failure function from which (26) is obtained. There is no reason to expect (39) to be more representative than (26). In [16], efforts were made to eliminate an independent strength property from (39) whilst keeping the manipulations all within the transverse plane.

5. Retrospection and Prospection

According to analytic geometry [9], the sufficient conditions for the failure envelope of transversely isotropic materials to be closed, i.e., to be an ellipsoidal surface, are:

$$A < 0;\ J > 0;\ D \neq 0 \text{ and } DI > 0 \tag{40}$$

where A, J and D are given in (12)–(14), respectively, and the following is another invariant of quadratic form (11):

$$I = F_{11} + 2F_{22} \tag{41}$$

Since I is always positive, given the expression of D in (14), and $F_{22} - F_{23}$ is always positive according to the fourth equation in (9) (because F_{44} is always positive), conditions $D \neq 0$ and $DI > 0$ can be re-written as:

$$\Delta = \frac{1}{2}F_{11}(3F_{22} - F_{44}) > 0 \text{ i.e., } 3F_{22} - F_{44} > 0 \text{ given } F_{11} > 0. \tag{42}$$

With F_{12} as given in (23), if the transverse shear strength τ_{23}^* (and hence F_{44}) was left as an independent property, (42) could not be satisfied in general. An ellipsoid could be guaranteed provided that τ_{23}^* was sufficiently large. In the case of isotropic materials obeying the von Mises criterion, as a special case of transversely isotropic materials, $D = \Delta = 0$, which satisfies (24) or (25) under the condition of (23), but violates (42). In other words, (42) would prohibit (11) to degenerate to the von Mises criterion as a special case for isotropic materials of equal tensile and compressive strengths, because $D = 0$ as a necessary condition is shared by cylinders and elliptic paraboloids, but not by ellipsoids, according to analytic geometry. It is clear now that having a closed failure envelope and having the von Mises criterion as a special case are two mutually exclusive propositions. Sometimes, it only requires a very small experimental or data processing error to nudge D from positive to negative; then, the nature of the failure envelope would change dramatically from an ellipsoid to an elliptic paraboloid or a hyperboloid. Whilst the imposition of (17) alone eliminates the possibility of the failure envelope being an ellipsoid or hyperboloid, errors in measured transverse strengths could still make the elliptic paraboloid imaginary according to analytic geometry. Relationship (25), in addition to (23), not only ensures that the failure envelope is a real and unique elliptic paraboloid, but also offers a way to mitigate experimental errors amongst measured transverse strengths. This illustrates a perfect example of how mathematical consistency brings insight into physical problems.

The conventional Tsai–Wu criterion could degenerate to the von Mises criterion for isotropic materials of equal tensile and compressive strengths [11] because the failure envelope in the original Tsai–Wu criterion was not limited to an ellipsoid, no matter how much it was desired; it could be ellipsoid, elliptic paraboloid and different hyperboloids depending on the value of the transverse shear strength τ_{23}^*, whilst satisfying the inequalities in (3). The necessary conditions of (3) were far too loose to impose such a restriction. The controlling factor of the shape of the failure envelope rests on the transverse shear strength τ_{23}^*, which affects the values of various invariants, in particular D. Narrowing the failure envelope down to an elliptic paraboloid, as proposed in this paper, can help to filter out many physically prohibitive scenarios whilst restoring the mathematical and logical consistency of the Tsai–Wu criterion. This is the spirit of rationalisation.

The fully rationalised Tsai–Wu criterion for general 3D stress states is presented in (26). For this criterion, the transverse shear strength is not a required strength property, but it can be predicted from the criterion. There will inevitably be a discrepancy between the predicted value and the measured value. The available independently measured transverse shear strength should be used to correct the experimental errors in other transverse strength properties, instead of being employed as an independent strength property for the failure criterion based on a quadratic failure function. It should be employed to correct the transverse tensile and compressive strengths in order to minimise systematic errors in these measured strengths in the same way that the experimentally measured ν_{21} should be employed to correct ν_{12}, E_1 and E_2, rather than being incorporated in the generalised Hooke's law. A practical method of such corrections will be addressed in a subsequent publication.

The 2D version of the fully rationalised Tsai–Wu criterion is given in (29). Although it looks identical to its familiar form, as presented in [11], the transverse tensile and compressive strengths required to evaluate F_{22} should be understood as their corrected values using the transverse shear strength, whenever available. The rationalisation still makes a difference to the 2D Tsai–Wu criterion but in an implicit way.

The failure criterion for transversely isotropic materials based on a single quadratic failure function can now be considered fully established, with a complete understanding of the nature of a quadratic failure function having been achieved mathematically. It can be fully and logically defined with five independent strength properties, namely tensile and compressive strengths in the directions along fibres and transverse to fibres, and the in-plane shear strength. All coefficients involved in the failure function can be expressed with mathematical rigour and logical consistency. Any additional strength property employed as an independent one will compromise the consistency of the theoretical framework, even if they were experimentally measured.

Having established the failure criterion (26) as the fully rationalised Tsai–Wu criterion, any significant deficiency or genuine discrepancy with experimental observations can now be confidently attributed to the lack of representativeness of the basic assumptions underlying the criterion:

(1) The transverse isotropy and homogeneity of the material;
(2) Failure function being a single quadratic function.

Any deviation from these ideal positions will inevitably have an effect on the outcomes and should be anticipated in assessing the accuracy of the predictions. The use of these assumptions brings convenience but also restrictions. The obtained criterion can only be as accurate as its underlying assumptions allow. The consistent use of a quadratic function for a failure criterion helps to eliminate undue errors and anomalies due to mathematical and logical oversights, but not the generic deficiency due to these assumptions themselves.

The limitations of using a single quadratic function as the failure function could be improved whilst remaining within the framework of a phenomenological approach in one of two ways:

(1) To use a higher order polynomial for the failure function;
(2) To partition the stress space into subspaces and to use a quadratic function in each subspace.

Either way, the number of independent strength properties will inevitably increase. One should always bear in mind how the newly introduced properties are to be determined, experimentally or otherwise. Abandoning phenomenological approaches completely or discontinuing their improvement before alternatives are established is partially responsible for the confused state of the art on the subject of composite failure criteria. With consistency established, one would be in a position to explore meaningful ways of determining strength properties as required for alternatives to experimental measurements, for instance, molecular dynamics as investigated in [23], although the material used in [23] was not transversely isotropic.

The first option above seems to pose formidable difficulties, as the understanding of the analytic geometry of polynomial functions higher than the second order in multidimensional spaces has been very limited and is certainly far less than that of quadratic functions. This is not an attractive direction forward practically.

The second option is likely to offer a practical way forward. The Hashin criterion [3] can be seen as an attempt in this direction, followed by a wide range of different attempts, including those in [15,24]. However, it is the authors' view that an appropriate rationalisation is necessary for the Hashin criterion before a fundamental breakthrough can be expected. This will be the objective of another one of our subsequent publications.

In the literature, there is lack of failure criteria for genuinely orthotropic materials, apart from the maximum stress/strain criteria. The Tsai–Wu criterion was nominally proposed for orthotropic materials. However, without the appropriate means to determine the coefficients to the interactive terms, it has never been applied seriously beyond transversely isotropic materials. This is a problem we will address in another subsequent publication.

6. Conclusions

Using a complete quadratic failure function, there is only one rational choice for the failure envelope—an elliptic paraboloid, as a single-sided open surface that allows infinite strength under a unique stress ratio in triaxial compression. Any alternative would lead to contradictions one way or another. This quadratic failure function can only accommodate five independent strength properties. Seven were initially introduced in the original Tsai–Wu criterion [1] after considerations of transverse isotropy. A relationship was established between the two interactive coefficients F_{12} and F_{23} based purely on analytic geometry. It ensures that the failure envelope is an elliptic paraboloid. A further relationship between these coefficients can be obtained from a logical consideration that the sensitivity of the failure criterion to these two coefficients should be unbiased. Thus, they can both be uniquely determined analytically as $F_{12} = -\frac{1}{2}\sqrt{F_{11}F_{22}}$ and $F_{23} = -\frac{1}{2}F_{22}$. Although the former is well known, it can only be fully justified in presence of the latter, which can be re-written as $3(\tau_{23}^*)^2 = \sigma_{2t}^*\sigma_{2c}^*$, given the transverse isotropy of the material. This introduces a relationship amongst transverse strengths as a natural consequence of the rationalisation; the relationship is an intrinsic property of the quadratic failure function for transversely isotropic materials. Failing to comply with this relationship compromises the consistency of the criterion based on a single quadratic failure function.

Due to the high variability in the measured transverse strengths, the obtained relationship amongst transverse strengths may not be perfectly represented in the available experimental data. In presence of all three transverse strengths independently obtained experimentally, the transverse shear strength can be employed to validate the relationship in (25), i.e., $3(\tau_{23}^*)^2 = \sigma_{2t}^*\sigma_{2c}^*$. The failure criterion can only be as accurate as the agreement between τ_{23}^* and $\sqrt{\frac{\sigma_{2t}^*\sigma_{2c}^*}{3}}$.

The fully rationalised Tsai–Wu criterion is given in (26) for 3D stress states. Its 2D version in (29) looks identical to its familiar form. Although no changes were made to (29), the rationalisation offers a firm basis for (29), and the determination of F_{12} no longer relies on experimental data fitting or other empiricism, but on a rigorous deduction from the mathematical and logical consequences of the basic assumptions introduced instead.

The required five independent strength properties are limited to conventional and widely available tensile and compressive strengths in the directions along and transverse to fibres and the in-plane shear strength for the failure predictions of both 2D and 3D stress states.

The most significant outcome of the present paper is the achievement of a thorough understanding of the nature of the quadratic failure function, with its intrinsic relationships having been revealed. Obeying these relationships ensures the self-consistency of the failure criterion. This should conclude a phase of investigations on the subject of failure criteria involved in recent publications [6,16,25] based on a quadratic failure function, as far as transversely isotropic materials are concerned. The failure criterion given in (26) can be employed in design and analysis with confidence within its applicability defined by its assumptions, viz. the transverse isotropy and homogeneity of the material, and the failure function being a single quadratic function.

Author Contributions: Conceptualization, S.L.; methodology, S.L. and E.S.; validation, S.L., M.X. and E.S; formal analysis, S.L.; investigation, S.L.; writing—original draft preparation, S.L.; writing—S.L. and E.S.; visualization, S.L. All authors have read and agreed to the published version of the manuscript.

Funding: This research received no external funding.

Institutional Review Board Statement: Not applicable.

Informed Consent Statement: Not applicable.

Data Availability Statement: There is no additional data.

Acknowledgments: The second author wishes to acknowledge the financial support from CSC, China and the scholarship from the Faculty of Engineering, the University of Nottingham.

Conflicts of Interest: The authors declare no conflict of interest.

References

1. Tsai, S.W.; Wu, E.M. A general theory of strength for anisotropic materials. *J. Compos. Mater.* **1971**, *5*, 58–80. [CrossRef]
2. Gol'denblat, I.I.; Kopnov, V.A. Strength of glass-reinforced plastics in the complex stress state. *Mekhanika Polim.* **1965**, *7*, 70. [CrossRef]
3. Hashin, Z. Failure criteria for unidirectional fiber composites. *J. Appl. Mech.* **1980**, *47*, 329–334. [CrossRef]
4. Li, S.; Sitnikova, E. A critical review on the rationality of popular failure criteria for composites. *Compos. Commun.* **2018**, *8*, 7–13. [CrossRef]
5. Li, S.; Sitnikova, E. *Representative Volume Elements and Unit Cells—Concepts. Theory, Applications and Implementation*; Woodhead Publishing Series in Composites Science and Engineering; Elsevier: Duxford, UK, 2019.
6. Li, S.; Sitnikova, E.; Liang, Y.; Kaddour, A.-S. The Tsai-Wu failure criterion rationalised in the context of UD composites. *Compos. Part A Appl. Sci. Manuf.* **2017**, *102*, 207–217. [CrossRef]
7. *BS EN ISO 527*; Plastics—Determination of Tensile Properties—Part 4: Test Conditions for Isotropic and Orthotropic Fibre-Reinforced Plastic Composites. The British Standards Institution: London, UK, 1997.
8. *ASTM D3039/D3039M-14*; Standard Test Method for Tensile Properties of Polymer Matrix Composite Materials. ASTM International: West Conshohocken, PA, USA, 2014.
9. Korn, G.A.; Korn, T.M. *Mathematical Handbook for Scientists and Engineers*; Sections 2.4, 3.5 & 4.7; McGraw-Hill: New York, NY, USA, 1968.
10. Kaddour, A.S.; Hinton, M.J. Input data for test cases used in benchmarking triaxial failure theories of composites. *J. Compos. Mater.* **2012**, *46*, 2295–2312. [CrossRef]
11. Tsai, S.W.; Hahn, H.T. *Introduction to Composite Materials*; Technomic Publishing Company: Westport, CT, USA, 1980.
12. Clarkson, E. *Hexcel 8552 AS4 Unidirectional Prepreg Qualification Statistical Analysis Report*; FAA Special Project no SP4614WI-Q, Report No NCP-RP-2010-008 Rev D; Wichita State University: Wichita, KS, USA, 2011.
13. Von Mises, R. *Mechanik der Festen Körper im Plastisch-Deformablen Zustand*; Nachrichten von der Gesellschaft der Wissenschaften zu Göttingen, Mathematisch-Physikalische Klasse; Gesellschaft der Wissenschaften zu Göttingen: Göttingen, Germany, 1913; pp. 582–592.
14. Puck, A.; Schurmann, H. Failure analysis of FRP laminates by means of physically based phenomenological models. *Compos. Sci. Tech.* **1998**, *58*, 1045–1067. [CrossRef]
15. Knops, M. *Analysis of Failure in Fiber Polymer Laminates. The Theory of Alfred Puck*; Springer: Berlin/Heidelberg, Germany, 2008.
16. Christensen, R.M. Completion and closure on failure criteria for unidirectional fiber composite materials. *ASME J. Appl. Mech.* **2013**, *81*, 011011. [CrossRef]
17. Raghava, R.; Caddell, R.M.; Yeh, G.S.Y. The macroscopic yield behaviour of polymers. *J. Mater. Sci.* **1973**, *8*, 225–232. [CrossRef]
18. Christensen, R.M. *The Theory of Materials Failure*; Oxford University Press: Oxford, UK, 2013.
19. Soden, P.D.; Hinton, M.J.; Kaddour, A.S. Lamina properties, lay-up configurations and loading conditions for a range of fibre-reinforced composite laminates. *Compos. Sci. Technol.* **1998**, *58*, 1011–1022. [CrossRef]
20. Kaddour, A.S.; Hinton, M.J.; Smith, P.A.; Li, S. Mechanical properties and details of composite laminates for the test cases used in the third world-wide failure exercise. *J. Compos. Mater.* **2013**, *47*, 2427–2442. [CrossRef]
21. Mohr, O. Welche Umstände bedingen die Elastizitätsgrenze und den Bruch eines Materials? *Civilingenieur* **1900**, *44*, 1572–1577.
22. Li, S. A reflection on the Mohr failure criterion. *Mech. Mater.* **2020**, *148*, 103442. [CrossRef]
23. Zho, Y.; Hu, M. Mechanical behaviors of nanocrystalline Cu/SiC composites: An atomistic investigation. *Comput. Mater. Sci.* **2017**, *129*, 129–136. [CrossRef]
24. Pinho, S.T.; Dávila, C.G.; Iannucci, L.; Robinson, P. *Failure Models and Criteria for FRP under in-Plane or Three-Dimensional Stress States including Shear Non-Linearity*; NASA/TM-2005-213530; NASA Langley Research Center: Hampton, VA, USA, 2005.
25. Christensen, R.M. Failure criteria for fiber composite materials, the astonishing sixty year search, definitive usable results. *Compos. Sci. Tech.* **2019**, *182*, 107718. [CrossRef]

Journal of
Composites Science

Article

The Direct Sandwich Composite Molding (D-SCM) Process: Sandwich Manufacturing and Characterization

Felix Behnisch [1,*], Jonathan Brütsch [1], Henrik O. Werner [2,3] and Frank Henning [1,3]

1 Fraunhofer-Institute for Chemical Technology ICT, Joseph-von-Fraunhofer Str. 7, 76327 Pfinztal, Germany; jonathan.bruetsch@gmail.com (J.B.); frank.henning@ict.fraunhofer.de (F.H.)
2 Institute for Applied Materials–Materials Science and Engineering (IAM-WK), Karlsruhe Institute of Technology (KIT), 76131 Karlsruhe, Germany; henrik.werner@kit.edu
3 Institute of Vehicle System Technology (FAST), Karlsruhe Institute of Technology (KIT), 76131 Karlsruhe, Germany
* Correspondence: felix.behnisch@ict.fraunhofer.de

Abstract: Sandwich structures benefit from the geometrical stiffening effect due to their high cross-sectional area moment of inertia. Transferred to carbon fiber-reinforced plastic (CFRP) components, the needed amount of carbon fiber (CF) material can be reduced and with it the CO_2 footprint. The combination of a light foam core with continuous fiber-reinforced face sheets is a suitable material combination for lightweight design. Traditionally, CFRP sandwich structures with a foam core are manufactured in a two-step process by combining a prefabricated foam core with fiber-reinforced face sheets. However, in addition to the reduction in the used CFRP material, manufacturing processes with a high efficiency are needed. The objective of this paper is the sandwich manufacturing and characterization by using the Direct Sandwich Composite Molding (D-SCM) process for the one-step production of CFRP sandwich structures. The D-SCM process utilizes the resulting foaming pressure during the reactive polyurethane (PUR) foam system expansion for the impregnation of the CF-reinforced face sheets. The results of this work show that the production of sandwich structures with the novel D-SCM process strategy is feasible in one single manufacturing step and achieves good impregnation qualities. The foam density and morphology significantly influence the core shear properties and thus the component behavior under a bending load.

Keywords: sandwich structures; lightweight structures; advanced composites; polyurethane foam

Citation: Behnisch, F.; Brütsch, J.; Werner, H.O.; Henning, F. The Direct Sandwich Composite Molding (D-SCM) Process: Sandwich Manufacturing and Characterization. *J. Compos. Sci.* **2022**, *6*, 81. https://doi.org/10.3390/jcs6030081

Academic Editor: Stelios K. Georgantzinos

Received: 9 February 2022
Accepted: 28 February 2022
Published: 5 March 2022

1. Introduction

Existing legal frameworks and social pressure set the requirements for material and production process developments. The environmental impact by harmful emissions is increasingly monitored, from manufacture to disposal (life cycle assessment). Carbon fiber-reinforced plastic (CFRP) components have to face the challenge of the high resource consumption during material extraction (greenhouse gas impact of 38.9 kg CO_2-equivalent per kg CF [1]) and component manufacture counteracting the weight and emission savings they offer during the use phase. Thus, the development focus must not be exclusively on mechanical performance and cycle times. The environmental impact of CFRP components from the beginning of the product life cycle must be reduced to improve the overall balance of the component during its whole life cycle compared to conventional materials.

One strategy for achieving this is to use a sandwich design for large and flat components. The geometric stiffening effect by increasing the cross-sectional area moment of inertia allows the reduction in the reinforcing carbon fiber material and thus the CO_2 footprint for the components. A suitable combination for lightweight design is, for example, the use of continuous fiber-reinforced face sheets with a light foam core [2–5].

In addition to the reduction in CFRP material, more resource-efficient manufacturing processes are needed. CFRP sandwich structures with a foam core are manufactured by combining a prefabricated foam core with fiber-reinforced face sheets in a press process. For the face sheets, either pre-impregnated materials (so-called prepregs) can be used, or they can be in situ impregnated during the component manufacturing process. Thus, at least a two-step process is needed since the pressure-sensitive foam core must be manufactured within a previous production step [6–8].

A processing strategy for utilizing the internal pressure generated by a foam core has already been used by M. Ware for the manufacturing of composite sandwich structures within the Thermal Expansion RTM (TERTM) process [9]. A polyurethane (PUR) or polyimide (PI) foam core, combined with dry fabric reinforcing material, is placed in a cold cavity and the resin system is injected. A thermal expansion of the foam core is induced by increasing the mold temperature. The resulting pressure on the infiltrated face sheets improves their impregnation [9,10]. An alternative strategy by using foamable films in-between two pre-impregnated face sheets is proposed by Burr et al. in a DOW Global Technologies patent from 2014 for epoxy-based thermosets [11] and by Beukers et al. for thermoplastic polymers [12].

A one-step process strategy is introduced by Hopmann et al., using a syntactic, epoxy-based foam system [13]. Hollow polymeric microspheres, loaded with a low-boiling liquid, are used as a blowing agent. The expansion of the blowing particles is induced at rising temperatures due to the heated cavity as the polymeric microsphere shell becomes flexible and the transition of the low-boiling liquid into the gas phase leads to a volume increase. Glass fiber textile reinforcements serve as filters and prevent the microspheres from penetrating the sandwich face sheets. The fabric impregnation is realized by the expansion pressure of the blowing particles, which forces the matrix system into the face sheets. After the curing of the matrix material, the solid polymer network fixates the expanded hollow spheres that remain as pores in the syntactic foam material [13].

Weißenborn et al. introduced another one-step manufacturing strategy by using the expansion reaction of foamable PUR in a closed cavity [14]. Both the foam core and the matrix system for the textile reinforcement of the sandwich face sheets are realized by one single reactive PUR foam system. By manipulating the pressure conditions during the process, the pore content in the face sheets can be influenced. A roving impregnation of 84% and thus a pore volume of 16% were achieved at 770 kg/m^3 foam density [14,15].

The one-step sandwich manufacturing method introduced by Hopmann et al. is based on two independent PUR material systems for the foam core and the face sheet impregnation [16]. The face sheet textile reinforcement is wetted by a non-foamable PUR material outside the mold and vacuum supports the pre-impregnation process. Combined with the reactive PUR foam material in the heated cavity, the temperature-induced expansion reaction of the foam system leads to an internal force that presses the face sheets against the mold wall during curing. The investigations showed that the penetration of foam pores into the face sheets cannot be prevented even if the vacuum-assisted wetting of the fiber structure is almost optimal [16,17].

Studies conducted by Hopmann et al. [16,17] and Weißenborn et al. [14,15] showed an uncontrolled penetration of foam pores into the face sheets that lower the mechanical performance. Furthermore, relatively high foam densities (e.g., 770 kg/m^3 [14,15]) are necessary for sandwich manufacturing, which decreases the potential of being light weight. This leads to the need for a more efficient and robust one-step manufacturing technology for a CFRP sandwich structure production. This paper contributes to the development of a one-step manufacturing method for sandwich structures by introducing and investigating the novel Direct Sandwich Composite Molding (D-SCM) process. The focus of this manufacturing strategy is on avoiding foam pore penetration into the face sheets and on achieving low foam core densities. The objectives of this paper are to prove the feasibility of the D-SCM production method and to characterize the achievable sandwich properties in terms of fiber volume content and mechanical behavior under a three-point bending load.

2. The Direct Sandwich Composite Molding (D-SCM) Process Strategy

The basic idea of the intended processing strategy within this paper is the use of the expansion pressure generated during the PUR foam reaction for the fabric impregnation and consolidation of the face sheets. Two independent material systems for the foam core and the fabric impregnation in the face sheets will be used. A thermoplastic PUR (TPU) film is used as an impermeable layer for separating the face sheet and the reactive foam materials to avoid the uncontrolled face sheet penetration of foam cells. Consequently, different resin types for the face sheets (e.g., use of epoxy-based systems) can be used to improve the process flexibility. Additionally, an optimized polyurethane foam system is used for ensuring a high lightweight potential of the structures by decreasing the necessary foam density and increasing the achievable fiber volume content (FVC) of the sandwich face sheets [18].

The focus of this paper is the realization of a one-step process—the Direct Sandwich Composite Molding (D-SCM) Process—for the manufacturing of CFRP sandwich structures by using an optimized polyurethane foam system [18]. The process strategy is described in Figure 1. The stack of a wetted textile fabric and a TPU film is introduced in the heated cavity (1), an optimized reactive PUR foam system [18] is applied into the mold (2) and a second wetted fabric stack plus TPU film forms the upper face sheet (3). After closing the mold to the final component thickness (4), the chemical reaction of the foam material results in an expansion pressure on the face sheets (5). This internal pressure leads to the impregnation of the textile reinforcement, followed by the consolidation (curing) of the face sheets (6). After curing, the cavity can be opened and the sandwich component demolded (7). The resulting sandwich structures are characterized within a three-point bending analysis.

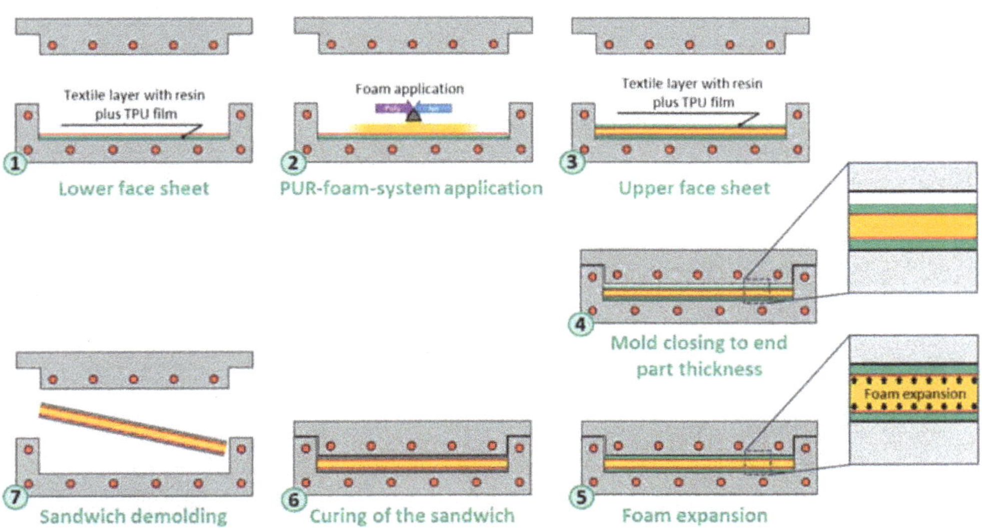

Figure 1. Schematic illustration of the ideal one-step sandwich process: (1) first wetted fabric stack is introduced in the cavity, (2) reactive PUR foam system is applied, (3) second wetted fabric stack is introduced in the cavity, (4) mold is closed to the final component thickness, (5) chemical PUR foam expansion, (6) consolidation (curing) of the face sheets, (7) demolding of the sandwich component.

3. Materials and Methods

3.1. Materials

The biaxial (0°/90°), 50 K-based carbon fiber non-crimp fabric (NCF) PX35MD030B-127T from Zoltek Corporation was used for the fiber reinforcement of the face sheets. The basic structure of the textile with a basis weight of 304 g/m^2 consists of Panex® 35 carbon fibers. These were divided proportionally with 150 g/m^2 between the biaxial orientations of the NCF. The carbon fiber filaments were coated with 1.5 wt.-% sizing. The material was tricot-stitched with 4 g/m^2 polyester thread [19]. Two layers of the fiber semi-finished product were combined per face sheet to form a symmetrical reinforcement structure of [(0°/90°), (90°/0°)].

Two thermosetting amine-based epoxy resin systems were used as matrix material for the face sheets. The Epikote™ TRAC 06170 system from Hexion is a low-viscosity resin system with a pot life of 42 min ± 5 at 23 °C, developed for liquid impregnation processes [20]. It will be referred to as "Epikote" in the following. The Araldite® LY 3031 system from Huntsman is a fast-curing resin system for automotive mass production and has a pot life of 15 to 25 min at 23 °C [21]. The system will be referred to as "Araldite" in the following.

The used PUR foam material for the D-SCM process is based on the two-component system EP 3587/2 from Rühl Puromer GmbH [22]. It consists mainly of polyether polyols and diphenylmethane diisocyanate (MDI). Further additives (e.g., distilled water, catalysts and nucleating agents) are mixed into the polyol component. The PUR system formulation is adjusted to the D-SCM process requirements and includes 3.1 parts of water and 1.1 parts of catalyst related to the polyol quantity [18].

Acmos 36-5238 from Acmos Chemie KG is used as an external mold release agent in the PUR foam manufacturing. For the epoxy resin systems, the semi-permanent solvent-based mold release agent Chemlease® 255 from Chem-Trend L.P. was used.

The impermeable barrier layer was formed by a TPU film Type 4110 from Saxonymed GmbH. This is transparent and contains no antiblocking agents. The film had a thickness of 0.1 mm.

3.2. Experimental Setup

A circular plate geometry with 150 mm diameter and adjustable height (shown in Figure 2a) was used for the one-step manufacturing of CFRP sandwich structures. An electric cylinder (Festo ServoPress YJKP-17-400-AX-SB-5) operated the trial setup and allowed the programming of closing profiles and highly accurate movements of the mold. The schematic cross section of the mold is shown in Figure 2b. The resin sealing prevents from leakage of the reactive materials and the vacuum sealing allows an evacuation of the cavity during the process. The mold was water-heated at 80 °C, a pressure sensor (Kistler type 6167A) was installed in the upper mold for online recording of the pressure values and the cavity height can be adjusted via distance rings.

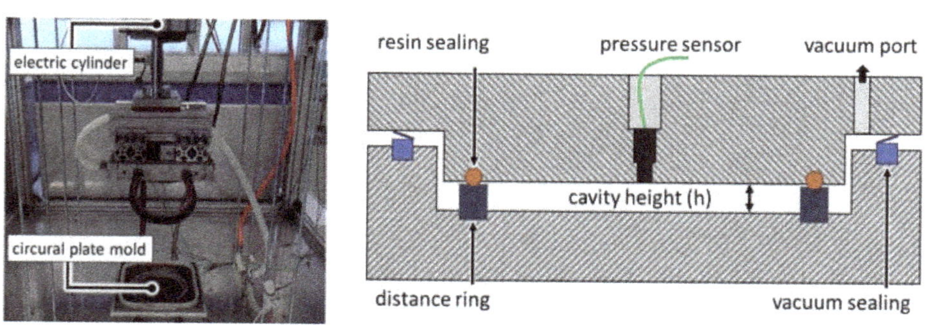

(a) (b)

Figure 2. (a) experimental setup for D-SCM sandwich manufacturing, (b) schematic cross section of the circular plate mold.

3.3. Sandwich Specimen Manufacturing

The one-step sandwich manufacturing follows the general processing steps from Figure 1. The respective matrix material (Epikote or Araldite) was applied on the stack of two biaxial CF fabric layers (148 mm in diameter) and the impermeable TPU film. The stack was introduced into the heated circular plate mold. Simultaneously, the foam system was manually mixed and applied on the TPU film that covered the wet fabric reinforcement of the lower face sheet. The necessary foam material amount was dependent on the aimed foam density. Table 1 summarizes the manufactured trial configurations. A foam thickness of 4 mm was aimed for for the core. The second face sheet, consisting of wetted reinforcement fabrics, and the TPU film were placed on top of the reactive foam material. Subsequently, the mold was closed to the final sandwich height of 5.7 mm in 11 s according to the programmed closing profile of the electric cylinder. At a mold opening of 3 mm, the vacuum was activated and applied throughout the entire process. From the point at which the resin seal was engaged at 2 mm, the vacuum no longer acted on the sandwich structure. The active vacuum time was 5.5 s. The target face sheet thickness was 0.75 mm, which corresponds to 44.2% FVC. The curing time was 300 s for the Araldite system and 600 s for the Epikote system. The specimens for detailed characterization were prepared by waterjet cutting out of the produced circular sandwich plates.

Table 1. Investigated trial configurations.

Epoxy System	Foam Core Density	Foam Core Material	Face Sheet Material	Curing Time
Araldite	150 g/L	10.6 g	9.5 g	300 s
Araldite	200 g/L	14.1 g	9.5 g	300 s
Araldite	250 g/L	17.7 g	9.5 g	300 s
Epikote	150 g/L	10.6 g	9.5 g	600 s
Epikote	200 g/L	14.1 g	9.5 g	600 s
Epikote	250 g/L	17.7 g	9.5 g	600 s

The TPU film provides high adhesion properties to the PUR core and the Epoxy face sheet material. To be able to determine the correct FVC, the face sheets must be produced to be detachable from the foam core. For this purpose, D-SCM trials were carried out and a PTFE film of identical thickness was inserted between the foam system and the face sheets instead of the TPU film. As there was adhesion between the used reactive materials and the PTFE film, this allowed the foam core and face sheets to be separated from each other after the process. The detailed trial configurations for the sandwich processing can be found in Table 1.

3.4. Characterization Methods

3.4.1. Thermogravimetric Analysis of the Face Sheets

The FVC was determined by a thermogravimetric analysis (TGA). In this characterization method, the polymeric content of the composite is pyrolyzed and the mass of the remaining carbon fibers is determined.

For the TGA, three specimens (22×30 mm^2) were cut out of each face sheet by water jet cutting and their density ρ_{Sample} was determined in a gas pycnometer (PYCNOMATIK ATC from Microtrac Retsch GmbH) according to DIN 66137-2, and their weights m_{Sample} were measured. The pyrolysis was carried out in a macro TGA of the type TGA701 from Leco Corporation. The test procedure was implemented according to the method of Bücheler et al. [23]:

- Drying of the specimens for 6 h at 100 °C.
- Pyrolysis of the specimens in nitrogen atmosphere:
    - Heating rate: 2 K/min.
    - Pyrolysis temperature: 430 °C.

○ Pyrolysis duration: 20 h.

Subsequently, the mass of the matrix-free fibers m_{CF} was determined. Referring to the investigations of Bücheler et al., the sizing was burned off from the fiber during the pyrolysis process. For the correct calculation of the fiber volume content, the mass of the sizing m_{sizing} (1.5 % · m_{CF}) was again added to the fiber mass. Thus, the combined density ρ_{CF} of the Panex® 35 carbon fiber of 1.81 g/cm^3 [19] can be used to determine the fiber volume V_{CF}. The fiber volume content was calculated as follows:

$$FVC = \frac{V_{CF}}{V_{sample}} \cdot 100\ \% = \frac{\left(m_{CF} + m_{sizing}\right) \cdot \rho_{CF}}{m_{sample} \cdot \rho_{sample}} \cdot 100\ \% \quad (1)$$

3.4.2. Tensile Characterization of the Face Sheets

The stiffness of the face sheets E_{FS} was determined in a tensile test based on DIN EN ISO 527-1. For this purpose, the face sheets were separated from the core structure and the test specimens (120 × 15 mm^2) with the corresponding fiber orientation of $[0/90]_S$ were prepared. A small amount of foam residue and the TPU film remained on the tensile specimen. Their influence on the face sheet stiffness was neglected in the following, since the mechanical properties were significantly lower compared to the CFRP specimen (e.g., Young's modulus of the TPU Elastollan®: 400 MPa at 20 °C [24]). The tests were carried out on a standard universal testing machine (Hegewald & Peschke, inspekt 50 table) with a test speed of 1 mm/min.

3.4.3. Three-Point Bending Characterization of Sandwich Structure

The test was performed in accordance with ASTM C393/C393M–16, one of the standard test methods for sandwich characterization [25,26]. The 3-point bending test fixture was installed in a Zwick/Roell Z2.5 universal testing machine, shown in Figure 3. The radius of the three supports was 2 mm. In accordance with the standard, an elastomer layer was introduced at the pressure fin (shore hardness: 60; width: 25 mm; thickness: 3 mm) for homogenized introduction of the test load into the specimen (dimensions: 120 × 15 × 5.7 mm^3). The fiber orientation for each face sheet was $[0/90]_S$. The material combinations of the D-SCM sandwich structures shown in Table 1 were tested.

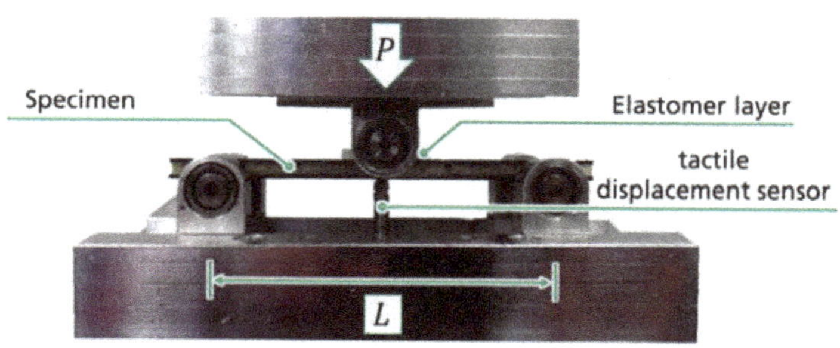

Figure 3. Three-point bending setup for sandwich characterization.

A 2.5 kN load cell records the test force required to deflect the specimen. The corresponding specimen deflection was recorded via a tactile displacement sensor centered underneath the specimen. To prevent settling effects, a preload of 10 N was applied before starting the test. The bending specimens were tested at a support span length (L) of 105 mm and at a test speed of 1 mm/min until failure. The test ended at a force reduction of 20% to the measured maximum force (force cut-off threshold). Three specimens of each configuration were tested.

The maximum shear stress τ_{max} corresponds to the stress state at the maximum test load P_{max}. Formula (2) describes the general calculation basis to determine τ_{max}. Since the evaluation according to the ASTM C393/C393M-16 standard only allows shear failure and such failure behavior can be observed in all specimens, the maximum core shear stress is equivalent to the core shear strength for the sandwich composite under consideration of a 3-point bending load.

$$\tau = \frac{dM}{dx}\frac{1}{b \cdot c} = \frac{P}{b \cdot c}$$
$$with\ M = bending\ moment,\ b = specimen\ width,$$
$$c = combined\ thickness\ of\ core\ and\ one\ face\ sheet \quad (2)$$

The bending stiffness D_S and the shear stiffness G_C can be determined by considering the compliance $C = \delta / P$ (δ = displacement) and the correlation in Formula (3) [27]:

$$C = \frac{L^3}{48 \cdot b \cdot D_S} + \frac{L}{4 \cdot t_C \cdot b \cdot G_C}$$
$$with\ L = support\ distance,\ t_C = core\ thickness \quad (3)$$

Due to the valid assumption of significantly stiffer and relatively thin face sheets compared to the core, the bending stiffness D_S was reduced to the bending stiffness D_0 of the face sheets respective to the centroidal axis of the entire sandwich cross section and can be determined directly using the face sheet stiffness E_{FS} measured in the tensile tests by applying Formula (4). The core shear stiffness G_C can then be derived by inserting the calculated bending stiffness D_S into Formula (3).

$$D_S = D_0 = \frac{E_D * b * c^2 * t_D}{2} \quad (4)$$

4. Results

4.1. The D-SCM Manufacturing Process

It is possible to manufacture sandwich structures (shown in Figure 4) using the D-SCM process strategy. All samples listed in Table 1 can be produced and result in testable components. The surface qualities indicate a good fiber impregnation in the face sheets. Starting from a centered resin application, the fiber reinforcement in the face sheets is impregnated by the fluid flow, induced through the PUR foaming pressure.

(a) (b)

Figure 4. D-SCM sandwich structures: in the lower cavity after mold opening (**a**); demolded, cured sandwich component (**b**).

4.2. Fiber Volume Content of the Face Sheets

Figure 5 shows the determined FVC in the face sheets for the D-SCM sandwich structures, according to the foam core density and the resin system. For both resin systems, the average fiber volume contents range between 30 and 35% and thus are at similar levels.

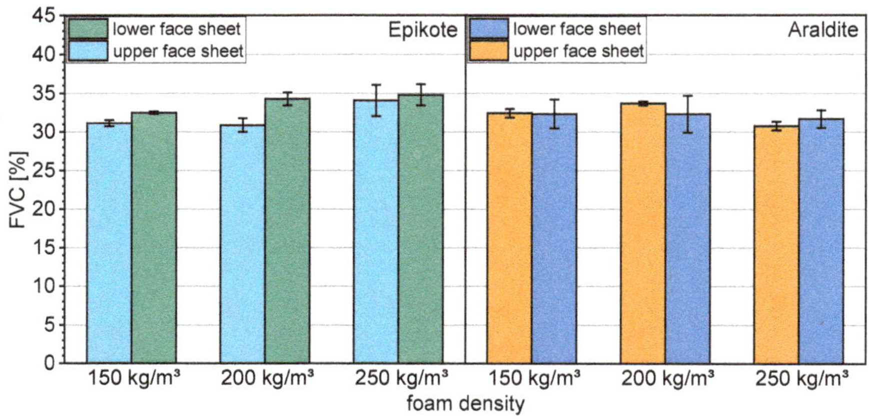

Figure 5. Fiber volume contents of the lower and upper face sheets according to the foam core density for the Epikote (**left**) and the Araldite resin systems (**right**).

For the Epikote resin system, the lower face sheet had a higher average FVC than the upper face sheet. This is due to the fact that the lower face sheet was inserted into the mold first and warmed up. As a result, the viscosity of the resin initially decreased due to the temperature, and the resin system could already start to flow over the surface and impregnate the fibers due to weight and capillary forces. Due to the comparatively slow reaction rate of the Epikote system, the material in both face sheets was still flowable at the time of the PUR pressure build-up in the closed mold; however, the flow front had already more progressed in the lower face sheet due to the above-mentioned temperature effect. The mean FVC increased with increasing foam density. The increasing foam density leads to a higher foam pressure and consequently to a higher fluid pressure in the face sheets. This allows the flow front to impregnate the dry fiber reinforcement further in the face sheet, resulting in higher FVC.

However, for the Araldite resin system, no such significant trend with respect to the FVC in the face sheets can be identified. The standard deviations for the upper and the lower face sheets always overlap and also do not show an analogous behavior like they do for the Epikote specimens. The Araldite system is more challenging to be processed in the manual procedure due to its significantly higher reactivity. In case of small time differences in the individual process steps during the manual specimen manufacturing, viscosity increases can occur during processing. This limits the flowability and no clear trend of the FVC can be identified with respect to the increasing foam density.

4.3. *Stiffness of the Face Sheets*

The results for the face sheets stiffness E_{FS} determined in the tensile test relating to the foam density are shown in Figure 6. In addition, the bending stiffness D_S determined on the basis of the face sheet stiffness using Formula (4) is also illustrated.

The face sheet stiffness determined in the tensile test are at similar levels between 40.4 and 45.4 GPa for all specimens tested. For both investigated resin systems, a moderate increase in the average stiffness with increasing foam density can be identified. This can be explained by the also slightly increasing FVC in the face sheets (shown in Section 4.2). However, the overlapping standard deviations do not indicate any significant effect. With regard to the values of the bending stiffness D_S, an analogous trend of the mean values can be observed. Since the face sheet reinforcement as well as the FVC do not change significantly, the results of the tensile tests are plausible.

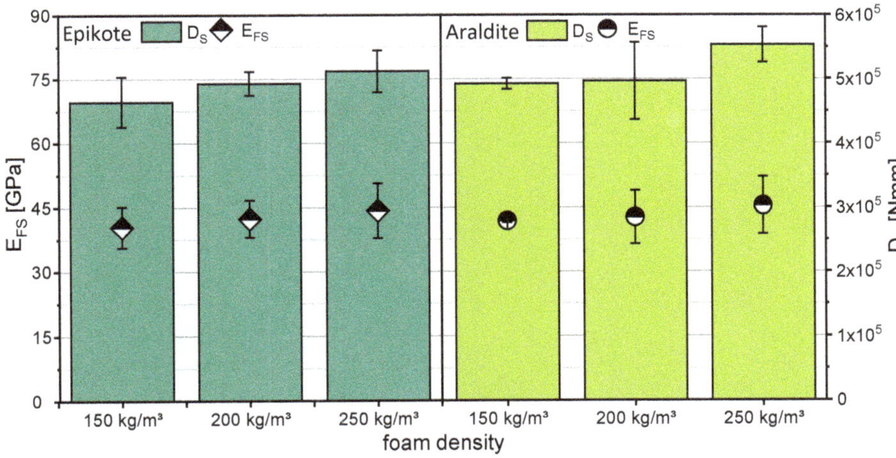

Figure 6. Face sheet stiffness, determined in tensile tests E_{FS} (dot plot) and the resulting bending stiffness D_S (bar graph) for the Epikote (**left**) and the Araldite (**right**) resin systems.

4.4. Three-Point Bending Behavior of the D-SCM Sandwich Structures

The specimen failure is comparable in all tests performed. A face sheet-parallel crack propagation in the foam structure on the specimen compression side near the face sheet/core interface occurred in the longitudinal specimen direction. The failure pattern resembles a classic delamination. However, the TPU interface is not the weak component of the sandwich structure. In all specimens, a continuous layer of foam residue remains adhered to the TPU film (shown in Figure 7). Consequently, the foam structure near the interface is the weak point (cohesive failure) of the composite.

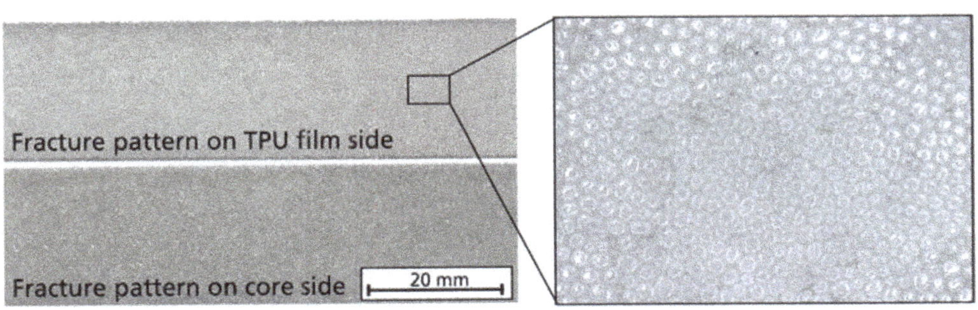

Figure 7. Fracture pattern of the cohesive failure of 3-point bending specimens.

The calculated maximum core shear strengths according to Formula (2) of the tested sandwich specimens are shown in Figure 8.

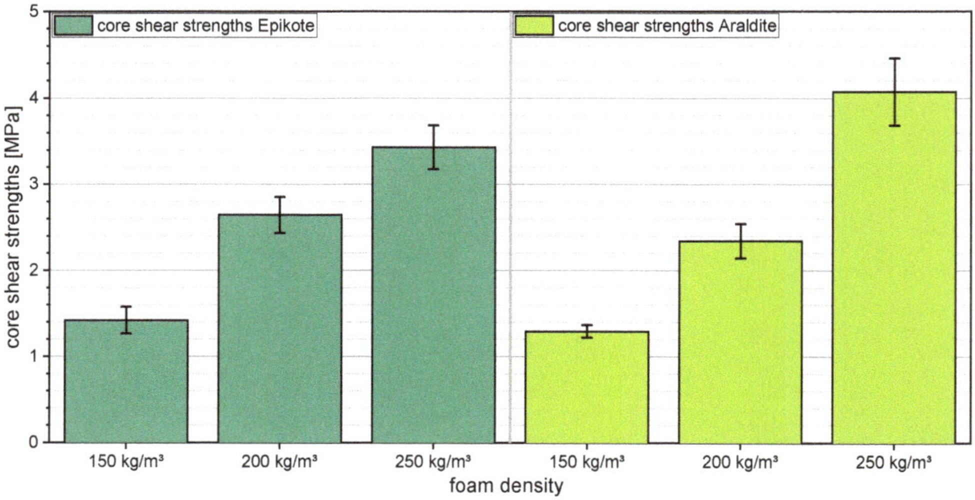

Figure 8. Core shear strength of the sandwich structures for the Epikote resin system (**left**) and the Araldite resin system (**right**).

For the specimens of the Epikote resin system in the face sheets, the core shear strength increased with the foam density from 1.4 MPa at a density of 150 kg/m^3 to 3.4 MPa for a density of 250 kg/m^3, representing an increase of 143 %. For the Araldite system, these values evolved from 1.3 MPa at a density of 150 kg/m^3 to 4.1 MPa at a density of 250 kg/m^3, which corresponds to an increase of 215 %. The determined values are compared to commercially available PUR foam core materials (e.g., LEOcore®: 1.2 MPa at 145 kg/m^3 density [28]). The results for the determined core shear moduli G_C by applying Formula (3), considering the bending face sheet stiffness D_S are given in Figure 9.

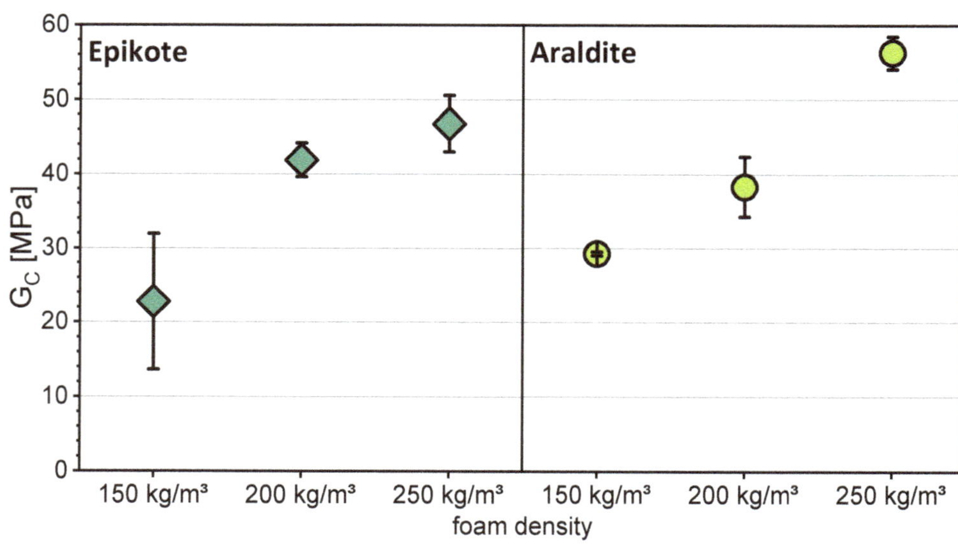

Figure 9. Core shear stiffness of the sandwich structures for the Epikote resin system (**left**) and the Araldite resin system (**right**).

For a 150 kg/m^3 foam density, an average shear stiffness between 22.8 (Epikote) and 29.3 MPa (Araldite) was found. At 250 kg/m^3 density, the stiffness was increased by 105% to 46.8 for Epikote, and by 92% to 56.3 MPa for the Araldite system. The core shear modulus increased with foam density for both resin systems. For the Araldite system in the face sheets, this effect is even statistically significant as the standard deviations do not overlap.

In addition, Figure 10 shows that the foam morphology also becomes more homogeneous with increasing density and includes less instabilities. The resulting increase in foam quality also leads to a higher load-bearing capacity of the core structure.

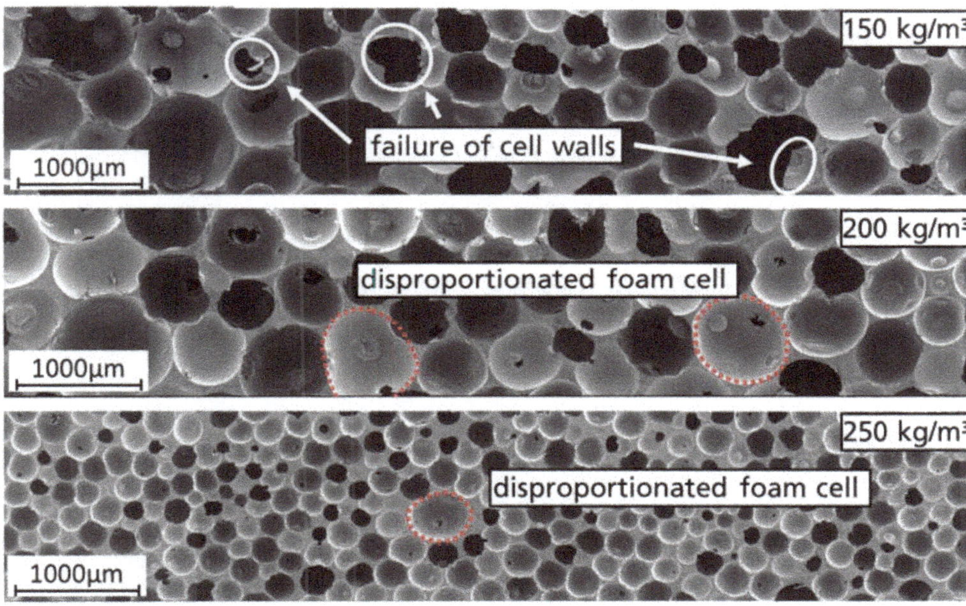

Figure 10. PUR foam morphology at 150 kg/m^3 (**top**), 200 kg/m^3 (**middle**) and 250 kg/m^3 (**bottom**) foam density.

Disproportionations of the foam cells occur due to the higher pressure in smaller pores during the gas bubble growing phase of the chemical foaming process. In this process, the generated CO_2 diffuses from small to larger cells to compensate for the local pressure difference [29,30]. Consequently, irregular cell sizes and open cell structures occur. Such foaming behavior is disadvantageous, since it impairs the mechanical properties of the foam structure.

This effect is further enhanced by the adapted foam system for the D-SCM process and the resulting higher pressure level. The comparatively low polymer mass at 150 kg/m^3 foam density is not able to sufficiently stabilize the cell walls during the foam formation.

The foam structure at 200 kg/m^3 foam density still shows instabilities. However, in contrast to the structures obtained at a density of 150 kg/m^3, these instabilities are much less pronounced. The cell sizes at 150 and 200 kg/m^3 foam density are comparable. Even at an increased density, disproportionation effects still seem to occur, preventing a highly distributed structure of small cells. The improved foam quality is due to the supporting effect of the increasing polymeric PUR material at a density of 200 kg/m^3. Although the process pressure increases with the increasing foam density, the polymer mass is able to stabilize the cell structure to a large extent and prevent open-celling. Thus, useful for their mechanical stablility, foam structures can be generated at higher process pressures.

Figure 10 (bottom) shows a representative section of a foam structure with a 250 kg/m^3 foam density. The cells are closed without exception and the distribution is largely homogeneous. The cell size is significantly smaller compared to the 150 and 200 kg/m^3 foam densities. Local disproportionation effects lead very occasionally to larger foam pores. In contrast to the foam structures for the 150 and 200 kg/m^3 foam densities, the increased PUR mass for the 250 kg/m^3 foam density is able to stabilize the cell walls during the entire foaming process. Due to the increased density, the D-SCM process pressure for a density of 250 kg/m^3 is increased. However, the cell morphology shows that this effect has no negative influence on the chemical foaming process.

5. Discussion

The determined fiber volume contents for the face sheets of the D-SCM sandwich structures are all within a range between 30 and 35% and thus below the target values of 44.2%. Comparable two-step process variants such as RTM in combination with a foam core allow FVC between 50 and 60%. For the investigated process window between 150 and 250 kg/m^3 foam density, the expected increase in the FVC with rising foam pressure has not been identified for the Araldite system in the face sheets. For the Epikote resin system in the face sheets, a slight increase in the FVC mean values can be observed. This is due to the characteristics of the D-SCM process, in that the foam pressure acts simultaneously on the dry textile and on the applied fluid. This results in a necessary resin flow through the dry textile in the non-impregnated areas, which are already compacted by the foam pressure. This compaction of the dry fabric areas leads to a decreased permeability and thus a higher flow resistance. As a result, the fluid flow in the face sheets is impeded. The higher the foam pressure, the higher the compaction of the dry areas and thus the flow resistance. This effect partially offsets the increased fluid pressure due to the higher foam pressure.

Following the relatively constant fiber volume content, the stiffness of the face sheets and the corresponding bending stiffness do not change significantly by increasing the foam density. This behavior is plausible due to the constant face sheet reinforcement.

The core shear properties influence the sandwich behavior under a three-point bending load significantly. Due to the increased amount of polymer material, the load-bearing capacity of the foam core is increased at higher densities. As solely shear failure in the foam core was observed during the three-point bending tests, the shear strength of the core is the failure-relevant parameter. An increased shear strength leads directly to an increase in the sandwich structure performance under bending load. Analogically to the shear strength, the shear modulus also increases as a result of the increased amount of polymer material in the core with higher foam densities. In general, the determined shear strengths (150 kg/m^3 density: 1.4 MPa for Epikote, 1.3 MPa for Araldite) are within a plausible range of values, since literature values of comparable rigid PUR foams are in the same order of magnitude (e.g., LEOcore® with 145 kg/m^3 foam density and 1.2 MPa shear strength [28]).

In addition to the foaming pressure [18], the foam morphology is also influenced by the PUR foam density. During the chemical foaming process, the cell structure must be stabilized by a sufficient amount of polymeric PUR material. However, an increasing foam density results in a higher component weight and must therefore be optimized on the application requirements. The increasing core shear properties with an increasing density represent a combined effect of a higher polymer mass and a more homogeneous foam structure.

The TPU film as an impermeable layer prevents foam penetration into the face sheets. The adhesion of both reactive systems (resin and foam) does not represent an explicit weak point of the sandwich structure. The TPU film thus fulfills its intended purpose in the D-SCM process. For a large-area and complex geometries, integration of the film within the process chain may be challenging. The TPU material used in this work has only limited stretchability, which makes it difficult to drape double-curved areas. For such an application, the film must be replaced by a more stretchable material.

6. Conclusions

The infiltration of the face sheets is possible with both resin systems investigated. The potentially time-critical processing with the more reactive Araldite system is not confirmed. For the relevant process steps, the viscosity ranges of both resin systems are not critical. The D-SCM process can be used for the one-step manufacturing of sandwich structures.

The processing conditions and the resulting structural properties of the sandwich components can be mainly controlled by the foam density when considering a constant fiber reinforcement structure in the face sheets.

In principle, the substitution of individual materials for the production of sandwich structures in the D-SCM process is possible. The required basic knowledge for the material substitution was determined within the investigations. In the case of material substitution, the effects of the changing, time- and temperature-dependent properties of the reactive components must be considered in the context of the overall process strategy. The face sheet resin system must have sufficient flowability corresponding to the pressure build-up time of the foam. For the TPU film as well as the fiber structure, basic compatibility with the reactive systems involved is essential. In addition, the necessary flexibility for molding the required part geometry must be ensured.

Author Contributions: Conceptualization, F.B. and H.O.W.; methodology, F.B., J.B. and H.O.W. validation, F.B., J.B. and H.O.W.; formal analysis, F.B., J.B. and H.O.W.; investigation, F.B., J.B. and H.O.W.; resources, F.B., J.B. and H.O.W.; data curation, F.B., J.B. and H.O.W.; writing—original draft preparation, F.B.; writing—review and editing, H.O.W., J.B. and F.H.; visualization, F.B.; supervision, F.H.; project administration, F.B. and H.O.W.; funding acquisition, F.B. and F.H. All authors have read and agreed to the published version of the manuscript.

Funding: Financial support by the project Profilregion Mobilitätssysteme Karlsruhe is gratefully acknowledged. In the state of Baden-Württemberg, the Profilregion Mobilitätssysteme is funded equally by the Ministry for Economy, Labor and Housing, the Ministry of Science, Research and the Arts as well as through contributions from the Fraunhofer Society, the Karlsruhe University of Applied Sciences (HsKA) and the Karlsruhe Institute of Technology (KIT). Furthermore, the Profilregion Mobilitätssysteme Karlsruhe is supported by the strategic industry partner Hengrui Group which is greatly appreciated.

Informed Consent Statement: Not applicable.

Data Availability Statement: Not applicable.

Acknowledgments: The authors would like to thank their colleagues from Fraunhofer ICT, Pfinztal: Cahit Arik, Markus Reeb and Christian Ress for technical and analytical support. The polyurethane foam supplier Rühl Puromer GmbH as well as the fiber textile supplier Zoltek Corporation and resin suppliers Huntsman Corporation and Hexion are also gratefully acknowledged for providing the materials used. Special thanks to the strategic industry partner Hengrui Group for the great support.

Conflicts of Interest: The authors declare no conflict of interest.

References

1. European Composites Industry Association. *Carbon Fiber LCA Data in EcoCalculator: An Essential Extension*; Composites Europe: Stuttgart, Germany, 2018.
2. Vinson, J.R. Sandwich Structures. *Appl. Mech. Rev.* **2001**, *54*, 201–214. [CrossRef]
3. Zenkert, D. *An Introduction to Sandwich Structures—Student Edition*; Engineering Materials Advisory Services Publishing: Birmingham, UK, 1995; ISBN 978-0947817770.
4. Zenkert, D.; Nordic Industrial Fund. *The Handbook of Sandwich Construction*; Engineering Materials Advisory Services Publishing: Birmingham, UK, 1997; ISBN 978-0947817961.
5. Thomsen, O.T.; Bozhevolnaya, E.; Lyckegaard, A. (Eds.) Sandwich Structures Technology in Commercial Aviation. In *Sandwich Structures 7: Advancing with Sandwich Structures and Materials, Proceedings of the 7th International Conference on Sand-wich Structures, Aalborg University, Aalborg, Denmark, 29–31 August 2005*; Springer: Dordrecht, The Netherlands, 2005; ISBN 978-1-4020-3444-2.
6. Gerstenkorn, J. Entwicklung einer Methodik zur Analyse und Vorhersage des Mechanischen Verhaltens Polymerer Hartschäume im RTM-Prozess. Ph.D. Thesis, Karlsruher Institute for Technology, Karlsruhe, Germany, 2021.

7. Rosenberg, P. Entwicklung einer RTM Prozessvariante zur Kavitätsdruckgeregelten Herstellung von Faserverbundstrukturbauteilen. Ph.D. Thesis, Karlsruher Institute for Technology, Karlsruhe, Germany, 2018.
8. Seibert, H.F. Applications for PMI foams in aerospace sandwich structures. *Reinf. Plast.* **2006**, *50*, 44–48. [CrossRef]
9. Ware, M. Thermal Expansion Resin Transfer Molding (TERTM)—A Manufacturing Process for RP Sandwich Core Structures. In Proceedings of the 40th Annual Conference, Atlanta, GA, USA, 28 January–1 February 1985; Reinforced Plastics/Composite Institute: New York, NY, USA; The Society of Plastic Industry, Inc.: New York, NY, USA, 1985.
10. Ware, M. Thermal Expansion Resin Transfer Molding(TERTM)—An Advanced Composite Mass Production Process. In *Society of Manufacturing Engineers, Proceedings of the 5th Conference on Composites in Manufacturing, Los Angeles, CA, USA, 13–16 January 1986*; Society of Manufacturing Engineers: Southfield, MI, USA.
11. Burr, S.T.; Tudor, J.M.; Athreya, S.R.; Parsons, T.J. Expanding Foam Core Prepreg. PCT/US2013/070033, 14 November 2013.
12. Beukers, A. A new technology for sandwich plates and structures based on the use of in-situ foamable thermoplastic films. In Proceedings of the 37th International SAMPE Symposium and Exhibition, Anaheim, CA, USA, 9–12 March 1992; pp. 172–185.
13. Hopmann, C.; Karatzias, C.; Böttcher, A. Syntaktische Schäume zur einstufigen Sandwichherstellung von FVK. *Lightweight Des.* **2016**, *9*, 44–51. [CrossRef]
14. Weißenborn, O. Entwicklung eines Neuartigen Imprägnierschäumverfahrens zur Herstellung Komplex Geformter Polyurethan-Sandwichverbundstrukturen. Ph.D. Thesis, Technische Universität Dresden, Dresden, Germany, 2020.
15. Weißenborn, O.; Geller, S.; Jaschinski, J.; Gude, M. Modelling the bending behaviour of novel fibre-reinforced sandwich structures with polyurethane foam core. *IOP Conf. Ser. Mater. Sci. Eng.* **2018**, *416*, 012073. [CrossRef]
16. Hopmann, C.; Wagner, R.; Fischer, K.; Böttcher, A. One Step Production of High-performance Sandwich Components. *Cell. Polym.* **2018**, *36*, 135–150. [CrossRef]
17. Hopmann, C.; Riedel, R.; Karatzias, C. Strukturelle Sandwichbauteile mit prozessintegriert ausgeschäumtem Schaumkern. *Fachmag. Für Die Polyurethanindustrie* **2015**, *91*, 11–12.
18. Behnisch, F.; Lichtner, V.; Henning, F.; Rosenberg, P. Development of a Reactive Polyurethane Foam System for the Direct Sandwich Composite Molding (D-SCM) Process. *J. Compos. Sci.* **2021**, *5*, 104. [CrossRef]
19. ZOLTEK Corporation (Toray Group). *ZOLTEK™ PX35 Multi-Directional Fabrics*; Technical Datasheet; ZOLTEK Corporation (Toray Group): Bridgeton, MO, USA, 2021.
20. Hexion™ Inc. *Datenblatt EPIKOTE™ Resin Trac 06170*; Hexion Inc.: Columbus, OH, USA, 2017.
21. Huntsman Corporation. *Datenblatt Araldite® LY 3031/Aradur® 3032*; Huntsman Corporation: The Woodlands, Texas, USA, 2015.
22. Rühl Puromer GmbH. *Datenblatt Rühl Puromer EP 3587/2*; Rühl Puromer GmbH: Friedrichsdorf, Germany, 2017.
23. Bücheler, D.; Kaiser, A.; Henning, F. Using Thermogravimetric Analysis to Determine Carbon Fiber Weight Percentage of Fiber-Reinforced Plastics. *Compos. Part B Eng.* **2016**, *106*, 218–223. [CrossRef]
24. BASF, SE. Elastollan®—Physikalische Eigenschaften. Available online: https://plastics-rubber.basf.com/emea/de/performance_polymers/products/elastollan.html?at_medium=sl&at_campaign=PM_BAW_EMEA_DE_Elastollan_TRA_CROSS&at_term=%2Btpu%20%2Bfolie&at_creation=Search_Google_SERP_Elastollan-General-EMEA-DE&at_platform=google&at_variant=Elastollan-General-EMEA-DE (accessed on 12 December 2021).
25. Xiong, J.; Ma, L.; Wu, L.; Liu, J.; Vaziri, A. Mechanical behavior and failure of composite pyramidal truss core sandwich columns. *Compos. Part B Eng.* **2011**, *42*, 938–945. [CrossRef]
26. Wesolowski, M.; Ludewicz, J.; Domski, J.; Zakrzewski, M. Shear properties evaluation of a truss core of sandwich beams. *IOP Conf. Ser. Mater. Sci. Eng.* **2017**, *251*, 012085. [CrossRef]
27. Allen, H.G. *Analysis and Design of Structural Sandwich Panels: The Commonwealth and International Library: Structures and Solid Body Mechanics Division*; Elsevier: Oxford, UK, 2013; ISBN 978-1483126623.
28. Gaugler & Lutz OHG. Produktübersicht: Ihr Starker Partner für den Leicht- und Sandwichbau. June 2019. Available online: https://www.gaugler-lutz.de/fileadmin/user_upload/Unternehmen/Produktuebersicht_LS_DE.pdf (accessed on 12 December 2021).
29. Minogue, E. An In-Situ Study of the Nucleation Process of Polyurethane Rigid Foam Formation. Ph.D. Thesis, Dublin City University, Dublin, Ireland, 2000.
30. Lander, R.; Modro, H.; Hubel, R. Influencing the Cell Structure of Flexible Polyurethane Foams by Additives. In Proceedings of the 2014 Polyurethane Technical Conference, Dallas, TX, USA, 22–24 September 2014; pp. 22–24.

Journal of Composites Science

Article

Fundamental Frequency Optimization of Variable Angle Tow Laminates with Embedded Gap Defects

João Carvalho [1], Abdolrasoul Sohouli [2,*] and Afzal Suleman [1,2]

[1] IDMEC—Instituto de Engenharia Mecânica, IST—Instituto Superior Técnico, Universidade de Lisboa, 1049-001 Lisboa, Portugal; joao.f.carvalho@tecnico.ulisboa.pt (J.C.); suleman@uvic.ca (A.S.)
[2] Department of Mechanical Engineering, University of Victoria, Victoria, BC V8P 5C2, Canada
* Correspondence: sohouli@uvic.ca

Abstract: Variable stiffness composite laminates can improve the structural performance of composite structures by expanding the design space. This work explores the application of variable stiffness laminated composite structures to maximize the fundamental frequency by optimizing the tow angle. To this end, an optimization framework is developed to design the fiber angle for each layer based on the maximization of the fundamental frequency. It is assumed that the design process includes the manufacturing constraints encountered in the automated fiber placement process and a linear fiber angle variation. The current study improves existing results by considering embedded gap defects within the optimization framework. The plates are assumed symmetric, with clamped and simply supported boundary conditions. The optimal results and a comparison between the non-steered and steered plates with and without gaps are presented. Results show that, although gaps deteriorate the structural performance, fiber steering can still lead to an increase in the fundamental frequency depending on the plate's geometry and boundary conditions.

Keywords: variable stiffness composite; variable angle tow; automated fiber placement; Defect Layer Method; natural frequency; genetic algorithm optimization

Citation: Carvalho, J.; Sohouli, A.; Suleman, A. Fundamental Frequency Optimization of Variable Angle Tow Laminates with Embedded Gap Defects. *J. Compos. Sci.* **2022**, *6*, 64. https://doi.org/10.3390/jcs6020064

Academic Editors: Stelios K. Georgantzinos and Francesco Tornabene

Received: 31 December 2021
Accepted: 17 February 2022
Published: 20 February 2022

1. Introduction

Nowadays, the use of composites materials is increasing in a wide range of sectors such as the aerospace, automotive, naval and others. This is the main motivation to develop innovative and cost effective composite manufacturing techniques which allow the production of composite laminates meeting certain requirements in a more effective way. In particular, the development of the Automated Fiber Placement (AFP) technique opens the possibility of manufacturing Variable Stiffness Composite Laminates (VSCLs), which allow the stiffness to vary spatially. VSCLs include, for instance, those made with variable fiber spacing (VFS) and those in which the fibers are deposited following curvilinear paths, denominated as variable-angle tow (VAT) laminates. They allow new design possibilities by enlarging the design space when compared to conventional composite laminates. Hence, several authors have been studying them with the objective of improving composite laminates characteristics. For instance, the studies presented in [1–3] show the influence of these kind of laminates in the buckling behaviour of laminated panels.

Although most studies for vibration analysis are performed on composite laminated plates with constant stiffness, some authors also considered the use of VAT laminates. Most of these studies aimed to study the influence of the curvature of the fibers on the fundamental frequency achieved by a plate. Honda et al. [4] considered splines to represent the fiber paths of curvilinear fibers. Honda et al. [5] developed a multi-objective approach in order to maximize both fundamental frequencies and in-plane strengths of VAT plates, while Pereira et al. [6] performed a multi-objective optimization considering the fundamental frequency and the first mode specific damping capacity (SDC). Serhat et al. [7], also maximized the fundamental frequency of variable stiffness panels but proposing a more efficient method to design variable stiffness composite panels using lamination parameters. Although the steering radius was not directily constrained during the optimization process, the design space is controlled to ensure a smooth changes in the layer angles. The design methodology efficiency developed in [7] is improved by Serhat et al. [8] study by combining the lamination parameters formulation with a spectral Tchebychev modeling approach. In this study, the fundamental frequencies of the variable stiffness designs were maximized for different boundary conditions and aspect ratios. It was verified that optimal variable stiffness designs provide higher frequencies (up to 10%) when compared to constant stiffness ones. More recently, Farsadi et al. [9] included plates with different aspect ratios, boundary conditions and skew angles. The fundamental frequency was also studied and optimized together with other composite characteristics.

The most common manufacturing constraint considered when designing curvilinear composite laminates is the maximum curvature constraint [10–14]. Akhavan and Ribeiro [10], for example, considered this constraint in the study of the natural frequencies and mode shapes of VAT with a linear curvilinear fiber path. They used a p-version finite element and a Third-order Shear Deformation Theory (TSDT). It was verified that the use of curvilinear fibers can be advantageous to adjust frequencies and mode shapes. Demir et al. [14] developed a lamination parameter optimization algorithm using a penalty parameter to enforce a maximum value for the fiber curvatures. The proposed methodology is applied to design for the minimum compliance in-plane and out-of-plane problems. In a recent study, Rashed and Demir [15] maximized the fundamental frequency of variable stiffness plates. Manufacturing constraints were considered and the method employed uses lamination parameters as the design variables. The optimum lamination parameters distribution found is in good agreement with other studies present in the literature.

The previous studies mentioned assumed that the VAT laminated plate is ideal, which means that the models used to obtain the natural frequencies and mode shapes did not take into account the manufacturing process, in particular, the AFP and the respective manufacturing induced imperfections. As a result of that, some models were developed in order to take defects into consideration and better predict the structural behaviour of VAT laminates. Blom [16] developed a methodology where each ply of the laminate is modeled including gaps and overlaps generated by tow drops. Afterwards, a 2-D mesh is generated on top of each ply and the centroids of all the elements in all the plies are checked for the tow drop region. If it is the tow drop region considered as a gap, then at that position resin properties are assigned to the respective element stack. In case of an overlap, the thickness assigned to the element stack is twice the thickness. It is important to emphasize that a tow drop defect is only identified if the respective element centroid is located over it. In consequence, a very refined mesh is required to have a good precision which is associated to a higher computational cost. A similar strategy was developed by Fayazbakhsh [17,18]. However, each element properties are assigned in accordance with its volume fraction of gaps and overlaps using micro-mechanics. In this way, it is not required such a refined mesh and computational cost is saved. Several authors considered this modeling methodology to include the effect of both gaps and overlaps [19–21]. Marouene et al. [19] study considered this last methodology to model variable stiffness plates with the goal of maximizing the in-plane stiffness and the buckling load. The numerical results obtained were then validated with experimental studies and a good accordance between them was found. Akbarzadeh et al. [20] study was focused on evaluating the impact that AFP manufacturing defects, namely, gaps and overlaps, have on the laminate properties of variable stiffness plates. Both static bending, buckling and free vibration have been studied. It also examined the role of shear deformation on those plates. On the one hand, for very thin plates, all equivalent single layer theories used provided similar results. On the other hand, the results of moderately thick plates showed some discrepancies, especially between those obtained using the Classical Laminate Theory (CLPT) and the Third Order Shear Deformation Theory (TSDT). It was verified an increase of the buckling load and fundamental frequency when compared to the steered plate with no defects, if overlaps were considered. When gaps were the plate manufacturing defect, a decrease in both mentioned parameters was verified.

In addition to the studies presented in the previous paragraph, other novel methods to include manufacturing defects were developed in [22,23]. Brooks and Martins [22] treated tow paths as the streamlines of a unit-vector field. In the developed mathematical formulation the gap/overlap formation is related with the divergence of a 2D vector field. A positive divergence indicates a gap formation and a negative value the occurrence of an overlap. Mishra et al. [23] explored the effect of gaps on the stiffness and buckling load of variable stiffness panels. The developed methodology used to account the effect of gaps showed to be significantly more efficient than the Defect Layer Method described in [17,18].

Other authors considered instead more detailed and comprehensive 3D analysis [24]. Marrouzé et al. [25], Heinecke et al. [26] and Willie et al. [27] approach models gaps and overlaps in both laminate, lamina and micro structure level. Li et al. [28] developed 3D meshing tools to automatically generate ply by ply models with gaps and overlaps. It automatically inserts intra and inter-ply cohesive elements in order to capture the influence of splitting and delamination. As a result, out-of-plane waviness and ply thickness variations are automatically modeled. These detailed 3D finite element models can capture effects like load redistribution and stress localization. Therefore, these type of analysis could be justifiable if higher levels of detail are necessary.

Regarding experimental studies, in Antunes et. al. study [29], which is the a continuation of Rodrigues et al. study [30], experimental modal analysis were performed on a rectangular plate considering various boundary conditions including one edge clamped and the remaining edges free (CFFF), two opposite edges clamped and the other two free (CFCF) and all the edges free (FFFF). Moreover, modal damping ratios were also identified in addiction to both natural frequencies and mode shapes. In this article, the theoretical model is based on two p-version type finite element models, one based on the Classical Plate Theory (CPT) and another based on the First-order Shear Deformation Theory (FSDT). Similar results were obtained using the two models, which can be justified by the use of a very thin plate. The authors reported that the discrepancies found between the experimental and numerical results were due to the occurrence of gaps and overlaps in the plate, which the FEM model did not take into account.

In this work, an optimization framework is developed with the objective of designing composite plates with curvilinear fibers for maximum fundamental frequency. Square and rectangular plates under fully clamped and fully simply supported boundary conditions are optimized using a Genetic Algorithm. The optimization is done first considering only straight fiber and then using curvilinear fibers with the purpose of evaluating the difference between the maximum frequencies obtained. The optimization of curvilinear fiber plates is performed considering both the absence and different values of the maximum curvature constraint of the AFP machine, thereby allowing to assess its effect on both maximum frequency and optimal designs. Moreover, the optimization of curvilinear fiber plates is performed considering the absence and the presence of embedded gap defects to evaluate its effect on both maximum frequencies and optimal designs. Gap imperfections are taken into consideration using the Defect Layer Method [17]. The authors consider that the consideration of defects within the optimization framework is the principal contribution of this work.

The remainder of this article is organized as follows: first, the variable stiffness plates considered are described in Section 2. This Section also includes the methodology for modeling the VATs and for including the effect of gaps in the FE model. The optimization problem is formulated in Section 3. Section 4 presents the developed framework used to solve the optimization problem. Section 5 discusses and analyses the results obtained. Finally, Section 6 presents the main conclusions drawn of this work.

2. Materials and Methods

2.1. Variable Angle Tow Laminates

In the AFP process, a band of fibers, denominated course, is composed by individual units named tows. The courses are deposited following a reference fiber path which can be curvilinear. Different curvilinear fiber paths have been studied by various authors. In this work, the fiber path presented in the following section is the one considered to solve the optimization problem.

2.1.1. Fiber Path Definition

The reference course fiber path considered was first introduced in [31] in 1993. In this formulation, the fiber angle (θ), measured with respect to the x^* axis, varies linearly along the axis x^*. The respective fiber angle variation is presented in Equation (1). The fiber path trajectory is presented in Equation (2) [32,33]. The variable T_0, represents the value of θ in the middle of the plate, while the variable T_1 represents the value of θ when $x^* = \pm a/2$. The variable ϕ represents the orientation of x^* relative to the global axis x. This fiber path is represented in Figure 1a.

In this work, ϕ is considered as zero. Thus, both axis x and x^* coincide and T_1 represents the fiber angle at both edges in x of a plate. The middle and edge fiber angles are written as $< T_0, T_1 >$. As an example, the fiber path correspondent to $< 60°, 15° >$ is presented in Figure 1b for a plate with length a in the x axis.

$$\theta(x^*) = \phi + (T_1 - T_0)2\frac{|x^*|}{a} + T_0 \tag{1}$$

$$y^*(x^*) = \begin{cases} \frac{a}{2(T_1-T_0)}[-\ln(\cos(T_0))+ \\ \ln\left(\cos\left(T_0 + \frac{2(T_0-T_1)}{a}x^*\right)\right)], -\frac{a}{2} \leq x^* \leq 0 \\ \frac{a}{2(T_0-T_1)}[-\ln(\cos(T_0))+ \\ \ln\left(\cos\left(T_0 + \frac{2(T_1-T_0)}{a}x^*\right)\right)], 0 \leq x^* \leq \frac{a}{2} \end{cases} \tag{2}$$

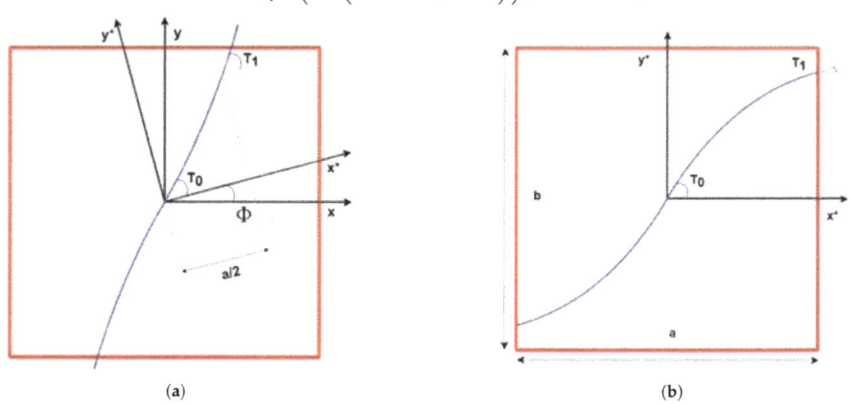

Figure 1. Reference fiber path: (**a**) Reference fiber path, with $\phi \neq 0°$; (**b**) Reference fiber path $< T_0, T_1 >=< 60°, 15° >$, with $\phi = 0°$.

2.1.2. Induced Defects

During the deposition of the course, the AFP machine's head is perpendicular to the local fiber angle. Thus, each point along the AFP head has the same orientation as the one corresponding to the reference path. Therefore, the points that form each tow of the course can be calculated using Equation (3). In this equation, n_t represents the number of tows in a course, t_w is the tow width and i is the tow index which range increases by one from $-n_t/2$ to $n_t/2$. Moreover, the superscript ref is used to denote the points in the reference path.

$$x = x^{ref} - i \cdot t_w \sin\left(\theta^{ref}\right)$$
$$y = y^{ref} + i \cdot t_w \cos\left(\theta^{ref}\right) \quad (3)$$

The fiber paths of the adjacent courses are obtained by moving the centerline fiber path in the y direction with the shift distance, which leads to the formation of manufacturing defects within the laminate. The shift distance (d_s) considered is the minimum vertical distance between the top and bottom boundaries of the reference course and it is calculated as presented in Equation (4). The points of the shifted course (y_s, x_s) are obtained as presented in Equation (5).

$$d_s = min(\frac{n_t t_w}{\cos\left(\theta^{ref}\right)}) \quad (4)$$

$$x_s = x^{ref} - i \cdot t_w \sin\left(\theta^{ref}\right)$$
$$y_s = y^{ref} + i \cdot t_w \cos\left(\theta^{ref}\right) + d_s \quad (5)$$

The deposition of both reference and adjacent courses is presented in Figure 2a considering a course with 8 tows. The reference course and its reference curve are represented in black and green, respectively. The adjacent course and its reference curve are represented in blue and red, respectively. In Figure 2a the intersection points between the reference and adjacent course are also visible. These intersections lead to the formation of defects which can be either gaps or overlaps depending on the way the tows are cut. If the entire course is cut or restarted when there is an intersection with the adjacent course, a large gap or overlap area is formed. So, in order to solve this problem and minimize the defect area, the tow cut and restart capability of the AFP machine is used. It allows to individually control each tow, to cut and restart its deposition and leads to a reduced defect area. In this work, an one-sided cut strategy is employed, which means that the tows are only cut on one side of each course. It is also used the zero percent coverage parameter in the tow drop strategy, which implies that the cut is performed as soon as one edge of a tow intersects the boundary of the adjacent course, also known as complete gap strategy or 0% coverage parameter. The gap formation which results from this cut strategy is presented in Figure 2b, where the gap set is represented in a gray colour.

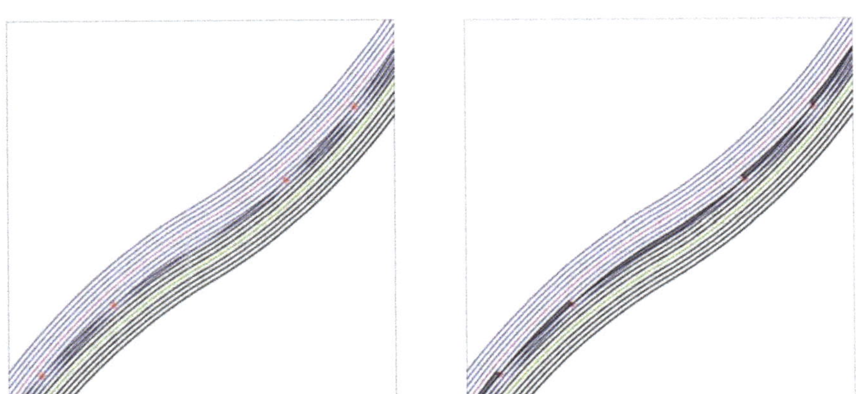

(a) (b)

Figure 2. (**a**) Tow cut and restart positions; (**b**) Gap pattern between consecutive courses.

The distance between two consecutive gap sets is the shift distance (Equation (4). So, the gap set is translated vertically with the shift distance across a ply. The gap pattern over an entire ply is represented in Figure 3a for a square plate, but considering that the AFP machine deposits 32 tows, each one with a width of 3.175 mm. The gap distribution is then intersected with the ply mesh in order to calculate the gap area fraction (A_m) within each element as represented in Figure 3b. Each element equivalent properties are then calculated using a 'modified' rule of mixtures (Equations (6)–(10)). In these equations the subscript c represents the non-defective composite and the subscript m the matrix or gaps. The composite area fraction, A_c, and the gap area fraction, A_m, are related by $A_m = 1 - A_c$. It is important to note that both area fractions are equivalent to volume fractions because it is assumed a constant thickness across the ply, with or without gaps.

$$E_1 = A_c E_{1c} + A_m E_m \tag{6}$$

$$E_2 = \frac{E_{2c} E_m}{A_m E_{2c} + A_c E_m} \tag{7}$$

$$G_{12} = \frac{G_{12c} G_m}{A_m G_{12c} + A_c G_m} \tag{8}$$

$$G_{23} = A_c G_{23c} + A_m G_m \tag{9}$$

$$\nu_{12} = A_c \nu_{12c} + A_m \nu_m \tag{10}$$

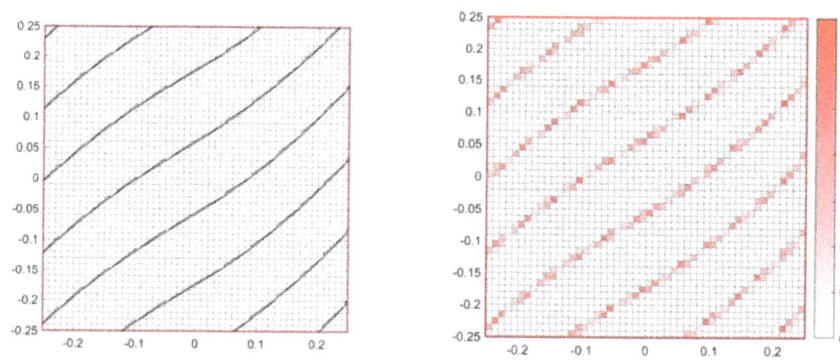

Figure 3. (**a**) Gap distribution on a plate (**b**) Gap area fraction in each element.

2.1.3. Manufacturing Constraints

The maximum curvature constraint or inverse, minimum curvature radius, depends on the AFP machine and limits the maximum value of curvature of the reference fiber path of each course. Typical curvature constraints are presented in Table 1. This constraint limits the design space by limiting the fiber angle distribution that each lamina possesses. This constraint is dependent on the AFP machine and must be evaluated before the production stage in order to ensure that the composite can be manufactured. For that reason, it is most often used as an optimization constraint when VAT laminates are optimized to maximize or minimize determined characteristics.

Table 1. Curvature Constraint, adapted from [34].

Tow Width	Typical Minimum Turning Radius	Typical Maximum Curvature
(mm)	(mm)	(m^{-1})
3.175	635	1.57
6.35	1778	0.56
12.7	8890	0.1125

The curvature equation was analytically obtained using the fiber reference path presented in Equation (1) and also the definition of curvature of a function with a a single variable presented in Equation (11).

$$K = \frac{\frac{d^2 f(x)}{dx^2}}{(1 + (\frac{df(x)}{dx})^2)^{3/2}} \tag{11}$$

$f(x)$ was substituted in Equation (11) by $y^*(x^*)$ of the reference fiber path (Equation (2)) treated here as $y(x)$, because $\phi = 0$. The first and second derivatives of y with respect to x are calculated in the following equations:

$$\frac{dy}{dx} = \tan(\theta) \tag{12}$$

$$\frac{d^2y}{dx^2} = \frac{d\tan\theta}{dx} = \frac{1}{\cos^2\theta}\frac{2(T_1 - T_0)}{a} \tag{13}$$

The trigonometric relation presented in Equation (14) is also considered:

$$1 + \tan^2\theta = \frac{1}{\cos^2\theta} \tag{14}$$

After substituting Equations (12)–(14) in Equation (11), the curvature equation for the reference fiber path considered is obtained (Equation (15)). This last equation was also considered in references [10,22].

$$K = 2\frac{T_1 - T_0}{a}\cos\left((T_1 - T_0)2\frac{|x^*|}{a} + T_0\right) \tag{15}$$

3. Optimization

The optimization problem that this work intends to solve is defined in Equation (16). The goal is to maximize the fundamental frequency, treated here as f_1, of a laminated plate. It is assumed that the laminate is symmetric and that the total number of plies (N_p) is equal to eight. The problem is solved considering different cases where the vector of design variables (**x**) depends on whether only rectilinear or curvilinear fibers (whose orientation is presented in Equation (1) are considered. The cases that only consider rectilinear fibers are treated as non-steered cases and the cases that allow curvilinear fibers are treated as steered cases. In the cases where only rectilinear fibers are considered ($T_0 = T_1$), the design variables correspond to the orientation of each layer, T_0 (Equations (17) and (18)) and, therefore, the number of design variables is four (one per ply for a half laminate). In the cases where steered fibers are considered, the design variables correspond to the angles T_0 and T_1 of each layer (Equations (17) and (18)). Consequently, there is a total of eight design variables (two per ply for a half laminate). Each design variable is allowed to continuously vary between $-90°$ and $90°$, except when gaps were considered. In those cases the domain of design variables is allowed to vary between $-89°$ and $89°$, due to singularities of the *Matlab* subroutine used. In some of the steered cases, the maximum curvature constraint, displayed in Equation (15), is imposed in order to guarantee that the laminate can be manufactured using an AFP machine. Thus, in some steered cases the curvature of each ply is constrained by the maximum value of curvature allowed (K_{max}). It is not necessary to impose the maximum curvature constraint in the non steered cases, because the fibers are rectilinear and therefore the curvature (K) is always zero. Thus, the optimization of the non-steered cases is an unconstrained problem. The boundary conditions considered are the fully clamped and fully simply supported boundary conditions.

$$\begin{aligned} \text{maximize} \quad & f_1(\mathbf{x}) \\ \text{by varying} \quad & \mathbf{x} \in \mathbb{R}^n \\ \text{subject to} \quad & K_n(\mathbf{x}) \leq K_{max}, \quad \text{if steered} \\ & n = 1, 2, \ldots\ldots, N_p \end{aligned} \tag{16}$$

$$\begin{aligned} \text{given} \quad & \mathbf{x} = [x^{(1)} \ \ldots \ x^{(i)} \ \ldots \ x^{(N_p)}]^T \\ \text{such that} \quad & -90^\circ < x^{(i)} < 90^\circ \end{aligned} \tag{17}$$

$$\text{where} \quad x^{(i)} = \begin{cases} [T_0^{(i)}], & \text{if non-steered} \\ [T_0^{(i)}, T_1^{(i)}], & \text{if steered} \end{cases} \tag{18}$$

The notation used to distinguish the different cases is orientation distribution—boundary condition—unconstrained/constrained. The orientation distribution notation can be NS (non-steered), if only non-steered plies are considered or LS (linearly steered), if the fiber path orientation varies linearly across x^* (Equation (1)). The boundary conditions fully clamped (CCCC) and fully simply suported (SSSS) are treated here as C and S, respectively. The unconstrained and constrained cases are denominated UN or CON, respectively. In the constrained cases, it is considered a maximum curvature constraint (K_{max}) with two different values: 1.57 m^{-1} [16], treated here as constraint A, and 3.28 m^{-1} [32], denominated here as constraint B. For instance, LS-C-CON-A is a plate with a linear fiber angle variation, fully clamped and with a maximum curvature constraint of 1.57 m^{-1}.

The plates considered are symmetric with a total of eight plies, each one with a thickness of 0.159 mm. The material properties considered are displayed in Table 2. The length in the x^* axis, a, is kept constant and equal to 0.5 m throughout all the plates. A square plate and two rectangular plates with aspect ratios (a/b) equal to 0.5 and 2.0 are considered.

Table 2. Carbon epoxy Cytec® G40-800/5276-1 and resin properties [17].

Property	E_1 [GPa]	E_2 [GPa]	G_{12} [GPa]	ν_{12}	ρ [kg/m^3]
G40-800/5276-1	143	9.1	4.8	0.3	1650
resin	3.7	3.7	1.4	0.3	1310

4. Implementation

The developed framework can be divided into two main components: the *Matlab* and the Finite Element Model (*Python*/ABAQUS) environments as it is represented in Figure 4. The optimization process starts in the *Matlab* where the optimizer, *Matlab*'s built-in genetic algorithm, is implemented and where each generation's population is created. After that, the population information is passed to a *Python* script where the geometry of the plate, property assignment, boundary and loading conditions are defined before running the structural analysis using *ABAQUS*. The input data generated depend if the case considered only non-steered, steered without imperfections and steered plates with imperfections. Following the structural analysis, its results are then again passed to the *Matlab* environment where they are post-processed and the resulting fitness is evaluated by the genetic algorithm optimizer. If there is convergence, the optimization stops, otherwise another population is created and the optimization process continues in the next generation.

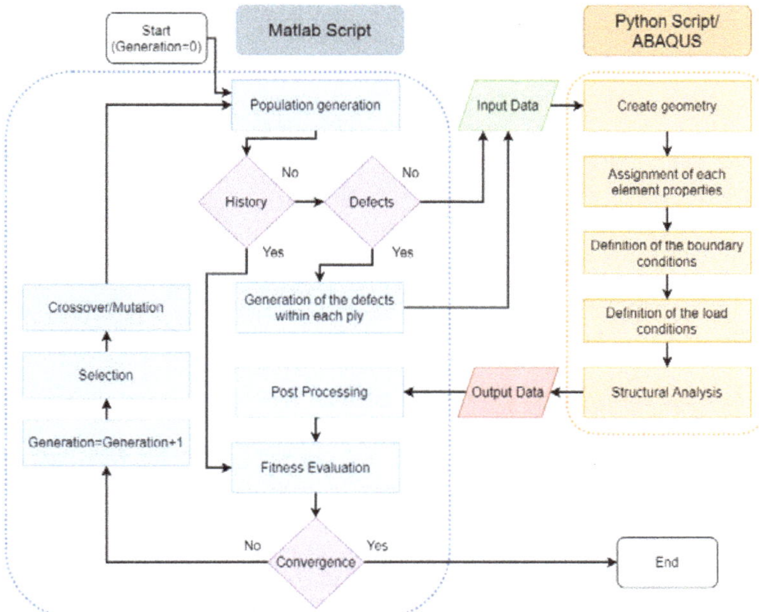

Figure 4. Framework scheme.

4.1. Finite Element Model

The geometry creation, element property assignment as well as the structural analysis is defined in *Python* scripts and performed by *ABAQUS*. The orientation of the fibers and material properties can differ in each iteration and it is generated as the input data by *Matlab*. In the non-steered cases all the elements of a ply have the same orientation, so the orientation is defined layer by layer and therefore *Matlab* only passes each layer orientation to the *Python Script*. In the steered cases, the element orientation in a ply is calculated in accordance with Equation (1), where the fiber orientation varies linearly along the x^*. As a consequence, the orientation can no longer be defined layer by layer and has to be defined element by element in each layer which leads to an increase of computational cost. *Matlab* generates each ply T_0 and T_1 for half laminate (due to its symmetry) which are read by the *Python* script. Then, each element orientation in a ply is assigned by substituting in Equation (1) each element centroid position in x^* and its respective ply T_0 and T_1. If defects are being considered in the steered case, each element property is dependent on its resin percentage. So, not only the orientation is defined element by element in each layer, but also the material properties. For that reason, the steered cases considering defects are those that require a higher computational cost.

It is chosen a mesh with a total of 2500, 5000, 1250, FSDT based shell S4R mesh elements [35] for the square plates and plates with $a/b = 0.5$ and $a/b = 2.0$, respectively. This choice was made after a mesh convergence study with the objective of simultaneously assuring a good compromise between the model convergence and the computational cost necessary to run a structural analysis.

4.2. Matlab Environment

The main script starts by generating a population. The problem is constrained and the generated population should satisfy the optimization constraints which are defined in a different script. After this, each individual is checked by a script that saves the optimization history. That history has saved all the individuals already analysed and their respective fitness evaluation. So, in case a certain individual has already been evaluated, then there is no need to run the structural analysis and this step can be skipped. This saves computational time during the optimization process. If the individual has not yet been evaluated, the respective input data needed to run the Python Script and ABAQUS CAE is created.

The creation of input data depends on whether the plate considered is non-steered or steered, and if defects are or are not being considered. It is important to note that only symmetrical composites were considered in all cases, so the input data is only referred to half of the laminate. When defects are considered, namely gaps, they are modeled in a *Matlab* subroutine in the same way as in Fayabazh et al. [18]. Afterwards, each element resin percentage is calculated and passed to the Python Script where the equivalent element properties are calculated [18].

Following the input data creation the next steps that include the generation of the plate geometry, property and fiber path assignment, boundary and loading conditions definition and structural analysis are performed by the *Python* script which used *ABAQUS* to run the structural analysis. After the structural analysis is finished, the output data, in this case, the plate natural frequency, is passed to the *Matlab* script that defines the objective function in order to evaluate its fitness. Finally, after all the individuals of a generation are evaluated, the optimizer checks if the convergence has yet been achieved. If it meets the convergence criterion, the optimization process is terminated. Otherwise, the optimizer generates the next population using genetic operators such as selection, crossover and mutation and the optimization process continues. A population size of 80, a function tolerance of 10^{-5}, a maximum number of stall generations equal to 25 and a constraint tolerance of 10^{-6} are considered. This choice was made after a study where these parameters were varied with the objective of minimizing the computational cost necessary during the optimization process while assuring a very high probability of finding the optimal solution.

4.3. Validation

The finite element model was verified by comparing its results with the linear natural frequencies obtained in the study performed by Akhavan and Ribeiro [10] and by Akbarzadeh et al. [20].

In the first study, imperfections are not considered and the fiberpath considered is also the one presented in Equations (1) and (2). The respective plate dimensions and material properties are presented in Table 3. The results, namely, the first three natural frequencies are shown in Table 4, where a fully clamped boundary condition was considered. In the same table it is possible to verify that the results obtained using the finite element model developed in this work are in accordance with those obtained by Akhavan and Ribeiro [10].

The plate properties considered in the second study are presented in Table 5, where square plates ($a = b = 1$ m) are considered. The comparison of results is presented in Table 6, where the dimensionless fundamental frequency ($\bar{\omega} = \frac{\omega a^2}{h}\sqrt{\frac{\rho}{E_2}}$) calculated using the developed framework and obtained by Akbarzadeh et al. [20] are displayed and compared. It is worth mentioning that the results were compared considering a defect free plate and the same one with complete gap manufacturing defects. It is also important to note that Akbarzadeh et al. [20] considered a plate with a constant curvature fiber path, which can be the cause of the slight difference in the results displayed in Table 6.

Table 3. Characteristics of plate 1 [10].

Characteristic	a [m]	b [m]	h [mm]	E_1 [GPa]	E_2 [GPa]	G [GPa]	ν_{12}	ρ [kg/m^3]
Value	1	1	0.01	173	7.2	3.76	0.29	1540

Table 4. Natural frequencies [10].

Layup [°]		f_1 [Hz]	f_2 [Hz]	f_3 [Hz]
[<0, 45>, <−45, −60>, <0, 45>]	Ref. [10]	579.398	821.532	1225.79
	Abaqus	579.790	822.538	1230.008
	Difference (%)	0.07	0.12	0.34
[<30, 0>, <45, 90>, <30, 0>]	Ref. [10]	667.177	862.919	1234.64
	Abaqus	667.492	863.899	1239.187
	Difference (%)	0.05	0.11	0.37
[<90, 45>, <60, 30>, <90, 45>]	Ref. [10]	710.771	912.183	1335.49
	Abaqus	710.597	912.246	1336.922
	Difference (%)	0.02	0.01	0.11

Table 5. Prepreg and resin properties [20].

Property	E_1 (GPa)	E_2 (GPa)	G_{12} (GPa)	G_{13} (GPa)	G_{23} (GPa)	ν_{12}	ρ kg/m^3
Prepreg	143	9.1	4.82	4.9	4.9	0.3	1500
Resin	3.72	3.72	1.43	1.43	1.43	0.3	1100

Table 6. Dimensionless fundamental frequency—$\bar{\omega}$.

a/h	Layup	Manufacturing Defects	ω—Literature Results FSDT ($c_f = 5/6$) [20]	ω—Present Results	Difference (%)
200	$[\pm<58\,|\,39>]_{4s}$	Defect free	16.4566	16.2358	1.34
		Complete gap	15.7916	15.7874	0.02

5. Results and Discussions

This section is dedicated to present and discuss the optimization results which aim to maximize the fundamental frequency.

5.1. Non Steered Plates

This section has the goal of presenting the optimal results of plates that only consider non steered plies. The first natural frequency is presented in Table 7 as well as the optimal orientation of each ply (T^*_{0i}). The computational cost is represented by the number of generations (Gen.) required to perform the optimization.

Focusing first on the square plates, it is observable in Table 7 that the fully clamped supported case achieves a higher first natural frequency (51.601 Hz) than the one obtained in the simply supported case (31.223 Hz). It is also noticeable that the optimal orientation of each ply is different in both cases. On the one hand, in the NS-C, the optimal orientation alternates between approximately −90° and nearly 0°. On the other hand, in the NS-S, the exterior ply and its respective symmetric one have an orientation of approximately 45°, while all the interior ones have an orientation near −45°. That again proves that the optimal orientation of each ply is dependent on the imposed boundary conditions. Comparing now both cases, but focusing on the last column of Table 7, it is possible to verify that both required the same number of generations to perform the optimization.

The non steered rectangular optimal results are also presented in Table 7. When the aspect ratio is equal to 0.5, the optimal orientations are near zero degrees for all plies and for both boundary conditions. A similar behaviour is observed for an aspect ratio of 2, where all plies have an orientation near 90° for both boundary conditions. So, for non steered plies, these two cases are equivalent, because all plies are oriented parallel to the smaller side of the plate for both boundary conditions considered. Regarding the computational cost, it is visible that all the non steered cases have a similar number of function evaluations due to the fact that all of them have the same number of design variables. The small discrepancy is explained by the randomness associated to the genetic algorithm used.

Table 7. Non Steered Results.

	Design	T^*_{01} [°]	T^*_{02} [°]	T^*_{03} [°]	T^*_{04} [°]	f^*_1 [Hz]	f^*_2 [Hz]	f^*_3 [Hz]	Gen.
$a/b = 0.5$	NS-C	0.0	0.0	−0.1	0.1	49.260	50.717	53.943	39
	NS-S	1.6	−1.0	−0.9	−2.4	22.042	23.963	28.207	44
$a/b = 1.0$	NS-C	−89.9	0.6	−89.5	0.5	51.601	92.924	119.100	55
	NS-S	45.0	−45.0	−45.0	−45.0	31.223	70.371	73.737	55
$a/b = 2.0$	NS-C	−90.0	89.9	89.9	−89.5	197.443	203.191	216.032	61
	NS-S	−89.9	89.9	−89.8	−89.5	88.275	95.840	112.720	43

5.2. Defect Free Steered Optimization Results

The optimal results for the ideal square steered plates are presented in Table 8 and shown in Figure 5. The optimization is performed considering unconstrained and constrained cases with different maximum curvature constraints ($K_{max} = 1.57$ m^{-1} (A) and $K_{max} = 3.28$ m^{-1} (B)) in order to assess the curvature effect on the final solution. All these cases are optimized for fully clamped and fully simply supported boundary conditions. Table 8 displays the optimal T_0 and T_1 for the first four plies (because the composite laminate is symmetric) and the respective first three natural frequencies for the cases considered. It is also visible the number of generations (Gen.) required to perform the optimization.

The optimization results for a square plate considering the fully clamped boundary condition are presented in columns four to six of Table 8. The unconstrained optimization (no maximum curvature considered), the constrained case A ($K_{max} = 1.57$ m^{-1}) and the constrained case B ($K_{max} = 3.28$ m^{-1}) are presented in the fourth, fifth and sixth columns, respectively. An observation that can be made is that the unconstrained case (LS-C-UN) is the one that achieves the highest first natural frequency (59.471 Hz) which represents an increase of 15.25% with respect to the square NS-C case (Table 7). However, six plies of this case have a maximum curvature ($K_{max} = 6.42$ m^{-1} for ply 1 and 8, $K_{max} = 6.52$ m^{-1} for ply 2 and 7, $K_{max} = 6.53$ m^{-1} for ply 3 and 6) higher than the K_{max} values considered here. Consequently, this laminate can not be manufactured by an AFP machine. In spite of that, the result of the unconstrained case is useful not only to show where the optimal solution is, but also as, with the development of new manufacturing techniques that enable to produce plates with higher fiber curvatures, it could become a feasible design in the close future.

Focusing now in both constrained cases, LS-C-CON-A ($K_{max} = 1.57$ m^{-1}) and LS-C-CON-B ($K_{max} = 3.28$ m^{-1}), Table 8 shows that LS-C-CON-B is the case with the highest fist natural frequency (56.814 Hz) between the two. That constitutes an increase of 10.10% with respect to the square NS-C case (Table 7), while the LS-C-CON-A increase is only 4.58%. LS-C-CON-B is also the constrained case that allows a higher maximum curvature, which, again, reinforces the influence of this parameter in the frequency obtained. It is also worth noting that for both cases, K_{max} is an active constraint for the first two layers and their respective symmetric ones. Therefore, it can be argued that lowering the maximum curvature allowed tends for the frequency to decrease. Regarding plies four and five of LS-C-CON-B, their maximum curvature approaches zero similarly to the unconstrained case. That is also the case for plies three and six in the LS-C-CON-A case.

The optimization results considering the fully simply supported boundary condition case are presented in columns seven to nine of Table 8. The unconstrained optimization (no maximum curvature considered), the constrained case A ($K_{max} = 1.57$ m^{-1}) and the constrained case B ($K_{max} = 3.28$ m^{-1}) are presented in the seventh, eighth and ninth columns, respectively. The unconstrained optimal solution (LS-S-UN) has a first natural frequency equal to 31.229 Hz which is very similar to the one achieved by both constrained cases. The same phenomena is verified with respect to the second and third natural frequencies. So, it is no surprise, that all the cases have similar T_0 and T_1 combinations and, consequently, also similar maximum fiber curvatures (K^*_{max}) in each ply. Regarding the K^*_{max} in each ply, it is important to note that its value in all cases and in all plies is near zero, which implies that the maximum curvature is not an active constraint. It also means that the unconstrained case is possible to manufacture using the AFP machine (because the maximum curvature of all plies is lower than both K_{max} considered here) and that the first maximum natural frequency achieved is near the one obtained considering only rectilinear fibers. This last affirmation can be proved by comparing the first natural frequency values obtained in Table 8 with the one considering only non steered fibers in Table 7 for a fully simply supported boundary condition. The respective increase of the first natural frequency caused by the introduction of steered fibers is only 0.02%. So, it can be concluded that the use of non steered fibers is almost as efficient as steered fibers in order to maximize the first natural frequency of a square plate with a fully simply supported boundary condition.

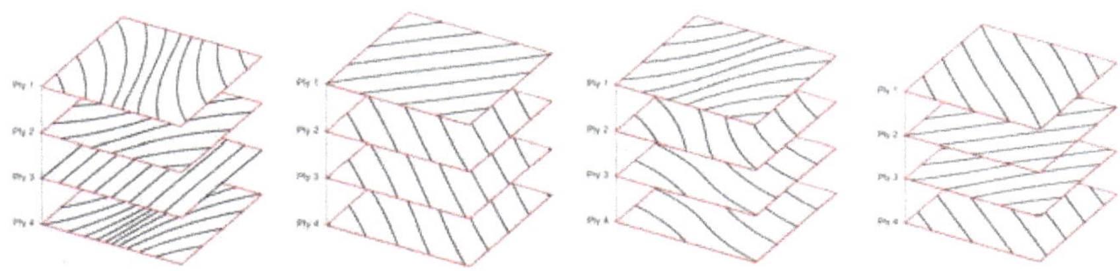

(a) (b) (c) (d)

Figure 5. Square plate optimal reference fiber paths layouts. (**a**)LS-C-CON-A defect free; (**b**) LS-S-CON-A defect free. (**c**) LS-C-CON-A with complete gap. (**d**) LS-S-CON-A with complete gap.

Focusing now on the plates with $a/b = 0.5$ (visible in Figure 6), columns two and three of Table 8, it is observable that the orientation of each ply is near 0° for the fully clamped boundary condition. Therefore, the optimal solution found is similar to the one obtained considering only rectilinear fibers (Table 7). So, it appears that the use of curvilinear fibers is not justifiable in this case. On the contrary, for the simply supported boundary condition, there is a slight improvement of 1.49% on the optimal solution with the introduction of curvilinear fibers. The influence of the curvature of the fibers is especially relevant because the maximum curvature is an active constraint in the first two plies and their symmetric ones ($K_{max} = 1.57$ m^{-1}). Regarding the plates with $a/b = 2.0$ (last two columns of Table 8), the maximum curvature is also an active constraint for the fully simply supported boundary condition. For this case, the orientation in the middle of all plies (T_0) is near ±90°, which is also the optimal orientation of all plies for the correspondent non steered case (Table 7). The orientation at the edge of the plies is clearly restricted by the maximum curvature constraint, that, as aforementioned, is an active constraint in all plies. Analyzing now the case with the fully clamped boundary condition, it is visible that all plies have almost zero curvature and an orientation near ±90°. So, the optimal solution obtained is very similar to the one achieved in the non steered case.

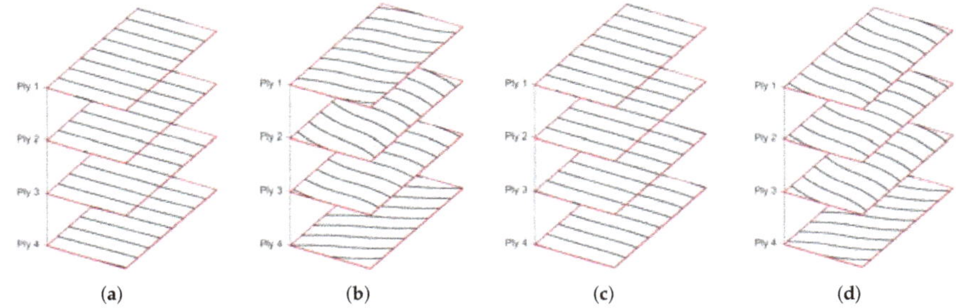

Figure 6. Rectangular plate ($a/b = 0.5$) optimal reference fiber paths optimal layouts. (**a**) LS-C-CON-A defect free; (**b**) LS-S-CON-A defect free; (**c**) LS-C-CON-A with complete gap; (**d**) LS-S-CON-A with complete gap.

Table 8. Optimal defect free steered results.

	$a/b = 0.5$		$a/b = 1.0$						$a/b = 2.0$	
Design	LS-C	LS-S	LS-C			LS-S			LS-C	LS-S
	CON	CON	UN	CON		UN	CON		CON	CON
	A	A	-	A	B	-	A	B	A	A
$<T_0^\circ, T_1^\circ>_1^*$	<0.0, 0.3>	<0.1, 22.6>	<90.0, −1.9>	<−74.9, −43.7>	<−81.8, −28.4>	<−42.8, −45.9>	<42.3, 46.0>	<42.1, 46.0>	<89.9, 90.0>	<−90.0, −52.9>
$<T_0^\circ, T_1^\circ>_2^*$	<0.2, 0.1>	<−5.4, −28.0>	<−89.7, 3.7>	<63.3, 35.7>	<74.9, 23.7>	<47.8, 44.0>	<−47.1, −43.8>	<−48.0, −43.7>	<−89.6, −89.5>	<90.0, 52.9>
$<T_0^\circ, T_1^\circ>_3^*$	<1.4, −0.5>	<2.5, −17.1>	<−90.0, 3.6>	<−90.0, −89.2>	<87.1, 32.4>	<47.9, 43.9>	<−48.5, −43.9>	<−49.1, −43.8>	<89.2, 88.6>	<88.8, 52.2>
$<T_0^\circ, T_1^\circ>_4^*$	<1.5, 2.7>	<20.4, 32.6>	<−5.4, −5.9>	<80.9, 48.2>	<2.0, 2.6>	<42.7, 45.3>	<−49.9, −43.0>	<−47.3, −43.9>	<89.4, 88.3>	<90.0, 52.9>
$f_1^*(Hz)$	49.258	22.370	59.471	53.962	56.814	31.229	31.229	31.229	197.436	90.146
$f_2^*(Hz)$	50.715	26.603	95.181	85.656	90.604	70.595	70.601	70.648	203.181	105.580
$f_3^*(Hz)$	53.943	33.571	129.029	126.412	129.466	73.518	73.463	73.445	216.032	138.403
Gen.	59	66	66	89	76	74	55	80	82	117

5.3. Defect Free Steered Results with Complete Gap

The previous section results do not consider the occurrence of any type of defect inside the plate, which is a limitation of the model used. Therefore, the results would be different from those shown. In order to better estimate that difference, the three first natural frequencies of the previous constrained optimal layouts are calculated considering the existence of gaps caused by the one sided cut with a zero percent coverage parameter of an AFP machine. The results are shown in Table 9 for square and rectangular plates considering the deposition of courses with 8, 16 and 32 tows. Each tow has a width of 3.175 mm. It is also visible the volumetric gap fraction ($v_G[\%]$) of each analysed case. As previously mentioned, in a one sided cut strategy the tows of each course are only cut and restarted on one side of the course, where the tow drop areas are formed. By fixing each tow width, if the number of tows increases, so does the total course width. In consequence, the number of course intersections inside the plate decreases, which leads to a reduction of the tow drop area. So, as expected, it is clear that the volumetric gap fraction increases in Table 9 with the decrease of the number of tows.

Another important aspect to notice in Table 9 is that all the first natural frequencies decreased when compared with those obtained considering defect free steered plates

(Table 8). It is also observable that with the increase of the volumetric gap fraction, the fundamental frequency diminishes. This could imply that the existence of gaps is responsible for the decrease of the first natural frequency. The stiffness reduction in the areas where gaps are located can be the explanation for that frequencies' decrease. It is also worth mentioning that, in spite of that, the use of steered fibers in a square plate with a fully clamped boundary condition considering 32 or 16 tows can still achieve a higher first natural frequency when compared to a plate with non steered fibers (Table 7) for both K_{max}. However, if the boundary condition used is a fully simply supported one, the results show that the opposite happens. The highest first natural frequency is instead achieved by the plate that uses non steered fibers. This may be explained by the fact that when considering ideally steered square plates, the obtained results were very similar to those obtained for non steered plates (only 0.02% increase). When further adding the effect of gaps, the first natural frequency decreases past the one obtained for non steered square plates. The same argument could be used regarding the fully clamped rectangular plate cases, where the achieved frequencies are lower than those obtained in the respective optimal non steered plates as expected. In the rectangular plates simply supported cases, it is observed that with the decrease of the number of tows, the gap area increases, and the fundamental frequency decreases until it is no longer advantageous to have curvilinear fibers considering this manufacturing process.

Table 9. Optimal defect free steered results with complete gaps introduced.

$a/b = 1.0$												
Design	**32 tows**				**16 tows**				**8 tows**			
	LS-C-CON		**LS-S-CON**		**LS-C-CON**		**LS-S-CON**		**LS-C-CON**		**LS-S-CON**	
	A	**B**	**A**	**B**	**A**	**B**	**A**	**B**	**A**	**B**	**A**	**B**
f_1^* [Hz]	53.280	55.970	31.020	31.008	52.634	54.889	30.789	30.776	51.447	52.747	29.933	30.007
f_2^* [Hz]	84.742	89.295	70.128	70.149	83.879	87.946	69.705	69.673	81.921	85.323	67.829	68.087
f_3^* [Hz]	124.606	127.393	72.946	72.912	122.9	124.727	72.291	72.296	120.084	119.229	70.243	70.35
v_G [%]	3.51	3.14	1.72	1.71	6.65	6.49	3.47	3.54	10.86	13.51	8.81	9.17

	$a/b = 0.5$						$a/b = 2.0$					
Design	**32 tows**		**16 tows**		**8 tows**		**32 tows**		**16 tows**		**8 tows**	
	LS-C	**LS-S**	**LS-C**	**LS-S**	**LS-C**	**LS-S**	**LS-C**	**LS-S**	**LS-C**	**LS-S**	**LS-C**	**LS-S**
	CON-A						**CON-A**					
f_1^* [Hz]	48.698	22.171	48.504	21.926	47.732	21.367	196.588	85.604	194.405	81.433	190.883	72.004
f_2^* [Hz]	50.113	26.364	49.919	26.087	49.088	25.460	202.152	101.618	199.172	100.132	195.686	96.127
f_3^* [Hz]	53.262	33.259	53.033	32.910	52.112	32.132	214.851	133.890	212.234	130.241	207.145	123.160
v_G [%]	2.94	1.99	4.30	4.31	8.57	9.51	1.66	7.01	4.26	12.39	9.67	22.26

5.4. Complete Gap Steered Optimization Results

It was shown in the latest section that the occurrence of gaps leads to a decrease of the fundamental frequency, which can be correlated to the reduction of the elastic properties in the areas where gaps are located. It was also shown that the higher the gap percentage is, or the lower the number of tows used, the lower the fundamental frequency is. However, the optimization has not yet been performed considering the Defect Layer Method included in the finite element model built, which is interesting in order to verify not only how the defects affect the fundamental frequency, but also to check how they affect the optimal laminated layout. With that in mind, the results presented in the following sections assume that the AFP machine deposits 32 tows, each one with a width of 3.175 mm for both fully clamped and fully simply supported boundary conditions. Only constrained cases are presented in order to assure that the plates could be manufactured.

The square plate results are displayed in Table 10 for both boundary conditions and maximum curvature constraints values considered. As expected, all the optimal fundamental frequencies achieved are lower than the respective optimal ones obtained with the absence of defects. For instance, regarding the square plate cases with a fully clamped boundary condition, fourth and fifth columns of Table 10, it is observable that there is a decrease of approximately 0.83% and 2.65% in cases LS-C-CON-A (Figure 5) and B, respectively, relative to the equivalent ones but without considering defects (Table 8). Nonetheless, it is still verified an increase of 3.71% and 7.19%, respectively, with respect to the NS-C solution (Table 7). Therefore, it can be argued that although a decrease of the frequency is verified by the occurrence of gaps, it is still advantageous to use curvilinear fibers to maximize the frequency for square plates with this boundary condition. It is also important to note that the optimal layout for both cases differs from the one obtained when no defects are considered. This evidence can be verified by comparing the optimal designs from Tables 8 and 10. Nevertheless, the maximum curvature constraint is still an active one in some of the plies and, when it is not, the maximum curvature obtained is near it. The reason why the optimal layout has changed when gaps are considered can be explained by comparing the results from Tables 9 and 10 when considering 32 tows. In Table 9 it is visible that LS-C-CON-A and B have a gap volumetric fraction ($v_G[\%]$) of 3.51% and 3.14%, respectively, while in Table 10 1.88% and 2.21% are the respective gap volume fractions. The reduction in the occurrence of gaps is one of the factors that explains that the optimal solution considering gaps is superior to the one found considering ideal plates with gaps inserted afterwards. So, when the layout is optimized with the occurence of gaps included it may be said that in order to maximize the fundamental frequency there is a trade-off between having the desired fiber curvature and the minimum gap volume fraction.

Regarding the square plate, but focusing now on the fully simply supported boundary condition (sixth and seventh columns of Table 10), it is again visible that the maximum curvature of all plies is near zero. In consequence, the maximum curvature is not an active constraint in both cases A and B. This phenomenon has been previously observed when analysing the same cases, but without defects being considered (Table 8). The plate layout is also similar to the one previously presented with the orientation of each layer tending to $\pm 45^{\circ}$, so the frequencies obtained are very similar to those obtained considering only non steered fibers (NS-S, Table 7) and steered fiber but without gaps included (Table 8). Since the optimal solutions found in this section still present some curvature, although tending to zero, some gaps occur. In consequence, the optimal fundamental frequency is 0.56% and 0.46% lower in cases LS-S-CON A and B, respectively, when compared with the respective non steered optimal solution (Table 7). So, because of the previously stated reasons, it can be argued that it is more advantageous to use only rectilinear fibers for square plates with this boundary condition.

The rectangular plate results are visible in Table 10 considering only $K_{max} = 1.57\ \mathrm{m}^{-1}$ for both boundary conditions and aspect ratios. As it could be expected, the optimal fundamental frequencies achieved are lower than those obtained when defects are not considered. Focusing first on the plates with $a/b = 0.5$ (visible in Figure 6), there is a decrease of 0.31% and 0.74% for the LS-C-CON-A and LS-S-CON-A, respectively, with respect to their equivalent cases where complete gaps are not included (Table 8). On the one hand, in the fully clamped case, it is observable that the maximum curvature of each ply tends to zero. On the other hand, in the simply supported boundary condition, the constraint is active in some plies, as in the respective defect free cases. It is also worth noting that there is a decrease for the fully clamped boundary condition of approximately 0.31% when compared to the optimal solution considering only non steered plies (Table 7). As a result is can be said that it is preferable to use rectilinear fibers for this case. On the contrary, when the BC is simply supported there is an increase near 0.74% when compared to the respective optimal NS case (Table 7), even taking into consideration the occurrence of gaps.

Table 10. Optimal steered results with complete gaps.

	$a/b = 0.5$		**$a/b = 1.0$**				**$a/b = 2.0$**	
Design	**LS-C-CON**	**LS-S-CON**	**LS-C-CON**		**LS-S-CON**		**LS-C-CON**	**LS-S-CON**
	A	**A**	**A**	**B**	**A**	**B**	**A**	**A**
$<T_0^\circ, T_1^\circ>_1^*$	<−0.3 1.0>	<1.2 −23.5>	<66.0, 37.7>	<−53.9, −6.6>	<−41.2, −46.9>	<47.6, 47.8>	<87.4, 87.3>	<−84.1, −59.2>
$<T_0^\circ, T_1^\circ>_2^*$	<1.0 −0.4>	<−1.1 −23.6>	<−56.2, −30.2>	<65.3, 16.4>	<44.7, 44.9>	<−51.2, −41.1>	<−84.9 −81.7>	<84.9, 51.4>
$<T_0^\circ, T_1^\circ>_3^*$	<0.0 −2.1>	<−6.1 −28.6>	<−36.5, −13.4>	<80.0, 27.3>	<52.3, 41.8>	<−43.2, −45.5>	<87.9 86.9>	<86.2, 51.2>
$<T_0^\circ, T_1^\circ>_4^*$	<1.7 −4.0>	<9.4 30.0>	<−36.5, −13.4>	<55.2, 8.2>	<−39.6, −39.7>	<89.0, 34.0>	<−88.7 −83.2>	<89.0, 65.5>
$f_1^*(Hz)$	49.107	22.205	53.517	55.311	31.047	31.080	195.988	88.151
$f_2^*(Hz)$	50.548	26.415	100.660	105.858	69.767	69.389	201.819	105.223
$f_3^*(Hz)$	53.739	33.315	111.872	112.376	73.204	74.660	215.154	139.443
$v_G[\%]$	3.078	3.77	1.88	2.21	1.64	2.24	0.96	1.87
Gen.	66	62	122	75	83	88	84	91

A similar phenomenon occurs for plates with $a/b = 2.0$, however in these cases the frequencies achieved are much higher. The plate's maximum curvature tends to zero in all plies for the fully clamped boundary condition and it is near the maximum curvature constraint in some plies for the simply supported boundary condition (Table 10). It is verified a reduction in frequency of 0.73% and 2.21% for LS-C-CON-A and LS-S-CON-A, respectively, in relation to their optimal equivalent defect free cases (Table 8). Since there is a reduction of 0.73% in the LS-C-CON-A with respect to its respective optimal non steered case (Table 7), the non steered optimal solution is preferable for this boundary condition just like in the previous case. Focusing now on the simply supported boundary condition, there is a slight decrease of 0.14% with respect to its corresponding non steered optimal solution (Table 7). In addition, the maximum curvatures especially in plies 2 and 4, and their respective symmetric ones, are near the maximum curvature constraint. Therefore, it can also be said that it is also preferable to use non steered plies for this boundary condition as opposed to the previously case with an $a/b = 0.5$. Since the maximum curvature of the plies does not tend to zero, it is possible to speculate that a better solution could be found if the maximum curvature constraint imposed was higher.

In a similar way as with square plates, the optimal layout when complete gap imperfections are considered during the optimization differs from the one obtained for ideal plates. These differences in the orientations can be observed by comparing the optimal layouts in Tables 8 and 10. When adding the complete gap imperfections to the ideal optimal solutions (Table 9), it is visible for a plate with $a/b = 0.5$ that the $v_G[\%]$ obtained is lower than the one obtained when defects are taken into consideration within the optimization framework (Table 10). On the contrary, for a plate with $a/b = 2.0$, $v_G[\%]$ is lower in Table 10 than on Table 9 considering a course with 32 tows. Therefore, when the layout is optimized with the occurrence of gaps included it is once again visible that there is a trade-off between having the desired fiber path and the minimum gap volume fraction depending on the plate's geometry and boundary condition.

6. Conclusions

An optimization framework has been implemented with the objective of tailoring the performance of laminated composite structures by maximizing the fundamental natural frequency. The optimization is performed for both the non-steered and steered fibers in order to evaluate the effects of curvilinear fibers on the structural performance. Furthermore, manufacturing issues such as tow drop gaps are also taken into account using the Defect Layer Method with the objective of assessing its effect on the fundamental frequency.

The steered plates were first optimized considering no manufacturing embedded defects. It was found that the fibers curvature effect on the fundamental frequency depends not only on the boundary condition considered, but also on the plate geometry. For square plates, it was observed that the use of steered fibers is advantageous when a fully clamped boundary condition is imposed, while the optimal fiber orientation is rectilinear when a fully simply supported boundary condition is imposed. However, the use of curvilinear fibers was found to be advantageous for rectangular plates with simply supported boundary conditions.

Before considering embedded defects within the optimization framework, complete gap defects were added to the optimal ideal steered plate designs. When tow drop defects such as gaps were considered, results show that these rich resin areas lead to a reduction in the fundamental frequency, which appears to be a consequence of a reduction of the material equivalent elastic properties. It was also verified that reducing the number of tows leads to an increase in the occurrence of rich resin areas and therefore to a higher decrease in the fundamental frequency.

The plates were then optimized considering embedded defects within the optimization framework. The results revealed that the optimal designs that consider ideal plates can differ from the ones where manufacturing defects are considered. It can also be verified that not considering defects within the optimization framework leads to a different (often worse) gap area percentage. In spite of the occurrence of complete gaps, it was observed that the use of steered fibers is still advantageous when a fully clamped boundary condition is imposed in square plates. Moreover, the optimal fiber orientation is also rectilinear when a fully simply supported boundary condition is imposed in square plates. Regarding rectangular plates, results show that curvilinear fibers are only advantageous for a rectangular plate with $a/b = 0.5$ and with a simply supported boundary condition.

It has been shown that the use of curvilinear fibers could be an effective way to tailor the natural frequency of laminated composite plates. Nonetheless, variable angle tow laminates are still a relatively recent field, therefore experimental validation and verification would significantly contribute to the improvement of the numerical models proposed in this paper. Future work would include critical buckling load and to account for the minimum cut length constraint of the AFP machines.

Author Contributions: J.C.: Methodology, Software, Validation, Formal Analysis, Writing—original draft. A.S. (Abdolrasoul Sohouli): Methodology, Software, Validation, Formal Analysis, Writing—original draft. A.S. (Afzal Suleman): Supervision, Conceptualization, Resources, Writing—review & editing. All authors have read and agreed to the published version of the manuscript.

Funding: This research received no external funding.

Institutional Review Board Statement: Not applicable.

Informed Consent Statement: Not applicable.

Data Availability Statement: Not applicable.

Acknowledgments: The authors acknowledge the Fundação para a Ciência e a Tecnologia (FCT), through IDMEC, under LAETA, project UIDB/50022/2020. A.S. (Afzal Suleman) also acknowledges the NSERC Canada Research Chair and Discovery funding programs.

Conflicts of Interest: The authors declare no conflict of interest.

Abbreviations

The following abbreviations are used in this manuscript:

AFP	Automated Fiber Placement
VAT	Variable Angle Tow
LS	Linearly Steered
NS	Non Steered
C	Boundary Condition Fully Clamped
S	Boundary Condition Fully Simply Supported
VSCL	Variable Stiffness Composite Laminate
SDC	Specific Damping Capacity
CPT	Classical Plate Theory
FSDT	First-Order Shear Deformation Theory
FEM	Finite Element Method
GA	Genetic Algorithm

References

1. Gürdal, Z.; Tatting, B.F.; Wu, C. Variable stiffness composite panels: Effects of stiffness variation on the in-plane and buckling response. *Compos. Part A Appl. Sci. Manuf.* **2008**, *39*, 911–922. [CrossRef]
2. Alhajahmad, A.; Mittelstedt, C. Design tailoring of curvilinearly grid-stiffened variable-stiffness composite cylindrically curved panels for maximum buckling capacity. *Thin-Walled Struct.* **2020**, *157*, 107132. [CrossRef]
3. Setoodeh, S.; Abdalla, M.M.; IJsselmuiden, S.T.; Gürdal, Z. Design of variable-stiffness composite panels for maximum buckling load. *Compos. Struct.* **2009**, *87*, 109–117. [CrossRef]
4. Honda, S.; Narita, Y.; Sasaki, K. Maximizing the fundamental frequency of laminated composite plates with optimally shaped curvilinear fibers. *J. Syst. Des. Dyn.* **2009**, *3*, 867–876. [CrossRef]
5. Honda, S.; Igarashi, T.; Narita, Y. Multi-objective optimization of curvilinear fiber shapes for laminated composite plates by using NSGA-II. *Compos. Part B Eng.* **2013**, *45*, 1071–1078. [CrossRef]
6. Pereira, D.; Guimarães, T.; Resende, H.; Rade, D. Numerical and experimental analyses of modal frequency and damping in tow-steered CFRP laminates. *Compos. Struct.* **2020**, *244*, 112190. [CrossRef]
7. Serhat, G.; Basdogan, I. Lamination parameter interpolation method for design of manufacturable variable-stiffness composite panels. *AIAA J.* **2019**, *57*, 3052–3065. [CrossRef]
8. Serhat, G.; Bediz, B.; Basdogan, I. Unifying lamination parameters with spectral-Tchebychev method for variable-stiffness composite plate design. *Compos. Struct.* **2020**, *242*, 112183. [CrossRef]
9. Farsadi, T.; Asadi, D.; Kurtaran, H. Fundamental frequency optimization of variable stiffness composite skew plates. *Acta Mech.* **2021**, *232*, 555–573. [CrossRef]
10. Akhavan, H.; Ribeiro, P. Natural modes of vibration of variable stiffness composite laminates with curvilinear fibers. *Compos. Struct.* **2011**, *93*, 3040–3047. [CrossRef]
11. Van Campen, J.M.; Kassapoglou, C.; Gürdal, Z. Generating realistic laminate fiber angle distributions for optimal variable stiffness laminates. *Compos. Part B Eng.* **2012**, *43*, 354–360. [CrossRef]
12. Dodwell, T.J.; Butler, R.; Rhead, A.T. Optimum fiber steering of composite plates for buckling and manufacturability. *AIAA J.* **2016**, *54*, 1146–1149. [CrossRef]
13. Parnas, L.; Oral, S.; Ceyhan, Ü. Optimum design of composite structures with curved fiber courses. *Compos. Sci. Technol.* **2003**, *63*, 1071–1082. [CrossRef]
14. Demir, E.; Yousefi-Louyeh, P.; Yildiz, M. Design of variable stiffness composite structures using lamination parameters with fiber steering constraint. *Compos. Part B Eng.* **2019**, *165*, 733–746. [CrossRef]
15. Rashed, A.; Demir, E. Design of variable stiffness composites for maximum fundamental frequency considering manufacturing constraints of tow steering. *Compos. Struct.* **2022**, *284*, 115151. [CrossRef]
16. Blom, A.W.; Lopes, C.S.; Kromwijk, P.J.; Gurdal, Z.; Camanho, P.P. A theoretical model to study the influence of tow-drop areas on the stiffness and strength of variable-stiffness laminates. *J. Compos. Mater.* **2009**, *43*, 403–425. [CrossRef]
17. Fayazbakhsh, K.; Nik, M.A.; Pasini, D.; Lessard, L. Defect layer method to capture effect of gaps and overlaps in variable stiffness laminates made by Automated Fiber Placement. *Compos. Struct.* **2013**, *97*, 245–251. [CrossRef]
18. Fayazbakhsh, K.; Nik, M.A.; Pasini, D.; Lessard, L. The effect of gaps and overlaps on the in-plane stiffness and buckling load of variable stiffness laminates made by automated fiber placement. In Proceedings of the 15th European Conference on Composite Materials, Venice, Italy, 24–28 June 2012.
19. Marouene, A.; Boukhili, R.; Chen, J.; Yousefpour, A. Effects of gaps and overlaps on the buckling behavior of an optimally designed variable-stiffness composite laminates—A numerical and experimental study. *Compos. Struct.* **2016**, *140*, 556–566. [CrossRef]
20. Akbarzadeh, A.; Nik, M.A.; Pasini, D. The role of shear deformation in laminated plates with curvilinear fiber paths and embedded defects. *Compos. Struct.* **2014**, *118*, 217–227. [CrossRef]

21. Arranz, S.; Sohouli, A.; Suleman, A. Buckling Optimization of Variable Stiffness Composite Panels for Curvilinear Fibers and Grid Stiffeners. *J. Compos. Sci.* **2021**, *5*, 324. [CrossRef]
22. Brooks, T.R.; Martins, J.R. On manufacturing constraints for tow-steered composite design optimization. *Compos. Struct.* **2018**, *204*, 548–559. [CrossRef]
23. Mishra, V.; Peeters, D.M.; Abdalla, M.M. Stiffness and buckling analysis of variable stiffness laminates including the effect of automated fibre placement defects. *Compos. Struct.* **2019**, *226*, 111233. [CrossRef]
24. Heinecke, F.; Willberg, C. Manufacturing-induced imperfections in composite parts manufactured via automated fiber placement. *J. Compos. Sci.* **2019**, *3*, 56. [CrossRef]
25. Marrouze, J.; Housner, J.; Abdi, F. Effect of manufacturing defects and their uncertainties on strength and stability of stiffened panels. In Procedings of the 19th International Conference on Composite Materials, Montreal, QC, Canada, 28 July–2 August 2013; p. 10.
26. Heinecke, F.; van den Brink, W.; Wille, T. Assessing the structural response of automated fibre placement composite structures with gaps and overlaps by means of numerical approaches. In Procedings of the 20th International Conference on Composite Materials, Copenhagen, Denmark, 19–24 July 2015; p. 12.
27. Heinecke, F.; Wille, T. In-situ structural evaluation during the fibre deposition process of composite manufacturing. *CEAS Aeronaut. J.* **2018**, *9*, 123–133. [CrossRef]
28. Li, X.; Hallett, S.R.; Wisnom, M.R. Modelling the effect of gaps and overlaps in automated fibre placement (AFP)-manufactured laminates. *Sci. Eng. Compos. Mater.* **2015**, *22*, 115–129. [CrossRef]
29. Antunes, A.M.; Ribeiro, P.; Rodrigues, J.D.; Akhavan, H. Modal analysis of a variable stiffness composite laminated plate with diverse boundary conditions: Experiments and modelling. *Compos. Struct.* **2020**, *239*, 111974. [CrossRef]
30. Rodrigues, J.D.; Ribeiro, P.; Akhavan, H. Experimental and finite element modal analysis of variable stiffness composite laminated plates. In Proceedings of the 11th Biennial International Conference on Vibration Problems (ICOVP-2013), Lisbon, Portugal, 9–12 September 2013; p. 30.
31. Gurdal, Z.; Olmedo, R. In-plane response of laminates with spatially varying fiber orientations-variable stiffness concept. *AIAA J.* **1993**, *31*, 751–758. [CrossRef]
32. Waldhart, C. Analysis of Tow-Placed, Variable-Stiffness Laminates. Master's Thesis, Faculty of the Virginia Plytechnic Institute and State University, 1996 .
33. Langley, P.T. Finite Element Modeling of Tow-Placed Variable-Stiffness Composite Laminates. Ph.D. Thesis, Virginia Tech, 1999.
34. Blom, A.W. Structural Performance of Fiber-Placed, Variable-Stiffness Composite Conical and Cylindrical Shelss. Ph.D. Thesis, Delft University of Technology, Delft, The Netherlands, 2010.
35. Smith, M. *ABAQUS/Standard User's Manual, Version 6.9*; Dassault Systèmes Simulia Corp: Providence, RI, USA, 2009.

Journal of *Composites Science*

Article

Mechanical Behaviour Evaluation of Porous Scaffold for Tissue-Engineering Applications Using Finite Element Analysis

Akesh Babu Kakarla [1], Ing Kong [1,*], Satya Guha Nukala [1] and Win Kong [2]

[1] School of Computing, Engineering and Mathematical Sciences, La Trobe University, Bendigo 3552, Australia; a.kakarla@latrobe.edu.au (A.B.K.); s.nukala@latrobe.edu.au (S.G.N.)
[2] BASF Corporation, 1609 Biddle Avenue, Wyandotte, MI 48192, USA; win.kong@basf.com
* Correspondence: I.Kong@latrobe.edu.au;

Abstract: In recent years, finite element analysis (FEA) models of different porous scaffold shapes consisting of various materials have been developed to predict the mechanical behaviour of the scaffolds and to address the initial goals of 3D printing. Although mechanical properties of polymeric porous scaffolds are determined through FEA, studies on the polymer nanocomposite porous scaffolds are limited. In this paper, FEA with the integration of material designer and representative volume elements (RVE) was carried out on a 3D scaffold model to determine the mechanical properties of boron nitride nanotubes (BNNTs)-reinforced gelatin (G) and alginate (A) hydrogel. The maximum stress regions were predicted by FEA stress distribution. Furthermore, the analysed material model and the boundary conditions showed minor deviation (4%) compared to experimental results. It was noted that the stress regions are detected at the zone close to the pore areas. These results indicated that the model used in this work could be beneficial in FEA studies on 3D-printed porous structures for tissue engineering applications.

Keywords: boron nitride nanotubes; porous scaffold; finite element analysis; representative volume elements; mechanical properties

Citation: Kakarla, A.B.; Kong, I.; Nukala, S.G.; Kong, W. Mechanical Behaviour Evaluation of Porous Scaffold for Tissue-Engineering Applications Using Finite Element Analysis. *J. Compos. Sci.* **2022**, *6*, 46. https://doi.org/10.3390/jcs6020046

Academic Editor: Stelios K. Georgantzinos

Received: 18 December 2021
Accepted: 29 January 2022
Published: 1 February 2022

1. Introduction

In recent years, 3D bioprinting has significantly boosted the research and development in tissue regeneration [1]. The technique can be used to create complex tissue structures according to patient-specific geometries and compositions. Compared to 3D bioprinting, traditional methods are restricted in producing scaffolds with an adequate pore size that enhances in vitro behaviour. For instance, the internal geometry of the scaffold greatly influences cell adhesion, proliferation, and nutrient transportation for tissue regeneration. Customising suitable scaffold geometry for creating biological environments is addressed by 3D-printing technology rather than traditional methods [1–4]. Scaffolds generated by 3D printing have lattice structures with various pore sizes and serve as a template for cell interaction and cell-extracellular matrix formation. These scaffolds are required to furnish structural assistance for the newly generated tissue. In addition to delivering the required biological properties, the scaffolds provide biomechanical properties during tissue regeneration and implantation [5]. Biomechanical properties, such as shear stress, deformation, and tensile or compressive stress, must match with natural healthy tissue or bone structure properties. Ideal scaffolds are produced with a well-regulated pore structure and can reproduce the shape of the implants [6–9]. Research studies have indicated that anisotropic porous structures with a combination of small and large pores in various shapes are advantageous for cell growth and can improve cell proliferation over time [7,10,11]. Therefore, characterising and predicting the biomechanical properties of 3D-printed scaffolds using different materials is essential. Identifying suitable biomaterials that support biomechanical and biological properties is a significant challenge in 3D bioprinting for tissue-engineering applications. Natural and synthetic polymers are widely used to produce scaffolds through 3D bioprinting [12–16]. By contrast, natural or synthetic polymers are limited in terms of their biomechanical properties. Thus, researchers have been focusing on hybrid polymer or polymer nanocomposite-based materials applicable in 3D bioprinting techniques for tissue engineering. However, most studies on these materials are conducted through experiments, which is time consuming and costly. Therefore, finite element modelling (FEM) and FEA provide the alternatives for determining the biomechanical properties of biomaterials without printing or performing extensive, time-consuming experiments. Additionally, FEM helps improve the design process and methodology to provide high accuracy in the geometric configuration of 3D-printed scaffolds [6,17–19].

Using FEM, tissue-specific or material-specific design analysis and prediction of biomechanical properties can be determined [20–24]. The analysis aids in accelerating the progression in choosing the materials or structures that are adequate in 3D bioprinting. Miranda et al. [25] simulated the mechanical behaviour of hydroxyapatite (HA) and beta-tricalcium phosphate (β-TCP) lattice scaffold structure by using FEM. The results predicted by the FEM were validated by comparing with experimental data, justifying the suitability of the 3D scaffold for bone tissue-engineering applications. Hashemi et al. [26] predicted the mechanical behaviour of the HA-wollastonite scaffold model with different porosity percentages by FEA. According to the simulation results, the increase in the percentage of porosity enhanced the strength of the scaffold. The findings satisfied prospects of being a bone scaffold material with suitable mechanical strength [26]. Ali et al. [27] designed high-porosity scaffolds with gyroid- and lattice-based structures, and they were analysed using FEA. The results demonstrated that the lattice-based structures showed high moduli and compressive strength, and the permeability was highly influenced by porosity and design [27]. Additionally, it was reported that lattice-based structures with high porosity can effectively mimic bone structure properties [27]. Bagde et al. [28] developed a 3D-printed bio-ceramic scaffold used in bone tissue engineering, and its mechanical properties were analysed using FEA. Thirty-six scaffolds with differing geometrical design parameters composed of β-TCP (matrix) reinforced with four different filler materials (zirconium dioxide, magnesium oxide, aluminium oxide, and hydroxyapatite) for extrusion-based 3D bioprinting were used in the simulation. The results indicated that β-TCP with hydroxyapatite scaffold presented the Young's modulus closely related to natural bone tissue [28]. Patel et al. [29] developed a scaffold of poly(3-hydroxybutyrate-co-3-hydroxyvalerate) with a porous architecture, and the mechanical properties were analysed using FEA. The results showed that the use of linear elastic material structures exhibited higher rigidity compared with bilinear models. The scaffolds demonstrated deformation at sharp corners and necked regions only [29]. The study illustrated the optimal predictions of mechanical behaviour of porous structures when subjected to peripheral loading [29]. Jiang et al. [30] constructed a 3D model of an auricle silicone scaffold to optimise the thickness and hardness. The results successfully validated the data taken from computed tomography scans. The auricle silicone scaffold displayed sufficient intensity and hardness to resist deformation [30]. Blázquez-Carmona et al. [31] designed a patient-specific ceramic scaffold model for bone regeneration. The FEA data indicated that the optimised porosity and pore size levels provided a more significant mechanical constraint [31].

Research studies have been conducted to improve the biomechanical properties of scaffolds by incorporating filler (nanoparticles, nanotubes, and nanosheets) materials, such as carbon nanotubes (CNTs) [32], graphene [33], titanium oxide [34], and HA [26,35], into the main polymer matrix. CNTs have been widely used as the reinforcement of advanced composites [36]. In biomedical applications, the toxicity of a material is a crucial factor when considering material for implants [37]. However, it has been reported that CNTs are cytotoxic than carbon black and quartz [38,39]. Hence, researchers are finding alternative nanomaterial for biomedical applications.

BNNTs are structure analogues of CNTs with distinctive physical properties. BNNTs possess excellent mechanical and thermal properties, making them a favourable nanomaterial to be incorporated in the polymer matrix [40]. Several previous experimental studies on BNNTs combined with aluminium [41], polycaprolactone [42], gelatin [43], β-TCP [44], and polyvinyl alcohol [45] have shown significant improvements in the mechanical properties of the polymers. However, the analytical approaches on predicting BNNTs with polymer matrix on 3D scaffolds are limited. Therefore, the present study aims to create a simulation model and perform a mechanical test on the BNNTs-reinforced gelatin and alginate. A customised library composed of material properties was created for the scaffold model to analyse the tensile properties. A random distribution of BNNTs in gelatin and alginate was developed in RVE, and a 3D model was generated using SolidWorks (Dassault Systèmes, USA). The geometry and engineering data generated through RVE were analysed using ANSYS (ANSYS, Inc., USA) software. The FEA of the scaffold model subjected to static loading was evaluated to predict the mechanical properties.

2. Materials and Methods

2.1. Basic Properties of BNNTs with Gelatin and Alginate Scaffold

The computer-aided design model of a quadrilateral lattice structure was designed using SolidWorks modelling software. The scaffold was designed with a pore size of 2×2 mm^2 and strands spaced 1 mm apart. The strands with 0° and 90° in the X and Z direction were considered for the scaffold design. Subsequently, the model was imported into ANSYS for simulation. In addition, a custom library of material properties, such as elastic modulus and Poisson's ratio, was created based on literature reports (Table 1) as properties of raw materials. The raw materials properties were furtherly used in RVE to generate a scaffold of BNNTs-reinforced alginate and gelatin properties. The BNNTs with gelatin and alginate scaffold was considered linearly elastic, isotropic, and homogeneous for the simulation.

Table 1. Properties of raw materials.

Properties	Alginate (A)	Gelatin (G)	Boron Nitride Nanotubes
Young's modulus	30 kPa [46]	39 kPa [47]	1300 kPa [44]
Poisson's ratio	0.4	0.33	0.35

2.2. Representative Volume Elements (RVE)

The random distribution of BNNTs in the matrix was developed using RVE through a random sequential algorithm (RSA). RSA was based on adding fibres (diameter of 85 nm) to a predefined space by randomly generating the coordinates (xy, yz, and xz planes) and orientation angles [44]. Through this technique, the fibres are not allowed to overlap with the former fibres, and the cycle lasts until the desired volume fractions (5%) of the fibres are obtained. BNNTs are assumed to be a solid cylindrical bar, as shown in Figure 1a. The polymer matrix is shown in Figure 1b. The combination of polymer matrix and BNNTs is shown in Figure 1c, which was isotropic, elastic, and homogenous. The aspect ratio of BNNTs was considered to be 40 based on previous studies [44], and contact between matrix and BNNTs was predicted to be perfectly bonded, and 100% load interchanging occurred. After generating the RVE, all the required data (Table 2) for an elastic analysis was attached to the engineering data of the analysis system. The generated RVE model was exported as static structural into ANSYS software for analysis of mechanical properties.

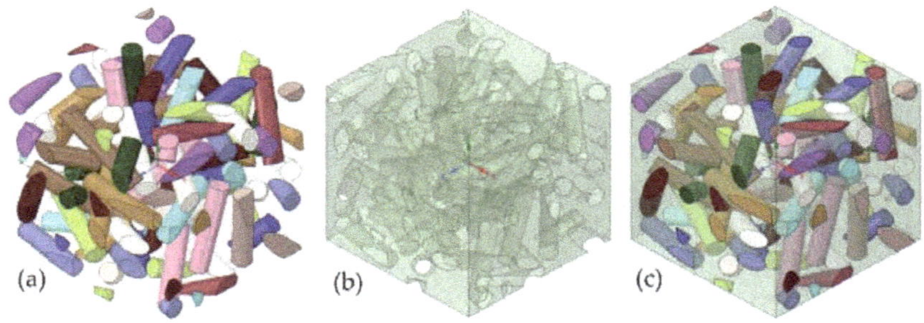

Figure 1. The RVE model and its details: (**a**) BNNTs; (**b**) polymer matrix; and (**c**) the combination of polymer matrix and BNNTs.

Table 2. Material properties of the scaffold.

Young's modulus X-direction	0.14 MPa
Young's modulus Y-direction	0.12 MPa
Young's modulus Z-direction	0.16 MPa
Poisson's ratio	0.33

2.3. Finite Element Analysis (FEA)

The RVE and scaffold model (Figure 2a) with characteristics listed in Table 3 were imported into the ANSYS as material designer and geometry. The material designer was connected to the engineering data in the FEA analysis system. The homogenous isotropic properties with the scaffold model were related to the geometry of the FEA analysis. Therefore, a scaffold featuring randomly distributed BNNTs was generated to analyse the mechanical properties.

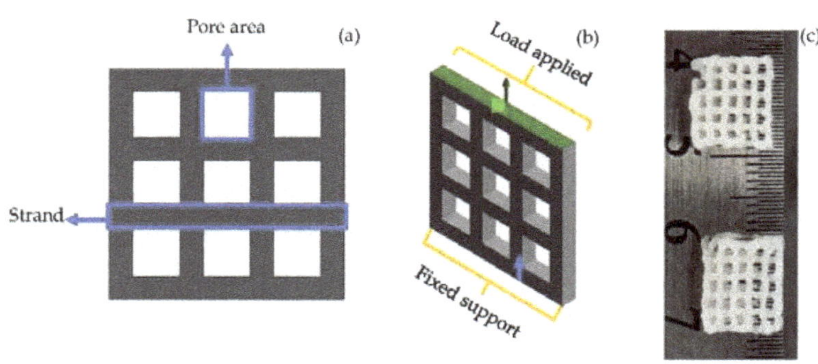

Figure 2. (**a**) 3D scaffold designed for analysis; (**b**) applied boundary condition on scaffold model; (**c**) 3D-bioprinted lattice structured scaffold of BNNTs reinforced alginate and gelatin.

2.4. Boundary Conditions

The investigation was carried in static load conditions with increase in load for each second up to 5 s on the scaffold. For the simulation, a transient plugin in ANSYS was used for the tension test. The displacement rate was kept at 2 mm/min based on a typical quasi-static loading rate for testing bone and biomaterials for tissue-engineering applications. The Poisson's ratio was kept constant throughout the FEA. To predict the tensile properties, one end of the scaffold was fixed (blue arrow, Figure 2b), and the other end (green arrow, Figure 2b) was applied with load. The maximum von Mises stress required for deformation of the scaffold was calculated.

Table 3. Characterisation of the scaffold.

Parameter	Scaffold
Cell size	L = 2 mm d = 2.82 mm
Pore area (mm^2)	8
Porous volume (mm^3)	4
Total volume (mm^3)	100
Surface area (mm^2)	240
Porosity (%)	84

L, length of a pore area; *d*, diameter of the square pore.

The porosity of the scaffold was calculated using Equation (1) as follows:

$$\text{Porosity} = (\text{porous volume}/\text{total volume}) \times 100\% \quad (1)$$

2.5. Experimental

According to previous report [48], the alginate, gelatin, and BNNTs hydrogel composite scaffolds were produced. Briefly, alginate (5 *w/v*%) and gelatin (6 *w/v*%) were mixed in deionised water through constantly stirring for 1 h. Later, BNNTs (1 *w/v*%) was slowly added into the alginate and gelatin solution and stirred for another 1 h at 60 °C. The obtained solution was loaded into a 3-mL syringe with attached 22-gauge nozzle. The syringe was fixed to a 3D bioprinter (Cellink INKREDIBLE+, Sweden). Afterwards, the scaffold was printed with 105 ± 5 kPa pressure (Figure 2c). The obtained grid-like, porous scaffold was crosslinked with 100 Mm calcium chloride solution and freeze-dried prior to the tension test. The freeze-dried samples were tested according to Oladapo et al.'s [49] described method for universal testing machine. The test was carried out in the tensile test method using INSTRON 5982 (INSTRON, USA) with a constant displacement of 0.5 mm/min at room temperature. The engineering stress and strain values obtained in the experiment were transformed into the true stress and true strain values corresponding to Equations (2) and (3).

$$\sigma_{true} = \sigma_{engineering} \times \left(1 + \epsilon_{engineering}\right) \quad (2)$$

$$\epsilon_{true} = ln\left(1 + \epsilon_{engineering}\right) \quad (3)$$

where σ and ∈ represent the stress and strain of the material. The deformation rate was kept to be the same as in the simulation. However, the mesh relevance and element size were modified to obtain the most precise results.

3. Results

3.1. Mesh Generation

Mesh convergence is one of the concerns in the simulation process, as it can affect the accuracy of the results. Therefore, Solid 95 soft mesh was selected to mimic the scaffold structure, as it creates a smooth mesh and avoids simulation stress-convergence errors. The soft mesh automatically generates higher node elements in higher curvature areas without the need for mesh control. The element size of 0.2 mm (Figure 3) was used to obtain accurate results closer to the mechanical test results.

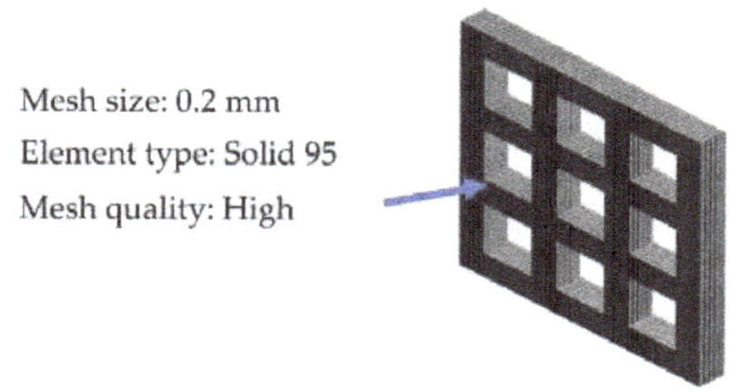

Figure 3. Meshing of the scaffold model.

3.2. Mechanical properties

The von Mises distribution was shown in Figure 4a. The maximum (red arrows) and minimum stress (yellow arrows) concentration on the scaffolds when it was stretched are shown in Figure 4a. It is vital to recognize the stress concentration of a structure, as it will aid in suggesting areas of failures within the structure. The FEM results demonstrated that the scaffold underwent maximum stress at 2.7 MPa (Figure 4a). The corresponding moderate equivalent strain (Figure 4b) in the scaffold was shown at a higher rate of 6%.

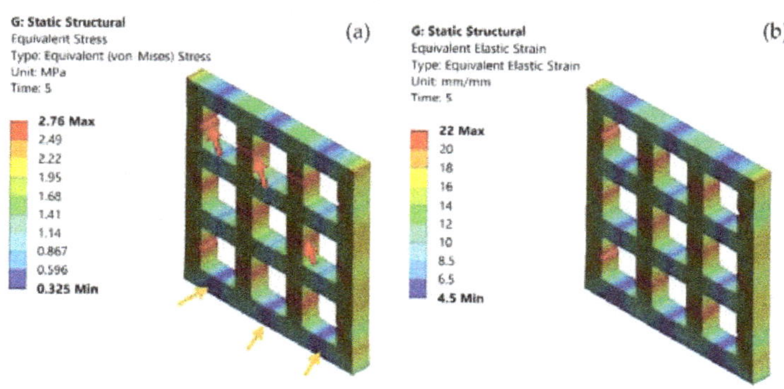

Figure 4. (**a**) von Mises stress distribution on the scaffold model; maximum stress (red arrows), minimum stress (yellow arrows); and (**b**) elastic strain on the scaffold model.

Additionally, the stress-strain curves were compared with the experimental data. The plot graph of stress and strain in Figure 5 demonstrated the maximum experimental stress compared to the FEA predictions. However, both practical and simulation showed the maximum stress rate was attained at approximately 6%. Both FEA and experimental results showed the elastic region corresponding to the pore edge bending or face stretching. The second region was a plastic region corresponding to the progressive pore collapse due to the load applied. The fracture region corresponds to the pore's failure at maximum stress of 2.7 MPa for FEA and 2.8 MPa for experimental data. The FEA and experimental plots showed a variation in plastic and fracture regions due to the variation of the load. Furthermore, in FEA, the material was considered isotropic. The percentage of error for FEA and experimental was approximately 4%. Li et al. [50] demonstrated sodium alginate, gelatin, and carbon nanotubes mechanical testing of circular scaffolds mechanical strength of 1.24 MPa. Serrano-Aroca et al. [51] reported alginate-graphene oxide composite hydrogel maximum stress at 8.98 ± 0.35 MPa. Similarly, the BNNTs-reinforced alginate and gelatin showed 2.8 MPa maximum stress. Additionally, the maximum strength of human soft tissues and hard tissues range between 0.01 MPa to 150 MPa [52–54]. Thus, it was evident that BNNTs-reinforced gelatin and alginate could be a potential scaffold for regenerating tissue with good pore interconnectivity as well as a good agreement between experimental and analysis.

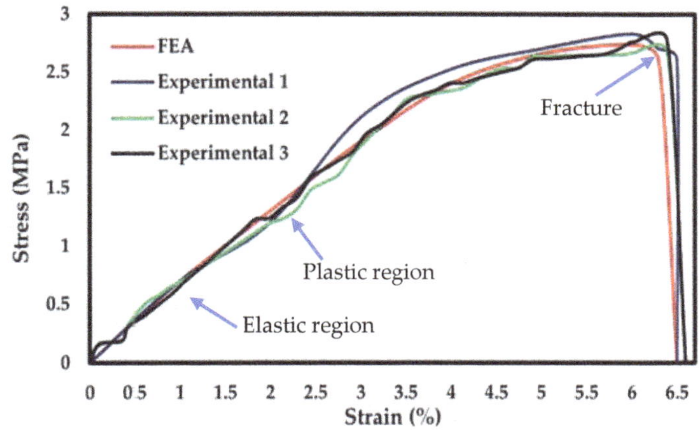

Figure 5. Stress-strain curve comparison of FEA with experimental.

According to studies conducted by Ambu et al. [55], Smith et al. [56], and Maskery et al. [57], the smaller pore areas are adequate for FEA simulation compared with larger pore size. Hence, in this study, 2×2 mm^2 pore areas were considered for simulation. The BNNTs-reinforced gelatin and alginate scaffold results indicated that the stress in the scaffold showed a homogeneous distribution at the fixed end and heterogeneous distribution at the pore area. The areas close to the pores have higher stress (Figure 6a, red arrows) concentration by default due to soft spots. The areas at the edges and corners have less stress concentration (Figure 6a, yellow arrows). The fracture occurred at the maximum stress point, as shown in Figure 6b (red arrow). The porous BNNTs-reinforced gelatin and alginate scaffold with the smallest pore size was an excellent combination to produce a scaffold with high mechanical properties while providing an excellent porosity (84%). Furthermore, an adequate pore size and porosity are important factors to the scaffold's properties because the pores aid in cell proliferation and differentiation as well as encourage development of tissue structures [17,58,59].

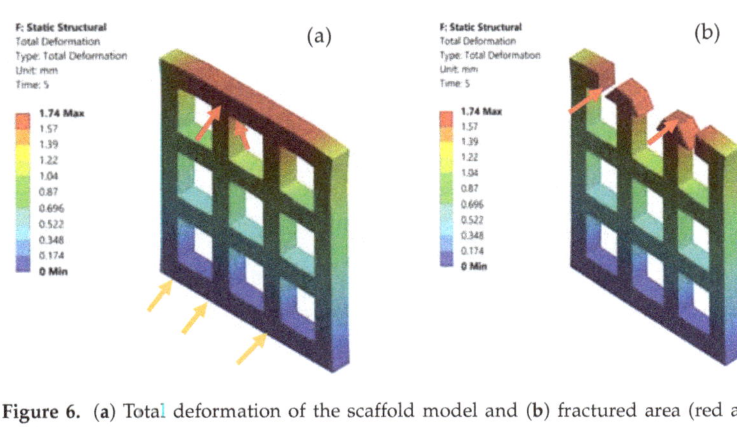

Figure 6. (**a**) Total deformation of the scaffold model and (**b**) fractured area (red arrow) of the scaffold model.

4. Conclusions

In this paper, a FEM model combined with geometry and RVE was developed to analyse the mechanical behaviour of the porous scaffold for tissue-engineering applications. The scaffold was designed with regular pore interconnectivity and strand distance. Additionally, the analysed data were validated with experimental results of a 3D-printed scaffold. The simulation results showed that $2 \times 2\ mm^2$ pore size was found to play a significant role in determining the maximum stress region. The higher stress concentration areas were observed at the soft zones close to the pore area, considered default stress regions. The analysed maximum strength was obtained at 2.7 MPa and experimental at 2.8 MPa. The FEA and experimental stress-strain curves corresponded to each other and displayed analogous slopes and trends within the range. In addition, the investigation of lattice models with a random distribution of BNNTs in gelatin and alginate is both novel and helpful to the designer of 3D bioprinting, particularly in discovering the biomechanical properties. However, the findings are limited to uniform strand and pore size. Further work will focus on evaluating a widening type of pore sizes, strand directions, and widths with FEA and experiments.

Author Contributions: Conceptualization, A.B.K.; methodology, A.B.K.; software, A.B.K.; validation, A.B.K. and I.K.; formal analysis, A.B.K. and S.G.N.; investigation, I.K and W.K.; resources, I.K.; data curing, A.B.K. and S.G.N.; writing—original draft preparation, A.B.K.; writing—reviewing and editing, I.K and W.K.; visualization, A.B.K. and I.K.; supervision, I.K.; funding acquisition, I.K.; project administration, I.K. All authors have read and agreed to the published version of the manuscript.

Funding: This research received no external funding.

Institutional Review Board Statement: Not applicable.

Informed Consent Statement: Not applicable.

Data Availability Statement: Not applicable.

Acknowledgments: Vipulkumar Ishvarbhai Patel is acknowledged for his support and valuable feedback on methodology and analysis.

Conflicts of Interest: The authors declare no conflict of interest.

References

1. Giannitelli, S.M.; Accoto, D.; Trombetta, M.; Rainer, A. Current trends in the design of scaffolds for computer-aided tissue engineering. *Acta Biomater.* **2014**, *10*, 580–594. [CrossRef] [PubMed]
2. Melchels, F.P.W.; Barradas, A.M.C.; van Blitterswijk, C.A.; de Boer, J.; Feijen, J.; Grijpma, D.W. Effects of the architecture of tissue engineering scaffolds on cell seeding and culturing. *Acta Biomater.* **2010**, *6*, 4208–4217. [CrossRef] [PubMed]
3. Pires, T.; Santos, J.; Ruben, R.B.; Gouveia, B.P.; Castro, A.P.G.; Fernandes, P.R. Numerical-experimental analysis of the permeability-porosity relationship in triply periodic minimal surfaces scaffolds. *J. Biomech.* **2021**, *117*, 110263. [CrossRef] [PubMed]
4. Ibañez, R.I.R.; do Amaral, R.J.F.C.; Reis, R.L.; Marques, A.P.; Murphy, C.M.; O'Brien, F.J. 3D-Printed Gelatin Methacrylate Scaffolds with Controlled Architecture and Stiffness Modulate the Fibroblast Phenotype towards Dermal Regeneration. *Polymers* **2021**, *13*, 2510. [CrossRef] [PubMed]
5. Karageorgi, V.; Kalpan, D. Porosity of 3D biomaterial scaffolds and osteogenesis. *Biomaterials* **2005**, *26*, 5474–5491. [CrossRef]
6. Soufivand, A.A.; Abolfathi, N.; Hashemi, S.A.; Lee, S.J. Prediction of mechanical behavior of 3D bioprinted tissue-engineered scaffolds using finite element method (FEM) analysis. *Addit. Manuf.* **2020**, *33*, 101181. [CrossRef]
7. Gómez, S.; Vlad, M.D.; López, J.; Fernández, E. Design and properties of 3D scaffolds for bone tissue engineering. *Acta Biomater.* **2016**, *42*, 341–350. [CrossRef]
8. Adachi, T.; Osako, Y.; Tanaka, M.; Hojo, M.; Hollister, S.J. Framework for optimal design of porous scaffold microstructure by computational simulation of bone regeneration. *Biomaterials* **2006**, *27*, 3964–3972. [CrossRef]
9. Oladapo, B.I.; Zahedi, S.A.; Ismail, S.O.; Olawade, D.B. Recent advances in biopolymeric composite materials: Future sustainability of bone-implant. *Renew. Sustain. Energy Rev.* **2021**, *150*, 111505. [CrossRef]
10. Childs, P.G.; Boyle, C.A.; Pemberton, G.D.; Nikukar, H.; Curtis, A.S.G.; Henriquez, F.L.; Dalby, M.J.; Reid, S. Use of nanoscale mechanical stimulation for control and manipulation of cell behaviour. *Acta Biomater.* **2016**, *34*, 159–168. [CrossRef]
11. Hao, M.; Wei, C.; Liu, X.; Ge, Y.; Cai, J. Quantitative evaluation on mechanical characterization of Ti6Al4V porous scaffold designed based on Weaire-Phelan structure via experimental and numerical analysis methods. *J. Alloys Compd.* **2021**, *885*, 160234. [CrossRef]

12. Jin, Y.; Chai, W.; Huang, Y. Printability study of hydrogel solution extrusion in nanoclay yield-stress bath during printing-then-gelation biofabrication. *Mater. Sci. Eng. C* **2017**, *80*, 313–325. [CrossRef] [PubMed]
13. Curti, F.; Stancu, I.-C.; Voicu, G.; Iovu, H.; Dobrita, C.-I.; Ciocan, L.T.; Marinescu, R.; Iordache, F. Development of 3D Bioactive Scaffolds through 3D Printing Using Wollastonite–Gelatin Inks. *Polymers* **2020**, *12*, 2420. [CrossRef] [PubMed]
14. Holzwarth, J.M.; Ma, P.X. Biomimetic nanofibrous scaffolds for bone tissue engineering. *Biomaterials* **2011**, *32*, 9622–9629. [CrossRef]
15. Müller, M.; Becher, J.; Schnabelrauch, M.; Zenobi-Wong, M. Nanostructured Pluronic hydrogels as bioinks for 3D bioprinting. *Biofabrication* **2015**, *7*, 35006. [CrossRef]
16. Gong, Y.; Wang, F.; Al-Furjan, M.S.H.; Shan, L.; He, J.; Bian, X.; Bi, Z.; Liu, H.; Li, W.; Shao, H.; et al. Experimental Investigation and Optimal 3D Bioprinting Parameters of SA-Gel Porous Cartilage Scaffold. *Appl. Sci.* **2020**, *10*, 768. [CrossRef]
17. Ostrowska, B.; Di Luca, A.; Moroni, L.; Swieszkowski, W. Influence of internal pore architecture on biological and mechanical properties of three-dimensional fiber deposited scaffolds for bone regeneration. *J. Biomed. Mater. Res. Part A* **2016**, *104*, 991–1001. [CrossRef]
18. Sun, K.; Li, R.; Li, H.; Fan, M.; Li, H. Analysis and Demonstration of a Scaffold Finite Element Model for Cartilage Tissue Engineering. *ACS Omega* **2020**, *5*, 32411–32419. [CrossRef]
19. Uth, N.; Mueller, J.; Smucker, B.; Yousefi, A.-M. Validation of scaffold design optimization in bone tissue engineering: Finite element modeling versus designed experiments. *Biofabrication* **2017**, *9*, 015023. [CrossRef]
20. Mustafa, N.S.; Akhmal, N.H.; Izman, S.; Ab Talib, M.H.; Shaiful, A.I.M.; Omar, M.N.B.; Yahaya, N.Z.; Illias, S. Application of Computational Method in Designing a Unit Cell of Bone Tissue Engineering Scaffold: A Review. *Polymers* **2021**, *13*, 1584. [CrossRef]
21. Rodríguez-Montaño, Ó.L.; Cortés-Rodríguez, C.J.; Uva, A.E.; Fiorentino, M.; Gattullo, M.; Manghisi, V.M.; Boccaccio, A. An Algorithm to Optimize the Micro-Geometrical Dimensions of Scaffolds with Spherical Pores. *Materials* **2020**, *13*, 4062. [CrossRef] [PubMed]
22. De Santis, R.; Russo, T.; Rau, J.V.; Papallo, I.; Martorelli, M.; Gloria, A. Design of 3D Additively Manufactured Hybrid Structures for Cranioplasty. *Materials* **2021**, *14*, 181. [CrossRef] [PubMed]
23. Mohonee, V.K.; Lim Goh, K.; Mishnaevsky, L.; Pasbakhsh, P. Capsule based self-healing composites: New insights on mechanical behaviour based on finite element analysis. *Comput. Mater. Sci.* **2021**, *192*, 110203. [CrossRef]
24. Mirtaghavi, A.; Luo, J.; Muthuraj, R. Recent Advances in Porous 3D Cellulose Aerogels for Tissue Engineering Applications: A Review. *J. Compos. Sci.* **2020**, *4*, 152. [CrossRef]
25. Miranda, P.; Pajares, A.; Guiberteau, F. Finite element modeling as a tool for predicting the fracture behavior of robocast scaffolds. *Acta Biomater.* **2008**, *4*, 1715–1724. [CrossRef]
26. Hashemi, S.A.; Esmaeili, S.; Ghadirinejad, M.; Saber-Samandari, S.; Sheikhbahaei, E.; Kordjamshidi, A.; Khandan, A. Micro-Finite Element Model to Investigate the Mechanical Stimuli in Scaffolds Fabricated via Space Holder Technique for Cancellous Bone. *ADMT J.* **2020**, *13*, 51–58.
27. Ali, D.; Sen, S. Finite element analysis of mechanical behavior, permeability and fluid induced wall shear stress of high porosity scaffolds with gyroid and lattice-based architectures. *J. Mech. Behav. Biomed. Mater.* **2017**, *75*, 262–270. [CrossRef]
28. Bagde, A.D.; Kuthe, A.M.; Nagdeve, S.R.; Dahake, S.W.; Sapkal, P.S.; Daronde, S.B.; Lande, N.H.; Sarode, B.D. Geometric Modeling and Finite Element Simulation for Architecture Design of 3D Printed Bio-ceramic Scaffold Used in Bone Tissue Engineering. *J. Indian Inst. Sci.* **2019**, *99*, 361–374. [CrossRef]
29. Patel, R.; Lu, M.; Diermann, S.H.; Wu, A.; Pettit, A.; Huang, H. Deformation behavior of porous PHBV scaffold in compression: A finite element analysis study. *J. Mech. Behav. Biomed. Mater.* **2019**, *96*, 1–8. [CrossRef]
30. Jiang, T.; Shang, J.; Tang, L.; Wang, Z. Thickness optimization of auricular silicone scaffold based on finite element analysis. *J. Mech. Behav. Biomed. Mater.* **2016**, *53*, 397–402. [CrossRef]
31. Blázquez-Carmona, P.; Sanz-Herrera, J.A.; Martínez-Vázquez, F.J.; Domínguez, J.; Reina-Romo, E. Structural optimization of 3D-printed patient-specific ceramic scaffolds for in vivo bone regeneration in load-bearing defects. *J. Mech. Behav. Biomed. Mater.* **2021**, *121*, 104613. [CrossRef] [PubMed]
32. Agrawal, R.; Nieto, A.; Chen, H.; Mora, M.; Agarwal, A. Nanoscale Damping Characteristics of Boron Nitride Nanotubes and Carbon Nanotubes Reinforced Polymer Composites. *ACS Appl. Mater. Interfaces* **2013**, *5*, 12052–12057. [CrossRef] [PubMed]
33. Chung, C.; Kim, Y.K.; Shin, D.; Ryoo, S.R.; Hong, B.H.; Min, D.H. Biomedical applications of graphene and graphene oxide. *Acc. Chem. Res.* **2013**, *46*, 2211–2224. [CrossRef]
34. Fiedler, T.; Belova, I.V.; Murch, G.E.; Roether, J.A.; Boccaccini, A.R. Tailoring elastic properties of PLGA/TiO2 biomaterials. *Comput. Mater. Sci.* **2012**, *61*, 283–286. [CrossRef]
35. Du, X.; Dehghani, M.; Alsaadi, N.; Nejad, M.G.; Saber-Samandari, S.; Toghraie, D.; Su, C.-H.; Nguyen, H.C. A femoral shape porous scaffold bio-nanocomposite fabricated using 3D printing and freeze-drying technique for orthopedic application. *Mater. Chem. Phys.* **2022**, *275*, 125302. [CrossRef]
36. Weidt, D.; Figiel, Ł. Finite strain compressive behaviour of CNT/epoxy nanocomposites: 2D versus 3D RVE-based modelling. *Comput. Mater. Sci.* **2014**, *82*, 298–309. [CrossRef]
37. Gaharwar, A.K.; Peppas, N.A.; Khademhosseini, A. Nanocomposite hydrogels for biomedical applications. *Biotechnol. Bioeng.* **2014**, *111*, 441–453. [CrossRef]

38. Warheit, D.B. Comparative Pulmonary Toxicity Assessment of Single-wall Carbon Nanotubes in Rats. *Toxicol. Sci.* **2003**, *77*, 117–125. [CrossRef]
39. Lam, C.-W. Pulmonary Toxicity of Single-Wall Carbon Nanotubes in Mice 7 and 90 Days after Intratracheal Instillation. *Toxicol. Sci.* **2003**, *77*, 126–134. [CrossRef]
40. Kostoglou, N.; Tampaxis, C.; Charalambopoulou, G.; Constantinides, G.; Ryzhkov, V.; Doumanidis, C.; Matovic, B.; Mitterer, C.; Rebholz, C. Boron Nitride Nanotubes Versus Carbon Nanotubes: A Thermal Stability and Oxidation Behavior Study. *Nanomaterials* **2020**, *10*, 2435. [CrossRef]
41. Sedigh, P.; Zare, A.; Montazeri, A. Evolution in aluminum applications by numerically-designed high strength boron-nitride/Al nanocomposites. *Comput. Mater. Sci.* **2020**, *171*, 109227. [CrossRef]
42. Akesh Babu, K.; Cin, K.; Wei, K.; Ing, K.; Kakarla, A.B.; Kong, C.; Kong, W.; Kong, I. Synthesis and Characterization of Boron Nitride Nanotubes-Polycaprolactone Nanocomposite. *Mater. Sci. Forum* **2019**, *951*, 39–44. [CrossRef]
43. Nagarajan, S.; Belaid, H.; Pochat-Bohatier, C.; Teyssier, C.; Iatsunskyi, I.; Coy, E.; Balme, S.; Cornu, D.; Miele, P.; Kalkura, N.S.; et al. Design of Boron Nitride/Gelatin Electrospun Nanofibers for Bone Tissue Engineering. *ACS Appl. Mater. Interfaces* **2017**, *9*, 33695–33706. [CrossRef] [PubMed]
44. Ali, D.; Sen, S. Finite element analysis of the effect of boron nitride nanotubes in beta tricalcium phosphate and hydroxyapatite elastic modulus using the RVE model. *Compos. Part B Eng.* **2016**, *90*, 336–340. [CrossRef]
45. Lu, X.; Nautiyal, P.; Bustillos, J.; Loganathan, A.; Zhang, C.; Chen, Y.; Boesl, B.; Agarwal, A. Hydroxylated boron nitride nanotube-reinforced polyvinyl alcohol nanocomposite films with simultaneous improvement of mechanical and thermal properties. *Polym. Compos.* **2020**, *41*, 5182–5194. [CrossRef]
46. Ceccaldi, C.; Fullana, S.G.; Alfarano, C.; Lairez, O.; Calise, D.; Cussac, D.; Parini, A.; Sallerin, B. Alginate Scaffolds for Mesenchymal Stem Cell Cardiac Therapy: Influence of Alginate Composition. *Cell Transplant.* **2012**, *21*, 1969–1984. [CrossRef]
47. Czerner, M.; Fellay, L.S.; Suárez, M.P.; Frontini, P.M.; Fasce, L.A. Determination of Elastic Modulus of Gelatin Gels by Indentation Experiments. *Procedia Mater. Sci.* **2015**, *8*, 287–296. [CrossRef]
48. Kakarla, A.B.; Kong, I.; Turek, I.; Kong, C.; Irving, H. Printable gelatin, alginate and boron nitride nanotubes hydrogel-based ink for 3D bioprinting and tissue engineering applications. *Mater. Des.* **2022**, *213*, 110362. [CrossRef]
49. Oladapo, B.I.; Ismail, S.O.; Adebiyi, A.V.; Omigbodun, F.T.; Olawumi, M.A.; Olawade, D.B. Nanostructural interface and strength of polymer composite scaffolds applied to intervertebral bone. *Colloids Surfaces A Physicochem. Eng. Asp.* **2021**, *627*, 127190. [CrossRef]
50. Li, L.; Qin, S.; Peng, J.; Chen, A.; Nie, Y.; Liu, T.; Song, K. Engineering gelatin-based alginate/carbon nanotubes blend bioink for direct 3D printing of vessel constructs. *Int. J. Biol. Macromol.* **2020**, *145*, 262–271. [CrossRef]
51. Serrano-Aroca, Á.; Iskandar, L.; Deb, S. Green synthetic routes to alginate-graphene oxide composite hydrogels with enhanced physical properties for bioengineering applications. *Eur. Polym. J.* **2018**, *103*, 198–206. [CrossRef]
52. Sakuma, I.; Nishimura, Y.; Chui, C.K.; Kobayashi, E.; Inada, H.; Chen, X.; Hisada, T. *In Vitro Measurement of Mechanical Properties of Liver Tissue under Compression and Elongation Using a New Test Piece Holding Method with Surgical Glue BT—Surgery Simulation and Soft Tissue Modeling*; Ayache, N., Delingette, H., Eds.; Springer: Berlin/Heidelberg, Germany, 2003; pp. 284–292.
53. Pervin, F.; Chen, W.W.; Weerasooriya, T. Dynamic compressive response of bovine liver tissues. *J. Mech. Behav. Biomed. Mater.* **2011**, *4*, 76–84. [CrossRef] [PubMed]
54. Holzapfel, G.A. Biomechanics of Soft Tissue. In *Handbook of Materials Behavior Models*; Elsevier: Amsterdam, The Netherlands, 2001; pp. 1057–1071.
55. Ambu, R.; Morabito, A. Porous Scaffold Design Based on Minimal Surfaces: Development and Assessment of Variable Architectures. *Symmetry* **2018**, *10*, 361. [CrossRef]
56. Smith, M.; Guan, Z.; Cantwell, W.J. Finite element modelling of the compressive response of lattice structures manufactured using the selective laser melting technique. *Int. J. Mech. Sci.* **2013**, *67*, 28–41. [CrossRef]
57. Maskery, I.; Aremu, A.O.; Parry, L.; Wildman, R.D.; Tuck, C.J.; Ashcroft, I.A. Effective design and simulation of surface-based lattice structures featuring volume fraction and cell type grading. *Mater. Des.* **2018**, *155*, 220–232. [CrossRef]
58. Bružauskaitė, I.; Bironaitė, D.; Bagdonas, E.; Bernotienė, E. Scaffolds and cells for tissue regeneration: Different scaffold pore sizes—Different cell effects. *Cytotechnology* **2016**, *68*, 355–369. [CrossRef]
59. Griffon, D.; Sedighi, M.; Schaeffer, D.; Eurell, J.; Johnson, A. Chitosan scaffolds: Interconnective pore size and cartilage engineering. *Acta Biomater.* **2006**, *2*, 313–320. [CrossRef] [PubMed]

Journal of
Composites Science

MDPI

Article

Microstructure-Free Finite Element Modeling for Elasticity Characterization and Design of Fine-Particulate Composites

Yunhua Luo

Department of Mechanical Engineering, University of Manitoba, Winnipeg, MB R3T 2N2, Canada; Yunhua.Luo@umanitoba.ca

Abstract: The microstructure-based finite element modeling (MB-FEM) of material representative volume element (RVE) is a widely used tool in the characterization and design of various composites. However, the MB-FEM has a number of deficiencies, e.g., time-consuming in the generation of a workable geometric model, challenge in achieving high volume-fractions of inclusions, and poor quality of finite element mesh. In this paper, we first demonstrate that for particulate composites the particle inclusions have homogeneous distribution and random orientation, and if the ratio of particle characteristic length to RVE size is adequately small, elastic properties characterized from the RVE are independent of particle shape and size. Based on this fact, we propose a microstructure-free finite element modeling (MF-FEM) approach to eliminate the deficiencies of the MB-FEM. The MF-FEM first generates a uniform mesh of brick elements for the RVE, and then a number of the elements, with their total volume determined by the desired volume fraction of inclusions, is randomly selected and assigned with the material properties of the inclusions; the rest of the elements are set to have the material properties of the matrix. Numerical comparison showed that the MF-FEM has a similar accuracy as the MB-FEM in the predicted properties. The MF-FEM was validated against experimental data reported in the literature and compared with the widely used micromechanical models. The results show that for a composite with small contrast of phase properties, the MF-FEM has excellent agreement with both the experimental data and the micromechanical models. However, for a composite that has large contrast of phase properties and high volume-fraction of inclusions, there exist significant differences between the MF-FEM and the micromechanical models. The proposed MF-FEM may become a more effective tool than the MB-FEM for material engineers to design novel composites.

Keywords: particulate composite; microstructure; finite element modeling; representative volume element; elastic property; microstructure-based; microstructure-free

Citation: Luo, Y. Microstructure-Free Finite Element Modeling for Elasticity Characterization and Design of Fine-Particulate Composites. *J. Compos. Sci.* **2022**, *6*, 35. https://doi.org/10.3390/jcs6020035

Academic Editor: Francesco Tornabene

Received: 28 December 2021
Accepted: 18 January 2022
Published: 20 January 2022

1. Introduction

Particulate composites are widely used in industrial products and engineering structures due to their merits, such as ease-of-manufacturing and great design flexibility. At the length scale of material representative volume element (RVE), the behavior of a particulate composite can be usually considered as homogeneous and isotropic if (1) the geometric aspect ratio of inclusions is small, (2) both the distribution and orientation of inclusions are statistically homogeneous, and (3) the characteristic dimension of inclusions is adequately small compared with the size of RVE. An important task in the design of composites is to predict their elastic properties based on phase properties and volume fractions. A number of methods is available for the prediction, and they are often classified into experimental, analytical, and numerical categories. Experimental methods are direct and reliable, but they are also expensive and time consuming. Analytical solutions developed from micromechanics models are convenient and efficient; however, they have various limitations in application because they are based on special assumptions regarding composite microstructure. Numerical methods, mainly represented by finite element modeling of RVE, have a number of advantages over the other two categories of methods. They are more efficient than experimental methods, and they do not make any special assumption about composite microstructure. The microstructure-based finite element modeling (MB-FEM) of material RVE has played an important role for material engineers to understand the micromechanics of composite cracking and debonding, which often initiate and develop in the phase materials or at their interfaces [1–3]. An accurate representation of the composite microstructure is critical in the finite element analysis of microscopic damage.

However, for the design of a novel composite, the main interest is in the elastic properties at the macroscopic level, rather than the interactions of phase materials at the microscopic level. The commonly considered design variables are elastic properties and volume fractions of phase materials. The design process is often iterative and involves a large number of finite element analyses of intermediate designs; the conventional MB-FEM has a number of deficiencies for this purpose. First, the creation of a workable geometric model for the RVE microstructure is often tedious and time-consuming. A number of algorithms has been developed to generate RVE of particulate composites, e.g., [4–13] among others. These algorithms either have difficulty to accommodate a high volume-fraction of inclusions [7–9,13], or have to use a time-consuming iterative process to achieve a high volume-fraction with the sacrifice of microstructure randomness [10–12]. It should be mentioned that the majority of the algorithms are developed for simple inclusion shapes such as sphere, ellipsoid, and cylinder. Inclusion shapes in actual composite materials are often irregular, which would further complicate the creation of valid RVE geometric models. Geometric models of such composites usually contain a large number of degenerated and small geometric entities. Therefore, even with the success of creating an RVE geometric model, there are still challenges in the generation of valid and high-quality finite element meshes, because the degenerated and small entities in the geometric models either prevent the generation of finite element meshes or result in poor-quality elements. Iterative mesh adaptation algorithms usually have to be used to improve mesh quality, but a satisfactory quality is still not guaranteed.

To eliminate the limitations of the conventional MB-FEM for the characterization and design of fine-particulate composites, we first demonstrate that if the size ratio of inclusion to RVE is adequately small, elastic properties of RVE computed by finite element modeling are independent of inclusion shape and size. Based on this fact, we propose microstructure-free finite element modeling (MF-FEM). We validate the MF-FEM against experimental data reported in the literature and compare the MF-FEM with the widely used micromechanical models.

2. Effect of Inclusion Shape and Size on Elastic Properties of Particulate Composites

To study the effect of inclusion shape and size on the elastic properties of particulate composites, the following assumptions are made: (1) the materials of the matrix and the inclusions are homogeneous, isotropic, and perfectly bond to each other; (2) the distribution and orientation of inclusions of small aspect ratio are statistically homogeneous; (3) the applied loading only introduces elastic deformation. Under the above assumptions, we hypothesize that for a fixed volume-fraction of inclusions, if the ratio of inclusion characteristic length to RVE size is adequately small, composite elastic properties such as Young's modulus, Poisson's ratio, and shear modulus are independent of inclusion shape and size, and also independent of loading orientation. To verify the above hypothesis, a series of particulate-composite RVE models, which are different from each other only by their inclusion shape and size, are constructed and analyzed. The models are described as follows:

- All RVE models are in a cubic shape, and the sides have a length of 100 units. Since inclusion-to-RVE size ratio is of interest, the unit can be in any length from nanometer to meter.

- The composite is a particulate-filled glassy polymer [14]. The matrix material has Young's modulus E_m = 2.68 GPa and Poisson's ratio V_m = 0.394, and the inclusion material has Young's modulus E_i = 70.0 GPa and Poisson's ratio V_i = 0.23.
- The volume fraction of inclusions in all models is fixed at 30%.
- The inclusions in a model have the same shape, i.e., either spheroid, almond-shaped, or pill-shaped.
- The inclusions in a model have the same size. The inclusion-to-RVE size ratio is gradually reduced from 0.8 to 0.02 in the series of models, by decreasing inclusion sizes.

A set of such models is shown in Figure 1.

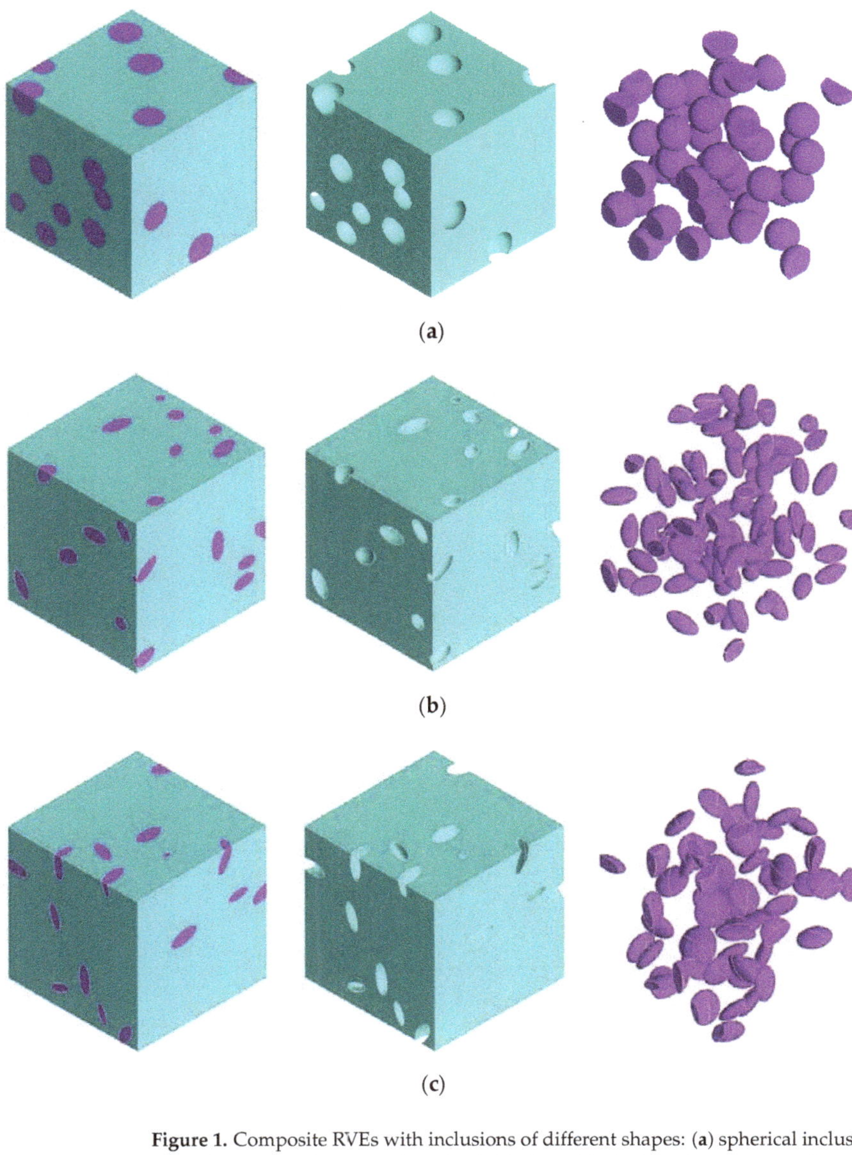

Figure 1. Composite RVEs with inclusions of different shapes: (**a**) spherical inclusions; (**b**) almond-shaped inclusions (aspect ratio = 2); (**c**) pill-shaped inclusions (aspect ratio = 2).

RVE Young's modulus, Poisson's ratio, and shear modulus are characterized by the conventional MB-FEM. The coordinate system used in the finite element modeling is illustrated in Figure 2. Boundary conditions for characterizing the elastic properties are listed in Table 1.

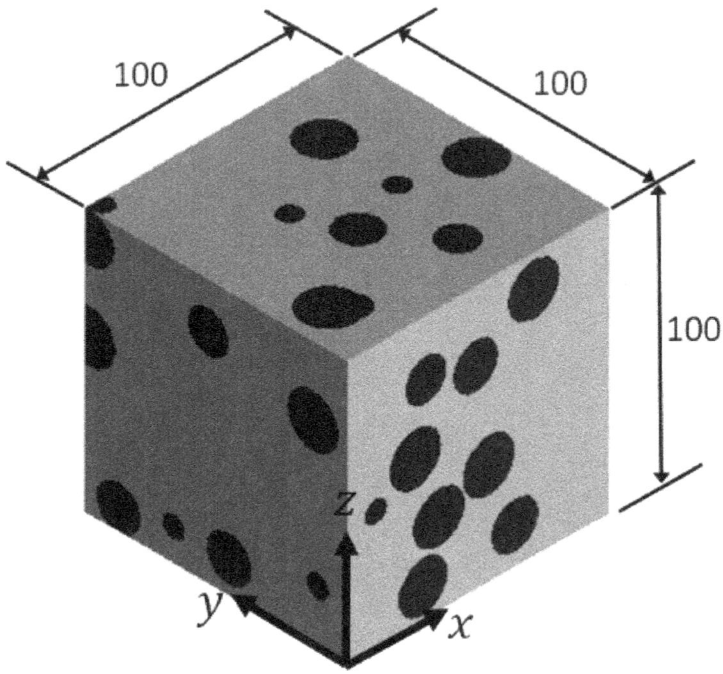

Figure 2. RVE of particulate composite.

Table 1. RVE boundary conditions for characterization of composite elastic properties.

RVE Surface	Young's Modulus (E_i, $i = x, y, z$) and Poisson's Ratio (v_{ij}, $i, j = x, y, z$)			Shear Modulus (G_{ij}, $i, j = x, y, z$)		
	E_x, v_{xy}, v_{xz}	E_y, v_{yx}, v_{yz}	E_z, v_{zx}, v_{zy}	G_{xy}	G_{yz}	G_{zx}
$x = 0$	$u_x = 0$	$u_x = 0$	$u_x = 0$	$u_x = u_y = u_z = 0$	Free	Free
$y = 0$	$u_y = 0$	$u_y = 0$	$u_y = 0$	Free	$u_x = u_y = u_z = 0$	Free
$z = 0$	$u_y = 0$	$u_y = 0$	$u_z = 0$	Free	Free	$u_x = u_y = u_z = 0$
$x = 100$	$u_x = 1$	u_x (coupled DOFs)	u_x (coupled DOFs)	$u_y = 1$, u_x (coupled DOFs)	Free	Free
$y = 100$	u_y (coupled DOFs) *	$u_y = 1$	u_y (coupled DOFs)	Free	$u_z = 1$, u_y (coupled DOFs)	Free
$z = 100$	u_z (coupled DOFs)	u_z (coupled DOFs)	$u_z = 1$	Free	Free	$u_x = 1$, u_z (coupled DOFs)

* Homogeneous boundary condition, i.e., all nodes on the surface $y = 100$ are forced to have the same displacement u_y.

Based on the mean-field homogenization theory [15], RVE Young's modulus (E_i), shear modulus (G_{ij}), and Poisson's ratio (ν_{ij}) are determined from RVE average stresses ($\overline{\sigma}_i$, $\overline{\tau}_{ij}$) and average strains ($\overline{\varepsilon}_i$, $\overline{\gamma}_{ij}$):

$$E_i = \frac{\overline{\sigma}_i}{\overline{\varepsilon}_i}, \quad (i = x, y, z) \tag{1}$$

$$G_{ij} = \frac{\overline{\tau}_{ij}}{\overline{\gamma}_{ij}}, \quad (i, j = x, y, z) \tag{2}$$

$$\nu_{ij} = -\frac{\overline{\varepsilon}_j}{\overline{\varepsilon}_i}, \quad (i, j = x, y, z) \tag{3}$$

The average stresses and strains are calculated as

$$\overline{\sigma}_i = \frac{1}{V}\int_V \sigma_i \, dV, \; \overline{\varepsilon}_i = \frac{1}{V}\int_V \varepsilon_i \, dV, \overline{\tau}_{ij} = \frac{1}{V}\int_V \tau_{ij} \, dV, \; \overline{\gamma}_{ij} = \frac{1}{V}\int_V \gamma_{ij} \, dV, \; (i, j = x, y, z) \tag{4}$$

In Equation (4), σ_i, ε_i, τ_{ij}, and γ_{ij} are, respectively, the normal stress, normal strain, shear stress, and shear strain that are determined by the finite element modeling; V is the volume of the RVE.

All finite element analyses in this study were conducted using commercial software, ANSYS Mechanical APDL (2020 R1). For each model, the mean value ($\overline{P}$) and standard deviation (σ) of Young's moduli, shear moduli, and Poisson's ratios characterized from the three loading orientations were calculated, i.e.,

$$\overline{P} = \frac{P_x + P_y + P_z}{3}, \quad \sigma = \sqrt{\frac{\left(P_x - \overline{P}\right)^2 + \left(P_y - \overline{P}\right)^2 + \left(P_z - \overline{P}\right)^2}{3}} \tag{5}$$

where P represents one of the three elastic properties. Variations of the mean values and standard deviations with the inclusion-to-RVE size ratio are shown in Figure 3.

The following observations can be made from the results shown in Figure 3:

- With the inclusion-to-RVE size ratio approximately smaller than 0.04, the RVE elastic properties are almost not affected by the inclusion shape and size.
- Anisotropy in the RVE elastic properties, which is measured by the error bars in Figure 3, is gradually reduced and then disappears with the decreased inclusion-to-RVE size ratio.

It is not surprising that the hypothesis is valid, because it is the fundamental assumption adopted in the continuum mechanics of composite materials, and it has been demonstrated to be true in a large number of previous studies, which were conducted on different inclusion shapes and phase properties, e.g., [16–23] among others. The threshold of inclusion-to-RVE size ratio referred in the above is equivalent to the critical RVE size used in other studies. Trias et al. [21] reported that for a typical unidirectionally carbon-fiber-reinforced polymer, to properly represent the random distribution of fibers in the transverse plane, the minimum size of RVE should be 50 times the fiber radius. Harper et al. [16] showed that for a planar model of discontinuous carbon fiber composites, the convergence of results from multiple realizations (i.e., models of the same volume fraction) is achieved at RVE edge lengths that are four times the fiber length, irrespective of fiber volume fraction; however, if only one realization is allowed, the ratio of fiber length to RVE size must be reduced to 0.03. The studies show that the averaging of multiple realizations has the effect of reducing anisotropy and thus underestimating the effect of RVE size or inclusion-to-RVE size ratio. For this reason, averaging of multiple realizations of the same volume fraction is not used in this study, and the results in Figure 3 were produced from one realization. It should be pointed out that the critical size of particulate-composite RVE can be significantly different in different regimes, for example, linear vs. nonlinear properties [22].

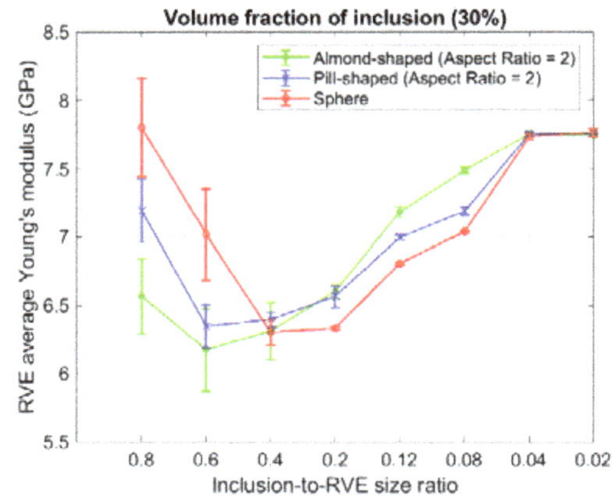

(a)

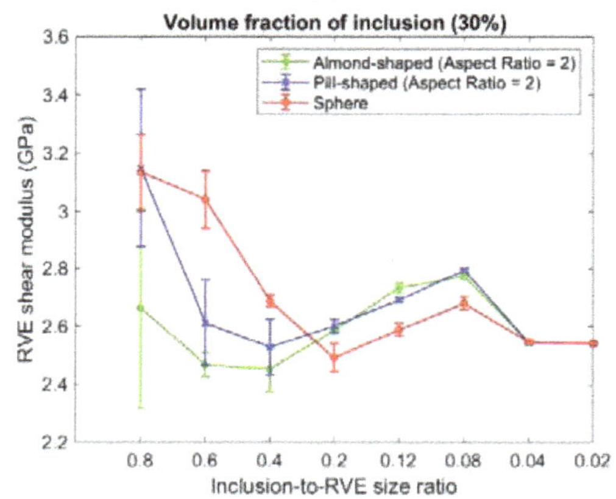

(b)

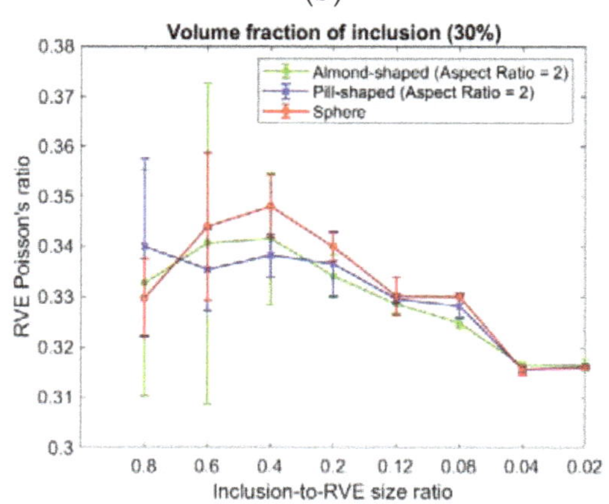

(c)

Figure 3. Variations of RVE elastic properties with inclusion-to-RVE size ratio: (**a**) Young's modulus; (**b**) shear modulus; (**c**) Poisson's ratio.

3. Microstructure-Free Finite Element Modeling of Particulate-Composites

The validity of the hypothesis demonstrated in the previous section justifies the idea of microstructure-free finite element modeling (MF-FEM) for particulate composites. Now that RVE properties are independent of inclusion shape and size if the inclusion-to-RVE size ratio is adequately small, theoretically we can consider any inclusion shape and size that are convenient for the finite element modeling. In this section, we propose a microstructure-free finite element modeling procedure for the characterization and design of two-phase particulate composites. The procedure is similar to the conventional MB-FEM. The only difference is in the representation of inclusions. A geometric model of the RVE microstructure is not needed; instead, the inclusions are represented by finite elements. The procedure is described below.

- A cubic RVE with side length of 100 units is constructed. The unit can be at any length scale from nanometer to meter, depending on the composite material to be studied.
- The RVE is meshed with brick elements of the same size, which is determined by the critical inclusion-to-RVE size ratio, i.e., 0.04 to 0.02 times the RVE side length, as discussed in the previous section. All the elements are first assigned with the properties of the matrix material.
- Then, a number of the elements is randomly selected and re-assigned with the properties of the inclusion material. The number of the selected elements is determined by the desired volume fraction of inclusions and the volume of each element. Samples of such microstructure-free finite element models are shown in Figure 4.
- Then, the boundary conditions described in Table 1 are applied, and finite element analyses are conducted.
- RVE properties such as Young's modulus, shear modulus, and Poisson's ratio are calculated using Equations (1)–(4).

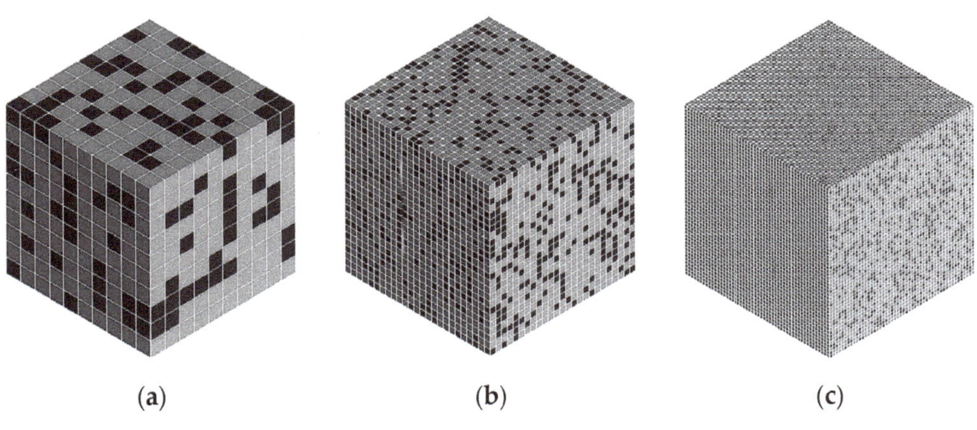

Figure 4. Microstructure-free finite element models of different element-to-RVE size ratios (**a**) 1/10; (**b**) 1/25; (**c**) 1/50 (The volume fraction of inclusions is 30%).

The effect of element-to-RVE size ratio on elastic properties predicted by the MF-FEM was studied using the same composite described in the previous section, with the volume fraction of inclusion elements set to 30%. The variations of RVE properties, averaged in the three loading orientations, with element-to-RVE size ratio, are shown in Figure 5. Similar to those computed by the MB-FEM, RVE properties predicted by the MF-FEM also show anisotropy if a large element-to-RVE size ratio is used; the degree of anisotropy is indicated by the error-bar length in the figure. With the element-to-RVE size ratio decreased, the anisotropy disappears, as shown in Figure 5. By comparing Figures 3 and 5, one can observe that the variations of properties with element-to-RVE size ratio shown in Figure 5 are much smaller than those in Figure 3, indicating that even with a relatively larger element-to-RVE size ratio, and thus a smaller number of elements, the MF-FEM is still able to predict RVE properties with reasonable accuracy.

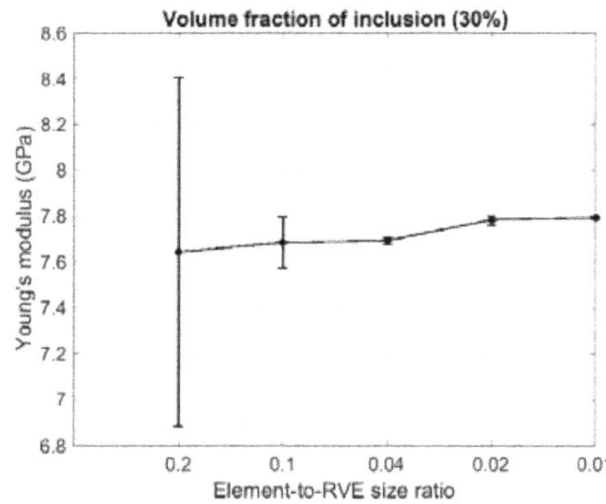

(**a**)

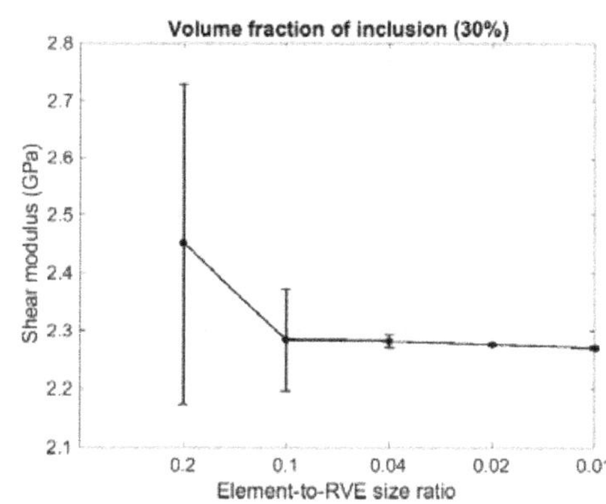

(**b**)

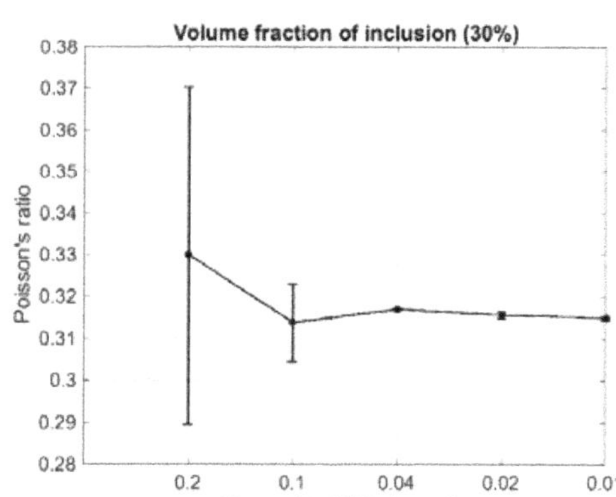

(**c**)

Figure 5. Variation of RVE properties with element-to-RVE size ratio: (**a**) Young's modulus; (**b**) shear modulus; (**c**) Poisson's ratio.

Comparison of RVE properties computed by the MF-FEM and the MB-FEM is displayed in Figure 6. With the inclusion-to-RVE (or element-to-RVE for the MF-FEM) size ratio of 0.02, differences in RVE properties computed by the two types of finite element models are negligible, suggesting that the MF-FEM is able to replace the MB-FEM in the prediction of effective properties of particulate composites.

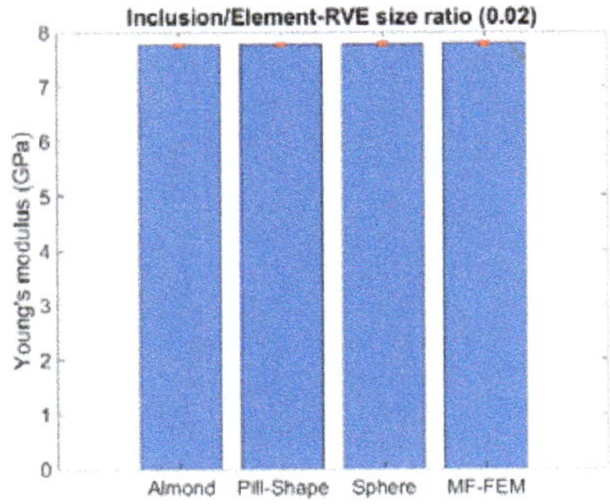

(**a**)

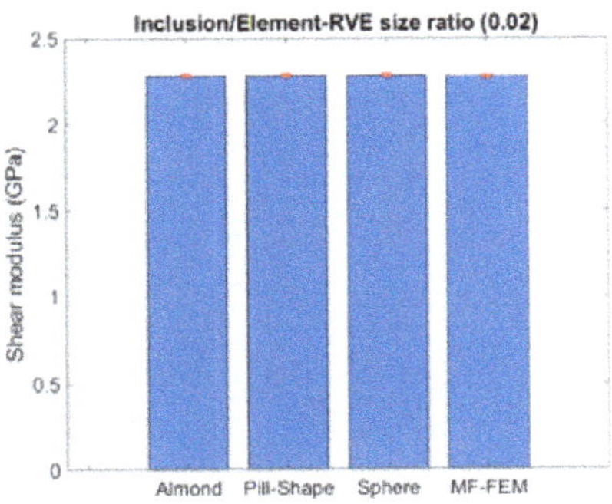

(**b**)

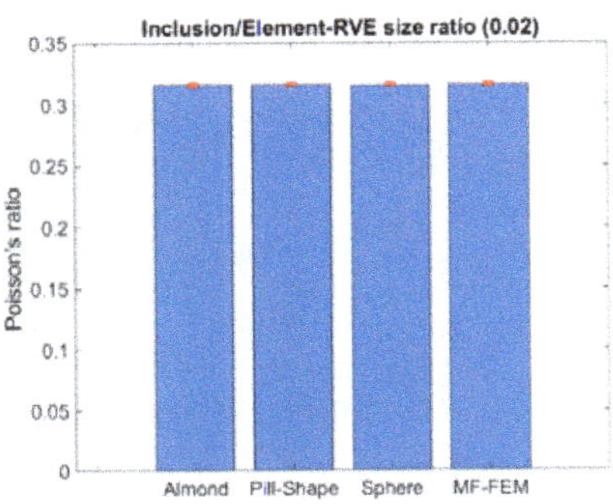

(**c**)

Figure 6. Comparison of RVE properties computed by the MF-FEM and the MB-FEM: (**a**) Young's modulus; (**b**) shear modulus; (**c**) Poisson's ratio.

4. Validation of MF-FEM against Experimental Data and Comparison with Popular Micromechanics Models

In this section, the MF-FEM was validated against experimental data previously reported in the literature and compared with popular micromechanics models. Two representative cases of experimental data were selected for the validation. One is WC–cobalt alloy, where the phase properties are close to each other [24], and the other is glass particulate reinforced polyester, where the phase properties have a large contrast [25]. Using a dynamic resonance method, Doi et al. [24] measured the Young's modulus, shear modulus, bulk modulus, and Poisson's ratio for a series of WC–cobalt alloys, with volume fractions of WC varying from 0.55 to 0.91. Properties of WC and cobalt used in the study are listed in Table 2.

Table 2. Properties of WC and cobalt phases [24].

Phase	Young's Modulus (GPa)	Shear Modulus (GPa)
Cobalt	206.99	79.00
WC	700.43	293.31

Richard [25] conducted a series of tensile tests on specimens of glass microsphere filled polyester composites to measure their Young's modulus and Poisson's ratio; the composites had volume fractions of glass microsphere varying from 0.0 to 0.464. The phase material properties used in Richard's study are listed in Table 3. Bulk modulus and shear modulus of the composites and phase materials were calculated using the elasticity relationships.

Table 3. Properties of glass microsphere and polyester [25].

Phase	Young's Modulus (GPa)	Poisson's Ratio
Polyester	1.69	0.45
Glass microsphere	68.95	0.21

Although there exists a large number of micromechanics models available for comparison [23,26,27], we selected the models that have explicit analytical solutions, explicitly developed for particulate composites and widely used in the current literature. Based on the above considerations, we chose Voigt–Reuss (V–R) bounds, Hashin–Shtrikman (H–S) bounds, the generalized self-consistent (GSC) scheme, the Mori–Tanaka (M–T) method, and the Voigt–Reuss–Hill (V–R–H) average for comparison.

Figures 7 and 8 show the results of validation and comparison. As it can be observed from Figure 7, for the WC–cobalt alloys, all the properties computed by the MF-FEM, except Poisson's ratio, had excellent agreement with the experimental data and the predictions by the GSC and the M–T methods.

For the glass–polyester composites, as can be seen from Figure 8a,b, at low volume fractions, the MF-FEM predictions were well congruous with the experimental data and the predictions of GSC and M–T methods. However, a deviation trend occurred among the curves with the increase of glass volume fraction. Unfortunately, experimental data of high volume-fraction of glass were not available for the validation. Experimental data of bulk modulus and Poisson's ratio, as displayed in Figure 8c,d, showed much greater scatterness, and the agreement between the experimental data and the MF-FEM predictions was not as good as in the Young's modulus or shear modulus.

It should be pointed out that the specific loading and constraint conditions used in the experiments are not available in the literature, which may be considerably different from those described in Table 1 and used in the MF-FEM. This may explain the differences between the experimental data and the MF-FEM predictions.

From Figures 7 and 8 it can be noticed that compared with the experimental data, micromechanics models such as GSC and M–T underestimate composite Young's modulus, shear modulus, and bulk modulus, which may become even more significant for composites with a large contrast of phase properties and a high volume-fraction of inclusions. On the other hand, the predictions of MF-FEM are generally higher than the experimental data; the possible reason is that porosity may exist in the experimental materials, while the MF-FEM considers the materials as non-porous.

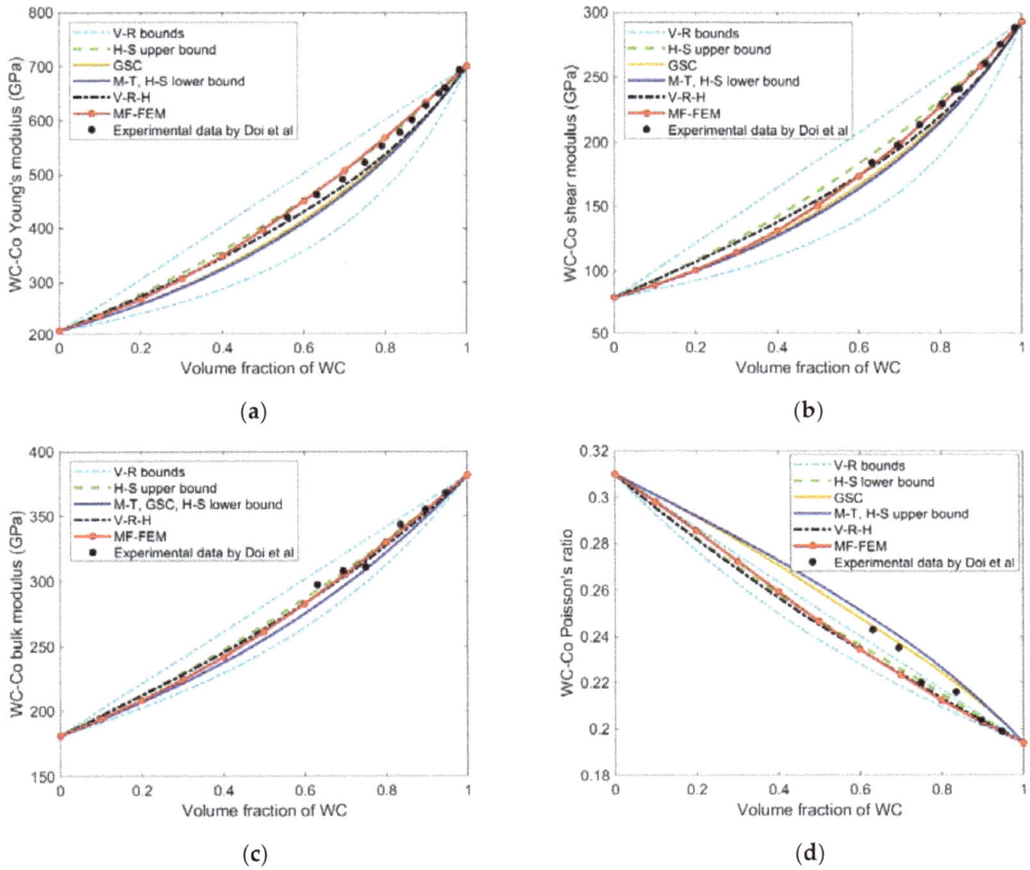

Figure 7. Validation and comparison against experimental data of WC–cobalt alloy [24]: (**a**) Young's modulus; (**b**) shear modulus; (**c**) bulk modulus; (**d**) Poisson's ratio.

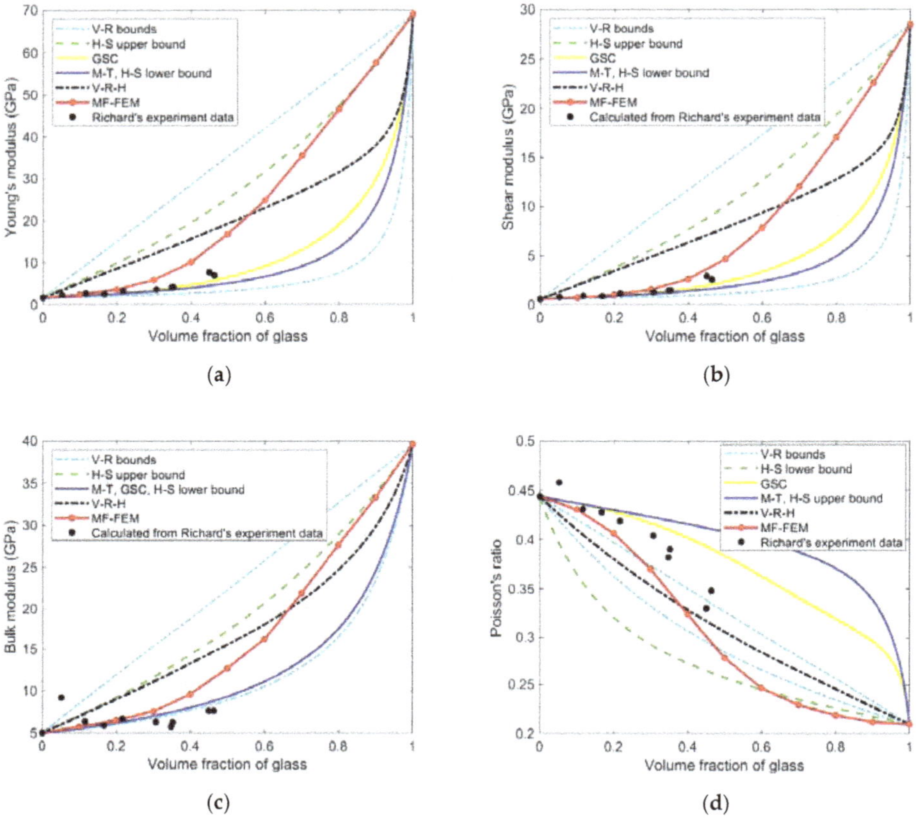

Figure 8. Validation comparison against experimental data of glass–polyester [25]: (**a**) Young's modulus; (**b**) shear modulus; (**c**) bulk modulus; (**d**) Poisson's ratio.

5. Discussion

Although there exist differences between the MF-FEM predictions and the experimental data, composite properties computed by the MF-FEM are in excellent agreement with those produced by the MB-FEM, as shown in Figure 6. The MF-FEM will become a more powerful and more efficient tool than the conventional MB-FEM in the design of particulate composites.

The MF-FEM has a number of advantages over the MB-FEM for the prediction of composite properties. There is no need to create a geometric model for the composite microstructure, which is usually tedious and time-consuming. There is no need to determine the size of RVE. Mesh quality is always guaranteed, and the number of elements does not change with the complexity of the composite microstructure. Once a universal element-to-RVE size ratio is determined, even a convergence study is not required, which is always necessary for the conventional MB-FEM. A high volume-fraction of inclusion can be easily achieved in the MF-FEM.

However, limitations of the MF-FEM are also obvious. The MF-FEM is not able to predict stress distribution in the phase materials, because the microstructure is not represented. If the inclusion-to-RVE size ratio is not adequately small, composite properties predicted by the MF-FEM will have low accuracy.

6. Conclusions

Based on the study results, it can be reasonably concluded that the MF-FEM is able to replace the conventional MB-FEM for the prediction of composite properties at the macroscopic level, but the MF-FEM is not effective for the study of composite damage at the microscopic level.

Funding: This research was funded by the Natural Sciences and Engineering Research Council (NSERC) via grant RGPIN-2019-05372 and the Research Manitoba via grant RM47950.The APC was funded by both NSERC and Research Manitoba.

Institutional Review Board Statement: Not applicable.

Informed Consent Statement: Not applicable.

Data Availability Statement: Not applicable.

Conflicts of Interest: The author has no conflict of interest to declare.

References

1. Raghavan, P.; Li, S.; Ghosh, S. Two scale response and damage modeling of composite materials. *Finite Elem. Anal. Des.* **2004**, *40*, 1619–1640. [CrossRef]
2. Gad, S.I.; Attia, M.A.; Hassan, M.A.; El-Shafei, A.G. Predictive computational model for damage behavior of metal-matrix composites emphasizing the effect of particle size and volume fraction. *Materials* **2021**, *14*, 2143. [CrossRef]
3. Ferdous, W.; Manalo, A.; Yu, P.; Salih, C.; Abousnina, R.; Heyer, T.; Schubel, P. Tensile fatigue behavior of polyester and vinyl ester based GFRP laminates—A comparative evaluation. *Polymers* **2021**, *13*, 386. [CrossRef]
4. Gusev, A.A. Representative volume element size for elastic composites: A numerical study. *J. Mech. Phys. Solids* **1997**, *45*, 1449–1459. [CrossRef]
5. Sherwood, J.D. Packing of spheroids in three-dimensional space by random sequential addition. *J. Phys. A Math. Gen.* **1997**, *30*, L839. [CrossRef]
6. Cooper, D.W. Random-sequential-packing simulations in three dimensions for spheres. *Phys. Rev. A* **1988**, *38*, 522. [CrossRef]
7. Adamczyk, Z.; Weronski, P. Random sequential adsorption of spheroidal particles: Kinetics and jamming limit. *J. Chem. Phys.* **1996**, *105*, 5562–5573. [CrossRef]
8. Buryachenko, V.A.; Pagano, N.J.; Kim, R.Y; Spowart, J.E. Quantitative description and numerical simulation of random microstructures of composites and their effective elastic moduli. *Int. J. Solids Struct.* **2003**, *40*, 47–72. [CrossRef]
9. Torquato, S.; Uche, O.U.; Stillinger, F.H. Random sequential addition of hard spheres in high Euclidean dimensions. *Phys. Rev. E* **2006**, *74*, 061308. [CrossRef]
10. Zhou, J.; Qi, L.; Gokhale, A.M. Microstructure model for discontinuously reinforced composite by modified random sequential absorption method. *J. Eng. Mater. Technol.* **2016**, *138*, 021001. [CrossRef]
11. Catalanotti, G. On the generation of RVE-based models of composites reinforced with long fibres or spherical particles. *Compos. Struct.* **2016**, *138*, 84–95. [CrossRef]
12. Segurado, J.; Llorca, J. A numerical approximation to the elastic properties of sphere-reinforced composites. *J. Mech. Phys. Solids* **2002**, *50*, 2107–2121. [CrossRef]
13. Yang, L.; Sycheva, A.V.; Black, D.M.; Eastell, R. Site-specific differential effects of once-yearly zoledronic acid on the hip assessed with quantitative computed tomography: Results from the HORIZON Pivotal Fracture Trial. *Osteoporos. Int.* **2013**, *24*, 329–338. [CrossRef]
14. Smith, J.C. Experimental values for the elastic constants of a particulate-filled glassy polymer. *J. Res. Natl. Bur. Stand. A Phys. Chem.* **1976**, *80*, 45–49. [CrossRef]
15. Genin, G.M.; Birman, V. Micromechanics and structural response of functionally graded, particulate-matrix, fiber-reinforced composites. *Int. J. Solids Struct.* **2009**, *46*, 2136–2150. [CrossRef]
16. Harper, L.T.; Qian, C.; Turner, T.A.; Li, S.; Warrior, N.A. Representative volume elements for discontinuous carbon fibre composites–Part 2: Determining the critical size. *Compos. Sci. Technol.* **2012**, *72*, 204–210. [CrossRef]
17. Drugan, W.J.; Willis, J.R. A micromechanics-based nonlocal constitutive equation and estimates of representative volume element size for elastic composites. *J. Mech. Phys. Solids* **1996**, *44*, 497–524. [CrossRef]
18. Chen, L.; Gu, B.; Zhou, J.; Tao, J. Study of the effectiveness of the RVEs for random short fiber reinforced elastomer composites. *Fibers Polym.* **2019**, *20*, 1467–1479. [CrossRef]
19. Kanit, T.; Forest, S.; Galliet, I.; Mounoury, V.; Jeulin, D. Determination of the size of the representative volume element for random composites: Statistical and numerical approach. *Int. J. Solids Struct.* **2003**, *40*, 3647–3679. [CrossRef]
20. Kari, S.; Berger, H.; Rodriguez-Ramos, R.; Gabbert, U. Computational evaluation of effective material properties of composites reinforced by randomly distributed spherical particles. *Compos. Struct.* **2007**, *77*, 223–231. [CrossRef]

21. Trias, D.; Costa, J.; Turon, A.; Hurtado, J. Determination of the critical size of a statistical representative volume element (SRVE) for carbon reinforced polymers. *Acta Mater.* **2006**, *54*, 3471–3484. [CrossRef]
22. Pelissou, C.; Baccou, J.; Monerie, Y.; Perales, F. Determination of the size of the representative volume element for random quasi-brittle composites. *Int. J. Solids Struct.* **2009**, *46*, 2482–2855. [CrossRef]
23. Pindera, M.-J.; Khatam, H.; Drago, A.S.; Bansal, Y. Micromechanics of spatially uniform heterogeneous media: A critical review and emerging approaches. *Compos. Part B* **2009**, *40*, 349–378. [CrossRef]
24. Doi, H.; Fujiwara, Y.; Miyake, K.; Oosawa, Y. A systematic investigation of elastic moduli of WC-Co alloys. *Metall. Mater. Trans. B* **1970**, *1*, 1417–1425. [CrossRef]
25. Richard, T.G. The mechanical behavior of a solid microsphere filled composite. *J. Compos. Mater.* **1975**, *9*, 108–113. [CrossRef]
26. Christensen, R.M. A critical evaluation for a class of micromechanics models. *J. Mech. Phys. Solids* **1990**, *38*, 379–404. [CrossRef]
27. Raju, B.; Hiremath, S.R.; Mahapatra, D.R. A review of micromechanics based models for effective elastic properties of reinforced polymer matrix composites. *Compos. Struct.* **2018**, *204*, 607–619. [CrossRef]

Journal of
Composites Science

Article

Assessing Failure in Steel Cable-Reinforced Rubber Belts Using Multi-Scale FEM Modelling

Siegfried Martin Frankl [1], Martin Pletz [1,*], Alfred Wondracek [2] and Clara Schuecker [1]

[1] Designing Plastics and Composite Materials, Department of Polymer Engineering and Science, Montanuniversitaet Leoben, 8700 Leoben, Austria; siegfried.frankl@unileoben.ac.at (S.M.F.); clara.schuecker@unileoben.ac.at (C.S.)
[2] Semperit Technische Produkte Gesellschaft m.b.H., 2632 Wimpassing, Austria; alfred.wondracek@semperitgroup.com
* Correspondence: martin.pletz@unileoben.ac.at

Abstract: This work introduces a finite element model of a steel cable-reinforced conveyor belt to accurately compute stresses in the splice. In the modelled test rig, the belt runs on two drums and is loaded with a cyclic longitudinal force. An explicit solver is used to efficiently handle the high number of elements and contact conditions. This, however, introduces some issues of dynamics in the model, which are subsequently solved: (a) the longitudinal load is applied with a smooth curve and damping is introduced in the beginning of the simulation, (b) residual stresses are applied in regions of the belt that are initially bent around the drums, and (c) supporting drums are introduced at the start of the simulation to hinder oscillations of the belt at low applied forces. To accurately capture the tensile and bending stiffness of the cables, they are modelled by a combination of solid and beam elements. The results show that numerical artefacts can be reduced to an acceptable extent. In the region of highest stresses, the displacements are additionally mapped onto a submodel with a smaller mesh size. The results show that, for the investigated belt, the local maximum principal stresses significantly increase when this region of highest stresses comes into contact with, and is bent by, the drum. Therefore, it is essential to also consider the belt's bending to predict failure in such applications.

Keywords: Finite Element Method; conveyor belts; steel cables; submodelling

Citation: Frankl, S.M.; Pletz, M.; Wondracek, A.; Schuecker, C. Assessing Failure in Steel Cable-Reinforced Rubber Belts Using Multi-Scale FEM Modelling. *J. Compos. Sci.* **2022**, *6*, 34. https://doi.org/10.3390/jcs6020034

Academic Editor: Francesco Tornabene

Received: 22 December 2021
Accepted: 18 January 2022
Published: 20 January 2022

1. Introduction

Conveyor belts are used in a wide range of applications such as supermarkets, logistic centres, and mining. The conveyor belts in mining are reinforced with steel cables to reach the high strengths required. Such conveyor belts can have lengths of several kilometres. They consist of belt segments that are connected on site. In this connection, which is called splice, the steel cables are arranged in a specific laying scheme. The strength of this splice limits the belt's strength in cyclic loading and is thus of great interest to engineers. The splice strength is experimentally determined in a test rig where a conveyor belt with one splice runs on two drums and is cyclically loaded [1], see Figure 1. The failure of the belt can be caused by cable–rubber debonding, tearing of the rubber, and the rupture of cables.

To certify a belt for a tensile force, a certain number of cycles must be reached in the test rig without failure. To better understand the failure mechanisms, some effort has been put into monitoring the evolution of damage during the test [2,3]. This could be an initial debonding of cables from the rubber, which, at higher cycle numbers, leads to the rupture of individual cables, which ultimately leads to failure of the belt. The large variety of belt damage mechanisms has been illustrated in a study of damage due to objects such as rocks falling on a conveyor belt [4]. This complex damage behaviour indicates that it is not trivial to predict this failure which depends on the cable and rubber properties, the splice geometry, and the debonding strength of the cable–rubber interface [5]. For conveyor belts of lower strength classes, which are reinforced with textiles, Finite Element Method (FEM) models that compute the stress-distribution in belts that are loaded in tension exist [6,7]. The steel cables, on the other hand, pose a challenge in such models due to their complex stiffness: steel cables have high tensile stiffness but very low bending stiffness, and can feature tension/torsion coupling [8]. This can play a role in the failure of splices.

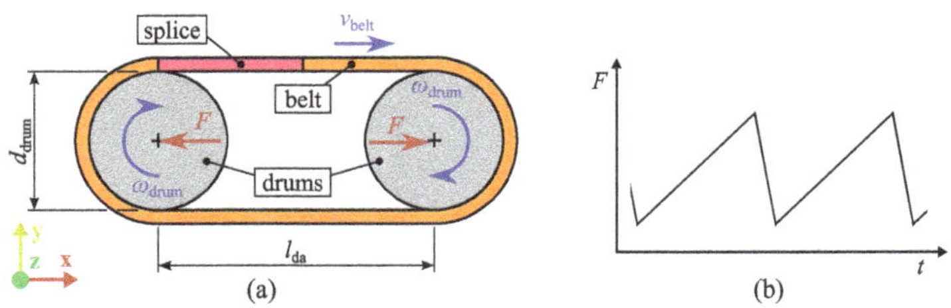

Figure 1. Belt test rig to determine the cyclic strength of a belt splice following [1]: (**a**) Setup of the test rig and (**b**) applied load curve.

Various research groups have developed numerical models to predict the strength of a splice, putting the emphasis on different aspects of the complex conveyor belt system. Nordell et al. [9,10] presented a very detailed FEM model of a splice, which even accounts for the tension/torsion coupling of cables. Similar to Keller [11], they argued that simpler specimens can be designed to show damage behaviour similar to that observed in the belt splice, which can then be used to study the damage behaviour in detail, where the influence of belt-drum contact and bending was also discussed. Another research group introduced similar FEM models to study a range of rubber materials and introduce an automated tool for investigating splice schemes [12,13]. In more recent work, other groups such as Li et al. [14] and Wheatley and Keipour [15] presented similar models, focusing on simplified specimens loaded in tension. None of those models account for bending of the belt on the drums in the test rig. Usually, the splice optimisation is carried out according to evaluated stresses in the FEM models. The bending of the belt, however, can affect the local stress fields in the splice in a nonlinear way. This can shift the damage mechanisms compared to a belt loaded purely in tension.

This work introduces a conveyor belt test rig model that fully accounts for contact and bending of the belt on the two drums. The computed stress fields can be used as an indicator of the belt's strength (Li et al. [16] use a stress-based criterion for damage initiation that agrees well with experimental results of failure). It can, thus, predict the influence of bending on the tested strength of a belt. Note that this work computes static stresses and could be extended towards fatigue models that are based on those stresses, as described by Carraro et al. [17] and Ferdous et al. [18]. Since all steel cables of a belt need to be modelled to capture the stress variations in the splice, a full 3-d model including drums contact is very challenging in terms of computational time. For this reason, a two-scale approach using an explicit full-scale test rig model and a finer-meshed implicit submodel of the region of highest stresses is taken. The models are used to assess whether the bending of the belt at the drums introduces additional stresses in the splice. This work, which is the first full 3-d splice model with bending, can thus answer the question of whether the bending loads are relevant for the mechanical design of conveyor belts. Challenges such as obtaining an initial state of movement and stresses in the belt and coping with initial dynamic artefacts are solved in the explicit model.

2. Methods

In this section, the modelling of the conveyor belt setup and the splice scheme used are explained. Additionally, the setup of the global test rig model and the submodel is presented. The belt is stress-free and flat after production. In the global test rig model, the initially-bent regions are applied with residual stresses. Furthermore, specific techniques are described to avoid unwanted dynamic effects in the explicit global test rig model.

2.1. Material Models and Splice Geometry

This work uses a simple conveyor belt with nine cables and only one rubber material. This belt, which can be simulated much faster, can be assumed to feature similar effects of the bending to realistic belts. The setup of a conveyor belt is shown in Figure 2a with the longitudinal direction, the out-of plane direction, and the lateral direction, defined as x, y, and z-directions, respectively. Such a steel cable-reinforced conveyor belt can consist of several reinforced and non-reinforced rubber layers. The belt considered in this work is made up of only one rubber material reinforced by parallel steel cables and no additional transverse reinforcement, see Figure 2b. The belt has a total thickness of $h = 42.5\,\text{mm}$ with $h_1 = 16.75\,\text{mm}$ and $h_2 = 25.75\,\text{mm}$. The rubber is modelled as a hyper-elastic material using the Mooney-Rivlin formulation. The strain energy function is given by

$$U = C_{10}(I_1 - 3) + C_{01}(I_2 - 3) + \frac{1}{D}(J - 1)^2, \tag{1}$$

where C_{10}, C_{01}, and D are material parameters, and I_1 and I_2 are the first and second invariant of the left Cauchy—Green deformation tensor, respectively. J is the determinant of the deformation gradient. The material parameters are listed in Table 1. The parameters C_{10} and C_{01} are taken from Froböse et al. [12] for a rubber material with a shear modulus of $G = 1.7\,\text{MPa}$. The rubber density is not given and is assumed as $\rho_{\text{rubber}} = 1000\,\text{kg/m}^3$. To use a typical Poisson's ratio for rubber of ν very close to 0.5 would introduce high-frequency noise in the explicit simulation, as described in [19], suggesting an upper limit of $\nu = 0.495$. Therefore, the parameter D is set to reach a Poisson's ratio of $\nu = 0.485$.

Table 1. Material parameters.

C_{10} [MPa]	C_{01} [MPa]	D [1/MPa]
0.7083	0.1417	0.03565

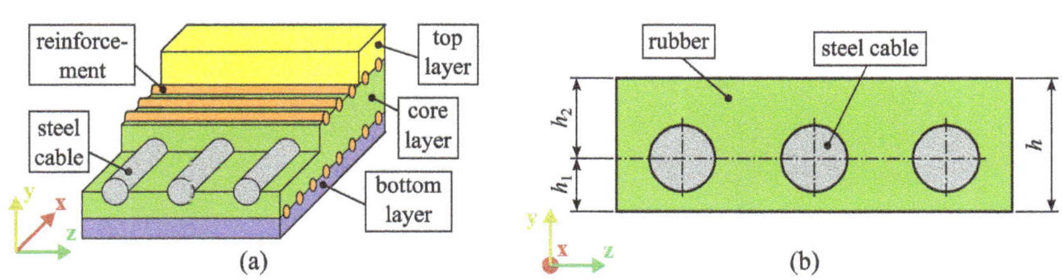

Figure 2. General structure of a conveyor belt with possible rubber layers and reinforcements (**a**) and cross-section of the conveyor belt used in this work (**b**).

Due to their inherent structure, the steel cables cannot be modelled accurately by a simple solid material model. Steel cables feature high tensile stiffness and low bending stiffness compared to a solid material of the same diameter. Also, they can feature a tension/torsion coupling, which requires special modelling in FEM. In this work, the steel cables are modelled so that they reach accurate tensile and bending stiffness. To this end, the volume of the cylindrical steel cables is meshed with solid elements and, in the cable's axis, additional beam elements are introduced. This setup is illustrated in Figure 3. The mechanical response is assumed to be linear elastic. The beam elements share nodes with the solid elements. It should be mentioned that the volume elements are not affected by the rotations of their connected nodes. The steel cables in this work have a radius of $r_{\text{cable}} = 6.75\,\text{mm}$.

A very small radius of the beam is chosen so that the solid elements account for the whole bending stiffness of the cable. To determine the Young's modulus for the solid elements E_{solid}, a three-point-bending test of a rubberised steel cable with a bending length $l_{\text{bend}} = 110\,\text{mm}$ has been performed. From the linear region at the beginning of the force-displacement curve, a bending stiffness $S_{\text{bend}} = 95\,\frac{\text{N}}{\text{mm}}$ is obtained and yields a Young's modulus for the solid elements as

$$E_{\text{solid}} = \frac{S_{\text{bend}}\, l_{\text{bend}}^3}{12\, r_{\text{cable}}^4\, \pi} = 1615.68\,\text{MPa}. \tag{2}$$

To determine the average Young's modulus of the cable, a tensile test of a single cable with a test length of $l_{\text{tensile}} = 200\,\text{mm}$ has been performed. A tensile stiffness $s_{\text{tensile}} = 25\,\text{kN/mm}$ is obtained from the linear region at the beginning of the force-displacement curve and yields a Young's modulus for the cable:

$$E_{\text{cable}} = \frac{s_{\text{tensile}}\, l_{\text{tensile}}}{r_{\text{vol}}^2\, \pi} = 34{,}931.13\,\text{MPa}. \tag{3}$$

In order to not induce any additional bending stiffness by the beam, the area ratio of the solid and the beam cross-sections is set to $\xi_{\text{volbeam}} = 1000$. This results in a beam radius of $r_{\text{beam}} = 0.214\,\text{mm}$. The Young's modulus of the beam E_{beam} needs to be adjusted to reach the total cable stiffness by the combination of the beam and solid elements. Therefore, E_{beam} has to account for the tensile stiffness that has not been accounted for by the solid elements as

$$E_{\text{beam}} = \xi_{\text{volbeam}}(E_{\text{cable}} - E_{\text{solid}}) = 33{,}300\,\text{GPa}. \tag{4}$$

The density of the steel cable ρ_{cable} is computed from the density of the rubber ρ_{rubber}, the density of steel $\rho_{\text{steel}} = 7850\,\text{kg/m}^3$, and an assumed volume ratio of steel in the rubberised cables $\xi_{\text{st}} = 0.6$. Therefore, the density of the steel cable is given by

$$\rho_{\text{cable}} = \xi_{\text{st}}\, \rho_{\text{steel}} + (1 - \xi_{\text{st}})\rho_{\text{rubber}} = 5110\,\frac{\text{kg}}{\text{m}^3}. \tag{5}$$

The Poisson's ratio of the solid and the beam elements is assumed to be 0.3.

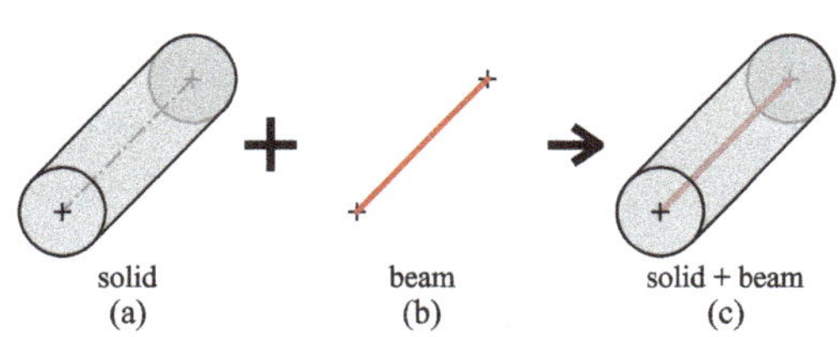

Figure 3. Steel cable modelling: (**a**) Solid elements that account for bending stiffness, (**b**) beam elements that account for tensile stiffness, and (**c**) the combination of solid and beam elements.

The splice scheme used for the splice is shown in Figure 4 and the corresponding parameters of the splice are listed in Table 2. Outside the splice, the steel cables are arranged in parallel with an equal distance between the cable axes of s_{belt}. Within the splice, the steel cables are rearranged to a lateral distance between the cables of s_{splice}. Cubic splines are used to describe the lateral change in the steel cable axes between outer regions and the splice.

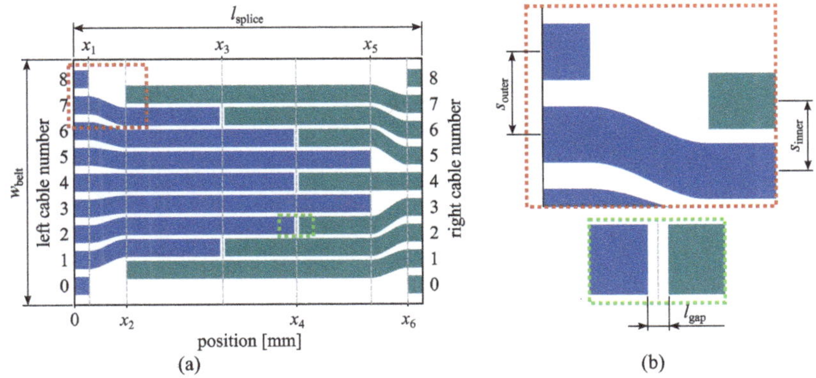

Figure 4. Geometry of the splice: (**a**) full view of the splice and (**b**) closer view of one part of the splice. Note that the lateral cable distance is smaller within the splice compared to outside the splice.

Table 2. Parameters of the splice.

Parameter Name	Symbol	Value	Unit
belt width	w_{belt}	190	mm
splice length	l_{splice}	4.7	m
cable positions	x_i	0.2, 0.7, 2, 3, 4, 4.5	m
outer steel cable spacing	s_{belt}	20	mm
inner steel cable spacing	s_{splice}	17	mm
gap between cable ends	l_{gap}	70	mm

2.2. *Test Rig Model*

In this section, the global FEM model of the conveyor belt test rig is described. In the test rig [1], the belt is cyclically loaded in tension as it is running on the drums. This work considers the quasi-static case of the conveyor belt running around the drums with a constant applied load F. For reasons of efficiency, an explicit solver is used. This means that dynamic effects in the model are considered as artefacts and have to be reduced to a small extent. To be able to start the simulation with a belt running on the two drums, the bending stresses of the belt on the drums have to be considered. This is reached by the application of residual stresses. Additionally, the application of the load is optimised and additional drums are introduced that avoid belt oscillations in the beginning of the simulation. The model is used to determine stress and strain fields for the conveyor belt running in the test rig. The validity for the quasi-static case is examined. The region of highest stresses in the belt running around the drum is determined and investigated further in a submodel.

2.2.1. Model Setup

The conveyor belt test rig shown in Figure 1 is represented as a quasi-static FEM model. The belt geometry, splice scheme, and materials of the model are taken from above. The left end of the splice is initially positioned right above the centre of the left drum. All model parameters are listed in Table 3. To handle the non-linearities in the material, the contact between the belt and drums and a high number of elements, an explicit solver is used, which is much more efficient for such models. Velocities in the model, thus, have no physical meaning and dynamic effects are considered as artefacts.

The drums are modelled as rigid discrete surfaces with reference points in each of their axes. In the x- and y-axes, the drum's rotation is fixed. The displacement of the drum's reference points is also constrained in y- and in the z-direction. An additional reference point P for applying the tensile load of the belt F is introduced, see Figure 5a. The reference point P is coupled to the reference points of the left drums P_{ld} and right drum P_{rd} in the x-direction:

$$u_x + \eta_c\, u_{ldx} = 0 \text{ and}$$
$$u_x - \eta_c\, u_{rdx} = 0, \tag{6}$$

with u_x as the displacement of P, u_{ldx} as the displacement of P_{ld}, u_{rdx} as the displacement of P_{rd}, and η_c as the weight factor for the coupling. Since F should be applied in the belt, the weight factor for the coupling is set to $\eta_c = 4$. The contact between the drums and the belt is modelled with penalty contact. The friction coefficient between rubber and steel μ_{rs} for the dry condition is between 0.6 to 1.1 according to Cruz Gómez et al. [20]. Therefore, $\mu_{rs} = 0.8$ is assumed.

Table 3. Parameters of the test rig.

Parameter Name	Symbol	Value	Unit
distance of the drum axes	l_{da}	7	m
diameter of the drums	d_{drum}	1.25	m
velocity of the belt Equation (7)	v_{belt}	6.454	m/s
number of circulations per load cycle [1]	n_{cplc}	18	1
time period of one load cycle [1]	t_{lc}	50	s
angular velocity of the drums Equation (8)	ω_{drum}	9.987	rad/s
point mass of the drum Equation (9)	m_{drum}	1502	kg
moment of inertia of the drums Equation (10)	$I_{z\,drum}$	429	kg m^2
tensile load in the belt Equation (12)	F_{belt}	655.2	kN

The explicit simulation starts with a running belt with the initial velocity of the belt v_{belt}, which can be calculated from the number of circulations per tensile load cycle n_{cplc} and the time period of one load cycle t_{lc}, as defined in [1]:

$$v_{belt} = \frac{n_{cplc}(d_d\, \pi + 2\, l_{da})}{t_{lc}}. \tag{7}$$

Note that this means that the belt is running with the actual speed stated in Table 3, even though the model is developed for a quasi-static analysis. The angular velocity of the drums ω_{drum} that corresponds to the belt speed calculates as

$$\omega_{drum} = \frac{2\, v_{belt}}{d_{drum} + h}, \tag{8}$$

where the neutral axis is assumed to be in the middle of the belt thickness. Since the right drum is the driving drum, this angular velocity is applied to the right drum continuously throughout the simulation and to the left drum as an initial condition only.

A point mass m_{drum} is applied at the reference points of the drums. For the mass, a hollow cylinder of steel with an external diameter of $d_{outer} = d_{drum}$, an internal diameter $d_{inner} = d_{drum} - 2 \cdot 200\,\text{mm}$, and a height of $h_{drum} = w_{belt} + 2 \cdot 50\,\text{mm}$ is assumed. The point mass m_{drum} is then given by

$$m_{drum} = \rho_{steel}\, \pi\, h_{drum} \frac{d_{outer}^2 - d_{inner}^2}{4}. \tag{9}$$

A moment of inertia $I_{z\,drum}$ for rotation in the drum axis is specified for the left drum, which is given for a hollow cylinder by

$$I_{z\,drum} = \frac{1}{2}\, m_{drum} \left(\frac{d_{outer}^2 + d_{inner}^2}{4} \right). \tag{10}$$

The tensile load F is applied smoothly enough and slowly enough so that dynamic effects do not alter the results and, therefore, the displacement caused by the load application increases smoothly. The tensile force on the belt is applied to point P in a way so that oscillations in the belt are minimised: it was discovered that starting with a force $F = F_{belt}/8$ at $t = 0$ and then using a polynomial with the smooth-step option of Abaqus to increase it to F_{belt} at $t = t_1$ works well, see Figure 5b. Additionally, the reference point P is connected to a damper in the x-direction, which has a damping coefficient d of $100\,\mathrm{Ns/mm}$. The time t_1 is chosen such that the constant applied force is reached before the splice runs into the bent region on the drum:

$$t_1 = \frac{l_{da} - l_{splice}}{v_{belt}} = 0.3564\,\mathrm{s}. \tag{11}$$

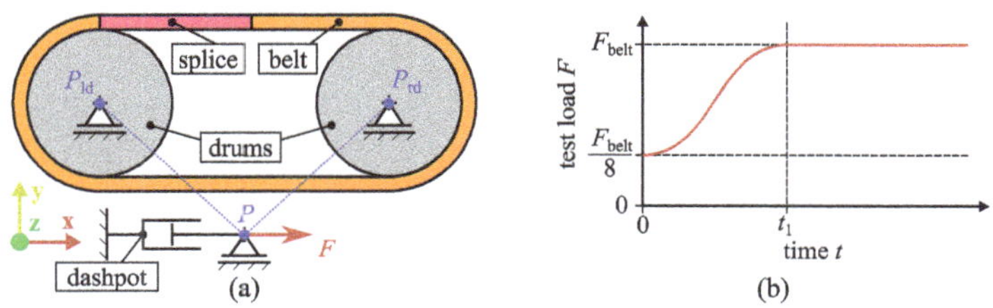

Figure 5. Setup of the conveyor belt test rig model (**a**) and applied test load vs. time curve (**b**).

The applied load is selected to be 60% of the nominal strength per width k_N of $7.8\,\frac{\mathrm{kN}}{\mathrm{mm}}$ of such a steel cable-reinforced belt. The width in the nominal strength is taken as the number of cables in the splice n_{sc} (here $n_{sc} = 7$) times their distance s_{belt}. The force F_{belt} thus calculates as

$$F_{belt} = 0.6\, n_{sc}\, s_{belt}\, k_N. \tag{12}$$

The belt is meshed with a global mesh size of 3 mm. Four elements are used over the gap length of the rubber gaps between cables. Outside the splice region, a mesh size of 10 mm is used along the running direction of the belt. To avoid hour glassing, eight-noded fully-integrated hexahedral elements are used for the volume except for the regions near the steel cable ends, where it is not possible to mesh this geometry change using purely hexahedral elements. Thus, those regions are meshed using six-noded wedge elements with reduced integration and four-noded tetrahedral elements. The beams of the steel cables are modelled using two-noded Thimoshenko beam elements. The cables and the rubber are perfectly connected (shared nodes), except for the end faces of the cables, where an initial debonding is assumed. This is because the necessary surface modification of the steel cables to increase the adhesion to the rubber is not present at the ends that have been cut. The drums are meshed with a global mesh size of 10 mm and four-noded rigid quad elements.

To solve the model, the nonlinear explicit solver of the commercial FEM code Abaqus [19] is used where large deformations are considered. A simulation time of 0.80638 s is sufficient for about 60% of the splice running into the bent region of the drum, and the region of highest stresses is located in the middle of the bent region. Mass scaling is used to achieve a stable time increment of $\Delta t = 1.75 \cdot 10^{-6}\,\mathrm{s}$ with a total change in the model's mass of 4.46%. The test rig model has a total of about 3,250,000 elements and takes 92 h to run on 72 cores.

2.2.2. Applying Initial Stresses in Bent Belt Regions

Whenever the conveyor belt is bent, it contains some bending stresses in the cables and the rubber regions. In FEM models, the geometry is usually considered stress-free at the start of the computation. For the regions of the belt that are initially not straight but are bent around the drums, the stresses that correspond to bending with a curvature of $2/d_{drum}$ need to be applied as initial stresses. Since the bending stress in a cable-reinforced rubber belt is not trivial, a small implicit simulation is carried out to obtain those stresses. This model contains the belt in its bent shape and straightens it, where the bent shape corresponds to the shape it will initially have in the test rig model (see Figure 6a,b). The stresses in the cable direction are evaluated and their inverse values are applied element-wise in the initial step of the test rig model. It has been verified that this procedure yields an approximately stress-free state in the segment, see Figure 6c,d.

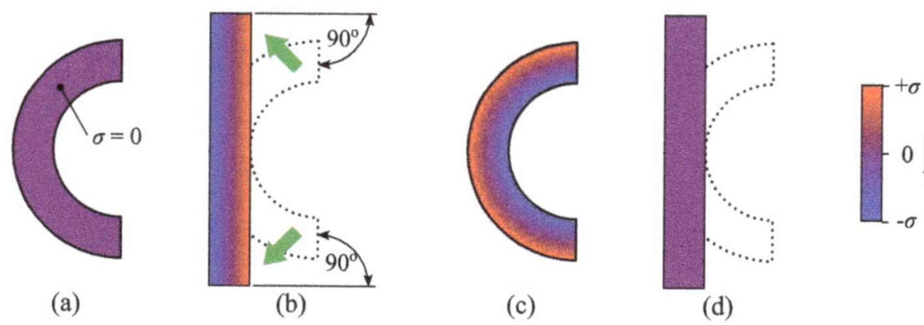

Figure 6. Procedure to determine and apply the residual stresses in the initially bent regions: (**a**) bent belt, (**b**) straightened belt and resulting stress distribution, (**c**) bent belt with applied residual stresses, and (**d**) approximately stress-free belt after straightening.

2.2.3. Avoiding Belt Oscillations Using Support Drums

Due to the internal stresses in the initially bent regions, the belt tends to deform as shown in Figure 7a, particularly if only low tensile forces are applied to the drums. In a dynamic model, this induces initial oscillations. To avoid such oscillations, additional support drums are used, but only during the first phase until the final loading state is reached, see Figure 7b. Note that these drums are only introduced to avoid numerical artefacts and are not part of the test rig. The horizontal distance of the support drum axes l_{sda} is set to 6.8 m, therefore they lie 100 mm closer together than the test rig drums. The diameter of the support drums is the same as the test rig drums, and the vertical distance of the support drum axes w_{sda} is set to 2.585 m. The support drums initially have the same angular velocity ω_{drum} as the test rig drums; however, their rotational velocity is free to change during the simulation. Apart from rotation around the z-axis, where an inertia of $I_{z\,drum}$ is applied as in the left drum of the test rig, all translations and rotations of the drums are constrained. The support drums use the same mesh and contact conditions to the belt as the test rig drums. Once the maximum applied load is reached at t_1, their contact condition between the support drums and belt is switched off in order to effectively remove them from the simulation.

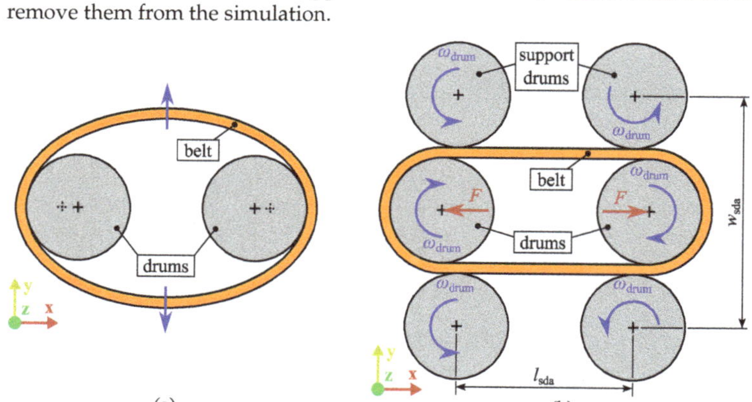

Figure 7. To avoid oscillations with belt deformations schematically shown in (**a**), four support drums are introduced (**b**).

2.3. Submodel

In this work, the region with the highest stresses in the rubber is modelled in the submodel with a finer mesh to investigate this region in more detail. This region is defined in Figure 8, where the same belt geometry and the same material models as the test rig model are used. The submodel length l_{sub} and the submodel width w_{sub} are set to 135 mm and 51.5 mm, respectively. The displacements of the test rig model are mapped to the lateral surfaces and the bottom surface of the submodel, see Figure 8. The displacements are mapped separately for the steel cable and the rubber nodes. The nodes at the steel cable/rubber interface use the displacement field of the cable elements in the global model. To avoid convergence issues, the rotational degree of freedom around the beam axes of the beam nodes belonging to the left lateral surface is fixed.

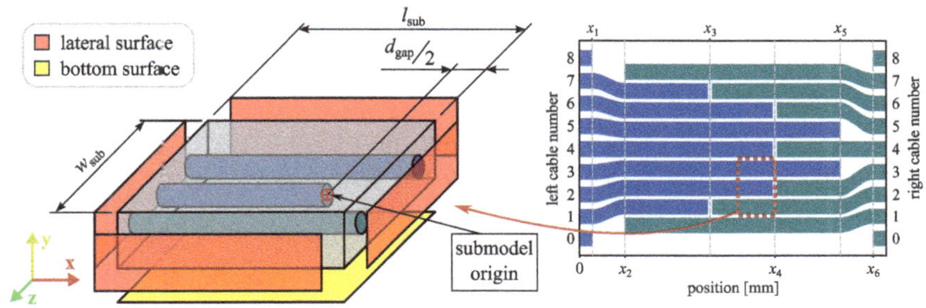

Figure 8. Sketch of the submodel, which is located in the splice. Additionally, the surfaces are defined for boundary conditions.

The submodel uses eight-noded fully-integrated hybrid hexahedral elements and is meshed with a global mesh size of 1.5 mm. Eight elements are used in the rubber gap between the steel cables. The beams of the steel cables are modelled using two-noded Thimoshenko beam elements, as in the global model. For computation, the nonlinear implicit solver of the commercial FEM code Abaqus [19] is used, considering large deformations. For verification of the submodelling approach, the stresses of the global model and the submodel are compared to see if the submodelling approach is valid for that case.

3. Results and Discussion

Before showing the results of the test rig model and the submodel, this section starts with an evaluation of the test rig model. It is discussed whether the strategies for avoiding dynamic artefacts have been successful in reaching a quasi-static loading in the model. Figure 9a shows the force applied to point P in the x-direction as well as the computed velocity and displacement at this point as a function of simulation time. After reaching a constant force ($t > 0.3564$ s), a small delay occurs before reaching a displacement plateau. Therefore, the used values for damping and applying the initial tensile force, as well as the used support drums, yield accurate steady-state stress fields for $t > 0.48$ s, where the deviation from the maximum value $u_{max} = 327.5$ mm is less than 1.5%. The residual stresses remaining after straightening the initially bent belt are shown in Figure 9b. Only slight artefacts are visible in the stresses, which confirms that the residual stresses have been applied in a valid manner. The increased stresses in the bent region for $t = 0.05$ s compared to $t = 0$ s occur due to the applied loading at this time in the simulation.

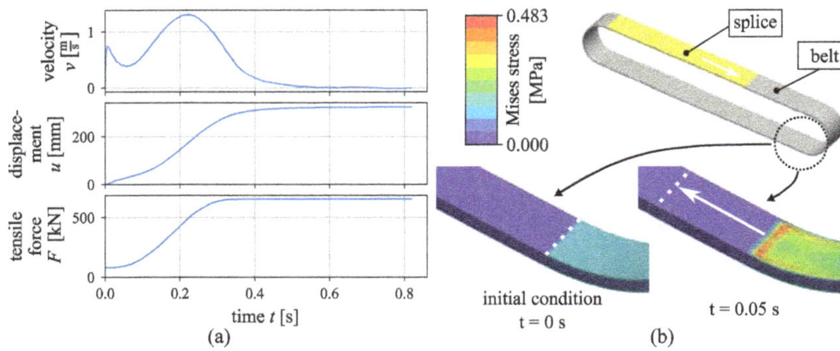

Figure 9. (**a**) Evolution of velocity, displacement, and tensile force in the x-direction of the reference point P and (**b**) validation of the applied residual stress in the initially bent regions of the conveyor belt.

The results of the global model for $t = 0.806$ s are shown in Figure 10. The maximum principal stress σ_I is plotted for the plane of the steel cable centre axes in Figure 10a, where σ_I is computed at the element centre. A small schema shows the position of the splice at the time of the stress evaluation. The highest maximum principal stress $\sigma_{I\,max}$ occurs at the cable end of the third and seventh cable from the left-hand side, counted from the bottom. These regions feature $\sigma_{I\,max}$ in all time steps. One of these regions is plotted in more detail and is indicated by a red frame. This region's stress field is plotted for $t = 0.556$ s, when it lies in the flat part of the test rig, in Figure 10b. The $\sigma_{I\,max}$ occur at the cable end since, in the model, the cable end face is detached from the rubber. This means that, at the front surface, all nodes except the outer nodes are duplicated. Here, $\sigma_{I\,max}$ occurs at the bottom of the cable due to shear load between a cable from the left and the right-hand side of the splice. The influence of bending at the drums can be seen in Figure 10c, where the same region is plotted for $t = 0.806$ s, where it lies in the bent region of the belt. In the flat and in the bent region $\sigma_{I\,max}$, values of 38.9 MPa and 45.7 MPa are reached, respectively. This indicates that the bending of a belt in the test rig can have a significant influence on arising stresses and is not negligible. It is obvious that the mesh in Figure 10 only roughly approximates the stress field; therefore, a study with a finer mesh is relevant.

The results of the submodel are shown in Figure 11, where the same times and regions as in Figures 10b,c are plotted. Due to the finer-meshed submodel, the stress field can be approximated better and the σ_I are higher. For the flat and the bent region $\sigma_{I\,max}$, values of 61.0 MPa and 69.7 MPa are obtained, respectively. In the submodel, the σ_I values at locations which are not close to the stress concentrations at the cable ends are similar to the σ_I values in the global model. This indicates the validity of this submodel.

The $\sigma_{I\,max}$ results of the global model and submodel are plotted over time in Figure 12. For $t < 0.65$ s, the investigated region is located in the flat region, and for $t > 0.65$ s, the investigated region is located in the bent region of the test rig. The σ_I values in the belt in the bent case are higher than in the flat case by 17.5% and 14.3% in the global model and the submodel, respectively. For $t < 0.65$ s, first a pronounced increase in $\sigma_{I\,max}$ occurs due to bending, followed by a slight increase in both models. This slight increase at the end of the curve can be observed more clearly in the submodel.

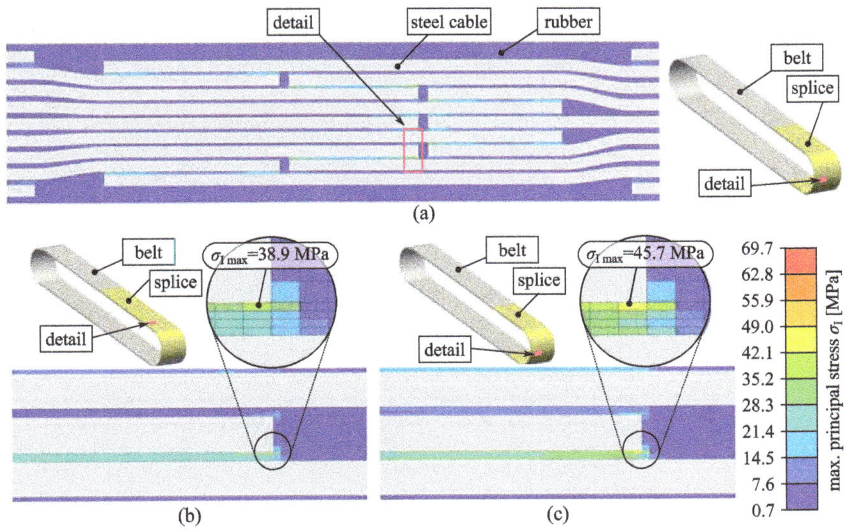

Figure 10. Stress results of the splice determined in the global model: (**a**) Top view of the splice with a general view where the splice location is shown, (**b**) detail of the splice for straight condition, and (**c**) detail for the splice for bent condition.

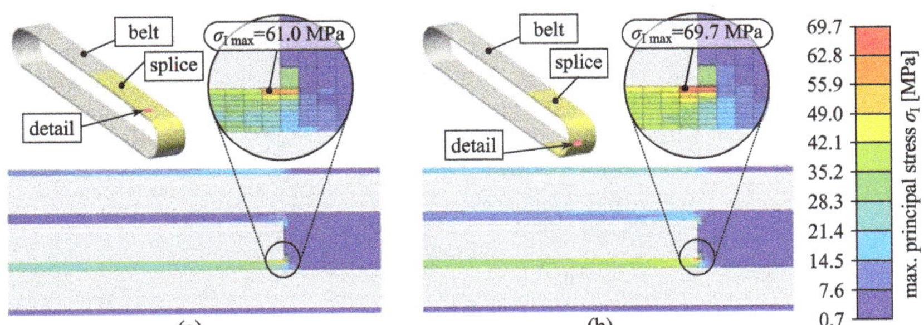

Figure 11. Stress results of the splice determined in the submodel: (**a**) Detail of the splice for straight condition, and (**b**) detail for the splice for bent condition.

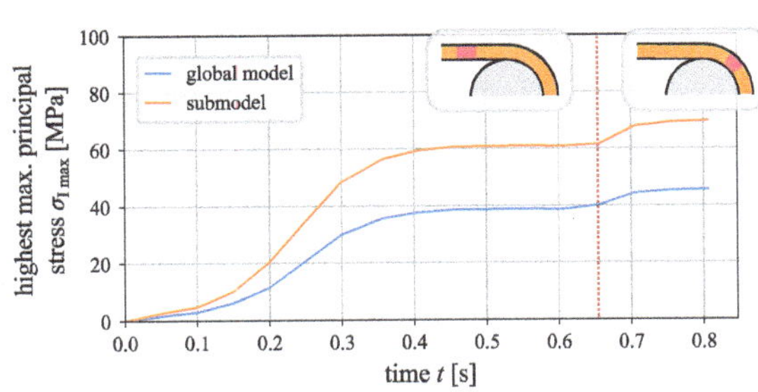

Figure 12. Time series of highest maximum principal stress values evaluated in the test rig model and the submodel. At $t = 0.65$ s, the point of highest stress runs into the bent region of the belt.

4. Conclusions

This work introduces a modelling approach using FEM to predict the stress and strain fields of a conveyor belt in a test rig, where the influence of the bending of the belt around the test rig drums is also considered. It is shown how a quasi-static loading can be achieved for an explicit FEM-simulation using suitable modelling techniques such as smooth load application with damping and added support drums. A computational cost of 3.8 days using 72 CPUs on a cluster is obtained for the investigated seven-cable splice belt. This means that larger belts with more steel cables can also be computed in an acceptable time frame. Existing conveyor belt models from the literature only consider tensile loading of the belt, therefore one key question addressed in this work is how much the stresses are influenced by the bending. The test rig model and an additional submodel of the region of highest stresses show that

- The region of highest stresses in the used splice scheme occurs at two cable ends due to shear stresses to the neighbouring cables;
- The test rig model computes by 17.5% higher maximum principal stresses while the critical position of the splice is bent at the drums compared to in the flat region;
- The submodel, where eight instead of four elements are used between the steel cables, computes higher stresses than the global model. The maximum principal stresses reach 14.3% higher values in the bent region than in the flat region.

The developed modelling approach shows a significant influence of bending on the local rubber stresses in the used test rig setup. Therefore, the influence of bending is essential for further fatigue modelling, where accurate stress fields are required. The submodel illustrates how, at a smaller scale, more accurate fracture mechanic concepts such as damage criteria can be used. Other researchers and engineers can benefit from this work by seeing the significance of bending loads on splices and being presented with a modelling approach that can compute this.

Author Contributions: Conceptualization, S.M.F., M.P. and C.S.; methodology, S.M.F. and M.P.; software, S.M.F.; validation, S.M.F.; formal analysis, S.M.F.; investigation, S.M.F.; resources, S.M.F., M.P., A.W. and C.S.; data curation, S.M.F.; writing—original draft preparation, S.M.F. and M.P.; writing—review and editing, S.M.F., M.P., A.W. and C.S.; visualization, S.M.F.; supervision, C.S.; project administration, M.P. and A.W.; funding acquisition, C.S. and M.P. All authors have read and agreed to the published version of the manuscript.

Funding: This research was supported by the Austrian Research Promotion Agency (FFG) within the "BRIDGE" framework as part of the project "Entwicklung einer Methodik zur Vorhersage des Versagens in elastomeren Gurten mittels Finite Elemente Simulation", grant agreement 864702.

Data Availability Statement: The data presented in this study are available on request from the corresponding author. The data is not publicly available due to size of the FEM result files.

Conflicts of Interest: The authors declare no conflict of interest.

References

1. *DIN 22110-3:2015-04, Prüfverfahren für Fördergurtverbindungen_- Teil_3: Ermittlung der Zeitfestigkeit für Fördergurtverbindungen (Dynamisches Prüfverfahren)*; Technical Report; Beuth Verlag GmbH: Berlin, Germany, 2015 . [CrossRef]
2. Kirjanów-Błażej, A.; Jurdziak, L.; Burduk, R.; Błażej, R. Forecast of the remaining lifetime of steel cord conveyor belts based on regression methods in damage analysis identified by subsequent DiagBelt scans. *Eng. Fail. Anal.* **2019**, *100*, 119–126. [CrossRef]
3. Kozłowski, T.; Wodecki, J.; Zimroz, R.; Błażej, R.; Hardygóra, M. A Diagnostics of Conveyor Belt Splices. *Appl. Sci.* **2020**, *10*, 6259. [CrossRef]
4. Fedorko, G.; Molnar, V.; Marasova, D.; Grincova, A.; Dovica, M.; Zivcak, J.; Toth, T.; Husakova, N. Failure analysis of belt conveyor damage caused by the falling material. Part II: Application of computer metrotomography. *Eng. Fail. Anal.* **2013**, *34*, 431–442. [CrossRef]
5. Bonneric, M.; Aubin, V.; Durville, D. Finite element simulation of a steel cable -rubber composite under bending loading: Influence of rubber penetration on the stress distribution in wires. *Int. J. Solids Struct.* **2019**, *160*, 158–167. [CrossRef]
6. Heitzmann, P.; Froböse, T.; Wakatsuki, A.; Overmeyer, L. *Optimierung von Textil-Fördergurtverbindungen Mittels Finite Elemente Methode (FEM)*; Medium: application/pdf; Wissenschaftliche Gesellschaft für Technische Logistik, Rostock-Warnemünde: Chemnitz, Germany, 2016; Volume 2016. [CrossRef]
7. Bajda, M.; Błażej, R.; Hardygóra, M. Impact of Selected Parameters on the Fatigue Strength of Splices on Multiply Textile Conveyor Belts. *IOP Conf. Ser. Earth Environ. Sci.* **2016**, *44*, 052021. [CrossRef]
8. Costello, G.A. *Theory of Wire Rope*; Mechanical Engineering Series; Springer: New York, NY, USA, 1997. [CrossRef]

9. Nordell, L.; Qiu, X.; Sethi, V. Belt conveyor steel cord splice analysis using finite element methods. *Bulk Solids Handl.* **1991**, *11*, 863–868.
10. Nordell, L. Steel cord belt and splice construction: Modernizing their specifications, improving their economics. *Bulk Solids Handl.* **1993**, *13, 685–693* .
11. Keller, M. Zur Optimierung Hochfester Stahlseilgurtverbindungen. Ph.D. Thesis, Universität Hannover, Hannover, Germany, 2001.
12. Froböse, T.; Heitzmann, P.; Overmeyer, L.; Wakatsuki, A. *Entwicklung eines FE-Modells zur Optimierung von Stahlseil-Fördergurtverbindungen*; Medium: Application/pdf; Wissenschaftliche Gesellschaft für Technische Logistik, Rostock-Warnemünde: Chemnitz, Germany, 2014; Volume 2014. [CrossRef]
13. Froböse, T. *Verfahren zur Ermittlung der Materialparameter für die Auslegung von Stahlseil-Fördergurtverbindungen mit Hilfe der FEM*; Number 2017, Band 01 in Berichte aus dem ITA; PZH Verlag: Hanover, Germany, 2017.
14. Li, X.; Long, X.; Shen, Z.; Miao, C. Analysis of Strength Factors of Steel Cord Conveyor Belt Splices Based on the FEM. *Adv. Mater. Sci. Eng.* **2019**, *2019*, 1–9. [CrossRef]
15. Wheatley, G.; Keipour, S. FEA of Conveyor Belt Splice Cord End Conditions. *UPB Sci. Bull. Ser. D Mech. Eng.* **2021**, *83, 205–216.*
16. Li, X.G.; Long, X.Y.; Jiang, H.Q.; Long, H.B. Finite element simulation and experimental verification of steel cord extraction of steel cord conveyor belt splice. In Proceedings of the IOP Conference Series: Materials Science and Engineering, Kitakyushu City, Japan, 10–13 April 2018; Volume 369, p. 012025. [CrossRef]
17. Carraro, P.; Maragoni, L.; Quaresimin, M. Prediction of the crack density evolution in multidirectional laminates under fatigue loadings. *Compos. Sci. Technol.* **2017**, *145*, 24–39. [CrossRef]
18. Ferdous, W.; Manalo, A.; Yu, P.; Salih, C.; Abousnina, R.; Heyer, T.; Schubel, P. Tensile Fatigue Behavior of Polyester and Vinyl Ester Based GFRP Laminates—A Comparative Evaluation. *Polymers* **2021**, *13*, 386. [CrossRef] [PubMed]
19. *ABAQUS Version 2020 User's Manual*; Dassault Systèmes Simulia Corp.: Providence, RI, USA, 2020.
20. Cruz Gómez, M.; Gallardo-Hernández, E.; Vite Torres, M.; Peña Bautista, A. Rubber steel friction in contaminated contacts. *Wear* **2013**, *302*, 1421–1425. [CrossRef]

Journal of
Composites Science

MDPI

Article

Modeling Fracture Formation, Behavior and Mechanics of Polymeric Materials: A Biomedical Implant Perspective

Quazi Md. Zobaer Shah [1,2], Md. Arefin Kowser [3], Mohammad Asaduzzaman Chowdhury [2,*], Muhammad Tariq Saeed Chani [4], Khalid A. Alamry [5], Nayem Hossain [6] and Mohammed M. Rahman [4,5,*]

1 Maintenance & Engineering Department, Healthcare Pharmaceuticals Ltd., Rajendrapur, Gazipur 1703, Bangladesh; qmzshah@engineer.com
2 Department of Mechanical Engineering, Dhaka University of Engineering and Technology, Gazipur 1707, Bangladesh
3 Department of Materials and Metallurgical Engineering, Dhaka University of Engineering and Technology, Gazipur 1707, Bangladesh; arefin@duet.ac.bd
4 Center of Excellent for Advanced Materials Research, King Abdulaziz University, Jeddah 21589, Saudi Arabia; mtmohamad@kau.edu.sa
5 Department of Chemistry, Faculty of Science, King Abdulaziz University, Jeddah 21589, Saudi Arabia; k_alamry@yahoo.com
6 Department of Mechanical Engineering, IUBAT—International University of Business Agriculture and Technology, Dhaka 1230, Bangladesh; nayem.hossain@iubat.edu
* Correspondence: asad@duet.ac.bd (M.A.C.); mmrahman@kau.edu.sa (M.M.R.)

Abstract: In industrial applications where contact behavior of materials is characterized, fretting-associated fatigue plays a vital role as a failure agitator. While considering connection, it encounters friction. Biomaterials like polytetrafluoroethylene (PTFE) and ultra-high-molecular-weight polyethylene (UHMWPE) are renowned for their low coefficient of friction and are utilized in sophisticated functions like the hip joint cup and other biomedical implants. In addition to the axial stresses, some degree of dynamic bending stress is also developed occasionally in those fretting contacts. This research investigated the fracture behavior of a polymer PTFE under bending fretting fatigue. Finite element analysis justified the experimental results. A mathematical model is proposed by developing an empirical equation for fracture characterization in polymers like PTFE. It was found that the bending stiffness exists below the loading point ratio (LPR) 3.0, near the collar section of the specimen. Along with fretting, the bending load forces the specimen to crack in a brittle-ductile mode near the sharp-edged collar where the maximum strain rate, as well as stress, builds up. For a loading point ratio of above 3, a fracture takes place near the fretting pads in a tensile-brittle mode. Strain proportionality factor, k was found as a life optimization parameter under conditional loading. The microscopic analysis revealed that the fracture striation initiates perpendicularly to the fretting load. The fretting fatigue damage characteristic of PTFE may have a new era for the biomedical application of polymer-based composite materials.

Keywords: biomedical implants; mathematical modeling; loading point ratio; strain proportionality factor; polymer materials

Citation: Shah, Q.M.Z.; Kowser, M.A.; Chowdhury, M.A.; Chani, M.T.S.; Alamry, K.A.; Hossain, N.; Rahman, M.M. Modeling Fracture Formation, Behavior and Mechanics of Polymeric Materials: A Biomedical Implant Perspective. *J. Compos. Sci.* **2022**, *6*, 31. https://doi.org/10.3390/jcs6010031

Academic Editors: Francesco Tornabene and Thanasis Triantafillou

Received: 16 November 2021
Accepted: 7 January 2022
Published: 17 January 2022

Publisher's Note: MDPI stays neutral with regard to jurisdictional claims in published maps and institutional affiliations.

1. Introduction

Fretting fatigue is a type of wear process that takes place in contact pairs, joints under static or dynamic cyclic stresses [1]. Based on shear stress distribution (uniformly/unidirectional) over the cross-sectional area, dynamic bending fatigue can be classified as reciprocating and rotating bending fatigue [2]. When bending fatigue is carried out under the effect of fretting action, wear as well as surface deterioration occurs. Polytetrafluoroethylene (PTFE) is a fluoropolymer or tetrapluoroethyelene consisting of carbon and fluorine. It is solid at room temperature and has one of the lowest coefficients of friction of any solid. Tanaka [3] studied the wear process of PTFE. Experimental and electron microscopy results revealed the wear mechanism. Instead of crystallizing, the width of bands affected the wear rate. However, both of the factors affected friction. The tensile deformation of PTFE by Fourier analysis determined the effect of crystallization in [4]. There exist distinguishable differences between the fatigue behavior of polymers and metals. Where the metal fatigue process includes crack initiation, propagation, and rupture, a polymer's fatigue behavior is significantly affected by viscoelastic effects. At the same time, frequency has a greater impact on polymer fatigue due to melting by generated heat, unlike the metals [5]. Blanchet and Kennedy [6] investigated PTFE under the mild-severe transition where sliding speed and temperature were the functions. A fracture-based model demonstrated severe wear and secondary changes in debris formation. It was concluded that fillers decrease wear by obstructing debris formation as well as crack propagation. Figure 1 shows the electron image (1A) and strip formation of PTFE (1B) under sliding friction. Aglan, H., et al. [7] proposed an MCL model to demonstrate the fatigue crack propagation (FCP) and fracture resistance behavior of PTFE. Figure 1C shows the micrographs for PTFE at 100× and 500×. Ductile tearing features are shown in Figure 1D. In both magnification's fracture surface is covered by micro fibrils and drawn ligaments.

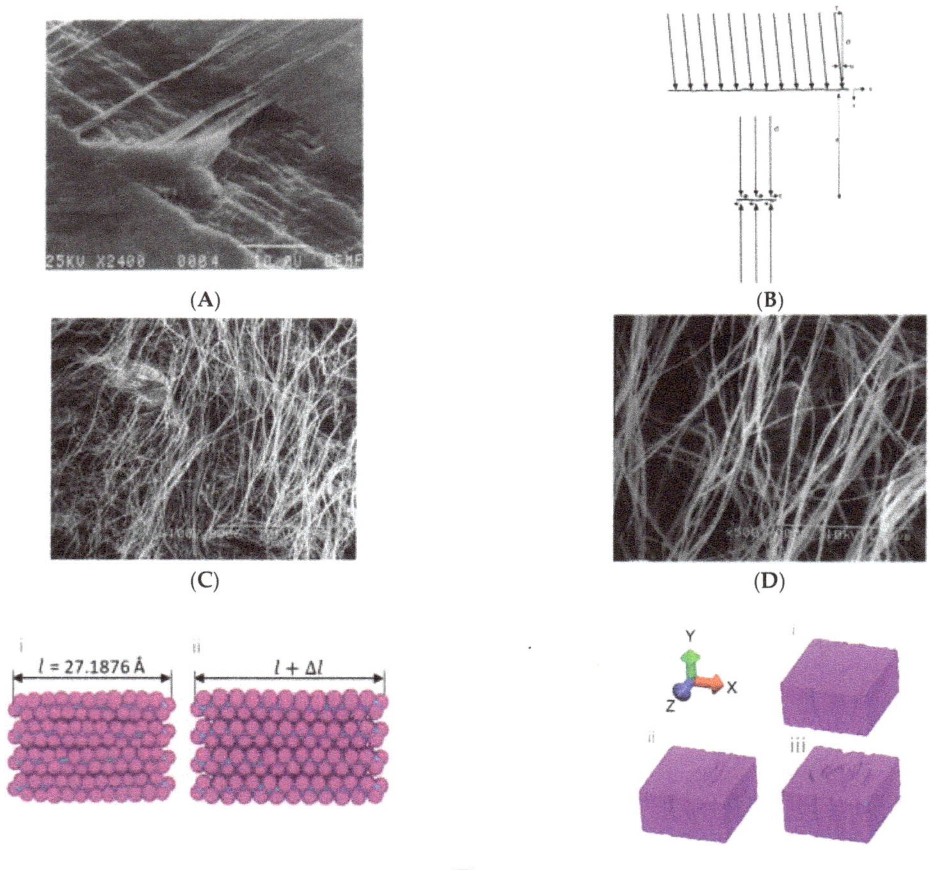

Figure 1. (**A**) Electron image of strip formation and (**B**) schematic drawing of crack [6], and micrograph of damage species on the fracture surface within the stage of stable crack propagation for the virgin polytetrafluoroethylene (PTFE) (**C**) 100×, (**D**) 500× [7], Deformation and indentation of PTFE [8] (**E**).

Joyce and James [9] showed that at slightly below ambient temperature (20 °C) quick fracture resistance degradation takes place. This can be compensated by rapid loading. Rae [10,11] demonstrated the effect of the relationship between crystallinity and microstructure with fatigue behavior, both in tension and compression. A temperature-induced transition based relationship was established between failure behavior and microstructure. A mixed-mode-I/II fracture in polytetrafluoroethylene (PTFE) was studied in [12]. Mode-I fracture in PTFE 7C showed strong phase dependence with a brittle-to-ductile transition. Nunes studied the mechanical behavior of PTFE in [13]. In this research, the authors showed the tensile properties of Teflon under different strain rates. They derived a mathematical model to determine the stress–strain curve of PTFE under different conditions. Gao, Yang, et al. [14] proved the higher the loading of PTFE fiber the better the tribological properties. The wear mechanism was derived from the thin film of PTFE formed on the contact-surfaces during sliding. A constitutive mathematical model was proposed in [15] that derived the compressive properties of PTFE at different strain rates. Sonne and Hattel [16] modeled a micro-scale deformation of PTFE. Frictional behavior between PTFE and steel on micro-scale proved the importance of PTFE stamps. Stress-cycle (S-N) behavior was determined by the crack growth mechanism in [17] for mechanical fatigue of a polymer. Brownell [8] investigated the deformation and mechanical failure of PTFE using molecular dynamics (MD). A coarse-grained model was developed to observe the mechanical properties at the micro-scale. Typical deformation and indentation of PTFE are shown in Figure 1E [8].

By using DIC and FEM, Sawada [18] characterized the mechanical properties of PTFE under bending and tensile tests. A unique fracture criterion was presented by an exponential function. Bending test results for a loading ratio of 1.0 mm/s are shown in Figure 2A. Shah et al. [19] studied the effect of friction on the fatigue strength of PTFE. However, there remain almost no data on the fretting fatigue behavior of Teflon. Depending on the friction types, both static and sliding friction is applicable in polymer design. Deviated angled eccentric loading components can be found in [20,21] where bending takes place in parallel with tension and compression. Biological implants like the hip joint and femur occasionally experience bending fatigue [22]. Along with the hip-cup joint, PTFE can also be used in bone/shaft implants. When it is in contact with bone, slow erosion occurs as no fibrous reaction seals it off from the bone [23]. Many more examples of PTFE use in biomedical implants are available in [24] where recent trends of such research (temperature/lubrication) can be found in [25–27]. In Figure 2B, green arrows show the compressive load while red arrows show bending loads on the bone as well as joint. In the cup-joint frictional zones, PTFE is used. However, continuous frictional bending stress causes failure (Figure 2C) that should be given importance for increasing the lifetime.

Polymeric materials have long repeating chains of molecules. They can be both natural such a hemp, shellac, amber, wool, silk, natural rubber and synthetic such as polyethylene, polypropylene, polystyrene, polyvinyl chloride, synthetic rubber, phenol formaldehyde resin, neoprene, nylon etc. Synthetic polymers are made by step-growth polymerization and chain polymerization methods. These materials show superior mechanical performance and that is why their applications are seen in transportation, medicine and construction [28–32].

PTFE is a high thermally stable material. It is hydrophobic, stable in most types of chemical environment, and generally considered to be inert inside the body. Study shows that PTFE-based composite material reinforced with glass fibers found good behavior in vitro but poor performance in vivo. After 1 year of implantation the composite developed a pasty surface that could easily be worn away. Moreover, the filler acted abrasively and lapped the metal counter face. Furthermore, compared to other material, the composite showed a higher rate of infection and loosening.

The materials those will be used as hip implant must be highly non-toxic, and should not cause any inflammatory or allergic reactions in the human tissues or cells, high corrosion- and wear resistant, should have excellent thermal conductivity, high strength, high fracture toughness, hardness, biocompatibility, and high stiffness [33].

In this research, a mathematical model is developed for describing the fracture mechanism of a polymer, PTFE. Based on the experimental observations, an empirical equation is proposed to predict the fracture location under bending fretting fatigue. A FE method is used to verify the model. Path stress analysis and the fractographic observation conclude some remarkable findings on the fretting fracture of PTFE.

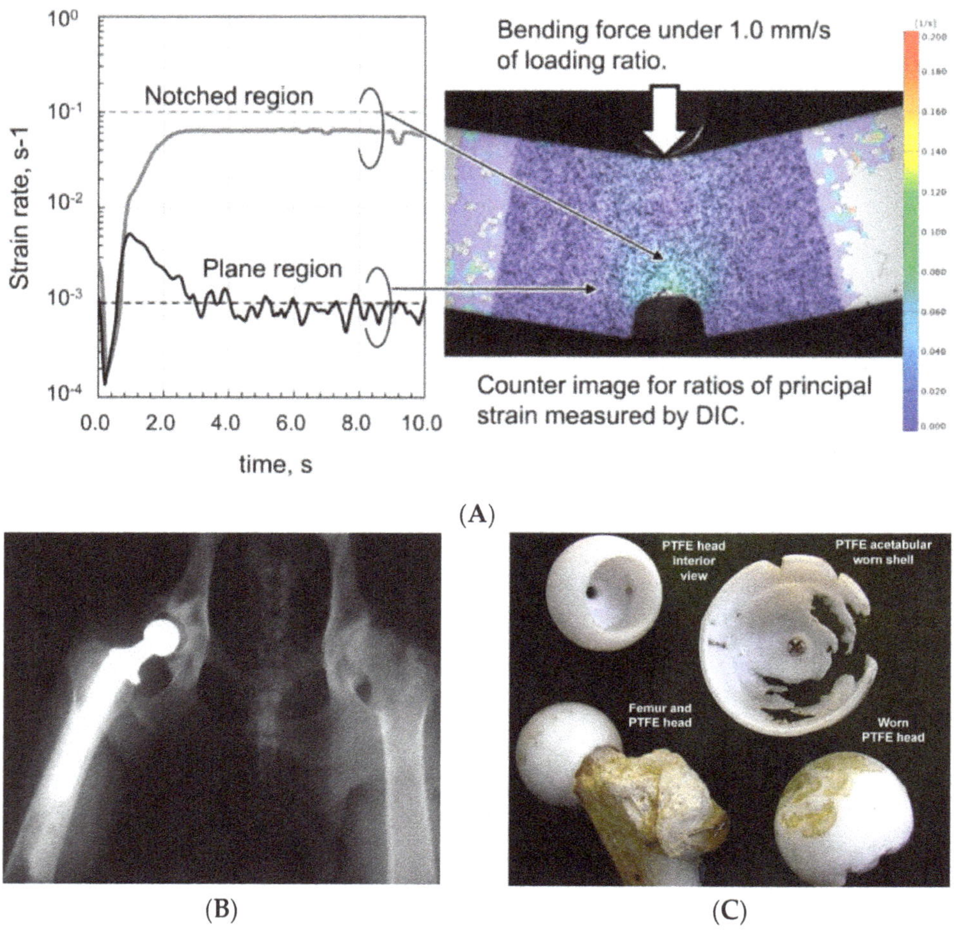

Figure 2. (**A**) Bending test results of PTFE under quasi-static loading [18], (**B**) hip joint cup and (**C**) biological implants showing PTFE components [22].

2. Mathematical Modeling

2.1. Loading Point Ratio

Figure 3A presents a cantilever beam of length L. It is loaded by a point load F at the free end and two fretting pads at l distance from the other end. The rest of this end is considered as a fixed support.

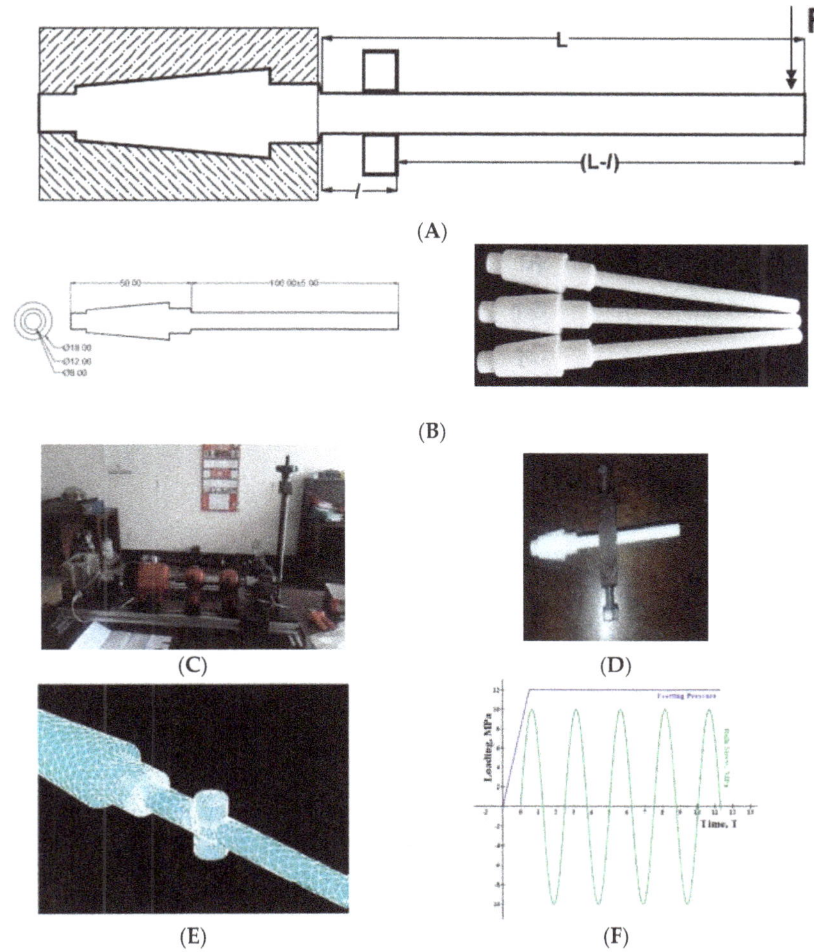

Figure 3. (**A**) cylindrical cantilever beam under bending fretting fatigue. (**B**) Designed specimen (dimensions are in mm). (**C**) Meshed geometry (**D**) conditional loading, (**E**) rotating bending fatigue setup, and (**F**) fretting proving ring.

Let us consider the first case i.e., the beam without fretting pads. From the curvature of the neutral surface, the relation can be expressed by:

$$\frac{M}{I} = \frac{E}{\rho} \quad (1)$$

Here,
M = Moment,
E = Young's modulus,
I = Inertia,
ρ = Radius of curvature,
This can be rewritten as,

$$\frac{M}{EI} = \frac{1}{\rho} \quad (2)$$

From the curvature of a point Q(x, y), the radius of curvature can be expressed as,

$$\frac{1}{\rho} = \frac{d^2y}{dx^2} \quad (3)$$

$$\frac{M}{EI} = \frac{d^2y}{dx^2} \quad (4)$$

Successive integration yields,

$$\frac{dy}{dx} = \frac{1}{EI}\int M dx + C_1 \quad (5)$$

$$EIy = \int\int M dx\, dx + C_1 x + C_2 \quad (6)$$

where, C_1 and C_2 are constants that can be calculated from the boundary conditions.

Applying boundary conditions (At the fixed end, $x = L$; $y = 0$) for a cantilever beam yields the values of C_1 and C_2, and thus putting values of C_1 and C_2 in the equation yields,

$$y_1 = \frac{FL^3}{3EI} \quad (7)$$

where, F and L are point load and the distance from the fixed end, respectively. ($M = FL$).

In the second case, when fretting ring is positioned about l distance from the fixed end, let us consider a new shorter cantilever beam at the fretting position where the new cantilever's length is $(L - l)$.

From Equation (7) we obtain,

$$y_2 = \frac{F(L - l)^3}{3EI} \quad (8)$$

thus, the deflection as well as the loading point ratio can be shown as the ratio of deflection of the original (L) cantilever beam to the deflection of the fretting ring shortened $(L - l)$ cantilever beam, i.e.,

$$\text{Loading Point Ratio, } LPR = \frac{y_1}{y_2} \quad (9)$$

$$= \frac{\frac{FL^3}{3EI}}{\frac{F(L-l)^3}{3EI}} \quad (10)$$

$$\therefore \text{Loading Point Ratio, } LPR = \frac{L^3}{(L-l)^3} \quad (11)$$

This is absolutely a geometry-dependent parameter. This aspect could be researched in greater scope in the future. It is just an interpretation of fretting failure characterization of biomedical implants under variable loading point and stain ratios. It is a new idea, and thus may require more research in the laboratory.

2.2. Strain Proportionality Factor

Let us consider the strains for bending and fretting point loading as ε_1 and ε_2 respectively. Strain due to the bending load,

$$\varepsilon_1 = \frac{\sigma_b}{E}$$

Here,

$$\sigma_b = Bendingstress$$

$$E = Young'smodulus$$

$$\text{Strain due to the fretting load, } \varepsilon_2 = \frac{(D-d)}{D} \tag{12}$$

$$[D = Undeformed\ diameter\ and\ d = deformed\ diameter]$$

Similarly, strain proportionality factor can be defined as the ratio of strain due by fretting to the strain by bending, i.e.,

$$k = \frac{\varepsilon_2}{\varepsilon_1} = \frac{\frac{(D-d)}{D}}{\frac{\sigma_b}{E}} \tag{13}$$

$$\therefore \text{Strain Proportionality Factor, } k = \frac{E(D-d)}{\sigma_b D} \tag{14}$$

This depends on the strain factor due to the fretting fatigue load, thus more attention should be given to the loading condition than the geometry/design.

3. Methodology

3.1. Experimental Design

As an experimental setup, a general rotating bending fatigue test rig was considered here. It mainly includes a motor, coupling, and control board with VFD, cycle counting scheme, chuck, specimen and load bearing shaft with wheel supports mounted on a fixed vibration free base. In order to imply normal bulk stress on the end of specimen, the top wheel is turned in the clockwise direction. A cut-off switch, located near the load bearing stops the motor once the specimen fails. A cycle can be counted from the frequency of the VFD and stop watch and, alternatively, from the auto cycle counting scheme, as well. Cylindrical Teflon bars were machined through a CNC lathe machine to shape into the designed specimens. Mechanical properties and technical dimensions are provided in Table 1 and Figure 3B, respectively.

Table 1. Mechanical properties.

Material	Density (g/cc)	Youngs Modulus, E (GPa)	Poisson's Ratio	Yield Tensile Strength, YTS (MPa)	Ultimate Tensile Strength, UTS (MPa)
Teflon (PTFE)	2.3	0.5	0.46	30	43

In order to create the fretting phenomenon, an 8 mm bolted proving ring was prepared as shown in Figure 3D. Screwing the flat bolt through the ring, on the cylinder surface induces fretting action. Details of such a comprehensive experimental procedure are available literally and can be referred in [34,35].

3.2. Numerical Model

An FE model was developed by using ANSYS 17. For contact pair surfaces, CONTA 174 and TARGE170 were used as contact and target elements, respectively. Adaptive mesh refinement was used for mesh convergence. In order to optimize mesh refinement by saving CPU runtime at the same time, the refining process was carried out unless the stress convergence arrives within a reasonable gap. Ideally, a consecutive stress level difference of 2% was found as an acceptable range. Refinement optimization yields element size-independent results. Displacement condition was determined as Ux = Uz = 0, Uy = free. Figure 3E,F shows mesh convergence and loading sequence for fretting fatigue.

4. Results and Discussion

As expected, fretting affects the fatigue life of PTFE considerably. Based on the loading point ratio, fracture behavior was characterized. However, from a different angle of view, results can be evaluated differently.

4.1. Effect of Loading Point Ratio

Two cases were observed for fracture failure locations. For the fretting point at the loading point above 70 mm of the bending load side, the specimen cracks near the collar that experiences few fatigue cycles due to the higher strain rate [18] by the combined effect of fretting and notch. For the loading point below 70 mm, it cracks near the fretting pad location that faces more cycles for the same identical conditions for the loading point at above 70 mm from the free end, as shown in [28].

From Equation (3),

For fretting at 75 mm from the free end, LPR = 2.744, fractures near the collar.

For fretting at 70 mm from the free end, LPR = 3.375, fractures near the fretting pads.

For fretting at 68 mm from the free end, LPR = 3.68, fractures near the fretting pads.

Thus, it can be said that, the bending stiffness of PTFE lies below the LPR 3.0, i.e., below 3.0, resistance to the bending is high and the stress and strain accumulate [18] near the non-filleted collar section.

Figure 4 shows that even for a diametric strain of $\frac{1}{2}$, it cracks near the collar.

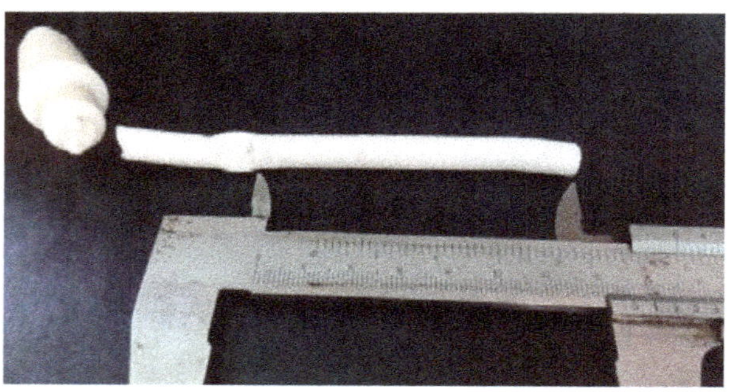

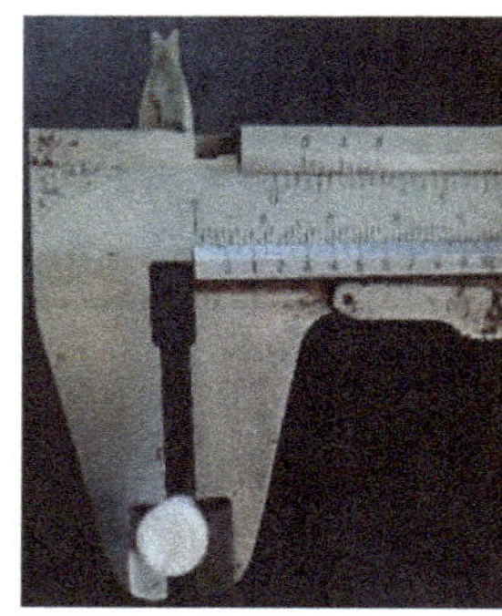

Figure 4. Cracked specimen for fretting ring distance above 70 mm (**left**), and deformed diameter (**right**).

4.2. Fractography Analysis

Figure 5A shows the typical collars of PTFE samples after a fracture. If it were metals, the fracture would occur just at the sharp edge collar with a plain surface. However, a small amount of debris is left at the collar. This suggests the tearing action of the polymer at the neck. An interesting observation is that the self-weight and rotation of the fretting pad induce or try to induce a moment that creates slippage over the specimen body. Once the fretting pad locates near the collar, due to the dual effect of fretting and sharp neck a striped surface is observed as shown in Figure 5B,C (blue colored). Such striped fibers can be compared at the microscopic level with the ductile tearing fibers [7] presented in Figure 1C,D. The PTFE samples fractured just near the collar section, which implies that the stress concentration factor does not have significant effects on the failure of polymers.

Figure 6 shows debris formation and striation on the fractured surface. PTFE shows both brittle and ductile manners of fracturing. A ductile fracture causes slant lip formation while a brittle fracture shows a plain cracked surface [36]. Due to the ductile shear near the collar neck of the specimen, a crescent-type fracture with slant shear lip at 45° occurs with prominent debris. On the other hand, for LPR above 3.0, a relatively plain surface (90° brittle, tensile mode) with striation is found as shown in Figure 6b.

In order to create a simple illustration, a cracked surface schematic is shown in Figure 7.

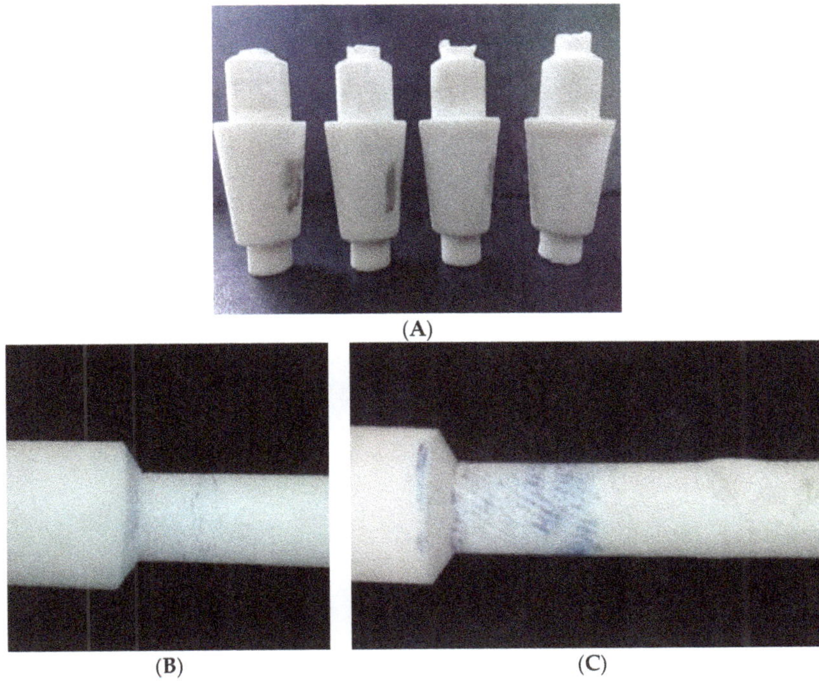

Figure 5. (**A**) Typical fractured heads showing debris of the cylindrical body, fresh specimen (**B**), tested specimen after certain cycle's period (**C**).

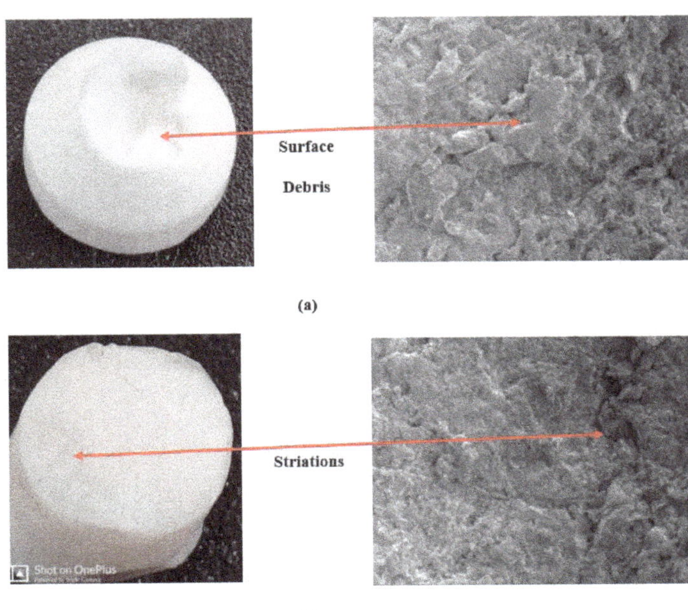

Figure 6. (**a**) Crescent shaped edge. (**b**) fractured surface for lower fretting depth.

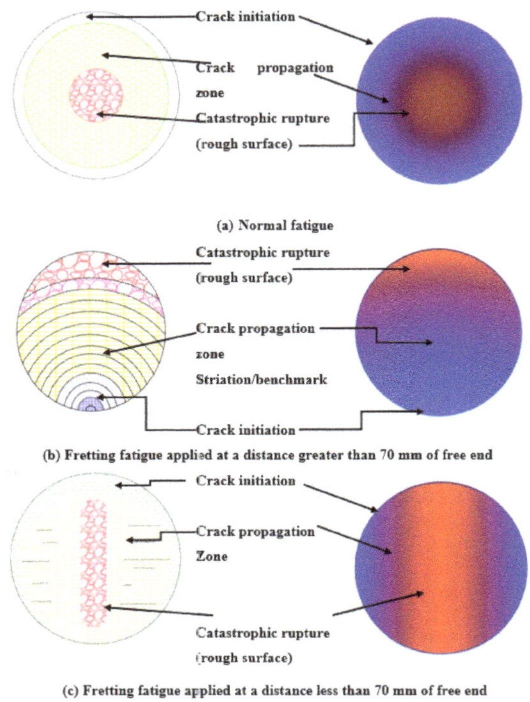

Figure 7. Schematic diagram of fractured PTFE surface.

For normal fatigue, no naked eye irregularities are found in the fractured surface. The core of the specimen exists in the middle surrounded by almost regular and smooth surfaces like a nucleus (Figure 7a). Gradual and comparatively slow crack initiation occurs at the external diameter of the body that steadily propagates towards the center. It is rough at the center because of the inability of the imbalanced body at the critical cross-sectional point to cope with the applied stress any more. However, the crack nucleates at the corner for loading greater than 70 mm and propagates like striation-benchmark/concentric rings outwards (Figure 7b). Relatively spaced striation suggests the up-gradation of stress that discontinues smoother propagation. Here, the granular-rougher rapid failed surface is found at the corner in a circular shape. Unlike normal fatigue, fretting fatigue (less than 70 mm from the free end) causes a nearly oval-shaped rough center that is perpendicular to the direction of the applied fretting load (Figure 7c). Constant rubbing of the cracked surface causes a smoother surface than the fast-growing rougher nucleus. Macroscopic lines, sometimes known as a chevron, are found in the propagation passage.

Typical fractured surfaces of PTFE, if the LPR is below 3.0, are shown in Figure 8A. Because of the fretting pressure, the polymer deforms, and is compressed (Figure 1E), hence becoming more solid through the central passage line. It becomes rougher and catastrophic failure occurs here. Striation and debris formed near the canal type rupture are shown in Figure 8B. Interestingly, this manner looks like the sliding wear condition as shown in Figure 1. If observed, it will be clear that cracks, as well as fractures, occur perpendicularly to the fretting action. A schematic drawing demonstrated the phenomenon as shown in Figure 9.

Figure 10 demonstrates a better distinguishable view for various mode of fractures on the thermal LUTs (look up tables). The maximum stressed zone is specified by the red zone. For plain fatigue, the cross-sectional area under the neck collar bears the maximum stress uniformly (Figure 10a). Under ductile fracture mode, tearing in a 45° crescent shape is obvious in Figure 10b. For brittle mode, the crack initiates at the corner in 90° that incorporates maximum stress (Figure 10c). The trough section, along with striations, is obvious for higher fretting loading when it is fractured near the fretting ring (Figure 10d).

(A)

(B)

Figure 8. (**A**) Typical fractured surface cracked near the fretting zone. (**B**) Debris and striations near canal type rupture.

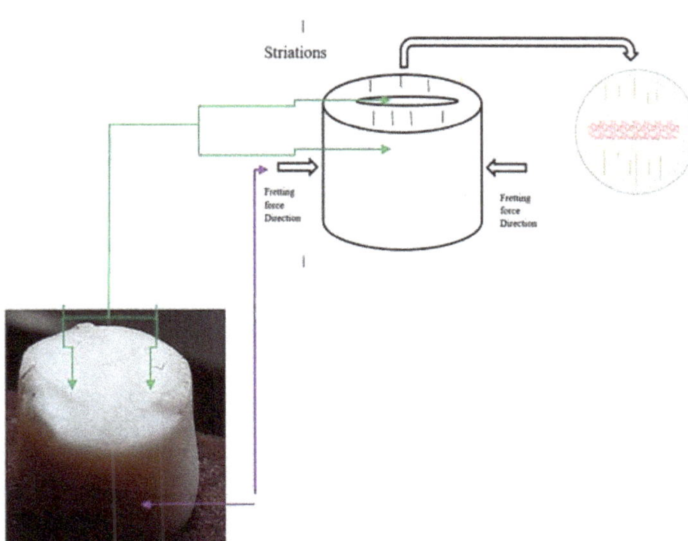

Figure 9. Schematic demonstration of fracture failure.

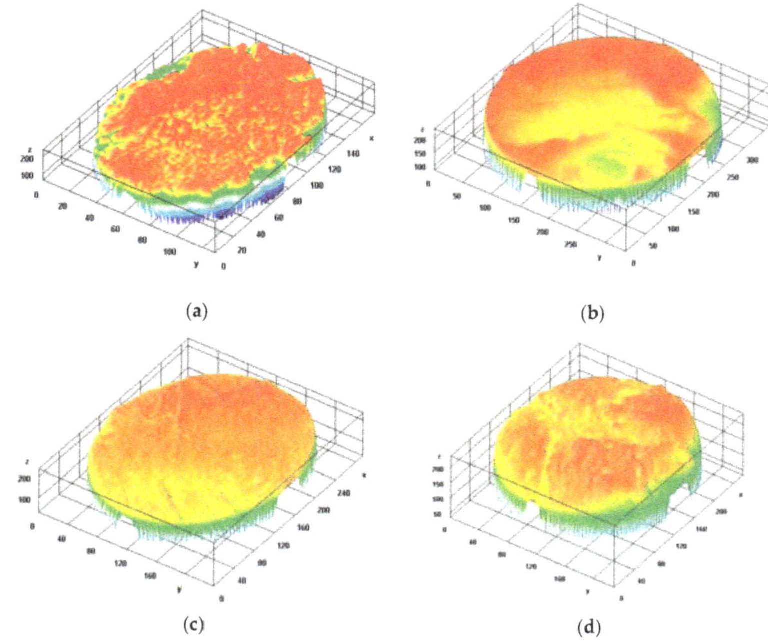

Figure 10. Typical thermal LUTs (look up tables) on fractured surfaces, (**a**) normal mode (plain fatigue), (**b**) crescent (ductile) mode, (**c**) brittle mode (initial order fretting), (**d**) trough mode (high order fretting).

4.3. Impact of Strain Proportionality Factor

Responses of strain proportionality factor, k with the diametric strain, and longitudinal strain are shown in Figure 11A,B. Here, a similarity was found between stress concentration factor and strain proportionality factor that acts linearly to the stress and strain, respectively [37]. Even so, it also shows similarity with the relation between K_I and displacement of the crack tip [38]. Interestingly, the significance of this factor may be observed for LPR > 3.0. Figure 11C shows that, for lower and middle order loading, fatigue life increases for incremental diametric strain up to a pre-intermediate level. After reaching the extreme point, it decreases again with the increment in diametric strain. Due to the mutually opposed force actions of fretting and bending loads, the shaft tries to be straightened to some degree. Thus, stress concentration, as well as strain range, is minimized that holds the fatigue life until the optimal point. After the optimal level, counterbalance of the opposite forces becomes imbalanced, hence, life degradation occurs. However, for higher-order bending load, the influence of the factor, k almost vanishes.

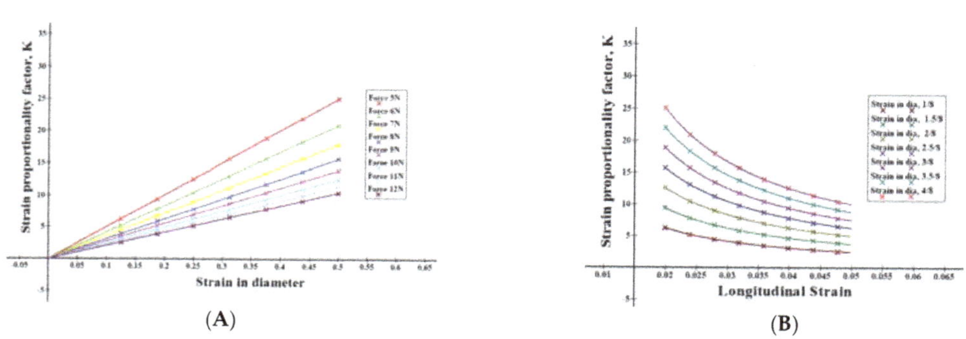

Figure 11. *Cont.*

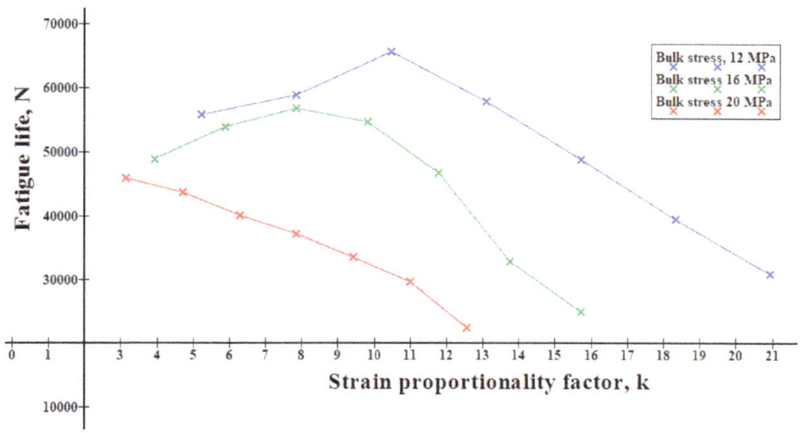

(C)

Figure 11. Relationship between (**A**) k and diametric strain, (**B**) k and longitudinal strain, (**C**) relationship between fatigue life and strain proportionality factor.

4.4. Finite Element Model (FEM) Analysis

As found earlier [28], when fretting pressure acted on more than 70 mm from the free end of the specimen, it failed at the collar section, not at the fretting zone. Stress distribution along the contact paths for both the initial and optimal fretting load steps [28] justified the way it failed. This shows that maximum higher stress concentrates at the collar edge side of the specimen. The stress distribution is shown for 30 mm distance along the contact path of the fretting pad and cylindrical specimen. Thus, the expected fracture occurs at the edge corner. However, when fretting pressure, as well as the penetration depth, is increased up to the optimal point, and fluctuating stress crest and trough limit extend at the fretting zones. On the other hand, stress upper range decreases down at the edge collar section, although it does not lie below the fretting zone side stress level to fail at that section. Thus, it develops a higher fatigue life by reducing the edge corner stress intensity/concentration.

To predict the consequence as well as observe the overall effect of variable parameters on the fatigue life of PTFE, surface graphs were created. As the frequency increases, negligible heat generation in bending fatigue is considered to accelerate the fracture failure of Teflon. Such concepts are demonstrated in the surface graphs of Figure 12. Fatigue life decreases with the increase of bulk stress as well as bending load. RPM increment shows the same results for fatigue life decrement. As found earlier, fretting pressure, as well as penetration depth, optimize the fatigue life up to a definite point.

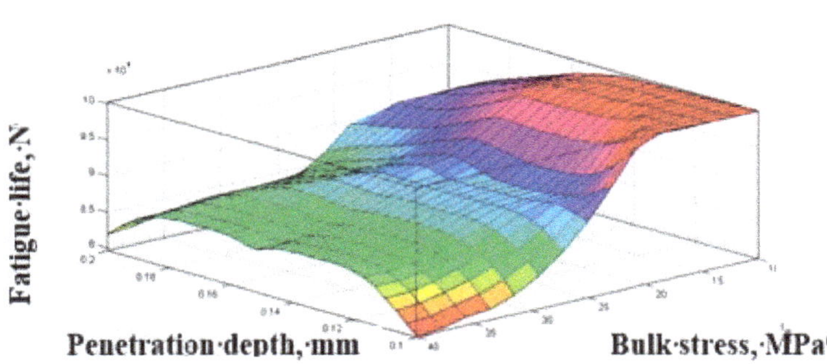

(A)

Figure 12. *Cont.*

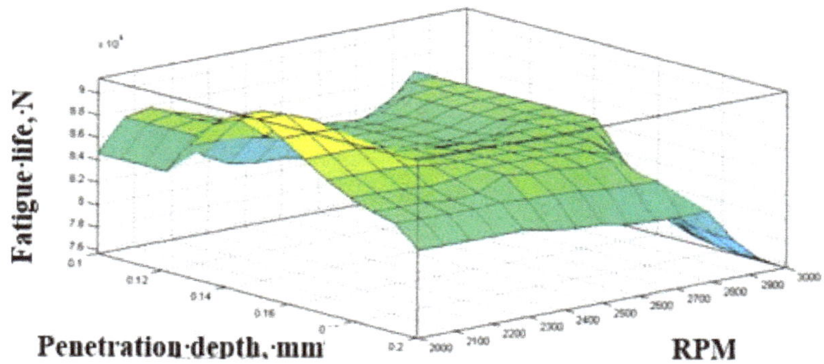

(B)

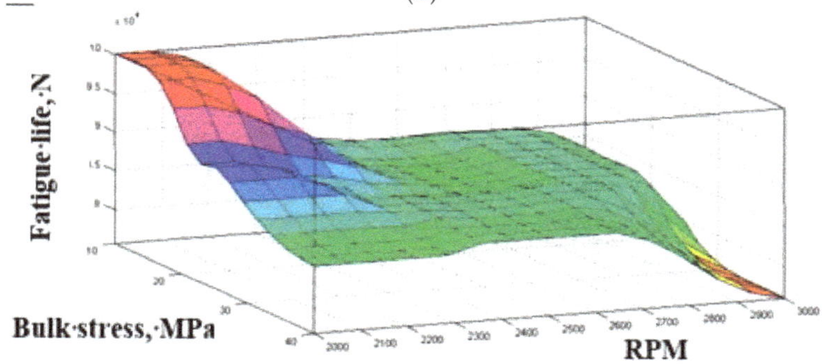

(C)

Figure 12. (**A**) Relationships for bending stress, (**B**) relationships for RPM, penetration and fatigue life and (**C**) fatigue life and penetration relationships for bending stress, RPM and fatigue life.

5. Conclusions

Although well known, about 90% of fatigue life is reduced by fretting action. The aspect of fretting phenomena on fatigue strength is still being researched as well as being discovered day by day. However, for a polymer, comparatively less attention has been paid to the fretting fatigue behavior of PTFE. In this research, a new horizon has been unveiled by introducing the fracture mechanism of a polymer, PTFE, under the bending fretting fatigue phenomenon. Some measurable outcomes can be summarized as follows:

(1) Fretting action depends on the loading point ratio. Location of fretting within 70 mm from free end causes the sample fracture near the fretting zone while for loading greater than 70 mm it causes failure near the neck point or collar due to the combined effect of fretting and notch. A mathematical model is developed and an empirical equation is derived for the loading point ratio. For LPR > 3.0, the specimen fractures at the fretting zone, otherwise at the collar section. Stress distribution along the fretting contact path supports the results obtained. Thus, it suggests a point of influence for fretting action that should be avoided while designing.

(2) The strain proportionality factor, k, increases up to a certain (optimal) value, and improves fatigue life due to the counterbalance of the reversed (or negative) contact pressure and opposite (positive) bending stress. From finite element stress distribution, it was found that fretting compensates the excess bulk stress at the edge corner that improves fatigue life. However, once it reaches that optimal value, fatigue life starts to yield because of the loss in strength due to imbalanced necking.

(3) From geometrical aspects and schematic demonstrations, fretting acts perpendicular to the formation of cracks. Rapid rupture occurs at an angle of 90° to the direction of

fretting pressure. PTFE shows both the ductile and brittle behavior in the fractured zone. For LPR < 3.0, slant lips at 45° were found for a crescent-type fracture near the collar section. Therefore, not only the fretting quantities but also fretting qualities affect fatigue life badly.

According to the above discussion, it can be said that along with the geometric aspects, failure characteristics of PTFE depends on the loading and strain ratio with location. This could help us to develop new polymer-based biomaterial composites.

Author Contributions: Q.M.Z.S.: Conceptualized the model and established the empirical theorem, conducted the experiment, analyzed and visualized the data, designed the FE simulation, wrote the paper. M.A.K., M.A.C.: Supervised the research as well as the FE analysis, arranged computing and laboratory resources. M.A.K. Introduced the idea of "Fretting fatigue" while M.A.C. Introduced the idea of "PTFE" in this project. M.A.C., M.M.R.: Supervised and financed the project, provided materials and laboratory resources. Review & submission, Collaborated in and Co-ordinated the research as the project head. M.M.R. suggested and chose the journal for submission. M.M.R., M.T.S.C., K.A.A.: Funding Acquisition, Project Administration, Supervision & Review, Resources, Collaboration. N.H.: Participated in the manuscript revision phase. All authors have read and agreed to the published version of the manuscript.

Funding: This project was funded by the Deanship of Scientific Research (DSR) at King Abdulaziz University, Jeddah, under grant no. G-131-130-1441. The authors, therefore, acknowledge with thanks DSR for technical and financial support.

Conflicts of Interest: The authors declare no conflict of interest.

References

1. Chowdhury, M.A.; Kowser, A.; Shah, Q.M.Z.; Das, S. Characteristics and damage mechanisms of bending fretting fatigue of materials. *Int. J. Damage Mech.* **2017**, *27*, 453–487. [CrossRef]
2. Shah, Q.M.Z.; Chowdhury, A.; Kowser, A. On the diversity in design for different bending fretting fatigue mechanism. *SN Appl. Sci.* **2019**, *1*, 1067. [CrossRef]
3. Tanaka, K.; Uchiyama, Y.; Toyooka, S. The mechanism of wear of polytetrafluoroethylene. *Wear* **1973**, *23*, 153–172. [CrossRef]
4. Wecker, S.M.; Davidson, T.; Cohen, J.B. Study of deformation in polytetrafluoroethylene by x-ray line broadening. *J. Appl. Phys.* **1974**, *45*, 4453–4457. [CrossRef]
5. Dao, K.C.; Dicken, D.J. Fatigue failure mechanisms in polymers. *Polym. Eng. Sci.* **1987**, *27*, 271–276. [CrossRef]
6. Blanchet, T.A.; Kennedy, F.E. Sliding wear mechanism of polytetrafluoroethylene (PTFE) and PTFE composites. *Wear* **1992**, *153*, 229–243. [CrossRef]
7. Aglan, H.; Gan, Y.; El-Hadik, M.; Faughnan, P.; Bryan, C. Evaluation of the fatigue fracture resistance of unfilled and filled polytetrafluoroethylene materials. *J. Mater. Sci.* **1999**, *34*, 83–97. [CrossRef]
8. Brownell, M.; Nair, A.K. Deformation mechanisms of polytetrafluoroethylene at the nano- and microscales. *Phys. Chem. Chem. Phys.* **2018**, *21*, 490–503. [CrossRef] [PubMed]
9. Joyce, J.A. Fracture toughness evaluation of polytetrafluoroethylene. *Polym. Eng. Sci.* **2003**, *43*, 1702–1714. [CrossRef]
10. Rae, P.; Dattelbaum, D. The properties of poly(tetrafluoroethylene) (PTFE) in compression. *Polymer* **2004**, *45*, 7615–7625. [CrossRef]
11. Rae, P.; Brown, E. The properties of poly(tetrafluoroethylene) (PTFE) in tension. *Polymer* **2005**, *46*, 8128–8140. [CrossRef]
12. Brown, E.N.; Rae, P.J.; Liu, C. Mixed-mode-I/II fracture of polytetrafluoroethylene. *Mater. Sci. Eng. A* **2007**, *468*, 253–258. [CrossRef]
13. Nunes, L.; Dias, F.; Mattos, H.D.C. Mechanical behavior of polytetrafluoroethylene in tensile loading under different strain rates. *Polym. Test.* **2011**, *30*, 791–796. [CrossRef]
14. Gao, Y.; Sun, S.; He, Y.; Wang, X.; Wu, D. Effect of poly(ethylene oxide) on tribological performance and impact fracture behavior of polyoxymethylene/polytetrafluoroethylene fiber composites. *Compos. Part B Eng.* **2011**, *42*, 1945–1955. [CrossRef]
15. Zhang, J.-F.; Ju, Y.-T.; Sun, C.-X.; Wang, P.-B. The Research on Compressive Properties of Polytetrafluoroethylene at High Strain Rate. *Def. Technol.* **2013**, *9*, 181–185. [CrossRef]
16. Sonne, M.R.; Hattel, J.H. Modeling the constitutive and frictional behavior of PTFE flexible stamps for nanoimprint lithography. *Microelectron. Eng.* **2013**, *106*, 1–8. [CrossRef]
17. Chandran, K.S.R. Mechanical fatigue of polymers: A new approach to characterize the SN behavior on the basis of macroscopic crack growth mechanism. *Polymer* **2016**, *91*, 222–238. [CrossRef]
18. Sawada, T. Fracture criterion considering notch brittleness of polytetrafluoroethylene under quasi-static loading. *Theor. Appl. Fract. Mech.* **2019**, *103*, 102285. [CrossRef]
19. Shah, Q.Z.; Chowdhury, M.A.; Kowser, M. Failure Mechanism of Polytetrafluoroethyleneunder Friction Fatigue. *J. Fail. Anal. Prev.* **2019**, *19*, 245–249. [CrossRef]

20. Hobbs, J.; Burguete, R.; Heyes, P.; Patterson, E. The effect of eccentric loading on the fatigue performance of high-tensile bolts. *Int. J. Fatigue* **2000**, *22*, 531–538. [CrossRef]
21. Meyer, R.R. Buckling of 45° Eccentric-Stiffened Waffle Cylinders. *J. R. Aeronaut. Soc.* **1967**, *71*, 516–520. [CrossRef]
22. Surgical Research. 2001, pp. 426–441. Available online: https://www.sciencedirect.com/topics/medicine-and-dentistry/artificial-hip-joint (accessed on 15 November 2021).
23. Charnley, J. Tissue reactions to polytetrafluoroethylene. *Lancet* **1963**, *282*, 1379. [CrossRef]
24. Merola, M.; Affatato, S. Materials for Hip Prostheses: A Review of Wear and Loading Considerations. *Materials* **2019**, *12*, 495. [CrossRef] [PubMed]
25. Zuo, J.; Xu, M.; Zhao, X.; Shen, X.; Gao, Z.; Xiao, J. Effects of the depth of the acetabular component during simulated acetabulum reaming in total hip arthroplasty. *Sci. Rep.* **2021**, *11*, 9836. [CrossRef]
26. Seo, Y.; Wang, Z.J. Measurement and evaluation of specific absorption rate and temperature elevation caused by an artificial hip joint during MRI scanning. *Sci. Rep.* **2021**, *11*, 1134. [CrossRef] [PubMed]
27. Cui, Z.; Tian, Y.-X.; Yue, W.; Yang, L.; Li, Q. Tribo-biological deposits on the articulating surfaces of metal-on-polyethylene total hip implants retrieved from patients. *Sci. Rep.* **2016**, *6*, 28376. [CrossRef]
28. Liu, M.; Luo, Y.; Jia, D. Synthesis of mechanically durable superhydrophobic polymer materials with roughness-regeneration performance. *Compos. Part A Appl. Sci. Manuf.* **2020**, *133*, 105861. [CrossRef]
29. Zhang, Z.P.; Rong, M.Z.; Zhang, M.Q. Mechanically robust, self-healable, and highly stretchable "living" crosslinked polyurethane based on a reversible CC bond. *Adv. Funct. Mater.* **2018**, *28*, 1706050. [CrossRef]
30. Zhong, B.; Jia, Z.; Luo, Y.; Jia, D. A method to improve the mechanical performance of styrene-butadiene rubber via vulcanization accelerator modified silica. *Compos. Sci. Technol.* **2015**, *117*, 46–53. [CrossRef]
31. Hamdia, K.M.; Silani, M.; Zhuang, X.; He, P.; Rabczuk, T. Stochastic analysis of the fracture toughness of polymeric nanoparticle composites using polynomial chaos expansions. *Int. J. Fract.* **2017**, *206*, 215–227. [CrossRef]
32. Talebi, H.; Silani, M.; Bordas, S.P.A.; Kerfriden, P.; Rabczuk, T. A computational library for multiscale modeling of material failure. *Comput. Mech.* **2013**, *53*, 1047–1071. [CrossRef]
33. Aherwar, A.; Singh, A.K.; Patnaik, A. Current and future biocompatibility aspects of biomaterials for hip prosthesis. *AIMS Environ. Sci.* **2015**, *3*, 23–43. [CrossRef]
34. Shah, Q.M.Z.; Chowdhury, M.A.; Kowser, A. Fretting & friction induced fatigue failure: Damage criterion of polytetrafluoroethylene. *Heliyon* **2020**, *6*, e04066. [CrossRef]
35. Zobaer, S.; Quazi, M.; Arefin, K.; Mohammad, A.C. A parametric investigation on the fretting fatigue behaviour of heat treated Al 6061-T6 under rotating bending phenomena. *Aust. J. Mech. Eng.* **2019**, 1–13. [CrossRef]
36. Perez, N. Fatigue Crack Growth. In *Fracture Mechanics*; Springer: Cham, Switzerland, 2017; pp. 327–372.
37. Perez, N. Linear-elastic fracture mechanics. In *Fracture Mechanics*; Springer: Cham, Switzerland, 2017; pp. 79–130.
38. Perez, N. Crack Tip Plasticity. In *Fracture Mechanics*; Springer: Cham, Switzerland, 2017; pp. 187–225.

Journal of
Composites Science

MDPI

Article

FEM-Validated Optimal Design of Laminate Process Parameters Based on Improved Genetic Algorithm

Xing Mou, Zhiqiang Shen, Honghao Liu, Hui Xv, Xianzhao Xia and Shijun Chen *

School of Mechanical Engineering, Hefei University of Technology, Hefei 230000, China; mouxing2021@163.com (X.M.); shenzhiqiang158@foxmail.com (Z.S.); 13700588738@163.com (H.L.); xuhuicslg@163.com (H.X.); xiaxz4557@126.com (X.X.)

* Correspondence: newbonchen@hotmail.com; Tel.: +86-136-85-519-410

Abstract: In tape placement process, the laying angle and laying sequence of laminates have proven their significant effects on the mechanical properties of carbon fibre reinforced composite material, specifically, laminates. In order to optimise these process parameters, an optimisation algorithm is developed based on the principles of genetic algorithms for improving the precision of traditional genetic algorithms and resolving the premature phenomenon in the optimisation process. Taking multi-layer symmetrically laid carbon fibre laminates as the research object, this algorithm adopts binary coding to conduct the optimisation of process parameters and mechanical analysis with the laying angle as the design variable and the strength ratio R as the response variable. A case study was conducted and its results were validated by the finite element analyses. The results show that the stresses before and after optimisation are 116.0 MPa and 100.9 MPa, respectively, with a decrease of strength ratio by 13.02%. The results comparison indicates that, in the iterative process, the search range is reduced by determining the code and location of important genes, thereby reducing the computational workload by 21.03% in terms of time consumed. Through multiple calculations, it validates that "gene mutation" is an indispensable part of the genetic algorithm in the iterative process.

Keywords: composite materials; genetic algorithm; important genes; strength ratio; finite element analysis

Citation: Mou, X.; Shen, Z.; Liu, H.; Xv, H.; Xia, X.; Chen, S. FEM-Validated Optimal Design of Laminate Process Parameters Based on Improved Genetic Algorithm. *J. Compos. Sci.* **2022**, *6*, 21. https://doi.org/10.3390/jcs6010021

Academic Editors: Stelios K. Georgantzinos and Francesco Tornabene

Received: 17 December 2021
Accepted: 4 January 2022
Published: 7 January 2022

Publisher's Note: MDPI stays neutral with regard to jurisdictional claims in published maps and institutional affiliations.

1. Introduction

Carbon fibre reinforced composite materials (CFRP) are widely used in the fields of aviation equipment structures [1], fuselages and wings manufacturing due to their high corrosion resistance, high strength and lightweight [2,3]. As a typical representative of the practical application of composite materials, laminates have a direct impact on the mechanical properties of CFRPs, which depend highly on their laying angle and laying sequence. Hence, many investigations have been conducted in-depth on exploring the influence of these two variables and searching for their optimal combinations.

A genetic algorithm (GA) is a random search method abstracted by simulating gene selection and genetic mechanisms in the natural environment. It can not only solve linear problems but also can be used to solve nonlinear problems; and genetic algorithm is sensitive to the problem of discretization, so it has attracted the attention of scholars [4], and it has been applied to many occasions. At present, scholars worldwide favor the use of genetic algorithms to optimise the design of the mechanical properties of laminates, and certain research results have been obtained. Wang et al. [5] applied a multi-island genetic algorithm to optimise the damage of different loads to the laminate during low-velocity impact and optimised the laminate sequence. Strain energy was taken as the optimisation objective and the laying angle as the design variable. Their results show that the impacted area was reduced by 42.1%, and the residual compression strength increased by 10.79%. It is notable that the laying angle was optimised as a continuous variable rather than discrete as in the other investigations mentioned above. Chen et al. [6] aimed at optimising the fibre laying angle to achieve the minimum stress of the laminate, and used a genetic algorithm to obtain the best laying angle and minimum stress of the laminate. Their method realised the optimisation of the laying angle of two-layer laminates and three-layer laminates by taking the design variable as a continuous variable.

In order to improve the reliability, some optimisation strategies were proposed by Xiu and Cui [7], which took the buckling load of laminate as the optimisation objective, and the laying angle of laminate as constraint condition, and used the neural network model and genetic algorithm to optimise the ratio of laying angle to the buckling load of the laminate. The optimisation method proposed in their investigation is capable of producing a better laying sequence with the same material weight under the condition that the laying angles are given and the number of layers is a constant; nevertheless, the critical process parameter, laying angle, cannot be optimised by this method and is just taken as a given constant. Feng et al. [8] took the strength of laminate as the optimisation objective, and the laying angle of laminate as the constraint condition. In order to obtain a reasonable laying angle and ply number, Jin et al. [9] proposed a two-step optimisation strategy and developed a three-level optimisation model with embedded genetic algorithm and Tsai-Wu failure criterion. The concept of sublayer was introduced into the optimisation method for the optimisation of the location of layers, geometric dimensions, number of layers, and laying sequence. The authors claimed that the introduction of this concept improved the manufacturability of laminates, however, the laying angle was not optimised in this work. By embedding the Tsai-Hill failure criterion, Park et al. [10] used genetic algorithms to apply different loads to the symmetrical composite laminates under different boundary conditions to perform mechanical optimisation design. Tournament selection and uniform crossover methods were adopted in the optimisation process. The laying angle was taken as the design variable as a continuous variable, and a low failure index was obtained. The results showed that the modified genetic algorithm was capable of identifying a global optimal solution for the parameter's optimisation of laid laminates.

Furthermore, new encoding approaches and operators have been developed for genetic algorithms to accelerate the convergence speed and computational accuracy. A genetic algorithm was used to calculate the fibre laying angle and validated that the calculation results can effectively improve the strength of the laminate. This work improved the genetic algorithm by modifying the encoding approach and the integer encoding was embedded in the iteration process for the laminate parameters optimisation. The advantage of this modification is that integer encoding has the potential to avoid unnecessary iterations of floating number, thereby improving the computational efficiency theoretically. The laying angle in this optimisation was specified to 0°, $\pm 45^\circ$, and 90°, leading to weakening the designability of laminates to some extent. By designing a new mutation operator and crossover operator, Wang et al. [11] adopted a symbol encoding method for the development of a self-adaptive genetic algorithm with the bending stiffness parameter as the fitness function. They deeply explored the influence of the adaptively changing crossover operator and the mutation operator on the genetic algorithm and further optimised the laying order. Compared with the traditional genetic algorithm, their modification enhanced the reliability, convergence speed, and computational efficiency of the optimisation process. Another self-adaptive genetic algorithm was proposed by Yang et al. [12] through the automatic change of the mutation rate and crossover rate according to fitness value. This algorithm took the maximum deformation of laminate as optimisation objective, the laminate angle as constraint condition, and called ABAQUS to analyse the stress of the laminate. The results show that this algorithm is easier to get convergence than the traditional genetic algorithm. Hwang et al. [13] designed an elite comparison operator for the genetic algorithm to accelerate the iteration process for the optimisation of laminates. This elite comparison operator was used to identify the differences in the design variables from two different generations, thereby maintaining the similar ones and changing the different ones to find their possible values. Two optimisation problems were solved by this modified genetic algorithm and the results indicate that the designed operator has the potential to speed up the convergence of the optimisation process.

The above investigations mainly focus on the change of optimisation parameters, design of encoding methods, and improvement of operators in the application of traditional genetic algorithms. However, in order to improve the precision of traditional genetic algorithms and eliminating the premature phenomenon in the optimisation process, this article improves the genetic algorithm (improved genetic algorithm, referred as IGA below) on the basis of the principles of traditional genetic algorithms through the application of binary coding to explore important gene codes and their locations. In this IGA, the laying angle is set as the constraint condition, and the minimum stress as the optimisation objective, which customised this algorithm for the optimisation of process parameters of laminates tape placement. Based on the Tsai-Wu failure criterion [9], the laying angle and laying sequence were optimised; and the obtained stress was compared with that of the traditional genetic algorithm to validate the IGA. In addition, the optimal solutions were further validated by a finite element analysis of a laminate with the same geometry, loading, and boundary conditions.

2. Mathematical Model of Laminate and the IGA

2.1. On- and off-Axis Stress Conversion of Laminates

Generally speaking, a multi-layer laminate can be formed by stacking multiple mono-layer boards at different laying angles. According to the designability of composite materials and the theoretical basis of elastic mechanics, the on-axis direction of the monolayer boards can be laid at different angles and in different orders according to the loading conditions to meet the requirements of strength and stiffness design of the structural parts under the load conditions.

Research for the classical laminate theory (CLT), the establishment and derivation of formulae are mainly based on the following assumptions:

- The layers of the laminate need to be firmly bonded so that the deformation between the layers is consistent, and relative slippage between the layers can be avoided;
- Laminates are thin plates with an invariable thickness, and their strength in the thickness direction is relatively small to negligible in respect to that in other directions;
- The bending deformation of the laminate needs to be in a small deflection range, and the straight line perpendicular to the midplane must be maintained before and after the bending deformation while the length of the straight line remains unchanged (viz. $\gamma_{xz} = \gamma_{yz} = 0$; $\varepsilon_z = 0$);
- It is considered that each monolayer of the laminate is in a plane stress state, that is, the theory of plane stress is suitable for the analysis of the laminate structure of thin planes, curved surfaces, or shells.

For simplification of the formula derivation process, the median plane of the laminate is taken as the reference plane, viz. the plane perpendicular to the z-direction at the middle point of the thickness of the laminate. According to the above assumptions, let the stress function x and y be φ, then the corresponding stress components σ_x, σ_y, τ_{xy}, are given in Equation (1) as below:

$$\sigma_x = \frac{\partial \varphi^2}{\partial y^2}, \; \sigma_y = \frac{\partial \varphi^2}{\partial x^2}, \; \tau_{xy} = \frac{-\partial \varphi^2}{\partial x \partial y} \tag{1}$$

Since the classic laminate theory only considers plane stress and ignores the existence of interlayer stress in the iterative process, the stiffness coefficient of each layer of the laminate is different so that the internal force of the laminate can only be integrated layer by layer. Assuming that the stress of the kth layer of the laminate is $\left\{\sigma^k\right\}$, the internal force of the laminate can be expressed as:

$$\begin{Bmatrix} N_x \\ N_y \\ N_{xy} \end{Bmatrix} = \sum_{k=1}^{n} \int_{z_{k-1}}^{z_k} \begin{Bmatrix} \sigma_x^k \\ \sigma_y^k \\ \tau_x^k \end{Bmatrix} dz, \quad \begin{Bmatrix} M_x \\ M_y \\ M_{xy} \end{Bmatrix} = \sum_{k=1}^{n} \int_{z_{k-1}}^{z_k} \begin{Bmatrix} \sigma_x^k \\ \sigma_y^k \\ \tau_x^k \end{Bmatrix} zdz \tag{2}$$

The stress–strain relationship of the kth layer of the laminate is as follows:

$$\begin{Bmatrix} \sigma_x^k \\ \sigma_y^k \\ \tau_{xy}^k \end{Bmatrix} = \begin{bmatrix} \overline{Q_{11}^k} & \overline{Q_{12}^k} & \overline{Q_{16}^k} \\ \overline{Q_{21}^k} & \overline{Q_{22}^k} & \overline{Q_{26}^k} \\ \overline{Q_{61}^k} & \overline{Q_{62}^k} & \overline{Q_{66}^k} \end{bmatrix} \begin{Bmatrix} \varepsilon_x^k \\ \varepsilon_y^k \\ \gamma_{xy}^k \end{Bmatrix} \tag{3}$$

Substituting Equation (2) into Equation (1), and through integration, the stress–strain relationship of the laminate can be obtained as follows:

$$\begin{Bmatrix} N_x \\ N_y \\ N_{xy} \\ M_x \\ M_y \\ M_{xy} \end{Bmatrix} = \begin{bmatrix} A_{11} & A_{12} & A_{16} & B_{11} & B_{12} & B_{16} \\ A_{21} & A_{22} & A_{26} & B_{21} & B_{22} & B_{26} \\ A_{61} & A_{62} & A_{66} & B_{61} & B_{62} & B_{66} \\ B_{11} & B_{12} & B_{16} & D_{11} & D_{12} & D_{16} \\ B_{21} & B_{22} & B_{26} & D_{21} & D_{22} & D_{26} \\ B_{61} & B_{62} & B_{66} & D_{61} & D_{62} & D_{66} \end{bmatrix} \begin{Bmatrix} \varepsilon_x^0 \\ \varepsilon_y^0 \\ \gamma_{xy}^0 \\ k_x \\ k_y \\ k_{xy} \end{Bmatrix} \tag{4}$$

where, A_{ij}, B_{ij}, D_{ij} are

$$\left.\begin{array}{r} A_{ij} = \sum_{k=1}^{n} \int_{z_{k-1}}^{z_k} \overline{Q_{ij}^k} dz = \sum_{k=1}^{n} \overline{Q_{ij}^k} (z_k - z_{k-1}) \\ B_{ij} = \sum_{k=1}^{n} \int_{z_{k-1}}^{z_k} \overline{Q_{ij}^k} zdz = \frac{1}{2} \sum_{k=1}^{n} \overline{Q_{ij}^k} ({z^2}_k - {z^2}_{k-1}) \\ D_{ij} = \sum_{k=1}^{n} \int_{z_{k-1}}^{z_k} \overline{Q_{ij}^k} z^2 dz = \frac{1}{3} \sum_{k=1}^{n} \overline{Q_{ij}^k} ({z^3}_k - {z^3}_{k-1}) \end{array}\right\} \tag{5}$$

where, i, j = 1, 2, 6; A_{ij} is the in-plane stiffness coefficient, B_{ij} is the coupling stiffness coefficient, and D_{ij} is the bending stiffness coefficient.

The flexibility matrix of the laminate is

$$S' = \begin{bmatrix} A' & B' \\ H' & D' \end{bmatrix} \tag{6}$$

where, $A' = A^{-1} + A^{-1}B(-BA^{-1}B + D)^{-1}BA^{-1}; B' = B^{-1} - A^{-1}BD;\ H' = B^{-1} - D^{-1}BA^{-1};\ D' = (-BA^{-1}B + D)^{-1}$.

As the presence of the matrix [B] indicated, the laminate is prone to the coupling of bending deformation and tension and compression deformation when subjected to force; in addition, in-plane deformation is relatively easy to occur when subject to bending stress, and further leads to the warpage of the laminate after curing (that is, $B_{ij} \neq 0$, which means it is hard for the calculation of laminates). For this reason, the composite material laminates generally used in engineering should be symmetrically laid ($B_{ij} = 0$) to avoid the calculation difficulties caused by the coupling stiffness [14]. Note that symmetry is assumed in this study to make sure the laminate only subject to tensile strain but no flexure [14], though it is difficult to always keep symmetry exactly in industry. Therefore, the flexibility matrix S of the laminate at this time can be simplified as:

$$S' = \begin{bmatrix} A^{-1} & 0 \\ 0 & D^{-1} \end{bmatrix} \tag{7}$$

At the same time, through the stress conversion matrix $[T_\sigma]$, the off-axis stress of each layer can be converted into the on-axis stress. The specific conversion formula is shown in Equation (8):

$$\left\{ \begin{array}{c} \sigma_1 \\ \sigma_2 \\ \tau_{12} \end{array} \right\} = \left[\begin{array}{ccc} m^2 & n^2 & 2mn \\ n^2 & m^2 & -2mn \\ -mn & mn & m^2 - n^2 \end{array} \right] \left\{ \begin{array}{c} \sigma_x \\ \sigma_y \\ \tau_{xy} \end{array} \right\} \tag{8}$$

where, $m = \cos\ \theta,\ n = \sin\theta$, and θ is the laying angle of each layer of fibres.

2.2. Failure Criterion for Laminates

After comprehensively considering the advantages and disadvantages of multiple strength criteria, Tsai and Wu [15] proposed a new strength criterion based on the tensor form. That is, the matrix breaks based on Tsai-Wu criterion, as below:

$$F_{11}\sigma_1^2 + F_{22}\sigma_2^2 + F_{66}\sigma_6^2 + 2F_{12}\sigma_1\sigma_2 + F_1\sigma_1 + F_2\sigma_2 = 1 \tag{9}$$

where, σ_1 is the longitudinal stress, σ_2 is the transverse stress, σ_6 is the shear stress, X_T is the longitudinal tensile strength, X_C is the longitudinal compressive strength, Y_T is the transverse tensile strength, Y_C is the transverse compressive strength, S is the shear strength; F_1, F_{11}, F_2, F_{22}, and F_{66} are strength parameters as shown below in Equation (10).

$$\left. \begin{array}{cc} F_1 = \frac{1}{X_T} - \frac{1}{X_C} & F_{11} = \frac{1}{X_C X_T} \\ F_2 = \frac{1}{Y_T} - \frac{1}{Y_C} & F_{22} = \frac{1}{Y_T Y_C} \\ \multicolumn{2}{c}{F_{66} = \frac{1}{S^2}} \end{array} \right\} \tag{10}$$

The value of F_{12} here has little effect on the calculation of the strength criterion, and this is validated by the investigation cited here [16]. Then let $F_{12} = 0$, the calculation result of Equation (9) and that of the Hoffman strength criterion are relatively close with an error of 10%. Hence, Equation (9) is improved as below in Equation (11).

$$F_{11}\sigma_1^2 + F_{22}\sigma_2^2 + F_{66}\sigma_6^2 + 2F_{12}\sigma_1\sigma_2 + 2F_{16}\sigma_1|\sigma_6| + 2F_{26}\sigma_2|\sigma_6| + F_1\sigma_1 + F_2\sigma_2 = 1 \tag{11}$$

The value of F_{16}, F_{26} here is obtained by a semi-empirical formula [17], and the value of F_{12} is the value when the Tsai-Wu strength criterion and the Hoffman criterion are the same. See Equation (12) for details.

$$\left. \begin{array}{cc} F_{16} = \frac{1}{5S\sqrt{X_T X_C}} & F_{26} = \frac{1}{5S\sqrt{Y_T Y_C}} \\ \multicolumn{2}{c}{F_{12} = -\frac{1}{2}F_{11}} \end{array} \right\} \tag{12}$$

Note that the value of F_{12} here is not $-1/2\sqrt{F_{11}F_{22}}$ because:

- The absolute value of Equation (12) is relatively not very large, and it is a negative value, which is reasonable;
- When part of the curves in Figure 1 [15] is under the 0 MPa line, there is a situation where the value of σ_1 is greater than $2X_C$; and this situation does not exist in reality because when greater than, fracture has already appeared for composite materials.

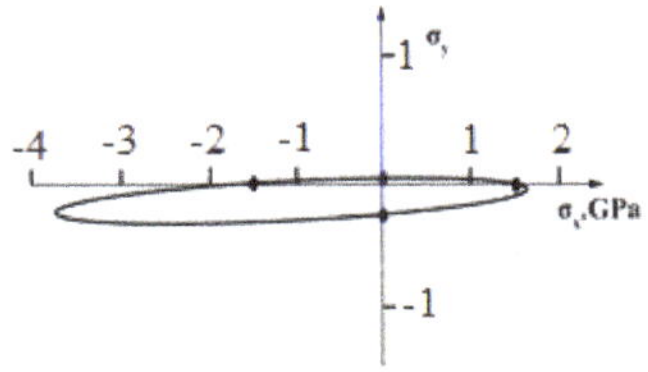

Figure 1. Strength curve of Tsai-Wu failure criterion [15].

From Equation (11), when the calculated value on the left side of the equation equates 1, it indicates that the material has just reached the critical failure state; when the left side is greater than 1, it indicates that damage has begun; when the left side is less than 1, it indicates that the material is in a safe state, and there is still a certain safety margin.

In order to obtain better calculation speed and expected values in genetic algorithms, a suitable. fitness value is generally specified. Here, the strength ratio is applied to the laminate, namely

$$R = \frac{\sigma_i}{\sigma} \tag{13}$$

where R is the ratio of a certain component of the ultimate stress to the corresponding component of loading stress, and i = 1, 2, 6.

The meaning of "correspondence" here [18] assumes that the loading is linear, that is, the stress components increase simultaneously in a certain proportion, and this situation is in line with the actual application in industrial production. Transforming Equation (13) to get $R\sigma = \sigma_i$, substitute this into Equation (11) and after simplification to produce Equation (14) below.

$$\begin{gathered}(F_{11}\sigma_1^2 + F_{22}\sigma_2^2 + F_{66}\sigma_6^2)R^2 + 2(F_{12}\sigma_1\sigma_2 + F_{16}\sigma_1|\sigma_6| + F_{26}\sigma_2|\sigma_6|)R^2 \\ +(F_1\sigma_1 + F_2\sigma_2)R = 1\end{gathered} \tag{14}$$

This is a quadratic equation in one variable. It is easy to get Equation (15) as below.

$$R = \frac{-b + \sqrt{b^2 + 4a}}{2a} \tag{15}$$

where:

$$\begin{aligned}a &= F_{11}\sigma_1^2 + F_{22}\sigma_2^2 + F_{66}\sigma_6^2 + 2(F_{12}\sigma_1\sigma_2 + F_{16}\sigma_1|\sigma_6| + F_{26}\sigma_2|\sigma_6|) \\ b &= F_1\sigma_1 + F_2\sigma_2\end{aligned} \tag{16}$$

Equation (15) is called the strength ratio equation. From the definition of the strength ratio, it can be seen that when $R > 1$, the material has not failed; when the applied stress is increased to $(R - 1)$ times, the material fails. R cannot be less than 1 because this is mathematically meaningless; on contrast, $R < 1$ in engineering applications still has a certain reference value, that is, it indicates that the applied stress must be reduced or the size of the structure must be increased to ensure that the structure is not damaged.

2.3. Genetic Algorithm and Its Improvement

Genetic algorithm is based on the natural law of survival of the fittest by simulating the theory of genetics and biological evolution proposed by Darwin, thereby retaining the excellent genes and individuals [19].

The core content of genetic algorithm mainly includes four parts, namely selection, crossover, mutation, and fitness evaluation. The optimisation process is shown in Figure 2.

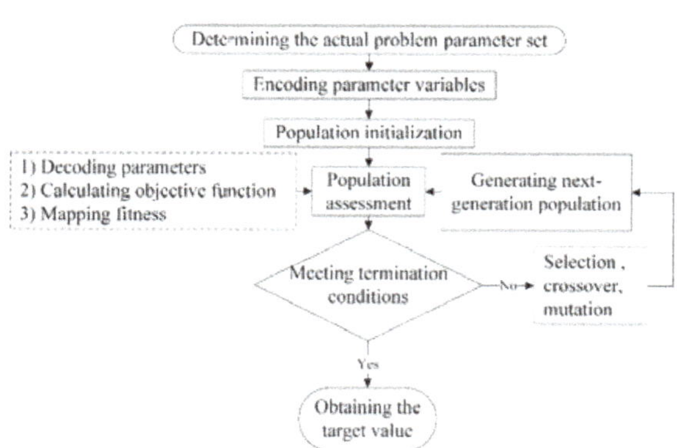

Figure 2. Iteration process of genetic algorithm.

At present, the improvement of genetic algorithm mainly focuses on self-adaptive genetic operators, that is, generating a certain number of individuals, improving the crossover rate and genetic probability between individuals during genetic operations, and using average fitness to improve calculation efficiency. The method proposed in this paper is to start from the population; when generating the population, only one individual is generated. Then the genes of this individual mutate later in the evolution process, and the gene mutation means gene transfer from one vertex to another according to the definition of the hyperplane. The calculation is performed by gene mutation; the better genes and positions are selected through fitness, and negative feedback adjustment is used to generate individuals with important genes, and then identify the corresponding laying angle through decoding. The basic mathematical theory mentioned above is:

For the full set space composed of individuals whose gene length is l [20], that is, the individual space is given in Equation (17)

$$S_{sp} = \{0,1\}^l \tag{17}$$

where we regard this as the spatial vertex set with the dimension l. When $l = 3$, it becomes a three-dimensional cube.

As Figure 3 indicated, for three-dimensional individual space, the genes at each vertex can be connected by 12 straight lines. It can also be seen that in a straight line, we only need to change the value of one of the genes, then the gene can be converted into the gene at the other end of the same line; in any plane, as long as the values of two genes are changed, the genes at the four vertices can be switched to each other. Therefore, in a three-dimensional individual space, the straight line and the plane are called the "hyperplane", and the interconnection between them constitutes the "subspace" in the space.

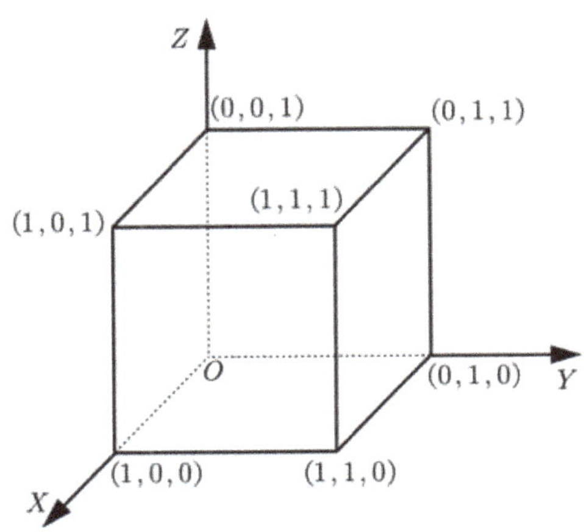

Figure 3. Three-dimensional individual space.

Therefore, according to the above definition of "hyperplane", for the *i-th* gene, if

$$f((x_1,\cdots x_{i-1},1,x_{i+1},\cdots x_l)) \geq f((x_1,\cdots x_{i-1},0,x_{i+1},\cdots x_l)) \tag{18}$$

for any $x_k = 0,\ 1(k \neq i)$ holds, or

$$f((x_1,\cdots x_{i-1},0,x_{i+1},\cdots x_l)) \geq f((x_1,\cdots x_{i-1},1,x_{i+1},\cdots x_l)) \tag{19}$$

for any $x_k = 0,\ 1(k \neq i)$ holds, the gene i is taken as an important gene. When Equation (18) is established, important gene 1 is the better choice; while when Equation (19) is established, important gene 0 is the better choice.

Once the "hyperplane" is determined, it is possible to locate the gene value of the corresponding position after confirming that the gene value of some gene position is valid, thereby reducing the search range and further reducing the computational workload. The specific process of the IGA is shown in Figure 4.

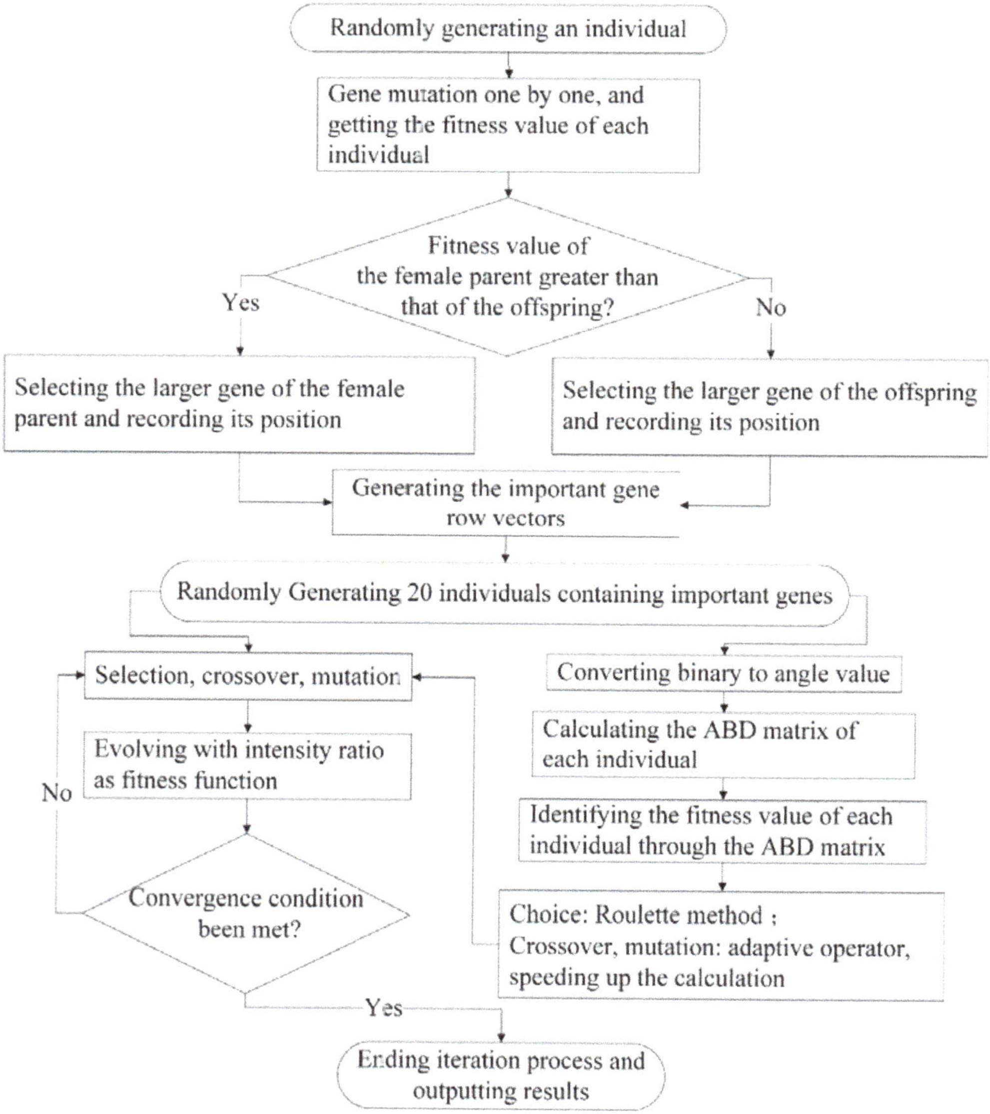

Figure 4. Flow chart of the IGA.

3. IGA Applied to Composite Laminate Design

3.1. Ply Design Requirements for Laminates

For the ply design of laminates, the coupling effect caused by stretching and bending during the curing process should be avoided since it causes resin cracking, warping deformation and even delamination for laminates. Therefore, a design generally complies with the following process design rules [21].

- Standard laid layers should be adopted, which is supposed to include four angle values of 0°, ±45°, and 90°, respectively;
- Adjacent 4 monolayers cannot be formed at the same laying angle so as to prevent the substrate from cracking, and simultaneously a full 90° ply should be avoided;
- The proportion of 0° laid layers is between 20% and 40%, ±45° layers between 40% and 60%, and 90° layers must be between 10% and 30%.

3.2. Genetic Algorithm Coding

In the process of searching optimal solutions, the genetic algorithm generally uses a finite length string code to define the solution to be solved, such as a binary code. The layup of a laminate is a matter of discrete variables, so the choice of variables affects the length of the code string to some extent. Since the other encoding schemes, such as integer encoding and floating-point number encoding, are not unique in the subsequent mutations, they are not conducive to genetic operations such as crossover and mutation. Therefore, traditional binary encoding is selected hereof. The concrete corresponding method is:

$$00\text{: } 0°;\ 01\text{: } -45°;\ 10\text{: } 45°;\ 11\text{: } 90°$$

Since the initial population generates randomly, there must be a certain number of initial individuals that do not meet the requirements of the laminate layup design; these individuals must be eliminated, and the others that meet the requirements should remain. The subsequent operations conduct all the above operations for each subsequent generation, thereby ensuring that the population individuals are suitable. The operation is like a "human intervention" in nature.

3.3. Fitness Function

In the genetic algorithm, fitness is used to evaluate the goodness of individuals adapting to the "environment", thereby further weighing the degree to which individuals can reach or approach the optimal solution in the process of genetic evolution [22]. Therefore, the fitness function selected for this calculation is

$$R_s = R - (5n_1 - 10n_2) \qquad i = 1,\ 2\ldots n \tag{20}$$

where, n is the total number of laminates, $R(i)$ is the strength ratio of each layer of fibre, n_1 is the number of adjacent four layers with the same laying angle, n_2 is the number of layers that do not meet the requirements of the layup ratio, $5n_1 + 10n_2$ is the penalty factor. Because the initial individuals generated may not all meet the design requirements of the laminate, as mentioned above, a certain high penalty is required for the fitness factor in the searching process. Hence, through reduction of the fitness function value of unsuitable. Individuals, the probability of being selected in selection replication is further reduced. In actual operations, there may be a phenomenon that a high penalty makes R_s a negative value, and such a value is not conducive to the calculation of the subsequent results. Therefore, when it is a negative value, set R_s = 0.05, thereby forcing it to become a small positive value by "human interference" to facilitate subsequent calculations. According to the equations above, the optimisation process of the strength of laminate by the IGA algorithm is the process of analysing and calculating the best strength ratio of the single layer in the population.

3.4. Relevant Parameters Selection

In this paper, T300/5228 epoxy carbon fibre with good specific strength is selected, and the material parameters are shown in Table 1.

Table 1. Mechanical properties of T300 epoxy resin carbon fibre.

E_1/Gpa	E_2/Gpa	X_T/MPa	X_C/MPa	G_{12}/Gpa	Y_T/MPa	Y_C/MPa	S/MPa	μ_1	ρ/(kg/m^3)
144	9.3	1633	1021	4.68	53.8	232	90	0.312	1610

In an optimisation process, the communication of individuals between populations is not a complete fusion, but on the basis of absorbing each other's excellent genes and simultaneously ensuring that their own excellent genes are preserved. Therefore, individuals need to use different crossover rates P_C and mutation rates P_m. Furthermore, the values of P_C and P_m also affect the global and local search capabilities and lead to different optimisation results; empirically, researchers recommend to adopt a large value of P_C and small value of P_m. Here in this article, the two probability factors are specified as 0.95 and 0.005, respectively. The specific parameter settings related to the genetic algorithm are detailed in Table 2.

Table 2. Parameter settings of genetic algorithm.

P_C	P_m	Population Size	Maximum Number of Iterations
0.95	0.005	20	500

At the same time, in order to speed up the optimisation process, the evaluation parameters of the adaptive genetic operator are introduced [23], namely

$$U = \frac{R_{max} - R_{avg}}{R_{avg} - R_{min} + \varepsilon} \tag{21}$$

where, R_{max} is the maximum fitness value in each generation of individuals; R_{avg} is the average fitness value in each generation of the population; R_{min} is the minimum fitness value in each generation; ε is an infinitesimal with the main purpose to prevent the extreme case of 0 in the denominator.

Through Equation (21), the value of P_C and P_m in the two cases of $U > 1$ and $U < 1$ is further derived so as to speed up the calculation speed.

4. A Case Study

4.1. Relevant Parameters Selection

Given that a rectangular composite laminate beam with four sides simply supported is designed, the size of the laminate is, length $b = 2l$ with the value of 1300 mm, width $a = 132$ mm, thickness $h = 2.25$ mm, and the bearing pressure perpendicular to the surface $q = 0.5$ MPa. The design adopts a symmetrical 18-layer layup scheme. The geometric model of the laminate is illustrated in Figure 5.

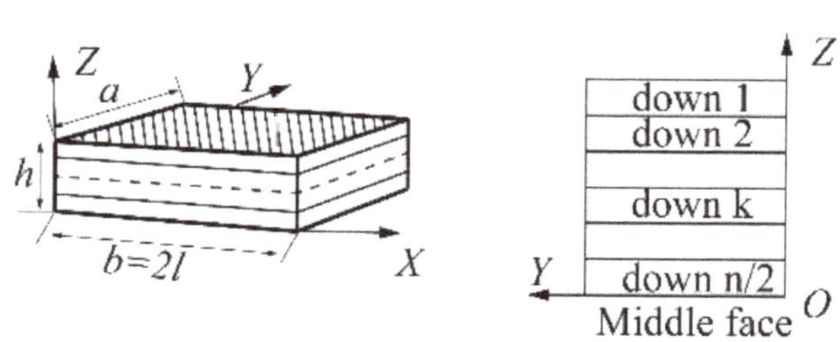

Figure 5. Structure and geometric dimensions of the laminate model.

The schematic diagram of the simply supported laminate beam is shown in Figure 6, as below.

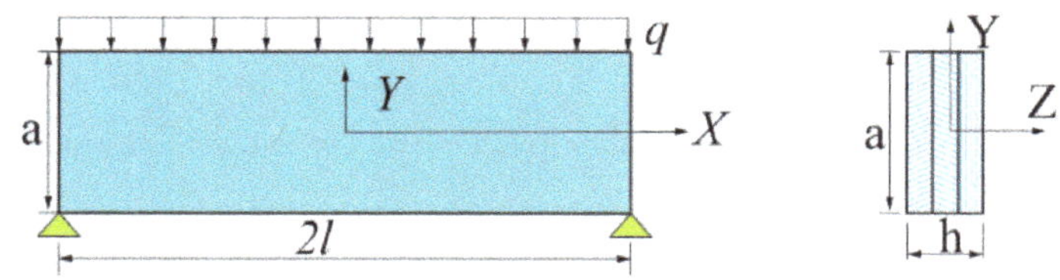

Figure 6. Loading diagram of simply supported laminate beam.

Since σ_y is caused by the lateral load q, and q is a fixed value, which means that σ_y is a function of y, here we can set (See the detailed derivation process of the equations in this section at the Appendix A.)

$$\sigma_y = f(y) = \frac{\partial^2 \varphi}{\partial x^2} \tag{22}$$

Then integrating Equation (22) it becomes

$$\sigma_y = \frac{1}{2}x^2 f(x) + x f_1(x) + f_2(y). \tag{23}$$

Substituting Equation (23) into compatibility Equation (26) it becomes

$$\begin{gathered}\frac{x^2}{2}\frac{d^4 f(y)}{dy^4}S_{11} + \left(\frac{d^4 f_1(y)}{dy^4}S_{11} - 2\frac{d^3 f(y)}{dy^3}S_{16}\right)x + (2S_{11}+S_{66})\frac{d^2 f(y)}{dy^2} - 2\frac{d^2 f_1(y)}{dy^2}S_{16} \\ + \frac{d^4 f_2(y)}{dy^4}S_{11} = 0\end{gathered} \tag{24}$$

where, the balance equation and compatibility equation are shown in Equations (25) and (26), and the stress function φ is shown in Equation (1) above

$$\left.\begin{aligned}\frac{\partial \sigma_x}{\partial x} + \frac{\partial \tau_{xy}}{\partial y} = 0 \\ \frac{\partial \sigma_y}{\partial y} + \frac{\partial \tau_{xy}}{\partial x} = 0\end{aligned}\right\} \tag{25}$$

$$\left.\begin{aligned}\frac{\partial \sigma_x}{\partial x} + \frac{\partial \tau_{xy}}{\partial y} = 0 \\ \frac{\partial \sigma_y}{\partial y} + \frac{\partial \tau_{xy}}{\partial x} = 0\end{aligned}\right\} \tag{26}$$

If for any x, Equation (23) holds, then all the coefficients of Equation (24) need to be 0, that is

$$\begin{gathered}\frac{S_{11}}{2}\frac{d^4 f(y)}{dy^4} = 0 \\ \frac{d^4 f(y)}{dy^4}S_{11} - 2\frac{d^3 f(y)}{dy^3}S_{16} = 0 \\ (2S_{11}+S_{66})\frac{d^2 f(y)}{dy^2} - 2\frac{d^2 f_1(y)}{dy^2}S_{16} + \frac{d^4 f_2(y)}{dy^4}S_{11} = 0.\end{gathered} \tag{27}$$

Hence, it can set $f(y)$ as

$$f(y) = \frac{A_1}{6}y^3 + \frac{A_2}{2}y^2 + A_3 y + A_4 \tag{28}$$

According to the principle of Saint-Venant it gives

$$\left.\begin{array}{ll}\tau_{xy}=0 & y=\pm\frac{h}{2}\\ \left.\begin{array}{ll}\sigma_y=q & y=+\frac{h}{2}\\ \sigma_y=0 & y=-\frac{h}{2}\end{array}\right\}\end{array}\right\}. \quad (29)$$

At the same time, the component force according to the shear stress is Q, namely

$$Q=-ql \quad x=l\ . \quad (30)$$

Substituting Equation (28) into boundary condition Equations (29) and (30), and according to the axial component force in the axial direction is 0, the stress expression of σ_x, σ_y, τ_{xy} can be derived as [24]

$$\begin{aligned}\sigma_x&=\frac{12q}{h^3}\left(\frac{2S_{12}+S_{66}}{6S_{11}}y^3-\frac{1}{2}x^2y\right)+\left(\frac{6ql^2}{h^3}-\frac{3q}{10h}\frac{2S_{12}+S_{66}}{S_{11}}\right)y\\ \sigma_y&=-\frac{2q}{h^3}y^3+\frac{3q}{2h}y+\frac{q}{2}\\ \tau_{xy}&=x\left(\frac{6q}{h^3}y^2-\frac{3q}{2h}\right)\end{aligned} \quad (31)$$

In order to conduct optimisation, R is first obtained by combining Equations (31) and (15) according to the above requirements, and then it is substituted into Equation (20) to produce the penalty fitness function; and the MATLAB script was written. For speeding up the convergence process, here the adaptive operator of Equation (21) was introduced, then the size of the crossover rate P_C was calculated and so was the mutation rate P_m between individuals, thereby further obtaining the target value. The compiled calculation program was conducted on a PC with the CPUs (AMD Ryzen 71700 Eight-Core Processor, 2.99 GHz); and the important gene was 0 with 2 important gene positions, which were the 1st and 5th genes, respectively. In order to avoid contingency caused by one or two cycles, the program for finding important genes was run repeatedly and the important genes generated after each cycle were counted and sorted, and then through counting the number of occurrences and position numbers, finally it got the important gene positions as the 1st and 5th genes. Then injecting the obtained important genes into new individuals randomly generated, the strength ratio R was obtained during the evolution process. It needs to run multiple times to prevent local convergence that may occur accidentally and fail to reach the global convergence value.

4.2. Results Discussion

The optimisation diagram of final strength ratio R is shown in Figure 7. It suggests that after multiple runs, almost all optimisation curves had fluctuated in the fitness value in the early stage. This is because it is necessary to determine whether the population individual meets the requirements of the layup design after each iteration; if not, eliminate them and introduce new individuals. Therefore, in the continuous genetic evolution process, individuals that did not meet the requirements were eliminated, and new individuals that meet the requirements were added in thereby ensuring that the original design layup criteria were always met.

This value always changed in the optimisation process in the later stage. As indicated by the first, fourth, fifth, and seventh optimisation curves in Figure 7, the best strength ratio appeared at a certain genetic stage, and the genetic generation number was kept short. The "best value" is produced by "gene mutation" and is not the global optimal solution. Simultaneously, the 4th and 7th optimisation curves gradually increase with the number of iterations. Although the optimal strength ratio was found due to the "gene mutation", the final optimisation curve converges locally, which also indicates that the genetic algorithm has the disadvantage of local convergence, and thus, unable to find the global optimal solution smoothly.

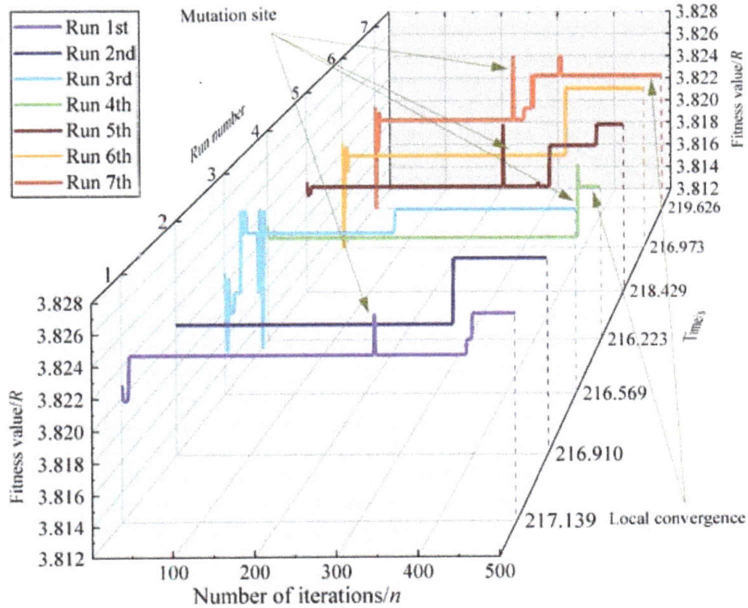

Figure 7. Ratio of number of iterations of the IGA (n) to strength ratio (R).

In order to determine whether the curve is reasonable and whether the calculation result meets the initial design requirements, the curve was processed through polynomial data fitting. The seven strength ratios were averaged in the fitting process, and the final fitting result is shown in Figure 8.

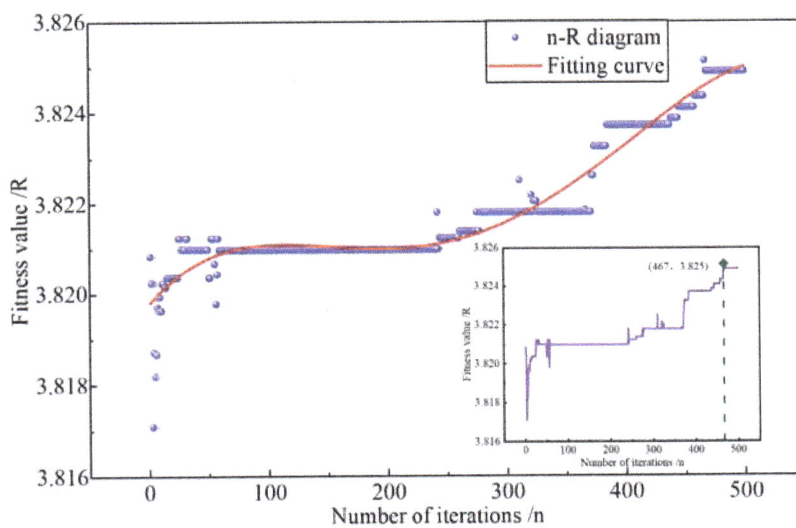

Figure 8. The fitted curve of the number of iterations of the IGA (n) to strength ratio (R).

The expression of the red curve in Figure 8 is given below as,

$$R = A_4 n^4 + A_3 n^3 + A_2 n^2 + A_1 n + A_0 \tag{32}$$

where

$$\begin{gathered} A_4 = -6.5604e-13 \quad A_3 = -7.3372e-10 \\ A_2 = -2.4327e-7 \quad A_1 = 3.0651e-5 \\ A_0 = 3.8198 \end{gathered} \tag{33}$$

According to statistics, Goodness of fitting T^2 is the indicator to evaluate the degree of fit [25]. It is expressed in Equation (34) below.

$$T^2 = \frac{\sum_{i=1}^{N} (\hat{R}_i - \overline{R})^2}{\sum_{i=1}^{N} (R_i - \overline{R})^2} \tag{34}$$

where, $\hat{R}_i$ is the fitted value; $\overline{R}$ is the average value of the initial data; R_i is the *i-th* initial data. Putting the average of the strength ratios of the IGA algorithm into Equation (34), it becomes T^2 = 0.943. The value range of T^2 is 0 to 1. The closer to 1, the better the fitting effect is. In order to further determine the reliability of this improved algorithm, the number of iterations was increased up to 550 generations, and the optimisation iteration was performed three times with their results averaged. At the same time, the calculation result of the fitted curve was substituted for comparison, as shown in Figure 9.

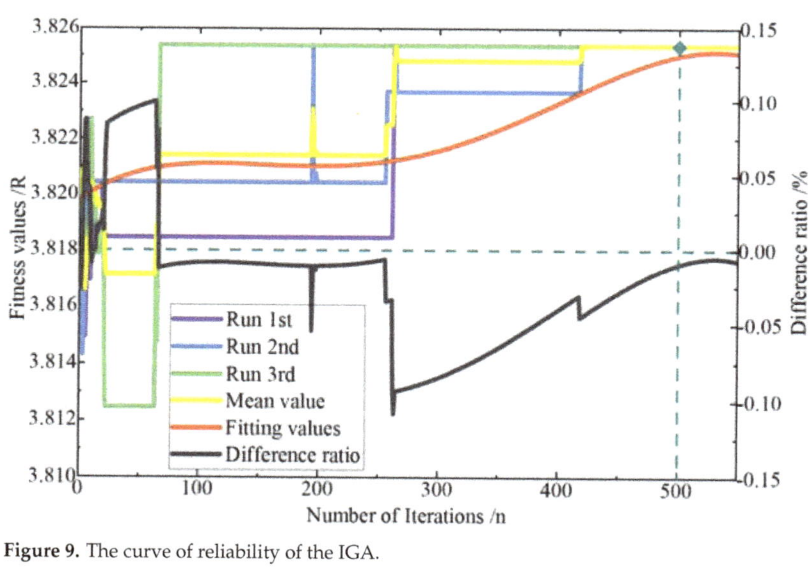

Figure 9. The curve of reliability of the IGA.

Figure 9 shows that after 550 generations with the maximum difference ratio of 0.1%, the minimum difference ratio of −0.18%, the maximum difference range of 0.208%, and T^2 of 0.943 at the same time; hence, it can be considered that the IGA algorithm and the fitted curve are reliable.

It can be seen from Tables 3 and 4 and Figure 7 that when the number of iterations evolves to 52 generations at the earliest stage, the strength ratio R reaches the maximum value of 3.825, suggesting a significant improvement with the increase rate of 48.83%, and further, in the iterative process, the laying angle ±45° gradually approaches the outermost layer, which can also effectively improve the impact resistance and stability of the laminate. At the same time, through decoding the optimised layering scheme inversely, it suggests that the number of genes 0 is gradually increasing. This gradual increase also shows that

0 is an important gene, which is consistent with the initial calculation results. Compared with the traditional genetic algorithm, in order to ensure a single variable, the population initially formed by the IGA was retained, and simultaneously ensuring that other initial settings remained unchanged; then optimisation iterations were performed. The calculation results are shown in Figure 10 and Table 5.

Table 3. Calculation data for searching for the optimal solution.

Earliest Number of Iterations	Converging Globally?	Strength Ratio/R
321	Y	
374	Y	
52	Y	
467	N	3.825
311	Y	
372	Y	
242	N	

Table 4. Comparison before and after layup optimisation.

	Strength Ratio/R	Layup Scheme/°
Before optimisation	2.57	$[0/+45/90/-45/0/+45/-45/90/-45]_S$ *
After optimisation	3.825	$[-45/+45/-45/+45/0/90/0/0/90]_S$ *

* The small "s" next to the right bracket denotes symmetric layup hereof.

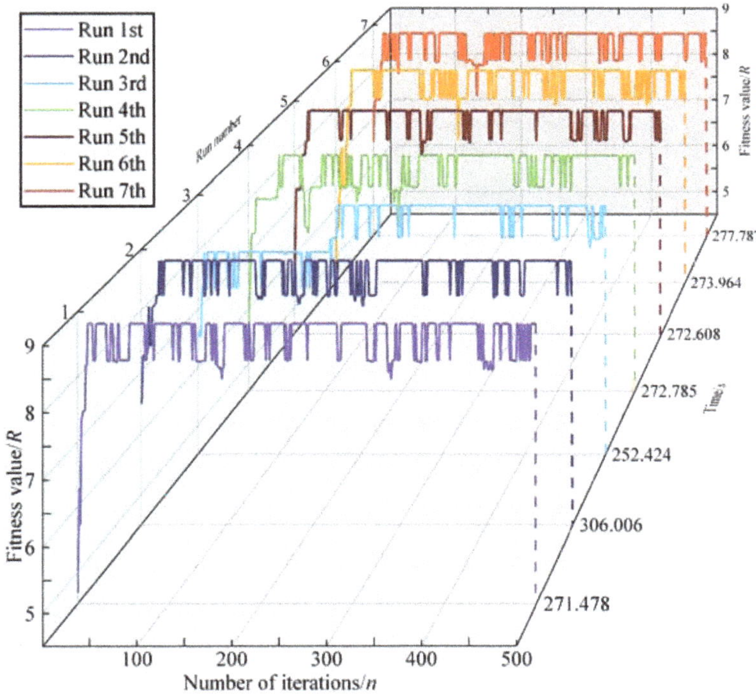

Figure 10. The curve of traditional genetic algorithm number of iterations N-strength ratio, R.

Table 5. Calculation results of traditional genetic algorithm.

Earliest Number of Iterations	Strength Ratio R	Layup Scheme/°
12/22/172/41/22/24/25	8.819	$[9]_S$ *

* "[9]" denotes there are nine layers with a laying angle of 0°.

Figure 10 presents that the strength ratio R obtained by the GA algorithm mainly shows an increasing trend at the beginning, and the maximum strength ratio R is 8.819. At the same time, Table 5 shows that the laying angles are all 0° at this stage, and then the strength ratio R fluctuates. This can be attributed to the "gene mutation" due to mutation during the evolution process. From Table 4, it is known that the "gene mutation" has produced Gene 1; however, as indicated by the layup scheme in Table 4, the laying angle does not meet the original design criteria, which led to the phenomenon of "prematurity".

At the same time, through the comparison between Figures 7 and 10 it results that:

- Due to the existence of "gene mutation", the IGA algorithm identifies the optimal strength ratio, but due to the existence of local convergence, it is necessary to rely on "gene mutation" to find the optimal strength ratio. Hence, "gene mutation" cannot be removed in actual operation. At the same time, in order to avoid contingency, it is necessary to run multiple times to ensure the unity of the results, thereby preventing local convergence produced by a single run. The GA algorithm has an "unreasonable" maximum strength ratio in the 149th generation, and this reversely indicates the occurrence of the "premature" phenomenon.
- The average running time of the IGA algorithm is 217.41 s, while the traditional genetic algorithm is 275.293 s. It is obvious that the IGA algorithm has a certain improvement in its iterative operation efficiency, with an increase of 21.03%.

5. FEM Validation

In order to validate the proposed IGA, a finite element model of a laminate was developed with a length of 1200 mm, a width of 130 mm, and thickness of 2.25 mm. The laminate has a total of 18 monolayers with a thickness of each layer of 0.125 mm. The Laminate plate was constrained by fixed the x, y, and z directions on all four sides (U1 = U2 = U3 = 0). That is to say, the laminate plate was set as a simply supported beam. Then the plate was meshed with hexagonal shell elements with eight nodes (SC8R) and reduced integration. A uniform load of 0.5 MPa was applied perpendicular to the x–y plane of the laminate plane. Finally, the finite element analysis was conducted through the commercial software ABAQUS and the simulation results before and after optimisation were obtained, as shown in Figure 11.

As indicated by Figure 11a,b, the maximum stress in the fibre direction before optimisation is 116.0 MPa, and the maximum stress after optimisation is 100.9 MPa. The stress value is greatly reduced, and the reduction ratio reaches 13.02%, indicating that the optimisation results meet the initial requirements. The stresses at the other two directions, namely s22 and s12 changed from 8.57 MPa and 8.80 MPa to 5.464 MPa and 11.11 MPa, respectively. These two directions are not the directions mainly subject to applied forces and the corresponding stresses are also relatively small to s11. At the same time, it can be found from the figures that the maximum stress is mainly concentrated on the two long sides of the laminate. This is because the use of four-sided simple support leads to stress concentration.

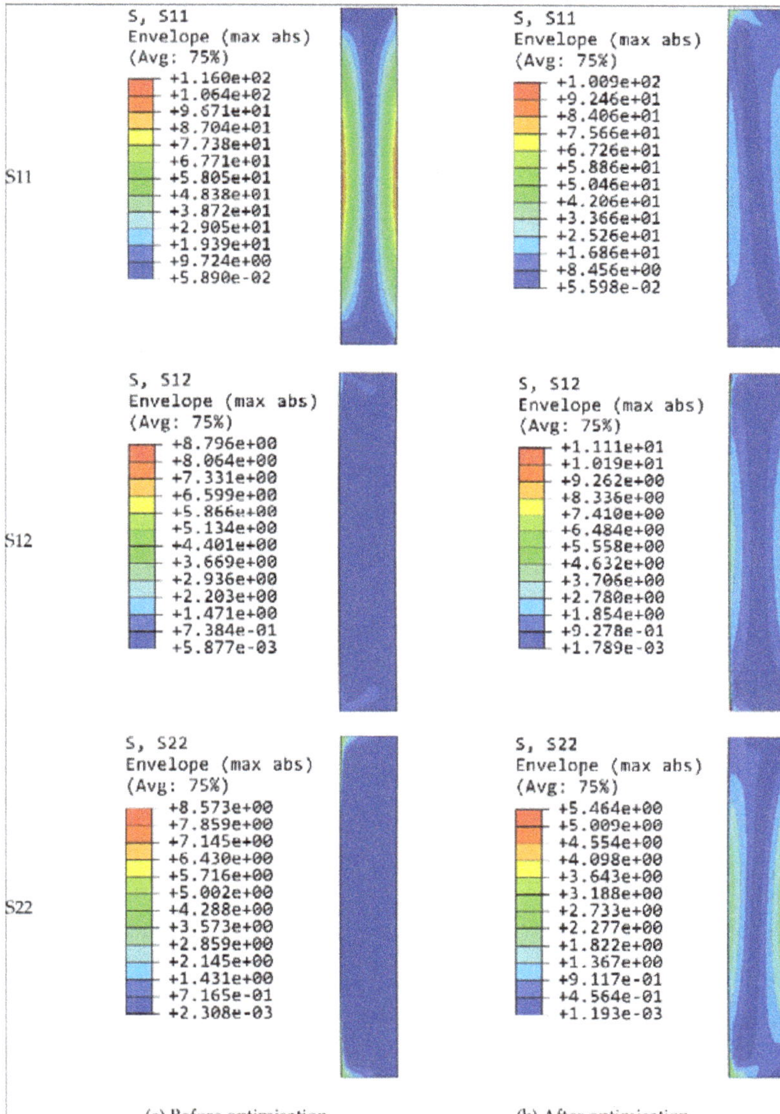

Figure 11. Stress contour of the laminate before and after optimisation: (**a**) Stress contour before optimisation; (**b**) Stress contour after optimisation.

6. Conclusions

Through the development of an IGA, the laying angle and laying sequence of laminates were optimised and the optimal results were compared with those of traditional genetic algorithms. The reliability of the optimisation was further validated by comparison with the finite element analysis results and the conclusions are drawn as below.

- Based on the genetic algorithm, this paper optimised the layering angle of laminate, and adopted the identification of important genes to determine the important genes and their positions and reduced the query range so that the calculation cost in terms of time was reduced by 21.03%.

- In the process of genetic evolution, the operation "gene mutations" is essential and indispensable, and it proves that the "gene mutation" has the potential to facilitate the identification of global optimal solutions.
- Through the development of MATLAB script, it found the important genes were the first and fifth genes. The optimisation results show that the improved strength ratio is 3.825, and the optimal laying angle is sequentially [−45°/+45°/−45°/ 45°/0°/90°/0°/ 0°/90°]s.
- The stresses before and after the optimisation were 116.0 MPa and 100.9 MPa, respectively, with a decrease of 13.02%. This comparison validates the IGA and the optimal results can provide a reference for engineering design.

Author Contributions: Conceptualization, X.M., Z.S. and S.C.; methodology, X.M. and H.L.; software, X.M.; validation, X.M., H.X. and X.X.; formal analysis, X.M.; investigation, X.M. and Z.S.; resources, X.X.; data curation, H.L. and H.X.; writing—original draft preparation, X.M. and Z.S.; writing—review and editing, X.M., Z.S. and S.C.; visualization, X.M.; supervision, S.C.; project administration, S.C.; funding acquisition, S.C. All authors have read and agreed to the published version of the manuscript.

Funding: This research is financially supported by the Fundamental Research Funds for the Central Universities (No.: JZ2021HGQB0274).

Institutional Review Board Statement: Not applicable.

Informed Consent Statement: Not applicable.

Data Availability Statement: Not applicable.

Conflicts of Interest: The authors declare no conflict of interest.

Nomenclature

$\overline{Q_{ij}^k}$	off-axis modulus of row i and column j
$\hat{R}_i$	the fitted value
σ_x^k	the stress of kth layer along x direction
σ_y^k	the stress of kth layer along y direction
τ_{xy}^k	the shear stress of kth layer in x–y plane
A'	in-plane flexibility matrix
A_i	polynomial coefficients, $i = 0 \ldots \ldots 4$
A_{ij}	in-plane stiffness of row i and column j in matrix
B'	coupling flexibility matrix
B_{ij}	coupling stiffness of row i and column j in matrix
D'	bending flexibility matrix
D_{ij}	bending stiffness of row i and column j in matrix
F_i, F_{ij}	strength parameter, $i, j = 1, 2, 6$ respresnting x, y, and x–y direction, respectively
$\overline{R}$	the average value of the initial data
R_{avg}	mean fitness values of individuals in each generation
R_i	the i-th initial data
R_{max}	maximum fitness in each generation of individuals
R_{min}	minimum fitness value in each generation of individuals
R_s	fitness function
S'	flexibility matrix
S'_{11}	flexibility coefficients
S'_{16}	flexibility coefficients
S'_{66}	flexibility coefficients
S'_{ij}	flexibility coefficient of row i and column j in matrix
S_{sp}	individual space

X_C	longitudinal compressive strength
X_T	longitudinal tensile strength
Y_C	transverse compressive strength
Y_T	transverse tensile strength
n_1	number of same layering angles for four consecutive layers
n_2	number of layers that do not meet the ratio requirement
x_i	the *kth* gene
P_C	crossover probability
P_m	mutation probability
γ_{xy}^0	the shear strain of middle plane in x–y plane
γ_{xy}^k	the shear strain of *kth* layer in x–y plane
ε_x^0	the strain of middle plane along x-direction
ε_x^k	the strain of *kth* layer along x-direction
ε_y^0	the strain of middle plane along y-direction
ε_y^k	the strain of *kth* layer along y-direction
σ_1	the stress of *kth* layer along fibre direction
σ_2	the stress of *kth* layer along the vertical fiber direction
σ_6	the shear stress of layer in-plane
τ_{12}	the shear stress of *kth* layer in-plane
a	width of laminate
b	length of laminate
E_1	longitudinal modulus of elasticity
E_2	transverse modulus of elasticity
G_{12}	shear modulus of elasticity in the plane of the lamina
h	thickness of laminates
k	*kth* layers
M_x	the bending moment of x-direction
M_{xy}	the torque of x–y plane
M_y	the bending moment of y-direction
n	total number of layers
N_x	the force of x-direction
N_{xy}	the shear force of x–y plane
N_y	the force of y-direction
Q	shear stress
q	uniform loading pressure
R	the ratio of a certain component of the ultimate stress to the corresponding component of loading stress.
S	the shear strength in x–y plane
T	the indicator to evaluate the degree of fitting
x	the fiber direction
y	the vertical fiber direction
Z_k	the z coordinate of *kth* layer
Z_{k-1}	the z coordinate of $k-1th$ layer
μ_1	poisson's ratio in the x-direction
ρ	density
φ	stress function
U	genetic operator evaluates parameters
l	half-length of laminate
ε	an infinitesimal value

Appendix A

Considering a composite beam with simply supported ends, it has

$$\begin{cases} \sigma_z = 0 \\ \tau_{yz} = \tau_{zy} = \tau_{zx} = \tau_{xz} = 0 \end{cases} \tag{A1}$$

Here, only $\sigma_x, \sigma_y, \tau_{xy}$ are not zero. Then according to statics, it obtains the following expressions.

- Balance equation (ignoring body stress)

$$\begin{cases} \frac{\partial \sigma_x}{\partial x} + \frac{\partial \tau_{xy}}{\partial y} = 0 \\ \frac{\partial \sigma_y}{\partial y} + \frac{\partial \tau_{xy}}{\partial x} = 0 \end{cases} \tag{A2}$$

Let stress $\varnothing$ be a function of x, y, the following relationship holds:

$$\sigma_x = \frac{\partial^2 \varnothing}{\partial y^2}, \sigma_y = \frac{\partial^2 \varnothing}{\partial x^2}, \tau_{xy} = -\frac{\partial^2 \varnothing}{\partial x \partial y} \tag{A3}$$

- Geometric compatibility equations

$$\frac{\partial^2 \varepsilon_x}{\partial y^2} + \frac{\partial^2 \varepsilon_y}{\partial x^2} = \frac{\partial^2 \gamma_{xy}}{\partial x \partial y} \tag{A4}$$

- Physical equations

$$S = A^{-1} \tag{A5}$$

$$\begin{cases} \varepsilon_x = S'_{11}\sigma_x + S'_{12}\sigma_y + S'_{16}\tau_{xy} \\ \varepsilon_y = S'_{12}\sigma_x + S'_{22}\sigma_y + S'_{26}\tau_{xy} \\ \gamma_{xy} = S'_{16}\sigma_x + S'_{26}\sigma_y + S'_{66}\tau_{xy} \end{cases} \tag{A6}$$

- Substituting (A3) into (A6) and then into (A4), it can obtain the equation below,

$$S'_{22}\frac{\partial^4 \varnothing}{\partial x^4} - 2S'_{26}\frac{\partial^4 \varnothing}{\partial x^3 \partial y} + \left(2S'_{12} + S'_{66}\right)\frac{\partial^4 \varnothing}{\partial x^2 y^2} - 2S'_{16}\frac{\partial^4 \varnothing}{\partial x \partial y^3} + S'_{11}\frac{\partial^4 \varnothing}{\partial y^4} = 0 \tag{A7}$$

The expressions of σ_x, σ_y, τ_{xy} from (A7) as below

① Since σ_y is caused by the constant shear force q, and its value does not depend on x, it is a function of y. Therefore, we can set $\sigma_y = f(y)$

$$\frac{\partial^2 \varnothing}{\partial x^2} = \sigma_y = f(y) \tag{A8}$$

Through integral formula (A8), it can produce:

$$\varnothing = \frac{1}{2}x^2 f(y) + x f_1(y) + f_2(y) \tag{A9}$$

② Therefore, substituting (A9) into the compatible equation, each item of (A7) can be obtained as follows:

$$\begin{gathered} \frac{\partial^4 \varnothing}{\partial x^4} = 0, \frac{\partial^4 \varnothing}{\partial x^3 \partial y} = 0 \\ \frac{\partial^4 \varnothing}{\partial x^2 y^2} = \frac{d^2 f(y)}{dy^2} \\ \frac{\partial^4 \varnothing}{\partial x \partial y^3} = x\frac{d^3 f(y)}{dy^3} + \frac{d^3 f_1(y)}{dy^3} \\ \frac{\partial^4 \varnothing}{\partial y^4} = x^2\frac{d^4 f(y)}{2dy^4} + x\frac{d^4 f_1(y)}{dy^4} + \frac{d^4 f_2(y)}{dy^4} \end{gathered} \tag{A10}$$

Substituting into the compatible Equation (A4), it becomes

$$\left(2S'_{12} + S'_{66}\right)\frac{d^2 f(y)}{dy^2} - 2S'_{16}\left(x\frac{d^3 f(y)}{dy^3} + \frac{d^3 f_1(y)}{dy^3}\right) + S'_{11}\left(x^2\frac{d^4 f(y)}{2dy^4} + x\frac{d^4 f_1(y)}{dy^4} + \frac{d^4 f_2(y)}{dy^4}\right) = 0 \tag{A11}$$

If it holds for any x of the above formulas, then it gets

$$\frac{S'_{11}}{2}\frac{d^4f(y)}{dy^4}=0 \tag{A12a}$$

$$S'_{11}\frac{d^4f_1(y)}{dy^4}-2S'_{16}\frac{d^3f(y)}{dy^3}=0 \tag{A12b}$$

$$(2S'_{12}+S'_{66})\frac{d^2f(y)}{dy^2}-2S'_{16}\frac{d^3f_1(y)}{dy^3}+S'_{11}\frac{d^4f_2(y)}{dy^4}=0 \tag{A12c}$$

If (a) holds, then

$$f(y)=\frac{A_1y^3}{6}+\frac{A_2y^2}{2}+A_3y+A_4 \tag{A13}$$

If (b) holds, then

$$f_1(y)=\frac{A_1}{12}\frac{S'_{16}}{S'_{11}}y^4+\frac{B_1}{6}y^3+\frac{B_2}{2}y^2+B_3y+B_4 \tag{A14}$$

Because B_4 is not used in next equation, it can be omitted.
If (c) holds, then

$$f_2(y)=\left[\frac{1}{30}A_1\left(\frac{S'_{16}}{S'_{11}}\right)^2-\frac{A_1}{120}\frac{(2S'_{12}+S'_{66})}{S'_{11}}\right]y^5+\left[\frac{1}{12}B_1\frac{S'_{16}}{S'_{11}}-\frac{A_2}{24}\frac{(2S'_{12}+S'_{66})}{S'_{11}}\right]y^4+\frac{C_1}{6}y^3+\frac{C_2}{2}y^2+C_3y+C_4 \tag{A15}$$

Because $(C_3\,y+C_4)$ is not used in next equation, it can be omitted.

Finally, through combining the Equations (A13)–(A15), it obtains stress function as below,

$$\begin{aligned}\varnothing&=\tfrac{1}{2}x^2f(y)+xf_1(y)+f_2(y)=\tfrac{1}{2}x^2\left(\tfrac{A_1y^3}{6}+\tfrac{A_2y^2}{2}+A_3y+A_4\right)+\\&x\left(\tfrac{A_1}{12}\tfrac{S'_{16}}{S'_{11}}y^4+\tfrac{B_1}{6}y^3+\tfrac{B_2}{2}y^2+B_3y\right)+\left[\tfrac{1}{30}A_1\left(\tfrac{S'_{16}}{S'_{11}}\right)^2-\tfrac{A_1}{120}\tfrac{(2S'_{12}+S'_{66})}{S'_{11}}\right]y^5+\\&\left[\tfrac{1}{12}B_1\tfrac{S'_{16}}{S'_{11}}-\tfrac{A_2}{24}\tfrac{(2S'_{12}+S'_{66})}{S'_{11}}\right]y^4+\tfrac{C_1}{6}y^3+\tfrac{C_2}{2}y^2\end{aligned} \tag{A16}$$

Through the expression of $\varnothing$, here it obtains its stress component $\sigma_x,\ \sigma_y,\ \tau_{xy}$

$$\begin{aligned}\sigma_x&=\tfrac{\partial^2\varnothing}{\partial y^2}=\left[\tfrac{2}{3}A_1\left(\tfrac{S'_{16}}{S'_{11}}\right)^2-\tfrac{A_1}{6}\tfrac{(2S'_{12}+S'_{66})}{S'_{11}}\right]y^3+\left[A_1\tfrac{S'_{16}}{S'_{11}}x+B_1\tfrac{S'_{16}}{S'_{11}}-\tfrac{A_2}{2}\tfrac{(2S'_{12}+S'_{66})}{S'_{11}}\right]y^2+\\&\left[\tfrac{A_1}{2}x^2+B_1x+C_1\right]y+\left[\tfrac{A_2}{2}x^2+B_2x+C_2\right]\\\sigma_y&=f(y)=\tfrac{A_1y^3}{6}+\tfrac{A_2y^2}{2}+A_3y+A_4\\\tau_{xy}&=-\left(\tfrac{\partial^2\varnothing}{\partial x\partial y}\right)=-\left[x\left(\tfrac{A_1}{2}y^2+A_2y+A_3\right)+\tfrac{A_1}{3}\tfrac{S'_{16}}{S'_{11}}y^3+\tfrac{B_1}{2}y^2+B_2y+B_3\right]\end{aligned} \tag{A17}$$

1. When the upper and lower boundaries $y=\pm h/2$, $\tau_{xy}=0$ (this holds for any x), through the above equation, the relational expressions of A_2, A_3, B_1, B_2, B_3, and A_1 can be derived.

$$\begin{cases}x\left(\tfrac{A_1}{2}\tfrac{h^2}{4}+A_2\tfrac{h}{2}+A_3\right)+\tfrac{A_1}{3}\tfrac{S'_{16}}{S'_{11}}\tfrac{h^3}{8}+\tfrac{B_1}{2}\tfrac{h^2}{4}+B_2\tfrac{h}{2}+B_3=0\\x\left(\tfrac{A_1}{2}\tfrac{h^2}{4}-A_2\tfrac{h}{2}+A_3\right)-\tfrac{A_1}{3}\tfrac{S'_{16}}{S'_{11}}\tfrac{h^3}{8}+\tfrac{B_1}{2}\tfrac{h^2}{4}-B_2\tfrac{h}{2}+B_3=0\end{cases} \tag{A18}$$

Because $\tau_{xy} = 0$ holds for any x, the first items of the Equation (A18) can produce:

$$\begin{cases} \frac{A_1}{2}\frac{h^2}{4} + A_2\frac{h}{2} + A_3 = 0 \\ \frac{A_1}{2}\frac{h^2}{4} - A_2\frac{h}{2} + A_3 = 0 \end{cases} \tag{A19}$$

Solving the Equation (A19), $A_2 = 0$ $A_3 = -\frac{h^2}{8}A_1$ can be obtained. The other items of the Equation (A18) can produce:

$$\begin{cases} \frac{A_1}{3}\frac{S'_{16}}{S'_{11}}\frac{h^3}{8} + \frac{B_1}{2}\frac{h^2}{4} + B_2\frac{h}{2} + B_3 = 0 \\ -\frac{A_1}{3}\frac{S'_{16}}{S'_{11}}\frac{h^3}{8} + \frac{B_1}{2}\frac{h^2}{4} - B_2\frac{h}{2} + B_3 = 0 \end{cases} \tag{A20}$$

Solving the Equation (A20), B_3 B_2 can be obtained

$$\begin{cases} B_3 = -\frac{h^2}{8}B_1 \\ B_2 = -\frac{h^2}{12}\frac{S'_{16}}{S'_{11}}A_1 = \frac{S'_{16}}{S'_{11}}\frac{q}{h} \end{cases} \tag{A21}$$

When the upper and lower boundaries the $y = \pm h/2$, the equations of $\begin{cases} y = +\frac{h}{2}, \sigma_y = q \\ y = -\frac{h}{2}, \sigma_y = 0 \end{cases}$, $\sigma_y = f(y) = \frac{A_1 y^3}{6} + \frac{A_2 y^2}{2} + A_3 y + A_4$ hold. Then the relational expression of A_2, A_3, A_4 and A_1 can be derived. The derivation process is shown as below

$$\begin{cases} \frac{A_1}{6}\frac{h^3}{8} + \frac{A_2}{2}\frac{h^2}{4} + A_3\frac{h}{2} + A_4 = q \\ -\frac{A_1}{6}\frac{h^3}{8} + \frac{A_2}{2}\frac{h^2}{4} - A_3\frac{h}{2} + A_4 = 0 \end{cases} \tag{A22}$$

Solving the Equation (A22), A_3 A_4 can be obtained

$$\begin{cases} A_3 = \frac{q}{h} - \frac{h^2}{24}A_1 \\ A_4 = \frac{q}{2} - \frac{h^2}{8}A_2 = \frac{q}{2} \end{cases} \tag{A23}$$

From the Equations (A19) and (A23), A_1 A_3 can be obtained

$$\begin{cases} A_1 = -12\frac{q}{h^3} \\ A_3 = \frac{3q}{2h} \end{cases} \tag{A24}$$

2. With the boundary of $x = l$, the shear force $Q = -ql$,

$$\int_{-\frac{h}{2}}^{\frac{h}{2}} \left\{ -\left[x\left(\frac{A_1}{2}y^2 + A_2 y + A_3 \right) + \frac{A_1}{3}\frac{S'_{16}}{S'_{11}}y^3 + \frac{B_1}{2}y^2 + B_2 y + B_3 \right] \right\} dy = -ql \tag{A25}$$

Then the relational expression of A_2, A_3, B_1, B_2, B_3 and A_1 can be derived. The derivation process is shown, as below

$$\left(-\frac{A_1}{2}l\frac{1}{3}y^3 - A_2 l\frac{1}{2}y^2 - A_3 ly - \frac{A_1}{3}\frac{S'_{16}}{S'_{11}}\frac{1}{4}y^4 - \frac{B_1}{2}\frac{1}{3}y^3 - B_2\frac{1}{2}y^2 - B_3 y \right)\Bigg|_{-\frac{h}{2}}^{\frac{h}{2}} = -ql \tag{A26}$$

Integration of the above formula, it produces

$$lA_1 + B_1 = \frac{24ql - 24hlA_3 - 24hB_3}{h^3} \tag{A27}$$

Because $A_3 = \frac{3q}{2h}$ as shown in Equation (A24) and $B_3 = -\frac{h^2}{8}B_1$ in Equation (A21), B_1 *and* B_3 can be obtained here as below.

$$\begin{cases} B_1 = 0 \\ B_3 = -\frac{h^2}{8}B_1 = 0 \end{cases} \tag{A28}$$

3. Since the σ_x at the x-axis direction is zero; the moment is also zero. Hence, it gives

$$\begin{cases} \int_{-\frac{h}{2}}^{\frac{h}{2}} (\sigma_x)_{x=l} dy = 0 & \text{(A29a)} \\ \int_{-\frac{h}{2}}^{\frac{h}{2}} (\sigma_x)_{x=l} y dy = 0 & \text{(A29b)} \end{cases}$$

Through this Equation (A29), the relational expression of $A_2, B_1, B_2, B_3, C_1, C_2,$ and A_1 can be derived. From (a) of (A29), it gets $C_2 = 0$; and from (b) of (A29), it gets $C_1 = \frac{6}{5}\frac{q}{h}\left(\frac{S'_{16}}{S'_{11}}\right)^2 + 6\frac{ql^2}{h^3} - \frac{3}{10}\frac{q}{h}\frac{(2S'_{12}+S'_{66})}{S'_{11}}$ Finally, here the coefficients, $A_1, A_2, A_3, A_4, B_1, B_2, B_3, C_1, C_2$ are all been obtained, the following equation holds:

$$\varnothing = \frac{1}{2}x^2\left(-\frac{2q}{h^3}y^3 + \frac{3q}{2h}y + \frac{q}{2}\right) + x\left(-\frac{q}{h^3}\frac{S'_{16}}{S'_{11}}y^4 + \frac{S'_{16}q}{2S'_{11}h}y^2\right) - \frac{2q}{5h^3}\left[\left(\frac{S'_{16}}{S'_{11}}\right)^2 - \frac{(2S'_{12}+S'_{66})}{4S'_{11}}\right]y^5 + \frac{1}{6}\left[\frac{6q}{5h}\left(\frac{S'_{16}}{S'_{11}}\right)^2 + \frac{6ql^2}{h^3} - \frac{3q}{10h}\frac{2S'_{12}+S'_{66}}{S'_{11}}\right]y^3 \tag{A30}$$

Finally, $\sigma_x, \sigma_y, \tau_{xy}$ can be obtained through Equation (A30) as below.

$$\begin{aligned} \sigma_x &= \frac{-12q}{h^3}\left\{\left[\frac{2}{3}\left(\frac{S'_{16}}{S'_{11}}\right)^2 - \frac{2S'_{12}+S'_{66}}{6S'_{11}}\right]y^3 + \frac{S'_{16}}{S'_{11}}xy^2 + \frac{1}{2}x^2y\right\} + \\ &\quad \left[\frac{6q}{5h}\left(\frac{S'_{16}}{S'_{11}}\right)^2 + \frac{6ql^2}{h^3} - \frac{3q}{10h}\frac{2S'_{12}+S'_{66}}{S'_{11}}\right]y + \frac{S'_{16}}{S'_{11}}\frac{q}{h}x \\ \sigma_y &= -\frac{2q}{h^3}y^3 + \frac{3q}{2h}y + \frac{q}{2} \\ \tau_{xy} &= -\left[x\left(-\frac{6q}{h^3}y^2 + \frac{3q}{2h}\right) - \frac{4q}{h^3}\frac{S'_{16}}{S'_{11}}y^3 + \frac{S'_{16}}{S'_{11}}\frac{q}{h}y\right] \end{aligned} \tag{A31}$$

References

1. Riccio, A.; Palumbo, C.; Acanfora, V.; Sellitto, A.; Russo, A. Influence of Failure Criteria and Intralaminar Damage Progression Numerical Models on the Prediction of the Mechanical Behavior of Composite Laminates. *J. Compos. Sci.* **2021**, *5*, 310. [CrossRef]
2. Hu, R.-X.; Zhou, J.Y. Optimal Design of Automobile Battery Shell Made of Carbon Fiber Reinforced Composites. *Mach. Des. Manuf.* **2017**, *12*, 229–233.
3. Wallentowitz, H. Innovation in Automotive Engineering:A look into the future. *J. Automot. Saf. Energy* **2013**, *4*, 95–108.
4. Feng, Z.; Yi, G.; Li, P.; Li, H.; Yu, D. Review of parallel genetic algorithm. *Comput. Appl. Softw.* **2018**, *35*, 1–7+80.
5. Wang, H.; Duan, Y.; Dilimulati, A.; Zhang, X. Design Optimization of CFRP Stacking Sequence Using a Multi-Island Genetic Algorithms under Low-velocity Impact Loads. *J. Wuhan Univ. Technol.* **2017**, *32*, 720–725. [CrossRef]
6. Chen, Y.T.; Xiang, S.; Zhao, W.P. Fiber Orientation Angle Optimization for Minimum Stress of Laminated Composite Plates. *Appl. Mech. Mater.* **2015**, *709*, 135–138. [CrossRef]
7. Xiu, Y.S.; Cui, D.G. Ply optimization design for stability of composite laminates. *Eng. Mech.* **2005**, *6*, 212–216.
8. Feng, X.B.; Huang, H.; Wang, W.; Zhang, H.D. Strength optimization of laminate based on the genetic algorithm. *Fiber Reinf. Plast. Compos.* **2012**, *3*, 7–13.
9. Jin, D.; Liu, Z.; Fan, Z. Ply optimization of composite laminate with ply drop based on genetic algorithm. *Acta Mater. Compos. Sin.* **2015**, *32*, 236–242.
10. Park, J.H.; Hwang, J.H.; Lee, C.S.; Hwang, W. Stacking sequence design of composite laminates for maximum strength using genetic algorithms. *Compos. Struct.* **2001**, *52*, 217–231. [CrossRef]
11. Wang, P.Y.; Chen, Z.; Geng, X.L.; Yue, Z.F. Stacking Sequence Optimization of Composite Laminates Based on a Modified Adaptive Genetic Algorithm. *Sci. Technol. Eng.* **2018**, *18*, 336–340.
12. Yang, Y.; Dai, W.; Nian, C. Optimum design of composite laminates based on adaptive genetic algorithms. *Mach. Build. Autom.* **2020**, *49*, 22–25.

13. Hwang, S.F.; Hsu, Y.C.; Chen, Y. A genetic algorithm for the optimization of fiber angles in composite laminates. *J. Mech. Sci. Technol.* **2014**, *28*, 3163–3169. [CrossRef]
14. Krishan, K.C. *Composite Materials*; Science and Engineering: London, UK, 2012.
15. Tsai, S.W.; Hahn, T.H. Introduction to Composite Materials. In *Fibrous and Textile Materials for Composite Applications*; Springer: Berlin/Heidelberg, Germany, 2016.
16. Adelman, M. Evaluation of the Tensor Polynomial and Hoffman Strength Theories for Composite Materials. *J. Compos. Mater.* **1977**, *11*, 366–377.
17. Wang, Z. *Composite Material Mechanics and Composite Material Structure Mechanics*; Machinery Industry Press: Xi'an, China, 1991.
18. Wang, Y. *Mechanics and Structural Design of Composite Materials*; East China University of Science and Technology Press: Shanghai, China, 2012.
19. Dassault. *Isight 5.9 User's Guide*; Dassault Systèmes Simulia Corp.: Johnston, RI, USA, 2014.
20. Zhang, W. *Mathematical Fundamentals of Genetic Algorithms*; Xi'an Jiaotong University Press: Xi'an, China, 2000.
21. Yi, X.; Du, S. *Handbook for Composite Materials*; Chemical Industry Press: Beijing, China, 2009.
22. Lei, Y.; Zhang, S. *MATLAB Genetic Algorithm Toolbox and Application*; Xidian University Press: Xi'an, China, 2014.
23. Lu, Z.; Zhang, E.Y.; Liu, B. Ply optimization of composite laminate based on improved adaptive genetic algorithm. *Fiber Reinf. Plast. Compos.* **2016**, *2*, 53–56.
24. Chen, L.; Zhao, X. Mechanical analysis of simply supported beam based on classical laminated plate theory. *Mech. Eng.* **2019**, *41*, 152–156.
25. Fan, Z. *Prediction and Analysis of Mechanical Properties of Carbon Fibre Reinforced Composite Laminates*; Dalian University of Technology: Dalian, China, 2018.

Journal of
Composites Science

MDPI

Article

A Cost Model for 3D Woven Preforms

James Clarke [1,*], Alistair McIlhagger [1], Dorian Dixon [1], Edward Archer [1], Glenda Stewart [2], Roy Brelsford [2] and John Summerscales [3]

[1] Engineering Composites Research Centre, Ulster University, Jordanstown BT37 OQB, UK; a.mcilhagger@ulster.ac.uk (A.M.); d.dixon@ulster.ac.uk (D.D.); e.archer@ulster.ac.uk (E.A.)
[2] Axis Composites Ltd., Airport Road, Belfast BT3 9DZ, UK; g.stewart@axiscomposites.co.uk (G.S.); r.brelsford@axiscomposites.co.uk (R.B.)
[3] Materials and Structures (MAST)/Composites Engineering Research Group, School of Engineering, Computing and Mathematics (SECaM), University of Plymouth, Plymouth PL4 8AA, UK; j.summerscales@plymouth.ac.uk
* Correspondence: clarkejames20@yahoo.ie

Abstract: Lack of cost information is a barrier to acceptance of 3D woven preforms as reinforcements for composite materials, compared with 2D preforms. A parametric, resource-based technical cost model (TCM) was developed for 3D woven preforms based on a novel relationship equating manufacturing time and 3D preform complexity. Manufacturing time, and therefore cost, was found to scale with complexity for seventeen bespoke manufactured 3D preforms. Two sub-models were derived for a Weavebird loom and a Jacquard loom. For each loom, there was a strong correlation between preform complexity and manufacturing time. For a large, highly complex preform, the Jacquard loom is more efficient, so preform cost will be much lower than for the Weavebird. Provided production is continuous, learning, either by human agency or an autonomous loom control algorithm, can reduce preform cost for one or both looms to a commercially acceptable level. The TCM cost model framework could incorporate appropriate learning curves with digital twin/multi-variate analysis so that cost per preform of bespoke 3D woven fabrics for customised products with low production rates may be predicted with greater accuracy. A more accurate model could highlight resources such as tooling, labour and material for targeted cost reduction.

Keywords: 3D woven fabrics; preform; complexity; cost model; learning; Weavebird; Jacquard

Citation: Clarke, J.; McIlhagger, A.; Dixon, D.; Archer, E.; Stewart, G.; Brelsford, R.; Summerscales, J. A Cost Model for 3D Woven Preforms. *J. Compos. Sci.* **2022**, *6*, 18. https://doi.org/10.3390/jcs6010018

Academic Editor: Stelios K. Georgantzinos

Received: 6 December 2021
Accepted: 31 December 2021
Published: 5 January 2022

1. Introduction

3D woven composites have promising growth prospects in a wide range of markets [1,2]. They possess superior mechanical properties in some respects compared with conventional 2D preforms, for example a composite made from a non-crimp 0, 90 2D reinforcing fabric of interlacing orthogonal sets of warp and weft tows, with the warp tows at 0 degrees running along the length of the weaving loom and the weft tows at 90 degrees to the warp tows [3]. However, acceptance of 3D woven composites has been difficult in sectors such as aerospace which increasingly demand lower-cost materials with mechanical performance at least the same or greater than for 2D laminates. Table 1 compares a 0, 90 2D non-crimp fabric composite and a 3D woven composite.

At high production rates, for most manufacturing operations, material cost dominates other resources such as tooling, capital and labour, while tooling and labour costs dominate for bespoke manufacturing [4,5]. Dry 3D preforms are highly complex materials. There are two types of 3D woven preform: multi-axial and interlock. Interlock preforms are multi-layered fabrics produced by interlacing three sets of fibre tows in a specialised weaving machine. A general definition of a 3D warp interlock fabric was proposed to better describe the position of the various yarns located inside the 3D woven structure [6].

Alternate layers of warp and weft are placed in cross-layers at 0 degrees and 90 degrees, respectively, in the plane of multi-layered weave. The warp and weft layers are interlocked/interlaced by a third set of tows called binder tows. The binder tows are also called warp weavers because interlocking is generally achieved through warp tows [5,7]. Interlock preforms can be categorised as follows:

a. Angle-interlock orthogonal composites: through-the-thickness interlock weave.
b. Layer-to-layer interlock or multi-layer weave found in both orthogonal interlock and angle interlock weaves.

Table 1. Comparison of 2D and 3D woven composites.

	2D Woven Composite	3D Woven Composite
Fabric Manufacture	Conventional loom for weaving a fabric with interlacing tows in X and Ydirections	Specialist loom for weaving a fabric with interlacing tows in X, Y and Z directions.
Fabric Structure	Warp tows run along the length of the weaving loom or X direction andweft tows run in the cross direction of the loom, or Y direction.	Warp, weft and binder tows run in X, Y and Z directions.
Properties	Higher in-plane-specific stiffness and strength. Lower delamination resistance. Lower out-of-plane stiffness andstrength.	Lower in-plane-specific stiffness and strength. Higher delamination resistance due to z-binder. Higher out-of-plane stiffness andstrength.

Examples of angle interlock layer-to-layer and orthogonal interlock architectures are shown in Figures 1 and 2.

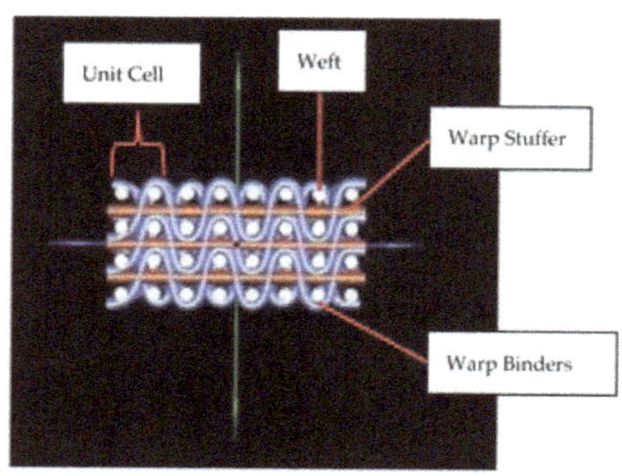

Figure 1. Layer-to-layer weave architecture.

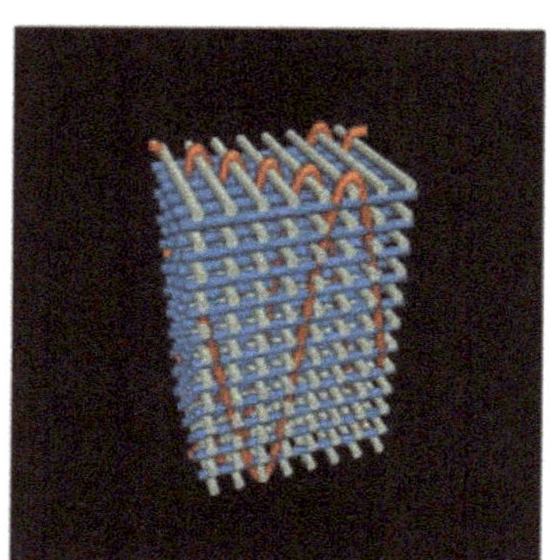

Figure 2. Orthogonal weave architecture.

Due to the inherent complexity of the preform manufacturing process, 3D woven textiles can be expensive [8]. Therefore, knowledge of fabric manufacturing cost is essential for devising strategies to reduce cost and enable successful competition with long-established 2D materials. However, while 3D preform commercial cost models are alluded to [9,10], they are not normally available in the public domain.

The aim of this study was to develop a model to estimate the cost of a hypothetical 3D preform. Data werederived from 17 bespoke 3D preforms manufactured on Weavebird and Jacquard weaving looms. A resource-based modelling approach [11] was developed that took account of the bespoke production of each preform. The model was based on the principles that cost is determined by resources such as material, tooling, labour and general overheads [4], and that manufacturing time, and therefore cost, increases with part complexity [12–14]. Data for resource inputs such as materials, equipment, labour and energy are approximate values for commercial sensitivity reasons. Technical cost modelling (TCM) was added as a further refinement in the form of a sub-model detailing how weaving equipment, labour and energy costs scale with part features such as part shape and complexity.

Manufacturing cost is not simply the addition of cost elements. How they interact with each other is a function of learning, which can be either by human agency (for example, the combination of textile designer and production operative) or an autonomous machine control algorithm The study included a description of how preform manufacturing cost can be reduced once a certain level of learning is attained. For future work, research into digital twin and/or multi-variate analysis for enhanced learning is proposed as a strategy for reducing the manufacture cost of bespoke preforms made singly or in small batches.

Two cost models were developed, one for a Weavebird loom (centre closed dobby shedding mechanism) and another for a Jacquard loom (mechanised production of patterned textiles). In both cases, there was a high correlation, measured by correlation coefficient r^2, between manufacturing time and preform complexity for preforms woven on each loom. Constants derived from time–preform complexity curves for both looms were input to the model to estimate and compare the cost of a large bespoke 3D preform to be woven on both looms. Weave tooling and labour accounted for approximately 80% of preform cost for the bespoke preform. The Jacquard loom is more automated and hence much more cost effective for large preforms compared to the Weavebird. Learning through experience will significantly reduce manufacturing weave time and cost per preform.

1.1. Literature Review

There are a wide variety of cost models for composite parts by market sector. Huber [11] proposed three categories for cost modelling of aerospace composites: analogous, parametric, and bottom-up cost estimation. Two possible cost estimation scenarios exist:

- Some historic cost data/experience exists for a top-down cost estimation.
- Design and process knowledge for a bottom-up, detailed cost calculation.

Essentially, all models are either one of these scenarios or a combination of them. This generalisation applies equally to cost models in other sectors such as automotive, marine and construction. The proprietary nature of fundamental data and equations leave most developed models unusable for third parties.

1.1.1. Manufacturing Cost Models

Esawi [4] provided a comprehensive summary of manufacturing cost model approaches. The required output of a model will depend on the context. In competitive bidding, the model must deliver a precise, absolute cost as an error of a few percent can make the difference between profit and loss. When predicting the approximate part cost where historical data arenot readily available, for example in the early stages of design, a cost accurate to within a factor of two is acceptable.

Function-costing or parametric methods extrapolate the cost of a part that is a variant of an existing family for which historic cost data already exist. In this case, two conditions must be met. The part must be a member of a closely related family. Secondly, the family must have many members with established historical cost data. Similar empirical or cost scaling methods can be used for part costing which are based on correlations using historical data for estimating the manufacturing cost of a part with given features. The cost of a new part having certain features made by a given process can be estimated by analysing cost correlations between previous parts with these features against their size, shape and complexity and then locating the new part in this cost field. Activity-based costing methods calculate and sum the cost of each unit operation involved in the manufacture of a part. However, a large amount of pre-existing input data arerequired.

Resource-based modellingaccounts for materials, energy, equipment and infrastructure capital, time, and information resources required for part manufacture. The method is approximate as values for these inputs are often unknown. TCM is a further refinement of resource modelling and includes sub-models for how equipment, tooling cost, and production rate scale with part features such as part mass, size and complexity. Costs can be approximate and are isolated, giving TCM flexibility, scalability and adaptability. As more data become available, detail can be added to the model to improve predictive power [4].

For TCM calculations based on established data and discussions with experts, Esawi [4] assumed a complexity factor varying from 1 (minimum complexity) to 5 (maximum complexity), with a value of 2 assumed for an average complexity factor in calculations of tooling cost, capital cost and production rate for injection moulding, extrusion and casting operations. For these operations, tooling cost, capital cost and production rate scale with part mass and complexity. Tooling and capital vary non-linearly, with exponents for mass and complexity varying between 0 and 1, implying an economy of scale with increasing part mass and complexity. Production rate decreases with increase in mass and complexity, so values of exponents for mass and complexity are negative for injection moulding, extrusion, and casting operations.

Hagnell and Akermo [15] describe a TCM for a generic aeronautical wing in which costs scale with part features for a given production method. An integrated top-down and bottom-up approach was employed, depending on available cost data. For a generic aeronautical wing, hand layup is normallythe most cost-effective method of those studied for annual volumes of less than 150 structures per year. For higher production volumes, automatic tape layup (ATL) followed by hot drape forming (HDF) are the most cost-effective choices. For all production methods, cost per part fell as production rates increased until material cost dominated at a minimum production rate.

Gutowski et al. [16] derived a series of cost equations incorporating variables and constants to estimate composite part manufacturing costs for an aircraft structure. The estimated results fit well with the Advanced Composite Cost Estimating Manual (ACCEM) model [11]. However, the system does not account for quality inspection processes. Verrey et al. [17] studied two resin transfer moulding (RTM) processes for automotive body-in-white (BIW) structures. An epoxy system was compared with a novel reactive polyamide 12 (PA12) via characterisation of reaction kinetics and the production of carbon thermoplastic (TP) fibre floor pan quadrant demonstrators incorporating typical geometrical features. Parametric TCM tools were used to compare the two RTM variants for full floor-pan production at volumes of 12,500–60,000 parts per year. TCM offered flexibility together with easy manipulation of processing and economic factors for sensitivity studies. A 22% increase in cost occurred for the standard TP-RTM cycle versus the epoxy system. In-mould cycle time was dominated by thermal cycling of the tool which was required to reduce component temperature below Tg before demoulding the thermoplastic part. A study of alternative strategies showed that a reduction in non-crimp fabric scrap gavemajor cost savings. Cost per part reduced with increase in production volume, with carbon non-crimp fibre (NCF) material cost accounting for 66% of part cost at a minimum production volume of 60,000 parts year. At this volume, carbon fibre becomes economic at a maximum price of €10/kg compared with glass fibre and steel.

Schubel [18] employed TCM to compare the cost of making a 40 m wind turbine blade by handlayup, prepreg, vacuum infusion and resin transfer moulding with automated manufacturing techniques such as automated tape laying (ATL), automated fibre placement

(AFP) and overlay braiding. ATL and AFP reduced manufacturing costs by up to 8% despite the high capital costs of automated equipment. Part size, production volume, material cost and tooling cost were accounted for. Cost centres were isolated and clearly indicated the dominance of materials and labour. For the manufacture of a large wind turbine blade, material deposition in the tool is only one of a string of labour-intensive processes. A holistic automated blade manufacturing approach is required to see true labour saving benefits.

3D woven fabrics are promising materials for growing market sectors, e.g., wind turbine blades for renewable energy generation [10]. However, high cost is still a major obstacle for uptake of high-performance materials such as carbon fibre. Ennis et al. [19] assessed the commercial viability of cost-competitive carbon fibre composites specifically suited for the unique loading conditions experienced by wind turbine blades. The wind industry is costdriven while carbon fibre materials have been developed for the performance-driven aerospace industry. Carbonfibre has known benefits for reducing wind turbine blade mass due to significantly improved stiffness, strength and fatigue resistance per unit mass compared to fibreglass. Novel carbon fibre reinforcements derived from the textile industry, and characterised using a validated material cost model and mechanical testing, were studied as potentially more optimal materials for wind turbine blades.

A novel heavy tow textile carbon fibre was compared [19] with commercial carbon fibre and fibreglass materials in representative land-based and offshore reference wind turbine blade models. Some advantages of carbon fibre spar caps are observed in reduced blade mass and improved fatigue life. The heavy tow textile carbon fibre has improved cost performance over the baseline carbon fibre and performed similarly to commercial carbon fibre in wind turbine blade design at a significantly reduced cost. The novel carbon fibre was observed to outperform fibreglass when comparing material cost estimates for spar caps optimised to satisfy design constraints. The study outlined a route for broader carbon fibre usage by the wind industry to enable larger rotors that capture more energy at a lower cost. Heavytow textile carbon fibre cost is estimated at €9.46/kg for an annual production volume of 2400 tonnes per year, reducing by 43% to €6.88/kg for an increased annual production volume of 6000 tonnes per year.

Fibre-reinforced composites play a key role in automotive applications because of their high strength to weight and stiffness to weight ratios compared with metals [20]. An integrated assessment of the durability, reliability and affordability of these materials is critical for facilitating their inclusion in new designs. A method to develop this assessment is described for fabricating sheet moulding compound (SMC) parts, together with the concept of Preform Insert Assembly for improved affordability in composite part manufacture.

A computer-aided material selection tool was developed for selecting the most suitable carbon fibre-reinforced composite configuration for aircraft structures [21]. The procedure is based on technical, economic and environmental performance objectives for a given design, in a multi-disciplinary and multi-objective optimisation scenario.

Carbon-fibre-based composite manufacturing processes have been considered for automotive body panel applications [22]. A full-scale front wing–fender component was produced using two composite manufacturing processes, a semi-impregnated (semi-preg) system and a novel directed fibre preforming–resin transfer moulding process. Both processes were compared with an existing stamped steel component for mechanical properties, weight saving and cost, using a TCM procedure. Mechanical testing demonstrated that the carbon fibre composite solutions provided 40–50% weight saving for an equivalent bending stiffness compared to steel panels and greatly improved dent resistance. For the part studied, carbon fibre semi-preg systems offered the lowest-cost process at approximately 500 parts/annum and directed fibre preforming technologies were cheaper, between 500 and 9000 parts/annum. The steel component was seen to be more cost effective at volumes above 9000 parts/annum.

A study was conducted to estimate the manufacture cost of a simple component in a number of different composite materials and by different manufacturing routes [23]. The materials and routes selected span the range of composites appropriate from general engineering to aerospace applications. A simple methodology is introduced for a comparison on the basis of cost-performance efficiency. It is demonstrated that more economic solutions may often be realised by the choice of 'expensive' carbon rather than 'cheaper' E-glass as the reinforcing fibre.

The majority of 3D woven preforms currently commercially available are formed by a 2D weaving process to build a preform with fibres oriented in three dimensions. Multiple insertion 3D differs from traditional weaving and involves 3D fabric formation with each process cycle, i.e., multi-layers at one time. The successful development and application of 3D woven composites will depend on an accurate understanding of the cost drivers in the manufacturing process. The costs associated with textile preforming are not as straightforward. A cost model was developed for multiple insertion 3D weaving [24] focusing on the effects of fabric design, fabric size (thickness and width) and fibresize (linear density) on setup cost, running production cost and conversion cost.

Despite the limited number of commercial 3D preform weave technologies, the design window for this class of materials is very broad. Even for one 3D weaving technology, and restricting fibre inputs to selected standard carbon and glass tows, design flexibility is still almost limitless. Process modelling, cost modelling, and performance modelling must all be applied to the design in terms of material, preform and performance in the final application so that development cycle times can be reduced. A concurrent engineering approach is described [25] for designing 3D woven fabrics that accounts for manufacturing and performance in addition to cost. A case study was presented to demonstrate that relatively minor design changes can result in very different performance and costs.

The cost-effective manufacture of carbon fibre-reinforced parts in high-wage economies is a major research goal for industry. An initiative is described [26] to develop a software tool for cost prediction in the early design stage to assist optimum process selection and highlight potential cost reductions.

While advanced composites can significantly reduce aircraft structural weight compared to conventional metal structures, the aerospace industry was reluctant to introduce them to new aircraft. The US Air Force Composites Affordability Initiative [27] found that the key to affordability in composites was to reduce assembly costs through the integration and bonding of parts. A partnership between various aerospace companies, the US Air Force Research Laboratory, and the US Office of Naval Research, was created to develop the materials and technologies required to fly large integrated and bonded structures. A multi-disciplinary approach was highlighted: maturation of materials and processes, an understanding of the structural behaviour of bonded joints, and quality assurance and non-destructive testing to ensure joints remain bonded throughout an aircraft's service life. The result was that technologies for large integrated and bonded composite structures were successfully developed across the fixed and rotary wing industrial base.

A design framework for cost analysis of a wind turbine blade made of variable stiffness composite laminates was outlined [28], consisting of design optimisation, time-variant reliability analysis, structural performance analysis, and life-cycle cost evaluation phases. Design optimisation will maximise stiffness via the material properties of the fibre-reinforced composites and correct orientation of the composite plies. Different volume constraints of carbon fibre-reinforced polymer (CFRP) are imposed on composite laminates in the load-carrying component. Structural performance and service lifetime of the blade designs were estimated based on a time-variant reliability assessment, which was evaluated using an out-crossing asymptotic method. Wind speed and material properties are considered as the random parameters during the reliability assessment. Maintenance cost of the various designs was determined by combining the estimated structural performance with an analytical method. The final designs are selected according to their cost-effectiveness using different discount rates and undiscounted costs.

1.1.2. Complexity, Organisational Learning

Organisational Learning is defined as a conscious attempt by organisations to improve productivity, effectiveness and innovation in complex economic and technological market conditions. Learning enables quicker and more effective responses to a complex and dynamic environment. Increasing complexity requires greater learning [29–32]. 3D woven preform manufacture is a highly complex process with numerous stepscarried out in a required sequence for successful manufacture [5]. If there is a delay in completing a given step, the time required to complete the overall preform will increase thereby increasing preform cost.

Wright [33] observed that as aircraft production increased, the cost in terms of direct labour hours fell. For a new component which has not been manufactured before, the learning required and therefore the cost to make the part will initially be high. As more units are made, there is a steep drop in direct labour hours per part until the rate of decrease in direct labour hours per part becomes smaller.

Klenow [34] and Baloff [35] reviewed various studies investigating learningby doing for a single defined production process across a variety of industrial sectors that showedestimates for a learning rate of approximately 20%, which is the rate at which productivity rises with a doubling of cumulative output. Lee [36] summarised learning rates from the literature by manufacturing sector and activity. Even in one overall activity, in this case manufacturing, learning rates will vary considerably by individual sector. Yelle [37] and ArgotteandEpple [38], observed that productivity rose across a variety of industries through a process of learningby doing.

A key assumption with learning in a manufacturing context is that production be continuous so that learning is reinforced and cost decreases. However, production may be discontinuous, leading to unlearning or forgetting [39]. Another assumption is the use of Wright's learning curve model [33] to estimate the cumulative number of preforms produced, based on the estimated time to make one preform and an assumed learning rate. The model yields production times equal to zero [40] after a high number of repetitions, which is impossible. Furthermore, it does not account for workers' prior experience [41], nor the influence of machinery in the learning process [42].

1.1.3. Jacquard and Dobby Looms

The cost model is based on the cost structure for 3D preform manufacture, which is split between loom setup and weaving (Section 3.3). In a Jacquard loom, Figure 3, harness cords extend down from a control head. Each harness cord is connected to one, two or sometimes four warp yarns which can be moved individually, allowing for weaving of much more intricate, complex and longer length 3D fabrics [43]. In the setup phase, fibre is wound onto bobbins. PTFE tubes glued to an eyeboard will prevent movement of tubing through the eyes and provides fibre tension. The bobbins with wound fibre are then mounted onto creels followed by fibre being thread through the tubing. Fabric is woven in a similar fashion to that described for a Weavebird except that each fibre is individually controlled by the Jacquard head.

The Weavebird (www.weavebird.com accessed on 20 September 2019), Figure 4, is a dobby loom. In setup, warp yarns are taken from a beam mounted on the back of the Weavebird loom and fed though the eyeboard. The eyeboard controls the warp ends as they enter eyelets on heddles sitting on loom shafts. The eyeboard houses PTFE tubing which protects the fibre and provides tension during the weaving process. The heddles are in a sequence determined by the required architecture. The heddles sit inside shafts or frames, which can lift the warp threads up or down, one warp thread for each heddle. During weaving, each time a group of heddles is lifted, a "shed" is created. The shed is the opening between the lifted and stationary warp threads. The weft is held in a shuttle or rapier, which passes the weft through the shed to the other side. The shed then closes, and a different set of heddles will be lifted, creating a new shed, effectively completing the interlacement of warp and weft.

Figure 3. Jacquard loom.

Figure 4. Weavebirdloom.

The warp ends are then threaded through the reed, a long, comb-like instrument that keeps the warp at the correct width and density and helps pack or beat the weft down into place. Beat up is the motion of weaving that compacts the weft/stuffer yarns with a consistent force ensuring an even density in the fabric. The woven fabric is wound on the take-up beam on the front of the loom until the warp on the back beam runs out.

Gurkan [7] notes that while dobby mechanisms work together with harnesses, there are harness cords for each warp yarn in a Jacquard loom. Therefore, the capability of Jacquard looms to make highly complex patterns is the highest among shedding mechanisms such as dobby, crank or cam.Stewart [43] observed that the main difference between a dobby and a Jacquard loom is how the warp yarns are moved up and down to form gaps or sheds through which the weft yarns are drawn by a shuttle to form the weave pattern. In the case of a dobby loom, the warp yarns can only be controlled in groups moved by harnesses attached to shafts or frames. When a harness goes up or down, all attached warp yarns move with the harness. As the loom can only hold a certain number of harnesses, this means that there is a limit on weave complexity. Dobby looms are best used for making simple geometric patterns and short fabric lengths because of harness limitations.

2. Methodology and Experimental

2.1. Methodology

Data for this study came from 17 bespoke 3D woven preforms manufactured by a Northern Ireland-based company. A resource-based modelling approach was developed that took account of the bespoke production of each preform utilising the principles that cost is determined by resources such as material, capital, tooling, energy and labour

Table 4. Preform 4 setup and weave manufacturing times.

Stage	Loom Setup, Design and Weave	Time Required (h)
1	Winding of bobbins	16
2	Bobbins insertion on creel	8
3	Tubing preparation time, 315 tubes	24
4	Passing 315 carbon tows through PTFE Tubing and loom	24
5	Weave time	3
6	Design on Scotweave	24
	Total Manufacturing Time:	99

3. Results

Each preform is unique in terms of complexity. In this study, the key metric for complexity is the product of the total number of fibre tows or warp stuffers, weft fillers and warp binders, any sub-features such as holes and the sum of preform structural elements $\sum SE_i$, which is a measure of the preform shape. Complexity is expressed by Equation (13).

Each preform complexity and manufacturing time is compared to a baseline preform complexity and manufacturing time, and expressed as the relative feature factor R_i/R_b and relative manufacturing time factor t_i/t_b, respectively. Tables 2 and 3 summarisepreforms woven on the Jacquard and Weavebird looms, respectively.

3.1. Calculation of t_i/t_b and R_i/R_b

The baseline fabric is the simplest in terms of woven architecture compared with the other fabrics and has the shortest manufacturing time t_b. R_b is complexity of Preform 1, (Equation (14)).

As the baseline is a single simple flat profile, $\sum SE_b$ is equal to 1, A_b is 360, the total number of fibre tows. As R_i is the same as R_b for Preform 1, R_i/R_b for Preform 1 is 1. Manufacturing time t_i for Preform 1 is 9 h. As t_i is the same as t_b for Preform 1, t_i/t_b for Preform 1 is 1. Values of t_i/t_b and R_i/R_b were found as follows for Preform 4, which is a fabric woven in the shape of a T-piece. A T-piece is assumed to be treated as 3 flat profiles (Figure 7), therefore the shape is given the number 3 or 3 structural elements. For Preform 4, the total number of fibre tows is 1380. Therefore, complexity R_4 for Preform 4 is: $R_4 = (1380)(3) = 4140$ so R_4/R_b is 4140/360 = 11.5 (Table 3). t_4 for Preform 4 is 99 h, while t_b is 9 h. Therefore, t_4/t_b is 99/9 = 11.

Values of t_i/t_b and R_i/R_b for the remaining profiles were calculated by the model, summarised in Tables 2 and 3 and plotted (Figure 10a–c) to validate Equation (12).

3.2. Data Analysis by Loom Type and Preform Architecture

Nine preforms were made on the Weavebird loom, and eight on the Jacquard loom. Figure 10a has 17 data points, one for each 3D woven preform, and shows a trend of increasing manufacturing time with increasing preform complexity. Each preform varies in complexity and architecture in terms of the number of weft and warp tows, preform shape and whether orthogonal or layerto layer. Figure 10a includes Preform 3 which took 252 h to produce a profile 20 m in length. The complexity value for Preform 3 was 36,901, the product of the total number of fibre tows (36,901) and one structural element as it is a flat profile with no extra features such as T sections. Production times for the remaining preforms ranged from 9 to 130 h. Correlation coefficient r^2 is 0.56.

In Figure 10b, Preform3 has been removed. Correlation coefficient r^2 is 0.51. Correlation between two variables will either be "weak" [45] or "well related" [46], depending on sector context. For example, correlation between two variables may be judged either "weak" in a manufacturing [45] context or "well related" in a public sector context [46]. Two outliers in Figure 10b are due to Preforms 7 and 16. If these are removed, Figure 10c for 14 profiles gives a significantly improved trend of increasing t_i/t_b with R_i/R_B, with r^2 = 0.62 compared with 0.51. Figure 10b,c show a tendency for preforms to separate out by loom type, with Jacquard preforms tending to group above the trendline and Weavebird preforms grouping below. Figure 11a is a plot of nine preforms from the Weavebird loom. Figure 11b is a plot of eight preforms from the Jacquard loom. Table 2 (Jacquard) and Table 3 (Weavebird) include the weave architecture for each preform.

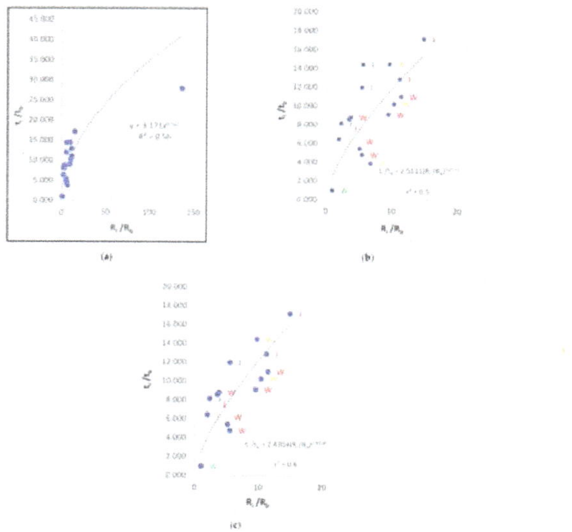

Figure 10. The (**a**) 17 preforms, (**b**) 16 preforms, and (**c**) 14 preforms.

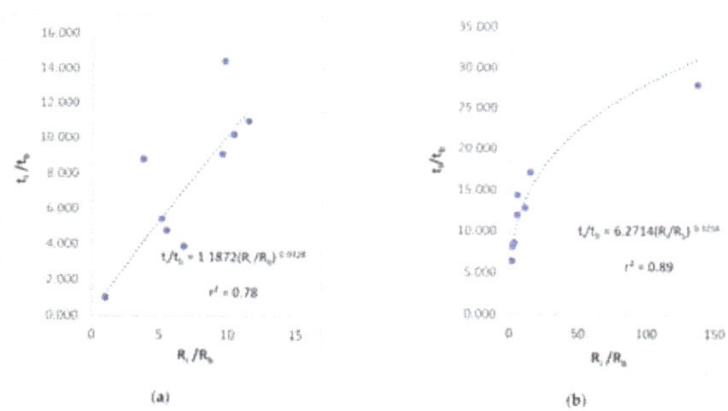

Figure 11. (**a**) The ninepreforms, Weavebirdloom. (**b**) The eightpreforms, Jacquard loom.

Figure 11b indicates a strong correlation between manufacturing time and preform complexity for the Jacquard preforms as shown by correlation coefficient $r^2 = 0.89$, while Figure 11a shows a moderately strong correlation with $r^2 = 0.78$ for nine Weavebird preforms. Additionally, Figure 11b shows a pronounced tendency for manufacturing time to level off with increasing preform complexity, i.e., the rate of change of t_i/t_b decreases with increasing complexity, R_i/R_B. For both Jacquard and Weavebird preforms, Tables 2 and 3 indicate that the relationship between t_i/t_b and R_i/R_B is independent of preform architecture.

3.3. Preform Cost Modelling for a Commercial Quote

The cost of a preform estimated by a local manufacturer was compared with the model-estimated preform cost. A Republic of Ireland-based manufacturer of resin transfer moulded 3D woven composites buys 3D fabrics from a US supplier, and requested a quote from the local manufacturer. Fabric profile data supplied by the ROI 3D woven composite manufacturer is shown in Table 5. The cost structure in 3D preform manufacturing is split between the proportion of costs due to loom setup and weaving, so Table 5 details the total number of warp and weft tows in the setup and weave phases. For example, fabric width is 1270 mm (both setup and weave). The number of warp layers is 3. The number of warp tows is 280/m, so the total number of warp tows in the preform will be 1.27 × 280 × 3 = 1067 tows. The preform length to be woven is 454 m. The number of weft tows is 190/m and the number of weft layers is 4, so the total number of weft tows in the woven preform will be 190 × 454 × 4 = 345,040 tows. The total number of tows, warp and weft, in the woven preform is 1067 + 345,040 = 346,107 tows.

Table 5. Hypothetical 3D woven preform fabric.

<table>
<tr><th colspan="2">Fabric Materials</th></tr>
<tr><td colspan="2">100% E-glass
Fibre content: warp stuffer: 98%, Weft filler, warp binder: 2%
Material cost, E-glass: £1/kg</td></tr>
<tr><td colspan="2">Warp tow: Setup and Weave</td></tr>
<tr><td colspan="2">Warp tows/cm/layer: 2.8
Warp tows/cm/layer/total: 2.8 × 127 = 356
Number of warp layers in preform: 3
Total number of warp tows in preform, set andweave = 356 × 3 = 1068</td></tr>
<tr><td>Warp andWeft: Setup</td><td>Warp andWeft: Weave</td></tr>
<tr><td>Setup length (cm): 2000
Width (cm): 127
Weft tows/cm/layer: 1.9
Number of preform weft layers: 4
Weft tows/cm/layer/total: 1.9 × 2000 = 3800
Weft tows, preform setup: 3800 × 3 = 15,200
Total number of tows: 1068 + 15,200 = 16,267</td><td>Length (cm): 45,400
Width (cm): 127
Weft tows/cm/layer: 1.9
Number of preform weft layers: 4
Weft tows/cm/layer/total: 1.9 × 45,400 = 86,260
Weft tows, weave: 86,260 × 4 = 345,040
Total number of tows: 1067 + 345,040 = 346,107</td></tr>
<tr><td>Material Cost</td><td></td></tr>
<tr><td>Setup fabric area (m^2): 1.27 × 20 = 25.4
Areal weight (g/m^2): 5200
Weight of woven fabric (kg): 5.2 × 25.4 = 132
Cost: £1/kg × 132 = £132</td><td>Weave fabric area (m^2): 1.27 × 454.27 = 577
Areal weight (g/m^2): 5200
Weight of woven fabric (kg): 5.2 × 577 = 3000
Cost: £1/kg × 3000 = £3000</td></tr>
<tr><td colspan="2">Total Material Cost: 3000 + 132 = £3132</td></tr>
</table>

Using constants y and m from Figure 10c (14 (Jacquard andWeavebird preforms), Figure 11a (9 Weavebird preforms) and Figure 11b (8 Jacquard preforms), manufacturing costs for the new preforms were estimated and compared for each set of constants. These plots were chosen as they have the highest correlation between manufacturing time and preform complexity as shown by correlation coefficient r^2.

Manufacturing cost based on constants derived from 14 preforms

The following equation was derived from Model—estimated data (Figure 10c):

$$\frac{t_i}{t_b} = 2.4704\left(\frac{R_i}{R_b}\right)^{0.6926}$$

where y is $\frac{t_i}{t_b}$ and x is $\left(\frac{R_i}{R_b}\right)$.

Feature Factor Calculation

Fabric manufacturing cost was estimated from Equation (16), which includes the relative feature factor since t_i for the fabric is unknown, with the feature factor is given by Equation (12).

Resource Cost Example: Proportion of Tooling Cost for Quoted 3D Fabric

The cost structure in 3D fabric manufacturing is split between the proportion of costs due to the loom setup and costs due to weaving. These costs are labour, capital and overheads. In the setup phase, 20 m of warp and weft tows will be woven while in the weave phase, 454 m will be woven. Two feature factors for complexity were calculated, one for setup, the other for weaving. The proportion of the capital tooling cost for setup and weaving is found from:

$$\text{Proportion of capital tooling } \cos t = 2.4704\left(\frac{R_i}{R_b}\right)^{0.6926}\frac{t_B}{T}\frac{C_t}{0.7t_{wo}}$$

where t_{wo} is the write-off time for capital equipment, 5 years, and C_t is capital tooling cost.

A load factor of 0.7 is assumed for the Jacquard loom on which the fabric would be woven.

Fabric complexity is a function of the total number of fibre tows and sub-features, e.g., holes, in the fabric and the shape of the fabric or number of structural elements (Equation (13)):

$$R_i = \left(\sum A_i + \text{sub} - \text{features}\right)\left(\sum SE_i\right)$$

where A_i is the total number of fibre tows and $\sum SE_i$ is the number of structural elements. This fabric is a flat profile with no sub-features, so $\sum SE_I = 1$ and

$$R_i = \left(\sum A_i\right)$$

Setup Feature Factor:
Fabric complexity R_i= 16,267 tows or 16k (Table 5). Baseline complexity R_B = 360 tows.
Therefore $\frac{R_i}{R_b}$ = 16,000/360 = 45
Weave Feature Factor
Fabric complexity I = 346,107 tows (Table 5). Baseline complexity R_b = 360 tows, so

$$\frac{R_i}{R_b} = 346,107/360 = 962$$

The company runs one shift per day, so total annual production time is 1840 h based on 8 h per day at 5 days per week for 46 weeks per year. The baseline setup time t_b is 9 h, which was the total time over two days, and the baseline weave time is 1 h, so t_B/T is 8/1840 = 0.004348 for the baseline setup time and 1/1840 = 0.000543 for the baseline weave time. The capital cost amortised over 5 years is £200,000/5. From Figure 10c, m is 2.4704 and n is 0.6926. A key variable is load factor L or machine utilisation. A load factor of 70% was agreed with the manufacturer. Using these values and approximate cost data for tooling, labour and overheads (Table 6), the proportion of annual capital tooling cost $C_{tooling}$ was calculated by the model for both setup and weave phases:

Loom setup: tooling cost = 24,704 × $45.2^{0.5209}$ × 0.00435 × 200,000/5 × 0.7 = £8595
Weaving: tooling cost = 24,704 × $962^{0.5209}$ × 0.000543 × 200,000/5 × 0.7 = £8934
Proportion of capital tooling cost, setup and weave = £8595 + £8934 = £17,529

Table 6. 3D preform resource costs.

Resource	Cost (£)
Material, 3132 kg, E-glass at £1/kg	3132
Labour, £30,000 for two operatives	60,000
Capital Tooling	
Jacquard loom	70,000
Four creels:	80,000
Bobbins:	20,000
Feed/Transport:	10,000
Other items:	20,000
Total	200,000
Overheads	25,000

In the same way, the proportion of labour and overhead costs for setup and weaving was calculated and summed to give an overall manufacturing cost of £46,736 for the fabric at a width of 127 cm and a total length of 454 m (Table 7). Using the same methodology, costs were estimated with constants derived from 8 Jacquard and 9 Weavebird preforms, as shown in Tables 8 and 9.

Table 7. Estimated cost of 3D woven fabric, 14 preforms.

Cost Element	Loom Setup	Weaving	Total	%
Capital Tooling (£)	8595	8934	17,529	37.5
Labour (£)	9025	9381	18,406	39.4
Overheads (£)	3760	3909	7669	16.4
3D woven fabric material (£)			3132	6.7
Total Cost (£)			46,736	100

Table 8. Estimated cost of 3D woven fabric, eight Jacquard preforms.

Cost Element	Loom Setup	Weaving	Total	%
Capital Tooling (£)	5393	1826	7218	34.2
Labour (£)	5662	1917	7579	35.9
Overheads (£)	2359	799	3158	15.0
3D woven fabric material (£)			3132	14.9
Total Cost (£)			21,087	100

Table 9. Estimated Cost of 3D woven fabric, nineWeavebird preforms.

Cost Element	Loom Setup	Weaving	Total	%
Capital Tooling (£)	10,317	22,355	32,671	38.4
Labour (£)	10,833	23,472	34,301	40.3
Overheads (£)	4514	9780	14,924	17.6
3D woven fabric material (£)			3132	3.7
Total Cost (£)			85,028	100

Table 10 shows the variation in quoted preform cost with values of constants n and m. Data for n and m from Figures 10c and 11a,b gave cost estimates of £46,736, £85,028 and £21,087, respectively. The biggest cost contributors are Labour and Tooling. The lowest value of exponent n is 0.3258 as all eight preforms in this case were made on the Jacquard, while n = 0.9328 when nine preforms were woven on the Weavebird. Figure 12 shows a steep rise in manufacturing cost for the quoted 3D fabric as weave manufacturing conditions change from those on the more efficient Jacquard to the less efficient Weavebird and complexity exponent n approaches 1, i.e., linearity.

Table 10. Quoted preform cost: cost breakdown (%) and total cost.

Numberof Preforms	n	m	Cost Breakdown (%)				Cost (£)
			Tooling	Labour	Overhead	Material	
8 J	0.3258	6.2714	34.2	35.9	15.0	14.9	21,087
14 J and W	0.6926	2.4707	37.5	39.4	16.4	6.7	46,736
9 W	0.9328	1.1872	38.4	40.3	17.6	3.7	85,028

J = Jacquard; W = Weavebird.

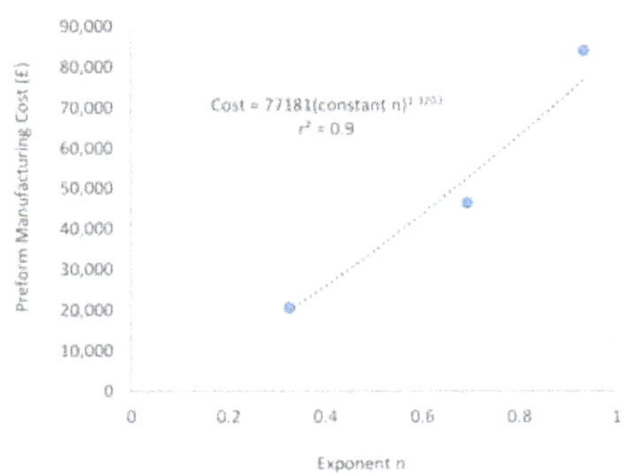

Figure 12. Manufacturing cost vs. complexity exponent.

The relationship between preform manufacturing cost C_i and exponent n (Figure 12) as a measure of decreasing loom efficiency from Jacquard to Weavebirdas n approaches 1 is:

$$C_i = 77,181n^{1.3203} \quad (17)$$

Although sample size is 3, r^2 = 0.90, implying a very strong correlation between preform manufacturing cost and complexity exponent n. More preform manufacturing data areneeded to fully validate this relationship and show whether the curve intercepts the *y*-axis or goes through the origin.

4. Discussion

4.1. Correlation of Preform Manufacturing Time and Complexity

Production of 17 individual preforms of varying fibre architecture and shape started in June 2017 and finished in June 2019. No data for other profiles wereavailable, so costs (capital, labour and overheads), profile complexity and time data for the 17 preforms wereemployed to develop a resource-based TCM to enable estimation of manufacturing cost for a new bespoke preform yet to be made. The working hypothesis is that preform production time, and therefore cost, scales with preform complexity. Increasingly accurate estimation of preform manufacturing cost is possible as more data become available [5].

Figure 10b shows a clear distinction between preforms manufactured on the Jacquard and Weavebird looms, with seven of the eight Jacquard preforms either on or above the trendline and seven of the nine Weavebird preforms below the line. In Figure 10c, which showed the same trend by loom, two outliers were removed resulting in an r^2 value of 0.62, indicating either a moderate or strong correlation between manufacturing time and complexity [46]. Although sample sizes were below 12, Figure 11a,b show strong and very strong correlations [45], 0.78 and 0.89, for nine and eight preforms made on the Weavebird and Jacquard looms, respectively (Table 11). The lower values of r^2 and greater scatter of data observed in Figure 10a–c can be explained by the presence of both Jacquard and Weavebird data points on the same plots. For a sample size comparison, Fingersh et al. [47] used sample sizes varying from 6 to 13 when determining r^2 for the dependence of wind turbine tower mass on blade swept area and blade hub height.

Significant variation across all 17 preforms was present according to profile shape and weave architecture. The Weavebird is suitable for weaving short length profiles while the Jacquard is suitable for longer preforms such as Preform 3, Table 2. The preforms were a variety of flat and T-section shapes with weave architecture varying from single layer, layer to layer and orthogonal. All 17 preforms were woven for the first time with loom setup issues such as fibre clumping, contact with loom framework and fibre breakage causing significant time delays. The time recorded for each preformincluded these time delays

(Tables 2 and 3, pp. 16–17). In Section 4.3, a reduction in manufacturing time as learning increases is discussed in detail (Equation (24), p. 29, Table 12, p. 26).

Table 11. Correlation coefficient and number of preforms by loom.

Number of Preforms	Correlation Coefficient r^2, $\frac{t_i}{t_b}$ vs. $\frac{R_i}{R_b}$
17 Jacquard and Weavebird	0.56
16 Jacquard and Weavebird	0.51
14 Jacquard and Weavebird	0.62
8 preforms woven on Jacquard	0.89
9 preforms woven on Weavebird	0.78

Table 12. Representative learning rates by industrial sector.

Sector Representative Learning Rates	
Aerospace	15%
Shipbuilding	15–20%
Machine Tools (new models)	15–20%
Electronics (repetitive)	5–10%
Electrical Wiring (repetitive)	15–25%
Machining	5–10%
75% Manual Assembly + 25% Machining	20%
50% Manual Assembly + 50% Machining	15%
25% Manual Assembly + 75% Machining	10%
Punch Press	5–10%
Raw Materials	5–7%
Purchased Parts	12–15%
Welding (repetitive)	10%

Jacquard looms can make long complex fabrics much more efficiently than dobby looms, e.g., the hypothetical fabric (454 m). This is shown clearly in Figure 11b for eight Jacquard preforms, in which manufacturing time increases at a decreasing rate with complexity and length, i.e., the loom becomes more efficient at weaving longer, more complex fabrics. Conversely, where all the preforms were made on a Weavebird loom, manufacturing time increases almost linearly with complexity, Figure 11a, and weaving efficiency does not increase with preform complexity.

4.2. Costing of the Hypothetical 3D Woven Preform

For nine Weavebird preforms, constants from Figure 11a gave a cost of £85,028, while for eight Jacquard preforms, constants from Figure 11b gave a preform cost of £21,087. Therefore, a less efficient Weavebird will give a much higher manufacturing cost for a large, complex preform compared to the cost when woven on a more efficient Jacquard. Values of complexity exponent n between 0 and 1 in Equation (12) for the feature factor imply an economy of scale for 3D woven preform manufacturing with increasing preform complexity:

$$\frac{t_i}{t_b} = m\left(\frac{R_i}{R_B}\right)^n$$
$$0 < n < 1$$

Esawi [4] found a similar relationship for injection moulding, extrusion and casting operations in which tooling cost and capital cost scale with complexity exponents y_t and y_c, respectively. Values for y_t and y_c vary between 0 and 1, implying greater economy of scale as tooling and capital equipment, for example a plastics injection press, become more complex:

$$0 < y_t < 1,\ 0 < y_c < 1$$

The analysis for preforms woven on the Jacquard and Weavebird looms indicates that as manufacturing conditions change from the more efficient Jacquard to the less efficient

Weavebird, as indicated by the increasing value of complexity exponent n, economy of scale will decrease and manufacturing cost increase for a large, complex preform (Table 10, Figure 12). In conclusion, Jacquard costs alone should be used to estimate costs for a large, complex preform intended to be made on the Jacquard loom. Therefore, £21,087 for the commercial preform is judged the most accurate estimate. Assuming a non-linear relationship between manufacturing time and complexity based on the available data and observed correlation coefficient for Jacquard and Weavebird manufactured preforms (Table 11), two feature factor sub-models are proposed for 3D preform weaving, one for the Jacquard and one for the Weavebird:

Jacquard feature factor: Weavebirdfeature factor:

$$\frac{t_i}{t_b} = 6.2714\left(\frac{R_i}{R_b}\right)^{0.3258} \tag{18}$$

$$\frac{t_i}{t_b} = 1.1872\left(\frac{R_i}{R_b}\right)^{0.9328} \tag{19}$$

which in turn leads to two cost models for preform manufacturing cost, C_i:

Jacquard cost model

$$C_i = \frac{mC_m}{(1-f)} + 6.27\left(\frac{R_i}{R_b}\right)^{0.3258}\frac{t_b}{T}\left(\frac{C_t}{Lt_{wo}} + \sum C_{\text{total salaries}} + \sum C_{\text{general overheads}}\right) \tag{20}$$

Weavebirdcost model

$$C_i = \frac{mC_m}{(1-f)} + 1.19\left(\frac{R_i}{R_b}\right)^{0.9328}\frac{t_b}{T}\left(\frac{C_t}{Lt_{wo}} + \sum C_{\text{total salaries}} + \sum C_{\text{general overheads}}\right) \tag{21}$$

The cost estimate of £21,087 is approximately three times that for the same preform supplied by a US manufacturer, or £7500. The lower cost may be due to a higher production rate coupled with a more efficient loom leading to lower preform cost and greater experience from embedded learning.

4.3. Cost Reduction by Learning

Organisational Learning is defined as a conscious attempt by organisations to improve productivity, effectiveness and innovation in complex economic and technological market conditions. Learning enables quicker and more effective responses to a complex and dynamic environment. The greater the complexity, the greater the need for learning [29–32]. 3D woven preform manufacture is a highly complex process with numerous stepscarried out in a required sequence for successful manufacture. If there is a delay in completing agiven step, the time required to complete the overall preform will increase thereby increasing preform cost. In this study, 17 preforms were made for the first time with no previous 3D preform manufacturing experience. Wright [33] observed that as aircraft production increased, the cost in terms of direct labour hours fell as shown in Figure 13, which is a learning curve (LC).

In general, learning curves (LC) can be described by Equation (22) [33]:

$$y = C_1x^b \tag{22}$$

where y is the average time (or cost) per unit required to produce x units;

C_1 is the time (cost) to produce the first unit; parameter b ($-1 < b < 0$), the slope of the LC, which describes the worker's learning rate.

For a new component not previously manufactured, the learning required and therefore the cost to make the part will initially be high as shown by the start of the slope on the left of Figure 11. As more units are made, there is a steep drop in labour hours per part until the rate of decrease in direct labour hours per part becomes smaller.

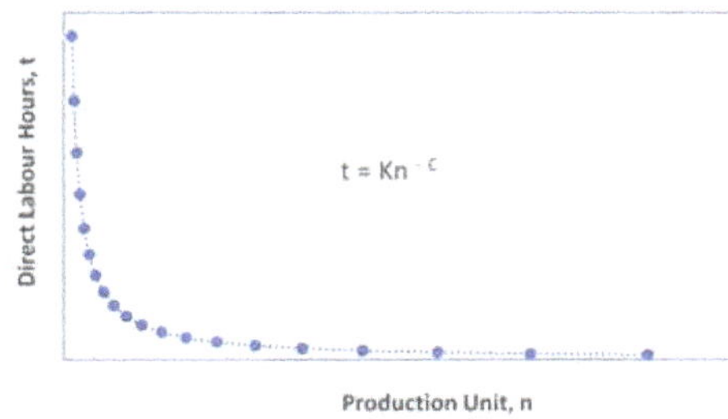

Figure 13. Generalrelationship, component cost and production volume.

Klenow [34] reviewed various studies investigating learning by doing for a single defined production process across a variety of industrial sectors, observing estimates of approximately 20% for the learning rate. Baloff [35], and Garg and Milliman [48] showed that 20% is the rate at which productivity rises with a doubling of cumulative output. Lee [36] summarised learning rates by manufacturing sector and activity (Table 12) and showed that even in one overall activity, in this case industrial manufacturing, learning rates will vary considerably by individual sector. In several studies, Yelle [37] and Argotte and Epple [38], observed that productivity rose across a variety of industries through a process of learning by doing.

In preform manufacture, direct labour hours are associated with activities such as bobbin winding and insertion, tube preparation time, loom maintenance and operation, and stoppage time due to issues encountered during weaving, e.g., damage to carbon and glass fibres from contact with the loom framework. Manufacturing time in this study is the time taken to complete these activities. With increased preform production, manufacturing time t_i and manufacturingcost should decrease with increased learning. The estimated manufacturing time t_i for one preform is found from Equation (12),

$$\frac{t_i}{t_b} = m\left(\frac{R_i}{R_B}\right)^n$$

from which

$$t_i = m\left(\frac{R_i}{R_b}\right)^n t_b$$

3D fabric manufacturing cost is split between the proportion of costs due to loom setup and costs due to weaving, therefore there are two manufacturing times for a given preform: $t_{i\ weave}$ and $t_{i\ setup}$. The estimated cost of the commercial preform was £21,087 (Table 10) for weaving on the Jacquard loom. The company has no experience of making this preform. Setup time and weave time t_b for the baseline preform was 8 and 1 h, respectively. Using values for constants m and n (Table 10), the estimated setup and weave times for the hypothetical preform are:

Setup time: $t_{i\ setup} = 6.2714(45.19)^{0.3258}$ 8 = 174 h
Weave time: $t_{i\ weave} = 6.2714(962)^{0.3258}$ 1 = 59 h
Total manufacturing time: $t_{i\ setup} + \text{I}t_{i\ weave}$ = 174 + 59 = 233 h

No data werepublicly available for 3D preform learning rates. The total estimated manufacturing time is 233 h. Setup time is the manual labour time involved in activities such as bobbin winding, bobbin placement on the creel and taking fibre tow onto the loom. The setup time is 174/233 or 74.7% while the weave or machine time is 25.3%. From Table 12, a learning rate of 20%, in which manufacturing time decreases by 20% for each doubling of cumulative production, corresponds to a manufacturing activity in which manual operations are 75% and machine time is 25% of total activity. Since manual setup time (74.7%) and weave time (25.3%) are closest to 75% manual assembly and 25% machining, 20% was the assumed learning rate for the new preform. Based on this rate, a learning curve (Figure 14) and an equation (Equation (23)) was derived by the model from Equation (22), to estimate a competitive manufacturing cost for the new preform.

$$t_i = 233b^{-0.32} \tag{23}$$

where t_i = preform manufacturing time; b = number of preforms.

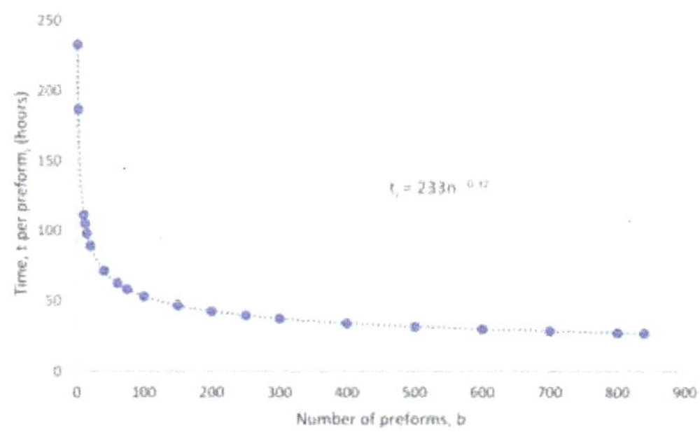

Figure 14. Learning curve for commercial preform i.

The model estimated a manufacturing time t_i of 98 h per preform after a cumulative production of 15 preforms. Therefore, setup time and weave time will have decreased with increased cumulative production. Insertion of this value for t_i in Equation (11) together with resource costs for tooling, salaries, overhead, write-off time (Table 6) and load factor 0.7 gave an estimated cost of £8002, approximately one third of the first-time preform cost of £21,087 and in line with the US supplier's cost of £7500. More manufacturing data will be required to clarify learning rates for 3D woven preforms to fully validate Equation (23) and provide a more accurate estimate of preform cost.

$$C_{p_i} = \frac{mC_m}{(1-f)} + \frac{t_i}{T}\left(\frac{C_t}{Lt_{wo}} + \sum C_{\text{total salaries}} + \sum C_{\text{general overhead}}\right)$$

$$= 3132 + \frac{98}{1840}\left(\frac{200,000}{0.7 \times 5} + 60,000 + 25,000\right) = £8002$$

Various issues were encountered during first-time preform manufacture:

- Fibre catching on the edges of the bobbin.
- Fibres splitting at the tensioning bars.
- Weight of bobbins causing tension problems.
- Damage to carbon and glass fibres due to contact with loom framework.
- Crossing fibres forming balls of carbon at the heddles.
- Weft insertion forming fibre clumps and splitting.

These issues accounted for the observed manufacturing times for each preform (Tables 2–4) due to low embedded learning and first-time preform manufacture. A key assumption underlying this analysis is that the commercial preform will be continuously manufactured so that unlearning or forgetting [49], due to discontinuous production is avoided. Another assumption is the use of Wright's learning curve model [33] to estimate the cumulative number of preforms produced from the estimated time to make one preform and an assumed learning rate. The model yields production times equal to zero [40] after a high number of repetitions, which is impossible. Furthermore, it does not account for workers' prior experience [41], nor the influence of machinery in the learning process [42]. However, workers' prior experience does not apply in this case as all seventeen preforms were made for the first time, while no data exist for the influence of weaving loom machinery on learning. Finally, Wright's model has been used successfully in various manufacturing sectors [41]. Therefore, the choice of this LC model is justified.

Irwin and Klernow [50] pointed out that productivity growth from learning by doing diminishes as experience accumulates with a technology. Even though learning by doing is largely specific for a given technology, a review of the literature showed that this same pattern holds for a wide variety of industries. An alternative visual representation of learning is a plot which shows learning increasing as a function of decreasing manufacturing time against the number of manufactured preforms. This relationship can be expressed as:

$$\text{Learning} = \left[1 - \left(\frac{t_i}{T}\right)\right] = n^c \tag{24}$$

Figure 15 represents Equation (24) using the same learning rate of 20% and a manufacturing time of 233 h for initial manufacture of the commercial preform.

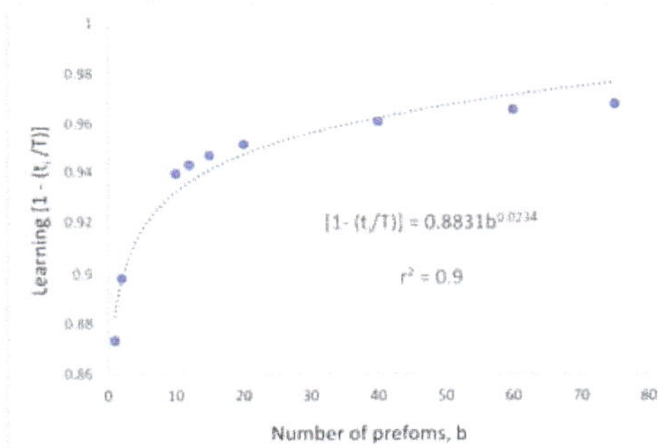

Figure 15. Learning as a function of manufacturing time and number of preforms.

Figure 15 shows a sharp initial increase in learning as a function of manufacturing time and the number of preforms produced, followed by a levelling off until there is no discernible increase in learning after a cumulative production of 80 preforms. Correlation coefficient r^2 is 0.96, denoting a strong correlation between manufacturing time t_i and b, the cumulative production of preforms. Exponent c has a value of 0.0234.

An alternative to continuous production of 15 preforms of the same complexity and size as the commercial preform is to acquire a more efficient Jacquard loom so that manufacturing time, and therefore cost, is reduced. Russell [51], and Pegels [52] observed that while productivity will initially fall with technology updates, it will gradually rise to overtake the level achieved with the old technology. However, Lee [36] and Hill [53] pointed out that reduced manufacturing cost through learning will not happen unless there is a willingness to learn, an ability to learn and, in many cases, an investment in learning. Many factors were identified that determine the learning curve for a given individual, team, factory or industry, including

Management styles and actions
Corporate culture
Organisation structure
Technology
Capital investment
Engineering
Product design
Direct and indirect labour efficiency
Economy of scale
Plant layout

Process improvement

To maximise learning, productivity and competitiveness, Skinner [54] emphasiseda limited, manageable set of products and markets for lowering costs, especially overhead. Therefore, to maximise learning and hence productivity in preform manufacture, the focus should be on a manageable set of 3D woven preform designs.

A summary of equations for cost, complexity and learning, from the literature and derived by the author, is presented in Table 13.

Table 13. Equation summary.

Equation	Attribution	No.
Cost = Material + Tooling Cost + Labour + Overheads	MA	1
$C_1 = C_m/(1-f)$	MA	2
$C_2 = C_t/n_r(1 + n_r/n_t)$	MA	3
$C_3 = 1/n_r(C_c/Lt_{wo})$	MA	4
$C_4 = C_{oh}/n_r$	MA	5
$C_{mc} = mC_m/(1-f) + C_t/n(1 + n_r/n_t) + 1/n_r(C_c/Lt_{wo} + C_{oh})$	MA	6
$C_{tooling} = t_i/T(C_t/t_{wo})$	JC	7
$C_{salaries} = t_i/T\sum C_{total\ annual\ salaries}$	JC	8
$C_{overhead} = \sum C_{general\ annual\ overhead}$	JC	9
$C_{overhead} = t_i/T\sum C_{general\ annual\ overhead}$	JC	10
$C_{p_i} = mC_m/(1-f) + t_i/T\left(C_t/Lt_{wo} + \sum C_{total\ salaries} + \sum C_{general\ overhead}\right)$	JC	11
$t_i/t_b = m(R_i/R_B)^n$	JC	12
$R_i = \sum(A_i + sub - features)(\sum SE_i)$	JC	13
$R_b = (A_b)(\sum SE_i)$	JC	14
$t_i = m(R_i/R_B)^n t_b$	JC	15
$C_{p_i} = mC_m/(1-f) + m(R_i/R_B)^n t_B/T\left(C_t/Lt_{wo} + \sum C_{total\ salaries} + \sum C_{general\ overhead}\right)$	JC	16
$C_i = 77,181n^{1.3203}$	JC	17
$t_i/t_b = 6.2714(R_i/R_b)^{0.3258}$	JC	18
$t_i/t_b = 1.1872(R_i/R_b)^{0.9328}$	JC	19
$C_i = mC_m/(1-f) + 6.27(R_i/R_b)^{0.3358} t_b/T\left(C_t/Lt_{wo} + \sum C_{total\ salaries} + \sum C_{general\ overheads}\right)$	JC	20
$C_i = \frac{mC_m}{(1-f)} + 1.19\left(\frac{R_i}{R_b}\right)^{0.9328} \frac{t_b}{T}\left(\frac{C_t}{Lt_{wo}} + \sum C_{total\ salaries} + \sum C_{general\ overheads}\right)$	JC	21
$y = C1x^b$	JC	22
$t_i = 233b^{-0.32}$	JC	23
$Learning = [1 - (t_i/T)] = n^c$	JC	24

JC—James Clarke.MA—Michael Ashby.

5. Conclusions and Recommendations for Further Work

3D woven preforms are promising materials for composite parts in numerous applications, for example wind turbine spar caps. They have unique mechanical properties and have the potential to reduce composite manufacturing costs due to near net-shape resin transfer moulding. At present, they are not widely used due to a perception of high cost and demanding safety protocols in market sectors such as aerospace. A predictive resource-based technical cost model (TCM) for bespoke manufacturing of 3D fabrics was developed based on the principles that cost is determined by resources such as tooling, labour and other overheads and that manufacturing time, and therefore cost, will scale with preform complexity. An expression for a cost scaling feature factor was introduced relating preform manufacturing time to preform architectural complexity, defined as a function of the number of fibre tows and preform shape.

The model is based on Equation (16) (Table 13) for costing the manufacture of 3D preforms and utilises two principles. Firstly, the manufacturing time for a single bespoke preform will depend on the unique complexity of that preform. Secondly, the resource cost for a given preform, for example tooling cost, will be a function of the time required to make that preform as a proportion of total annual production time (Equations (7), (8) and (10), Table 13). Loom tooling is not dedicated for a given preform. Plotting manufacturing time against preform complexity for seventeen preforms enabled derivation of constants m and n in Equation (12). Inserting these values into Equation (16) enables estimation of the manufacturing cost of a new 3D preform with a given architecture R_i:

$$C_{pi} = \frac{mC_m}{(1-f)} + m\left(\frac{R_i}{R_b}\right)^n \frac{t_B}{T}\left(\frac{C_t}{Lt_{wo}} + \sum C_{\text{total salaries}} + \sum C_{\text{general overheads}}\right)$$

Approximate resource costs for tooling, labour and overheads together with manufacturing times for seventeen unique preforms with varying architectures such as single layer, layerto layer and orthogonal were provided by a 3D preform manufacturer. The preforms were made on either a Weavebird handloom or a Jacquard loom. A hypothesis was proposed that preform manufacturing time will increase non-linearly with preform complexity.

Table 10 summarised model-estimated costs based on differing values for constants derived from separate plots for Jacquard and Weavebird preforms, and a plot with both Jacquard and Weavebird preforms. Eight preforms were woven on the Jacquard and nine on the Weavebird. Manufacturing time was plotted against preform complexity to derive separate plots for the Jacquard and Weavebird looms. For the separate looms, manufacturing time for a preform was shown to have a strong correlation with preform complexity. Analysis of the plots (Table 11) showed that those with nine Weavebird and eight Jacquard woven preforms gave the strongest positive correlation with preform complexity, as measured by correlation coefficient r^2, 0.78 and 0.89, respectively. Therefore, the hypothesis of preform manufacturing time increasing non-linearly with preform complexity is considered valid based on the cost information and preform data provided. More data from a wider range of preforms of varying complexity arerequired to fully validate the non-linear relationship between manufacturing time and complexity.

A composite parts manufacturer received a quote for a single large, complex preform currently made by a US manufacturer for £7500. The cost of the new preform was compared with cost estimates for the preform based on data from preforms made on either the Jacquard or Weavebird looms. Analysis of the plots showed that the Jacquard weaves large, complex preforms more efficiently than the Weavebird. Therefore, the estimated cost for the preform, £21,087 based on data derived from the plot for eight Jacquardwoven preforms, was judged the most realistic although almost three times that of the US-supplier's cost of £7500. Based on a weight of 3132 kg, the cost per kg for the US fabric is £2.4/kg. The raw fibre (E-glass) cost is £1/kg, so the material cost proportion for the fabric is 42%. For one-off preform manufacture, the material cost proportion varied from 3.7% to 4.9%, depending on whether the preform is to be made on a Weavebird or Jacquard loom. A recent cost modelling study of mass-produced wind turbine spar caps made with glass fibre composite {5} showed material proportions ranging from 35% to 52%, similar to a cost proportion of 42%, implying that the US-supplied fabric has been similarly mass produced, resulting in a cost per preform of £7500 compared to a one-off manufacturing cost of £21,087. A learning curve was derived based on a learning rate from the estimated labour and machine time proportions of total preform manufacturing time. From the learning curve, continuous production of 15 preforms resulted in a cost per preform of £8001, assuming no reduction in other costs such astooling and infrastructure.

In concluding, the results imply that it is possible to make unique 3D woven preforms competitively on a suitable loom machine, provided that sufficient learning is embedded in the manufacturing organisation coupled with greater automation. Although studies have indicated that cost per part will initially increase following machine installation, cost per part will fall below the level present before machine installation as new learning is embedded for successful machine operation. This should encourage increased uptake of suitably designed 3D woven composites in a wider range of applications.

Further work could investigate mass customisation, where short manufacturing runs for a part of given size and complexity are coupled with fast turnaround times and tool changes for another part. This can be very expensive owing to a lack of embedded learning in a fast-changing production environment. Short runs of bespoke 3D woven preforms could be modelled more accurately if extensions of traditional learning curve models incorporating multi-variate analysis can be developed. Multi-variate learning curves are based on two or more independent variables and are required when quantitative and qualitative factors run in tandem, e.g., when fast tool changes are required for a run of new preforms. To date, development of multi-variate analysis tools for constantly changing scenarios has been sparse, a key issue being lack of real time manufacturing process data. However, recent advances in digital twin technology enabling real time imaging of a manufacturing operation based on worker performance and process data should encourage the development of multi-variate learning curves for improving worker learning so that the cost of short production runs of 3D woven preforms is reduced.

Author Contributions: Conceptualisation, J.C.; data curation, J.C.; formal analysis, J.C.; investigation, J.C.; methodology, J.C.; project administration, A.M. and E.A.; resources, A.M. and E.A.; software, J.C.; validation, G.S. and R.B.; visualisation, J.C.; writing—original draft, J.C.; writing—review andediting, J.C., A.M., J.S. and D.D. All authors have read and agreed to the published version of the manuscript.

Funding: This research received no external funding.

Acknowledgments: The author would like to thankTomas Flanagan or Eire Composites for providing valuable advice and guidance in cost modelling for wind turbine spar cap manufacture.

Conflicts of Interest: The authors declare no conflict of interest.

References

1. The Aerospace Composites Market, 2012–2022. 2012. Available online: www.visiongain.com (accessed on 20 September 2019).
2. O'Dea, N. *Global Outlook for the Composites Industry, Presentation, Advanced Engineering, Composites Forum*; NEC: Birmingham, UK, 2018.
3. Stig, F.; Hallström, S. Assessment of the mechanical properties of a new 3D woven fibre composite material. *Compos. Sci. Technol.* **2009**, *69*, 1686–1692. [CrossRef]
4. Esawi, A.M.K.; Ashby, M.F. Cost estimates to guide pre-selection of processes. *Mater. Des.* **2003**, *24*, 605–616. [CrossRef]
5. Clarke, J. A Cost Model Framework for 3D Woven Composites. Ph.D. Thesis, Ulster University, Belfast, UK, 2020.
6. Boussu, F.; Cristian, I.; Neumann, S. Generaldefinition of 3D warp interlock fabric architecture. *Compos. Part B Eng.* **2015**, *81*, 171–188. [CrossRef]
7. Gurkan, P.; Namak, U. 3D Woven Fabrics. *Woven Fabrics* **2012**, *4*, 91–120.
8. Van Mourik, A. Why Conversion Costs of Composites in Aerospace are Still Way too High. LinkedIn Article. 16 July 2018. Available online: https://www.linkedin.com/pulse/why-conversion-costs-composites-aerospace-still-way-too-mourik-van/ (accessed on 20 September 2019).
9. McClain, M.; Goering, J. Rapid assembly of fiber preforms using 3D wovencomponents. *Sampe J.* **2013**, *49*, 24–31.
10. Mohamed, M.; Wetzel, K. 3D Woven Carbon/Glass Hybrid Spar Cap for Wind Turbine Rotor Blade. *J. Sol. Energy Eng.* **2006**, *128*, 562–573. [CrossRef]
11. Hueber, K.; Schledjewski, R. Review of cost estimation: Methods and models for aerospace composite manufacturing. *Adv. Manuf. Polym. Compos. Sci.* **2016**, *2*, 1–13. [CrossRef]
12. Fagade, A.A.; Kazmer, D.O. Early cost estimation for injection moldedparts. *J. Inject. Molding Technol.* **2000**, *3*, 97–106.
13. Fagade, A.A.; Kazmer, D.O. Modelling the effects of complexity on manufacturing costsand time-to-market of plastic injection molded products. In Proceedings of the Tenth Annual Conference of the Production and Operations Management Society, POM 99, Charleston, SC, USA, 20–23 March 1999.
14. Fagade, A.; Kazmer, D.; Kapoor, D. *A Discussion of Design and Manufacturing Complexity*; Department of Mechanical and Industrial Engineering, University of Massachusetts: Amherst, MA, USA, 2000. Available online: http://citeseerx.ist.psu.edu/viewdoc/download?doi=10.1.1.495.2899&rep=rep1&type=p (accessed on 20 September 2019).

15. Hagnell, M.K.; Akermo, M. A Composite Cost Model for the Aeronautical Industry: Methodology and Case Study. *Compos. Part B Eng.* **2015**, *79*, 254–261. [CrossRef]
16. Gutowski, T.G.; Neoh, E.T.; Polgar, K.C. *Adaptive Framework for Fabrication Time of Advanced Composite Manufacturing Processes*; Technical Report; Laboratory for Manufacturing and Productivity, Massachusetts Institute of Technology: Cambridge, MA, USA, 1995.
17. Verrey, J.; Wakeman, M.D.; Michaud, V.; Manson, J.A.E. *Manufacturing Cost Comparison of Thermoplastic and Thermoset RTM for an Automotive Floor Pan*; École Polytechnique Fédérale de Lausanne (EPFL), Laboratoire de Technologie des Composites etPolyme'res (LTC): Lausanne, Switzerland, 2005.
18. Schubel, P. Cost Modelling in polymer composite applications. Case study:Analysis of existing and automated manufacturing processes for a large windturbine blade. *Compos. Part B* **2012**, *43*, 953–960. [CrossRef]
19. Ennis, B.L.; Kelley, C.L.; Naughton, B.T.; Norris, R.E.; Das, S.; Lee, D.; Miller, D.A. *Optimized Carbon Fibre Composite in a Wind Turbine Blade Design*; Sandia Report, SAND 2019-14173; Sandia National Laboratories: Albuquerque, NM, USA, 2019.
20. Cabrera-Ríos, M.; Castro, J.M. The balance between durability, reliability and affordability in structural composites manufacturing. *Polym. Compos.* **2007**, *28*, 233–240. [CrossRef]
21. Calado, E.; Leite, M.; Silva, A. Selecting composite materials considering cost and environmental impact in the early phases of aircraft structure design. *J. Clean. Prod.* **2018**, *186*, 113–122. [CrossRef]
22. Turner, T.; Harper, L.; Warrior, N.; Rudd, C. Low-cost carbon-fibre-based automotive body panel systems: A performance and manufacturing cost comparison. *Proc. Inst. Mech. Eng. Part D J. Automob. Eng.* **2008**, *222*, 53–63. [CrossRef]
23. Bader, M. Selection of composite materials and manufacturing routes for cost-effective performance. *Compos. Part A Appl. Sci. Manuf.* **2002**, *33*, 913–934. [CrossRef]
24. Dickinson, L.; Mohamed, M.; Lienhart, B. Cost modeling for 3D woven preforming process. In Proceedings of the International SAMPE Symposium "Bridging the Centuries", Long Beach, CA, USA, 21–25 May 2000; Volume 45, pp. 127–140.
25. Dickinson, L.; Salama, M.; Stobbe, D. Design approach for 3D woven composites: Cost vs. performance. In Proceedings of the International SAMPE Symposium "2001: A Materials and Processes Odyssey", Long Beach, CA, USA, 6–10 May 2001; Volume 46, pp. 765–778.
26. Horejsi, K.; Noisternig, J.; Koch, O.; Schledjewski, R. Cost-based process selection for CFRP aerospace parts. *JEC Compos. Mag.* **2013**, *81*, 60–62.
27. Russell, J. Composites Affordability Initiative: Successes, failures—Where do we go from here? *Sampe J.* **2007**, *43*, 26–36.
28. Sohouli, A.; Yildiz, M.; Suleman, A. Cost analysis of variable stiffness composite structures with applications to a wind turbine blade. *Compos. Struct.* **2018**, *203*, 681–695. [CrossRef]
29. Fredendall, L.D.; Gabriel, T. Manufacturing Complexity: A Quantitative Measure. In Proceedings of the POMS Conference, Savannah, GA, USA, 4–7 April 2003.
30. Klir, G.J. Complexity: Some General Observations. *Syst. Res.* **1985**, *2*, 131–140. [CrossRef]
31. Simon, H. The architecture of complexity. *Proc. Am. Philos. Soc.* **1962**, *106*, 467–482.
32. Gell-Mann, M. What is complexity? Remarks on simplicity and complexity by the Nobel Prize-winning author of The Quark and the Jaguar. *Complexity* **1995**, *1*, 16–19. [CrossRef]
33. Wright, T.P. Factors Affecting the Cost of Airplanes. Presented at the Aircraft Operations Session, Fourth Annual Meeting, Curtiss-Wright Corporation. *J. Aeronaut. Sci.* **1936**, *1*, 122–128. [CrossRef]
34. Klenow, P.J. Learning Curves and the Cyclical Behaviour of Manufacturing Industries. *Rev. Econ. Dyn.* **1997**, *1*, 531–550. [CrossRef]
35. Baloff, N. Startups in machine-intensive production systems. *J. Ind. Eng.* **1966**, *14*, 25–32.
36. Lee, Q. Learning & Experience Curves in Manufacturing. *Strategos* **2014**, *1*, 1–15.
37. Yelle, L.E. The learning curve: Historical review and comprehensive survey. *Decis. Sci.* **1979**, *10*, 302–328. [CrossRef]
38. Argotte, L.; Epple, D. Learning curves in manufacturing. *Science* **1990**, *24*, 920–924. [CrossRef]
39. Argote, L. *Organizational Learning: Creating, Retaining and Transferring Knowledge*; Springer: New York, NY, USA, 1999.
40. Hurley, J.W. When are we going to change the learning curve lecture? *Comput. Oper. Res.* **1996**, *23*, 509–511. [CrossRef]
41. Teplitz, C.J. *The Learning Curve Deskbook: A Reference Guide to Theory, Calculations and Applications*; Quorum Books: New York, NY, USA, 1991.
42. De Jong, J.R. The Effects of Increasing Skill on Cycle Time and its consequences for time standards. *Ergonomics* **1957**, *1*, 51–60. [CrossRef]
43. Stewart, G. *Introduction to Weaving*; Presentation; NIACE Centre: Belfast, UK.
44. Ashby, M.; Shercliff, H.; Cebon, D. *Materials: Engineering, Science, Processing & Design*, 2nd ed.; Butterworth-Heinemann: Burlington, MA, USA, 2010; pp. 438–441.
45. Kent, R. *Energy Management in Plastics Processing: Strategies, Targets, Techniques and Tools*, 2nd ed.; Plastics Information Direct: Bristol, UK, 2013; pp. 28–29.
46. Hawkes, P.; Svensson, C. *Joint Probability: Dependence Mapping and Best Practice R&D*; Interim Technical Report FD2308/TR1; Defra/Environment Agency, Flood and Coastal Defence R&D Programme.Seacole Building, 2 Marsham Street: London, UK, 2003.
47. Fingersh, L.; Hand, M.; Laxson, A. *Wind Turbine Design Cost and Scaling Model*; National Renewable Energy Laboratory: Golden, CO, USA, 2006; pp. 1–43.
48. Garg, A.; Milliman, P. The aircraft progress curve modified for design changes. *J. Ind. Eng.* **1961**, *12*, 23–27.

49. Jaber, M.; Kher, H.; Davis, D. Countering forgetting through training and deployment. *Int. J. Prod. Econ.* **2003**, *85*, 33–46. [CrossRef]
50. Irwin, D.A.; Klenow, P.J. Learning-by-doing spillovers in the semiconductor industry. *J. Polit. Econ.* **1994**, *102*, 1200–1227. [CrossRef]
51. Russell, J.H. Progress function models and their deviations. *J. Ind. Eng.* **1968**, *19*, 5–11.
52. Pegels, C.C. On startup or learning curves: An expanded view. *AIIE Trans.* **1969**, *1*, 216–223. [CrossRef]
53. Hill, T. *Manufacturing Strategy*; Macmillan: London, UK, 1985.
54. Skinner, W. The Focused Factory. *Harv. Bus. Rev.* **1974**, *52*, 113–122.

Journal of
Composites Science

MDPI

Article

Modeling Stiffness Degradation of Fiber-Reinforced Polymers Based on Crack Densities Observed in Off-Axis Plies

Matthias Drvoderic, Martin Pletz and Clara Schuecker *

Chair of Designing Plastics and Composite Materials, Department of Polymer Engineering and Science, Montanuniversitaet Leoben, 8700 Leoben, Austria; matthias.drvoderic@unileoben.ac.at (M.D.); martin.pletz@unileoben.ac.at (M.P.)
* Correspondence: clara.schuecker@unileoben.ac.at

Abstract: A model that predicts the stiffness degradation in multidirectional reinforced laminates due to off-axis matrix cracks is proposed and evaluated using data from fatigue experiments. Off-axis cracks are detected in images from the fatigue tests with automated crack detection to compute the crack density of the off-axis cracks which is used as the damage parameter for the degradation model. The purpose of this study is to test the effect of off-axis cracks on laminate stiffness for different laminate configurations. The hypothesis is that off-axis cracks have the same effect on the stiffness of a ply regardless of the acting stress components as long as the transverse stress is positive. This hypothesis proves to be wrong. The model is able to predict the stiffness degradation well for laminates with a ply orientation similar to the one used for calibration but deviates for plies with different in-plane shear stress. This behavior can be explained by the theory that off-axis cracks develop by two different micro damage modes depending on the level of in-plane shear stress. It is found that besides influencing the initiation and growth of off-axis cracks, the stiffness degradation is also mode dependent.

Keywords: crack detection; fiber-reinforced polymers; fatigue damage model; composite fatigue; off-axis cracks

Citation: Drvoderic, M.; Pletz, M.; Schuecker, C. Modeling Stiffness Degradation of Fiber-Reinforced Polymers Based on Crack Densities Observed in Off-Axis Plies. *J. Compos. Sci.* **2021**, *6*, 10. https://doi.org/10.3390/jcs6010010

Academic Editor: Stelios K. Georgantzinos

Received: 7 December 2021
Accepted: 23 December 2021
Published: 29 December 2021

Publisher's Note: MDPI stays neutral with regard to jurisdictional claims in published maps and institutional affiliations.

1. Introduction

Components made from multidirectional fiber-reinforced composite laminates experience several distinct damage mechanisms when exposed to fatigue loads. The macroscopic damage mechanisms are matrix cracks, delamination, and fiber failure. This sequence of damage mechanisms during fatigue loading can be categorized into characteristic damage states [1]. The first fatigue-damage state of multi-axial laminates is matrix cracking in off-axis plies where multiple matrix cracks develop and grow in number and length. These so-called off-axis cracks typically span the whole thickness of a ply and propagate along the fiber direction. One of the main effects of off-axis cracks is a significant stiffness reduction of the laminate but not immediate failure of the component [2–8].

Since multiple similar cracks form under fatigue loading, the crack density is used as a measure for the amount of damage in the material in many progressive damage models [9–17]. These models describe the evolution of off-axis cracks as well as their effect on the stiffness of a laminate. Therefore, off-axis crack densities are often used in the development and calibration of these models or to compare their predictions to experimental data. Transmitted or transilluminated white light imaging (TWLI) can be used for transparent composites like glass fiber-reinforced polymers (GFRPs). It is an efficient, reliable and relatively simple method to capture off-axis cracks [2,10,18–20]. TWLI uses a light source placed behind a transparent specimen and a camera on the other side. An undamaged specimen appears bright as it only absorbs a small portion of the light. Cracks, on the other hand, scatter the light and therefore appear as dark lines in the images. For non-transparent laminates, more sophisticated techniques like computed tomography may be used [21]. Up to now, a few methods have been developed to compute the crack density from TWLI images. The simplest is to count all cracks along a straight line normal to the fiber direction and divide the number of counted cracks by the length of the line. This method only takes the number of counted cracks and not their length into account and the results are influenced by the selection of the path. As shown in Refs. [17,22,23], including the crack length results in an improved description of the average damage state of the material. A better approach that includes some information of the crack length is the weighted crack density used by Quaresimin et al. [7]. It clusters the cracks into eight groups of crack lengths and then computes a weighted average. Still, manually counting and categorizing the cracks is labor intensive and prone to human errors. An automated algorithm that takes the images as input and detects all cracks in a given direction has been developed by Glud et al. [24]. Based on this algorithm, we have developed *CrackDect*, an open-source package for evaluating crack densities [25]. With this package, even large fatigue test series can be evaluated efficiently. Figure 1 qualitatively shows the evolution of off-axis cracks and the associated stiffness degradation as well as examples of images taken during fatigue tests. CrackDect takes these images and computes the crack density.

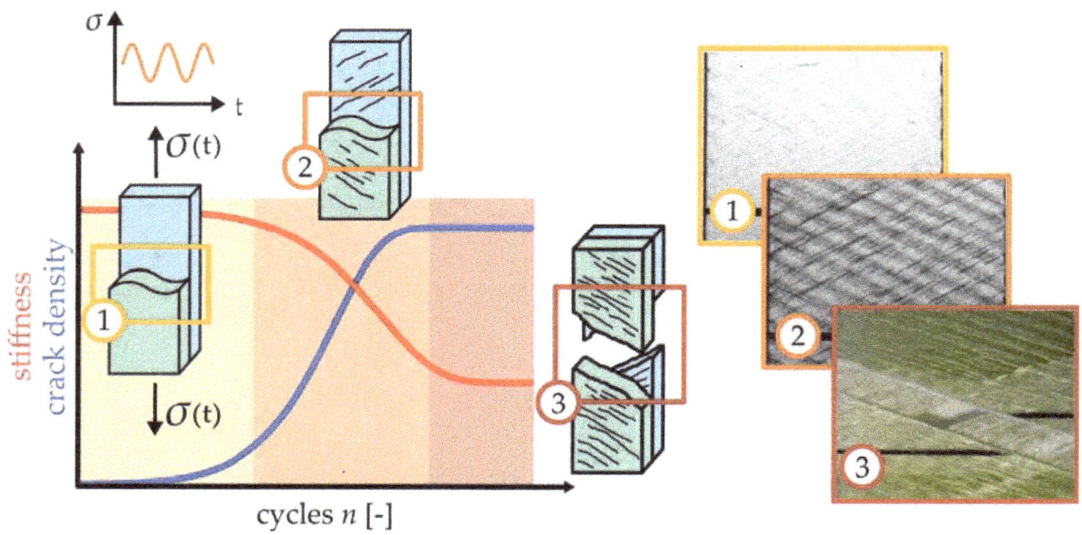

Figure 1. Evolution of off-axis cracks during fatigue tests. The stiffness stays constant as long as there are no cracks (1). After the onset of matrix cracking (2), the cracks grow in number and size and the stiffness starts to decrease due to the damage. (3) The last stage of matrix damage in composites is crack saturation and finally total failure usually due to other damage mechanisms like fiber failure and delamination.

Many fatigue models have been developed to describe the effect of fatigue damage in composite laminates. Degrieck and Van Paepegem [26] sorted them into three categories: fatigue life models, phenomenological models, and progressive damage models. To accurately describe the effect of distinct damage modes, progressive damage models are the most promising candidates since they take the actual cause of the degradation of a mechanical property—the damage—into account. Usually, this is done in a two step approach. In the first step, a damage model describes the evolution of a damage variable with respect to the undamaged material. The second step computes the effect of this damage on mechanical properties. Many models have been developed that establish a connection between off-axis cracks and laminate stiffness [9–17,27]. Often, the Finite Element Method (FEM) is used to compute the elastic response of a laminate. One widely-used approach is to model a representative volume element of the laminate with cracks in one or more plies [17,19].

Resolving the laminate into its plies and computing the effect of cracks on the ply-level has the advantage that the elastic response of any laminate can be computed. The drawback of FEM is that it is time-consuming when implemented as a sub-model for evaluation of large composite structures. Carraro et al. [9] further compared several modeling techniques and developed an analytical model based on shear lag analysis with the capability to compute the elastic response of any symmetric laminate. One drawback of this model is that the laminate must be symmetric and the elastic response is computed for cracks in both symmetric plies. This means that bending loads which yield unsymmetrical stress distributions in the laminate cannot be accounted for. Schuecker et al. [27] proposed a static damage model that computes the stiffness degradation based on the Mori-Tanaka method. The Mori-Tanaka method is a micro-mechanical approach to compute homogenized material properties of a material consisting of a matrix and an embedded inclusion [28]. This allows to compute the stiffness degradation for each ply separately and then combine all plies using classical laminate theory to compute the overall stiffness of the damaged laminate. In Schuecker's approach, the effect of crack-like void inclusions on the stiffness of a ply is computed. Even though this model has been developed for static load cases, it should also be applicable to fatigue since the stiffness degradation is only dependent on the amount of cracks. The main advantages of this model are:

1. The Mori-Tanaka homogenization scheme computes the effect of damage on the stiffness. The resulting stiffness tensor is positive, definite and symmetric. Therefore it meets the thermodynamic limits of the engineering constants of the damaged material without having to develop individual correlations for all the independent engineering constants [29].
2. The model can be calibrated easily to a new material. All data to calibrate the model can be obtained with standard static and fatigue tests.
3. The stiffness degradation is ply-based. Classical laminate theory is used to compute the overall stiffness of the laminate. Therefore, stress-redistribution to other plies is automatically accounted for. The model builds on well-established methods and focuses on efficiency.

Naturally, the model has limitations and prerequisites arising from classical laminate theory and the Mori-Tanaka method, but it has proven to provide an overall relatively simple yet effective approach to compute stiffness degradation of laminates due to cracks. Also, it does not consider delaminations or other damage mechanisms of composite laminates [27,30].

In this work, Schuecker's degradation model is combined with crack detection by replacing the evolution function for static loads by the crack evolution detected from experimental data. Based on the off-axis crack density from experiments, the stiffness degradation computed by the degradation model is compared to experimental stiffness data. Opposed to other works like [31], where crack densities of similar specimens have been averaged, we take the crack density of each individual test and compute the resulting stiffness degradation. This enables a comparison of experimental stiffness data and predictions based on experimental crack density data for individual specimens. Also, a mostly-automated procedure to fit experimental crack density data from the automated crack detection is presented.

2. Methods

2.1. Experimental Fatigue Data

Experimental data from fatigue tests of $\pm\theta_s$ GFRP laminates from [32] is used for comparison with computations of the damage model. The specimens with a gauge length of 100 mm, a width of 20 mm, and a thickness of 2 mm (12 layers) had been cut from GFRP laminate plates produced from a unidirectional glass fiber weave and epoxy resin. The plates had been produced by vacuum pressing manually-impregnated glass fiber layers. The stacking sequence of the laminates is $[+\theta_3/-\theta_3]_s$. For the fatigue tests, stress-controlled sinusoidal load cycles with an R-ratio of 0.1 and a frequency of 5 Hz had been periodically interrupted to perform displacement-controlled quasi-static tensile tests. The servo-hydraulic material testing system MTS 810 by MTS Systems Corporations had been used for all tests with an optical displacement measurement system (CV-X100 by Keyence) to provide a free field of few for the images taken during the tests. This procedure allows to track the change in stiffness as a function of the number of cycles and take images for the crack detection. For a more detailed description of the experimental setup, the reader is refered to [32]. In addition to the stiffness of the laminates, the crack density is evaluated from TWLI images with crack detection. For this work, the results of the fatigue tests for ±45°, ±60°, and ±75° laminates are used. Laminates with a ply orientation of less than ±45° show delamination as the main damage mode and only little off-axis cracks. The unidirectional 90° laminates show hardly any cracks before final failure.

For each laminate type, two fatigue tests had been conducted. Table 1 lists the static stiffness, transverse strength R_2, and in-plane shear strength R_{12} of the material. In the referenced data from [32], a miscalculation had happened in the evaluation of the in-plane shear strength and in-plane shear modulus G_{12}, which is corrected in this work. Additionally, the static ply properties are corrected with respect to the fiber volume fraction of the individual specimens. The procedure is described in Appendix A. The aforementioned miscalculation does not effect the validity of the tests since only the evaluation had to be redone. The load level of the fatigue tests, which is the ratio of the maximum load in the fatigue test to the static strength of the laminate, is computed by the Puck failure criterion [33]. The exact computation is given in Appendix B. The load levels of the tests are 75% for ±45°, 78% for ±60°, and 74% for ±75° laminates. The slight differences between load levels arise from the corrections done in the evaluation.

Table 1. Elastic constants of the composite ply material from static tests.

E_1 [GPa]	E_2 [GPa]	ν_{12} [−]	G_{12} [GPa]	R_2 [MPa]	R_{12} [MPa]
35.6	10.9	0.27	3.2	57.9	58.3

2.2. Crack Detection

The Python package *CrackDect* is used to to automatically evaluate the crack density from the TWLI images [25]. This package includes a sightly modified crack detection algorithm compared to [24]. Example pictures of a specimen at the beginning, during, and at the end of the fatigue test prior to image processing are shown in Figure 2. The processing pipeline of the images is as follows:

1. **Shift correction:** Since the individual images from a fatigue test are not aligned perfectly due to increasing strain and unavoidable inaccuracies of the test rig (see Figure 2), the shift of the specimen in the images must be corrected.
2. **Region of interest:** Only the area of the specimen without edges or other features like the black line that is used for optical strain measurement (see Figure 2), is evaluated by the crack detection since they might cause false detections.
3. **Crack detection:** Cracks are detected in a cumulative way. Cracks detected in the nth image are added to the n + 1st image.

The exact procedure of this processing pipeline is explained in [25] where the open-source code of all functions can be obtained. The input parameters are listed in Table 2 for each test series. It is observed that the crack width of the first major visible cracks varies slightly with the fiber orientation. Cracks in the ±45° specimens appear to be thinner than in ±75° specimens. Therefore, the average width of cracks that should be detected by the crack detection is set individually for each test series to get comparable results between the test series. To avoid artifacts in the crack detection or false detection due to inherent noise in the images, cracks of less than 50 pixels (0.7 mm) in length are excluded from the evaluation. The crack density is defined as

$$\rho_c = \frac{\sum_{i=1}^{n} L_i}{A} \quad , \tag{1}$$

with L_i as the length of crack i and A as the area of the region of interest (evaluation area). Since the crack detection only computes the crack density based on pixels, the conversion from pixel to millimeter is also listed in Table 2.

Table 2. Input parameters for CrackDect. The coordinates of the region of interest x_0 to y_1 are given in pixel.

Test	Ply Angle [°]	Crack Width [px]	Pixel per mm	x_0	x_1	y_0	y_1
±45° T1	45	8	69.2	200	1500	0	900
±45° T2	45	8	70.3	200	1500	0	950
±60° T1	60	8	68.8	100	1400	0	850
±60° T2	60	10	70.2	100	1400	0	1000
±75° T1	75	15	69.6	200	1400	0	1000
±75° T2	75	12	70.2	200	1450	0	900

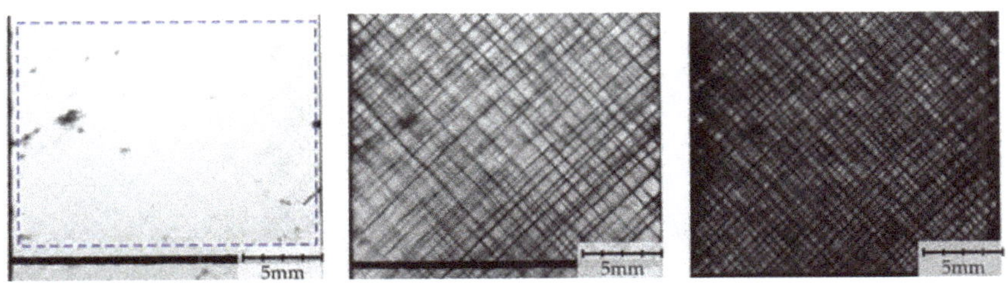

(**a**) (**b**) (**c**)

Figure 2. Example of TWLI images taken from a ±45° specimen with a load level of 75% before the test (**a**), after 254 cycles (**b**), and and after 4013 cycles (**c**). In (**a**), the evaluation area is shown for the crack density (region of interest) marked with the blue rectangle. In (**b**), a typical crack pattern for a ±45° is laminate shown and (**c**) shows the last image taken before failure. The shift of the specimen can easily be observed by the drift of the black line at the bottom from (**a**–**c**). In (**c**), delamination between the plies can be spotted as dark areas.

Instead of using the extracted densities directly, a crack density function is defined by fitting a cumulative Weibull distribution function to the experimental crack density data. The Weibull function is used since it has a form similar to the experimental crack density plotted over the number of cycles in logarithmic space. A direct fit of such a crack density function to the experimental data resulted in convergence problems, even with non-linear least squares algorithms (scipy.optimize.curve_fit) [34]. Therefore, a two-step approach was used to achieve a satisfactory quality of the fit. This fitting process is qualitatively illustrated in Figure 3. The first step is a linear regression in the region where the crack density increases linearly. For experiments that reach crack saturation, the linear fit is done from 30% to 85% of the maximal crack density. If crack saturation is not reached because the specimen fails prior to that, the region for the linear regression extends to 100%. In the second step, the following three-parameter cumulative Weibull distribution function is fitted to the linear regression

$$\rho_c^{fit}(n) = \left[1 - exp\left(-\left(\frac{n - n_0}{\lambda}\right)^k\right)\right] \cdot \rho_c^{sat} \quad , \tag{2}$$

with λ and k as scale and shape parameters respectively, and n_0 to shift the fitted curve along the x-axis. Note that n_0 is a fitting parameter and does not correlate with the cycles to damage initiation n^{init} defined later in Section 3.1. Since the Weibull distribution function has a span of 0–1, the fitted crack density function is scaled with the saturation crack density ρ_c^{sat}.

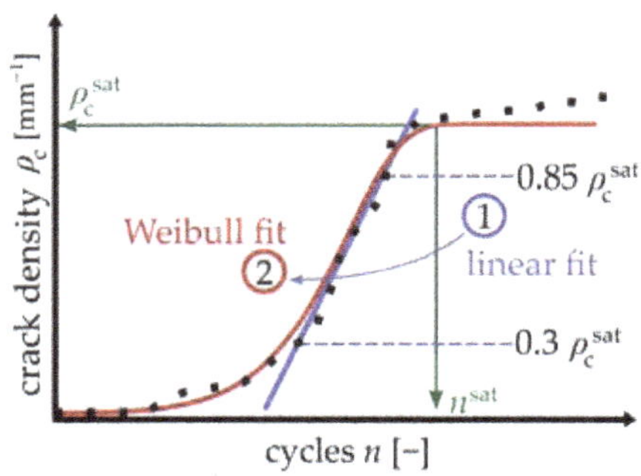

Figure 3. The experimentally obtained crack densities are fitted with a two-step approach. The first step is a linear regression from 30% to 85% of the saturation crack density. In the second step, a cumulative Weibull distribution function is fitted to the linear regression.

2.3. Damage Model

To compute the stiffness degradation of a laminate for a given damage state, Schuecke's damage model is used [27]. Like most progressive damage models, it consists of a part that describes the evolution of a damage variable and a part that computes the effect of this damage on the stiffness of the material. The first part computes the damage variable as a function of the loads for each ply. Then, in the second part, the degradation of stiffness based on the damage variable is computed using the Mori-Tanaka method. Here, only the second part of the model is used since the evolution of the damage variable is obtained by calibration to the fitted crack density functions. The damage variable in the model represents the volume fraction of crack-like inclusions in the Mori-Tanaka formulation, which for void inclusions, is given by

$$\boldsymbol{E}_{MTM,void} = \boldsymbol{E}^{(m)} \left[\boldsymbol{I} + \frac{V}{1-V} (\boldsymbol{I} - \boldsymbol{S})^{-1} \right]^{-1}, \tag{3}$$

where V is the inclusion volume fraction, $\boldsymbol{I}$ is the identity tensor, $\boldsymbol{E}^{(m)}$ is the initial stiffness tensor of the ply, and $\boldsymbol{S}$ is the Eshelby tensor [35,36]. To compute the Eshelby tensor for transversely isotropic materials, the numerical computation scheme by Gavazzi et al. [37] is used.

The Eshelby tensor for an inclusion depends on the surrounding material and the shape of the inclusion. It is assumed that the inclusion geometry is the same for all cracks and independent of the orientation of the ply. Often, an extremely sharp or disk-like inclusion geometry is used when the effect of cracks in a material is computed by the Mori-Tanaka method. Experimental evidence shows that off-axis cracks are often not straight but have a crooked path since the crack has to find the way of least resistance between the fibers. Cracks sometimes even split and merge on the way through the ply [20,38]. Here, the introduction of an orientation tensor to give idealized penny shaped cracks an orientation distribution similar to the crooked paths of the real cracks is avoided. Instead, one oblate ellipsoidal pore that represents the homogenized effect of these cracks qualitatively is used. It has been reported in [39,40] that this approach gives satisfactorily results. In this work, an aspect ratio of 100,1,10 in the 1,2,3-direction of the ply is chosen. The reasoning is as follows: The cracks are substantially longer in fiber direction than in out of plane direction so the 1-direction is set to 10 times the 3-direction. Also, cracks are relatively flat compared to the thickness of the ply so the 3-direction is 10 times the 2-direction. A schematic representation of this idealized inclusion is shown in Figure 4a and its effect on the ply properties is shown in Figure 4b. The curves of E_1, E_2, ν_{12} and G_{12} as function of to the inclusion volume show that E_2 is reduced the most relative to its initial value. The stiffness in fiber direction E_1 is reduced only slightly up to an inclusion volume fraction of 0.1. This behavior qualitatively agrees well with the stiffness degradation of a ply due to off-axis cracks. The degradation of E_1, E_2, ν_{12} and G_{12} could be fine-tuned by adjusting the aspect ratios if additional data from static tensile tests were available that allows to relate the engineering constants of a ply directly to off-axis cracks.

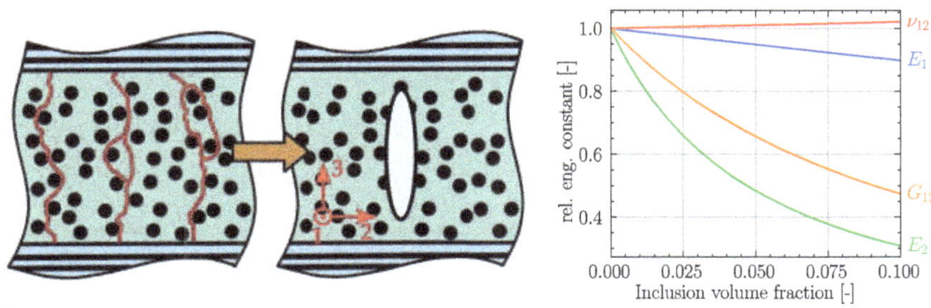

(a)

(b)

Figure 4. Representation of off-axis cracks in one ply (**a**) looking in fiber direction. The crack path is often not straight and can branch or merge. Therefore, an oblate ellipsoid is chosen as a single inclusion shape to compute the homogenized effect of these cracks. The effect of this inclusion is shown in (**b**) with the aspect ratios of 100,1,10 in 1,2,3-direction, respectively.

2.4. Calibration

The Mori-Tanaka method uses inclusion volume fractions to compute the effect of an inclusion in a surrounding matrix material, but since the crack density is analyzed, the model needs to be calibrated to experimental data. A link between the crack density and the inclusion volume fraction for the analyzed stiffness degradation must be established. For this, the $\pm 45°$ tests are used to calibrate the model. The following equation links the crack density (ρ_c) to the inclusion volume fraction (V) with a simple correlation factor (μ).

$$V = \mu \cdot \rho_c \tag{4}$$

The calibration process is illustrated in Figure 5. First, the experimental stiffness data is smoothed to reduce scatter using a lowess filter [41] with a window length of 40% the range of cycles. Then, the stiffness degradation relative to its initial value up to crack saturation (see Table 3) is calculated from this smoothed curve. Parallel to this, a model of the laminate is built with classical laminate theory. For each ply, the stiffness is reduced according to the damage model (see Equation (3)) as a function of the inclusion volume fraction. The inclusion volume fraction to reach the experimental stiffness degradation is optimized with the minimization algorithm from SciPy (*scipy.optimize.minimize*) [34]. From this, the correlation factor μ can easily be computed from Equation (4). This procedure is carried out for the two $\pm 45°$ tests and the average is used as the global calibration constant for the material.

Table 3. Results of the crack detection. The cycle number to damage initiation n^{init}, saturation n^{sat}, and crack density growth rate in semi-logarithmic space $d\rho_c/d(log(n))$ are listed to compare the laminates.

Test	ρ_c^{sat} [mm^{-1}]	n^{init}	n^{sat}	$d\rho_c/d(log(n))$ [mm^{-1}]
±45° T1	3	140	4000	2.8
±45° T2	3	30	1500	2.7
±60° T1	2.1	304	2100	3.2
±60° T2	1.7	487	2200	2.3
±75° T1	1.3	93,086	-	2.9
±75° T2	1.3	68,314	-	2.7

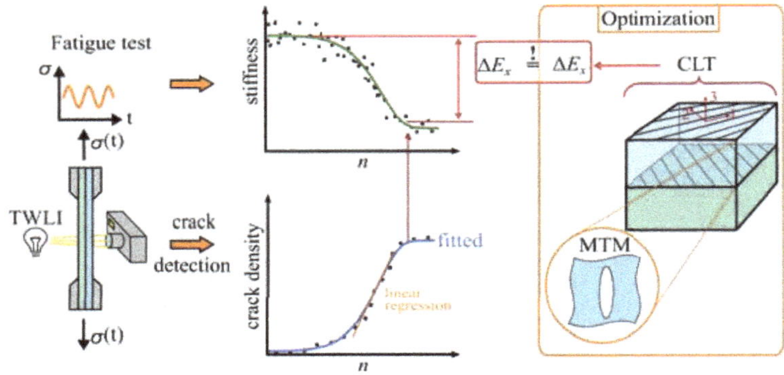

Figure 5. A schematic representation of the correlation process. The stiffness and crack density is evaluated in fatigue tests. The needed inclusion volume fraction is optimized to yield the same stiffness degradation as observed in the experiment. With Equation (4), the correlation constant is then computed from the experimental saturation density.

3. Results and Discussion

3.1. Crack Detection Results

The results of the crack detection are shown in Figure 6. An example of the cracks detected in +60° direction for the ±60° T1 specimen after 759 load cycles is shown in Figure 6a. At this state, the crack density is 0.8 mm/mm^2. The bigger cracks are detected well while cracks smaller than 50 pixel or 0.7 mm in length are filtered out. Since the laminate consists of 12 plies, cracks in the bottom layer will appear fainter and not as sharp as cracks in the top layer. Therefore, the detection is less reliable for cracks in the bottom layer. Cracks in negative fiber-direction are not included because the noise, blur and overlap with cracks in positive fiber direction from the top plies resulted in too many false detections. Since the plies in negative direction develop approximately the same amount of cracks as in positive direction (see Figure 2), only the positive direction is analyzed.

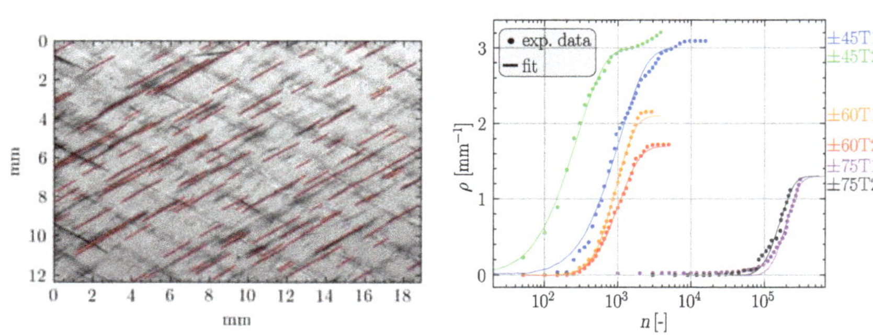

(**a**) (**b**)

Figure 6. Results of the crack detection. An example of the detected cracks for ±60° T1 is shown at 759 cycles (**a**). Cracks are only detected in the chosen direction of +60°. (**b**) shows the crack density results for all tests and the crack density functions.

Figure 6b shows the evolution of the crack density over loading cycles and the corresponding fitted crack density functions for all specimens. In the ±45° laminates, off-axis cracks initiate earlier than in ±60° and ±75° laminates. The ±45° laminates also show the highest crack saturation density. The ±75° laminates do not reach the state of crack saturation because they fail before. The growth rate of the crack density seems to be approximately constant for all tests when plotted on a logarithmic scale.

The results of the crack detection are listed in Table 3. The point of damage initiation n^{init}, crack saturation n^{sat} and the crack density growth rate of the linear region are used to compare the crack detection results. The saturation density ρ_c^{sat} is taken manually at the point where the crack accumulation reaches a plateau. Since the ±75° laminates do not reach crack saturation, the maximum crack density of ±75° T1 is used for the fitted curves. In the ±45° tests, the crack density increases again after reaching a first plateau due to delaminations that cause problems with the crack detection. Therefore, the saturation density is taken at the first plateau. The point of damage initiation n^{init} is defined as the first point with a crack density above 0.1 mm^{-1} and the point of crack saturation n^{sat} are the cycles needed to reach the saturation crack density ρ_c^{sat}. For ±75° tests, crack saturation is not reached. The crack density growth rate is the slope of the linear regression, the first step of the fitting process.

These results show that the automated crack detection is suited to efficiently obtain off-axis crack densities of multidirectional GRFP laminates. Damage initiation is easily detected and the crack density function can be modeled by a three-parameter cumulative Weibull distribution by fitting it to the linear regression of the region of constant crack density increase in semi-logarithmic space (see Figure 6b). The number of cycles until the saturation density is reached are also captured by the crack detection. It should be noted that the accuracy of the crack detection decreases when approaching saturation due to delamination and merging of cracks (see Figure 2c). From our experience, almost all cracks are detected up to approximately 80 percent of the saturation level. From there on, the merging of side-by-side cracks into one black line and delamination leads to misses and false detections. To improve the accuracy of the crack detection near saturation, image differencing techniques that allow to see only changes from one image to the next could be used (see [24]). The prerequisite to this is an extremely precise shift correction. This can be achieved with position markers on the images that allow a precise tracking of image shift and distortion. Since our images did not have these markers, a simpler version of the shift correction had to be used.

3.2. Stiffness Degradation Model

The calibration of the damage model for the material from [32] yields a correlation constant of 0.011 mm for the chosen inclusion aspect ratios of 100,1,10. Note that the correlation constant depends on the chosen aspect ratios. The results of the computed stiffness degradation are shown in Figure 7 along with experimental data for all laminates. The comparison between model and experiments for the ±45° laminates and ±60° T2 shows good agreement. Since the calibration constant is computed from the ±45° tests, good agreement of the stiffness drop is expected for ±45° specimens. Nonetheless, the curves follow the same shape as the experiments which is determined by the inclusion geometry. This indicates that the chosen inclusion geometry describes the effect of off-axis cracks for this laminate and our assumptions on the inclusion geometry are reasonable. For ±60° T1, the detected crack density is higher than for ±60° T2. Therefore, the computed stiffness degradation is also higher compared to ±60° T2. The trend of experimental data and the shape of the curves from the degradation model for ±45° and ±60° laminates is similar. Contrary to the ±45° tests, the experimentally observed stiffness degradation does not stop at the saturation of the off-axis cracks for the ±60° laminates (see Figure 7). As listed in Table 3, crack saturation is reached at around 2000 cycles for the ±60° tests. The computed stiffness of the ±75° laminates drops earlier than the experimental curves. It seems that off-axis cracks in ±75° laminates do not have the same effect at the local ply coordinate system as for ±45° laminates.

The error of the model increases from the ±45° laminates, where the calibration is carried out to the ±75° laminates. Carraro et al. [42] have shown that the macroscopic damage initiation is driven by two damage modes that depend on the level of in-plane shear stress. For plies with mostly in-plane shear stresses σ_{12} in the ply coordinate system shown in Figure 4a, the driving force for damage evolution at the micro scale is local maximum principal stress (LMPS). For plies with mostly positive in-plane transverse stress σ_{22}, local hydrostatic stress (LHS) is the driving force. The shift between LMPS and LHS in GFRP occurs at a fiber direction of around 60° [42]. Fractographic images also show different crack patterns for off-axis cracks depending on the in-plane shear stress. It has also been found that shear stress significantly reduces the number of cycles for damage initiation [7,43]. With differences in the micro-structure of the fracture plane between the two damage types, the effect of cracks on the ply stiffness can also be expected to be different for the LMPS/LHS damage types. For our model, this would require a separate correlation for the ±45° and ±75° tests, since these tests correspond to LMPS and LHS type damage, respectively. The presence of these two separate damage types would also explain the large difference in the number of cycles to damage initiation from ±45° to ±75° laminates (see Figure 6). The ±45° and ±60° tests show damage initiation at less than 500 cycles with ±45° being a bit lower than ±60°. On the other hand, the ±75° tests show damage initiation at more than 50,000 cycles, although the fatigue load level for all tests has been set to about 75% of the static strength. These findings back the theory of two distinct microscopic damage types controlled by in-plane shear stresses.

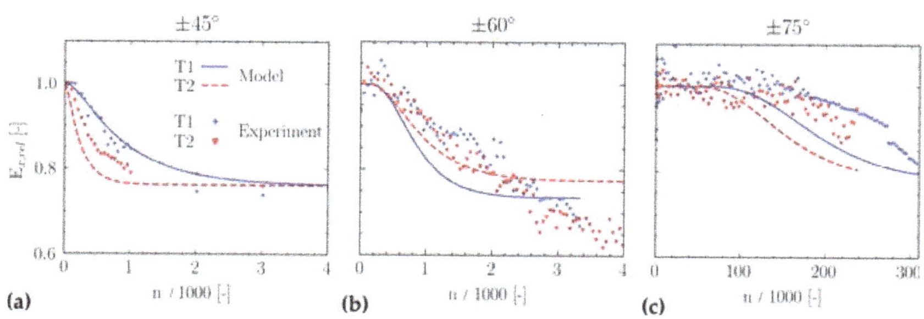

Figure 7. Stiffness degradation for the individual tests and damage model results. Comparison for (**a**) ±45° tests, (**b**) ±60° tests, and (**c**) ±75° tests.

In Figure 8, the experimental stiffness is plotted over crack density up to saturation. At the beginning, a small drop in stiffness without an increase in the crack density is visible for some specimens. After this initial drop in stiffness, all curves except ±75° T1 show a linear correlation up to saturation. Note that this does not conclude that the degradation of stiffness is linear since the crack density follows a S-shaped curve. For ±75° T1, a distinctive kink compared to the linear regression is visible. Also, the scatter of the experimental data for ±75° T1 (see Figure 7) is higher at the beginning. At around 200,000 cycles, this scatter nearly vanishes. This could be an indication of a problem in the evaluation or experimental procedure for this specimen. The linear relation between crack density and stiffness shows that the crack density is a good choice as a damage parameter.

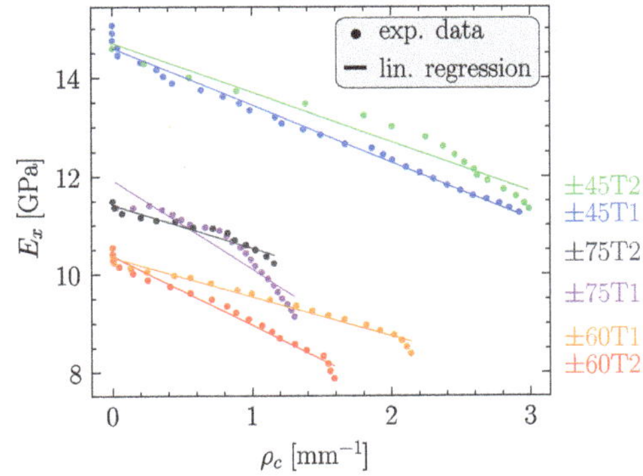

Figure 8. Relationship of the experimentally determined stiffness and crack density up to saturation. For the ±45° and ±60° specimen, the stiffness degradation correlates linearly with the crack density. Note that the fact that the stiffness for the ±75° specimen is higher compared to ±60° is a result of the given ply material properties and in accordance with classical laminate theory.

4. Conclusions

In this work, the effect of off-axis cracks in GFRP laminates is studied by using test results of the crack density and computing the effect of these cracks on the stiffness of the laminate. It is shown that crack detection can be used to efficiently evaluate images of off-axis cracks from fatigue tests and the fatigue crack density function can be modeled by a three-parametric cumulative Weibull distribution function. A mostly-automated scheme is presented for the calibration of the crack density functions from experimental data. Furthermore, the stiffness degradation model for multidirectional fiber-reinforced polymer laminates from Schuecker, which uses Mori-Tanaka homogenization on the ply-level, is tested against experimental fatigue data. The results suggest that it is necessary to distinguish the effect of off-axis cracks on the stiffness of a ply depending on the microscopic crack type. This requires a separate calibration for cracks formed under LMPS and LHS conditions, respectively. This observation agrees well with the theory by Carraro and Quaresimin which also distinguishes the evolution of off-axis fatigue cracks based on the micro-damage mechanisms driven by LMPS and LHS. A distinct jump in cycles to damage initiation is also found where the micro-damage mechanisms change. A new test campaign is under way to further test the damage model for both microscopic damage types.

Author Contributions: Conceptualization, C.S., M.P. and M.D.; methodology, M.D., C.S. and M.P.; software, M.D., validation, M.D.; formal analysis, M.D.; investigation, M.D.; data curation, M.D. and C.S.; writing—original draft preparation, M.D.; writing—review and editing, C.S. and M.P.; visualization, M.D.; supervision, C.S. and M.P.; project administration, C.S.; funding acquisition, C.S. All authors have read and agreed to the published version of the manuscript.

Funding: Part of this work has been performed within the COMET-project Experimental and numerical analysis of the damage tolerance behavior of manufactured induced defects and bonded repairs in structural aerospace composite parts (project-no.: VI-3.04) at the Polymer Competence Center Leoben GmbH (PCCL, Austria) within the framework of the COMET-program of the Federal Ministry for Transport, Innovation and Technology and the Federal Ministry for Digital and Economic Affairs with contributions by Montanuniversität Leoben (Chair of Designing Plastics and Composite Materials) and MAGNA Powertrain Engeneering Center Steyr GmbH CO KG. The PCCL is funded by the Austrian Government and the State Governments of Styria, Lower Austria and Upper Austria.

Data Availability Statement: The data presented in this study are available upon request from the corresponding author. The algorithms of the crack detection are available at GitHub (https://github.com/mattdrvo/CrackDect, accessed on 28 December 2021).

Acknowledgments: Special thanks go to René Rieser and the Chair of Materials Science and Testing of Polymers, Montanuniversitaet Leoben for providing experimental data and images for the crack detection.

Conflicts of Interest: The authors declare no conflict of interest. The funders had no role in the design of the study; in the collection, analyses, or interpretation of data; in the writing of the manuscript, or in the decision to publish the results.

Abbreviations

The following abbreviations are used in this manuscript:

GFRP	Glass fiber-reinforced polymer
TWLI	Transilluminated white light imaging
FEM	Finite element method
LMPS	Local maximum principal stress
LHS	Local hydrostatic stress

Appendix A. Quasi-Static Material Parameters

The basic characterization of the quasi-static material parameters is done with UD0° specimens for E_1 and ν_{12}, UD90° specimens for E_2 and ±45° specimens for G_{12}. In the referenced data set [32], a miscalculation has happened in the evaluation of the in-plane shear modulus G_{12} which is corrected here according to DIN EN ISO 14129. Additional measurements of the fiber volume fraction revealed differences between the specimens. Therefore, the ply stiffness of each laminate is corrected to a fiber volume fraction of 45%. For this, the semi-empirical Chamis model is used to approximate the engineering constants as a function of the fiber volume fraction since it showed good agreement with experimental data for GRFP [44]. This modified rule of mixture (ROM) replaces the fiber volume fraction in the iso-stress model (Reuss) for E_2^c and G_{12}^c with the root of the fiber volume fraction. The iso-strain model (Voigt) for E_1^c and ν_{12}^c is not altered. For isotropic matrix and fibers, this leads to the following set of equations:

$$\begin{aligned} E_1^c &= (1 - V^f)E^m + V^f E_1^f \\ E_2^c &= \left[\frac{1 - \sqrt{V^f}}{E_m} + \frac{\sqrt{V^f}}{E_2^f} \right]^{-1} \\ \nu_{12}^c &= (1 - V^f)\nu^m + V^f \nu^f \\ G_{12}^c &= \left[\frac{1 - \sqrt{V^f}}{G_m} + \frac{\sqrt{V^f}}{G_f} \right]^{-1} \end{aligned} \tag{A1}$$

For the correction, the elastic constants of the matrix E^m and ν^m must be known. Since the elastic constants for each laminate at a certain fiber volume fraction are tested, the elastic constants of the fibers can be estimated. This estimate is then reinserted in the same equations to obtain the elastic constants of the composite as a function of the fiber volume fraction. In Table A1, the fiber volume fractions for each laminate and the tested elastic constants are listed. With this data, the elastic constants at a fiber volume fraction of 45% in Table 1 are computed. In the degradation model, the stiffness is corrected for each laminate individually.

Table A1. Fiber volume fractions of the laminates and the quasi-static elastic constants in ply-coordinates determined from tests.

Laminate	Fiber Volume Fraction [−]	Elastic Constant
0°	42.2	E_1^c: 33.6 GPa, ν_{12}^c: 0.28
90°	42.4	E_2^c: 10.3 GPa
±45°	52.8	G_{12}^c: 3.7 GPa
±60°	41.8	-
±75°	45.7	-
Matrix	-	E^m: 3.55 GPa, ν^m: 0.43 GPa

Appendix B. Fatigue Load Level

The fatigue load level of laminates is computed by Puck´s failure criterion [33]. A load level of 75% means that the laminate is loaded in the fatigue tests up to 75% of its static strength. In Figure A1, the stress space for Puck mode A with the stress vectors of the tests is shown. In the ±45° laminates used to test the in-plane shear strength according to DIN EN ISO 14129, significant transverse stresses σ_{22} are present. The shear strength R_{12} for the material is therefore corrected using Puck´s failure criterion based on the strength of the ±45° and UD90° laminates. With this corrections, the fatigue level of the tests are 75% for ±45°, 78% for ±60° and 74% for ±75° laminates. As a comparison, the stress vector for a typical carbon fiber-reinforced laminate is also shown. The higher ratio of fiber stiffness to transverse stiffness yields nearly no transverse stress so no correction is necessary for carbon fiber-reinforced composites.

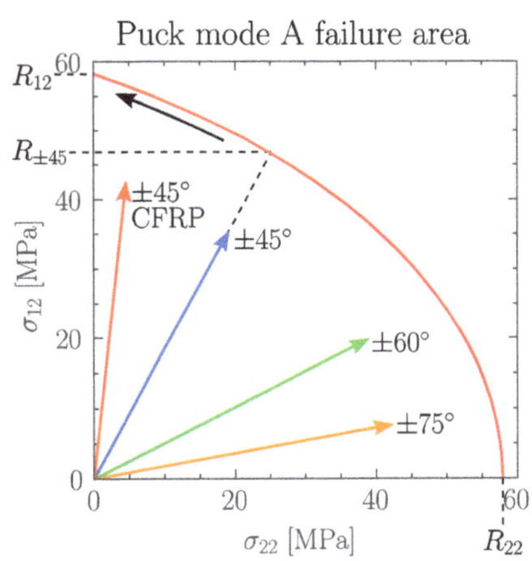

Figure A1. Puck mode A failure area (red) with stress vectors of the tests. In the ±45° laminate, significant transverse stress σ_{22} is present. Therefore, the in-plane shear strength R_{12} is corrected with Pucks failure criterion (black arrow). As a comparison, the stress vector for a typical ±45° CFRP laminate is also shown (red arrow).

References

1. Reifsnider, K.L. (Ed.) *Fatigue of Composite Materials*; Number 4 in Composite Materials Series; Elsevier: Amsterdam, The Netherlands; New York, NY, USA, 1991.
2. Tong, J.; Guild, F.J.; Ogin, S.L.; Smith, P.A. Off-axis fatigue crack growth and the associated energy release rate in composite laminates. *Appl. Compos. Mater.* **1997**, *4*, 349–359. [CrossRef]
3. Bartley-Cho, J.; Gyu Lim, S.; Hahn, H.; Shyprykevich, P. Damage accumulation in quasi-isotropic graphite/epoxy laminates under constant-amplitude fatigue and block loading. *Compos. Sci. Technol.* **1998**, *58*, 1535–1547. [CrossRef]
4. Wharmby, A.; Ellyin, F. Damage growth in constrained angle-ply laminates under cyclic loading. *Compos. Sci. Technol.* **2002**, *62*, 1239–1247. [CrossRef]

5. Wharmby, A. Observations on damage development in fibre reinforced polymer laminates under cyclic loading. *Int. J. Fatigue* **2003**, *25*, 437–446. [CrossRef]
6. Tohgo, K.; Nakagawa, S.; Kageyama, K. Fatigue behavior of CFRP cross-ply laminates under on-axis and off-axis cyclic loading. *Int. J. Fatigue* **2006**, *28*, 1254–1262. [CrossRef]
7. Quaresimin, M.; Carraro, P. Damage initiation and evolution in glass/epoxy tubes subjected to combined tension–torsion fatigue loading. *Int. J. Fatigue* **2014**, *63*, 25–35. [CrossRef]
8. Shen, H.; Yao, W.; Qi, W.; Zong, J. Experimental investigation on damage evolution in cross-ply laminates subjected to quasi-static and fatigue loading. *Compos. Part B Eng.* **2017**, *120*, 10–26. [CrossRef]
9. Carraro, P.; Quaresimin, M. A stiffness degradation model for cracked multidirectional laminates with cracks in multiple layers. *Int. J. Solids Struct.* **2015**, *58*, 34–51. [CrossRef]
10. Glud, J.; Dulieu-Barton, J.; Thomsen, O.; Overgaard, L. Fatigue damage evolution in GFRP laminates with constrained off-axis plies. *Compos. Part A Appl. Sci. Manuf.* **2017**, *95*, 359–369. [CrossRef]
11. Zhang, D.; Ye, J.; Lam, D. Ply cracking and stiffness degradation in cross-ply laminates under biaxial extension, bending and thermal loading. *Compos. Struct.* **2006**, *75*, 121–131. [CrossRef]
12. Thionnet, A.; Renard, J. Laminated composites under fatigue loading: A damage development law for transverse cracking. *Compos. Sci. Technol.* **1994**, *52*, 173–181. [CrossRef]
13. Nuismer, R.J.; Tan, S.C. Constitutive Relations of a Cracked Composite Lamina. *J. Compos. Mater.* **1988**, *22*, 306–321. [CrossRef]
14. Varna, J. Modelling mechanical performance of damaged laminates. *J. Compos. Mater.* **2013**, *47*, 2443–2474. [CrossRef]
15. Joffe, R.; Varna, J. Analytical modeling of stiffness reduction in symmetric and balanced laminates due to cracks in 90° layers. *Compos. Sci. Technol.* **1999**, *59*, 1641–1652. [CrossRef]
16. Jagannathan, N.; Gururaja, S.; Manjunatha, C. Probabilistic strength based matrix crack evolution model in multi-directional composite laminates under fatigue loading. *Int. J. Fatigue* **2018**, *117*, 135–147. [CrossRef]
17. Singh, C.V.; Talreja, R. A synergistic damage mechanics approach for composite laminates with matrix cracks in multiple orientations. *Mech. Mater.* **2009**, *41*, 954–968. [CrossRef]
18. Quaresimin, M.; Carraro, P. On the investigation of the biaxial fatigue behaviour of unidirectional composites. *Compos. Part B Eng.* **2013**, *54*, 200–208. [CrossRef]
19. Adden, S.; Horst, P. Stiffness degradation under fatigue in multiaxially loaded non-crimped-fabrics. *Int. J. Fatigue* **2010**, *32*, 108–122. [CrossRef]
20. Jespersen, K.M.; Glud, J.A.; Zangenberg, J.; Hosoi, A.; Kawada, H.; Mikkelsen, L.P. Uncovering the fatigue damage initiation and progression in uni-directional non-crimp fabric reinforced polyester composite. *Compos. Part A Appl. Sci. Manuf.* **2018**, *109*, 481–497. [CrossRef]
21. Sket, F.; Enfedaque, A.; Alton, C.; González, C.; Molina-Aldareguia, J.; Llorca, J. Automatic quantification of matrix cracking and fiber rotation by X-ray computed tomography in shear-deformed carbon fiber-reinforced laminates. *Compos. Sci. Technol.* **2014**, *90*, 129–138. [CrossRef]
22. Sørensen, B.F.; Talreja, R. Analysis of Damage in a Ceramic Matrix Composite. *Int. J. Damage Mech.* **1993**, *2*, 246–271. [CrossRef]
23. Lafarie-Frenot, M.; Hénaff-Gardin, C. Formation and growth of 90° ply fatigue cracks in carbon/epoxy laminates. *Compos. Sci. Technol.* **1991**, *40*, 307–324. [CrossRef]
24. Glud, J.; Dulieu-Barton, J.; Thomsen, O.; Overgaard, L. Automated counting of off-axis tunnelling cracks using digital image processing. *Compos. Sci. Technol.* **2016**, *125*, 80–89. [CrossRef]
25. Drvoderic, M.; Rettl, M.; Pletz, M.; Schuecker, C. CrackDect: Detecting crack densities in images of fiber-reinforced polymers. *SoftwareX* **2021**, *16*, 100832. [CrossRef]
26. Degrieck, J.; Van Paepegem, W. Fatigue damage modeling of fibre-reinforced composite materials: Review. *Appl. Mech. Rev.* **2001**, *54*, 279–300. [CrossRef]
27. Schuecker, C.; Pettermann, H. A continuum damage model for fiber reinforced laminates based on ply failure mechanisms. *Compos. Struct.* **2006**, *76*, 162–173. [CrossRef]
28. Mori, T.; Tanaka, K. Average stress in matrix and average elastic energy of materials with misfitting inclusions. *Acta Metall.* **1973**, *21*, 571–574. [CrossRef]
29. Benveniste, Y.; Dvorak, G.; Chen, T. On diagonal and elastic symmetry of the approximate effective stiffness tensor of heterogeneous media. *J. Mech. Phys. Solids* **1991**, *39*, 927–946. [CrossRef]
30. Schuecker, C.; Pettermann, H. Constitutive ply damage modeling, FEM implementation, and analyses of laminated structures. *Comput. Struct.* **2008**, *86*, 908–918. [CrossRef]
31. Carraro, P.; Maragoni, L.; Quaresimin, M. Prediction of the crack density evolution in multidirectional laminates under fatigue loadings. *Compos. Sci. Technol.* **2017**, *145*, 24–39. [CrossRef]
32. Rieser, R. Damage Mechanics of Composites under Fatigue Loads. Ph.D. Thesis, Montanuniversität Leoben, Leoben, Austria, 2016.
33. Puck, A. Failure analysis of FRP laminates by means of physically based phenomenological models. *Compos. Sci. Technol.* **2002**, *62*, 1633–1662. [CrossRef]

34. SciPy 1.0 Contributors; Virtanen, P.; Gommers, R.; Oliphant, T.E.; Haberland, M.; Reddy, T.; Cournapeau, D.; Burovski, E.; Peterson, P.; Weckesser, W.; et al. SciPy 1.0: Fundamental algorithms for scientific computing in Python. *Nat. Methods* **2020**, *17*, 261–272. [CrossRef]
35. Eshelby, J.D. The determination of the elastic field of an ellipsoidal inclusion, and related problems. *Proc. R. Soc. London Ser. Math. Phys. Sci.* **1957**, *241*, 376–396. [CrossRef]
36. Parnell, W.J. The Eshelby, Hill, Moment and Concentration Tensors for Ellipsoidal Inhomogeneities in the Newtonian Potential Problem and Linear Elastostatics. *J. Elast.* **2016**, *125*, 231–294. [CrossRef]
37. Gavazzi, A.C.; Lagoudas, D.C. On the numerical evaluation of Eshelby's tensor and its application to elastoplastic fibrous composites. *Comput. Mech.* **1990**, *7*, 13–19. [CrossRef]
38. Miskdjian, I.; Hajikazemi, M.; Van Paepegem, W. Automatic edge detection of ply cracks in glass fiber composite laminates under quasi-static and fatigue loading using multi-scale Digital Image Correlation. *Compos. Sci. Technol.* **2020**, *200*, 108401. [CrossRef]
39. Vasylevskyi, K.; Drach, B.; Tsukrov, I. On micromechanical modeling of orthotropic solids with parallel cracks. *Int. J. Solids Struct.* **2018**, *144–145*, 46–58. [CrossRef]
40. Drach, B.; Tsukrov, I.; Trofimov, A. Comparison of full field and single pore approaches to homogenization of linearly elastic materials with pores of regular and irregular shapes. *Int. J. Solids Struct.* **2016**, *96*, 48–63. [CrossRef]
41. Seabold, S.; Perktold, J. Statsmodels: Econometric and Statistical Modeling with Python. In Proceedings of the 9th Python in Science Conference, Austin, TX, USA, 28 June– 3 July 2010; pp. 92–96. [CrossRef]
42. Carraro, P.; Quaresimin, M. A damage based model for crack initiation in unidirectional composites under multiaxial cyclic loading. *Compos. Sci. Technol.* **2014**, *99*, 154–163. [CrossRef]
43. Plumtree, A. Fatigue damage evolution in off-axis unidirectional CFRP. *Int. J. Fatigue* **2002**, *24*, 155–159. [CrossRef]
44. Younes, R.; Hallal, A.; Fardoun, F.; Hajj, F. Comparative Review Study on Elastic Properties Modeling for Unidirectional Composite Materials. In *Composites and Their Properties*; Hu, N., Ed.; InTech: London, UK, 2012. [CrossRef]

Journal of
Composites Science

Article

Numerical and Experimental Evaluation of Mechanical and Ring Stiffness Properties of Preconditioning Underground Glass Fiber Composite Pipes

Mohamed K. Hassan [1,2,*], Ahmed F. Mohamed [1,3,*], Khalil Abdelrazek Khalil [4,5,*] and Mohammed Y. Abdellah [1,6,*]

1 Mechanical Engineering Department, College of Engineering and Islamic Architecture, Umm Al-Qura University, Makkah 21421, Saudi Arabia
2 Production Engineering & Design Department, Faculty of Engineering, Minia University, Minia 61111, Egypt
3 Mechanical Engineering Department, Faculty of Engineering, Sohag University, Sohag 82524, Egypt
4 Department of Mechanical and Nuclear Engineering, College of Engineering, University of Sharjah, Sharjah 27272, United Arab Emirates
5 Materials Department, Faculty of Energy Engineering, Aswan University, Aswan 81521, Egypt
6 Mechanical Engineering Department, Faculty of Engineering, South Valley University, Qena 83521, Egypt
* Correspondence: mkibrahiem@uqu.edu.sa (M.K.H.); afmohamed@uqu.edu.sa (A.F.M.); kabdelmawgoud@sharjah.ac.ae (K.A.K.); myahmad@uqu.edu.sa (M.Y.A.)

Citation: Hassan, M.K.; Mohamed, A.F.; Khalil, K.A.; Abdellah, M.Y. Numerical and Experimental Evaluation of Mechanical and Ring Stiffness Properties of Preconditioning Underground Glass Fiber Composite Pipes. *J. Compos. Sci.* **2021**, *5*, 264. https://doi.org/10.3390/jcs5100264

Academic Editor: Stelios K. Georgantzinos

Received: 8 September 2021
Accepted: 8 October 2021
Published: 10 October 2021

Abstract: The mechanical and ring stiffness of glass fiber pipes are the most determining factors for their ability to perform their function, especially in a work environment with difficult and harmful conditions. Usually, these pipes serve in rough underground environments of desert and petroleum fields; therefore, they are subjected to multi-type deterioration and damage agents. In polymers and composite materials, corrosion is identified as the degradation in their properties. In this study, tension and compression tests were carried out before and after preconditioning in a corrosive agent for 60 full days to reveal corrosion influences. Moreover, the fracture toughness is measured using a standard single edge notch bending. Ring stiffness of such pipes which, are considered characteristic properties, is numerically evaluated using the extended finite element method before and after preconditioning. The results reported that both tensile and compressive strengths degraded nearly more than 20%. Besides the fracture toughness decrease, the stiffness ring strength is reduced, and the finite element results are in good agreement with the experimental findings.

Keywords: composite; pipe; fracture toughness; damage; ring stiffness infrastructure industries

1. Introduction

Polymer-based composite materials have several important properties, such as lightweight, specific Young's modulus, and specific impact resistance, making them suitable for use in a wide range of industrial and aerospace applications [1,2]. In various environments, the stress corrosion and failure caused by these reinforced plastics were investigated [3–6]. Glass fiber reinforced plastic is a structural material made of short or long glass fibers and a thermosetting resin. Quartz particles are sometimes used as reinforcement by combining them with epoxy resin, forming the composite matrix. This matrix and numerous laminates of glass fiber reinforced polymer are used to make glass fiber reinforced pipes. Plastic composite pipes are a cost-effective alternative to metallic pipes. Metallic pipes are not permitted in applications where corrosion, weight, and environmental impact are critical. Fiberglass reinforced pipes have many applications, including pressure piping and water transfer above and below ground [7–10].

Farshad [11] performed compression tests on the fiberglass pipes and conditioned the pipe materials before running the tests in water at room temperature. Throughout the test, sample deflection was measured in relation to time. The findings indicated that the conditioned environment reduces the strength of these materials. In a related study by Nishizaki [12], the effect of water on durability was investigated. It was determined that the flexural strength of GRP is lower when immersed than when only at atmospheric humidity. Bergman [13] was able to handle plastic corrosion for chlorine dioxide, sodium chlorate, and chlorine in various environments. The differences between metal and plastic corrosion were compared to obtain typical reasons for using plastics in the processing industry and specific environments for plastic applications. Furthermore, after reviewing the corrosion behavior of GRP pipes, Hojo et al. [14] identified three types of corrosion: surface reaction, formation of a corrosive layer, and penetration-induced corrosion. Farshad and Nicola [15] tested unsaturated fiberglass reinforced polyester (GRP) pipe rings for long-term stress corrosion. In the lower region of the ring, these specimens were treated with 5% sulfuric acid. Compared to the non-acidic section of the same sample, the deformability of the tested section in the acidic environment was reduced by approximately 75%.

It was discovered that there is a direct relationship between specimen densities and dimensional accuracy with orientation [16]. In addition, the angular orientation caused significant anisotropy in the fracture toughness of single edge notch samples. The effects of seawater aging and curing on polymer composite cylinders used as marine structures were studied. Fully cured composite cylinders outperformed partially cured composite cylinders in terms of mechanical properties. Furthermore, aging in seawater improved hoop strength and stiffness but had no effect on radial strength. Similarly, the hardness and density of the same cylinders were investigated in the presence of salty water and found to change [17]. In addition, the effect of seawater on steel pipes rehabilitated with layers of fiberglass and epoxy was studied [18]. The hoop strength of the pipes being repaired increased before and after immersion, indicating that a thicker composite repair is required for complete rehabilitation. The association between the structural properties of the shell and the end bearing capacity (UBC) of fiberglass-reinforced mortar pipes (GRP) was simulated [19]. In a growing pipeline, the proportion of the size of the fibers and the ratio of the volume of the spirally wound layer increased, but UBC increased with the layers of tubules. It has been reported that the fracture and mechanical properties of GRP pipes degrade after immersion in a corrosive medium (solution of sodium chloride in water) [20,21]. This deterioration was attributed to corrosion caused by the use of a corrosive medium. On the other hand, the effect of thermal aging on the compression behavior of interlocking polymer network composites reinforced with fiberglass was discussed. As the temperature rose and the aging period progressed, the strength of these pipes decreased slightly.

It is observed that all previous work did not deal with fracture toughness and impact strength, besides the ring stiffness properties under a harsh underground environment. Therefore, the main objectives of this study are to investigate and fully comprehend the behavior of glass fiber reinforced epoxy pipes during service in the harsh underground environment of the petroleum field. Consequently, tensile, compression, single edge notch bending, and Charpy impact tests were performed before and after preconditioning in drag water of a petroleum environment field. The ring stiffness properties were numerically measured using an extended finite element method and compared to other experimental work to validate the model with one before preconditioning.

2. Experimental Work

It is well established that standard tests are performed on samples with standard dimensions when determining mechanical or fracture properties. When evaluating the behavior and properties of materials under specific operational conditions, tests are carried out under the same conditions and concerning previous similar experiments. In our case, the test was performed on a portion of the material composed under the same harmful and/or harsh operating conditions to demonstrate the extent of the condition's effect on those materials. entire domain [15,22–24].

2.1. Materials and Methods

According to the values shown in Figure 1 and listed in Table 1, the present study used glass fiber-reinforced polymer pipes with a heterogeneous structure consisting of random matt, roving, unsaturated polyester resin, and sand. Unsaturated polyester resin is environmentally and chemically resistant, and it also bonds the fiber in the pipe structure. Unsaturated polyesters are less expensive than other resins used to produce GRP pipes, but they offer only a slight increase in strength and chemical resistance. It is frequently a cost-effective option for less demanding, low-pressure service. The glass fiber composite pipes were created using the widely used filament winding technology. GRP has a complicated structure with inner and outer surface layers. The barrier and chop layers are followed by the structure layers in the outer and inner surfaces, with quartz sand placed between for a complete description of the manufacturing process, refer to [18]. The composite material is primarily determined by fiber geometry, lay-up thickness, fiber pretension, and manufacturing process quality. These constituent compositions were obtained using the ASTM D3171-99 standard [25] and the ignition removal technique. These pipes are used in chemical wastewater pipelines in the petroleum industry. The elastic properties of the glass fiber composite pipes (GRP) are listed in Table 2. the pipes used in the present study; are taken from situ and were in service for some time.

In Figure 1b, it is noted that the percentage of chlorine gas in petroleum wastewater is greater than that of solid impurities (S, Ca, Br, and Ag). The presence of Cl gas in wastewater increases corrosion [14] and erosion because it acts as an active agent by chemically reacting with the organic components of the composite pipe, such as resin, glass fiber, and quartz particles.

Table 1. Composition of GFR pipes [20,26].

Constituents	Average %	Viscosity/cp.25 °C	Thermal Deformation Temperature/°C	Tensile Strength /MPa	Specific Density (g/cm^3)
Thermosetting unsaturated polyester (Matrix)	30.2%	400	70	65	1.12
Roving	11.8	———	———	3100–3400	2.5
matt	13.5	———	———		
sand	44.5	———	———	———	2.66

Table 2. Elastic properties of GRP [24].

Properties	E_1 (GPa)	E_2 (GPa)	ν_{12}	G_{12} (GPa)	G_{13} (GPa)	G_{23} (GPa)
value	100	9	0.3	3.2	3.2	4

2.2. Preconditioning Procedures

The conditioning procedures are performed by immersion the test samples into a corrosive agent of wastewater taken from the petroleum field. The specimens were put in a wide container, as shown in Figure 2, in a field with sewage. The composition, illustrated previously in Figure 1, of a high amount of chlorine gas, is responsible for corrosion. They had a lousy effect on the pipe's material in a harsh environment. The specimens were left in the corrosive solution for 60 days at a room temperature range. Then the sample was left to dry from water then testing was carried out.

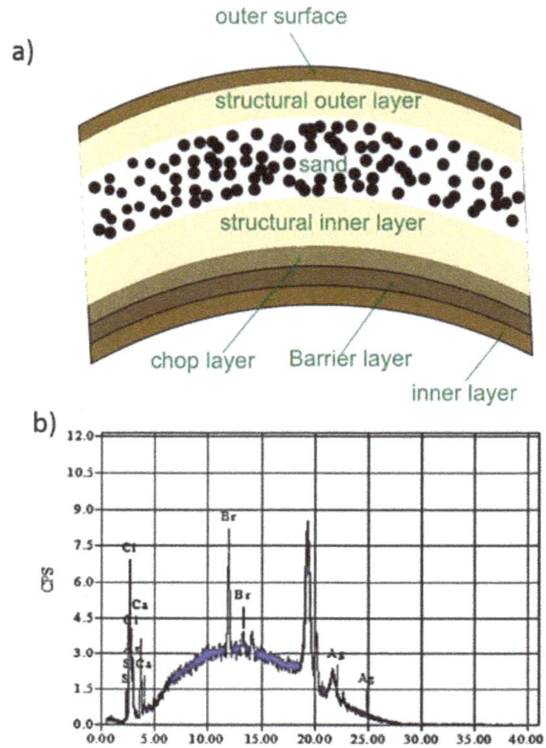

Element	Line type	Energy	ms%	mol%	K	net	Error %
S	K	2.31	4.7234	6.16665	0.0018651	334	1.2631
Cl	K	2.62	58.794201	69.4201	0.0194530	7868	1.3922
Ca	K	3.69	15.0785	15.7482	0.0029557	4484	0.4507
Br	K	11.91	2.6441	1.3852	0.0024013	10392	0.0488
Ag	L	2.98	18.7594	7.2800	0.0065527	1685	2.0120

Figure 1. GRP wastewater system; (**a**) GRP schematic [20] and (**b**) chemical analysis of a wastewater petroleum sample.

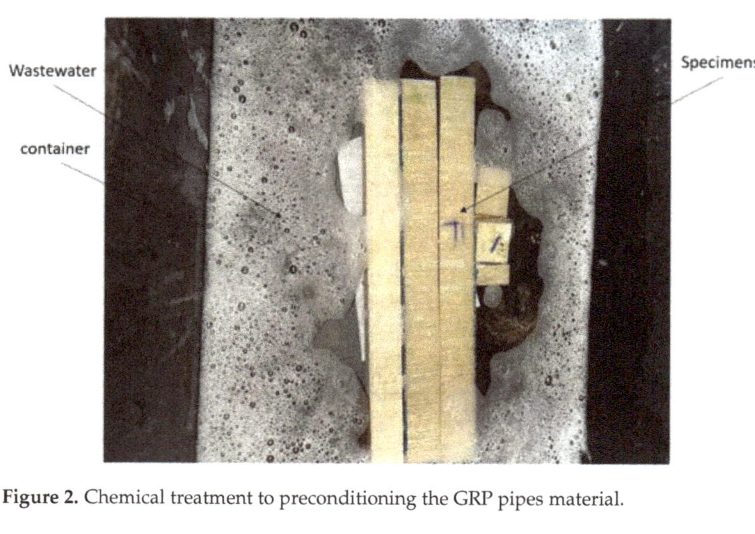

Figure 2. Chemical treatment to preconditioning the GRP pipes material.

2.3. Mechanical Testing

Tension and compression tests were carried out on specimens made from glass fiber reinforced polymer pipe wall thicknesses (DN = 520.7 mm and 13 mm). These tests were performed in accordance with ASTM D3039 [27] and ASTM D3410 [28] standards for tension and compression, respectively, on a computer control electromechanically universal testing machine (machine model WDW-100- Jinan Victory Instrument Co. Ltd, China) [29] with a load capacity of 100 kN and a controlled speed of 2 mm/min. The standard specimen geometry for tension and compression is shown in Figures 3 and 4, respectively. The compressive load was applied in both the longitudinal and transverse directions (see Figure 4a) and (see Figure 4b), respectively). These tests were repeated for specimens immersed into a petroleum field of wastewater for 60 days at room temperature. Each test required the use of five samples for measuring their average.

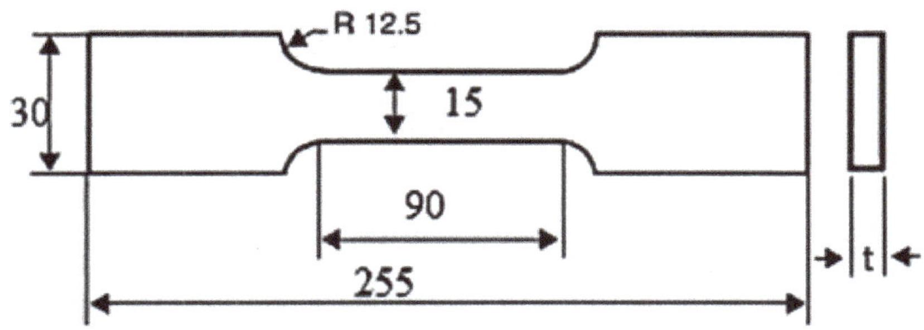

Figure 3. Geometry of tensile test specimens.

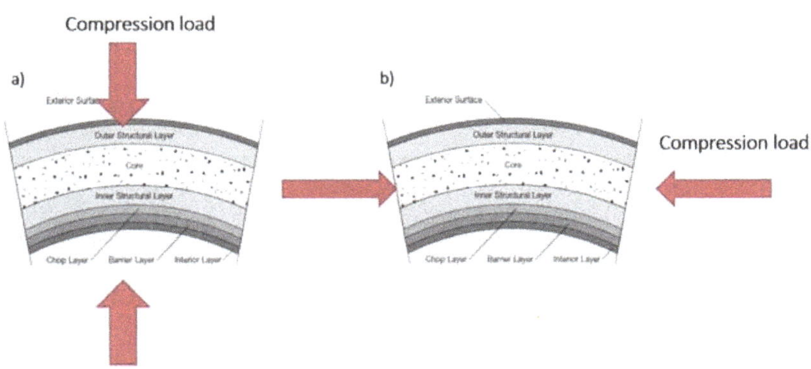

Figure 4. Compression loading direction is either (**a**) transverse or (**b**) longitudinal.

2.4. Fracture Toughness Test

The fracture toughness of these composite materials is emphasized because, in the event of a crack, it allows water to infiltrate and absorb into GRP pipes, causing harmful corrosion and deterioration of mechanical properties. As a result, fracture toughness has emerged as an essential and determining factor of pipe properties, and it must be measured. According to Anderson et al. [30], only one type of fracture toughness test is used to accomplish this. Measuring crack resistance is critical to halt deterioration and achieve good fracture toughness values. The single edge notch test specimen, shown in Figure 5b, was cut from the wall according to the dimensions specified in the ASTM D5045-14 standard [31], as shown in Figure 5a. The specimen has a thickness of 13 mm and a scaled mark on the face of the crack. A notch of about 45 percent of the beam width is formed in the middle of the beam span. After sharpening the root of the notch with a razor blade, the samples were pre-prepared to produce typical incisions 3–5 mm in length. The load-point is plotted against displacement as well.

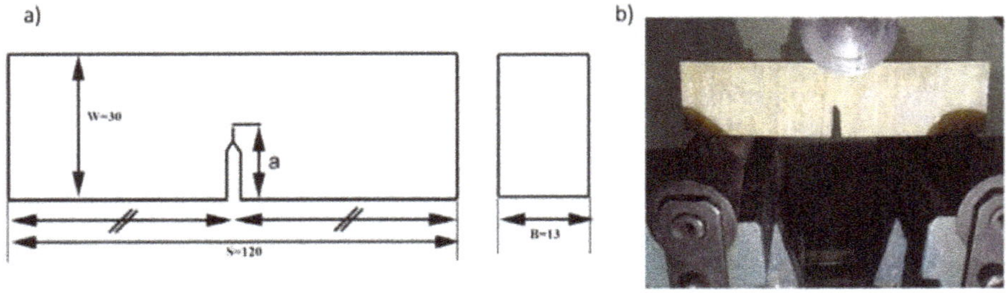

Figure 5. (**a**) specimen geometry, (**b**) test setup for single edge notch bending (SENB).

After the maximum load record, the fracture toughness (K_{IC}) can be measured [32] using the following Equations;

$$K_{IC}\frac{PS}{BW^{3/2}}f(a/w) \tag{1}$$

where $f(a/w)$ is the correction factor and can be calculated as following [32];

$$f(a/w) = 2.9\left(\frac{a}{w}\right)^{0.5} - 4.6\left(\frac{a}{w}\right)^{\frac{3}{2}} + 21.8\left(\frac{a}{w}\right)^{\frac{5}{2}} - 37.6\left(\frac{a}{w}\right)^{\frac{7}{2}} + 38.7\left(\frac{a}{w}\right)^{\frac{9}{2}} \tag{2}$$

where (a) is a crack length, (w) is specimen width, (p) is bending load, (B) is specimen thickness, and (s) is beam span.

Dynamic fracture toughness or impact fracture toughness can be measured using an ASTM D6110 [33] Charpy standard test pattern consisting of a metal (or other material) bar ($55 \times 10 \times 10$ mm) with a notch driven through one of the larger dimensions (see Figure 6a). There are five specimens used. The pendulum impact tester is shown in Figure 6b.

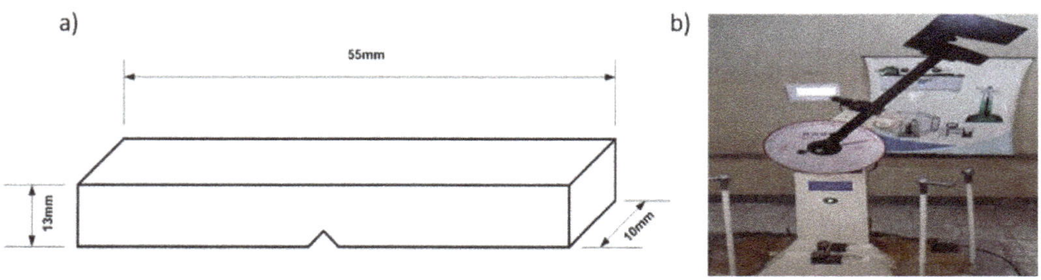

Figure 6. Charpy Impact test: (**a**) specimen geometry, (**b**) test setup.

2.5. Finite Element Method of Compression Stress (Stiffness Ring of Cylinder Test)

The stiffness ring of a cylinder is a physical characteristic of the pipe; it measures the resistance to ring deflection under external load. The following expression defines the stiffness ring of a pipe:

$$S = \frac{EI}{d_m^3} \tag{3}$$

where (S) is ring stiffness, (E) is young modulus, (I) moment of inertia of pipes, and (d_m) is mean pipe diameter.

Belytschko and Black [34] recently developed XFEM, the main idea of which is based on Melenk and Babuska [35], who used the concept of division of finite element units and the enrichment function. XFEM is distinguished because no network update is required to trace the crack path [36], reducing the need to interlace and re-interlace complex interruption surfaces. Because the crack extends around the tip of the fracture without reconnection or refinement, fracture analysis can be performed with high numerical precision; for a detailed description, see [37].

The 3D nonlinear elastic FEM is based on the primary, extended fracture method. A cylindrical domain (520.7 mm diameter × 500 mm) was created as a solid part (see Figure 7a). The reaction force and corresponding deflection are measured using extended finite element analysis. As shown in Figure 7a, the load is used as a constant displacement control for the upper surface of the cylinder while it is fixed at the two asymmetric ends of the cylinder. The 3D stress element type has an 8-node linear brick linear shape, reduced integration, and hourglass control (C3D8R) elements with an approximately hexagonal shape with a global size of 10 for region B were used in the domain. The total number of elements is 38,412; this is the best number chosen from three mesh densities investigated for mesh convergence (see Figure 7b). The maximum principal stress theory of failure was implemented with the mean stress of 30 MPa and 8 MPa for composite pipes without and with preconditioning in wastewater for 60 days. The damage evaluation criterion is the maximum fracture energy and the independent mixed mode that is applicable. The material used for the composite cylinder was based on DN 500 and SN 1000 [24].

Moreover, to create the failure propagation during the loading, a planar crack of (10 mm length) is inserted in the center of the cylinder region (A) (see Figure 7c). The element size in this region is defined as 5. These two values are derived from the standard tensile test results of the specimen of the curved pipe's layers before and after preconditioning. The equivalent young moduli are 35 GPa and 10.5 GPa, and the fracture energy with linear softening was 185 kJ/m^2 and 129 kJ/m^2 for after and before, respectively. The study considers the load only over the outer surface, and there are no internal pressures applied. This study is to simulate the ring stiffness test introduced by Ref. [24].

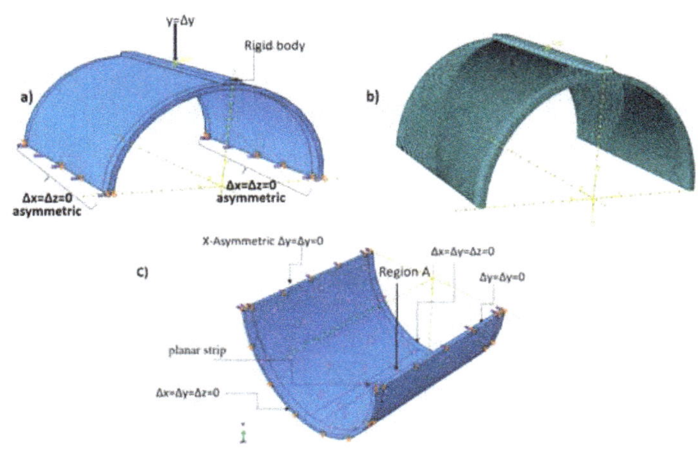

Figure 7. (**a**) Boundary condition, (**b**) mesh shape finite element domain, (**c**) domain with crack.

3. Results

3.1. Mechanical Properties

Figure 8 depicts the stress–strain relationship for tension test specimens before and after preconditioning. After preconditioning in corrosive water with increasing amounts of chlorine gas, the average tensile strength drops by nearly 50%. It is approximately 18 MPa with STDV equals 2 MPa; CV equals 11% before preconditioning and almost 8 MPa with STDV 1 MPa and CV equals 11.7% after preconditioning. This is due to hydro stress caused by moisture expansion coefficients in the polymer composite. Because of the material's thermoelastic and hydroelastic inhomogeneity and anisotropy, interlaminar and intralaminar stresses form. The net tension failure mode is depicted in Figure 9.

Preconditioning also affects the material's Young's modulus, resulting in a decrease of nearly 70%. The fracture behavior, as seen in microstructure examination, includes fiber pullout (Figure 10a) and fiber breakage, as well as matrix carking (Figure 10b). The fracture is ductile, as evidenced by the presence of pores in the matrix. A uniform corrosion attack may be used to attack the performance of such fiber-reinforced polymer (FRP) in a chlorine dioxide environment.

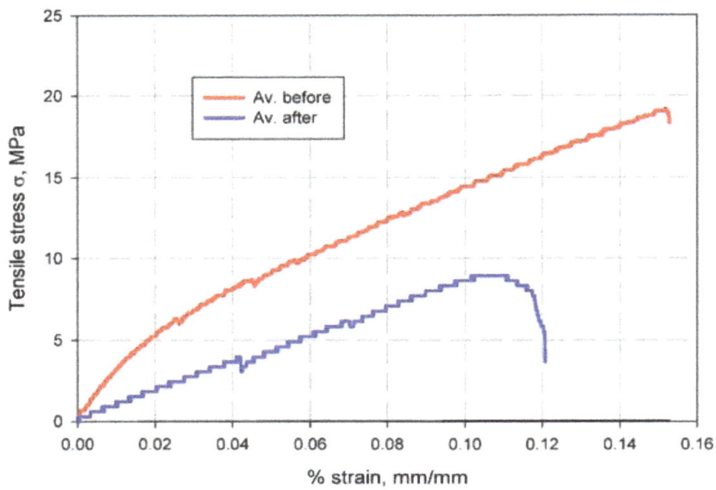

Figure 8. Effect of preconditioning on the stress and strain curves of composite pipes.

Figure 9. Tensile specimen fracture mode.

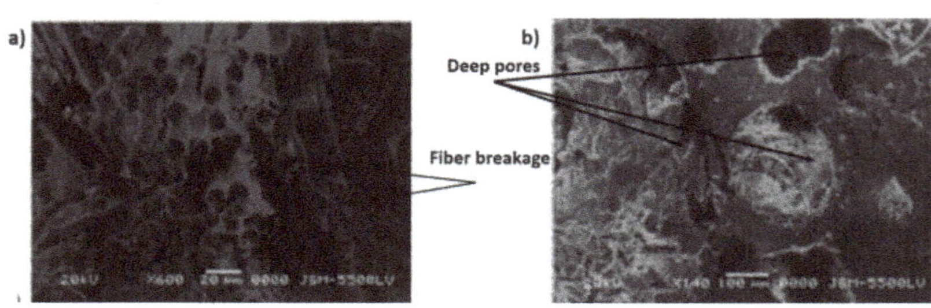

Figure 10. GRP SEM images: (**a**) fiber, (**b**) matrix.

As shown in Figure 11, the precondemning and corrosive agents affect the average compressive strength. A focus observation is made on the compressive behavior of glass fiber composite pipes in the transverse direction (see Figure 11a); it is discovered that there are two line slopes. Initially, the curve slope is nearly equal to 463.2 MPa and 501 MPa for after and before preconditioning in a chlorine dioxide solution, respectively, before changing to 1.4 GPa and 1.2 GPa. This might be attributed to the compressive load being resisted at the first stage by the exterior and interior curved glass fiber layers. The curvature acts like a spring to overcome compressive stress. The entire specimen resists the compressive load after this curvature of the internal layer straightness over the compressive lower platen; thus, the slope and stiffness increase until failure occurs (see Figure 12a). After preconditioning, compressive strength drops by nearly 5.9 percent. Figure 11b depicts the compressive stress and strain curve for compression loading in the longitudinal direction. It is demonstrated that the effect of preconditioning reduces compressive strength and young compressive modulus by nearly 21.5 percent, owing to the hygrothermal impact of moisture and the chemical reaction that occurs between chlorine gas CL and the polymer matrix [14]. Compressive stress steps correspond to the movement of quartz particles under compressive load; these steps were shifted back in the case of preconditioning, which could be attributed to the amount of water absorbed through the sand. Failure modes observed included serve damage and deeply cross cracks in the quartz phase, with no cracks in the exterior or interior layers. Delamination was also observed between glass fiber layers and quartz sand, which occurred because of the sand particles debonding from one another (see Figure 12b).

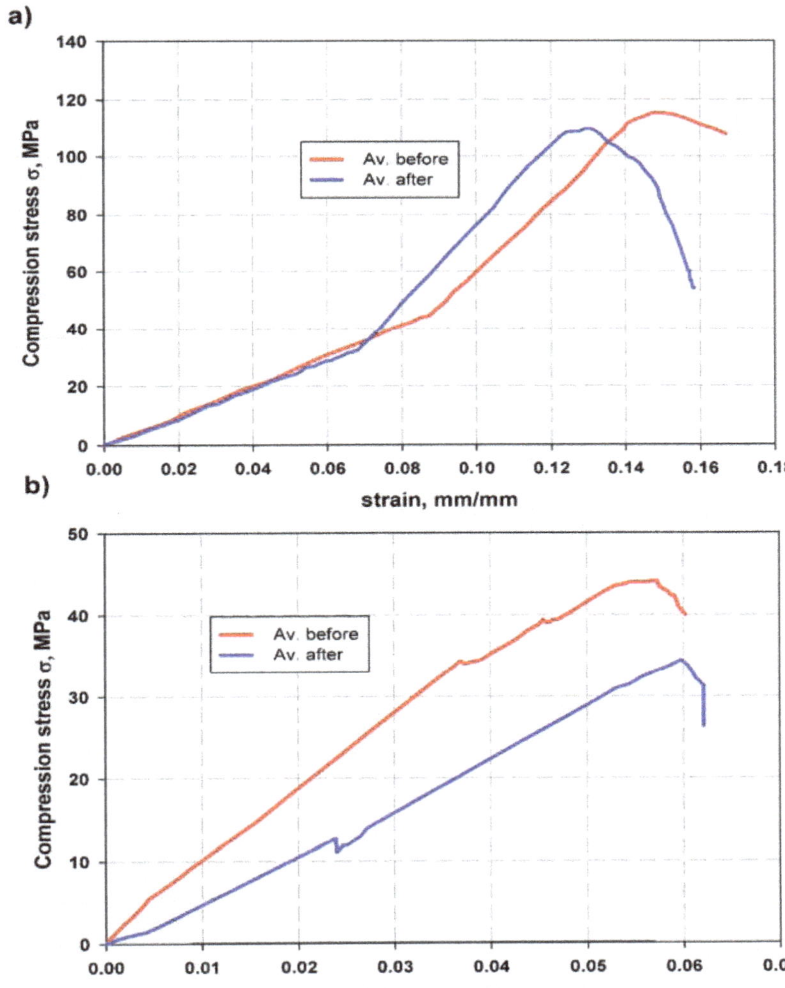

Figure 11. Composite pipe compressive behaviors; (**a**) transverse, (**b**) longitudinal.

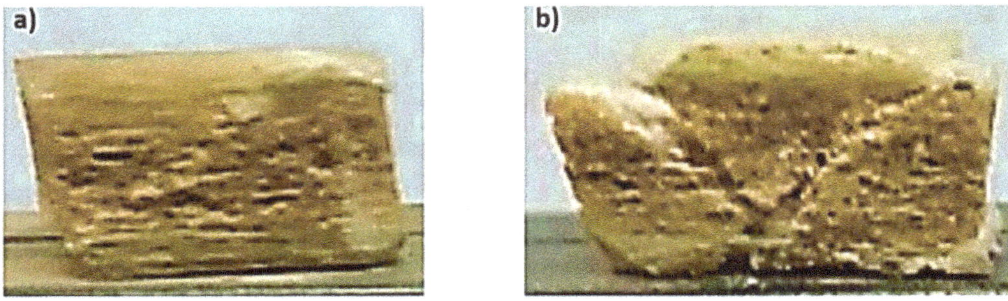

Figure 12. Compression failure mode; (**a**) transverse, (**b**) longitudinal.

Figure 13 depicts the load and displacement curve of a SENB specimen. It is observed that no softening occurred (red color line), and the curve was smooth, owing to the stress concentration near the blunt crack face, which is more sensitive. The load increases as the crack propagate through the material [32] due to the glass fiber layer resisting the load and the crack tip facing sand particles, which increases the crack tip and relieves stress; this explanation can be seen in the fracture mode. Figure 14 depicts the fiber bridging and the straight direction of the crack. When the maximum load (P) is read and then entered Equation (1). The fracture toughness is 283.4 $MPa\sqrt{m}$ with a standard deviation of 21.5 $MPa\sqrt{m}$. The material fracture behavior changed dramatically (see blue line), as the behavior softened after the critical value was reached, then decreased as the crack propagated through the material. This can be attributed to the amount of corrosive water with chlorine gas deeply absorbed by the sand and reach with the polymer, resulting in weak debonding between the sand particles, resulting in a decrease in fracture toughness, which is measured at 182.5 $MPa\sqrt{m}$. with a standard deviation of 15.5 $MPa\sqrt{m}$. (red color line). This is because the second specimen's crack takes a long time to propagate through the material due to the high debonding strength between the particles, whereas the first specimen's debonding strength is low.

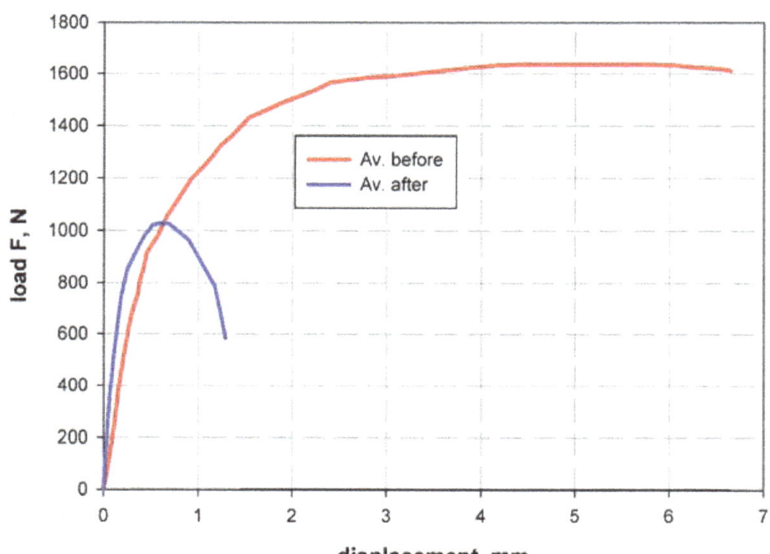

Figure 13. SENB specimen load-displacement curve.

Figure 14. SENB's fracture mode.

3.2. Charpy Impact Test

The shape of the sample significantly influences the fracture toughness of existing materials during the impact test. After immersion in chemically wasted petroleum water, the loading of the sample is much lower (see Figure 15), and the material softens. In contrast, the sample before immersion solidifies, and the quartz particles can absorb water through the thickness. As a result, any fine cracks or crevices on the surface are magnified. The inside of the GRP pipes can cause water leakage on the quartz particles, resulting in dangerous degradation, as illustrated in Figure 15. The corrosion effect resulted in a deterioration between the sand and epoxy layers in a proportion of approximately 50%, and Table 3 lists the impact toughness per joule before and after immersion.

Table 3. Impact energy of Charpy test.

Impact Toughness of Specimen	Unit (Joule)
Before immersion in wastewater	8.5
After immersion in wastewater	5

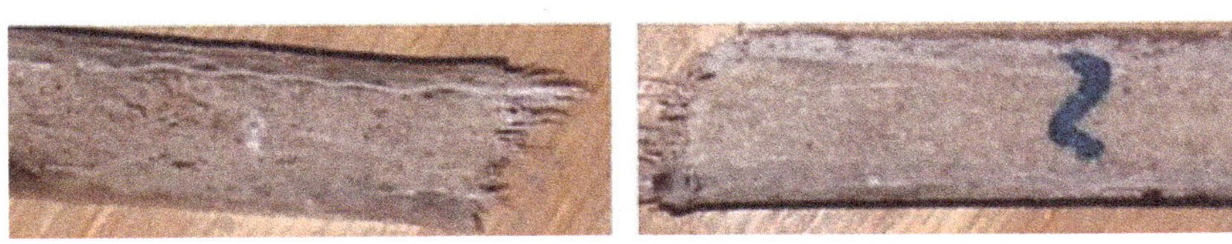

Figure 15. Within the photo of an impact fracture, the mode of failure can be seen.

3.3. Ring Stiffness Characteristic

Figure 16 depicts the initial ring stiffness of composite pipes, based on the values of three specimens from the experimental work of [24]. The average reaction force was 23 kN, while the FE analysis required 26.5 kN; thus, the prediction error was 15.2 percent, acceptable [38–41]. This error was a cumulative number of errors due to damage through used pipes, manufacturing defects, and so on. The numerical FE model assumes no delamination or damage between the glass fiber composites layers themselves or with the sand at interfaces, so the FE model data fit the initial portion of deflection very closely. As a consequence, the elastic behavior is prolonged until a sudden failure occurs. Moreover, this difference is most likely due to an incomplete adhesion in experimental specimens, resulting in higher stress concentration factors and premature failure [39]. Figure 17 depicts the stress distribution through the pipe wall; it is observed that a region of increasing stress concentration (red and orange color) around the crack (see Figure 17a,c), which corresponds to the fracture planes in experimental image Figure 17b [24], where surface cracks are located around the fracture plan. The precondemning affects the initial ring stiffness, as shown in Figure 16. (red dash lines). The hygrothermal effect of water absorption by the sand in the pipe's core reduces the strength of the pipe's material. The experimental findings will be used to validate the FE model.

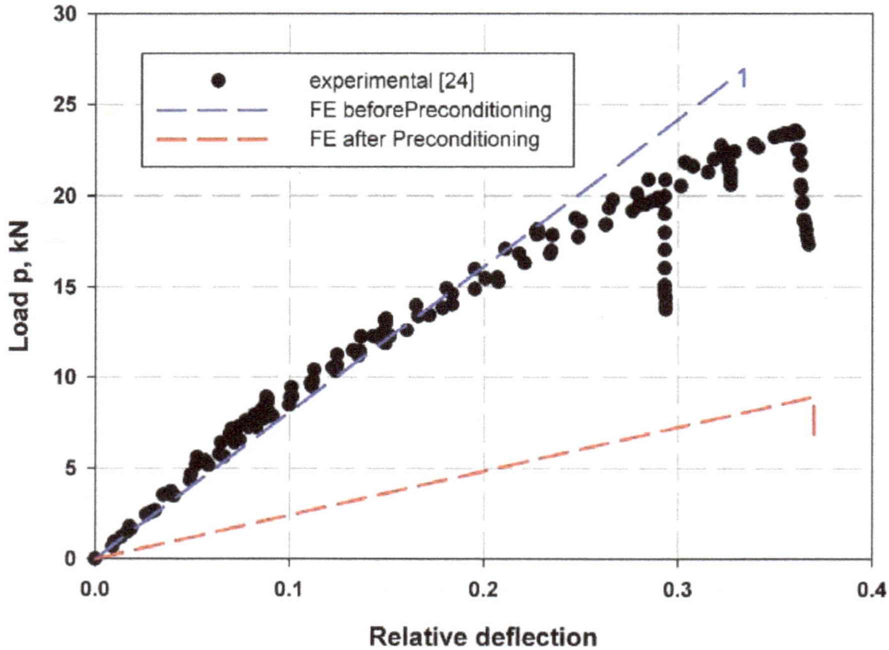

Figure 16. Validation of prediction ring stiffness load with experimental results of Ref. [24].

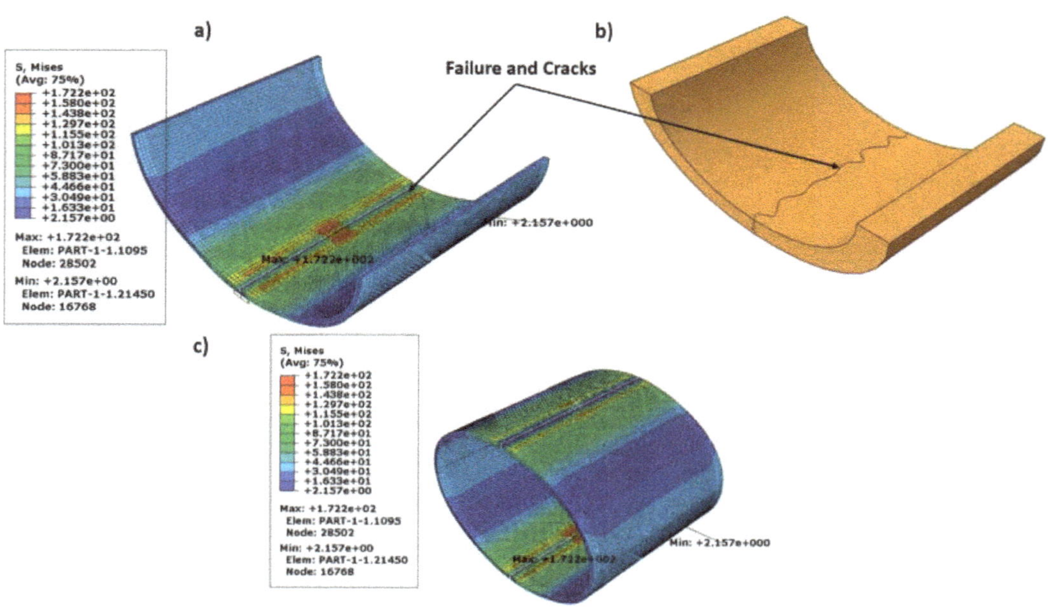

Figure 17. Failure mode prior to preconditioning; (**a**) FE prediction, (**b**) drawing based on experimental [24], (**c**) Von-Mises stress.

4. Conclusions

The mechanical properties of glass fiber composite pipes and the initial ring stiffness were investigated before and after preconditioning in a petroleum wastewater chlorine dioxide corrosive solution. Pipes of this type are considered the primary infrastructure in the petroleum field industry. Current assessments show that the preconditioning action harms the mechanical properties studied, such as compression and tensile strengths. These pipes were typically subjected to preconditioning steps in the aggressive environment of the petroleum field while transporting wastewater saturated with corrosive chlorine dioxide, which is the primary cause of the material's high degradation. Furthermore, both dynamic and static fracture toughness are significantly reduced by this action. These corrosive solutions also affect the ring stiffness properties. The compressive strength of pipe wall material is considerably higher in the transverse direction than in the longitudinal direction.

Author Contributions: Conceptualization, M.Y.A., M.K.H., and A.F.M.; methodology, M.Y.A. and A.F.M.; software, M.Y.A. and M.K.H.; validation, K.A.K. and M.K.H.; formal analysis, M.Y.A., M.K.H., and K.A.K.; investigation, M.Y.A., M.K.H., and A.F.M.; resources, K.A.K.; writing—original draft preparation, M.Y.A., M.K.H., and A.F.M.; writing—review and editing, M.Y.A. and A.F.M.; supervision, M.K.H., and M.Y.A.; project administration, M.Y.A. and K.A.K.; funding acquisition, K.A.K. All authors have read and agreed to the published version of the manuscript.

Funding: This research was funded by the Saudi Basic Chemical Industries (SABIC) and the Deanship of Scientific Research at Umm Al-Qura University, Grant Code: (20-UQU-0071-DSR).

Acknowledgments: The authors would like to thank the Saudi Basic Chemical Industries (SABIC) and the Deanship of Scientific Research at Umm Al-Qura University for supporting this work by Grant Code: (20-UQU-0071-DSR).

Conflicts of Interest: The authors declare no conflict of interest.

References

1. Fukushima, K.; Cai, H.; Nakada, M.; Miyano, Y. Determination of time-temperature shift factor for long-term life prediction of polymer composites. In Proceedings of the ICCM-17, 17th International Conference on Composite Materials, Edinburgh, UK, 27–31 July 2009.
2. Plota, A.; Masek, A. Lifetime Prediction Methods for Degradable Polymeric Materials—A Short Review. *Materials* **2020**, *13*, 4507. [CrossRef]
3. Julius, M.J. *Time, Temperature and Frequency Viscoelastic Behavior of Commercial Polymers*; West Virginia University: Morgantown, WV, USA, 2003.
4. Wang, J.Z.; Parvatareddy, H.; Chang, T.; Iyengar, N.; Dillard, D.A.; Reifsnider, K.L. Physical aging behavior of high-performance composites. *Compos. Sci. Technol.* **1995**, *54*, 405–415. [CrossRef]
5. Yao, J.; Ziegmann, G. Equivalence of moisture and temperature in accelerated test method and its application in prediction of long-term properties of glass-fiber reinforced epoxy pipe specimen. *Polym. Test.* **2006**, *25*, 149–157. [CrossRef]
6. Barbero, E.J.; Julius, M.J. Time-temperature-age viscoelastic behavior of commercial polymer blends and felt filled polymers. *Mech. Adv. Mater. Struct.* **2004**, *11*, 287–300. [CrossRef]
7. Feng, C.-W.; Keong, C.-W.; Hsueh, Y.-P.; Wang, Y.-Y.; Sue, H.-J. Modeling of long-term creep behavior of structural epoxy adhesives. *Int. J. Adhes. Adhes.* **2005**, *25*, 427–436. [CrossRef]
8. Chen, M. Accelerated Viscoelastic Characterization of E-Glass/Epoxy Composite. Ph.D. Thesis, Northwestern University, Evanston, IL, USA, 1991.
9. Goertzen, W.K.; Kessler, M. Creep behavior of carbon fiber/epoxy matrix composites. *Mater. Sci. Eng. A* **2006**, *421*, 217–225. [CrossRef]
10. Miyano, Y.; Nakada, M.; Sekine, N. Accelerated testing for long-term durability of FRP laminates for marine use. *J. Compos. Mater.* **2005**, *39*, 5–20. [CrossRef]
11. Farshad, M.; Necola, A. Effect of aqueous environment on the long-term behavior of glass fiber-reinforced plastic pipes. *Polym. Test.* **2004**, *23*, 163–167. [CrossRef]
12. Nishizaki, I.; Meiarashi, S. Long-term deterioration of GFRP in water and moist environment. *J. Compos. Constr.* **2002**, *6*, 21–27. [CrossRef]
13. Bergman, G. Managing corrosion on plastics-an analysis of experience from industrial applications. In Proceedings of the Corrosion-National Association of Corrosion Engineers Annual Conference, NACE, Orlando, FL, USA, 26–31 March 2000.
14. Hojo, H.; Tsuda, K.; Kubouchi, M.; Kim, D.-S. Corrosion of plastics and composites in chemical environments. *Met. Mater.* **1998**, *4*, 1191–1197. [CrossRef]

15. Farshad, M.; Necola, A. Strain corrosion of glass fibre-reinforced plastics pipes. *Polym. Test.* **2004**, *23*, 517–521. [CrossRef]
16. Stoia, D.I.; Marsavina, L.; Linul, E. Mode I Fracture Toughness of Polyamide and Alumide Samples obtained by Selective Laser Sintering Additive Process. *Polymers* **2020**, *12*, 640. [CrossRef]
17. Günöz, A.; Kepir, Y.; Memduh, K. The investigation of hardness and density properties of GFRP composite pipes under seawater conditions. *Turk. J. Eng.* **2022**, *6*, 34–39.
18. Shi, H.; An, Z.; Gao, R. Simulation of Mechanical Behavior and Structural Analysis of Glass Fiber Reinforced Mortar Pipes. *Rev. Romana Mater.* **2020**, *50*, 198–204.
19. Srinivasan, T.; Suresh, G.; Ramu, P.; Vignesh, R.; Vijay Harshan, A.; Vignesh, K.P. Effect of hygrothermal ageing on the compressive behavior of glass fiber reinforced IPN composite pipes. *Mater. Today Proc.* **2020**, *45*, 1354–1359. [CrossRef]
20. Abdellah, M.Y.; Hassan, M.K.; Alsoufi, M.S. Fracture and Mechanical Characteristics Degradation of Glass Fiber Reinforced Petroleum epoxy Pipes. *J. Manuf. Sci. Prod.* **2016**, *16*, 33–40. [CrossRef]
21. Seleem, A.-E.H.A. Failure and Corrosion Analysis of Composite Glass Fiber Reinforced Pipe Lines, in Mechaincal Engineering Department. Ph.D. Thesis, South Valley University, Qena, Egypt, 2015.
22. Abdellah, M.Y. Delamination Modeling of Double Cantilever Beam of Unidirectional Composite Laminates. *J. Fail. Anal. Prev.* **2017**, *17*, 1011–1018. [CrossRef]
23. Hassan, M.K.; Abdellah, M.Y.; Azabi, S.K.; Marzouk, W.W. Fracture Toughness of a Novel GLARE Composite Material. *Int. J. Eng. Technol.* **2015**, *15*.
24. Faria, H.Q.D. Failure Analysis of GRP Pipes under Compressive Ring Loads. Master's Thesis, Universidade do Porto, Porto, Portugal, 2005.
25. Standard, A. *D3171-99. Standard Test Methods for Constituent Content of Composite Materials*; ASTM International: West Conshohocken, PA, USA, 1999.
26. Abdellah, M.Y.; Alfattani, R.; Alnaser, I.A.; Abdel-Jaber, G.T. Stress Distribution and Fracture Toughness of Underground Reinforced Plastic Pipe Composite. *Polymers* **2021**, *13*, 2194. [CrossRef] [PubMed]
27. ASTM Standards. *ASTM D3039/D3039M-17, Standard Test Method for Tensile Properties of Polymer Matrix Composite Materials*; ASTM International: West Conshohocken, PA, USA, 1995; pp. 99–109.
28. ASTM Standards. *D3410/D3410M-03. Standard Test Method for Compressive Properties of Polymer Matrix Composite Materials with Unsupported Gage Section by Shear Loading*; ASTM International: West Conshohocken, PA, USA, 2008.
29. 100kN Computerized Universal Testing Machine. Available online: http://www.victorytest.com/products/wdw-50100-computerized-electromechanical-universal-testing-machine/ (accessed on 8 June 2020).
30. Anderson, T.L.; McHenry, H.I.; Dawes, M.G. *Elastic-Plastic Fracture Toughness Tests with Single-Edge Notched Bend Specimens, in Elastic-Plastic Fracture Test Methods: The User's Experience*; ASTM International: West Conshohocken, PA, USA, 1985.
31. ASTM International. *D5045-14. Standard Test Methods for Plane-Strain Fracture Toughness and Strain Energy Release Rate of Plastic Materials, Annual Book of ASTM Standards*; ASTM International: West Conshohocken, PA, USA, 2014; Volume 8.
32. Chen, Z.; Adams, R.; da Silva, L.F. Fracture toughness of bulk adhesives in mode I and mode III and curing effect. *Int. J. Fract.* **2011**, *167*, 221–234. [CrossRef]
33. ASTM International. *Standard Test Method for Determining the Charpy Impact Resistance of Notched Specimens of Plastics. ASTM D6110-10*; ASTM International: West Conshohocken, PA, USA, 2010.
34. Belytschko, T.; Black, T. Elastic crack growth in finite elements with minimal remeshing. *Int. J. Numer. Methods Eng.* **1999**, *45*, 601–620. [CrossRef]
35. Melenk, J.M.; Babuška, I. The partition of unity finite element method: Basic theory and applications. *Comput. Methods Appl. Mech. Eng.* **1996**, *139*, 289–314. [CrossRef]
36. Datta, D. Introduction to eXtended Finite Element (XFEM) Method. *arXiv* **2013**, arXiv:1308.5208.
37. Montasser Dewidar, N.S.K.; Mohammed, Y.; Abdellah Ayman, M.M. Finite element modeling of mechanical properties of titanium foam and dental application. In Proceedings of the Third International Conference on Energy Engineering (ICEE), Aswan, Egypt, 28–30 December 2015.
38. Khashaba, U. Tensile and flexural properties of randomly oriented gfrp composites. In *1st International Conference on Mechanical Engineering Advanced Technology for Industrial Production*; Assiut University: Assiut, Egypt, 1994; Volume 1, pp. 131–143.
39. Berbinau, P.; Filiou, C.; Soutis, C. Stress and failure analysis of composite laminates with an inclusion under multiaxial compression-tension loading. *Appl. Compos. Mater.* **2001**, *8*, 307–326. [CrossRef]
40. Soutis, C.; Curtis, P. A method for predicting the fracture toughness of CFRP laminates failing by fibre microbuckling. *Compos. Part A Appl. Sci. Manuf.* **2000**, *31*, 733–740. [CrossRef]
41. Abdellah, M.Y. Comparative Study on Prediction of Fracture Toughness of CFRP Laminates from Size Effect Law of Open Hole Specimen Using Cohesive Zone Model. *Eng. Fract. Mech.* **2018**, *191*, 277–285. [CrossRef]

Journal of
Composites Science

MDPI

Article

Qualitative Investigation of Damage Initiation at Meso-Scale in Spheroidized C45EC Steels by Using Crystal Plasticity-Based Numerical Simulations

Muhammad Umar [1,2,*], Faisal Qayyum [2], Muhammad Umer Farooq [1], Sergey Guk [2] and Ulrich Prahl [2]

[1] Department of Mechanical Engineering, Khwaja Fareed University of Engineering and Information Technology, Rahim Yar Khan 64200, Pakistan; umer.farooq@kfueit.edu.pk

[2] Institut für Metallformung, Technische Universität Bergakademie Freiberg, 09599 Freiberg, Germany; faisal.qayyum@imf.tu-freiberg.de (F.Q.); sergey.guk@imf.tu-freiberg.de (S.G.); Ulrich.Prahl@imf.tu-freiberg.de (U.P.)

* Correspondence: muhammad.umar@kfueit.edu.pk

Abstract: This research uses EBSD data of two thermo-mechanically processed medium carbon (C45EC) steel samples to simulate micromechanical deformation and damage behavior. Two samples with 83% and 97% spheroidization degrees are subjected to virtual monotonic quasi-static tensile loading. The ferrite phase is assigned already reported elastic and plastic parameters, while the cementite particles are assigned elastic properties. A phenomenological constitutive material model with critical plastic strain-based ductile damage criterion is implemented in the DAMASK framework for the ferrite matrix. At the global level, the calibrated material model response matches well with experimental results, with up to ~97% accuracy. The simulation results provide essential insight into damage initiation and propagation based on the stress and strain localization due to cementite particle size, distribution, and ferrite grain orientations. In general, it is observed that the ferrite–cementite interface is prone to damage initiation at earlier stages triggered by the cementite particle clustering. Furthermore, it is observed that the crystallographic orientation strongly affects the stress and stress localization and consequently nucleating initial damage.

Keywords: spheroidized steel; local ductile damage; crystal plasticity; numerical simulation; local deformation behavior; DAMASK

Citation: Umar, M.; Qayyum, F.; Farooq, M.U.; Guk, S.; Prahl, U. Qualitative Investigation of Damage Initiation at Meso-Scale in Spheroidized C45EC Steels by Using Crystal Plasticity-Based Numerical Simulations. *J. Compos. Sci.* **2021**, *5*, 222. https://doi.org/10.3390/jcs5080222

Academic Editors: Stelios K. Georgantzinos and Francesco Tornabene

Received: 31 July 2021
Accepted: 18 August 2021
Published: 20 August 2021

1. Introduction

The critical factor in the exponential growth of the mobility, mechanization, and infrastructure sectors in the last six to seven decades is the improvement in the existing and development of novel steel materials [1]. On average, about 1.8 billion tons of raw carbon steel is produced annually around the world. More than 83 million tons (approx. 5%) of steel are used by the automobile industry as a raw material in the form of different grades [1]. Therefore, the demand for high strength and lightweight steel is inevitable for technological, environmental, and economic progress. For instance, on average, 60% of the fuel in the automobile industry is consumed due to the vehicle's body weight itself [2]. Better steels can not only improve fuel consumption, minimize the detrimental carbon footprint, and assist in making aesthetically lucrative vehicle body shapes, but also increase collision safety. This motivates researchers and manufacturers to explore and use lightweight, highly deformable, and extended energy-absorbing steels in the front and rear sections of the vehicle body [3].

The industrial applications of carbon steels with multiple phases have expanded in the last decade due to their versatile microstructural configurations and consequent mechanical response [4]. Pearlitic-ferritic steels after spheroidization annealing with ferrite as the ductile phase and cementite particles as reinforcement elements have made their place in the automotive industry for components that need extended elongation during forming [5–7]. However, harder and stronger lamellar cementite present within the pearlite phase is a major limitation during the forming process, which undergoes brittle fracture without considerable plastic deformation [8]. A critically designed spheroidization heat treatment process can convert these challenging 3D cementite plates into globular cementite particles [9–11]. The resulting microstructure presents a remarkable combination of strength and ductility suitable for cold forming techniques [12]. In addition to the better formability, it also offers a margin in weight reduction by down-gauging the allowable thickness of the sheets in the automotive industry [13–15].

Researchers have used macroscopic techniques to explore the plasticity of these materials with simplifications and approximations which ignore the actual underlying plasticity mechanisms [16–19]. The difference in the microstructural attributes of ferrite matrix and cementite particles in these steels produces heterogeneous local stress and strain response [20]. To confidently predict the deformation behavior of spheroidized medium carbon steels, it is critical to understand the influence of cementite particles' size and distribution in the ferrite matrix at the microstructural level [15,21].

Virtually constructed RVEs used by Qayyum et al. [22] for the numerical simulations provided a clear correlation for stress and strain localization for 2D and 3D RVEs. However, both possess assumptions and limitations, and therefore lack an outlook of the actual material. Tasan et al. [20] published their work on the full phase numerical simulations of multi-phase steels and provided detailed insights influenced by the microstructure based on individual grain size, position, orientation, and morphology. This work provided an admirable comparison of numerical simulation and experimental results, yet it lacked ductile damage criteria.

Diehl et al. [23] comprehensively studied the influence of microstructural features of the second phase in a ductile ferrite matrix. They concluded that crystallographic orientation of the individual grains also contributes towards stress and strain localization. The absence of damage criteria resulting in an overestimation of plastic behavior of ferrite was reported by Umar et al., and Qayyum et al. [8,24–27]. It was concluded by Zhang et al., that the material insight is relatable to the actual material behavior at low strains with virtual RVEs, while for higher strains, the damage consideration is inevitable [28].

The present study continues the previously published work [8,25,29], and it is performed to overcome the described challenges. In this research, the real RVE consideration with actual crystallographic orientation and grain morphological attributes is carried out with EBSD-based data. To avoid the overestimation of global plastic flow of the material, the ductile damage criterion based on critical plastic strain energy dissipation is incorporated into the phenomenological material model with the Düsseldorf Advanced Material Simulation Kit (DAMASK) [30] to gain a comprehension of material behavior influenced by actual microstructure. Furthermore, a recently developed EBSD-based microstructural data cleaning technique is used to observe the actual interactive behavior of crystalline grains with distinct phases in C45EC medium carbon steel. This work is focused on implementing the improved material model to make the response of simulations more relatable.

For easier understanding, the work is divided into sections, where Section 1 contains the background of the study, the need for advanced steels, and the problem addressed in this study. Section 2 deals with the details of the methodology adopted to evaluate the numerical simulations modeling. It also contains a brief description of the phenomenological material model with ductile damage criterion implemented in this study. Section 3 contains results and their discussion to obtain meaningful inferences from them. Finally, Section 4 contains the conclusions of the study and the future outlook.

2. Methodology and Numerical Simulation Modelling

2.1. Experimental Data

To carry out the detailed study on the effect of microstructural constituents distributed in the ferrite matrix, thermo-mechanically processed C45EC steel samples with two spheroidization states, i.e., 83% and 97%, were prepared. The chemical composition of the steel sample in weight percentage is shown in Table 1.

Table 1. Chemical composition of the investigated C45EC steel in weight percentage.

Element	C	Si	Mn	P	S	Cr	Mo	Ni	Al	Cu
Percentage	0.44	0.10	0.75	0.007	0.017	0.04	0.011	0.04	0.03	0.03

The input steel samples with 0.44% carbon with notable traces of silicon and manganese were produced at an industrial scale. The steel then passed through hot rolling and was spheroidized in two different regimes. For S-83, the hot rolled steel sample was subjected to cold rolling and then spheroidization annealing. The elongated ferrite grain in the X-direction can be attributed to the cold rolling. The resulting microstructure converted the lamellar cementite from pearlitic microstructure into the globular particles, but they are not well distributed; rather, it can be seen in the form of bands at various places in the RVE.

For S-97, the hot rolled steel sample first goes went the first stage of spheroidization annealing to convert the pearlitic-interlayer cementite into free globular particles. The samples were then subjected to cold rolling with heavy mechanical deformation, resulting in the elongation of the ferrite grains with cementite particles in close vicinity. At the final stage, this material went through the spheroidization annealing process. As a result, the material attained very fine recrystallization with new ferrite grains and demonstrated a homogenous distribution of the cementite particles in the whole RVE instead of clustering.

The EBSD based microstructural analysis was performed using a Scanning Electron Microscope (SEM) Merlin Gemini II (ZEISS), at International center of electron microscopy for material science, AGH Krakow, Poland. Necessary crystallographic, morphological, and statistical grain information was recorded for appropriately prepared material samples according to the standard procedure in ASTM E3-11 [30]. In addition, the standard procedure of grinding and polishing the surface before etching with 3% nital solution was followed. Further details about the process of sample production, performance for EBSD analysis, and workflow of activities have been given in detail elsewhere [31].

To determine the global stress–strain curves, monotonic tensile tests were carried out at quasi-static state on dog-bone shaped samples according to ASTM E8 [30] standard test methodology. The integrated material testing system was used for applying a strain rate of 0.001 per second. The output data were recorded and presented in the form of engineering stress–strain curves in global results section. Spheroidization states of steel and ferrite matrix grain data considered for micromechanical crystal plasticity finite element method (CPFEM) simulations are mentioned in Table 2.

Table 2. The nomenclature and ferrite phase information for two spheroidization cases of C45EC steels.

Sample Case	Spheroidization Degree, %	Mean Ferrite Equivalent Diameter, μm	Max Ferrite Equivalent Diameter, μm	Global Strain at Damage Initiation, %
S-83	83	15.03	69.86	3.78
S-97	97	11.96	38.43	5.69

Inverse pole figure (IPF) maps of both samples at different spheroidization degrees are displayed in Figure 1, showing the ferrite grains and their crystallographic orientation distribution function (ODF) with reference to the directions given in legends.

2.2. Numerical Model Definition

Two-dimensional EBSD data obtained from SEM were usually in the raw form, and they needed some intelligent post-processing. In the present study, MTEX [32,33] was used for post-processing of the EBSD data as used by other researchers previously [34]. The clean EBSD data were used to attain the geometry description files on MATLAB to serve as an input for the material model implemented in DAMASK [30]. More details about the model are provided in Section 2.3, with very prominent local and global governing equations. The 2D RVE considered for the study had dimensions of 85 μm × 85 μm. The size and dimensions of real RVE depend upon the region of the sample selected for microstructural analysis. These values of region and resolution are chosen based on the particles to be studied, i.e., the cementite in this case. This specific resolution is chosen to ensure that the magnification and step size are large enough to assign multiple pixels to the individual cementite particles. The size is big enough to comprise multiple ferrite grains necessary for a polycrystalline RVE. It should not be large enough to make the simulations computationally too expensive to run with the available resources.

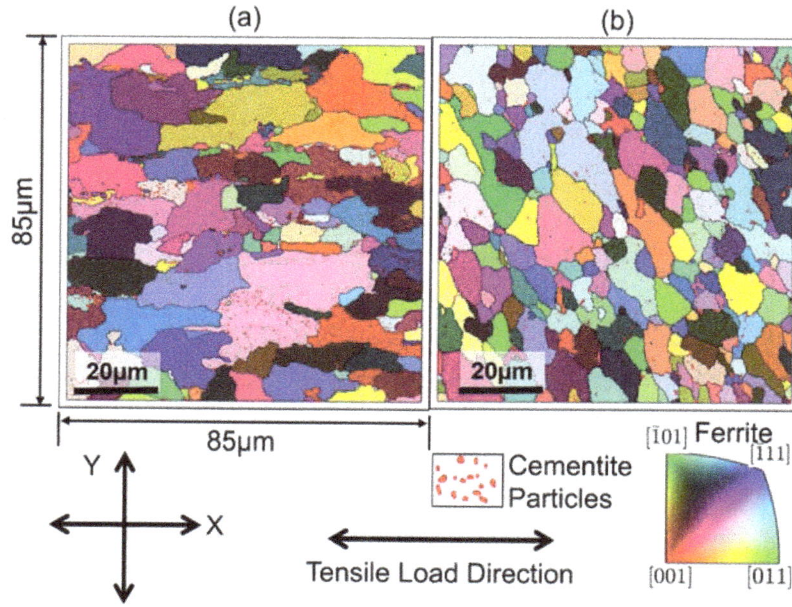

Figure 1. Post-cleaning EBSD-based RVEs with different grain morphologies depending upon the degree of spheroidization and thermo-mechanical processing route. The cementite particles are shown with the help of red color distributed heterogeneously in S-83 (**a**). After intermittent spheroidization treatment in S-97, the cementite particles are homogenously observable as red color in (**b**).

Researchers have shown various flow charts of activities performed during a numerical analysis for optimal usage or resources [35–39]. Similarly, the hierarchy of activities performed while implementing the micromechanical simulation model is in Figure 2 as a flow chart. It is shown that a complete description of the material to be simulated is fed in the form of input values containing information comprising phase positions, crystallographic orientation, elastic and/or plastic phase parameters, and damage description of all the phases present in the RVE.

The first step during numerical simulation is the implementation of the boundary conditions for all the calculation points defined in the material configuration file for a 2D RVE. Next, coupled crystal plasticity equations are solved numerically to reach an approximate solution. Due to computational complexity, if the solution is not converged under the error threshold of 1.0, the solution is reiterated until the value of the maximum allowed iterations is reached. In this case, a primary cut back is applied to the deformation gradient, and the process of finding an approximate solution is repeated. Once the solution converges and the result output file is obtained, the next step is to use this major repository of results into meaningful presentable outcomes, and various frames of RVE at various intervals during deformation are recorded and visualized using Paraview software [40].

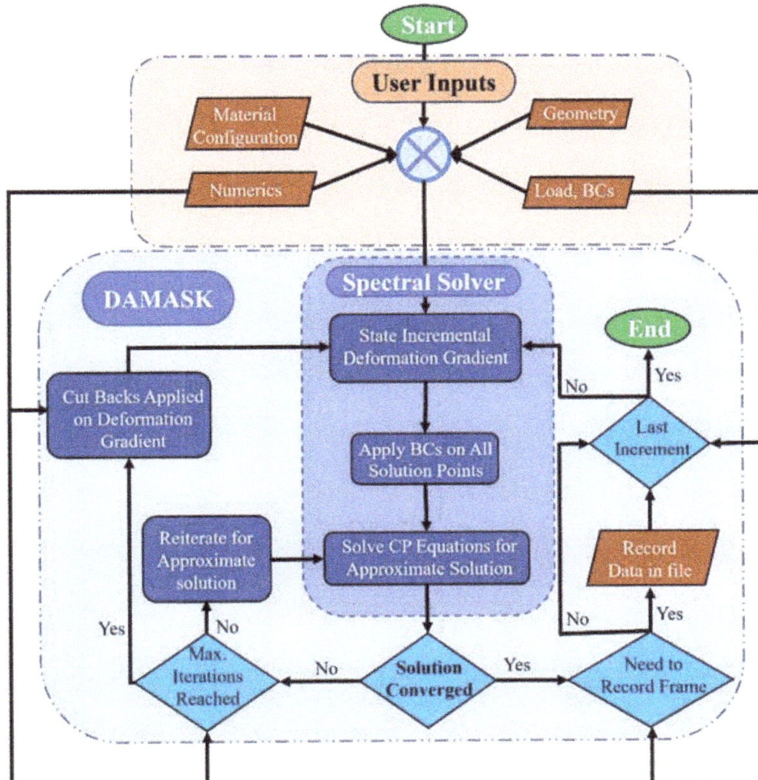

Figure 2. Flow diagram of the activities performed when solving the crystal plasticity-based phenomenological material model implemented in DAMASK. The input, output, process, and terminal points are shown as a parallelogram, rhombus, rectangle, and ellipse, respectively.

2.3. Numerical Model Definition

To observe the micromechanical response during deformation in the spheroidized medium carbon steels, the crystal plasticity and continuum mechanics-based numerical simulation model was used. The elastic, plastic, and ductile damage criteria were implemented in the phenomenological crystal plasticity model in DAMASK [30]. The ferrite phase is given elastic and calibrated plastic properties, while the hard cementite phase is assigned elastic properties due to its comparatively higher strength. The details about the process of calibrating the material model in DAMASK have been given elsewhere [41,42].

$\dot{\mathbf{F}}_{ij}$ and $\mathbf{P}_{ij}$ represent (Equations (1) and (2)) the deformation gradient tensor and first Piola–Kirchhoff stress tensor, respectively. "0" and "*" in the 3×3 matrix for both tensors show the complementary boundary conditions as "fix" and "free,", respectively.

$$\dot{\mathbf{F}}_{ij} = \begin{bmatrix} 1 & 0 & 0 \\ 0 & * & 0 \\ 0 & 0 & * \end{bmatrix} \times 10^{-3} \, . \, s^{-1} \quad (1)$$

$$\mathbf{P}_{ij} = \begin{bmatrix} * & * & * \\ * & 0 & * \\ * & * & 0 \end{bmatrix} Pa \tag{2}$$

Quasi-static uniaxial tensile load is applied during numerical simulation in the X-direction. The strain rate during the application of load was set to be 0.001 s^{-1}.

The most fundamental concept in CP phenomenological constitutive law is the slip resistance on every slip plane (Equation (3)). The plastic deformation is calculated based on plastic velocity gradient $\mathbf{L}_p$, which is also dependent on the relative resistance of the slip planes (Equation (4)). The localized deformation at any point is calculated based on the availability of slip planes, ease of slip, and critically resolved shear stress.

$$\dot{\gamma}^{\alpha} = \dot{\gamma}_0 \left| \frac{\tau^{\alpha}}{S^{\alpha}} \right|^{n} \mathrm{sgn}(\tau^{\alpha}),\ \alpha = 1, 2, 3 \ldots, N_{slip} \tag{3}$$

The value of N_{slip} is 24 for bcc ferrite

$$\mathbf{L}_p = \sum_{\alpha=1}^{N_{slip}} \dot{\gamma}^{\alpha}\, \mathbf{m}^{\alpha} \otimes \mathbf{n}^{\alpha} \tag{4}$$

The multistate model implemented in the DAMASK framework [30] calculates the global plastic behavior of the C45EC steel during virtually applied load using the mutual relationship between local and global stresses and strains, as given in Equations (3) and (4) [43]. The global flow curves are obtained based on the mean stress and strain of the entire field at each time interval during deformation in response to the applied load. The given model uses Equations (5) and (6) to calculate the average response of the stress and strain.

$$\varepsilon_{RVE} = \frac{\sum_{i=1}^{n} \varepsilon_i v_i}{\sum_{i=1}^{n} v_i} \tag{5}$$

$$\sigma_{RVE} = \frac{\sum_{i=1}^{n} \sigma_i v_i}{\sum_{i=1}^{n} v_i} \tag{6}$$

where

ε_{RVE} = equivalent plastic strain of the whole RVE
ε_i = equivalent plastic strain of an element
v_i = volume of an ith element in the RVE
n = total number of elements in the RVE
σ_{RVE} = equivalent stress of the whole RVE
σ_i = equivalent stress of an element.

The full-phase numerical simulation with the phenomenological model implemented in DAMASK [30] is a computation-intensive model. Furthermore, the complex damage phenomena make the simulation process crash at specific tensile load increments due to multiple unsuccessful attempts to converge the numerical solution to a pre-defined minimum threshold of error, i.e., 1.0. Therefore, the local results are observed and discussed after damage initiation but before the final rupture of the sample.

The critical plastic strain-dependent ductile damage criterion is also defined for the ferrite phase (refer to Equation (7)). As the primary mode of plastic deformation in the ferrite phase is dense plane slip, the phenomenological material model is employed to calculate the local deformation behavior of the spheroidized medium carbon steel at hand.

CPFEM uses individual elastic and/or plastic phase parameters to calculate the interactive response in a polycrystalline aggregate. The influence of the grain size, morphology, crystallographic orientation, and the respective position of the matrix grains and reinforcement-intended hard phase particles can be studied well using this methodology. The elastic and plastic phase parameters for ferrite are defined, while for cementite, elastic properties are used in this simulation, as detailed in Table 3. Strain energy-based ductile damage parameters are utilized in this model as given in Table 3 to observe the stress and strain evolution, consequently reaching damage. A minimum threshold of critical plastic strain (ε_{crit}) is set as 0.5. $\varphi_l = 0$ represents the material point with 100% degradation, while $\varphi_l = 1$ represents 0% degradation of the material at a local point.

$$\varphi_l = min\left(1, \frac{\epsilon_{crit}}{\sum_{a=1}^{n_{ss}} \gamma^{\alpha}}\right) \tag{7}$$

Table 3. Elastic, plastic, and damage parameters for the ductile ferrite matrix and elastic phase parameters for the hard cementite phase. Data adapted from previously published work [8,25] with the kind permission of Wiley and MDPI, respectively.

Elastic and Plastic Phase Parameters [8]		
Parameter Description	**Values for Ferrite**	**Unit**
C11, C12, C44	233.3, 235.5, 128.0	GPa
C11, C12, C44 (Cementite)	375.0, 161.0, 130.0	GPa
$\dot{\gamma}_0$	5.6×10^{-4}	ms^{-1}
$S_{0[111]}$, $S_{s[111]}$	95, 222	MPa
$S_{0[112]}$, $S_{s[112]}$	96, 412	MPa
h_o, $h_{\alpha\beta}$	1, 1	GPa
n, w	3, 2.0	-
N_{slip}	12, 12	-
N_{twin}	0	-
Ductile Damage Parameters [25]		
Interface Energy (g^0)	1.0	Jm^{-2}
Damage mobility coefficient (M)	0.001	s^{-1}
Critical plastic strain (ε_{crit})	0.5	-
Damage rate sensitivity coefficient (P)	10	-
Damage diffusion (D)	1.0	-
Damage type	Local	-

3. Results and Discussion

3.1. Global Results

The real microstructure-based RVEs of two differently spheroidization-annealed C45EC steel samples are modeled and subjected to monotonic quasi-static tensile load. The global material flow behavior is calculated by taking the local stress and strain average for each increment of load during numerical simulation (refer to Figure 3) as per Equations (5) and (6).

Generally, the plastic flow trends for both cases during simulation match well with the experimental results. For example, in Figure 3 (zone A), the experimental results show the actual material behavior. After the elastic region, both materials experience elastic to plastic transition where dislocation movement and pinning take place until a point is reached where the material starts to flow plastically. During simulation, this process of flow behavior is not observed similarly because the material model is meant to predict the plastic behavior, especially after zone A.

The trend for S-83 is obtained with less than 1% error in Figure 3, zone B involving the plastic flow region. For the S-97 sample, there is a maximum of 3.2% difference compared to experimental results. This accurate match suggests that the material model implemented is appropriately calibrated with experimental results globally. Therefore, it can be used to analyze and predict the local results of the RVE.

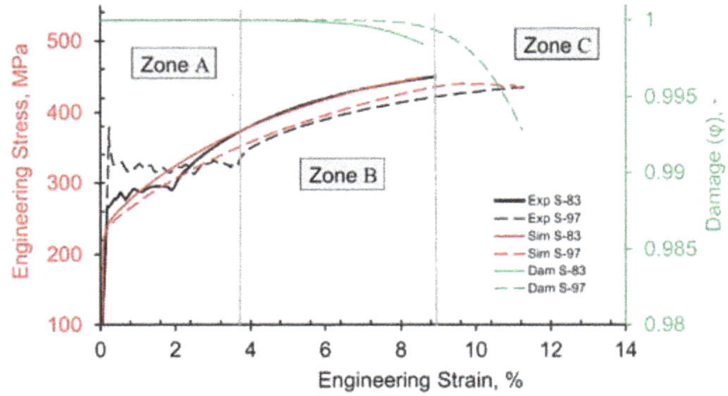

Figure 3. Comparison of experimental and simulation flow curves and damage trend lines of S-83 and S-97 samples with different spheroidization states based on actual RVEs. The three zones in the graph, A, B, and C, represent elastic, plastic, and extended plastic regions, respectively.

Figure 3, zone C has global results for the S-97 sample only, which undergoes comparatively more plastic flow during numerical simulations because of cementite particles' fine distribution due to the extended intermittent spheroidization-annealing.

It is observed that the global material response for the S-83 sample with damage exhibits more strain hardening. A similar trend can be seen in the case of S-97 at the beginning, with comparatively reduced strength. Local ductile damage initiates in both the cases represented by Dam S-83 and Dam S-97, at a difference of ~2% global strain. Slight differences at the starting point of damage can be attributed to the unique microstructural architecture, which will be discussed in detail later.

After the first incidence of damage, strain hardening continues with experimental results from 3.7% to about 9% global strain for the S-83 material sample. For S-97, due to the homogenous distribution of hard and fine cementite particles, the damage is initiated later, i.e., 5.7% global strain, and continues until ~10% (Figure 3, zone B). Stiffness degradation is observed in the S-97 case after 9% of global strain in zone C, and damage incidents increase, represented in Figure 3, with Dam S-97 dashed green trend line.

3.2. Local Results during Mechanical Deformation of S-83

Local maps for IPF, strain, stress and damage for the S-83 case are shown from top to bottom, respectively, in Figure 4 at increasing global strain, i.e., 2.5%, 5.3%, and 8.5%. Analyzing the local results at various critical points during the evolution of plastic deformation is very interesting to predict the initiation and high local concentration of stress and strain regions. Furthermore, the influence of cementite particle size, distribution, and clustering in the ductile ferrite matrix has a decisive effect on the overall mechanical behavior of the material. Therefore, this information is essential to understand and to improve the working of spheroidized C45EC steels.

Only ferrite phase is shown in all the local maps, while the cementite particles are represented by white dots (empty spaces) in local damage maps and by black dots in all the maps otherwise. This is because the substantial difference in the stresses and strains in both phases develops during deformation. For instance, the local stress points in cementite are observed as high as ~10GPa, while the average local stress shown for ferrite in the RVE for the S-83 sample is ~500 MPa. This difference distorts the scale and makes the simultaneous representation very challenging.

IPF map of the RVE can be seen to evolve due to applied strains. This evolution of the crystallographic orientation during deformation is interesting to note because the adjacent crystals shift their behavior according to the modified crystals' orientation after every step of load increment. Not all ferrite grains change their orientation simultaneously, yet their individual locality contributes very highly to deciding their modified orientation.

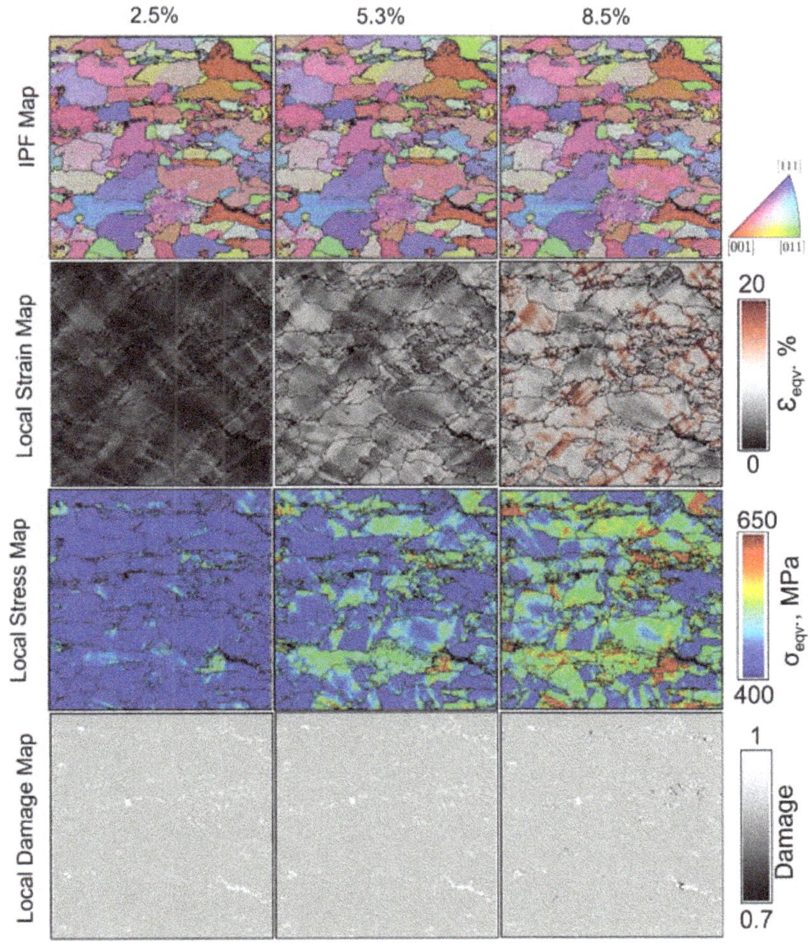

Figure 4. Local maps (from top to bottom) for IPF, strain, stress, and damage for the S-83 sample with respective legends/scales. Local maps at global strains, i.e., 2.5%, 5.3%, and 8.5%, are shown from left to right for local behavior insight during the evolution of mechanical deformation.

The local strains are heterogeneously distributed in the RVE due to varying cementite particle sizes and distributions and the random orientations of the ferrite grains. It is worth mentioning here that the local strain distribution map of S-83 at 8.5% global strain shows substantial heterogeneity at various positions. This difference in heterogeneity is attributed to the considerable difference in the strength of both phases. The local strains start to develop first at the ferrite–cementite interface. Then, they continue to grow in the oblique directions compared with the direction of application of the quasi-static tensile load.

Although local stress distribution maps show lower stresses at the 2.5% global strain, a significant contrast is observed during evolution up to 8.5% global strain. A few ferrite–ferrite grain interfaces are observed where local stresses start to appear at 2.5% and continue to grow until 8.5% global strain. Even the individual grains at some points develop local equivalent stresses near to 650MPa. Damage incidents, on the other hand, are not observed until 5.3% global strain. The initial micro cracks captured in the local damage map at 8.3% global strain demonstrate that the multiple sites have their unique history and circumstances for the initiation of the local damage. Further explanation and investigation of micro-cracks indeed demand postmortem analysis of the RVEs to predict the rationale of the specific observation.

A profound analysis of the local stress, strain, and damage maps of S-83 at 8.5% global strain is presented in Figure 5. It gives an obvious idea about the dependence of the stress and strain localization influenced by crystallographic orientations of the ferrite grains. The resulting damage maps suggest that initial micro-cracks in the RVEs develop due to noticeable orientation differences in ferrite grains. The influence was studied by highlighting three zones on the RVE as shown in Figure 5, i.e.,

- Zone A is defined to study the local stress, strain, and damage behavior of similarly orientated adjacent ferrite grain without any cementite particle inclusion.
- To analyze the effect of the different orientations of one ferrite grain on its neighboring grain, zone B is marked on three local maps in Figure 5. This zone does not contain cementite particles; therefore, the significant behavior difference is influenced by the orientation of the ferrite grains.
- The different orientations and the clustered cementite particles are highlighted in zone C, which is of utmost importance.

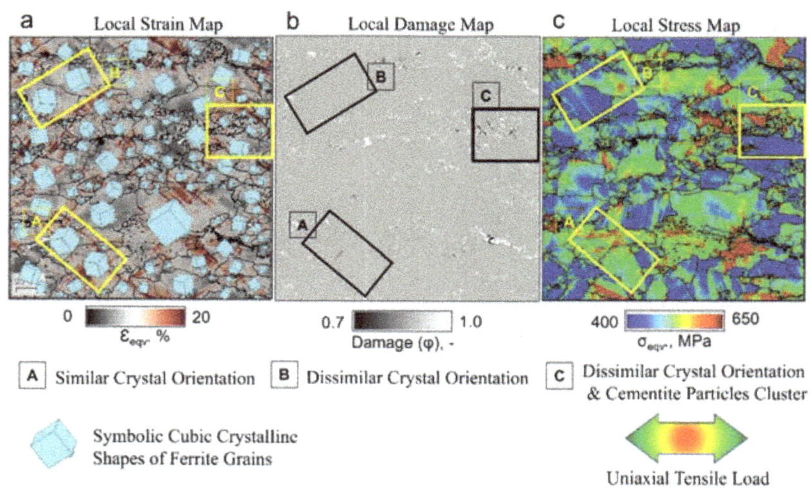

Figure 5. Comparison of local stress, strain, and damage maps of the S-83 sample at 8.3% global strain with zone A, B, and C pointing out crystallographic orientation that is similar, dissimilar, and dissimilar to the cementite particle cluster, respectively. (**a**) Local strain map with overlaid ferrite grain actual crystal orientation recorded during EBSD analysis, (**b**) local damage map of the S-83 RVE, and (**c**) local stress map.

To understand the influence of the crystallographic orientation, all three zones need to be discussed one by one. Zone A, with similar crystallographic orientations of adjacent ferrite grains, shows similar local strain distribution across the grain boundary. This suggests the idea that a similar effect of local strain is shifted to the adjacent ferrite grain. A limitation of the material model must be considered here because grain boundary consideration is neglected in the present model. Zone A, having almost no cementite grain, provides the perfect area to study the complete phenomenon observation due to variation in crystallographic orientation. A local stress map can be seen in zone A with no substantial local stresses across the grain boundary. In the damage map, there can be seen a thin black-colored line showing local damage developed due to local strains. Ferrite, a ductile phase, undergoes extended deformation in the large grains, and similarly, oriented ferrite grains further aggravate this plastic flow.

Conversely, in Zone B, the orientation of the adjacent ferrite grains is dissimilar, giving rise to the local stress development across the grain boundary due to blockage of slip planes. On the other hand, local strains are not observed with a substantial difference, and consequently, no local damage incident is observed in zone B in the local damage map.

In Figure 5, Zone C is of utmost importance because the combination of different orientation and cementite clustering effects in the RVE with 83% spheroidization degree creates multiple damage sites in the close vicinity. It is also observed in Figure 5b zone C that the cementite clusters are even more prone to the generation of multiple damage initiation sites as compared with the cementite bands present in the right lower corner of the RVE. This synergetic effect of the microstructure combined with highly dissimilar crystallographic orientations and cementite particle clusters gives rise to the sharp oblique strain band formation. Once the initial cracks appear at the ferrite–cementite interface and they start to coalesce in proximity to each other, the sharp damage lines start to appear at various locations in the very close vicinity, making this region a highly local damage-affected zone.

3.3. Local Results during Mechanical Deformation of S-97

The local results in the form of evolving IPF, local strain, stress, and damage maps for the case of S-97 are displayed in Figure 6. Generally, the evolution of the crystallographic orientation shows a slight difference, as in the case of S-83, due to random orientation.

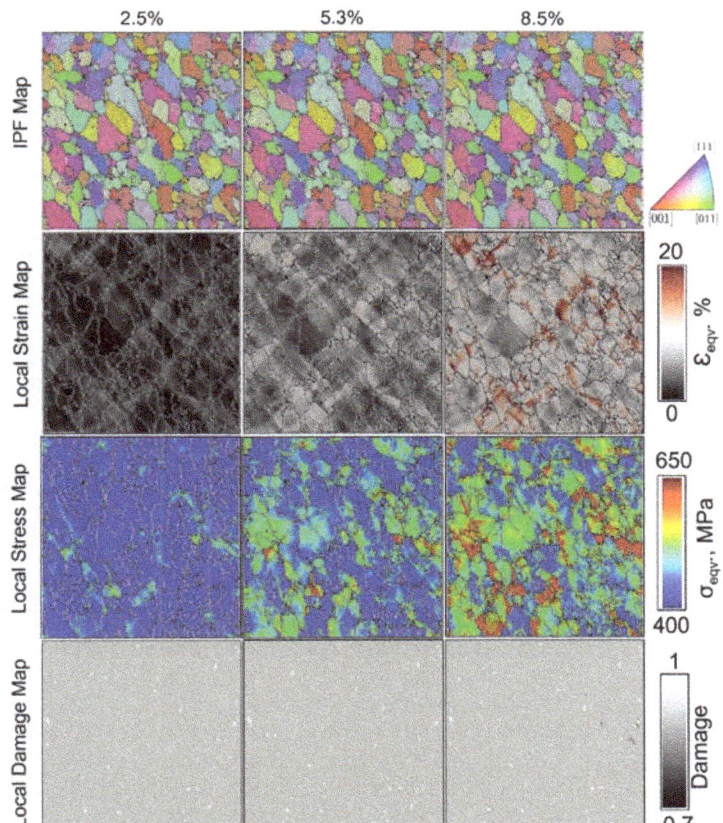

Figure 6. Local maps (from top to bottom) for IPF, strain, stress, and damage for the S-97 sample with respective legends/scales. Local maps at global strains, i.e., 2.5%, 5.3%, and 8.5%, are shown from left to right for local behavior insight during the evolution of mechanical deformation.

The evolution of local strains in the microstructure of S-97 with finely distributed cementite particles is comparatively more homogenous, and the stresses are also distributed in the whole RVE at 8.5% global strain. Due to extended spheroidization, fine and well-distributed cementite particles in the whole RVE, the damage incidents are less frequent than the S-83 samples at the same global strain. The first damage incident in the S-97 sample is observed at about 5.5% global strain, i.e., 2% later than S-83.

The effect of evolving crystallographic orientation with reference to crystal shapes is displayed in Figure 7. It is observed that the crystal oriented towards the [001] direction initially tends to shift its orientation slightly towards the [011] direction during deformation. This evolution in orientation from 2.5% to 11.2% global strain is shown in the top inset images labeled as Figure 7a. Furthermore, the changing texture of the specific grain represented by 1 in Figure 7b in the unique neighborhood of all the differently oriented ferrite grains numbered from 2 to 6 in the same image suggests the strong influence of the orientation of surrounding grains during deformation.

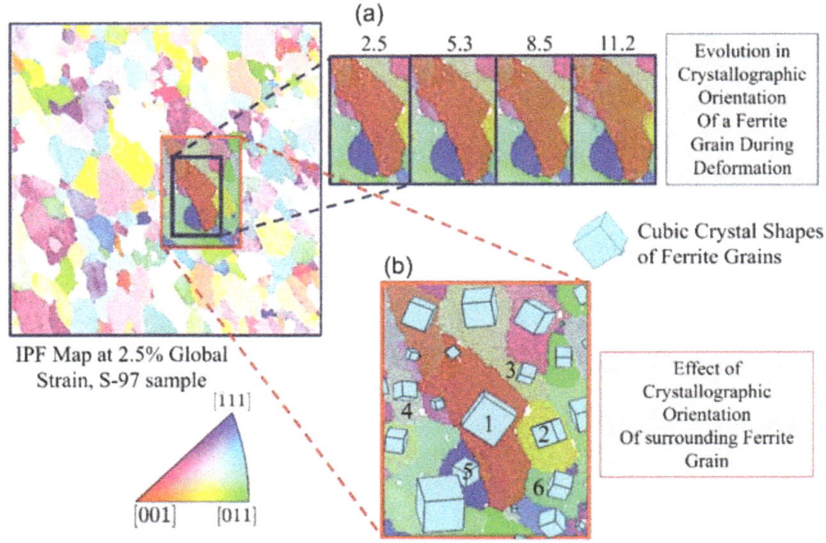

Figure 7. Crystallographic orientation evolution of ferrite grains for the S-97 sample shown (**a**) at multiple global strain values, i.e., 2.5%, 5.3%, 8.5%, and 11.2%. (**b**) Close look at the ferrite grain labeled with 1 and orientation distribution of neighboring grains labeled with numbers 2 to 6. Corresponding legends for the IPF map and crystal shapes are given for clear understanding and reference.

The evolution in crystallographic orientation during deformation largely depends on the alignment feasibility of adjoining ferrite grains. A higher conformity in the form of similar orientation can lead to greater ease in plastic flow, and a lower similarity can lead to enhanced slip resistance in the form of slip locking. This can further give rise to the local stress concentration at a local material point. The triple points also play a role here by enhancing the vulnerability of crack initiation and the strain concentration in local regions.

A profound analysis of the damage growth in the spheroidized medium carbon steel sample after 97% spheroidization is explained with the help of various constructions in Figure 8. During mechanical deformation evolution, local damage maps are shown from 9.5% to 11.2% in Figure 8a–c, respectively, in the S-97 sample during micromechanical simulation. It is observed in Figure 8a that local cracks first appear as independent sites with a single element in damage at various places. They start to grow immediately as the deformation progresses from 9.5% to 11.2% global strain.

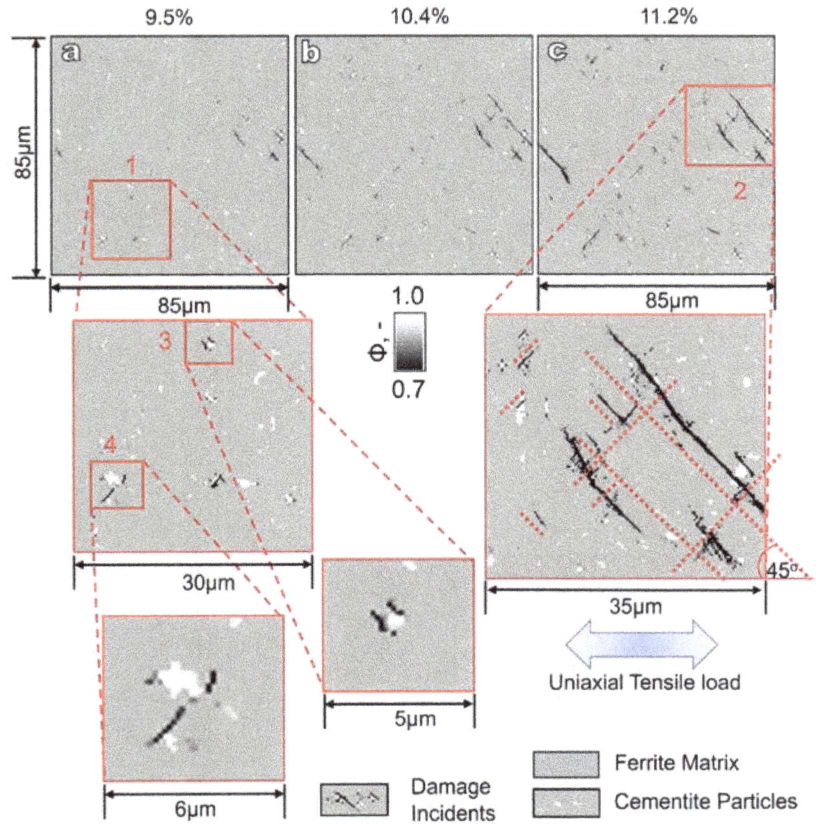

Figure 8. Local damage maps for the S-97 sample at different global strains, i.e., (**a**) 9.5%, (**b**) 10.4%, and (**c**) 11.2%. Zone 1 is selected to show the damage initiation sites at the ferrite–cementite interface magnified in zone 3 and 4. Zone 2 is marked and magnified to show the damage propagation lines at 45° with the direction of the application of the tensile load.

Zone 1 contains the sample area to study the initiation of the micro-cracks, with a magnified zone 3 and 4. The micromechanical model implemented in this study predicted the damage initiated at the ferrite–cementite interface, as shown in magnified zone 3 and zone 4. The process of the initiation of the micro-crack in areas with multiple cementite particles in close proximity causes severe strain localization, and the threshold of damage (i.e., $\varepsilon_{crit} = 0.5$) is reached at multiple points in close vicinity.

Various local damage incidents join together, and the crack line appears to propagate in the oblique direction compared to the load direction. Moreover, it is observed that the damage incidents occur at the ferrite–cementite interface. For example, in zone 3, the damage incidents occur at the interface on both sides of the cementite particle, and afterwards, the damage propagates in a specific direction. Similarly, for zone 4, the hard cementite phase particles do not carry on the usual deformation in the ductile ferrite matrix, and they start to dissociate from the matrix.

Upon further investigating the propagation of the cracks, it is observed in Figure 8, zone 2, that the angle of the crack growth lines is approximately equal to the 45° compared with the uniaxial tensile load direction.

4. Discussion

The micromechanical material behavior of spheroidized medium carbon steels is evaluated using two cases of differently heat-treated samples with the same ferrite and cementite phase fraction. The varying microstructure-based real RVEs were subjected to the virtual monotonic quasi-static tensile load. After implementing the calibrated phenomenological material model with DAMASK, the global and local results are presented and analyzed to gain valuable insights. These insights are found to be in compliance with the previously published findings and conclusions for similar materials.

The numerical simulation for the S-83 case terminated during computation at ~9% of global strain. This is probably due to computational complexities involved due to multiple damage sites initiated within the clustered cementite particles at various places in the RVE (refer to Figures 1 and 2).

Local damage initiation with a 2% delay in global strain for the S-97 sample suggests that the greater the homogeneous distribution of cementite particles, the lower the probability of starting damage incidents earlier. The S-83 sample is 83% spheroidized. It contains newly formed heterogeneously clustered cementite particles (~15 μm mean diameter). Within these clusters, the inter-particle spacing is small. During deformation, due to high local strains (refer to Figures 4 and 6, local strain maps), these zones are more susceptible to interface dissociation. It is supported by the already reported similar findings for ferrite and martensite aggregates by Habibi et al. [44] and Pathak et al. [45]. In addition, it has been reported by Tasan et al. [20] that the hard-phase particles in the ductile ferrite matrix lead to the matrix–particle decohesion and local micro-crack initiation.

Contrary to this, the S-97 sample, having the same volumetric phase fraction and instead more homogeneously distributed fine (~11 μm mean diameter) cementite particles, performs better in delaying the damage incident. Umar et al. [8] previously reported similar results for ferrite and cementite aggregate with a virtual microstructure subjected to tensile load under crystal plasticity-based numerical simulation.

This study has provided significant insights into the changing orientations of the ferrite grains during mechanical deformation (refer to Figures 5 and 7). It is reported that the evolving texture during mechanical deformation potentially changes the material behavior in crystalline materials [27,46]. It is observed that damage initiation depends on the second particle grain size and their distribution. At the same time, the crack propagation and void coalescence in the ductile ferrite matrix are governed by the orientation distribution of the surrounding grains. Similar observations have already been reported for the ferrite and martensite combination in steels [44,47]. As shown in Figure 7, grain rotation can have detrimental effects on local formability. Panin et al. [48] showed that the extent of rotation energy accumulation determines the starting point for crack opening in the heterogeneously distributed aggregate. It has been observed that microstructural heterogeneity may influence the crystallographic texture evolution and plastic strains during deformation. Sidor reported a similar observation for aluminum alloys [49].

The propagation of the damage lines at a ±45° angle to the applied tensile load is observed in this ferrite–cementite aggregate of medium carbon C45EC steel. This suggests of the generation of strain bands or lüder bands in the ductile ferrite matrix. The information obtained from the present study can be implemented on the dual-phase aggregates, with a comparable difference in flow properties in the phases. The mechanism of the local stress concentration in two-phase systems—evaluated by crystal plasticity models—has already been reported by researchers to behave in a more or less similar manner. The experimental results presented with the help of micro digital image correlation (μDIC) by Weidner et al. [50] and Tasan et al. [51] also present a detailed overview of similar growth line propagation in a martensite–ferrite combination during tensile loading. This information and development in model calibration and implementation with micro-scale material input can be used for modern generations of advanced high strength steels. Qayyum et al. [52] concluded their work on the heterogeneous aggregated steels with ductile austenite and hard martensite phase as the damage in ductile matrix propagates at 45 degrees to the loading direction.

An important observation in Figure 8 is that the most susceptible damage site in the whole aggregate is the ferrite–cementite interface. Therefore, it is critical to strengthen the weakest point in the RVE for better local formability of the steels. Similarly, it was concluded by Weidner et al. [53] that in multi-phase steels, the interface of the ductile ferrite and the hard second-phase particles is the most susceptible to damage initiation. This work also shows that the early start of damage at ~6% global strain in S-97 restricts the load shift from ductile ferrite to the cementite particles, not fully utilizing the latter's strength.

The following limitations are incorporated in the material model used for this study.

- The grain boundary is considered ideally akin to the grains, and its intrinsic effect has been ignored.
- The absence of a definition of ferrite–cementite interface-specific effects and properties.

In the near future, a consideration of the grain boundary and ferrite–cementite interface effects needs to be implemented in the material model. These additions can better predict stress and strain localization during crystallographic plane slip during adjacent grains interaction.

5. Conclusions

The microstructure dependence of differently spheroidized medium carbon steels is analyzed using calibrated phenomenological and micromechanics-based material model implemented in DAMASK. A methodology is presented which can be used to predict stress–strain localization and damage dynamics based on ferrite grain orientation and cementite particle size and distribution. After careful observation of results and correlating them with the similarly processed material behavior, as discussed in Section 4, the following conclusions are drawn for this study:

- It is shown that the presented methodology and numerical simulation model can be used to study the global flow trends that match, with greater than 99% and 97% accuracy between the simulation and experimental results for the S-83 and S-97 sa mples, respectively, achieved for plastic flow region in zone B. Furthermore, the minimum error suggests that the local results presented in this study can be attributed to the actual material's behavior.
- The S-83 sample with cementite particle clusters has shown multiple damage incidents earlier than the S-97 sample, having fine and homogeneously distributed cementite particles. In addition, the first damage incident in the same material is delayed until 2% global strain only because of the well-designed heat-treatment process.
- Contrary to the engineering (macroscopic) scale, the crystal plasticity-based simulation methodology can help us to understand the texture development and evolution during mechanical deformation. The crystallographic orientation of neighboring grains influences the grain orientation during thermo-mechanical deformation, changing the material behavior at the micro-level, and consequently at the component scale.
- Within the presently taken 2D RVEs under monotonic tensile load, the damage initiation occurs at the ferrite–cementite interface and propagates at $\pm 45^{\circ}$ in relation to the direction of the applied load.

Author Contributions: Conceptualization, M.U. and F.Q.; methodology, M.U., F.Q. and M.U.F.; software, M.U.; validation, F.Q., M.U. and S.G.; formal analysis, M.U.; investigation, F.Q.; resources, S.G. and M.U.F.; writing—original draft preparation, M.U., F.Q., M.U.F., S.G. and U.P.; writing—review and editing, M.U., F.Q., M.U.F., S.G. and U.P.; visualization, M.U. and F.Q.; supervision, S.G., M.U.F. and U.P.; project administration, U.P.; funding acquisition, F.Q. and U.P. All authors have read and agreed to the published version of the manuscript.

Funding: This research was funded by Deutsche Forschungsgemeinschaft (DFG, German Research Foundation) within the framework of the collaborative research group "TRIP Matrix Composites" project number 54473466–SFB 799.

Institutional Review Board Statement: Not applicable.

Informed Consent Statement: Not applicable.

Data Availability Statement: The MTEX algorithms for EBSD data cleaning and numerical simulation are not available online but can be shared upon request.

Acknowledgments: The authors acknowledge the DAAD Faculty Development for Candidates (Balochistan), 2016 (57245990)-HRDI-UESTP's/UET's funding scheme in cooperation with the Higher Education Commission of Pakistan (HEC) for sponsoring the stay of Faisal Qayyum at IMF TU Freiberg. This work is conducted within the DFG funded collaborative research group TRIP Matrix Composites (SFB 799). The authors gratefully acknowledge the German Research Foundation (DFG) for the financial support of the SFB 799. Freunde und Förderer der TU Bergakademie Freiberg e.V. is acknowledged for providing financial assistance to Muhammad Umar. The authors also acknowledge the support of Martin Diehl and Franz Roters (MPIE, Düsseldorf) for their help regarding the functionality of DAMASK. The authors acknowledge the support provided by AGH Krakow, Poland to perform EBSD analysis using Merlin Gemini II (ZEISS), at international center of electron microscopy for material science. Finally, the competent authorities at Khwaja Fareed University of Engineering and Information Technology, (KFUEIT) Rahim Yar Khan, Pakistan, and TU BAF Germany are greatly acknowledged for providing research exchange opportunities to Muhammad Umar at the Institute of Metal forming TU BAF Germany under a memorandum of understanding (MoU).

Conflicts of Interest: The authors declare no conflict of interest.

References

1. Rana, R.; Sing, S.B. *Automotive Steels, Design, Metallurgy, Processing and Applications*; Woodhead Publishing: Cambridgeshire, UK, 2017.
2. Pan, H. Development and Application of Lightweight High-strength Metal Materials. *MATEC Web Conf.* **2018**, *207*, 03010. [CrossRef]
3. Wilhelm, M. Materials used in automobile manufacture—current state and perspectives. *J. Phys.* **1993**, *3*, 7–31. [CrossRef]
4. Wong, S.L.; Madivala, M.; Prahl, U.; Roters, F.; Raabe, D. A crystal plasticity model for twinning- and transformation-induced plasticity. *Acta Mater.* **2016**, *118*, 140–151. [CrossRef]
5. Prasad, C.; Bhuyan, P.; Kaithwas, C.; Saha, R.; Mandal, S. Microstructure engineering by dispersing nano-spheroid cementite in ultrafine-grained ferrite and its implications on strength-ductility relationship in high carbon steel. *Mater. Des.* **2018**, *139*, 324–335. [CrossRef]
6. Harisha, S.R.; Sharma, S.; Kini, U.A.; Shankar, M.C.G. Study on Spheroidization and Related Heat Treatments of Medium Carbon Alloy Steels. *MATEC Web Conf.* **2018**, *144*, 2008. [CrossRef]
7. Joo, H.S.; Hwang, S.K.; Baek, H.M.; Im, Y.-T.; Son, I.-H.; Bae, C.M. The effect of a non-circular drawing sequence on spheroidization of medium carbon steel wires. *J. Mater. Process. Technol.* **2015**, *216*, 348–356. [CrossRef]
8. Umar, M.; Qayyum, F.; Farooq, M.U.; Khan, L.A.; Guk, S.; Prahl, U. Investigating the Effect of Cementite Particle Size and Distribution on Local Stress and Strain Evolution in Spheroidized Medium Carbon Steels using Crystal Plasticity-Based Numerical Simulations. *Steel Res. Int.* **2021**, *92*, 1–13. [CrossRef]
9. Song, W.; Choi, P.-P.; Inden, G.; Prahl, U.; Raabe, D.; Bleck, W. On the Spheroidized Carbide Dissolution and Elemental Partitioning in High Carbon Bearing Steel 100Cr6. *Met. Mater. Trans. A* **2013**, *45*, 595–606. [CrossRef]
10. Lv, Z.; Wang, B.; Wang, Z.; Sun, S.; Fu, W. Effect of cyclic heat treatments on spheroidizing behavior of cementite in high carbon steel. *Mater. Sci. Eng. A* **2013**, *574*, 143–148. [CrossRef]
11. Amos, P.K.; Bhattacharya, A.; Nestler, B.; Ankit, K. Mechanisms of pearlite spheroidization: Insights from 3D phase-field simulations. *Acta Mater.* **2018**, *161*, 400–411. [CrossRef]
12. Guk, S.; Augenstein, E.; Zapara, M.; Kawalla, R.; Prahl, U. Effect of Spheroidization Annealing on Pearlite Banding. *Mater. Sci. Forum* **2019**, *949*, 40–47. [CrossRef]
13. Tasan, C.; Diehl, M.; Yan, D.; Bechtold, M.; Roters, F.; Schemmann, L.; Zheng, C.; Peranio, N.; Ponge, D.; Koyama, M.; et al. An Overview of Dual-Phase Steels: Advances in Microstructure-Oriented Processing and Micromechanically Guided Design. *Annu. Rev. Mater. Res.* **2015**, *45*, 391–431. [CrossRef]
14. Mohammed, B.; Park, T.; Pourboghrat, F.; Hu, J.; Esmaeilpour, R.; Abu-Farha, F. Multiscale crystal plasticity modeling of multiphase advanced high strength steel. *Int. J. Solids Struct.* **2018**, *151*, 57–75. [CrossRef]
15. Ghadbeigi, H.; Pinna, C.; Celotto, S. Failure mechanisms in DP600 steel: Initiation, evolution and fracture. *Mater. Sci. Eng. A* **2013**, *588*, 420–431. [CrossRef]
16. Li, Z.-X.; Li, C.-S.; Zhang, J.; Qiao, B.; Li, Z.-Z. Effects of Annealing on Carbides Size and Distribution and Cold Formability of 1.0C-1.5Cr Bearing Steel. *Metall. Mater. Trans. A Phys.* **2015**, *46*, 3220–3231. [CrossRef]
17. Storojeva, L.; Ponge, D.; Kaspar, R.; Raabe, D. Development of microstructure and texture of medium carbon steel during heavy warm deformation. *Acta Mater.* **2004**, *52*, 2209–2220. [CrossRef]

18. Saha, A.; Mondal, D.K.; Maity, J. Effect of cyclic heat treatment on microstructure and mechanical properties of 0.6wt% carbon steel. *Mater. Sci. Eng. A* **2010**, *527*, 4001–4007. [CrossRef]
19. Matusiewicz, P.; Augustyn-Nadzieja, J.; Czarski, A.; Skowronek, T. Kinetics of pearlite spheroidization. *Arch. Metall. Mater.* **2017**, *62*, 231–234. [CrossRef]
20. Tasan, C.; Hoefnagels, J.; Diehl, M.; Yan, D.; Roters, F.; Raabe, D. Strain localization and damage in dual phase steels investigated by coupled in-situ deformation experiments and crystal plasticity simulations. *Int. J. Plast.* **2014**, *63*, 198–210. [CrossRef]
21. Maruschak, P.O.; Panin, S.V.; Stachowicz, F.; Danyliuk, I.M.; Vlasov, I.; Bishchak, R.T. Structural levels of fatigue failure and damage estimation in 17Mn1Si steel on the basis of a multilevel approach of physical mesomechanics. *Acta Mech.* **2015**, *227*, 151–157. [CrossRef]
22. Qayyum, F.; Guk, S.; Kawalla, R.; Prahl, U. On Attempting to Create a Virtual Laboratory for Application-Oriented Microstructural Optimization of Multi-Phase Materials. *Appl. Sci.* **2021**, *11*, 1506. [CrossRef]
23. Diehl, M.; An, D.; Shanthraj, P.; Zaefferer, S.; Roters, F.; Raabe, D. Crystal plasticity study on stress and strain partitioning in a measured 3D dual phase steel microstructure. *Phys. Mesomech.* **2017**, *20*, 311–323. [CrossRef]
24. Umar, M.; Qayyum, F.; Farooq, M.U.; Khan, L.A.; Guk, S.; Prahl, U. Analyzing the cementite particle size and distribution in heterogeneous microstructure of C45EC steel using crystal plasticity based DAMASK code. In Proceedings of the 2021 International Bhurban Conference on Applied Sciences and Technologies (IBCAST), Islamabad, Pakistan, 12–16 January 2021; pp. 15–20. [CrossRef]
25. Qayyum, F.; Umar, M.; Guk, S.; Schmidtchen, M.; Kawalla, R.; Prahl, U. Effect of the 3rd dimension within the representative volume element (Rve) on damage initiation and propagation during full-phase numerical simulations of single and multi-phase steels. *Materials* **2020**, *14*, 42. [CrossRef]
26. Qayyum, F.; Chaudhry, A.A.; Guk, S.; Schmidtchen, M.; Kawalla, R.; Prahl, U. Effect of 3d representative volume element (Rve) thickness on stress and strain partitioning in crystal plasticity simulations of multi-phase materials. *Crystals* **2020**, *10*, 944. [CrossRef]
27. Diehl, M.; Groeber, M.; Haase, C.; Molodov, D.A.; Roters, F.; Raabe, D. Identifying Structure–Property Relationships Through DREAM.3D Representative Volume Elements and DAMASK Crystal Plasticity Simulations: An Integrated Computational Materials Engineering Approach. *JOM* **2017**, *69*, 848–855. [CrossRef]
28. Zhang, C.; Li, H.; Eisenlohr, P.; Liu, W.; Boehlert, C.; Crimp, M.; Bieler, T. Effect of realistic 3D microstructure in crystal plasticity finite element analysis of polycrystalline Ti-5Al-2.5Sn. *Int. J. Plast.* **2015**, *69*, 21–35. [CrossRef]
29. Prahl, U.; Qayyum, F.; Guk, S.; Kawalla, R. Effect of Sulphur Content on Formability of a Low Alloyed Steel. In Proceedings of the NEMU 2019, Vitoria-Gasteiz, Spain, 10 May 2019.
30. Roters, F.; Diehl, M.; Shanthraj, P.; Eisenlohr, P.; Reuber, C.; Wong, S.; Maiti, T.; Ebrahimi, A.; Hochrainer, T.; Fabritius, H.-O.; et al. DAMASK—The Düsseldorf Advanced Material Simulation Kit for modeling multi-physics crystal plasticity, thermal, and damage phenomena from the single crystal up to the component scale. *Comput. Mater. Sci.* **2019**, *158*, 420–478. [CrossRef]
31. Guk, S.; Kawalla, R.; Prahl, U. Mathematical Description of the Microstructural Modifications and Changes in the Mechanical Properties during Spheroidization of Medium-Carbon Steel. *Steel Res. Int.* **2019**, *90*, 1–8. [CrossRef]
32. Hielscher, R.; Silbermann, C.B.; Schmidl, E.; Ihlemann, J. Denoising of crystal orientation maps. *J. Appl. Crystallogr.* **2019**, *52*, 984–996. [CrossRef]
33. Bachmann, F.; Hielscher, R.; Schaeben, H. Grain detection from 2d and 3d EBSD data—Specification of the MTEX algorithm. *Ultramicroscopy* **2011**, *111*, 1720–1733. [CrossRef] [PubMed]
34. Niessen, F.; Nyyssönen, T.; Gazder, A.A.; Hielscher, R. Parent grain reconstruction from partially or fully transformed microstructures in MTEX. *J. Chem. Inf. Model.* **2013**, *53*, 1689–1699.
35. Khan, U.A.; Hussain, A.; Shah, M.; Shuaib, M.; Qayyum, F. Investigation of mechanical properties based on grain growth and microstructure evolution of alumina ceramics during two step sintering process. In *IOP Conference Series: Materials Science and Engineering, Porceedings of 14th International Symposium on Advanced Materials, Islamabad, Pakistan, 12–16 October 2015*; IOP Publishing: Bristol, UK, 2016; Volunme 146, pp. 8–15. [CrossRef]
36. Najabat Ali, M.; Ansari, U.; Sami, J.; Qayyum, F.; Mir, M. To Develop a Biocompatible and Biodegradable Polymer-Metal Composite with Good; Mechanical and Drug Release Properties. *J. Mater. Sci. Eng.* **2016**, *5*, 1–4. [CrossRef]
37. Tsumura, K. Hierarchically Aggregated Optimization Algorithm for Heterogeneously Dispersed Utility Functions. *IFAC PapersOnLine* **2017**, *50*, 14442–14446. [CrossRef]
38. Chen, J.; Sun, Y. A new multiplexed optimization with enhanced performance for complex air conditioning systems. *Energy Build.* **2017**, *156*, 85–95. [CrossRef]
39. Różyło, P.; Wrzesińska, K. Numerical analysis of the behavior of compressed thin-walled elements with holes. *Adv. Sci. Technol. Res. J.* **2016**, *10*, 199–206. [CrossRef]
40. Kačeniauskas, A.; Pacevič, R.; Bugajev, A.; Katkevičius, T. Efficient visualization by using ParaView software on BalticGrid. *Inf. Technol. Control.* **2010**, *39*, 108–115. [CrossRef]
41. Qayyum, F.; Guk, S.; Schmidtchen, M.; Kawalla, R.; Prahl, U. Modeling the local deformation and transformation behavior of cast X8CrMnNi16-6-6 TRIP steel and 10% Mg-PSZ composite using a continuum mechanics-based crystal plasticity model. *Crystals* **2020**, *10*, 221. [CrossRef]

42. Qayyum, F.; Guk, S.; Prüger, S.; Schmidtchen, M.; Saenko, I.; Kiefer, B.; Kawalla, R.; Prahl, U. Investigating the local deformation and transformation behavior of sintered X3CrMnNi16-7-6 TRIP steel using a calibrated crystal plasticity-based numerical simulation model. *Int. J. Mater. Res.* **2020**, *111*, 392–404. [CrossRef]
43. Roters, F.; Eisenlohr, P.; Kords, C.; Tjahjanto, D.D.; Diehl, M.; Raabe, D. DAMASK: The Düsseldorf Advanced MAterial Simulation Kit for studying crystal plasticity using an FE based or a spectral numerical solver. *Procedia IUTAM* **2012**, *3*, 3–10. [CrossRef]
44. Habibi, N.; Vajragupta, N. Deformation and Damage Assessments of Two DP1000 Steels Using a Micromechanical Modelling Method. *Crystals* **2021**, *11*, 805. [CrossRef]
45. Pathak, N.; Butcher, C.; Worswick, M.J.; Bellhouse, E.; Gao, J. Damage evolution in complex-phase and dual-phase steels during edge stretching. *Materials* **2017**, *10*, 346. [CrossRef] [PubMed]
46. Raabe, D. Contribution of {123} ⟨111⟩ slip systems to deformation of b.c.c. metals. *Phys. Status Solidi.* **1995**, *149*, 575–581. [CrossRef]
47. Marteau, J.; Haddadi, H.; Bouvier, S. Investigation of Strain Heterogeneities Between Grains in Ferritic and Ferritic-Martensitic Steels. *Exp. Mech.* **2012**, *53*, 427–439. [CrossRef]
48. Panin, S.; Moiseenko, D.; Maksimov, P.; Vlasov, I.; Byakov, A.; Maruschak, P.; Berto, F.; Schmauder, S.; Vinogradov, A. Influence of energy dissipation at the interphase boundaries on impact fracture behaviour of a plain carbon steel. *Theor. Appl. Fract. Mech.* **2018**, *97*, 478–499. [CrossRef]
49. Sidor, J.J. Crystal plasticity and continuum mechanics-based modelling of deformation and recrystallization textures in aluminum alloys. *IOP Conf. Ser. Mater. Sci. Eng.* **2018**, *375*, 12028. [CrossRef]
50. Weidner, A.; Biermann, H. Review on Strain Localization Phenomena Studied by High-Resolution Digital Image Correlation. *Adv. Eng. Mater.* **2021**, *23*, 2001409. [CrossRef]
51. Tasan, C.; Diehl, M.; Yan, D.; Zambaldi, C.; Shanthraj, P.; Roters, F.; Raabe, D. Integrated experimental–simulation analysis of stress and strain partitioning in multiphase alloys. *Acta Mater.* **2014**, *81*, 386–400. [CrossRef]
52. Qayyum, F.; Guk, S.; Prahl, U. Studying the Damage Evolution and the Micro-Mechanical Response of X8CrMnNi16-6-6 TRIP Steel Matrix and 10% Zirconia Particle Composite Using a Calibrated Physics and Crystal-Plasticity-Based Numerical Simulation Model. *Crystals* **2021**, *11*, 759. [CrossRef]
53. Weidner, A.; Berek, H.; Segel, C.; Aneziris, C.G.; Biermann, H. In Situ Tensile Deformation of TRIP Steel/Mg-PSZ Composites. *Mater. Sci. Forum* **2013**, *738*, 77–81. [CrossRef]

Journal of
Composites Science

Article

Numerical Study of Soil-Thawing Effect of Composite Piles Using GMsFEM

Petr V. Sivtsev [1,*], Piotr Smarzewski [2] and Sergey P. Stepanov [1]

[1] Department of Computational Technologies, Institute of Mathematics and Information Science, Ammosov North-Eastern Federal University, 58 Belinskogo, 677000 Yakutsk, Russia; Cepe2a@inbox.ru
[2] Department of Structural Engineering, Faculty of Civil Engineering and Architecture, Lublin University of Technology, 20-618 Lublin, Poland; p.smarzewski@pollub.pl
* Correspondence: sivkapetr@mail.ru; Tel.: +7-924-169-17-47

Abstract: During construction works, it is advisable to prevent strong thawing and an increase in the moisture content of the foundations of engineering structures in the summer. Since the density of water and ice differ, due to the difference bulging of the foundation sections can occur when it freezes back in winter. In this work, the effect of fiber-reinforced piles on the thermal field of the surrounding soil is investigated numerically; that is, the study of the influence of aggregates with high and low thermal-physical properties on the temperature of frozen soils is conducted. Basalt and steel fiber reinforcement are compared. The difficulty of this work is that the inclusions inside piles are too small compared to the pile itself. Therefore, to solve the Stefan problem, a generalized multiscale finite element method (GMsFEM) was used. In the GMsFEM, the usual conforming partition of the domain into a coarse grid was used. It allowed reducing problem size and, consequently, accelerating the calculations. Results of the multiscale solution were compared with fine-scale solution, the accuracy of GMsFEM was investigated, and the optimal solution parameters were defined. Therefore, GMsFEM was shown to be well suited for the designated task. Collation of basalt and steel fiber reinforcement showed a beneficial effect of high thermal conductive material inclusion on freezing of piles in winter.

Keywords: Stefan problem; multiscale; generalized multiscale finite element method; composite pile; thermal conduction

Citation: Sivtsev, P.V.; Smarzewski, P.; Stepanov, S.P. Numerical Study of Soil-Thawing Effect of Composite Piles Using GMsFEM. *J. Compos. Sci.* **2021**, *5*, 167. https://doi.org/10.3390/jcs5070167

Academic Editor: Stelios K. Georgantzinos

Received: 22 May 2021
Accepted: 18 June 2021
Published: 28 June 2021

1. Introduction

Thermal calculations are important in the construction of engineering geotechnical structures and buildings in the permafrost zone. The temperature regime (a set of sequential temperature fields in the soil mass corresponding to any given points in time from the beginning of the calculation) is calculated as the forecast of the thermal effects on the upper and lower boundaries of the structure foundation set for the entire calculation period [1]. The most characteristic feature of these processes is the previously unknown ("free") boundaries between various thawed and frozen states of the soil. Due to this feature, their mathematical models are non-linear and difficult to analyze. The results of calculations of climatic phenomena, which are critical for geotechnical structures on permafrost foundations (in the permafrost zone), show a continuous repetition of the same constant seasonal cycle. This determines periodic harmonic oscillations of all quantities that determine the state of the foundation of a geotechnical structure, including temperature. However, cyclic harmonic changes in the foundation state caused by constantly repeating influences also occur during other processes. The course of temperature change during one period can have a different character. For example, the temperature can change abruptly, continuously increase or decrease, etc.

During construction works, it is advisable to prevent strong thawing and an increase in the moisture content of the foundations of engineering structures in the summer. Since the density of water and ice differ, due to the difference bulging of the foundation sections can occur when it freezes back in winter. Repetition of this phenomenon for many years causes the destruction of engineering structures and buildings. In the regions of the Far North, it leads to catastrophic emergencies.

In this work, the effect of fiber-reinforced piles on the thermal field of the surrounding soil is numerically investigated [2], that is, the study of the influence of aggregates with high and low thermal-physical properties on the temperature of frozen soils is conducted [3,4].

Basalt fibers are produced from molten basalt rock. They have very low thermal conductivity, good strength properties and, at the same time, are relatively cheap. Due to these characteristics, they are used in concrete [5–7]. However, in addition to the chemical and mechanical properties of basalt fibers, their cost varies, depending on the type and quality of the raw material and the production process of these fibers [8]. Nevertheless, the characteristics mentioned above and an environmentally friendly manufacturing process [9] might determine their application in high performance concrete structures instead of the most commonly used steel and polypropylene fibers. The length-to-diameter aspect for basalt fibers is about 1000.

As for steel fiber hybridization, industrial or recycled steel fibers can be used [10–13]. According to results available in the scientific literature [14,15], they show similar mechanical responses, both in terms of tensile strength and matrix-to-fiber bond. Recycled steel fibers are derived from waste tires. Their geometrical characterization can be highly variable: they are generally characterized by a nominal diameter ranging between 0.1 and 2 mm with a corresponding average aspect ratio (i.e., length-to-diameter ratio) ranging between 20 and 150. These variations mainly depend on both the original source (i.e., tires typology) and recycling processes.

The volume concentration of fibers is equal to 10%. In our article, we use the length-to-diameter ratio for both types of fibers equal to 32 for structured and 10 for random distribution of fibers for calculation convenience. Fibers' mechanical properties are presented in Table 1.

Table 1. Mechanical properties of fibers.

Fiber Type	Steel	Basalt
Density (kg/m^3)	7800	2700
Volume content (kg/m^3)	780	270
Elastic modulus (GPa)	200	70
Tensile strength (MPa)	>1060	>1700

One of the topical problems of mathematical physics is the Stefan problem, which is an initial–boundary value problem for a parabolic differential equation with discontinuous coefficients, and serves as a mathematical model of the change in the phase state of a substance with unknown interfaces [16,17]. The phase transition boundary is represented as a smearing zone when the phase transition occurs in a given temperature range [16,18].

The difficulty of this work lies in the pile inclusions that are too small compared to the solution area. For such geometrically complex problems, reduction techniques such as multiscale methods are preferred [19,20]. Multiscale methods are becoming very popular at this time [21–23]. Paper [24] develops a multiscale eigenvalue method based on multiscale substructure technology for multiscale analysis of periodic composite structures. This provides a comparative study from a user's viewpoint for multiscale methods, including the mathematical homogenization method, heterogeneous multiscale method, and multiscale finite element method. The article [25] studies the multiscale finite element method for solving elliptic problems arising from composite materials. The influence of production porosity, diameter [26], and density of structures on the thermal conductivity of a composite are considered using multiscale finite element modeling [27]. In [28,29], mathematical modeling is applied for the thermodynamic analysis and design of composite structures.

This article is organized as follows. Section 2 presents the formulation of a mathematical model that describes the dynamics of the temperature distribution with the phase transitions taken into account. In the third, finite element approximation is given. In the fourth, a generalized multiscale finite element method (GMsFEM) is considered. Section 5 presents the numerical results in a two-dimensional setting.

2. Mathematical Model

Let us consider a mathematical model describing the dynamics of the temperature distribution, taking into account the phase transitions of pore moisture into ice and back, at a certain given temperature of the phase transition T^* in area $\Omega = \Omega^- \cup \Omega^+$. Where $\Omega^+(t) = \{x | x \in \Omega,\ T(x,t) > T\}$ is the area occupied by the thawed soil, where the temperature exceeds the phase transition temperature and $\Omega^-(t) = \{x | x \in \Omega,\ T(x,t) < T\}$ is an area occupied by the frozen ground. A phase transition occurs at the interface between thawed and frozen soils $S = S(t)$ (Figure 1).

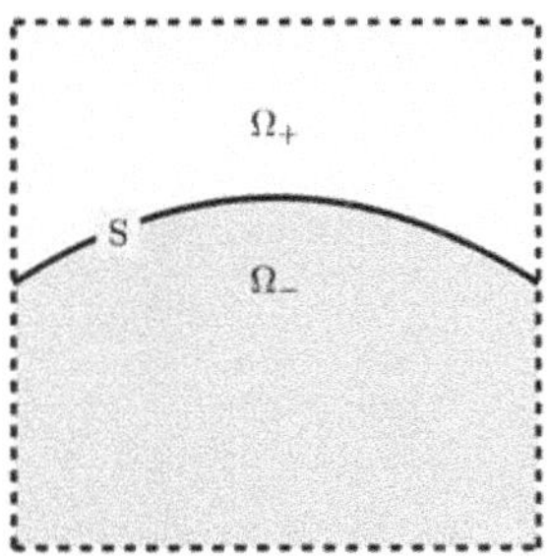

Figure 1. Phase transition.

To simulate heat transfer processes with phase transitions, the Stefan model is used. It describes thermal processes accompanying phase transformations of the medium with absorption and release of latent heat:

$$(c\rho(T) + m\rho_l L\delta(T - T^*))\frac{\partial T}{\partial t} - \mathrm{div}(\lambda(T)T) = f,\ x \in \Omega,\ t \in (0, t_{max}], \tag{1}$$

where L is the specific heat of phase transition, m is a porosity, $\delta(T - T^*)$ is Dirac delta function.

For the coefficients of the equation, there are following relations:

$$c\rho(T) = \begin{cases} c^-\rho^-, & T < T^*, \\ c^+\rho^+, & T \geq T^*, \end{cases} \quad \lambda(T) = \begin{cases} \lambda^-, & T < T^*, \\ \lambda^+, & T \geq T^*, \end{cases}$$

where $c^+,\ \rho^+, \lambda^+$ and $c^-,\ \rho^-, \lambda^-$ are the specific heat, density, and thermal conductivity of thawed and frozen soil, respectively.

Since the process of heat propagation is considered in a saturated porous medium, there are following thermal-physical characteristics:

$$c^-\rho^- = (1 - m)c_{sc}\rho_{sc} + mc_i\rho_i,\ c^+\rho^+ = (1 - m)c_{sc}\rho_{sc} + mc_w\rho_w,$$

where m is a porosity. The subscripts $sc,\ w,\ i$ denote the skeleton of the porous medium, water, and ice, respectively. For the coefficients of the thermal conductivity in the thawed and frozen zones, there are similar relations:

$$\lambda^- = (1 - m)\lambda_{sc} + m\lambda_i,\ \lambda^+ = (1 - m)\lambda_{sc} + m\lambda_w,$$

In practice, phase transformations do not take place instantly and can proceed in a small temperature range $[T-\Delta,\ T+\Delta]$. The discontinuous coefficients of Equation (1) are replaced by sufficiently smooth functions of temperature:

$$(c\rho)_\Delta(T) = c^-\rho^- + m(c^+\rho^+ - c^-\rho^-)\frac{1}{2}\left(1+\operatorname{erf}\left(\frac{T-T^*}{\sqrt{2\Delta^2}}\right)\right) + \frac{1}{\sqrt{2\pi\Delta^2}}e^{-\frac{(T-T^*)^2}{2\Delta^2}},$$

$$\lambda_\Delta(T) = \lambda^- + m(\lambda^+ - \lambda^-)\frac{1}{2}\left(1+\operatorname{erf}\left(\frac{T-T^*}{\sqrt{2\Delta^2}}\right)\right).$$

Thus, the following equation for the temperature is obtained:

$$(c\rho)_\Delta(T)\frac{\partial T}{\partial t} - \operatorname{div}(\lambda_\Delta(T)T) = f,\ x\in\Omega,\ t\in(0,t_{max}], \tag{2}$$

Thus, Equation (2) is a multidimensional quasilinear parabolic equation with smooth coefficients.

For piles and fibers the following is true: $(c\rho)_\Delta = (c\rho)_{p,f}$ and $\lambda_\Delta = \lambda_{\mathrm{p,f}}$, where $c_{p,f}, \rho_{p,f}, \lambda_{\mathrm{p,f}}$ are specific heat, density, and thermal conductivity of piles and fibers. That is, there is no phase transition in the pile and fiber areas, respectively.

Equation (2) is supplemented with the initial and boundary conditions

$$T(x,0) = T_0,$$

$$-\lambda\frac{\partial T}{\partial \eta} = \alpha(T - T_{air}),\ x\in\Gamma_t,$$

$$-\lambda\frac{\partial T}{\partial \eta} = 0,\ x\in\frac{\Gamma}{\Gamma_t}.$$

Here η—normal vector.

3. Fine-Scale Approximation

The quasilinear parabolic Equation (2) with the corresponding boundary and initial conditions is approximated using the finite element method in combination with a purely implicit linearized finite difference approximation in time. This means that during discretization, the coefficients depending on the desired function are taken from the previous time layer. Let us write down the variational problem statement for each time layer: find $T\in H^1$ such that:

$$a\left(T^{n+1},v\right) = L(v),\ \forall\in H_0^1,$$

where

$$a(T^{n+1},v) = \tfrac{1}{\tau}\int_\Omega(c\rho)_\sigma(T^n)T^{n+1}w dx + \int_\Omega\lambda_\sigma(T^n)T^{n+1}\cdot w dx + \int_{\Gamma t}\alpha T^{n+1}ds,$$

$$L(v) = \tfrac{1}{\tau}\int_\Omega(c\rho)_\sigma T^n w dx + \int_{\Gamma t}\alpha T_{air}ds.$$

To solve the problem numerically, it is necessary to pass from a continuous variational problem to a discrete one. For this, finite-dimensional spaces $V_\mathrm{h}\in H^1$, $\hat{V}_h\in H_0^1$ are introduced and the following problem is defined on them: find $T_h\in V_h$ such that:

$$a\left(T_h^{n+1},v\right) = L(v),\ \forall\in\hat{V}_h,$$

where

$$a(T^{n+1},v) = \tfrac{1}{\tau}\int_\Omega(c\rho)_\sigma(T^n)T^{n+1}w_h dx + \int_\Omega\lambda_\sigma(T^n)T_h^{n+1}\cdot w_h dx + \int_{\Gamma t}\alpha T^{n+1}ds,$$

$$L(v) = \tfrac{1}{\tau}\int_\Omega(c\rho)_\sigma T_h^n w_h dx + \int_{\Gamma t}\alpha T_{air}ds.$$

4. Generalized Multiscale Finite Element Method (GMsFEM)

Construction local reduction of a model on the snapshot space is described by solving several local spectral problems using GMsFEM. In the GMsFEM the usual conforming partition of domain into finite elements is used. This partition is called the coarse grid $\mathcal{T}_H$. The coarse grid is partitioned into coarse-grid blocks. That is, $\mathcal{T}_H$ is a coarse grid in domain Ω, such that $\mathcal{T}_H = \cup_{i=1}^{N_c} K_i$, where K_i is coarse cell. The $\mathcal{T}_h$ is a fine grid with $H \geq h \geq 0$, where h is size of fine grid. Each coarse-grid block (K_i) consists of the connected union of fine-grid blocks. Furthermore, a certain domain for this is constructed and denoted by N_c the coarse nodes number, by $\{x_i\}_{i=1}^{N_c}$ the vertices of the coarse mesh and define the neighborhood of the node x_i:

$$\omega_i = \cup\{K_i \in \mathcal{T}_H;\ x_i \in \overline{K_i}\}$$

The following steps need to be implemented for GMsFEM realization:

1. Coarse grid generation $\mathcal{T}_H$;
2. Offline space construction;
3. Construction of snapshot space that will be used to compute an offline space;
4. Construction of a small dimensional offline space by performing dimensional reduction in the space of local snapshots;
5. Solution of a coarse-grid problem for any force term and boundary condition.

In the first step of GMsFEM (Offline stage), the "snapshots" space must be constructed, a large dimensional snapshots space of local solutions. In the "snapshots" space, the following is considered:

$$-\operatorname{div}(k_\alpha \nabla \psi_j) = 0,\ x \in \omega_i,$$
$$\psi_j = \delta_j(x),\ x \in \partial\omega_i,$$

where k_α are the coefficients of the thermal conductivity, $\delta_j(x)$ are certain set of function defined on $\partial\omega_i$, here $j = 1,\ J_\omega$. The J_ω is a number of fine grid edges on ω. Therefore, following is defined:

$$V_{\text{snap}} = \text{span}\{\psi_j^{\text{snap}} : 1 \leq \text{j} \leq J_\omega\},\ \text{and}\ R_{\text{snap}} = \left[\psi_1^{\text{snap}}, \ldots, \psi_{J_\omega}^{\text{snap}}\right]$$

This allow reducing the snapshot space to offline space via some spectral procedure. Offline space is constructed using the following local spectral problems in the snapshots space:

$$\overline{A}_{\text{off}}\,\overline{\Psi}_k^{\text{off}} = \lambda_k \overline{S}_{\text{off}}\,\overline{\Psi}_k^{\text{off}},$$

where $\overline{A}_{\text{off}} = R_{\text{snap}} A R_{\text{snap}}^T$, $\overline{S}_{\text{off}} = R_{\text{snap}} S R_{\text{snap}}^T$, and

$$A = [a_{mn}] = \int_{\omega_i} (k_\alpha \nabla \varphi_m, \nabla \varphi_n) dx,\ S = [s_{mn}] = \int_{\omega_i} (k_\alpha \varphi_m, \varphi_n) dx.$$

To generate the offline space, the smallest $M_{\text{off}}^{\omega_i}$ eigenvalues are chosen and the corresponding eigenvectors $\psi_k^{\text{off}} = \sum_m \overline{\Psi}_{mk}^{\text{off}} \varphi_m^{\text{off}}$ are found for k = 1, 2, ... , $M_{\text{off}}^{\omega_i}$. The found eigenvectors must be multiplied by the partition of unity functions χ_i:

$$\varphi_k^{\omega_i} = \chi_i \psi_k^{\text{off}} \text{ for } 1 \leq i \leq N \text{ and } 1 \leq k \leq M_{\text{off}}^{\omega_i},$$

where $M_{\text{off}}^{\omega_i}$ denotes the number of eigenvectors that are sampled for each local ω_i.

After this procedure, conforming basis functions can be obtained in the space:

$$V_{\text{off}} = \text{span}\{\varphi_k^{\omega_i},\ 1 \le k \le M^{\omega_i},\ 1 \le i \le N\}.$$

Further, the projection matrix $R^T = \left[\varphi_1^{\omega_i}, \ldots, \varphi_{M_{\omega_i}}^{\omega_i}\right]$ is defined.

Using constructed multiscale space, the coarse-scale system is solved:

$$M\left(\frac{\partial T_c}{\partial t}\right) + A_c T_c = F_c,$$

where $A_c = RA_f R^T$ and $F_c = RF_f$.

After solving the coarse-scale solution, a solution on the fine grid $T_{ms} = R^T T_c$ can be calculated.

5. Numerical Results

In this section, the effect of fibers in a pile made of different materials on the temperature field of the soil is calculated. Thus, the melting effect of the pile is numerically simulated.

Conditions for the impact of piles on the temperature regime of the soil:

- Cement-sand mortar for filling the sinuses between the soil and the pile with a temperature of $+20\ ^\circ$C;
- Piles with a cross section of 40×40 cm, pile deepening is 10 m from the ground surface.

The computational domain consists of several soil layers; it has one composite pile (Figure 2).

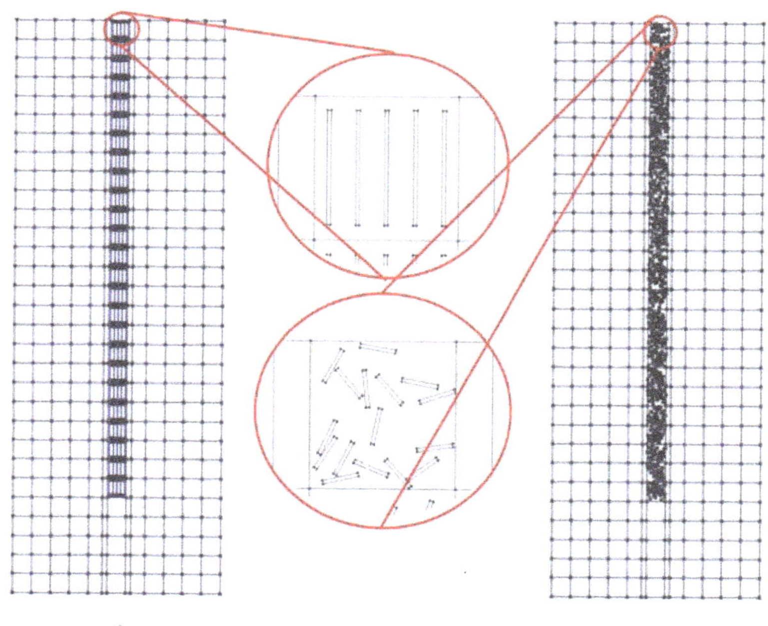

Figure 2. Computational domain (**a**) is structured, (**b**) is randomized.

There are 4 cases which were considered:

1. Fibers from basalt arranged in a structured manner;
2. Fibers made of steel arranged in a structured manner;
3. Fibers from basalt located randomly;
4. Fibers made of randomly located steel.

Soil temperature is 1.5 °C. Thermal-physical characteristics are taken from SP 25.13330.2012 and presented in Table 2. Calculations were performed for 365 days with a time step $\tau = 1$ day (24 h).

Table 2. Thermal-physical characteristics of soils.

Elements	Volumetric Heat Capacity $c\rho * 10^{-6}$ (J/m³/K)		Thermal Conductivity k (W/m/K)		Phase Transition Heat $L * 10^{-3}$ (J/m³)
	Thawed	Frozen	Thawed	Frozen	
Clay loam	3.17	2.41	2.67	2.84	101,600
Sand	2.31	2.14	2.15	2.37	114,800
Sand	2.78	2.26	2.26	2.62	101,600
Concrete	2.22		1.86		-
Basalt fiber	1.4		0.033		-
Steel fiber	368.8		53		-

The temperature of the daylight surface of the structure base was set considering the amplitude of the air temperature fluctuations taken from the data of the Yakutsk meteorological station. Figure 3 demonstrates the changes in the air temperature. It shows the distribution of the air temperature during one year.

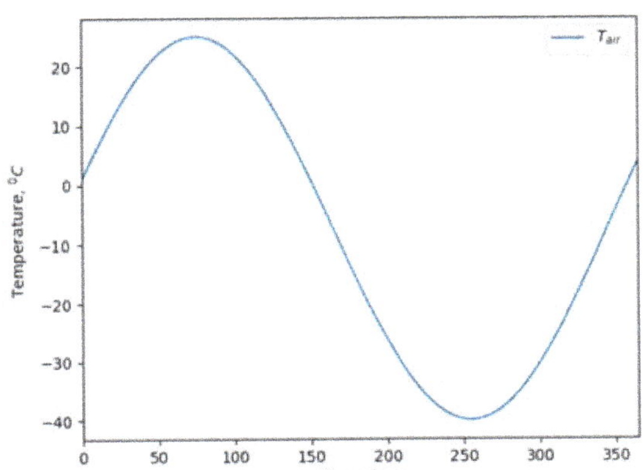

Figure 3. Air temperature.

Here are the results of numerical calculations. Figures 4 and 5 show the soil temperature distributions for different points in time. These results were obtained by the GMsFEM method for eight basis functions.

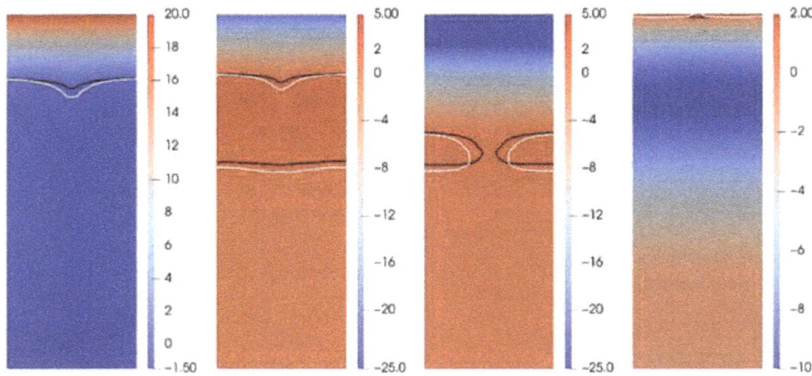

Figure 4. Temperature distribution for different time layers (t = 50, 200, 250, 365 days) when the fibers are structurally located. From left to right. (white line is isocline of zero of steel, black is isocline of zero of basalt).

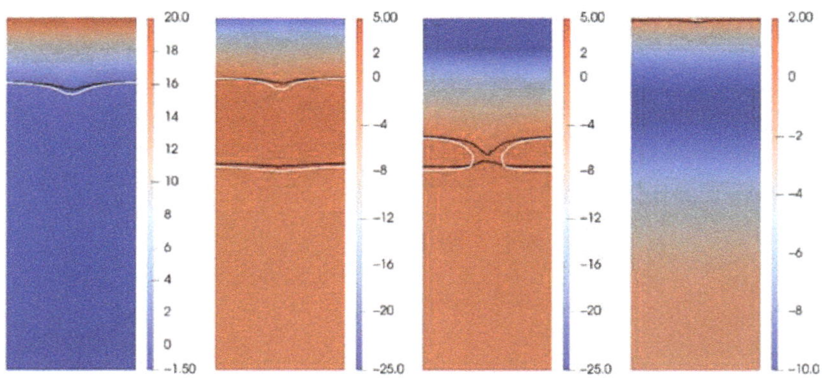

Figure 5. Temperature distribution for different time layers (t = 50, 200, 250, 365 days) when the fibers are randomly located. From left to right. (white line is isocline of zero of steel, black is isocline of zero of basalt).

For numerical comparison of the fine-scale and multiscale solutions, relative L_2 and energy errors are used.

$$\| e \|_{L_2} = \sqrt{\frac{\int_\Omega (T_h - T_{ms})^2 dx}{\int_\Omega {T_h}^2 dx}}, \| e \|_a = \sqrt{\frac{a_\phi(T_h - T_{ms}, T_h - T_{ms})}{a_\phi(T_h, T_h)}},$$

where T_h and T_{ms} are the fine-scale and multiscale solutions.

Tables 3 and 4 show the relative errors of L_2 and energies for different number of multiscale basis functions and corresponding degrees of freedom (DOF). Degrees of freedom for fine scale mesh is denoted as DOF_f. Here are the uncertainties for fibers arranged in structured order. When using basalt fibers, you can limit yourself to using 2 multiscale basis functions, and when using steel fibers, you need to use at least 4 multiscale basis functions.

Table 3. Relative L_2 and energy errors (%) for different number of multiscale basis functions (DOFf = 172,681). Case 1: structured basalt fibers.

M	DOF	$\|e\|_{L_2}$	$\|e\|_a$
		t = 50 days	
2	744	0.86	7.48
4	1488	0.38	6.04
8	2976	0.15	3.21
		t = 200 days	
2	744	0.79	8.28
4	1488	0.39	6.63
8	2976	0.12	3.32
		t = 250 days	
2	744	0.27	5.64
4	1488	0.17	4.93
8	2976	0.05	2.84
		t = 365 days	
2	744	0.58	12.55
4	1488	0.28	10.35
8	2976	0.11	4.91

Table 4. Relative L_2 and energy errors (%) for different number of multiscale basis functions (DOF_f = 172,681). Case 2: structured steel fibers.

M	DOF	$\|e\|_{L_2}$	$\|e\|_a$
		t = 50 days	
2	744	2.86	22.43
4	1488	0.92	9.97
8	2976	0.47	7.28
		t = 200 days	
2	744	2.89	20.76
4	1488	0.86	9.41
8	2976	0.40	6.64
		t = 250 days	
2	744	2.16	21.31
4	1488	0.45	7.04
8	2976	0.21	5.34
		t = 365 days	
2	744	2.95	23.65
4	1488	0.42	13.82
8	2976	0.20	8.77

Tables 5 and 6 show the relative errors of L_2 and energies for different number of multiscale basis functions for case, when fibers are located randomly. It can be seen that using four multiscale basis functions leads to good accuracy.

Table 5. Relative L_2 and energy errors (%) for different number of multiscale basis functions (DOF_f = 207,599). Case 3: random distribution of basalt fibers.

M	DOF	$\|e\|_{L_2}$	$\|e\|_a$
	t = 50 days		
2	744	1.56	39.13
4	1488	0.57	12.36
8	2976	0.30	10.05
	t = 200 days		
2	744	1.66	39.74
4	1488	0.59	12.71
8	2976	0.31	10.27
	t = 250 days		
2	744	1.06	37.96
4	1488	0.32	10.65
8	2976	0.17	8.79
	t = 365 days		
2	744	1.07	42.13
4	1488	0.35	16.69
8	2976	0.17	12.96

Table 6. Relative L_2 and energy errors (%) for different number of multiscale basis functions (DOF_f = 207,599). Case 4: random distribution of steel fibers.

M	DOF	$\|e\|_{L_2}$	$\|e\|_a$
	t = 50 days		
2	744	2.40	14.64
4	1488	0.81	9.62
8	2976	0.37	6.13
	t = 200 days		
2	744	2.38	14.28
4	1488	0.80	9.49
8	2976	0.35	5.89
	t = 250 days		
2	744	1.30	11.41
4	1488	0.48	8.40
8	2976	0.18	4.78
	t = 365 days		
2	744	1.09	18.42
4	1488	0.49	13.36
8	2976	0.18	7.94

A comparison was also made for different fiber arrangements (Figures 6 and 7). As the figures show, random arrangement of the fibers is better. However, this may be due to the fact that they are located horizontally.

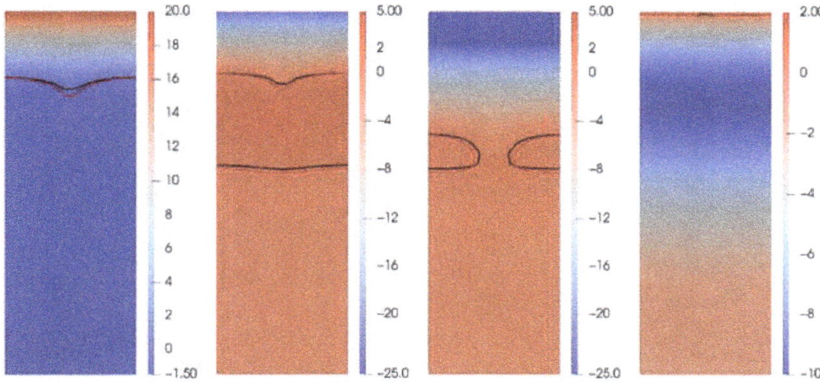

Figure 6. Temperature distribution for different time layers (t = 50, 200, 250, 365 days) in case of steel fiber reinforcement. From left to right. (red line is isocline of zero for structure locations, black is isocline of zero for random locations).

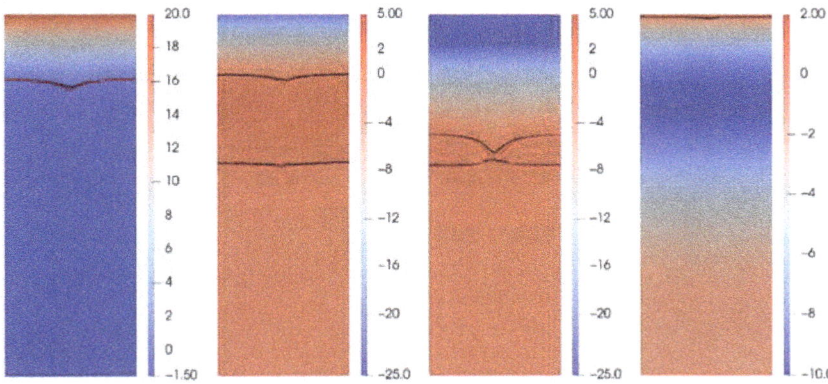

Figure 7. Temperature distribution for different time layers (t = 50, 200, 250, 365 days) in case of basalt fiber reinforcement. From left to right. (red line is isocline of zero for structure locations, black is isocline of zero for random locations).

6. Discussion

For the solution of the Stefan problem related to the soil thawing effect of composite piles, GMsFEM was used. According to results showing the accuracy of the method, the usage of four basis functions leads to the results with error in L_2 norm lower than 1%. The errors in energy norms are also very respectable. Thus, usage of GMsFEM for such problems is recommended.

Results show that the process of total freezing of piles with steel fiber inclusion in winter runs faster. Analysis shows that the length between isoclines of phase change is about 47 cm for structured and 37 for random distribution of fibers. Freezing of pile with steel fibers comes on the 239th day, and a pile with basalt inclusion freezes on the 250th day in the structured case. For random distribution, similar days are 245th and 252nd. Therefore, there are 11 and 7 days' delay between freezing of piles with steel and fiber inclusions, depending on the distribution.

7. Conclusions

In this work, the effect of a composite pile on the temperature field of the surrounding frozen soil using GMsFEM is considered. Results of calculations using GMsFEM show good accuracy. It can be said that when using materials with higher thermal conductivity, the thermal regime restores faster. In the future, we plan to consider the problems in the 3D statement and work with the piles with more complex compositions.

Author Contributions: Conceptualization, P.V.S. and S.P.S.; methodology, P.V.S.; software, S.P.S.; investigation, P.V.S.; resources, P.S.; data curation, P.S.; writing—original draft preparation, P.V.S.; writing—review and editing, P.S.; visualization, S.P.S.; supervision, P.S.; project administration, P.V.S.; funding acquisition, P.S. All authors have read and agreed to the published version of the manuscript.

Funding: This research was funded by grant RSCF 20-71-00133 and the mega-grant of the Russian Federation Government N14.Y26.31.0013.

Data Availability Statement: Data supporting reported results are available from the corresponding author by request.

Acknowledgments: The authors would like to thank every person/department who helped thorough out the research work. The careful review and constructive suggestions by the anonymous reviewers were gratefully acknowledged.

Conflicts of Interest: The authors declare no conflict of interest.

References

1. Vasiliev, V.I.; Sidnyaev, N.I.; Fedotov, A.A.; Ilyina, Y.u.S.; Vasilieva, M.V.; Stepanov, S.P. *Modeling the Distribution of Non-Stationary Temperature Fields in the Permafrost Zone in the Design of Geotechnical Structures*; Kurs: Moscow, Russia, 2017.
2. Nagy, B.; Nehme, S.G.; Szagri, D. Thermal Properties and Modeling of Fiber Reinforced Concretes. *Energy Procedia* **2015**, *78*, 2742–2747. [CrossRef]
3. Zhussupbekov, A.; Shin, E.C.; Shakhmov, Z.; Tleulenova, G. Experimental study of model pile foundations in seasonally freezing soil ground. *Int. J. GEOMATE* **2018**, *15*, 85–90. [CrossRef]
4. Montayeva, A.; Zhussupbekov, A.; Kaliakin, V.N.; Montayev, S. *IOP Conference Series: Earth and Environmental Science*; Analysis on Technological Features of Pile Foundations Construction in Frozen and Seasonal Thawing Soils, No. 1; IOP Publishing: Bristol, UK, 2021; p. 012006.
5. Jiang, C.; Fan, K.; Wu, F.; Chen, D. Experimental study on the mechanical properties and microstructure of chopped basalt fibre reinforced concrete. *Mater. Des.* **2014**, *58*, 187–193. [CrossRef]
6. High, C.; Seliem, H.M.; El-Safty, A.; Rizkalla, S.H. Use of basalt fibers for concrete structures. *Constr. Build. Mater.* **2015**, *96*, 37–46. [CrossRef]
7. Smarzewski, P. Study of Bond Strength of Steel Bars in Basalt Fibre Reinforced High Performance Concrete. *Crystals* **2020**, *10*, 436. [CrossRef]
8. Fiore, V.; Scalici, T.; Di Bella, G.; Valenza, A. A review on basalt fibre and its composites. *Compos. Part. B Eng.* **2015**, *74*, 74–94. [CrossRef]
9. Branston, J.; Das, S.; Kenno, S.Y.; Taylor, C. Mechanical behaviour of basalt fibre reinforced concrete. *Constr. Build. Mater.* **2016**, *124*, 878–886. [CrossRef]
10. Caggiano, A.; Folino, P.; Lima, C.; Martinelli, E.; Pepe, M. On the mechanical response of Hybrid Fiber Reinforced Concrete with Recycled and Industrial Steel Fibers. *Constr. Build. Mater.* **2017**, *147*, 286–295. [CrossRef]
11. Leone, M.; Centonze, G.; Colonna, D.; Micelli, F.; Aiello, M.A. Experimental Study on Bond Behavior in Fiber-Reinforced Concrete with Low Content of Recycled Steel Fiber. *J. Mater. Civ. Eng.* **2016**, *28*, 04016068. [CrossRef]
12. Mastali, M.; Dalvand, A. Use of silica fume and recycled steel fibers in self-compacting concrete (SCC). *Constr. Build. Mater.* **2016**, *125*, 196–209. [CrossRef]
13. Sengul, O. Mechanical behavior of concretes containing waste steel fibers recovered from scrap tires. *Constr. Build. Mater.* **2016**, *122*, 649–658. [CrossRef]
14. Caggiano, A.; Xargay, H.; Folino, P.; Martinelli, E. Experimental and numerical characterization of the bond behavior of steel fibers recovered from waste tires embedded in cementitious matrices. *Cem. Concr. Compos.* **2015**, *62*, 146–155. [CrossRef]
15. Aiello, M.A.; Leuzzi, F.; Centonze, G.; Maffezzoli, A. Use of steel fibres recovered from waste tyres as reinforcement in concrete: Pull-out behaviour, compressive and flexural strength. *Waste Manag.* **2009**, *29*, 1960–1970. [CrossRef]
16. Samarskii, A.A.; Vabishchevich, P.N. *Computational Heat Transfer*; Wiley: Chichester, UK, 1995.
17. Bernauer, D.I.M. Motion Planning for the Two-Phase Stefan Problem in Level Set Formulation. Ph.D. Thesis, Chemnitz University of Technology, Chemnitz, Germany, 2010. Available online: https://core.ac.uk/download/pdf/153228887.pdf (accessed on 25 June 2021).
18. Vasil'ev, V.; Vasilyeva, M. An Accurate Approximation of the Two-Phase Stefan Problem with Coefficient Smoothing. *Mathematics* **2020**, *8*, 1924. [CrossRef]
19. Vasilyeva, M.; Stepanov, S.; Spiridonov, D.; Vasil'Ev, V. Multiscale Finite Element Method for heat transfer problem during artificial ground freezing. *J. Comput. Appl. Math.* **2020**, *371*, 112605. [CrossRef]

20. Stepanov, S.; Vasilyeva, M.; Vasil'Ev, V.I. Generalized multiscale discontinuous Galerkin method for solving the heat problem with phase change. *J. Comput. Appl. Math.* **2018**, *340*, 645–652. [CrossRef]
21. Efendiev, Y.; Galvis, J.; Hou, T.Y. Generalized multiscale finite element methods (GMsFEM). *J. Comput. Phys.* **2013**, *251*, 116–135. [CrossRef]
22. Efendiev, Y.; Hou, T.Y. *Multiscale Finite Element Methods: Theory and Applications*; Springer Science & Business Media: Berlin, Germany, 2009; Volume 4.
23. Efendiev, Y.; Ginting, V.; Hou, T.Y. Multiscale Finite Element Methods for Nonlinear Problems and Their Applications. *Commun. Math. Sci.* **2004**, *2*, 553–589. [CrossRef]
24. Xing, Y.F.; Yang, Y.; Wang, X.M. A multiscale eigenelement method and its application to periodical composite structures. *Compos. Struct.* **2010**, *92*, 2265–2275. [CrossRef]
25. Hou, T.Y.; Wu, X.H. A multiscale finite element method for elliptic problems in composite materials and porous media. *J. Comput. Phys.* **1997**, *134*, 169–189. [CrossRef]
26. Alghamdi, A.; Alharthi, H.; Alamoudi, A.; Alharthi, A.; Kensara, A.; Taylor, S. Effect of Needling Parameters and Manufacturing Porosities on the Effective Thermal Conductivity of a 3D Carbon–Carbon Composite. *Materials* **2019**, *12*, 3750. [CrossRef] [PubMed]
27. Tomkova, B.; Sejnoha, M.; Novak, J.; Zeman, J. Evaluation of effective thermal conductivities of porous textile composites. *Int. J. Multiscale Comput. Eng.* **2008**, *6*, 153–167. [CrossRef]
28. Ai, S.; Fu, H.; He, R.; Pei, Y. Multi-scale modeling of thermal expansion coefficients of C/C composites at high temperature. *Mater. Des.* **2015**, *82*, 181–188. [CrossRef]
29. Zhao, Y.; Song, L.; Li, J.; Jiao, Y. Multi-scale finite element analyses of thermal conductivities of three dimensional woven composites. *Appl. Compos. Mater.* **2017**, *24*, 1525–1542. [CrossRef]

Journal of
Composites Science

Article

A Multi-Scale Method for Designing Hybrid Fiber-Reinforced Composite Drive Shafts with Carbon Nanotube Inclusions

Stelios K. Georgantzinos [1,2,*], Panagiotis A. Antoniou [2] and Stylianos I. Markolefas [2]

[1] Department of Aerospace Science and Technology, National and Kapodistrian University of Athens, 34400 Psachna, Greece
[2] General Department, National and Kapodistrian University of Athens, 34400 Psachna, Greece; panosant@core.uoa.gr (P.A.A.); stelmarkol@uoa.gr (S.I.M.)
* Correspondence: sgeor@uoa.gr

Abstract: In this paper, the modal and linear buckling analysis of a laminated composite drive shaft reinforced by 11 multi-walled carbon nanotubes (MWCNTs) was carried out using an analytical approach, as well as the finite element method (FEM). The theoretical model is based on classical laminated theory (CLT). The fundamental frequency and the critical buckling torque were determined for different fiber orientation angles. The Halpin–Tsai model was employed to calculate the elastic modulus of composites having randomly oriented nanotubes. The effect of various carbon nanotube (CNT) volume fractions in the epoxy resin matrix on the material properties of unidirectional composite laminas was also analyzed. The fundamental frequency and the critical buckling torque obtained by the finite element analysis and the analytical method for different fiber orientation angles were in good agreement with each other. The results were verified with data available in the open literature, where possible. For the first time in the literature, the influence of CNT fillers on various composite drive shaft design parameters such as the fundamental frequency, critical speed, and critical buckling torque of a hybrid fiber-reinforced composite drive shaft is finally predicted.

Keywords: buckling; fundamental frequency; finite element method; laminated composites; drive shaft; carbon nanotubes

Citation: Georgantzinos, S.K.; Antoniou, P.A.; Markolefas, S.I. A Multi-Scale Method for Designing Hybrid Fiber-Reinforced Composite Drive Shafts with Carbon Nanotube Inclusions. *J. Compos. Sci.* **2021**, *5*, 157. https://doi.org/10.3390/jcs5060157

Academic Editor: Francesco Tornabene

Received: 29 May 2021
Accepted: 8 June 2021
Published: 10 June 2021

1. Introduction

Drive shafts play important roles in the transmission vehicle system. They are used to transmit motion from the differential to the wheels. As a result, torsional, bending, and normal forces occur during its operation [1,2]. The demand for more powerful engines with higher torques and the new domains of application with increased stresses, as well as the improved materials and the new production processes, have imposed completely new operational requirements on drive shafts as a mechanical component [3–5]. Meanwhile, major progress has been made from the automotive industry in the field of drive shafts design, using new manufacturing processes, to conform to the constantly growing demands of various customers [6–11].

Over recent years, a rapid development has been observed concerning drive shafts made of composite materials that, in addition to their low weight, are able to respond effectively to their functional demands. Shinde et al. [12] designed a glass-epoxy composite drive shaft for light motor vehicles and investigated it for torsional strength, natural frequency, and critical speed. The results obtained were compared with a steel drive shaft. The developed composite shaft has proved to be the best alternative to the conventional steel shaft as its replacement resulted in a 73% weight reduction of the drive shaft. Nadeem et al. [13] carried out a review on the design and analysis of the performance of composite drive shafts made of different materials, such as carbon, glass, Kevlar, and boron with epoxy resin. They concluded that the fiber orientation angle and stacking sequence have great influences on the buckling torque and on the dynamic characteristics. Yefa et al. [14]

implemented a new mechanical analytical solution of torsional stiffness for a composite drive shaft with balance laminate, based on CLT and mechanical analysis. Finite element analysis, CLT, and experiments were carried for carbon-fiber-reinforced polymers (CFRP) drive shafts to predict the torsional stiffness. Other valuable works on both the FEM and CFRP composite materials can be found in [15,16].

Fiber-reinforced polymer matrix composites have emerged as a major class of structural materials and are being considered for use as a substitute for many traditional metallic materials in a large number of weight-critical components in aerospace, automotive, and other industries, due to their high strength/weight ratio and stiffness/weight ratio [17,18]. The simultaneous development of nanotechnology has given a new perspective on the use of composite materials, as the enhanced characteristics of nanostructures are able to improve significantly the already important mechanical properties of classic composite materials.

Nanocomposites have the potential of becoming the future structural material not only in terms of greater mechanical properties, but also in terms of superior thermal, electrical, optical, and other properties [19–22]. Polymer nanocomposites are polymer matrix composites in which the reinforcement has at least one of its dimensions in the nanometer range. The mechanical properties of the nano-reinforcement are considerably high and the ratio of their surface area to volume is high as well, which means that a great interfacial interaction with the matrix can be provided. These are the leading reasons for such highly improved properties in nanocomposites [23]. The usage of polymer nanocomposites as a material for primary structures is in its early stage, but their potential in future aircraft structures has been realized [24]. Lockheed Martin had revealed that the F-35 Lightning II will be the first mass-produced aircraft to integrate structural nanocomposites in non-load-bearing airframe components. A thermoset epoxy reinforced by CNTs will replace carbon fiber as the material used to produce F-35 wingtip fairings, beginning with low-rate initial production (LRIP)-4 aircraft [25]. Shakil et al. [26] carried out a review on the properties and fabrication techniques of fiber-reinforced polymer nanocomposites subjected to simulated accidental ballistic impacts. High-specific-surface-area nanoparticles were used for matrix modification to induce nano-scale toughness mechanisms, with a focus on the ballistic performance of composite structures. Tüzemen et al. [27] investigated the effects of nanoclay (NC), CNT, and a hybrid of both inclusions on the bending, tensile, and bearing strengths of nanocomposite plates. Mechanical properties were investigated by applying bending and tensile testing, as well as tensile testing, with bolted joints on the nanocomposite plates.

Khoramishad et al. [28] investigated the effect of adding MWCNTs on the high-velocity impact behavior of fiber metal laminates (FMLs). They concluded that incorporating 0.5 wt% of MWCNTs into the composite laminate of FML resulted in the maximum reduction of 29.8% in projectile residual velocity and the maximum increase of 18.9% in the absorbed energy during projectile perforation compared to the unreinforced FMLs. Chiou et al. [29] studied the synergistic effects on the mechanical properties of composites materials by adding graphene nanoplatelets (GNPs) and nanocarbon aerogel (NCA) in the epoxy resins. The experimental results demonstrated that the mechanical properties of GNP/NCA/epoxy nanocomposites and GNP/NCA/CFRP laminates were optimized via reinforcement upon the addition of GNP/NCA hybrids.

In the last two decades, several exotic forms of low-dimensional carbon inclusions have been discovered, including fullerenes (zero dimension), CNTs (one dimension), and graphene (two dimension). All these materials share graphite's sp2 carbon bonding. Graphene, the mother of all graphitic forms, is a monolayer of carbon atoms tightly packed into a two-dimensional hexagonal lattice sheet that can be wrapped up to fullerenes, rolled into CNTs, and stacked into graphite [30]. Since the discovery of CNTs by Iijima in 1991 [31], research on their growth, characterization, and application development has exploded [32,33]. Their extraordinary physical properties have drawn the interest of the scientific community studying for numerous applications in nanoelectromechanical systems and nanoelectronics [34].

Ahmadipour et al. [35] studied the synthesis of calcium copper titanate (CaCu3Ti4O12)/ MWCNT composites, using the ultrasonic technique. The effects of MWCNT content on the structural, dielectric, and mechanical properties of copper calcium titanate (CCTO) were investigated by dielectric measurement, as well as tensile strength and flexural strength tests. An important observation was that the CCTO mixed with 0.2 wt% MWCNT exhibited the highest dielectric permittivity and the lowest dielectric loss. With the addition of 0.2 wt% MWCNT, the values of load, tensile, and flexural strength increased to 10.38 kN, 101.88 MPa, and 275.07 MPa, respectively, due to the improvement in densification. These results have values for the fabrication of CCTO and the optimization of its performance for electronic devices such as capacitors and antennas.

Recently, a lot of attention has been also given by researchers to graphene nanosheets (GNS) and CNT/polymer matrix composite materials. Georgantzinos et al. studied the mechanical elastic behavior [36,37] and vibration response [38] of laminated polymer composite plates with carbon nanostructure inclusions, using a multilevel framework, starting from the nanoscale, up to the laminated hybrid composite plates. Rafiee et al. [39] compared the mechanical properties of epoxy resin nanocomposites with GNS, single-walled CNTs, and multi-walled CNTs at the same nanofiller weight fraction. In an excellent recent work, Tas and Soykok [40] determined theoretically the engineering constants of CNT-based composite laminas. Bending analysis was performed on a composite plate under concentrated and distributed load. The results showed that elastic constants increased with the added CNT fraction. Additionally, the flexural modulus of the laminated composite plate showed significant improvement and the maximum deflection decreased.

Various techniques and approaches on the modeling of the mechanical properties of composites and nanostructures have been developed. Stamoulis et al. [41] dealt with the numerical modeling of the low-velocity impact damage of laminated composites, which have increasingly important applications in aerospace primary structures. Their purpose was to present and validate a computationally efficient approach in order to explore the effect of critical parameters on the impact damage characteristics. Barretta et al. [42] focused on modeling the bending of armchair CNTs by means of gradient elasticity theory. The estimation of Young's modulus depending on the armchair CNT diameter and length was investigated. According to the results presented, Young's modulus was found to be dependent on small size effects and ranged between 816.87 and 1067.24 MPa.

Georgantzinos et al. [43] investigated the thermomechanical buckling of single-walled CNTs (SWCNTs) by a structural mechanics method, providing theoretical predictions, concerning the compressive buckling response of SWCNTs of different chiralities and sizes under thermal conditions. Aghdam et al. [44] proposed a new micromechanics approach for predicting the elastic modulus of randomly oriented and distributed wavy CNTs-reinforced polymer nanocomposites. The results revealed that the proposed method has a good accuracy according to the experimental data as compared. Mehar and Panda [45] investigated the deflection behavior of a composite plate reinforced with CNTs, numerically using FEM, and the result accuracy was verified via three-point experimental bending test data. The Young's modulus for plates consisting of randomly oriented MWCNTs obtained by using the Mori–Tanaka method showed an increase with the volume fraction of MWCNTs. Georgantzinos [46,47] developed a novel computational model for the prediction of the mechanical behavior of graphene and graphyne monolayers, according to their atomistic structure. The harmonic approximation was utilized for describing the interaction potential energies based on molecular theory. Spanos et al. [48] described a micromechanical finite element approach for the estimation of the elastic mechanical properties of graphene-reinforced composites. The load transfer conditions between the graphene and the matrix were modeled using joint elements connecting the two materials, simulating the interfacial region.

Feng et al. [49] investigated the mechanical properties of a graphene/PMMA nanocomposite system by using the molecular dynamics simulations. The results showed that the Young's and shear moduli increased with the graphene volume fraction and decreased as the temperature rose from 300 to 500 K, while the efficiency of the reinforcement reduced with the graphene content. Rui et al. [50] presented an atomistic study on the mechanical behaviors of a polymer nanocomposite reinforced with defective graphene using molecular dynamics simulations, with a particular focus on the influences of temperature changes and atom vacancy defects. The Halpin–Tsai model was then modified based on the results to enable the effects of temperature and graphene defects to be considered in determining the Young's modulus of graphene-reinforced nanocomposites.

Even though a comprehensive investigation, in general, concerning polymer nanocomposites has been performed in the open literature, as well as their possible applications in automotive industry already being noticed [51], the potential performance improvement of specific components due to the use of nanomaterials has not been extensively reported yet. In this study, the development of suitable computational procedures, based on theoretical and finite element analysis, for the prediction of the mechanical behavior of automotive drive shafts manufactured from laminated composite materials, reinforced with CNTs, was presented. To the author's best knowledge, this study is the first one in the open literature concerning the effect of CNTs on the mechanical performance of composite drive shafts, providing novel results. It aims to offer a relatively simple and complete procedure for the prediction of the critical design parameters helping the design process and development of drive shafts of enhanced performance.

2. Theoretical Approach

2.1. Unidirectional Composite Lamina with CNT Inclusions

The first step to determine the mechanical response of nanocomposites under certain loads is the estimation of their material properties such as Young's modulus, Poisson's ratio, and the shear modulus. These elastic constants of nano-reinforced composite laminas may be calculated theoretically using the combination of simple models such as the Halpin–Tsai model, the Tsai–Pagano equation, and the Rule of mixtures. The models can be implemented considering certain assumptions. Here, CNTs were considered as randomly and homogeneously oriented into the matrix, while the CNTs-polymer (hybrid) matrix exhibited isotropic behavior. The unidirectional fiber-reinforced composite lamina with CNT inclusions is shown in Figure 1.

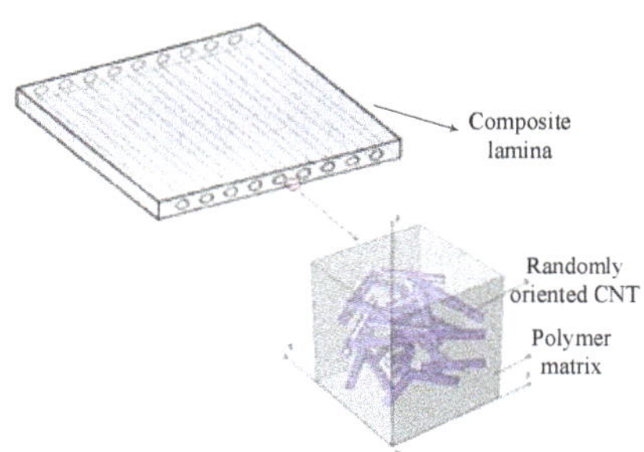

Figure 1. Schematic illustration of a unidirectional fiber-reinforced composite lamina, with the inclusion of CNTs in the polymer matrix. CNTs are considered as randomly oriented.

2.2. Hybrid CNT-Polymer Matrix Elastic Constants

The well-established Halpin–Tsai model for fiber-reinforced composites was utilized to predict the Young's modulus of the nanocomposites. The CNTs were considered as randomly oriented long discontinuous fiber laminas. The Young's modulus of a CNT-added matrix can be calculated from the following equation [39]:

$$E_{m-cnt} = E_m \left(\frac{3}{8} \frac{1 + 2\left(\frac{l_{cnt}}{d_{cnt}}\right)\beta_1 V_{cnt}}{1 - \beta_1 V_{cnt}} + \frac{5}{8} \frac{1 + 2\beta_2 V_{cnt}}{1 - \beta_2 V_{cnt}} \right) \quad (1)$$

where

$$\beta_1 = \frac{\left(\frac{E_{eq}}{E_m}\right) - 1}{\left(\frac{E_{eq}}{E_m}\right) + 2\left(\frac{l_{cnt}}{d_{cnt}}\right)} \text{ and } \beta_2 = \frac{\left(\frac{E_{eq}}{E_m}\right) - 1}{\left(\frac{E_{eq}}{E_m}\right) + 2} \tag{2}$$

E_m is the Young's modulus of the matrix, V_{cnt} is the volume content of CNTs, while d_{cnt} and l_{cnt} are the average outer diameter and length of the nanotube, respectively. E_{eq} is the equivalent modulus of nanotubes considering the hollow tube as a solid cylinder and can be expressed as:

$$E_{eq} = \left(\frac{2t}{r}\right) E_{cnt} \tag{3}$$

where t, r, and E_{cnt} are the wall thickness, radius, and Young's modulus of the nanotube, respectively.

The shear modulus of the CNT-polymer matrix can be calculated by using the Tsai–Pagano equation [40]:

$$G_{m-cnt} = E_m \left(\frac{1}{8} \frac{1 + 2\left(\frac{l_{cnt}}{d_{cnt}}\right)\beta_1 V_{cnt}}{1 - \beta_1 V_{cnt}} + \frac{1}{4} \frac{1 + 2\beta_2 V_{cnt}}{1 - \beta_2 V_{cnt}} \right) \tag{4}$$

The Poisson's ratio for the CNT-polymer matrix exhibiting isotropic behavior can be calculated from the following equation:

$$v_{m-cnt} = \left(\frac{E_{m-cnt}}{2G_{m-cnt}}\right) - 1 \tag{5}$$

The density of the nanocomposite (hybrid) matrix can also be calculated from the following equation:

$$\rho_{m-cnt} = \rho_{cnt} V_{cnt} + \rho_m V_m \tag{6}$$

where ρ_{cnt} is the density of the nanotube, V_m is the volume fraction of the matrix, and ρ_m is the matrix density.

2.3. Unidirectional Composite Lamina Elastic Constants

The material properties of the unidirectional composite lamina, adopting the rule of mixtures, may be calculated from the following equations [23,52]:

$$\text{Longitudinal Young's modulus } E_1 = E_f V_f + E_{m-cnt} V_{m-cnt} \tag{7}$$

$$\text{Transverse Young's modulus } E_2 = \frac{E_f E_{m-cnt}}{E_f V_{m-cnt} + E_{m-cnt} V_f} \tag{8}$$

$$\text{In-plane Shear modulus } G_{12} = \frac{G_f G_{m-cnt}}{G_f V_{m-cnt} + G_{m-cnt} V_f} \tag{9}$$

$$\text{Out-of-plane Shear modulus } G_{23} = \frac{E_2}{2(1 + v_{23})} \tag{10}$$

$$\text{Major Poisson's ratio } v_{12} = v_f V_f + v_{m-cnt} V_{m-cnt} \tag{11}$$

$$\text{Transverse Poisson's ratio } v_{23} = v_{12} \frac{(1 - v_{21})}{(1 - v_{12})} \tag{12}$$

where E_f, V_f, v_f, G_f, and $V_{m\text{-}cnt}$ are Young's modulus of the fiber, the volume fraction of the fiber, Poisson's ratio of the fiber, the shear modulus of the fiber, and the volume fraction of the hybrid matrix, respectively.

Alternatively, semi-empirical models may be adopted for the evaluation of the material properties of the unidirectional composite lamina. An efficient model is the Halpin–Tsai [53] model, which can be used over a wide range of elastic properties and fiber volume fractions.

In this approach, the longitudinal Young's modulus E_1 and major Poisson's ratio ν_{12} are evaluated by Equations (7) and (11), respectively, while the other engineering constants are provided by the following equation:

$$\frac{P}{P_{m-cnt}} = \left(\frac{1+\xi\eta V_f}{1-\eta V_f}\right) \tag{13}$$

where $P_{m\text{-}cnt}$ are the related properties of the CNT-polymer matrix, and P can be considered as the transverse Young's modulus E_2, transverse Poisson's ratio ν_{23}, in-plane shear modulus G_{12}, and out-of-plane shear modulus G_{23}. η is an experimental factor computed by using the next expression:

$$\eta = \frac{\left(\frac{P_f}{P_{m-cnt}}\right)-1}{\left(\frac{P_f}{P_{m-cnt}}\right)+\xi} \tag{14}$$

The value of the reinforcing factor ξ depends on the fiber geometry, packing geometry, and loading conditions. For circular fibers in a square array, ξ = 2 for E_2, and ξ = 1 for ν_{23}, G_{12}, and G_{23}, as referred in [52].

2.4. Theoretical Modeling of Composite Drive Shaft

CLT was used to analyze the behavior of the composite laminated drive shaft. The following assumptions must be considered in order to make the problem solvable [54,55]:

- The plate consists of orthotropic laminas bonded together, with the principal material axes of the orthotropic lamina orientated along arbitrary directions with respect to the x–y axes.
- The thickness of the plate, t, is much smaller than any characteristic dimension.
- The displacements u, v, and w are small compared with t.
- The in-plane strains ε_x, ε_y, and γ_{xy} are small compared with unity.
- Normal stress σ_z is negligible.
- Transverse shear strain is negligible, $\gamma_{xz} = \gamma_{yz} = 0$ (Kirchhoff hypothesis).
- Displacements u and v are assumed to be linear functions of the thickness coordinate z.
- Transverse normal strain ε_z is negligible.
- Each ply obeys Hooke's Law.
- The plate thickness is constant throughout the laminate.
- Transverse shear stresses τ_{xz} and τ_{yz} vanish on the laminate surfaces $z = \pm t/2$.

Given the previous assumptions, a lamina can be analyzed as a plane stress problem and the stress–strain relations can take the following matrix form [55]:

$$\begin{Bmatrix} \varepsilon_{11} \\ \varepsilon_{22} \\ \gamma_{12} \end{Bmatrix} = [S] \begin{Bmatrix} \sigma_{11} \\ \sigma_{22} \\ \tau_{12} \end{Bmatrix} \tag{15}$$

The square matrix on the right-hand side of Equation (15), denoted by $[S]$, is called the reduced compliance matrix. The elements of $[S]$ are given by:

$$S_{11} = \frac{1}{E_1}, S_{12} = -\frac{\nu_{12}}{E_1}, S_{22} = \frac{1}{E_2}, S_{66} = \frac{1}{G_{12}}, S_{16} = S_{26} = 0 \tag{16}$$

The inverse form of Equation (15) is as follows:

$$\begin{Bmatrix} \sigma_{11} \\ \sigma_{22} \\ \tau_{12} \end{Bmatrix} = [Q] \begin{Bmatrix} \varepsilon_{11} \\ \varepsilon_{22} \\ \gamma_{12} \end{Bmatrix} \tag{17}$$

The square matrix on the right-hand side of Equation (17), denoted by [Q], is called the reduced stiffness matrix, in which:

$$[Q] = \begin{bmatrix} Q_{11} & Q_{16} & 0 \\ Q_{12} & Q_{22} & 0 \\ 0 & 0 & Q_{66} \end{bmatrix} \tag{18}$$

and the elements of [Q] are given by:

$$Q_{11} = \frac{E_1}{1-\nu_{12}\nu_{21}}, Q_{12} = \frac{\nu_{12}E_2}{1-\nu_{12}\nu_{21}}, Q_{22} = \frac{E_2}{1-\nu_{12}\nu_{21}}, Q_{66} = G_{12}, \nu_{21} = \frac{E_2}{E_1}\nu_{12} \tag{19}$$

Unidirectional laminae are weak in transverse mechanical properties; to overcome this, they are usually placed in different orientations in a laminate. It is necessary to know the stress–strain relations for a generally orthotropic lamina [55].

We can relate global stresses to global strains as:

$$\begin{Bmatrix} \sigma_{11} \\ \sigma_{22} \\ \tau_{12} \end{Bmatrix} = [T]^{-1}[Q][R][T][R]^{-1} \begin{Bmatrix} \varepsilon_{11} \\ \varepsilon_{22} \\ \gamma_{12} \end{Bmatrix} \tag{20}$$

The matrix $\left[\overline{Q}\right] = [T]^{-1}[Q][R][T][R]^{-1}$ is called the transformed reduced stiffness matrix and it is obtained by multiplying the five 3 × 3 matrices on the right-hand side of Equation (20).

By carrying out the necessary multiplication and simplification, it can be shown that the elements of $\left[\overline{Q}\right]$ are given by:

$$\overline{Q_{11}} = Q_{11}\cos^4\theta + 2(Q_{12}+2Q_{66})\sin^2\theta\cos^2\theta + Q_{22}\sin^4\theta \tag{21}$$

$$\overline{Q_{12}} = (Q_{11}+Q_{22}-4Q_{66})\sin^2\theta\cos^2\theta + Q_{12}(\sin^4\theta\cos^4\theta) \tag{22}$$

$$\overline{Q_{16}} = (Q_{11}-Q_{12}-2Q_{66})\sin\theta\cos^3\theta + (Q_{12}-Q_{22}+2Q_{66})\sin^3\theta\cos\theta \tag{23}$$

$$\overline{Q_{22}} = Q_{11}\sin^4\theta + 2(Q_{12}+2Q_{66})\sin^2\theta\cos^2\theta + Q_{22}\cos^4\theta \tag{24}$$

$$\overline{Q_{26}} = (Q_{11}-Q_{12}-2Q_{66})\sin^3\theta\cos\theta + (Q_{12}-Q_{22}+2Q_{66})\sin\theta\cos^3\theta \tag{25}$$

$$\overline{Q_{66}} = (Q_{11}+Q_{22}-2Q_{12}-2Q_{66})\sin^2\theta\cos^2\theta + Q_{66}(\sin^4\theta\cos^4\theta) \tag{26}$$

The elements of $\left[\overline{Q}\right]$ are functions of stiffness elements Q_{11}, Q_{12}, Q_{22}, and Q_{66} and the lamina angle θ. As the elements of [Q] are functions of four engineering constants, the elements of $\left[\overline{Q}\right]$ are consequently functions of four engineering constants and the lamina angle θ [55].

The next step is to construct the extensional stiffness matrix [A], which is the summation of the products of the transformed reduced stiffness matrix $\left[\overline{Q}\right]$ of the layer and the respective thicknesses [56].

$$A_{ij} = \sum_{k=1}^{n} \{Q_{ij}\}_n (z_k - z_{k-1}) \tag{27}$$

where i,j = 1,2,6, the summation is carried over all n plies of the laminate, and z_k, z_{k-1} are the upper and lower z coordinates of the k_{th} ply, respectively.

Equation (21) describes the membrane deformations of a laminate under in-plane loads and incorporates the Kirchhoff plate theory assumption that plane sections remain planar and perpendicular to the neutral axis.

The matrix [A] is used to calculate E_x and E_h, which are the average moduli in the axial and hoop directions, respectively, and t is the total thickness of drive shaft.

$$E_x = \frac{1}{t}\left[A_{11} - \frac{A_{12}^2}{A_{22}}\right] \tag{28}$$

$$E_h = \frac{1}{t}\left[A_{11} - \frac{A_{12}^2}{A_{11}}\right] \quad (29)$$

2.5. Theoretical Calculation of Design Parameters

A drive shaft can be idealized as a simply supported beam. To determine the bending natural frequencies for a composite tube, Euler's equation for the lateral vibration of beams can be implemented [57,58]:

$$\omega_n = C_n\sqrt{\frac{E_x I}{mL^4}} \quad (30)$$

or

$$f_n = \frac{C_n}{2\pi}\sqrt{\frac{E_x I}{mL^4}} \quad (31)$$

where ω_n (rad/s) is the angular velocity, f_n (Hz) is the fundamental (rotational) frequency, m is the mass of a drive shaft per unit length, L is the length of a drive shaft, and I is the second moment of inertia given for a thin-walled tube as:

$$I_x = \frac{\pi}{4}(r_0^4 - r_i^4) \quad (32)$$

where r_0 is the outer radius, r_i is the inner radius, $(E_x I)$ represents the bending stiffness for a composite tube, and the number C_n depends on the boundary conditions of the drive shaft and the number of natural frequency n. Numerical values of C_n for simply supported boundary conditions [59] are $(n\pi)^2$.

The vibration of parts and assemblies increases as operating conditions become more demanding. Critical speed is defined as the angular rotating shaft speed that is equal to the lowest frequency of its natural vibration. Equations (30) and (31) are used to calculate the natural shaft oscillation (critical speeds) [60].

The main loading of a drive shaft is torsion. Instability may occur at a certain amount of loading. This parameter is critical in the design of drive shafts, and it is important to have an order of magnitude solution before starting the designing [61]. There are various empirical equations in the literature that are based on experimental studies and are used to calculate the torsional buckling load of long thin-walled shafts. The expression of the critical buckling torque for thin-walled orthotropic tubes is given as in [56]:

$$T_{cr} = (2\pi r^2 t)(0.272)\left[E_x E_h^3\right]^{1/4}\left(\frac{t}{r}\right)^{3/2} \quad (33)$$

where r and t are the mean radius and overall thickness of the composite drive shaft, respectively.

3. Finite Element Modeling

In this study, the numerical simulation of a hybrid drive shaft was performed using FEM. The composite drive shaft is considered as a thin-walled tube, balanced, while the stress–strain relationship is linear and elastic. The analysis consists of the following four stages.

3.1. Defining Material Properties

In the current study, epoxy resin, multi-walled CNTs, and carbon/glass fiber were chosen as matrix material, nano material, and reinforcing material, respectively. Material properties of the components constituting the composite material are shown in Table 1.

Table 1. Materials properties of the components.

e-Glass Fiber [62]		Carbon Fiber [40]		Epoxy Resin [23]		MWCNTs [39]	
$E_{f_{11}}$ (GPa)	74	$E_{f_{11}}$ (GPa)	230	E_m (GPa)	3.45	E_{cnt} (GPa)	450
$E_{f_{22}}$ (GPa)	74	$E_{f_{22}}$ (GPa)	15.41	v_m	0.36	l_{cnt} (nm)	20,000
$v_{f_{12}}$	0.25	$v_{f_{12}}$	0.29	G_m (GPa)	1.26	d_{cnt} (nm)	20
$v_{f_{23}}$	0.25	$v_{f_{23}}$	0.46			t_{cnt} (nm)	1.5
$G_{f_{12}}$ (GPa)	30	$G_{f_{12}}$ (GPa)	10,04			d_{icnt} (nm)	17
$G_{f_{23}}$ (GPa)	30	$G_{f_{23}}$ (GPa)	5.28			ρ_{cnt} (g/cm3)	0.6244

The material properties of CNTs-reinforced unidirectional composite laminas were calculated with two different approaches. In the first approach, concerning e-glass fiber/epoxy resin/CNTs, Equations (7)–(12) were used to calculate its orthotropic mechanical properties. In the second approach, concerning carbon fiber/epoxy resin/CNTs laminas, Equations (7), (11) and (13) were used to calculate the corresponding properties. Additionally, volume fractions of fiber and matrix were considered as 57.65% and 42.35%, respectively. In this way, the composite lamina properties without CNTs inclusions were close to the values referred in [56,58]. Following the approach presented in the previous section, the engineering constants are shown in Table 2. The CNTs volume fractions were considered as 0.00, 0.05, 0.10, 0.20, 0.35, and 0.50.

Table 2. Material properties of unidirectional composite laminas.

e-Glass Fiber/Epoxy Resin/CNTs Lamina						
V_{cnt}	0.00	0.05	0.10	0.20	0.35	0.50
E_1 (GPa)	40.312	41.619	42.943	45.655	49.939	54.647
E_2 (GPa)	6.874	11.871	16.570	25.204	36.651	46.890
G_{12} (GPa)	2.538	4.313	5.958	9.038	13.245	17.146
G_{23} (GPa)	2.434	4.176	5.815	8.928	13.231	17.212
v_{12}	0.3027	0.3164	0.3270	0.3355	0.3388	0.3377
v_{23}	0.4117	0.4211	0.4246	0.4115	0.3850	0.3621
ρ (g/cm^3)	1.9100	1.8457	1.7814	1.6529	1.4600	1.2672
Carbon fiber/epoxy resin/CNTs lamina						
V_{cnt}	0.00	0.05	0.10	0.20	0.35	0.50
E_1 (GPa)	134.056	135.215	136.389	138.795	142.594	146.770
E_2 (GPa)	8.079	10.579	12.335	15.080	18.639	22.200
G_{12} (GPa)	3.319	4.811	5.889	7.478	9.222	10.703
G_{23} (GPa)	2.654	3.593	4.242	5.216	6.391	7.509
v_{12}	0.3198	0.3320	0.3414	0.3489	0.3518	0.3508
v_{23}	0.4147	0.4284	0.4387	0.4467	0.4497	0.4486
ρ (g/cm^3)	1.6100	1.5607	1.5114	1.4129	1.2650	1.1172

Elastic constants E_1, E_2, G_{12}, and G_{23} were found to increase with the volume fraction of CNTs in the polymer matrix. Specifically, in the first approach, 35%, 582%, 575%, and 607% increases in E_1, E_2, G_{12}, and G_{23}, respectively, were observed for V_{cnt} = 0.50 compared with the laminated composite drive shaft without CNTs inclusions. In the second approach, 9%, 174%, 222%, and 182% corresponding increases were observed. In the first approach, the Poisson's ratio ν_{12} seemed to increase as V_{cnt} increased up to 0.35 and then remain stable, while in the second approach, it increased as the V_{cnt} increased up to 0.20 vol% and then remained stable. Finally, in the first approach, the Poisson's ratio ν_{23} increased as the V_{cnt} increased up to 0.10 and then decreased, while, in the second approach, it decreased with V_{cnt}.

3.2. *Configuration and Finite Element Type*

A composite drive shaft was considered with a length of 1730 mm and mean radius of 50.8 mm, which consists of four layers, stacked up as a $[\pm45^\circ{}_{\text{glass}}/0^\circ{}_{\text{carbon}}/90^\circ{}_{\text{glass}}]$ layup. The selected thicknesses for each layer were: $+45^\circ{}_{\text{glass}}$ = 0.1905 mm, $-45^\circ{}_{\text{glass}}$ = 0.1905 mm, $0^\circ{}_{\text{carbon}}$ = 0.635 mm, $90^\circ{}_{\text{glass}}$ = 1.016 mm.

The finite element modeling was implemented in ABAQUS CAE code. The drive shaft was discretized by using the S4R element type. This is a 3D 4-node, quadrilateral, stress/displacement shell element with reduced integration and a large-strain formulation. These elements allow transverse shear deformation and account for finite membrane strains and arbitrarily large rotations. Reduced integration usually provides more accurate results and significantly reduces the running time, especially in three dimensions. The total number of elements with a global element size l^e = 10 mm was calculated equal to 5536 elements.

3.3. Boundary Conditions

Appropriate boundary conditions must be applied to the drive shaft finite element model according to the analysis type, i.e., modal or elastic buckling analysis. In the modal analysis, boundary conditions are considered to be pinned-pinned, i.e., the drive shaft is simply supported in order to obtain the mode shapes and the corresponding natural frequencies.

In the buckling analysis, the model is fully restrained (fixed) at one end, while the drive shaft is subjected to a torsion load (moment) at the other end. In this way, the buckling modes and loads can be evaluated using a linear buckling analysis. The lowest value is the critical torsional buckling load.

3.4. Meshing Sensitivity Analysis

The effect of global size l^e on the natural frequency and buckling torque is presented in Figure 2a,b, respectively. It is obvious that a relatively coarse mesh was enough to reach the first 10 natural frequencies with reasonable accuracy, while a denser mesh was required in order for the model to converge in a specific magnitude of the critical buckling torque. Specifically, in Figure 2b, the solution seemed to converge with a global element size equal to 10 mm. This is the reason that, here, the global size l^e of the element in the finite element analysis was chosen to be l^e = 10 mm.

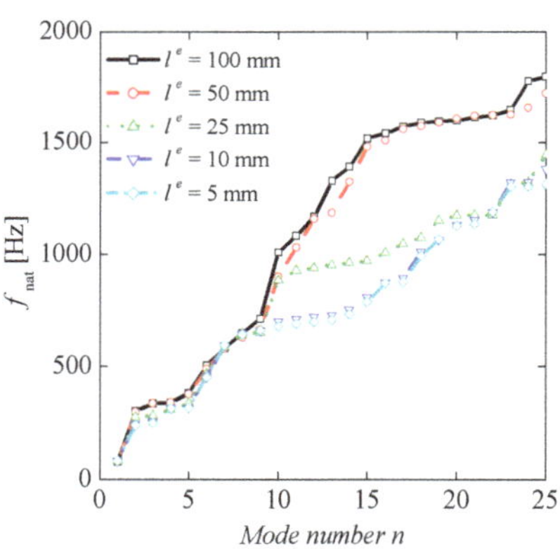

(a)

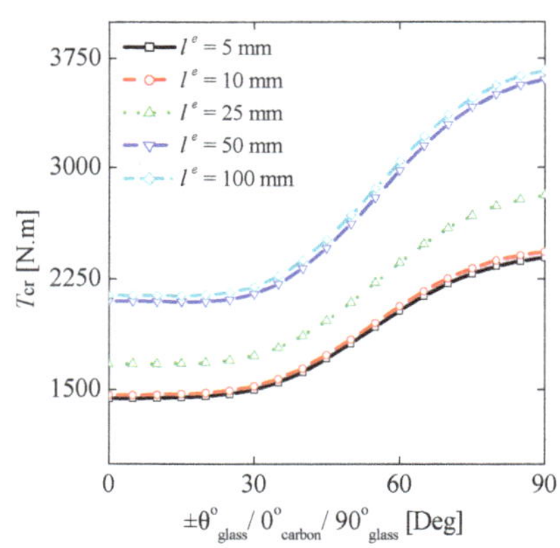

(b)

Figure 2. Mesh sensitivity analysis diagram of a hybrid composite driveshaft consisting of 4 layers, stacked up as a $[\pm45^\circ_{glass}/0^\circ_{carbon}/90^\circ_{glass}]$ layup, computed for lg = 1730 mm, r = 50.8 mm, and w.thk = 2.032 mm, for (**a**) modal analysis, and (**b**) linear bucking analysis, where the control parameter l^e was increased. To detect the appropriate global size, the model was discretized by using the S4R type of element, with the total No of 5536 elements used for l^e = 10 mm.

4. Results

4.1. Validation

To evaluate the validity of the proposed method in terms of natural frequency and torsional buckling load, a comparison with the analytical and numerical results available in [56] was performed concerning the drive shaft of the configuration presented in Section 3.2. The comparison is presented in Table 3 and shows a reasonable agreement between the results of the different methods. Note that the present model used engineering constants computed by the approach presented in the previous section that slightly differ with the corresponding values of [58]. Nevertheless, the proposed finite element model has also been tested using identical properties used in [58], and the deviations on the results were calculated to be lower than 1%.

Table 3. Comparison in the first three natural frequencies and critical torsional buckling load.

Mode Shape	FEA (Present)	FEA [58]	Equation (31)	Torsional Buckling Load	FEA Present	FEA [58]	Equation (33)	Analytical [58]
1st mode	91.65	90.46	96.33	Critical (Nm)	1730	1830	2149	2030
2nd mode	326.88	331.19	385.33					
3rd mode	635.73	663.35	866.99					

Figure 3 illustrates a set of eight natural frequencies and their corresponding mode shapes. The results also agree with the results presented in [58] in terms of both mode shapes and natural frequencies values. The greatest discrepancies between the two methods were observed in the breathing vibration modes; however, these modes were not involved in the basic (first three) mode shapes of vibration of the specific configuration of the composite shaft.

Furthermore, the fundamental natural frequency of a drive shaft consisting of four layers stacked as $[\pm 45^\circ_{glass}/0^\circ_{carbon}/90^\circ_{glass}]$ for various fiber orientations was evaluated. The variation is presented in Figure 4. The results were also compared with the results of the study [58] presenting, again, a reasonable agreement. It is observed that the fundamental frequency (Figure 4a) increased as the fiber angle of the first two $\pm\theta^\circ_{glass}$ layers tended to 0°. However, the optimum configuration may be determined taking into consideration the other design parameters as well. In Figure 4b, it is observed that the fundamental frequency decreased, as the fiber angle of the third 0° carbon layer tended to 90°. The results indicate that the drive shaft lost 55% of its natural frequency when the carbon fibers oriented at 90° in the hoop direction instead of 0°. In order to increase the modulus of elasticity in the longitudinal direction of the drive shaft, the carbon fibers layer, with its high modulus, must be oriented at zero angle.

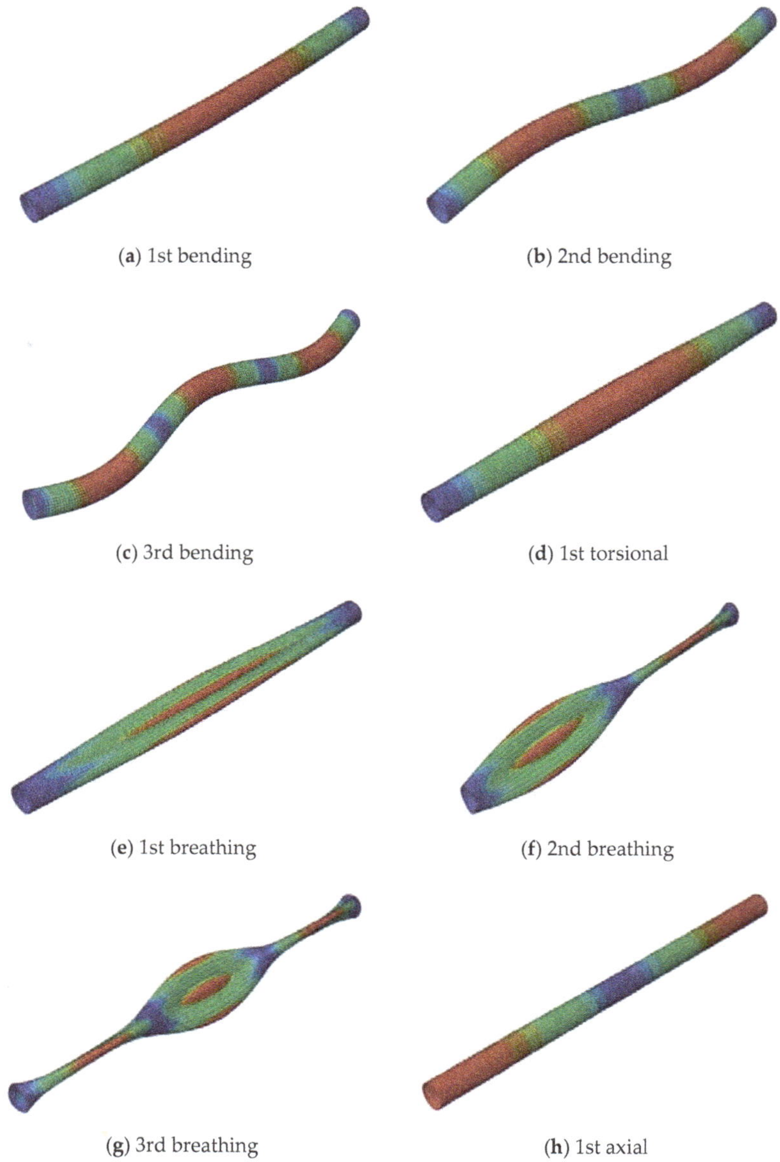

(**a**) 1st bending (**b**) 2nd bending (**c**) 3rd bending (**d**) 1st torsional (**e**) 1st breathing (**f**) 2nd breathing (**g**) 3rd breathing (**h**) 1st axial

Figure 3. Set of the first eight mode shapes of natural frequency of a simply supported composite drive shaft. To detect deformation and natural frequencies, FEM analysis was conducted in Abaqus CAE. The corresponding frequency values were as follows: (**a**) 91.65, (**b**) 326.88, (**c**) 635.73, (**d**) 442.34, (**e**) 334.69, (**f**) 334.91, (**g**) 402.28, and (**h**) 1485.60 Hz.

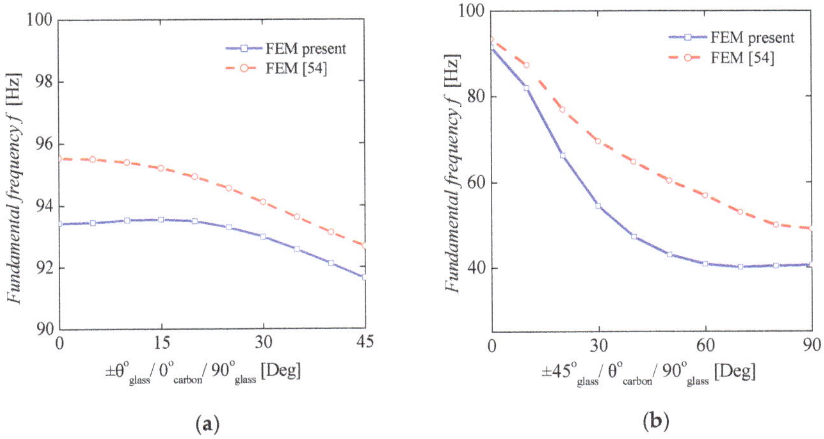

Figure 4. Fundamental frequency as a function of fiber orientation angle (θ°). A hybrid 372 composite drive shaft having four layers stacking of (**a**) [±θ°glass/0°carbon/90°glass] and (**b**) [±45°glass/θ°carbon/90°glass] was analyzed for length = 1730 mm, radius = 50.8 mm, and wall thickness = 2.032 mm, for modal analysis, where the fiber orientation and the control parameter θ° [deg] were transformed. Figures (**a**,**b**) show the effect of fiber orientation angle on f [Hz] by changing the first two layers and the third (carbon fiber) layer, respectively. To ensure the validity of the process, identical material properties were employed. The present approach and FEM [56] are presented by blue solid lines and dashed red lines, respectively.

4.2. Effect of CNTs Inclusion on Natural Frequencies and Critical Speed of the Composite Drive Shaft

In order to determine the effect of CNTs on the composite drive shaft on natural frequency, a parametric finite element analysis was conducted. The results are depicted in Figure 5.

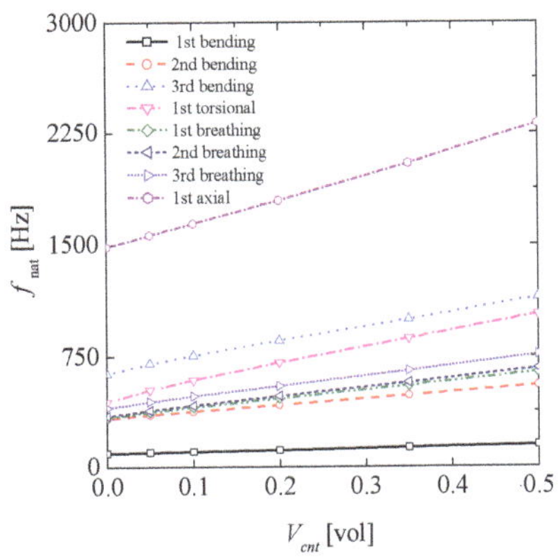

Figure 5. Natural frequencies as a function CNTs volume fraction. A hybrid composite drive shaft of stacking [$\pm45^{\circ}{}_{glass}/0^{\circ}{}_{carbon}/90^{\circ}{}_{glass}$] was analyzed for length = 1730 mm, radius = 50.8 mm, and wall thickness = 2.032 mm, for modal analysis, where the volumetric concentration of nanoparticles was increased. The gradient of the lines shows the effect of CNTs volume fraction on the natural frequencies computed for the following ratios: 0, 0.05, 0.10, 0.20, 0.35, and 0.50.

It is observed that as the volume fraction of CNTs increased, the natural frequency tended to increase. Specifically, the enhancement of the values of this set of natural frequencies reached 59%, 69%, 80%, 133%, 92%, 94%, 89%, and 55% for the 1st bending,

2nd bending, 3rd bending, 1st torsional, 1st breathing, 2nd breathing, 3rd breathing, and 1st axial natural frequency, respectively, concerning a 0.50 volume fraction of CNTs.

Figure 6 shows the variation in the critical speed, for different fiber orientation angles, and for various CNT volume fractions in the epoxy resin matrix. For volume fractions of CNTs equal to 0.50 into the hybrid composite drive shaft, the critical speed presented a significant increase in its value, approximately 60%, as depicted in Figure 6a,b, respectively.

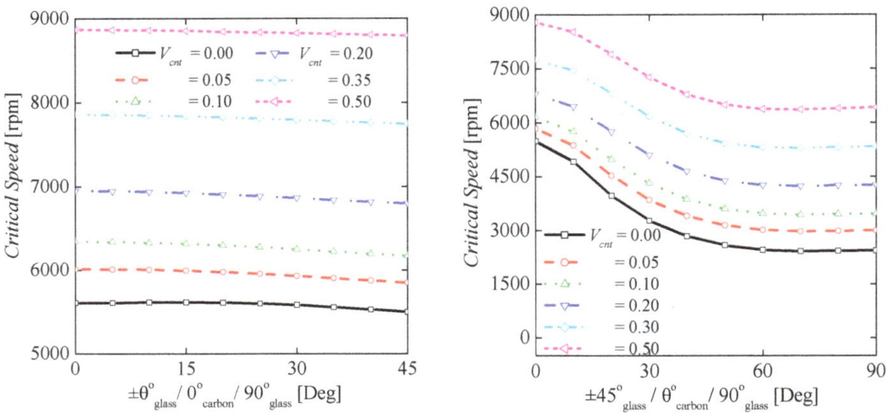

(a) (b)

Figure 6. Critical speed as a function of fiber orientation angle (θ°) for various CNT volume fractions. A hybrid composite drive shaft having four layers stacking of (**a**) $[\pm\theta^\circ_{glass}/0^\circ_{carbon}/90^\circ_{glass}]$ and (**b**) $[\pm45^\circ_{glass}/\theta^\circ_{carbon}/90^\circ_{glass}]$ was analyzed for length = 1730 mm, radius = 50.8 mm, and wall thickness = 2.032 mm, where the fiber orientation and volumetric concentration of nanoparticles were changed. Figures (**a**,**b**) show the effect of fiber orientation angle on the critical speed by changing the first two layers and the third (carbon fiber) layer, respectively. The curves show the effect of CNTs volume fractions on the critical speed, computed for CNT volume fractions: 0, 0.05, 0.10, 0.20, 0.35, and 0.50.

4.3. Effect of CNTs Inclusion on Buckling Torque

The first mode shape of buckling torque, as derived from the finite element analysis, is presented in Figure 7. The first buckling mode was similar to the corresponding one predicted in [61]. In this analysis, the computed load was close to the collapse load. The output from the analysis is a factor (eigenvalue) that is multiplied by the actual magnitude of the applied load in order to compute an estimation of the critical torque. To determine the variation in the composite drive shaft under buckling torque due to the presence of CNTs into the composite shaft, a parametric finite element investigation was also performed. The results are presented in Figure 8. It was observed that the performance of the critical buckling torque significantly improved with the amount of added CNTs in the matrix of the composite drive shaft.

This may be expected due to the increase in the mechanical properties of the composite lamina as the CNTs volume fraction is increased. Moreover, the variation in the critical buckling torque, for different fiber orientation angles, is also depicted. With the addition of CNT into the pure composite drive shaft, the critical buckling torque showed an enhancement in its value. Specifically, in Figure 8a, the critical buckling torque of the lamination [±0, 0, 90] increased up to 27%, 50%, 89%, 140%, and 187% for V_{cnt} = 0.05, 0.10, 0.20, 0.35, and 0.50, respectively. The critical buckling torque of the lamination [±45, 0, 90] increased up to 19%, 35%, 66%, 107%, and 145% for V_{cnt} = 0.05, 0.10, 0.20, 0.35, and 0.50, respectively. The critical buckling torque of the lamination [±90, 0, 90] increased 12%, 21%, 37%, 58%, and 79% for V_{cnt} = 0.05, 0.10, 0.20, 0.35, and 0.50, respectively. Moreover, it seems that the critical buckling torque was not considerably affected by the fiber orientation angle, as the amount of added CNTs increased. Finally, in Figure 8b, the critical buckling torque of the lamination [±45, 0, 90] increased up to 19%, 35%, 66%, 107% and 145% for V_{cnt} = 0.05, 0.10, 0.20, 0.35, and 0.50, respectively. Additionally, we notice that the critical buckling torque increased as the fiber angle of the third ply (carbon fiber) tended to 0° or 90°. On the contrary, the lowest value was observed at 45°.

Figure 7. First buckling mode shape of a hybrid composite driveshaft, stacked as $[\pm45°_{glass}/0°_{carbon}/90°_{glass}]$, computed for length = 1730 mm, radius = 50.8 mm, and wall thickness = 2.032 mm. To detect the deformation and critical buckling load, eigenvalue buckling analysis was conducted, where the model was fully fixed at one end and subjected to torsion load at the other end. The corresponding buckling torque value from Equation (33) was 2149 N·m and from FEA was 1730 N·m.

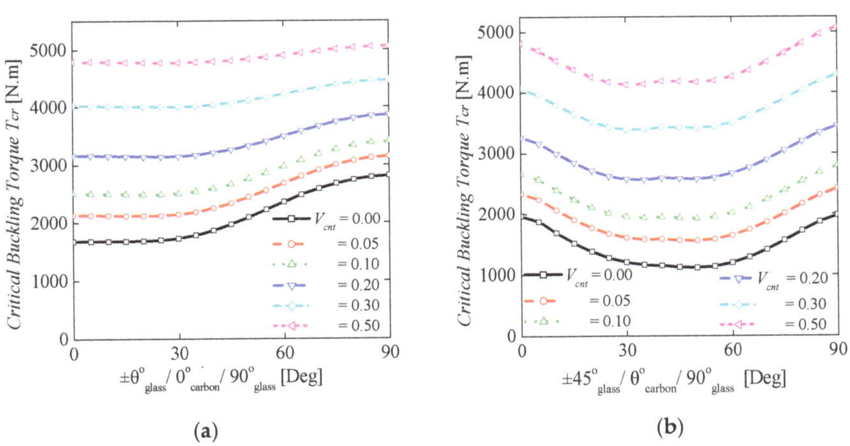

Figure 8. Critical buckling torque as a function of fiber orientation angle (θ°) for various 433 CNT volume fractions. A hybrid composite drive shaft having four layers stacking of (**a**) $[\pm\theta°_{glass}/0°_{carbon}/90°_{glass}]$ and (**b**) $[\pm45°_{glass}/\theta°_{carbon}/90°_{glass}]$ was analyzed with length = 1730 mm, radius = 50.8 mm, and wall thickness = 2.032 mm. To obtain the critical buckling torque eigenvalue, buckling analysis was conducted.

5. Conclusions

In this study, the effect of MWCNTs on a hybrid fiber-reinforced composite automotive drive shaft was investigated. The modified epoxy resin matrix with different CNTs volume fractions was theoretically modeled. The CNTs-modified matrix presented enhanced material properties compared to the pure epoxy matrix system, affecting the overall behavior of the drive shaft. According to the results, the following findings can be noticed:

- The elastic constants E_1, E_2, G_{12}, and G_{23} could be increased up to 35%, 582%, 575%, and 607% (e-glass-reinforced polymer lamina) and 9%, 174%, 222%, and 182% (carbon fiber-reinforced polymer lamina), respectively, due to the 0.50 volume fraction of CNTs in the polymer matrix for the specific drive shaft configuration considering uniform dispersion.

- Adding CNTs into the conventional lamina, the natural frequency and critical buckling torque of the laminated composite drive shaft showed significant improvements up to 60% and 145%, respectively.
- The method can be used for design purposes of hybrid fiber-reinforced composite automotive drive shafts with CNTs inclusions.

Eventually, the enhanced characteristics of CNTs-based nanocomposites have the potential to be employed for the design of composite structures in the areas of automotive, aerospace, construction, military, etc., in terms of improving stiffness or strength, in addition to its low weight. This computational study did not investigate the effects of CNT dispersion and distribution into the polymer matrix. Issues such as agglomeration are going to be modeled/studied theoretically and/or experimentally in future work. The obtained results can be considered as an upper bound limit of the potential enhanced mechanical performance. It seems that CNTs can play an important role as a significant secondary filler and the proposed approach may be used as a design tool.

Author Contributions: Conceptualization, S.K.G.; methodology, S.K.G.; software, P.A.A.; validation, P.A.A. and S.I.M.; formal analysis, S.K.G.; investigation, P.A.A.; resources, P.A.A.; data curation, P.A.A.; writing—original draft preparation, S.K.G.; writing—review and editing, S.K.G.; visualization, S.I.M.; supervision, S.K.G.; project administration, S.K.G. All authors have read and agreed to the published version of the manuscript.

Funding: This research received no external funding.

Conflicts of Interest: The authors declare no conflict of interest.

References

1. Godec, M.; Mandrino, D.; Jenko, M. Investigation of the fracture of car's drive shaft. *Eng. Fail. Anal.* **2009**, *16*, 1252–1261. [CrossRef]
2. Heisler, H. *Vehicle and Engine Technology*, 2nd ed.; SAE International: London, UK, 1999.
3. Swanson, S.R. *Introduction to Design and Analysis with Advanced Composite Material*; Prentice Hall: Hoboken, NJ, USA, 1997.
4. Seherr-Thoss, H.C.; Schmelz, F.; Aucktor, E. *Universal Joints and Driveshafts-Analysis, Design, Applications*; Springer: Berlin/Heidelberg, Germany, 2006.
5. Sun, Z.; Xiao, J.; Yu, X.; Tusiime, R.; Gao, H.; Min, W.; Tao, L.; Qi, L.; Zhang, H.; Yu, M. Vibration characteristics of carbon-fiber reinforced composite drive shafts fabricated using filament winding technology. *Compos. Struct.* **2020**, *241*, 111725. [CrossRef]
6. Cho, D.H.; Gil Lee, D.; Choi, J.H. Manufacture of one-piece automotive drive shafts with aluminum and composite materials. *Compos. Struct.* **1997**, *38*, 309–319. [CrossRef]
7. Elanchezhian, C.; Ramnath, B.V.; Raghavendra, K.S.; Muralidharan, M.; Rekha, G. Design and Comparison of the Strength and Efficiency of Drive Shaft made of Steel and Composite Materials. *Mater. Today Proc.* **2018**, *5*, 1000–1007. [CrossRef]
8. Nehe, S.R.; Ghogare, A.; Vatsa, S.; Tekade, L.; Thakare, S.; Yadav, P.; Shool, P.; Wankhade, P.; Dhakane, V.; Charjan, S. Design, Analysis, Simulation and Validation of Automobile Suspension System Using Drive-Shaft as a Suspension Link. *SAE Int. J. Passeng. Cars Mech. Syst.* **2018**, *11*, 129–138. [CrossRef]
9. Gong, L.; Gao, X.; Yang, H.; Liu, Y.; Yao, X. Design on the driveshaft of 3D 4-Directional carbon fiber braided composites. *Compos. Struct.* **2018**, *203*, 466–473. [CrossRef]
10. Karimi, S.; Salamat, A.; Javadpour, S. Designing and optimizing of composite and hybrid drive shafts based on the bees algorithm. *J. Mech. Sci. Technol.* **2016**, *30*, 1755–1761. [CrossRef]
11. Rastogi, N. *Design of Composite Driveshafts for Automotive Applications*; SAE Technical Paper Series: Warrendale, PA, USA, 2004. [CrossRef]
12. Shinde, R.M.; Sawant, S.M. Investigation on glass-epoxy composite drive shaft for light motor vehicle. *Int. J. Des. Eng.* **2019**, *9*, 22–35. [CrossRef]
13. Nadeem, S.S.; Giridhara, G.; Rangavittal, H. A Review on the design and analysis of composite drive shaft. *Mater. Today Proc.* **2018**, *5*, 2738–2741. [CrossRef]
14. Hu, Y.; Yang, M.; Zhang, J.; Song, C.; Hong, T. Effect of stacking sequence on the torsional stiffness of the composite drive shaft. *Adv. Compos. Mater.* **2016**, *26*, 537–552. [CrossRef]
15. Debski, H.; Rozylo, P.; Teter, A. Buckling and limit states of thin-walled composite columns under eccentric load. *Thin-Walled Struct.* **2020**, *149*, 106627. [CrossRef]

16. Liu, P.; Gu, Z.; Peng, X.; Zheng, J. Finite element analysis of the influence of cohesive law parameters on the multiple delamination behaviors of composites under compression. *Compos. Struct.* **2015**, *131*, 975–986. [CrossRef]
17. Galos, J. Thin-ply composite laminates: A review. *Compos. Struct.* **2020**, *236*, 111920. [CrossRef]
18. Bajpai, P.K.; Singh, I. *Reinforced Polymer Composites: Processing, Characterization and Post Life Cycle Assessment*; John Wiley & Sons: Hoboken, NJ, USA, 2019.
19. Kumar, S.K.; Krishnamoorti, R. Nanocomposites: Structure, Phase Behavior, and Properties. *Annu. Rev. Chem. Biomol. Eng.* **2010**, *1*, 37–58. [CrossRef] [PubMed]
20. Camargo, P.H.C.; Satyanarayana, K.G.; Wypych, F. Nanocomposites: Synthesis, structure, properties and new application opportunities. *Mater. Res.* **2009**, *12*, 1–39. [CrossRef]
21. Zhao, J.; Wu, L.; Zhan, C.; Shao, Q.; Guo, Z.; Zhang, L. Overview of polymer nanocomposites: Computer simulation understanding of physical properties. *Polymer* **2017**, *133*, 272–287. [CrossRef]
22. Isaac, C.W.; Ezekwem, C. A review of the crashworthiness performance of energy absorbing composite structure within the context of materials, manufacturing and maintenance for sustainability. *Compos. Struct.* **2021**, *257*, 113081. [CrossRef]
23. Mallick, P.K. *Fiber-Reinforced Composites: Materials, Manufacturing, and Design*; CRC Press: Boca Raton, FL, USA, 2007.
24. Siochi, E.J.; Harrison, J.S. Structural nanocomposites for aerospace applications. *MRS Bull.* **2015**, *40*, 829–835. [CrossRef]
25. Trimble, S. Lockheed Martin Reveals F-35 to Feature Nanocomposite Structures. Available online: https://www.flightglobal.com/lockheed-martin-reveals-f-35-to-feature-nanocomposite-structures/100174.article (accessed on 26 May 2011).
26. Shakil, U.A.; Bin Abu Hassan, S.; Yahya, M.Y.; Mujiyono; Nurhadiyanto, D. A review of properties and fabrication techniques of fiber reinforced polymer nanocomposites subjected to simulated accidental ballistic impact. *Thin-Walled Struct.* **2021**, *158*, 107150. [CrossRef]
27. Tüzemen, M.Ç.; Salamcı, E.; Avcı, A. Enhancing mechanical properties of bolted carbon/epoxy nanocomposites with carbon nanotube, nanoclay, and hybrid loading. *Compos. Part B Eng.* **2017**, *128*, 146–154. [CrossRef]
28. Khoramishad, H.; Alikhani, H.; Dariushi, S. An experimental study on the effect of adding multi-walled carbon nanotubes on high-velocity impact behavior of fiber metal laminates. *Compos. Struct.* **2018**, *201*, 561–569. [CrossRef]
29. Chiou, Y.-C.; Chou, H.-Y.; Shen, M.-Y. Effects of adding graphene nanoplatelets and nanocarbon aerogels to epoxy resins and their carbon fiber composites. *Mater. Des.* **2019**, *178*, 107869. [CrossRef]
30. Novoselov, K.S.; Geim, A.K.; Morozov, S.V.; Jiang, D.; Katsnelson, M.I.; Grigorieva, I.V.; Dubonos, S.V.; Firsov, A.A. Two-dimensional gas of massless Dirac fermions in graphene. *Nature* **2005**, *438*, 197–200. [CrossRef]
31. Iijima, S. Helical microtubules of graphitic carbon. *Nat. Cell Biol.* **1991**, *354*, 56–58. [CrossRef]
32. De Volder, M.F.; Tawfick, S.H.; Baughman, R.H.; Hart, A.J. Carbon nanotubes: Present and future commercial applications. *Science* **2013**, *339*, 535–539. [CrossRef] [PubMed]
33. Kim, G.; Nam, I.; Yang, B.; Yoon, H.; Lee, H.; Park, S. Carbon nanotube (CNT) incorporated cementitious composites for functional construction materials: The state of the art. *Compos Struct.* **2019**, *227*. [CrossRef]
34. Hu, Z.; Lu, X. *Mechanical Properties of Carbon Nanotubes and Graphene-Carbon Nanotubes and Graphene*; Elsevier: Amsterdam, The Netherlands, 2014.
35. Ahmadipour, M.; Arjmand, M.; Le, A.T.; Chiam, S.L.; Ahmad, Z.A.; Pung, S.-Y. Effects of multiwall carbon nanotubes on dielectric and mechanical properties of $CaCu_3Ti_4O_{12}$ composite. *Ceram. Int.* **2020**, *46*, 20313–20319. [CrossRef]
36. Georgantzinos, S.; Stamoulis, K.P.; Markolefas, S. Mechanical Response of Hybrid Laminated Polymer Nanocomposite Structures: A Multilevel Numerical Analysis. *SAE Int. J. Aerosp.* **2020**, *13*, 1–13. [CrossRef]
37. Georgantzinos, S.K.; Markolefas, S.I.; Mavrommatis, S.A.; Stamoulis, K. Finite element modelling of carbon fiber carbon nanostructure polymer hybrid composite structures. *MATEC Web Conf.* **2020**, *314*, 02004. [CrossRef]
38. Georgantzinos, S.K.; Giannopoulos, G.I.; Markolefas, S.I. Vibration Analysis of Carbon Fiber-Graphene-Reinforced Hybrid Polymer Composites Using Finite Element Techniques. *Materials* **2020**, *13*, 4225. [CrossRef]
39. Rafiee, M.; Rafiee, J.; Wang, Z.; Song, H.; Yu, Z.-Z.; Koratkar, N. Enhanced Mechanical Properties of Nanocomposites at Low Graphene Content. *ACS Nano* **2009**, *3*, 3884–3890. [CrossRef]
40. Taş, H.; Soykok, I.F. Effects of carbon nanotube inclusion into the carbon fiber reinforced laminated composites on flexural stiffness: A numerical and theoretical study. *Compos. Part B Eng.* **2019**, *159*, 44–52. [CrossRef]
41. Stamoulis, K.; Georgantzinos, S.K.; Giannopoulos, G. Damage characteristics in laminated composite structures subjected to low-velocity impact. *Int. J. Struct. Integr.* **2019**, *11*, 670–685. [CrossRef]
42. Barretta, R.; Brčić, M.; Čanađija, M.; Luciano, R.; de Sciarra, F.M. Application of gradient elasticity to armchair carbon nanotubes: Size effects and constitutive parameters assessment. *Eur. J. Mech. A Solids* **2017**, *65*, 1–13. [CrossRef]
43. Georgantzinos, S.K.; Giannopoulos, G.I. Thermomechanical buckling of single walled carbon nanotubes by a structural mechanics method. *Diam. Relat. Mater.* **2017**, *80*, 27–37. [CrossRef]
44. Hassanzadeh-Aghdam, M.; Ansari, R.; Darvizeh, A. A new micromechanics approach for predicting the elastic response of polymer nanocomposites reinforced with randomly oriented and distributed wavy carbon nanotubes. *J. Compos. Mater.* **2017**, *51*, 2899–2912. [CrossRef]
45. Mehar, K.; Panda, S.K. Elastic bending and stress analysis of carbon nanotube-reinforced composite plate: Experimental, numerical, and simulation. *Adv. Polym. Technol.* **2017**, *37*, 1643–1657. [CrossRef]

46. Georgantzinos, S.K. A new finite element for an efficient mechanical analysis of graphene structures using CAD/CAE techniques. *J. Comput. Theor. Nanosci.* **2017**, *14*, 5347–5354. [CrossRef]
47. Georgantzinos, S.K.; Siampanis, S.G. Size-dependent elastic mechanical properties of γ-graphyne structures: A comprehensive finite element investigation. *Mater. Des.* **2021**, *202*, 109524. [CrossRef]
48. Spanos, K.; Georgantzinos, S.; Anifantis, N. Mechanical properties of graphene nanocomposites: A multiscale finite element prediction. *Compos. Struct.* **2015**, *132*, 536–544. [CrossRef]
49. Lin, F.; Xiang, Y.; Shen, H.S. Temperature dependent mechanical properties of graphene reinforced polymer nanocomposites—A molecular dynamics simulation. *Compos. Part B Eng.* **2017**, *111*, 261–269. [CrossRef]
50. Sun, R.; Li, L.; Zhao, S.; Feng, C.; Kitipornchai, S.; Yang, J. Temperature-dependent mechanical properties of defective graphene reinforced polymer nanocomposite. *Mech. Adv. Mater. Struct.* **2021**, *28*, 1010–1019. [CrossRef]
51. Chandra, A.K.; Kumar, N.R. Polymer Nanocomposites for Automobile Engineering Applications. In *Properties and Applications of Polymer Nanocomposites*; Springer: Berlin/Heidelberg, Germany, 2017; pp. 139–172.
52. Kaw, A.K. *Mechanics of Composite Materials*; CRC Press: Boca Raton, FL, USA, 2006.
53. Halpin, J.C.; Kardos, J.L. The Halpin-Tsai equations: A Review. *Polym. Eng. Sci.* **1976**, *16*, 344–352.
54. Shokrieh, M.M. *Residual Stresses in Composite Materials*; Woodhead Publishing Ltd.: Amsterdam, The Netherlands, 2014.
55. Hyer, M.W.; Scott, W. *Stress Analysis of Fiber-Reinforced Composite Materials*; DEStech Publications Inc.: Lancaster, PA, USA, 2009.
56. Talib, A.R.; Aidy, A.; Badie, M.A.; Lah, C.N.A.; Golestaneh, A.F. Developing a hybrid, carbon/glass fiber- reinforced epoxy composite automotive drive shaft. *Mater. Des.* **2010**, *31*, 514–521. [CrossRef]
57. William, T.T.; Dahleh, M.D. *Theory of Vibration with Applications*; Pearson Education Inc.: London, UK, 1998.
58. Badie, M.; Mahdi, E.; Hamouda, A.M. An investigation into hybrid carbon/glass fiber reinforced epoxy composite automotive drive shaft. *Mater. Des.* **2011**, *32*, 1485–1500. [CrossRef]
59. Rao, S.S. *Vibration of Continuous Systems*; John Wiley & Sons: New York, NY, USA, 2007; Volume 464.
60. Sannikov, A.A.; Vitvinin, A.M. Calculating the critical speed of papermaking machine shafts, considering the stiffness of supports and screens. *Chem. Pet. Eng.* **1976**, *12*, 215–217. [CrossRef]
61. Shokrieh, M.M.; Hasani, A.; Lessard, L.B. Shear buckling of a composite drive shaft under torsion. *Compos. Struct.* **2004**, *64*, 63–69. [CrossRef]
62. Gay, D.; Hoa, S.V.; Tsai, S.W. *Composite Materials-Design and Applications*; CRC Press: Boca Raton, FL, USA, 2003.

Journal of
Composites Science

MDPI

Article

Thermal and Mechanical Properties of Green Insulation Composites Made from *Cannabis* and Bark Residues

Konstantinos Ninikas [1], Andromachi Mitani [1], Dimitrios Koutsianitis [1], George Ntalos [1], Hamid R. Taghiyari [2] and Antonios N. Papadopoulos [3,*]

[1] Department of Forestry, Wood Science and Design, University of Thessaly, GR-431 00 Karditsa, Greece; kninikas@uth.gr (K.N.); amitani@uth.gr (A.M.); dkoutsianitis@uth.gr (D.K.); gntalos@uth.gr (G.N.)
[2] Wood Science and Technology Department, Faculty of Materials Engineering & New Technologies, Shahid Rajaee Teacher Training University, Tehran 16788-15811, Iran; htaghiyari@sru.ac.ir
[3] Laboratory of Wood Chemistry and Technology, Department of Forestry and Natural Environment, International Hellenic University, GR-661 00 Drama, Greece
* Correspondence: antpap@for.ihu.gr

Abstract: The objective of this paper was to investigate the technical feasibility of manufacturing low density insulation particleboards that were made from two renewable resources, namely hemp fibers (*Cannabis sativa*) and pine tree bark, which were bonded with a non-toxic methyl cellulose glue, as a binder. Four types of panels were made, which consisted of varying mixtures of tree bark and hemp fibers (tree bark to hemp fibers percentages of 90:10, 80:20, 70:30, and 60:40). An additional set of panels was made, consisting only of bark. The results showed that addition of hemp fibers to furnish improved mechanical properties of boards to reach an acceptable level. The thermal conductivity unfavorably increased as hemp content increased, though all values were still within the acceptable range. Based on cluster analysis, board type 70:30 (with 30% hemp content) produced the highest mechanical properties as well as the optimal thermal conductivity value. It is concluded that low density insulation boards can be successfully produced using these waste raw materials.

Keywords: insulation composites; thermal properties; mechanical properties; *Cannabis*; hemp; bark

Citation: Ninikas, K.; Mitani, A.; Koutsianitis, D.; Ntalos, G.; Taghiyari, H.R.; Papadopoulos, A.N. Thermal and Mechanical Properties of Green Insulation Composites Made from *Cannabis* and Bark Residues. *J. Compos. Sci.* **2021**, *5*, 132. https://doi.org/10.3390/jcs5050132

Academic Editor: Stelios K. Georgantzinos

Received: 25 April 2021
Accepted: 14 May 2021
Published: 17 May 2021

1. Introduction

Building owners have become interested in a sustainable and healthy environment, which is a trend favoring ecological materials with outstanding performance. In addition, nowadays, thermal insulation can be considered to be a hot issue for civil engineering that tries to reduce cooling and heating costs and, at the same time, eliminate CO_2 emissions [1,2]. Insulating materials are produced for a variety of applications and with specific properties, based on their end use. Thermal conductivity is the most crucial property, followed by fire behavior and compressive strength [3]. The main products, which are available in the market for heat insulation materials, can be categorized, as follows: (i) synthetic materials, like polyurethane and polystyrene; (ii) inorganic materials, like mineral wool and glass. However, the recycling of such products is problematic since their degradation is quite slow and generates toxic substances [4]. In addition, the standards for contaminant emission as far as the building materials are concerned are significantly extended and, therefore, the impact of human's exposure to unhealthy materials is an important parameter that has to be taken into consideration [5]. A third, and perhaps more attractive option, is materials produced from renewable resources, which have gained increased popularity in recent years [6–12]. These may include wood residues, agricultural residues, and tree barks.

Various lignocellulosic materials have been used to manufacture low density insulation composites. Panyakaew and Fotios [1] produced low density binderless thermal insulation boards made from coconut husk and bagasse. They found that the bagasse insulation boards provided mechanical properties that were superior to those of coconut husk boards; on the other hand, the binderless coconut husk insulation boards showed greater stability against water. It was also reported that thermal conductivity values of boards that were made from this type of raw material were close to those of conventional insulation materials. Doost-hosseini et al. and Taghiyari et al. [2,13] manufactured insulation boards from sugar cane bagasse, and reported on the correlation between the physicomechanical properties and the permeability of the boards and the sound absorption coefficients. Low correlations were found between the sound absorption coefficients and physicomechanical properties, and it was concluded that sound absorption coefficients cannot be considered to be a reliable criterion to predict the board performance. Ibraheem et al. [14] developed insulation boards that were fabricated from polyurethane reinforced with kenaf fibers, at three different weigh contents. They reported the optical performance of boards at a weight of 50% kenaf fibers. In addition, it was stated that the thermal conductivity decreased with an increasing fiber content.

Tree bark, which is a byproduct of the timber industry, has also been applied for the manufacture of low density insulation properties. Bark is used mainly for low-value applications, such as a soil covering material in agriculture or as a fuel. It was reported that the thermal conductivity of bark is approximately 20% lower than that of solid wood [15]. Furthermore, bark is suggested as an insulation material due to its flame retardant properties, its favorable internal structure, and its low density [16,17]. Kaiņ et al. [8] made insulation boards from particles of larch bark, which were bonded with a tannin resin. It was found that the resin amount did not significantly influence the mechanical properties, but the panel density was reported to be the most important variable. It was also reported that light boards had a low thermal conductivity value. Recently, thermal insulation panels were made from various bark species (larch, pine, spruce, fir, and oak tree bark) and bonded with different resin systems (urea formaldehyde, melamine formaldehyde, Quebracho, Mimosa) [18]. It was found that all bark species were suitable for insulation panel production, while, at the same density, panels from barks with a low bulk density (i.e., pine and larch) are advantageous, because their compression ratios are higher, which improves the mechanical characteristics.

Tree bark has already been used within a wood-based sandwich panel, thereby proving its insulation properties [19]. Single layer bark insulation boards have also been constructed in laboratories, demonstrating that bark is a promising new insulation material [18]. On the other hand, hemp has been used as insulation material, together with wood, where the hydrothermal performance was studied. The thermal performance of this board was also competitive with an average thermal transmittance of 0.30 W/mK when compared to existing commercial oil based insulation panels [20]. Ninikas et al. studies the thermal properties of insulation boards that were made of hemp residues and tree bark [21]. However, their mechanical properties were not addressed, and the determination of the thermal conductivity was based on a different measurement setup, as described later in the paper.

Consequently, the objective of this paper was to investigate the technical feasibility of manufacturing low density insulation particleboards that were made from two renewable resources, namely hemp fibers (*Cannabis sativa*) and pine tree bark, bonded with a non-toxic methyl cellulose glue, as a binder.

2. Materials and Methods

2.1. Raw Material

The raw material that was used in this study, namely hemp fibers and pine tree bark, was collected from forest and cropland in Karditsa city, Central Greece. The bark was chipped by a mechanical hammer-mill chipper with a 18 and 20 mm round hole screen (Figure 1a). At least 92% of the bark particles were below 19 mm. The hemp fibres were manually cut with a pair of scissors into approximately 0.10 m length stripes to better bond with the bark (Figure 1b). At least 95% of the hemp fibers were 0.10 mm in length. The orientation of the hemp fibres was random at all three axes (X, Y, Z) when mixed with the bark. It was hypothesised that, for the composite boards, the geometry of the hemp fibres should be such to assist in bonding the two materials together due to the absence of a hot press. The bark went through a 20 mm diameter sieve (Figure 2a). The aim was to have bark flakes at a size that would accommodate the bonding with the hemp fibres with the minimum gap between the two materials that result in a robust final composite. The average bark consisted of an iregular shaped flake approximately of 0.013–0.018 m (Figure 2b). Bark and hemp fibers were both dried at 105 °C for 24 h to reach 6.5–7% moisture content.

(a) (b)

Figure 1. Pine tree bark (**a**) and hemp fibers (**b**).

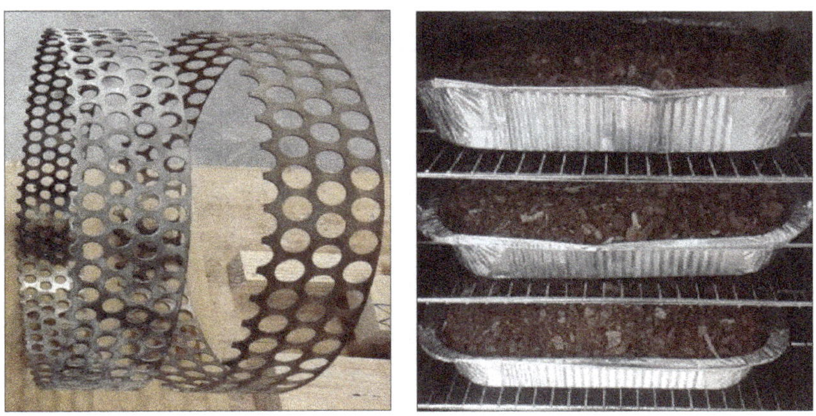

(a) (b)

Figure 2. The 18 mm (**left**) and the 20 mm (**right**) sieves used for the bark flakes (**a**) and the bark flakes after the sieving at the moisture chamber (**b**).

2.2. Board Manufacture

A non-toxic methyl cellulose glue, 4% as a percentage of the oven dry weight of raw material, was applied for single layer board manufacture. The glue (Glutolin N standard wallpaper adhesive, density 0.31 g/cm^3, and pH 7–8) was a hydrophilic white powder that dissolved in water at room temperature. According to the supplier, the solution was of 1.25% glue (dry powder). The solution was stirred for 15 min. and then sprayed into the two ingredients while they were mixed together before being placed in to 0.40 m × 0.40 m cast, as depicted in Figure 3. The casts were covered with flat fibreboards and tightened with hand clamps without any mechanical pressure in order to form the final board thickness of 0.047 m. The composites stayed under these conditions for 48 h at an average room temperature of 23 °C before being opened. Four types of panels were made, which consisted of varying mixtures of tree bark and hemp fibers (tree bark to hemp fibers percentages of 90:10, 80:20, 70:30, and 60:40); an additional set of panels was made, consisting only of bark, as shown in Figures 4 and 5. Three replicates were made for each board type. Target board density was 0.25 Kg/m^3.

Figure 3. The resinated raw material.

Figure 4. Particleboards made from varying mixtures of tree bark and hemp fibers (70:30 **left** and 60:40 **right**).

2.3. Determination of Mechanical Properties

The cboards were conditioned one week at 20 °C and 65% relative humidity prior to testing mechanical properties, namely the modulus of rupture (MOR) and modulus of elasticity (MOE) [22]. The 50 by 350 mm long beams were tested in third-point loading at a span of 320 mm at a loading rate of 3 mm per minute. The load and deflection were continuously recorded, and the resulting data were used to calculate the modulus of rupture (MOR) and modulus of elasticity (MOE).

Figure 5. Particleboards made from varying mixtures of tree bark and hemp fibers (80:20 **left** and 90:10 **center**) and panels consisting only of bark (**right**).

2.4. Determination of Thermal Conductivity

After taking out of the casts, the boards were directly forwarded to the thermal conductivity apparatus without being placed in a furnace to further reduce their moisture content. This was based on a more realistic approach, where the insulation boards do not usually have the ideal moisture content during the installation. This approach was different as compared to the one followed in a previous publication [21], where the thermal conductivity was measured in very dry conditions having the boards in a furnace, at a temperature of 103 °C for several hours, and, therefore, resulted in low values.

For the determination of the thermal conductivity of the boards, the 'box method', based on EN ISO 12,667, was applied [23]. For this purpose, a single box of EI-700 unit was used [24], as depicted in Figure 6, which measures the thermal characterisitics of homogeneous or heterogeneous, solid, or liquid materials with a comparatively low conductivity ($\lambda < 3$ W/mK). The specimens were 0.27 m × 0.27 m × 0.047 m (length × width × thickness). Two temperature probes have been applied, one for the upper (T_{uf}) and one for the lower (T_{lf}) board's surface temperatures, respectively. The room temperature was also recorded (21 °C). Having the samples within the 'box' for approximately 2 h, the temperature was stabilised and the readings for each probe were noted. The 'box method' produced readings with regard to the thermal conductivity value of all the insulation boards keeping each of them under stable thermal conditions for 24 h.

Figure 6. The EI-700 unit "box method" for the thermal conductivity measurements.

The thermal conductivity value was calculated based on the following equation:

$$\lambda = \frac{q_{in} * \Delta_x}{A * (T_{uf} - T_{lf})} \quad (1)$$

where

λ the thermal conductivity value (W/m*K)

q_{in} the heat flow from the indoor environment to the 'box' (W)

Δ_x the board's thickness (m): (0.047 m)

A the board's surface (m^2): ($0.27 \times 0.27 = 0.0729$ m^2)

T_{uf} the temperature at the upper surface of the board (°C)

T_{lf} the temperature at the lower surface of the board (°C)

The heat flow at a steady state is:

$$q_{in} = q_{out} + q_{spl} \quad (2)$$

where

q_{out} the heat flow from the "box" back to the environment (W)

q_{spl} the heat flow through the composite (W)

The heat flow (q_{in}) derives from the equation: [25]

$$q_{in} = \frac{V^2}{R} \quad (3)$$

where

q_{in} the heat flow emitted by Joule effect (W)

V the electric voltage of the device $V = 39.8$ Volt

R the device's resistance (Ω-Ohm) – $R = 1160$ Ω

All the above parametres are displayed in Figure 7. The grey rectangle at the bottom of the Figure, represents the insulation board, the "blue Π" illustrate the Unit's box.

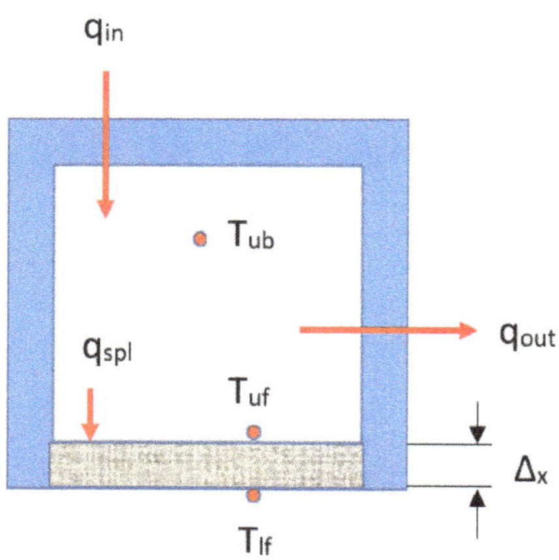

Figure 7. Schematic representation of the 'box'.

2.5. Statistical Analysis

Statistical analysi was conducted using the SPSS software program, version 24.0 (IBM, Armonk, NY, USA, 2018). One-way ANOVA was performed for identifying significant difference at the 95% level of confidence. Duncan's multiple range test grouping was carried out at 95% level of confidene, for each and every property measured, in order to discern significant difference among the five different panels produced and studied here. Contour and surface plots were designed using Minitab statistical software (version 16.2.2;

2010). Hierarchical cluster analysis was carried out based on all of the properties measures in this study, using Ward's method, in order to sum up similarities and dissimilarities among the five types of boards. In this analysis, the number of treatments (here, the number of five different panels) studied are shown as "Num". Subsequently, they are connected based on a scale-bar on top of the anlysis, showing numbers from 0 to 25. If treatments are connected by vertical lines on the lower numbers (in terms of the scale-bar), which means that they have more similarities. Going further away on the scale-bar upto 25 means the treaments have more dissimilarities. Cluster analysis demonstrates similarities and dissimilarites among treatments based on all properties that have been studied, giving a better overall outlook on all treatments.

3. Results and Discussion

3.1. Mechanical Properties

The moisture contents of the five types of insulation boards varied from 5.40 to 6.15%. Table 1 shows the mechanical properties of the single layer particleboards made from various tree bark/hemp fibers combinations. At this point, it has to be mentioned that preliminary tests revealed that it was not feasible to manufacture boards with higher hemp fiber content (tree bark to hemp fibers percentages of 50:50, 40:60), since this attempt lead to non-consistent boards. From the data that are presented in Tables 1 and 2, it can be seen that higher hemp fiber levels resulted in two opposing effects. From one side, the thermal conductivity increased as hemp content increased, although all of the values still remained within the acceptable range. From the other side, lack or low hemp contents (0, 10, and 20%) resulted in very low mechanical strength, so that the MOR values of these boards were lietrary unacceptable by the industry. Therefore, these board types (100:0, 90:10, and 80:20, as described in Table 1) cannot be recommended to the industry. The reduction in bending properties in boards, as the content of the hemp fibers is increased, can be attributted to the fact that hemp fibers is mainly comprised of relatively thin, short-walled, and weak cells [26]. As a consequense, hemp fibers are relatively weak and vulnerable to 'critical defects' inside the board structure and, therefore, a deterioration in bending properties is observed.

Table 1. Mechanical properties of various board types. Standard deviations in parentheses. Different letters show which values are statistically different at the 5% level.

Board Type (Tree Bark: Hemp Fibers)	Density [1] (Kg/m^3)	Weight of the Raw Material (g)		MOR [1] (N/mm^2)	MOE [1] (N/mm^2)
		Bark	Hemp		
100:0	0.24 A [3] (0.02) [2]	1660	0	0.01 C (0.01)	0.02 B (0.21)
90:10	0.24 A (0.02)	1440	160	0.01 C (0.01)	0.03 B (0.31)
80:20	0.23 A (0.02)	1280	320	0.09 B (0.01)	1.00 B (0.11)
70:30	0.22 A (0.02) [2]	1120	480	0.18 A (0.02)	2.25 A (0.31)
60:40	0.24 A (0.02)	960	640	0.15 A (0.01)	1.90 A (0.19)

[1] Each value is the mean of eight replicates. [2] Standard deviation. [3] Values followed by the same letter do not differ significantly from each other by a Duncan's multiple range test ($\alpha' = 0.05$).

The mechanical properties of the present particleboards were comparable to those of other low-density particleboards (100–500 Kg/m^3) produced from renewable resources (e.g., kenaf core, bagasse), reporting MOR values between 0.85 and 7 N/mm^2 and IB (Internal Bond Strength) values between 0.02 and 0.17 N/mm^2 [27,28]. A potential increase in the bending properties can be achieved through the increase of panel density. A recent study reported on thermal insulation panels made from various types of bark bonded with a variety of resins, with density values ranging from 0.25 to 0.50 Kg/m^3 [18]. The results of the physical-mechanical testing were analyzed using a multivariate ANOVA, and the panel density was considered as a covariate in the statistical model. It was found that bending properties, namely MOR and MOE, are highly affected by the panel density, type of the bark, and type of the resin. It is further reported that almost 60% of the variation in bending properties were attributed to the difference in density values, and that resin content did not show a significant effect on bending properties. It is interesting to mention that MOR and MOE were increased by 0.7 and 140 N/mm^2, respectively, with a density increase of 100 Kg/m^3 [18]. According to the results that were reported by Kain et al. [18], an approach to increase the bending properties of the boards made in this study, is to increase the density of the board.

Table 2. Thermal conductivity values of various board types. Standard deviations in parentheses. Different letters show which values are statistically different at the 5% level.

Board Type (Tree Bark: Hemp Fibers)	Thermal Conductivity Value [1] (λ) (W/m*K)
100:0	0.076 A [3] (0.04) [2]
90:10	0.081 A (0.02)
80:20	0.087 A (0.02)
70:30	0.094 B (0.02)
60:40	0.111 B (0.03)

[1] Each value is the mean of three replicates. [2] Standard deviation. [3] Values followed by the same letter do not differ significantly from each other by a Duncan's multiple range test ($\alpha' = 0.05$).

3.2. Thermal Conductivity

Table 2 depicts the thermal conductivity (λ) values of the produced insulation boards. From this, it can be seen that boards made only from bark demonstrated better (lower) thermal transmittance value as compared to the boards containing various amounts of hemp fibers. The bark-based insulation boards showed a thermal conductivity value of 0.076 W/m*K, which is in accordance with values that were reported in the literature [18] and higher than those of very light insulation boards (e.g., polystyrene, mineral wools with approximately 0.03 W/m*K). This disadvantage is compensated for by the low thermal diffusivity of bark [16,18]. This makes the material suitable for use as insulation layers that need to prevent quick cooling or overheating during summer.

A closer inspection of the data that are depicted in Table 2 reveals that the increase in hemp content resulted in higher thermal transmittance values and, furthermore, it is worth to be mentioned that the incorporation of hemp fibers up to 20%, did not significantly affect the thermal transmittance of the boards. This is in agreement with Ibraheem et al. [14], who developed insulation boards that were fabricated from polyurethane reinforced with kenaf fibers. They reported optimal performance of boards at a weight of 30% kenaf fibers. In addition, it was stated that the thermal conductivity decreased with increasing fiber content. The increased thermal conductivity values, as the content of the hemp fibers is increased, can be attributed to the high void content in the final panel [29]. Small pores are advantageous in this respect, because the air in such voids is static, and heat convection has a minor effect [30,31]. The existence of plenty of voids was also reported to improve sound absorption coefficients in insulating boards made from bagasse [32,33].

The different results for the thermal conductivity (for the 70:30 and 60:40 boards) that were obtained in this study in comparison with the previous study [21] are due the different measurement setup. In that study, the thermal conductivity was the one measured in very dry conditions having the boards at a furnace, at a temperature of 103 °C for several hours. In this study, it was chosen to identify the thermal conductivity readings without positioning the boards to a furnace that would reduce the moisture content even more as it happened herein. This means that the boards in this study were not so "dry" as in the previous study [21]. This explains the higher values (worst performance) as compared to the results reported by Ninikas et al. [21] for these two types of boards (70:30 and 60:40). This process was followed to all boards at this study and was based on a more realistic approach where the insulation boards do not have the ideal moisture content during the installation.

All types of boards resulted in acceptable thermal conductivity values, based on the value λ < 1.15 W/m*K, which is considered to be the limit for an appropriate insulation material [34]. In this connection, contour and surface plots between thermal conductivity versus different properties studied here demonstrated a clear relationship among properties within the acceptable range, although some minor discrepancies were also seen (Figure 8A,B). The discrepancies were attributed to the opposing effects of the addition of hemp on different properties. That is, the hemp content tended to increase thermal conductivity, as explained earlier; however, its delicate inter-connecting texture helped to ensure better integrity within the furnish, and ultimately the mechanical strength improved with higher hemp contents.

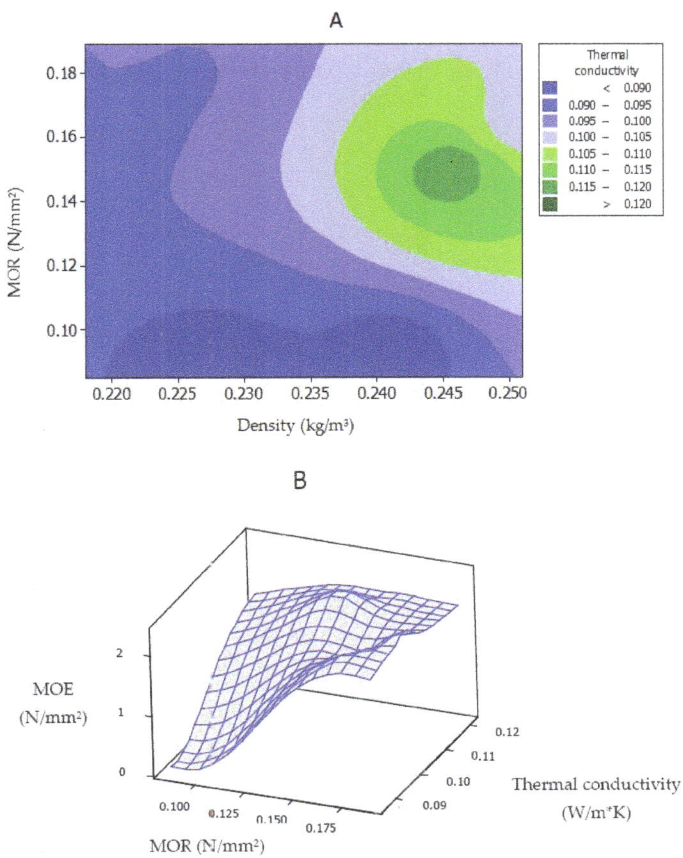

Figure 8. Contour (**A**) and surface (**B**) plots among different properties (thermal conductivity in W/m*K, MOR in N/mm^2, MOE in N/mm^2, and density in kg/m^3 values) within the acceptable range in the insulating boards made from *Cannabis* and bard residues.

A major issue addressed in this paper was to produce low density insulation boards from renewable resources, namely hemp fibers (*Cannabis sativa*) and pine tree bark. The use of a non-toxic methyl cellulose glue, formaldehyde-free, was a further challenge. The energy requirements (energy input) for constructing the five board types in this study, were kept relatively low due to the little energy input during the production procedure. The energy input for these boards was due to (a) the drying procedure (chamber) that was used for 24 h to reduce the moisture of the raw materials and (b) chipping procedure for the mechanical hammermill. Usually, the manufacturing procedure for a typical insulation board, with regards to the energy input during the production line, is immense due to the nature of the basic ingredients (petrol-based materials) [35]. The dimensional stability and biological durability of the panels produced in this study was not assessed, since the aim of the work was to produce low density insulation boards with an environmental friendly-non toxic adhesive. In such types of panels, these two properties are of minor importance. However, studies that are related to the manufacture of boards with these waste materials (bark and hemp), with higher density value and bonded with conventional formaldehyde resins, are under way and the results will be reported in due course.

Cluster analysis based on all of the properties measured in the present study categorized the five types of insulating boards, as depicted in Figure 9. It was demonstrated that all board types with hemp contents of lower than 20% were clustered very closely together; they are connected by vertical lines of less than digit "5" in terms of the scale-bar on the top of the graph. The other two board types with hemp contents of 30% and 40% (board types 70:30 and 60:40, as defined in Table 1) clustered very remotely from the other three types; as illustrated, they are connected by a vertical line on digit "25" in terms of the scale-bar. This was in close agreement with the low and unacceptable mechanical strength of the first three board types, as explained earlier. Cluster analysis also illustrated different clustering of the two board types with hemp contents of 30% and 40% (connecting to each other on digit "10" in terms of scale-bar). By taking the mechanical properties and thermal conductivity values into consideration (Tables 1 and 2), it can be deduced that board type 70:30 produced the optimal properties that can be recommended to the industry.

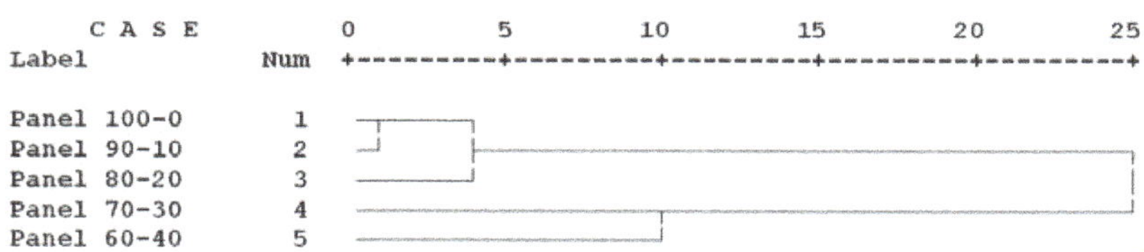

Figure 9. Cluster analysis of the five insulating board types, based on all properties measured (Num = number of five panels studid, based on the label column).

Low density insulation boards have been successfully produced using these waste raw materials. Their potential application can contribute to the reduction of cooling and heating costs and, at the same time, eliminate CO_2 emissions. The limited energy demand for the construction of these board types is expected to improve the carbon footprint of the insulation board and address a financially viable solution for producers who currently direct the residues in landfills with an additional cost.

4. Conclusions

This paper examined the technical feasibility of manufacturing low density insulation particleboards that were made from two renewable resources, namely hemp fibers (*Cannabis sativa*) and pine tree bark, which were bonded with a non-toxic methyl cellulose glue, as a binder. It was found that higher hemp fiber levels resulted in an increasing trend in thermal conductivity of boards, although all board types were still within the acceptable thermal conductivity range when compared to the value $\lambda < 1.15$ W/m*K which is considered to be the limit for an appropriate insulation material. Mechanical properties of boards with no or low hemp contents (0, 10, and 20% hemp) were not acceptable for the industry, although these produced the lowest thermal conductivity values. Based on the cluster analysis, it was concluded that board type 70:30 produced the boards with the highest mechanical properties and the optimal thermal conductivity. Therefore, using these waste raw materials for the production of insulating boards can be recommended.

Author Contributions: Methodology, K.N., A.M., D.K., G.N.; Validation, K.N., A.M., D.K., G.N.; Investigation, K.N., A.M., D.K., G.N., A.M.; Writing—Original Draft Preparation, K.N., G.N., A.N.P.; Writing—Review and Editing, G.N., H.R.T., A.N.P.; Visualization, K.N., A.M., D.K., G.N.; Supervision, G.N., H.R.T., A.N.P. All authors have read and agreed to the published version of the manuscript.

Funding: This research received no external funding.

Data Availability Statement: The data presented in this study are available on request from the corresponding author.

Conflicts of Interest: The authors declare no conflict of interest.

References

1. Panyakaew, S.; Fotios, S. New thermal insulation boards made from coconut husk and bagasse. *Energy Build.* **2011**, *43*, 1732–1739. [CrossRef]
2. Taghiyari, H.R.; Majidi, R.; Esmailpour, A.; Sarvari Samadi, Y.; Jahangiri, A.; Papadopoulos, A.N. Engineering composites made from wood and chicken feather bonded with UF resin fortified with wollastonite: A novel approach. *Polymers* **2020**, *12*, 857. [CrossRef] [PubMed]
3. Abdou, A.A.; Budaiwi, I.M. Comparison of Thermal Conductivity Measurements of Building Insulation Materials under Various Operating Temperatures. *J. Build. Phys.* **2005**, *29*, 171–184. [CrossRef]
4. Hanson, J.L.; Kopp, K.B.; Yesiller, N.; Cooledge, C.M.; Klee, E. (Eds.) *The Use of Recycled Materials as Thermal Insulation in Underground Construction*; Humboldt State University: Arcata, CA, USA, 2016.
5. Wi, J.H.; Park, Y.U.; Kim, S. Evaluation of environmental impact on the formaldehyde emission and flame-retardant performance of thermal insulation materials. *J. Hazard. Mater.* **2020**, *402*, 123463. [CrossRef]
6. Antov, P.; Krišťák, L.; Réh, R.; Savov, V.; Papadopoulos, A.N. Eco-Friendly Fiberboard Panels from Recycled Fibers Bonded with Calcium Lignosulfonate. *Polymers* **2021**, *13*, 639. [CrossRef] [PubMed]
7. Tsalagkas, D.; Börcsök, Z.; Pásztory, Z. Thermal, physical and mechanical properties of surface overlaid bark-based insulation panels. *Eur. J. Wood Prod.* **2019**, *77*, 721–730. [CrossRef]
8. Kain, G.; Guttler, V.; Barbu, M.C.; Petutschnigg, A.; Richter, K.; Tondi, G. Density related properties of bark insulation boards bonded with tannin hexamine resin. *Eur. J. Wood Prod.* **2014**, *72*, 417–424. [CrossRef]
9. Pizzi, A.; Papadopoulos, A.; Policardi, F. Wood Composites and Their Polymer Binders. *Polymers* **2020**, *12*, 1115. [CrossRef]
10. Igaz, R.; Kristak, L.; Ruziak, I.; Gajtanska, M.; Kucerka, M. Thermophysical properties of OSB boards versus equilibrium moisture content. *BioResources* **2017**, *12*, 8106–8118.
11. Antov, P.; Savov, V.; Krišťák, L'.; Réh, R.; Mantanis, G.I. Eco-Friendly, High-Density Fiberboards Bonded with Urea-Formaldehyde and Ammonium Lignosulfonate. *Polymers* **2021**, *13*, 220. [CrossRef]
12. Papadopoulos, A.N.; Taghiyari, H.R. Innovative Wood Surface Treatments Based on Nanotechnology. *Coatings* **2019**, *9*, 866.
13. Taghiyari, H.R.; Esmailpour, A.; Majidi, R.; Morrell, J.J.; Mallaki, M.; Militz, H.; Papadopoulos, A.N. Potential use of wollastonite as a filler in UF resin based medium-density fiberboard (MDF). *Polymers* **2020**, *12*, 1435. [CrossRef]
14. Ibraheem, S.A.; Ali, A.; Khalina, A. Development of Green Insulation Boards from Kenaf Fibres and Polyurethane. *Polym. Plast. Technol. Eng.* **2011**, *50*, 613–621. [CrossRef]
15. Martin, R.E. Thermal properties of bark. *For. Prod. J.* **1963**, *13*, 419–426.
16. Bauer, G.; Speck, T.; Blömer, J.; Bertling, J.; Speck, O. Insulation capability of the bark of trees with different fire adaptation. *J. Mater. Sci.* **2010**, *45*, 5950–5959. [CrossRef]
17. Tudor, E.M.; Dettendorfer, A.; Barbu, M.C.; Reh, R.; Kristak, L. Sound absorption coefficient of bark insulation panels. *Polymers* **2020**, *12*, 1012. [CrossRef] [PubMed]
18. Kain, G.; Tudor, E.M.; Barbu, M.C. Bark thermal insulation panels: An explorative study on the effects of bark species. *Polymers* **2020**, *12*, 2140. [CrossRef] [PubMed]
19. Kawasaki, T.; Kawai, S. Thermal insulation properties of wood-based sandwich panel for use as structural insulated walls and floors. *J. Wood Sci.* **2006**, *52*, 75–83. [CrossRef]
20. Latif, E.; Ciupala, M.A.; Tucker, S.; Wijeyesekera, D.C.; Newport, D.J. Hygrothermal performance of wood-hemp insulation in timber frame wall panels with and without a vapour barrier. *Build. Environ.* **2015**, *92*, 122–134. [CrossRef]
21. Ninikas, K.; Ntalos, G.; Hytiris, N.; Skarvelis, M. Thermal properties of insulation boards made of tree bark and hemp residues. *J. Sustain. Ach. Civ. Eng.* **2019**, *1*, 71–77. [CrossRef]

22. *Wood Based Panels-Determination of Modulus of Elasticity in Bending and of Bending Strength*; EN 310; Comite Europeen de Normalisation: Brussels, Belgium, 1993.
23. *Thermal Performance of Building Materials and Products—Determination of Thermal Resistance by Means of Guarded Hot Plate and Heat Flow Meter Methods*; BS EN 12667:2001; Products of high and medium thermal resistance; British Standards Institution: London, UK, 2001.
24. Koutsianitis, D.; Ninikas, K.; Mitani, A.; Ntalos, G.; Miltiadis, N.; Vasilios, A.; Taghiyari, H.R.; Papadopoulos, A.N. Thermal Transmittance, Dimensional Stability, and Mechanical Properties of a Three-Layer Laminated Wood Made from Fir and Meranti and Its Potential Application for Wood-Frame Windows. *Coatings* **2021**, *11*, 304. [CrossRef]
25. Available online: https://drive.google.com/file/d/1vgzuKkSfVhwXCjAaVTSaL5jD2SEXGvOJ/view?usp=sharing (accessed on 14 April 2021).
26. Shakel, A.; Chaudhery, M.H. *Green and Substainable Advanced Materials*; Scrivener Publishing LLC: Salem, MA, USA, 2018; Volume 1, pp. 191–212.
27. Xu, J.; Sugawara, R.; Widyorini, R.; Han, G.; Kawai, S. Manufacture and properties of low-density binderless particleboard from kenaf core. *J. Wood Sci.* **2004**, *50*, 62–67. [CrossRef]
28. Liao, R.; Xu, J.; Umemura, K. Low density sugarcane bagasse particleboard bonded with citric acid and sucrose: Effect of board density and additive content. *Bioresources* **2016**, *11*, 2174–2185. [CrossRef]
29. Brombacher, V.; Michel, F.; Niemz, P.; Volkmer, T. Untersuchungen zu Wärmeleitfähigkeit und Feuchteverhalten von Holzfaserplatten und Materialkombinationen. *Bauphysik* **2012**, *34*, 157–169. [CrossRef]
30. Kain, G.; Lienbacher, B.; Barbu, M.C.; Plank, B.; Richter, K.; Petutschnigg, A. Evaluation of relationships between particle orientation and thermal conductivity in bark insulation board by means of CT and discretemodeling. *Case Stud. Nondestruct. Test. Eval.* **2016**, *6*, 21–29. [CrossRef]
31. Arabi, M.; Faezipour, M.; Gholizadeh, H. Reducing resin content and board density without adversely affecting the mechanical properties of particleboard through controlling particle size. *J. For. Res.* **2011**, *22*, 659–664. [CrossRef]
32. Doost-hoseini, K.; Taghiyari, H.R.; Elyasi, A. Correlation between sound absorption coefficients with physical and mechanical properties of insulation boards made from sugar cane bagasse. *Compos. Part B* **2014**, *58*, 10–15. [CrossRef]
33. Taghiyari, H.R.; Elyasi, A.; Doost-hoseini, K.; Hosseinpourpia, R. Correlation between gas and liquid permeability with noise reduction coefficient in insulation boards made from sugar cane bagasse. *Bulg. J. Agric. Sci.* **2017**, *23*, 647–681.
34. Japanese Standards Association. *JIS Japanese Industrial Standard, Method for Thermal Transmission Properties of Thermal Insulations*; A1412-1994; Kain, G., Barbu, M.C., Petutschnigg, A., Eds.; Japanese Standards Association: Tokyo, Japan; Unasylva, Rome, 1994; Volume 67, pp. 67–75.
35. Zampori, L.; Dotelli, G.; Vernelli, V. Life cycle assessment of hemp cultivation and use of hemp-based thermal insulator materials. *Environ. Sci. Technol.* **2013**, *47*, 7413–7420. [CrossRef] [PubMed]

Journal of
Composites Science

MDPI

Article

Assessment of Analytical Orientation Prediction Models for Suspensions Containing Fibers and Spheres

Bastien Dietemann [1,*], Fatih Bosna [1], Harald Kruggel-Emden [2], Torsten Kraft [1] and Claas Bierwisch [1]

[1] Fraunhofer IWM, Wöhlerstraße 11, 79108 Freiburg, Germany; bosna@iwf.mavt.ethz.ch (F.B.); torsten.kraft@iwm.fraunhofer.de (T.K.); claas.bierwisch@iwm.fraunhofer.de (C.B.)

[2] Mechanical Process Engineering and Solids Processing (MVTA), TU Berlin, Ernst-Reuter-Platz 1, 10587 Berlin, Germany; kruggel-emden@tu-berlin.de

* Correspondence: bastien.dietemann@iwm.fraunhofer.de

Abstract: Analytical orientation models like the Folgar Tucker (FT) model are widely applied to predict the orientation of suspended non-spherical particles. The accuracy of these models depends on empirical model parameters. In this work, we assess how well analytical orientation models can predict the orientation of suspensions not only consisting of fibers but also of an additional second particle type in the shape of disks, which are varied in size and filling fraction. We mainly focus on the FT model, and we also compare its accuracy to more complex models like Reduced-Strain Closure model (RSC), Moldflow Rotational Diffusion model (MRD), and Anisotropic Rotary Diffusion model (ARD). In our work, we address the following questions. First, can the FT model predict the orientation of suspensions despite the additional particle phase affecting the rotation of the fibers? Second, is it possible to formulate an expression for the sole Folgar Tucker model parameter that is based on the suspension composition? Third, is there an advantage to choose more complex orientation prediction models that require the adjustment of additional model parameters?

Keywords: Folgar Tucker model; multicomponent suspensions; smoothed particle hydrodynamics

Citation: Dietemann, B.; Bosna, F.; Kruggel-Emden, H.; Kraft, T.; Bierwisch, C. Assessment of Analytical Orientation Prediction Models for Suspensions Containing Fibers and Spheres. *J. Compos. Sci.* **2021**, *5*, 107. https://doi.org/10.3390/jcs5040107

Academic Editor: Stelios K. Georgantzinos

Received: 23 March 2021
Accepted: 12 April 2021
Published: 13 April 2021

1. Introduction

1.1. Motivation

Mechanical properties of fiber-reinforced engineering materials often depend on their local orientation of fibers [1,2]. For example, specimens are reported to be stronger in the direction of fiber alignment [3], or the alignment of the fibers influences the thermal conductivity in sheet layers [4]. Consequently, there is a desire to predict the orientation for a specific process to optimize a part with respect to a specific mechanical property. The problem set up for investigation is illustrated in Figure 1, which shows an exemplary extrusion process in which a paste filled with fibers and spheres is extruded through a nozzle. The final orientation of the fibers in the extruded filament is complex to predict as it arises from an interplay of experienced hydrodynamic forces and interactions between the particles during the printing process.

Several strategies exist to predict the orientation of fibers. One obvious strategy is to perform experiments and measure the orientation of the fibers [5]. This approach is, however, tedious and time-consuming, and therefore numerical tools may facilitate collating information. Among the numerical tools, there exist two fundamental approaches working on different scales. On the one hand, Direct Numerical Simulations (DNS) can be applied on a particle scale, which means that the particles and the flow around them is fully resolved. Methods on this scale mainly differ in the choice whether the momentum of the particle phase is mapped back onto the fluid phase [6,7] (so called two-way coupling (TWC)), or not [2,8] (so called one-way coupling (OWC)), but they all provide the advantage that the orientation of particles can be measured. On this scale, it is also possible to include complex interactions such as surface chemistry [9], magnetism [10], or drying [11]. The disadvantage of these methods is that they are computationally too expensive to simulate the whole specimen being filled with particles with todays computational resources in reasonable time [12]. Therefore, on the other hand, a more pragmatic approach does not resolve the particles any more, and for this reason, it can be applied on the scale of the specimen [13,14]. This approach combines a fluid solver with an analytical prediction model, where the fluid solver predicts the velocity profile and provides this information to the analytical model that predicts the orientation. The quality of the orientation prediction thereby depends on the resolution of the fluid solver and the applicability of the prediction model for the given suspension and process conditions.

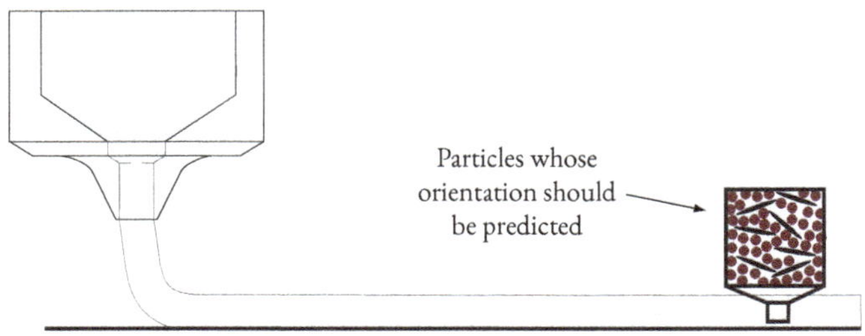

Figure 1. Illustration of an extrusion process with a zoom into the suspension showing round and elongated particles within a homogeneous matrix material.

1.2. Measuring Fiber Orientation States by Orientation Tensors

We assign a unit orientation vector $\mathbf{p}$ to each fiber that coincides with its principle axis. The ensemble average of all orientation vectors provide a concise description of the local orientation state in terms of a second order orientation tensor $\mathbf{A}$, with [15]

$$\mathbf{A} = \oint_p \mathbf{pp}\,\psi(\mathbf{p})\,\mathrm{d}\mathbf{p}, \tag{1}$$

where $\psi(\mathbf{p})$ is the orientation distribution function.

1.3. Orientation Prediction by Jeffery and Folgar Tucker

Multiple analytical prediction models exist, two famous ones are the models of Jeffery [16] and Folgar Tucker (FT) [17]. Jeffery's model became famous for its correct orientation prediction within a suspension of arbitrarily shaped particles in all kinds of flow fields and the only limitation of the model is that the concentration must be dilute (i.e., filling fraction $\phi < (1/a_\mathrm{r})^2$, with aspect ratio a_r) [17] to ensure insignificant particle interactions. The model of Jeffery is given by

$$\frac{\mathrm{d}\mathbf{A}}{\mathrm{d}t} = (\mathbf{W}\cdot\mathbf{A} - \mathbf{A}\cdot\mathbf{W}) + \lambda(\mathbf{D}\cdot\mathbf{A} + \mathbf{A}\cdot\mathbf{D} - 2\mathbf{D} : \mathbb{A}) \tag{2}$$

with a material derivate $\mathrm{d}/\mathrm{d}t$, a vorticity tensor $\mathbf{W} = \frac{1}{2}(\nabla\mathbf{v} - \nabla\mathbf{v}^T)$, a rate of strain tensor $\mathbf{D} = \frac{1}{2}(\nabla\mathbf{v} + \nabla\mathbf{v}^T)$, a velocity gradient $\nabla\mathbf{v}$, a form factor $\lambda = (a_\mathrm{r}^2 - 1)/(a_\mathrm{r}^2 + 1)$ to consider the shape of the fibers, a unit tensor $\underline{\mathbf{1}}$, and a fourth order orientation tensor $\mathbb{A}$. The fourth order tensor $\mathbb{A}$ is an unknown quantity in Equation (2) that is usually approximated as a function of the second order tensor by means of a closure approximation; in this work, the commonly applied hybrid closure approximation is used [18]. In this approximation, the elements A_{ijkl} of $\mathbb{A}$ are derived based on a combination of the linear [19] and the quadratic closure approximation [20,21] by [15,18] $A_{ijkl} = (1-f)\hat{A}_{ijkl} + f\tilde{A}_{ijkl}$, where the linear closure approximation is given by $\hat{A}_{ijkl} = c_1(\delta_{ij}\delta_{kl} + \delta_{ik}\delta_{jl} + \delta_{il}\delta_{jk}) + c_2(a_{ij}\delta_{kl} + a_{ik}\delta_{jl} + a_{il}\delta_{jk} + a_{kl}\delta_{ij} + a_{jl}\delta_{ik} + a_{jk}\delta_{il})$ with $c_1 = -1/24, c_2 = 1/6$ in two-dimensional (2D) and $c_1 = -1/35, c_2 = 1/7$ in three-dimensional (3D) formulations, δ_{ij} as the Kronecker operator and a_{ij} being the element of the second order tensor $\mathbf{A}$, the quadratic closure approximation is given by $\tilde{A}_{ijkl} = a_{ij}a_{kl}$, and a blending factor f, which in 2D is given by $f = 2\,a_{ij}a_{ji} - 1$ and in 3D by $f = 3/2\,a_{ij}a_{ji} - 1/2$. Additionally, we also apply the Invariant-based optimal fitting (IBOF) of Du Chung et al. that is exactly applied as explained in reference [22] for the 3D case and in reference [23] for the 2D case.

An orientation model that can also be applied to dense suspensions (i.e., $\phi > (1/a_\mathrm{r})^2$) is the FT model that provides an extension to the Jeffery model to consider the particles' interaction in dense suspensions. This extension introduces a phenomenological model parameter—the FT parameter C_I—to quantify the particles' interactions. The FT model belongs to the class of Isotropic Rotary Diffusion (IRD) models as the C_I parameter equally affects the orientation prediction in all dimensions. The FT model in tensor notation is given by [18]

$$\frac{\mathrm{d}\mathbf{A}}{\mathrm{d}t} = (\mathbf{W}\cdot\mathbf{A} - \mathbf{A}\cdot\mathbf{W}) + \lambda(\mathbf{D}\cdot\mathbf{A} + \mathbf{A}\cdot\mathbf{D} - 2\mathbf{D} : \mathbb{A}) + 2\,C_\mathrm{I}\,\dot{\gamma}\,(\underline{\mathbf{1}} - d\,\mathbf{A}), \quad (3)$$

with a scalar shear rate $\dot{\gamma} = \sqrt{2\,\mathbf{D} : \mathbf{D}}$. From all variables in Equation (3), the FT parameter C_I is the only parameter free to choose and all other parameter either arise from flow conditions or from the shape of the particles or the underlying closure approximation.

Many authors have determined the FT parameter, as shown in Figure 2, and often the parameter is provided as a function of the product of filling fraction of fibers ϕ_f and the aspect ratio of the fibers a_r. Despite the importance of the FT parameter, there exist contradicting trends and ideas about the FT parameter that is reported to range over at least five orders of magnitude. For example, Bay [24] and Mezher et al. [25] assume that C_I should decrease with $\phi_\mathrm{f}\,a_\mathrm{r}$; while Phan Thien et al. [26], Folgar and Tucker [17], Meyer et al. [7], and Fan et al. [27] report it to increase. A different dependency is published by Ranganathan and Advani [28], who suggest that when the model parameter is describing the particle–particle interactions, then it must depend on the local particle orientation and therefore change over time (decreasing from 0.6 to 0.05 in their work). They summarized their findings into an equation that does not depend on suspension parameter directly, but on a mean distance between the fibers and a fitting parameter that should, however, be independent of the suspension. The mean distance between the fibers cannot be mapped to $\phi_\mathrm{f}\,a_\mathrm{r}$, and therefore the shaded area in Figure 2 is a free interpretation of their statement to only visualize the order of magnitude in which they reported their FT parameter. All in all, although explanations for the variation in trends exist [29], we find a strong deviation of reported parameters among most authors, which shows that the knowledge towards an optimal FT parameter has not yet fully converged.

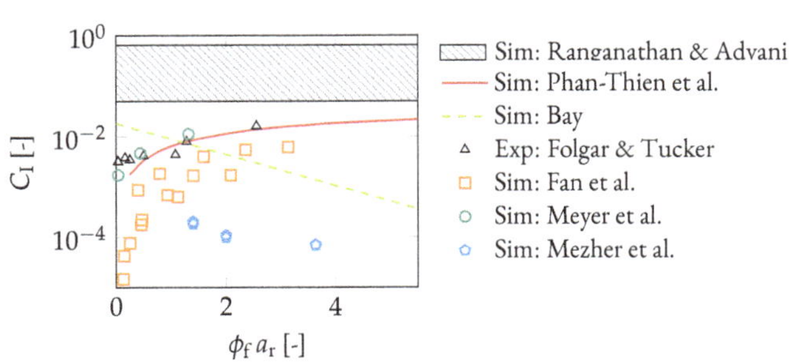

Figure 2. Comparison of C_I as published by various authors.

1.4. Further Analytical Orientation Prediction Models

Further prediction models have been introduced, most of which can be divided into two classes. One class [30,31] slows down the orientation rate to handle the phenomena that the Jeffery and FT models reach the steady-state within less strain than observed in corresponding experiments. These models have little influence on the predicted steady-state. A second class, Anisotropic Rotary Diffusion (ARD) [32], allows one to control the steady state within each dimension individually. Models that combine elements from both classes exist to fine-tune both the steady-state itself and the speed with which this steady state is reached. A detailed presentation of all existing models is beyond the scope of this manuscript, and we refer the interested reader to references [33,34] for a detailed comparison of all models or anisotropic models in particular. Table 1 shows an incomplete list of some commonly used prediction models from both classes. While each of the models is motivated by and designed for a different physical phenomena, we reduce them to their number of model parameter in this work. What all models again have in common is the non-existence of a concrete equation to compute the model parameters; instead, they have to be fitted against experimental data. Therefore, models that are more powerful might be available but the problem that we have been confronted with in the case of the FT model—not knowing the model parameter—still persists.

Table 1. Commonly used orientation prediction models sorted by the number of model parameters.

Model Name	Abbreviation	Parameters
Jeffery [16]	-	0
Folgar Tucker [17]	FT	1
Reduced Strain Closure [30]	RSC	2
improved Anisotropic Rotary Diffusion Retarding Principal Rate [31]	iARD-RPR	3
principle Anisotropic Rotary Diffusion [35]	pARD	3
Principal Anisotropic Rotary Diffusion Retarding Principal Rate [35]	pARD-RPR	3
Moldflow Rotational Diffusion [36]	MRD	4
Anisotropic Rotary Diffusion [32]	ARD	5
Anisotropic Rotary Diffusion Reduced Strain Closure [32]	ARD-RSC	6

In this work, we mainly address two research questions. First, to what degree can the FT model predict the orientation within various suspensions although the model provides only one fitting parameter? Second, should more powerful orientation prediction models be preferred? The novelty of this work is that the suspension under consideration consists of two different particle shapes, i.e., round particles and elongated particles that we relate to as disks and fibers, respectively. We varied the suspension parameter in terms of (a) the filling fraction of disks and fibers, (b) the size of the disks and (c) the aspect ratio of the fibers. The data on which this work is based on are openly accessible from reference [37].

This work is structured as follows. Section 2 briefly describes the fundamentals behind the numerical simulations of the suspension simulations that are used as a basis to fit the FT parameter. Section 3 shows the results of the individually fitted FT parameters, and it compares the accuracy of various prediction models to predict our simulated orientation. Section 4 discusses the results with the help of snapshots from selected simulations. Section 5 concludes this work.

2. Method

2.1. Suspension Simulations

Non-Brownian suspension simulations are applied to provide the orientation data against which we fitted the FT parameter. The algorithm and setup for the suspension simulations are precisely described and validated in reference [38]. A validation to an experiment showing that the model is able to predict the orientation accurately is presented in reference [39]. For the sake of brevity, we only summarize the fundamentals and refer the reader to the references for a deeper understanding of the scheme. We perform 2D simulations with Smoothed Particle Hydrodynamics (SPH) that is used to discretize the particle phase and the surrounding fluid phase with a direct coupling between both phases, which means that the local orientation of the particles has a direct influence on the local fluid phase and vice versa. The particles interact with each other by repulsion and friction forces, and the fluid is described by a Newtonian rheology model.

Some examples of the numerical setup are shown in Figure 3, in which we show the fibers in black, the disks in red and the surrounding fluid phase in blue. Figure 3 shows examples of the variation of four parameters of the present study: aspect ratio a_r of the fibers, the disk size and thereby the ratio of disk to fiber size A_d/A_f, and the filling fraction of fibers ϕ_f and disks ϕ_d.

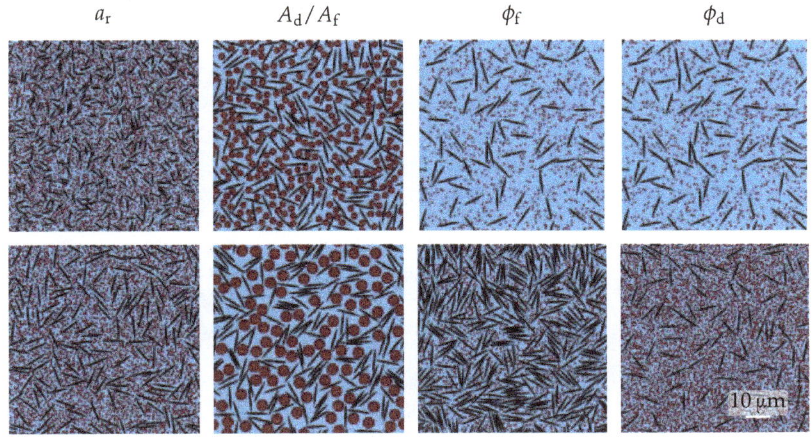

Figure 3. Micro-structures examples where in each column we show one variation of the quantity being denoted on top of the column.

Three types of deformation are applied within an Representative Volume Element (RVE) as illustrated in Figure 4 in terms of shear (subfigure a), elongation (subfigure b) or a combination of both (subfigure c). The product of $\dot{\gamma}\,t$ is used to quantify the amount of deformation.

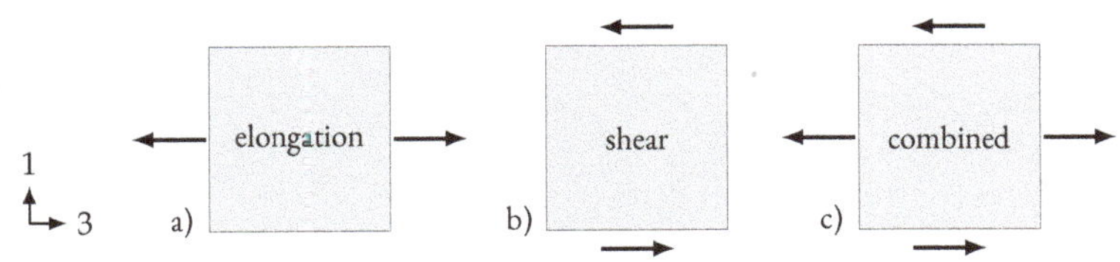

Figure 4. Schematic drawing of RVEs under deformation: (**a**) planar elongation, (**b**) simple shear, (**c**) combined shear and elongation.

2.2. Computation of FT Parameters

An FT parameter is retrieved by the following procedure. First, the orientation tensor is computed for a simulation. Second, we predict an orientation applying Equation (3) with exactly the same velocity gradient as used in the simulation and an initial orientation tensor as present in the simulation at $t = 0\,\mathrm{s}$. The FT parameter was used as a free fitting parameter to minimize the deviation between numerical simulation and FT prediction by a least mean square error method. An exemplary result is shown in Figure 5. Thereby, the FT parameter is chosen such that the whole transient evolution of the whole orientation tensor is most accurately described. The result of this procedure is therefore one FT parameter for each simulation.

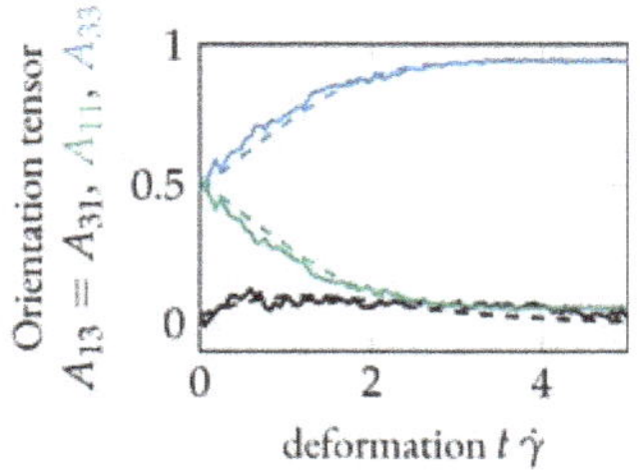

Figure 5. Exemplary Folger Tucker (FT) curves (dashed lines) being fitted to the simulation curves (solid lines).

3. Results

3.1. Influence of the FT Model Formulation

While the governing equation of the FT model is precisely described, its implementation allows a certain degree of freedom when it comes to the closure approximation (see Section 1.3). In this section, we want to study the effect of the FT model formulation on the resulting FT parameter. In particular, we want to compare the quadratic, hybrid. and IBOF closure approximation as well as the influence of whether the 2D or 3D formulation is applied. The simulations are performed in 2D, and therefore the default formulation applied in this work is the 2D formulation of the FT model in combination with the hybrid closure approximation if not stated differently.

We study the influence of the 2D and 3D formulation of the FT model on the orientation prediction. Figure 6 shows the A_{11} component of the orientation tensor of the 2D (left) and 3D (right) formulation of the FT model for the case of shear (top) and elongation (bottom), for five different FT parameters (color coded according to colorbar), and two different aspect ratios (solid and dashed line). A comparison between the left and right subfigures shows that there is a clear difference in orientation prediction between the 2D and 3D formulation that becomes more obvious with increasing C_I (see arrows). The black line in Figure 6 denotes the solution according to the Jeffery model (Equation (2)). We find the FT model with $C_I \leq 10^{-3}$ to be barely distinguishable from the Jeffery model in the 3D shear case and indistinguishable in all other cases. It should be noted that we performed our simulation in 2D, and the conclusion being drawn in this work might be different based on 3D simulations.

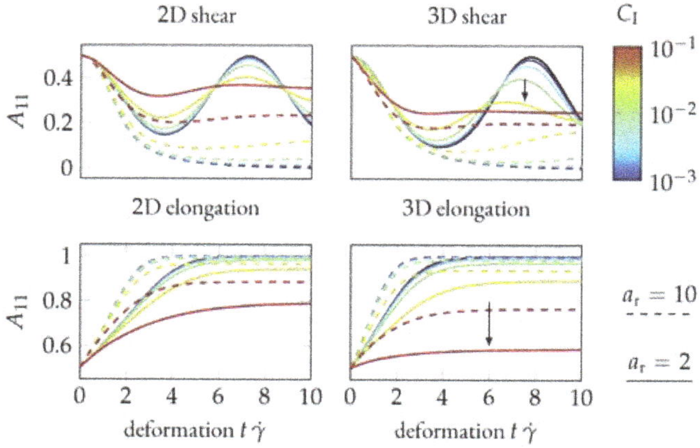

Figure 6. Influence of the dimensionality of the FT formulation for the case of shear and elongation for various FT parameters and two aspect ratios. The black line denotes the solution according to the Jeffery model.

Up to this point, we have studied the orientation prediction of the FT model purely based on the model formulation. Now we move on to the results of the particle simulation for the remaining part of this results section. From here onward, we relate to the FT parameters that have been fitted as explained in Section 2.2.

We study the influence of the FT model formulation (dimensionality and closure approximation) on the FT parameter. As we gave learned, the dimensionality and the closure approximation affect the orientation prediction, and the fitting routine towards the numerical data yields different parameters for each formulation. A histogram of FT parameters among all simulations for each FT model formulation is shown in Figure 7, in which each column corresponds to one specific closure approximation. The top row shows the histogram of parameters contrasting the effect the of dimensionality for a specific closure, and the bottom row shows a scatter plot in which each point denotes one pair of C_I from the 2D and 3D formulation to show the correlation between both.

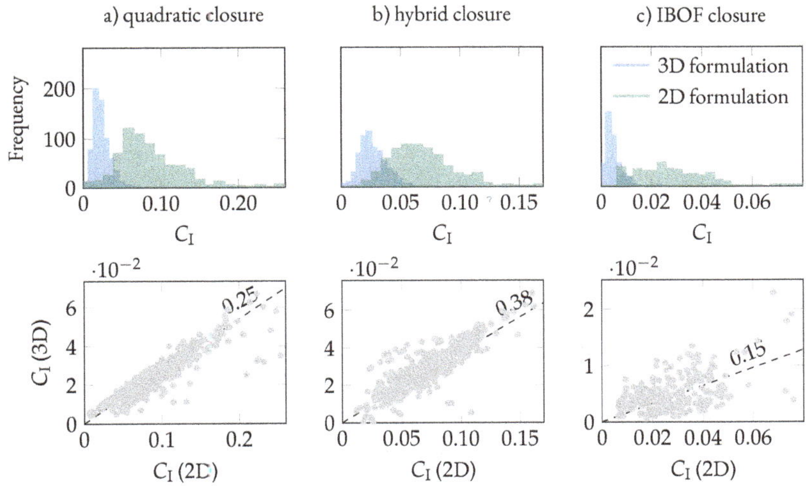

Figure 7. Distribution of Folgar Tucker parameters among all simulations for the 2D and 3D formulation of the Folgar Tucker model.

We see that the 3D formulation in general yields significantly smaller FT parameters than the 2D formulation. We also find the IBOF closure to yield smaller C_I than the hybrid closure, whose C_I parameters are again smaller than for the quadratic closure. There is a correlation (Pearson correlation test) of 0.93 and 0.91 for the quadratic and hybrid closure, and only of 0.6 for the IBOF closure. The reason for this difference in correlation among the closures in discussed in Section 4.1.

3.2. *Influence of the Flow Type*

The FT model predicts the orientation for a flow of all kinds of velocity gradients. In our simulations, we applied three different kinds of velocity gradient (shear, elongation, or a combination of both), and in this section, we study to what degree the FT parameters vary with flow type. Figure 8 shows a histogram of FT parameters originating from all suspensions from one specific flow type. We find all flow types to yield similar distribution of FT parameters with respect to peak position ($C_I = 0.06$) as well as minimum and maximum FT parameters.

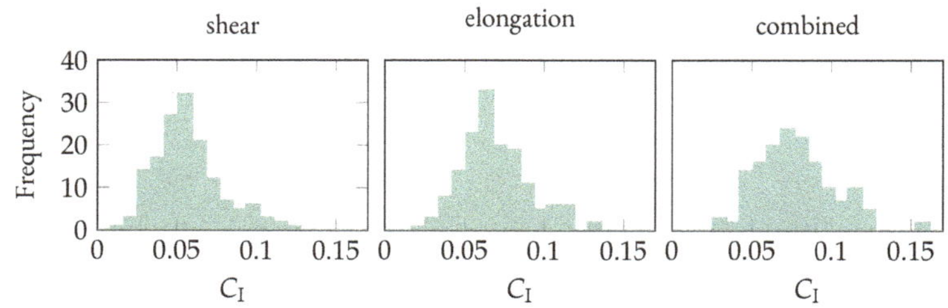

Figure 8. Histogram of FT parameters under consideration of the governing flow type.

3.3. Influence of Disk Size

The main purpose of this paper is to assess to what degree the FT model can predict the orientation in suspensions including a secondary disk phase. We performed simulations with varying fiber aspect ratio, disk size, and filling fraction of fibers and disks, and the resulting FT parameters are shown in Figure 9. The abscissa denote the area ratio A_d/A_f of one single disk to one single fiber, while all disks and all fibers have the same size in one suspension. As such, the abscissa describes the transition from very small to large disks in relation to the fibers. Every simulation yields one single FT parameter that is represented by a point where the size and the color of the point describes the suspension composition: the size of the point represents the aspect ratio of the fibers where small points represent small and large points large aspect ratios, and the color of the point shows the underlying filling fraction of disks in the suspension where a dark green relates to a high filling fraction of disks.

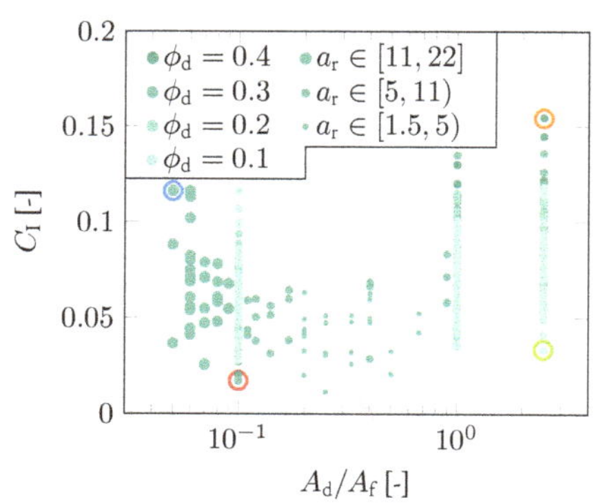

Figure 9. C_I over area ratio of disks to fibers A_d/A_f.

The point cloud originates from three classes: first, the region with $A_d/A_f \leq 0.2$ describes suspensions with long fibers and disks small in size compared to the fibers; second, for $0.2 < A_d/A_f < 1.0$, the suspensions are composed of short fibers and disks that are still smaller than the fibers; third, for $0.8 \leq A_d/A_f$, we have suspensions with long fibers and disks of equal or larger size. The FT parameters are generally widely distributed among all suspensions and classes. The highest FT parameters are generally found to be highest in the third region (large disks), followed by the first region (long fibers), and the smallest maximum FT parameter is found in the second region (small fibers). The majority of points in Figure 9 originate from a suspension with filling fraction of fibers $\phi_f = 0.2$ and disks $\phi_d = 0.3$, and both filling fractions have only been varied for $A_d/A_f \in \{0.1, 1, 2.5\}$; hence, lighter green colors appear only at these positions. For $A_d/A_f = 0.1$, the highest FT parameter is found among the suspensions with the highest amount of long fibers (please note that the coloring of the point describes the filling fraction of disks and not of fibers, and the filling fraction of fibers cannot be extracted from the figure) and is lowest for the suspension with a low amount of long fibers. For $A_d/A_f = 1$, and $A_d/A_f = 2.5$, contrariwise, the FT parameter is highest for those suspensions with many large disks and lowest for those suspensions with few large disks. Four FT parameters in Figure 9 are marked by a colored ring, and the underlying simulation yielding this parameter is discussed in Section 4.3.

3.4. Comparison of Existing Data

We compare our FT parameters in Figure 10 with those from the literature, which are shown in Figure 2. Here, we exceptionally use the parameters obtained from the 3D formulation of the FT model, as this is the formulation applied by all authors. The data being shown in Figure 2 include only those suspension simulations with no or only small disks. The data themselves are found in the same order of magnitude as those published by Phan-Thien et al., Folgar and Tucker, and Bay. A detailed discussion of the comparison with other authors is provided in Section 4.4.

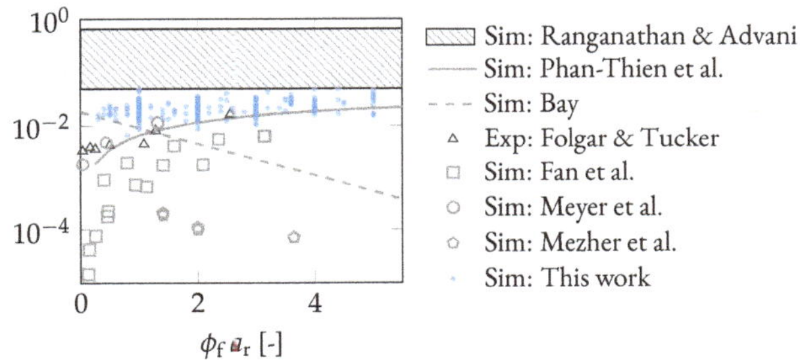

Figure 10. Literature data of Figure 2 shown in gray with an additional blue layer showing the FT parameters of this work.

3.5. Accuracy of FT Model in Contrast to Other Analytical Orientation Prediction Models

Various analytical prediction models exis,t and four of them are compared to the FT model in this section with regard to their accuracy to predict the orientation provided by the simulation. The first model, the Jeffery model, is a special case because it does not have any fitting parameters, nor is it designed to be applied for dense suspensions. The second model is the FT model with only one fitting parameter. The third model, the RSC model, is similar to the FT model, but it provides an additional parameter to fine-tune the speed with which the steady-state orientation is reached. The fourth model, the MRD model, has a structure similar to the FT model, but it involves three additional parameters to fine-tune the orientation alignment rate with the current orientation tensor. The fifth model, the ARD model, applies an additional anisotropic rotary tensor, which is given as a function of five parameters. We chose these models as there is a parameter set for which all models restore the FT model, or Jeffery model, respectively, and should be able to outperform this model. At this point, each model is characterized only by the number of model parameters.

In Figure 11, we assess the accuracy of all models to predict the orientation of suspensions, which means we compare the deviation between the analytical prediction and the simulation. Two cases are distinguished: the blue bars describe the case in which the analytical model has been fitted to every of our simulations individually with all fitting parameters being available, and the orange bars describe the case in which we have applied a fixed set of model parameters to predict the orientation among all suspensions independent of their composition. The parameters chosen for the second case are thereby extracted from the parameters obtained by the fitting routine. We always chose the parameter which in a histogram describes the peak position, as this value is expected to best describe all suspensions (for example, for FT, we used $C_I = 0.06$ as described in Section 3.2). The orange bars represent the case in which a perfect model parameter is unknown and has been retrieved by a fitting routine towards available data. Studying only the blue bars, we see that the total deviation reduces with increasing number of model parameters, and the largest reduction in deviation is obtained from moving from the Jeffery model to the FT model. When we focus on the orange bars with fixed parameters, we see that the deviation among all models with fitting parameters is similar.

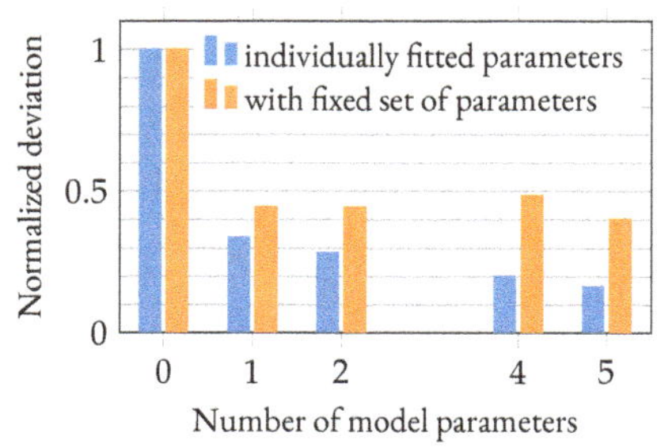

Figure 11. Total deviation between the numerical data and the Jeffery model (no parameter), the FT model (1 parameter), the Reduced Strain Closure (RSC) model (2 parameters), the MOldflow Rotational Diffusion (MRD) model (4 parameter) and the Anisotropic Rotary Diffusion (ARD) model (5 parameters). The total deviation is normalized by the deviation of the Jeffery model.

4. Discussion

The FT model is a widely applied analytical orientation prediction model. The equation itself is designed in such a way that all surrounding process conditions, i.e., the local velocity field, are fully considered, and what remains is only the FT parameter that, in theory, should only depend on the suspension parameters. From a mathematical point of view, the FT parameter regulates a dissipation term, which means that increasing the FT parameter either lowers the final steady-state orientation in elongation flow or it lowers the magnitude of the oscillation under shear flow. The value of the FT parameter itself has no direct meaning, but an increase in the FT parameter can be interpreted as showing that the particles are more hindered in their rotation. This link should be kept in mind for the remainder of this discussion, as we will use this analogy to explain why the FT parameter increases or decreases in some simulations.

4.1. Formulation

While the FT model is precisely formulated, its implementation adds a degree of freedom towards the choice of the closure approximation. In this work, we started with the influence of the choice of this closure approximation and, additionally, if the 2D or 3D formulation of the model is applied. We found that both the formulation and closure approximation significantly affect the FT parameter. For the same closure approximation, the 3D model yields smaller FT parameters than the 2D. Basically, this means that the same FT parameter has a stronger influence on the orientation prediction in 3D than in 2D. The reason for this behavior is unintuitive but understandable when we have a closer look at the term $2\,C_{\mathrm{I}}\,\dot{\gamma}\,(\underline{\mathbf{1}} - d\,\mathbf{A})$ that has been added to the Jeffery model. Let us consider an orientation tensor that describes the steady state in which the A_{11} entry of the orientation tensor is approaching unity. In this example, the term $\mathbf{1} - d\,\mathbf{A}$ (looking only at A_{11}) yields $1 - 3 \times 1 = -2$ or $1 - 2 \times 1 = -1$ in 3D, or 2D, respectively. This means that the material derivative is larger in 3D than in 2D, and, therefore, the corresponding FT parameter must be chosen smaller in 3D than in 2D to give the same orientation prediction. Hence, the FT parameter has a stronger influence on the orientation prediction in 3D than in 2D, which simultaneously means that the 2D model needs significantly larger values to adjust the orientation curves. Even more interesting than the influence of the 2D vs. 3D formalism is the influence of the closure approximation. We find FT parameters with the quadratic and hybrid closure to be significantly larger than when applying the IBOF closure. This behavior that the FT parameters with the IBOF closure are significantly smaller is unsurprising if we follow the strategy of Du Cheng et al. [22] to formulate this closure approximation. They provide a set of 63 fitting parameters, all of which have been fitted to numerical data that have been created assuming $C_I = 0$. As such, the closure is designed in such a way that the hydrodynamic part (i.e., the Jeffery part without the FT extension) is already capable of predicting a reasonable orientation within the suspension. Therefore, the magnitude with which the FT extension has to correct the prediction by the hydrodynamic part is obviously smaller and this correlates to a small FT parameter. There exist other closure methods (e.g., NAT [23], OWR [40], OWR3 [41]) that have not been tested in this but in other works (e.g., [34]) coming to the same conclusion that the choice of closure has a significant impact on the orientation prediction. We also find a high correlation between the C_I values for the 2D and 3D formulation for the hybrid and quadratic closure, while the correlation was found lower for the IBOF closure. This finding is expected for all three closures. The quadratic and hybrid closure apply the same equation to compute the fourth order tensor; the IBOF closure, in contrast, applies different equation for the 2D and 3D case, which differ in many details—63 model parameters vs. no model parameter and 8 equations vs. 3 equations for the 3D in contrast to the 2D case. It follows, we conclude, that a comparison of FT parameters among different authors is only meaningful when they apply the same closure approximation.

4.2. Underlying Flow Profile

A study of FT parameters among three flow types (shear, elongation, and combination) has shown that all of them yield a similar distribution. This behavior is expected as the governing flow type is already considered within the hydrodynamic part of the FT model, and the FT parameter itself should only represent the particle-particle interaction and, therefore, should explicitly not be a function of the governing flow type. Still, to the best of our knowledge, this is the first work to compare the FT parameters among different flow types, as usually simulation data are only available for shear. The confirmation of the independence of the FT parameter on the flow type is satisfying especially considering that Figure 6 shows a large influence of the flow type on the actually predicted orientation. This mean that, while the underlying flow profile has no influence on the FT parameter itself, the flow profile has a major influence on the actual result of the prediction.

4.3. Suspension Composition

The FT parameter is said to account for particle–particle interactions, and as such, we initially expected to find a clear relationship between the FT parameter and any of the suspension parameters. The main purpose of this work is to understand to what degree the FT model can predict a suspension including a second sphere phase by according adjustments to the FT parameter. Our results from Figure 9 are clearly illustrated in Figure 12 with the kind of composition that has led to the corresponding FT parameter including the approximate size of the disks, the aspect ratio of the fibers, and the filling fractions of both. Please note that Figure 12 might not show a generally valid picture of all possible parameters but only summarizes the elements of the parameter study being conducted in this work and any untested combination might contribute to FT parameters outside our result space. From a broad perspective, it seems possible to formulate an equation of state describing the whole parameter space, but then again, this equation would only provide a false impression of accuracy for three reasons. First, we have found that the exact values strongly depend on the FT formalism (dimensionality and closure). Second, our investigated parameter space is not encompassing. Third, and most important, the distribution of C_I parameters as shown in Figure 9 does not indicate that there exists a simple formulation for C_I as a function of suspension properties.

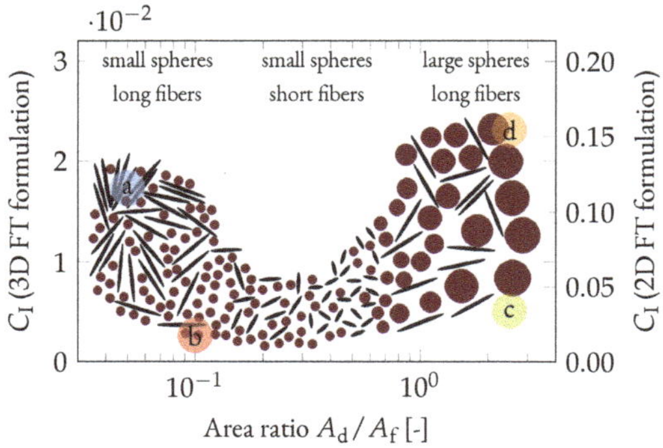

Figure 12. Generalization of Figure 9 illustrating the assumed suspension composition yielding the corresponding FT parameter. The colored points are drawn at the positions of the colored points in Figure 9.

In Figure 9, why do certain area ratios A_d/A_f yield both large and small FT parameters? To answer this question, we look onto some selected simulations yielding the FT parameters as marked by the colored points in Figure 12 to understand the mechanism dominating the suspension. Therefore, Figure 13 shows cutouts of these simulations showing some fibers that we assume show the dominant mechanism leading to the corresponding FT parameter. All sequences are oriented in such a way that the fibers in steady state should orient horizontally, which is the direction of the yellow double-ended arrow in the top left image. We discuss each simulation individually. In subfigure a, we see large fibers, three of which are highlighted by arrows, surrounded by a high number of small disks. The fibers are limited in their rotation due to the length of the fibers in combination with their large filling fraction but not because of the high number of disks. These fibers have formed a local fiber bundle that behaves like a rigid body. Obviously, this bundle rotates as one single object, and they have a different dynamic than one single fiber. The formation of these bundles for a high filling fraction of long fibers is also observed in experiments [42] and, therefore, is assumed reasonable. After some deformation, we still find the fibers orthogonal to the steady-state orientation, while other fibers obviously have managed to orient horizontally. Please remember that a large FT parameter correlates to a hindrance in rotation, and in this example, it is the combination of long fibers and a high fiber filling fraction that leads to the large FT parameter. Making a prediction when these bundles form is complex as they arise from an interplay between many local conditions that cannot be deduced from global suspension parameters. These bundles may not form in one-way coupled simulations in which the velocity gradient is continuously acting on each fiber, and there is no flow that might prevent the fiber bundles from moving as one single quantity. In subfigure b, in contrast, we again see a large number of small disks, but this time, the fibers are fewer in number and shorter in length. The rotation of the fibers is hindered neither by surrounding fibers nor by the small disks that simply flow around the fibers. It comes as no surprise that the corresponding FT parameter (red point) is smaller than in the simulation of subfigure a (blue point). In subfigure c, we have exactly the same fiber geometry and filling fraction as in b, but the disks are larger in size and smaller in number. The three fibers chosen as a reference (marked by arrow) can rotate freely without being affected by the surrounding ensemble of disks, and the FT parameter remains small. In subfigure d, we have a similar situation as in c but with a higher filling fraction of large disks. In this simulation, we see that the three fibers marked by arrows are stuck within the surrounding disk phase, and consequently the fibers cannot rotate freely, which becomes obvious if we compare the three fibers to the other surrounding fibers that are already oriented in steady-state orientation. In conclusion, we find the FT parameter to be high whenever the rotation of fibers is hindered, but the origin of this hindrance is multifaceted.

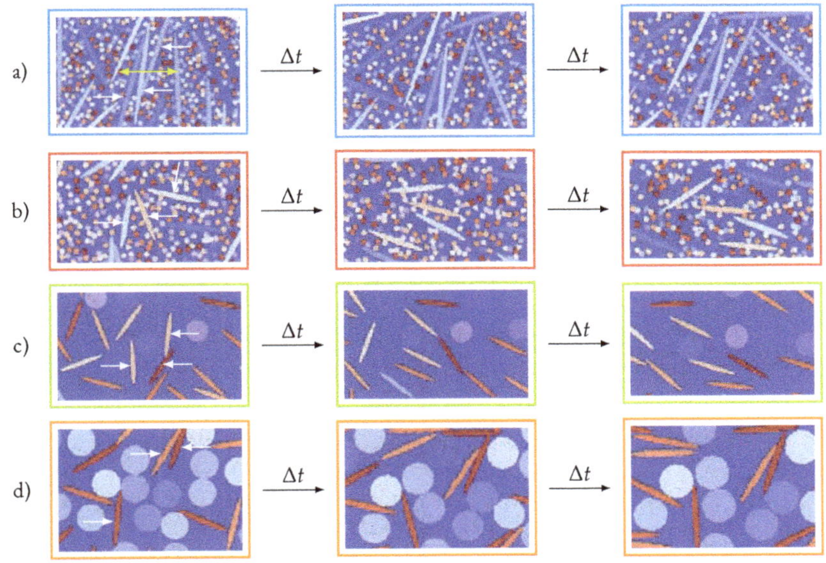

Figure 13. Cutouts of selected snapshots from the simulations marked by colored points in Figures 9 and 12. The yellow double-ended arrow points into the direction of steady-state orientation for the fibers.

4.4. Comparison with Literature

In Figure 10, we compared our FT parameters with those published in the literature and the figure includes a long list of authors because it serves two purposes. On the one hand, we validate our data, and, on the other hand, we want to point to the broad confusion on a good choice for the FT parameter. All in all, FT parameters are published within a broad window between 1×10^{-7} and 0.6 (values below 1×10^{-5}, e.g., from reference [43] are not shown in Figure 10). Comparing our parameters with those of other authors, we find that they match those from Bay for $\phi_f a_r < 1$ and those from Folgar and Tucker and Phan-Thien et al. for $\phi_f a_r \geq 1$, while there is a small positive offset to all of them. From all of the listed authors in Figure 10, the equation provided by Phan-Thien et al. is often used as a reference to compute the FT parameter by other authors [44–46], and, therefore, finding our parameters close to theirs is interpreted as validation of the parameters. We assume two possible reasons for this positive offset, both of which might act simultaneously. Phan-Thien et al. applied simulations with OWC in which shear forces are acting on all particles. In our simulations, we applied TWC for which we saw the formation of fiber bundles that act like rigid bodies with different dynamics because only the outer fibers of the fiber bundles experience the shear force and therefore the center fibers only start rotating because of the initiated rotation of the outer fibers. This bundle of particles must therefore rotate slower than individual fibers, and the FT parameters of TWC simulation should be larger than in OWC simulations. The second reason is the application of only a 2D instead of a 3D simulation, which is discussed in detail in Section 4.6. The presence of the disks is not expected to contribute to the positive offset as only simulations with small disks are included within Figure 10, and in these simulations, the disks have been shown to not hinder the rotation of the fibers. A deviation to the experimental data from Folgar and Tucker can also be explained by different measurement methods between simulation and experiment. Simulations allow using the full transient curve of all orientation tensor elements. Experiments require one to selectively measure the orientation from samples that are assumed in steady-state. This steady-state, however, does not exist in the oscillatory behavior of shear flow, which is the kind of flow most often applied. Additionally, pictures to compute the orientation tensor are often taken in 2D, which makes it difficult to measure the orientation in depths direction of the image [47].

Above our FT parameters, we find the broad window of Ranganathan and Advani that is never shown, as it cannot be visualized as a function $\phi_f\, a_r$, and, therefore, is only indicated by the shaded area in Figure 10. Ranganathan and Advani state that the FT parameter should actually change over the deformation as the particle–particle interaction in a fully oriented ensemble must be different than in a fully random ensemble. According to the authors, the parameter should decrease over one order of magnitude from the initial to the final orientation. Figure 10 includes also values below 1×10^{-4}, which are usually considered as too small [26]. We also find such small values to have insignificant influence on the prediction by the FT model. What all authors reporting $C_I \leq 1 \times 10^{-4}$ have in common is again that they apply a different formulation of the FT model, which is either not based on the orientation tensor but on an orientation distribution function like Fan et al., or is one for which they apply a different closure approximation like IBOF in the case of Mezher. We have already discussed in Section 4.1 that especially the IBOF approximation yields significantly smaller FT parameters, which explains the strong deviation between the authors applying different closure approximations. Every cited author reports that the FT model can be used to predict the orientation for their suspension, but disagreement exists among all authors on the actual value for the FT parameter. We conclude that the FT model can predict the orientation within many suspensions (including all considered in this work), as it is the hydrodynamic that mostly affects the orientation. The model also provides a tool to fine-tune the orientation prediction for a given setup, but providing an expression for the FT parameter as a function of suspension composition parameters appears not achievable.

4.5. Accuracy of the FT Model in Comparison to Other Prediction Models

If it appears impossible to include all physical effects within one parameter, do more complex prediction models include more fitting parameters than the solution? These models could include further terms, each of which is particularly designed to account for certain effects. A long list of such parameters is possible: deformation of particles, their Young's modulus, the size distribution, the filling fraction or its local variation, the dependency on the local fiber orientation, adhesion between particles, chemical reactions, and local changes in rheology due to additives, to name just a few.

In Section 3.5, we compared the FT model with other models in terms of their applicability to predict the orientation. Usually, experiments are first performed to which the model parameters are fitted—we call this a calibrated model. This calibrated model is then used to predict the orientation in a similar suspension assuming that the parameters can also be used for the next suspension. We applied this strategy and used the numerical data to first predict an individual set of model parameter for each simulation. For the assessment of the models, we tested for the deviation between the numerical data and the calibrated prediction model. It comes as no surprise that models with more fitting parameters outperform models with fewer fitting parameters, as the fitting parameter would be useless otherwise. More interesting is the case in which we applied one fixed set of parameters and then again screened for the deviation between the data. This fixed set of parameters is chosen to be the mode value among all fitted parameters, as this value is expected to best represent the full dataset, and it should therefore comes closest to a well-calibrated model. We find that prediction models with more parameters do not significantly outperform the simple FT model with only one single fitting parameter. The RSC and MRD model have a worse performance and only the ARD model has a better performance than the FT model, while the difference in deviation among all models is insignificant in our opinion. This finding is understandable in the present context of our applied 2D simulation for three reasons. First, direct numerical simulations do not show the slow orientation kinetics seen in physical experiments, and therefore, slow orientation kinetics models (i.e., RSC) have no intrinsic benefit in the present case [34]. Second, a 2D simulation does not require models to capture the difference in the missing third direction and anisotropic diffusion models (i.e., ARD and MRD) are particularly designed for capturing these effects [34]. Third, we consider a short stiff fiber system for which the FT model is reported to yield adequate results [32], in contrast to long flexible fiber, for which the aforementioned models might provide an advantage.

Obviously, the prediction models used in this study have not been designed for this suspension including differently shaped and sized particles. As such, one could argue that the underlying dataset is not suited for some models. However, the general problem we are addressing persists. We calibrate parameter to data where the number of dates is limited by project time and financial resources. From this limited dataset, we derive a set of parameters, which is now used continuously for a large study assuming the model terms can represent all relevant physical effects. How should we know? As we have not found any published dependency of any model parameter on suspension parameters for models other than the FT model, we assume that this finding holds for current analytical orientation prediction models in general.

We find the FT model to greatly predict the orientation in all kinds of different suspensions that originate from the fact that underlying hydrodynamics have such a strong influence on the orientation. With respect to the FT parameter, there is evidence that one fixed parameter is enough, but this parameter, however, should be calibrated once (e.g., by the mechanism by Willems et al. [48]), taking care of the exact FT model formulation including the closure approximation.

4.6. Limitation of This Work

The fitted parameters in this work rely on data from 2D simulations. This introduces a deviation from the real experiment as fibers in 2D cannot avoid collision with the same degree of freedom. We have not quantified to what degree this directly affects the resulting FT parameter, but some thought experiments can be made on this topic. In 2D, especially in highly filled systems, the fibers can rotate less freely than in 3D, as they will find themselves entangled more frequently without the possibility to escape in the missing third dimension. The overall rotation is therefore reduced, and this will increase the FT parameters in a 2D simulation compared to a 3D simulation. Our FT parameter will therefore be higher than in corresponding 3D simulations. However, the main result of this paper is not an equation of state linking all suspension parameters to one single equation of state, but rather that the FT model can predict the orientation for a huge variety of different suspensions. As such, even for a fixed suspension parameter set, we found a relatively large fluctuation among the FT parameters, even though we applied only 2D simulations in which, with fibers having less degree of freedom, simulation results should have less statistical variations than in 3D. Therefore, we assume that the main finding—that the FT model can also be used for suspensions with additional disk—can be transferred to the case of additional spheres in the 3D world.

5. Conclusions

Fiber orientations from non-Brownian suspension simulations have been used to fit the model parameters of analytical prediction models like the Folgar Tucker model for a wide range of different suspension compositions. The novelty in this work is a secondary suspended particulate phase that affects the rotation of the fibers, and which is not reflected in any orientation prediction model. Still, we find the FT to predict the orientation for all suspensions to a satisfying degree despite the additional second phase. This prediction, we suggest, should be made with one fixed FT parameter and not with an equation that is over-fitted to suspension composition parameters, such as the fiber aspect ratio or filling fraction. If this parameter is taken from the literature, we recommend checking that the authors apply the same FT formulation with respect to the closure approximation.

More complex prediction models than the FT model exist, and they can predict the orientation more accurately at the expense of requiring more fitting parameters. For one specific suspension composition, we also find the more complex models to reproduce the underlying orientation data more accurately than the simple FT model. However, we also tested the case in which the model parameter are fitted once and then applied for further prediction within a broad variety of suspension compositions. Within the limitation of having applied 2D simulations, we find that the more complex models do not necessarily lead to a more accurate description compared to the FT model with only one parameter.

Different flow types (i.e., shear flow, elongation flow, or a combination) have led to the similar FT parameters, but we want to stress that the orientation predicted under these flow types greatly differs. As such we emphasize that for an accurate orientation prediction, it is more important that the velocity profile is predicted correctly, which sheds the light on details like the choice of the rheology model or boundary condition (e.g., applying no-slip boundary condition or not). This means that, if there is an unexpected mismatch between experimentally measured and numerically predicted orientation, then checking the assumption of the flow simulation (e.g., boundary condition, rheology model, temperature dependency) is more likely to dissolve the mismatch than switching to a different orientation prediction model that might only change the prediction slightly and leaves the user again with unknown model parameters. In this regard, we refer the reader to more sophisticated rheology models [49–51] that allow coupling to the local orientation states.

Author Contributions: B.D.: Conceptualization, methodology, software, validation, formal analysis, investigation, writing—original draft preparation, visualization; F.B.: Methodology, software, investigation; H.K.-E.: writing—review and editing; T.K.: resources, writing—review and editing, project administration, funding acquisition; C.B.: Conceptualization, software, resources, writing—review and editing, supervision. All authors have read and agreed to the published version of the manuscript.

Funding: This research was funded by the German Research Foundation (DFG) grant number KR 1729/13-2.

Institutional Review Board Statement: Not applicable.

Informed Consent Statement: Not applicable.

Data Availability Statement: The data presented in this study are openly available from Fraunhofer Fordatis at http://dx.doi.org/10.24406/fordatis/117 (uploaded and accessed 21 January 2021).

Conflicts of Interest: The authors declare no conflict of interest.

Abbreviations

The following abbreviations are used in this manuscript:

2D	two-dimensional
3D	three-dimensional
ARD	Anisotropic Rotary Diffusion
ARD-RSC	Anisotropic Rotary Diffusion Method with Reduced Strain Closure
DNS	Direct Numerical Simulations
FT	Folgar Tucker
iARD	improved Anisotropic Rotary Diffusion
IBOF	Invariant-based optimal fitting
IRD	Isotropic Rotary Diffusion
MRD	Moldflow Rotational Diffusion
NAT	Natural Closure Approximiation
OWR	Orthotropic Closure Approximation
OWC	one-way coupling
pARD	principle Anisotropic Rotary Diffusion
RPR	Retarding Principal Rate
RSC	Reduced Strain Closure
RVE	Representative Volume Element
SPH	Smoothed Particle Hydrodynamics
TWC	two-way coupling

References

1. Wang, Z.; Smith, D.E. Rheology Effects on Predicted Fiber Orientation and Elastic Properties in Large Scale Polymer Composite Additive Manufacturing. *J. Compos. Sci.* **2018**, *2*, 10. [CrossRef]
2. Simon, S.A.; Bechara Senior, A.; Osswald, T. Experimental Validation of a Direct Fiber Model for Orientation Prediction. *J. Compos. Sci.* **2020**, *4*, 59. [CrossRef]
3. Fu, S.Y.; Hu, X.; Yue, C.Y. Effects of Fiber Length and Orientation Distribution On The Mechanical Properties of Short-Fiber-Reinforced Polymers. *J. Soc. Mater. Sci. Jpn.* **1999**, *48*, 74–83. [CrossRef]
4. Terao, T.; Zhi, C.; Bando, Y.; Mitome, M.; Tang, C.; Golberg, D. Alignment of Boron Nitride Nanotubes in Polymeric Composite Films for Thermal Conductivity Improvement. *J. Phys. Chem. C* **2010**, *114*, 4340–4344. [CrossRef]
5. Yang, D.; Zhang, H.; Wu, J.; McCarthy, E.D. Fibre flow and void formation in 3D printing of short-fibre reinforced thermoplastic composites: An experimental benchmark exercise. *Addit. Manuf.* **2020**, 101686. [CrossRef]
6. Hashemi, M.R.; Fatehi, R.; Manzari, M.T. SPH simulation of interacting solid bodies suspended in a shear flow of an Oldroyd-B fluid. *J. Non–Newton. Fluid Mech.* **2011**, *166*, 1239–1252. [CrossRef]
7. Meyer, N.; Saburow, O.; Hohberg, M.; Hrymak, A.N.; Henning, F.; Kärger, L. Parameter Identification of Fiber Orientation Models Based on Direct Fiber Simulation with Smoothed Particle Hydrodynamics. *J. Compos. Sci.* **2020**, *4*, 77. [CrossRef]
8. Do-Quang, M.; Amberg, G.; Brethouwer, G.; Johansson, A.V. Simulation of finite-size fibers in turbulent channel flows. *Phys. Rev. E Stat. Nonlinear Soft Matter Phys.* **2014**, *89*. [CrossRef]
9. Lu, J.; Das, S.; Peters, E.; Kuipers, J. Direct numerical simulation of fluid flow and mass transfer in dense fluid-particle systems with surface reactions. *Chem. Eng. Sci.* **2018**, *176*, 1–18. [CrossRef]
10. Zhang, J.; Wang, C. Numerical Study of Lateral Migration of Elliptical Magnetic Microparticles in Microchannels in Uniform Magnetic Fields. *Magnetochemistry* **2018**, *4*, 16. [CrossRef]
11. Hwang, H.; Son, G. Direct numerical simulation of 3D particle motion in an evaporating liquid film. *J. Mech. Sci. Technol.* **2016**, *30*, 3929–3934. [CrossRef]
12. Polfer, P.; Kraft, T.; Bierwisch, C. Suspension modeling using smoothed particle hydrodynamics: Accuracy of the viscosity formulation and the suspended body dynamics. *Appl. Math. Model.* **2016**, *40*, 2606–2618. [CrossRef]
13. Gudžulić, V.; Dang, T.S.; Meschke, G. Computational modeling of fiber flow during casting of fresh concrete. *Comput. Mech.* **2018**, *31*, 751. [CrossRef]
14. Li, M.; Zhang, Y.; Zhang, S.; Hou, B.; Zhou, H. Experimental investigation and modeling study of the fiber orientation behavior. *Eng. Comput.* **2020**, *367*. [CrossRef]
15. Advani, S.G.; Tucker, C.L. The Use of Tensors to Describe and Predict Fiber Orientation in Short Fiber Composites. *J. Rheol.* **1987**, *31*, 751–784. [CrossRef]
16. Jeffery, G.B. The Motion of Ellipsoidal Particles Immersed in a Viscous Fluid. *Proc. R. Soc. A: Math. Phys. Eng. Sci.* **1922**, *102*, 161–179. [CrossRef]
17. Folgar, F.; Tucker, C.L. Orientation Behavior of Fibers in Concentrated Suspensions. *J. Reinf. Plast. Compos.* **1984**, *3*, 98–119. [CrossRef]
18. Advani, S.G.; Tucker, C.L. A numerical simulation of short fiber orientation in compression molding. *Polym. Compos.* **1990**, *11*, 164–173. [CrossRef]
19. Hand, G.L. A theory of anisotropic fluids. *J. Fluid Mech.* **1962**, *13*, 33–46. [CrossRef]
20. Doi, M. Molecular dynamics and rheological properties of concentrated solutions of rodlike polymers in isotropic and liquid crystalline phases. *J. Polym. Sci. Polym. Phys. Ed.* **1981**, *19*, 229–243. [CrossRef]
21. Lipscomb, G.G.; Denn, M.M.; Hur, D.U.; Boger, D.V. The flow of fiber suspensions in complex geometries. *J. Non–Newton. Fluid Mech.* **1988**, *26*, 297–325. [CrossRef]
22. Du Chung, H.; Kwon, T.H. Invariant-based optimal fitting closure approximation for the numerical prediction of flow-induced fiber orientation. *J. Rheol.* **2002**, *46*, 169–194. [CrossRef]
23. Verleye, V.; Couniot, A.; Dupret, F. Numerical prediction of fibre orientation in complex injection-moulded parts. *Eng. Sci.* **1994**, *4*. [CrossRef]
24. Bay, R.S. Fiber Orientation in Injection-Molded Composites: A Comparison of Theory and Experiment. Ph.D. Thesis, University of Illinois, Champaign, IL, USA, 1991.
25. Mezher, R.; Perez, M.; Scheuer, A.; Abisset-Chavanne, E.; Chinesta, F.; Keunings, R. Analysis of the Folgar & Tucker model for concentrated fibre suspensions in unconfined and confined shear flows via direct numerical simulation. *Compos. Part A Appl. Sci. Manuf.* **2016**, *91*, 388–397. [CrossRef]
26. Phan-Thien, N.; Fan, X.J.; Tanner, R.I.; Zheng, R. Folgar–Tucker constant for a fibre suspension in a Newtonian fluid. *J. Non–Newton. Fluid Mech.* **2002**, *103*, 251–260. [CrossRef]
27. Fan, X.; Phan-Thien, N.; Zheng, R. A direct simulation of fibre suspensions. *J. Non–Newton. Fluid Mech.* **1998**, *74*, 113–135. [CrossRef]
28. Ranganathan, S.; Advani, S.G. Fiber–fiber interactions in homogeneous flows of nondilute suspensions. *J. Rheol.* **1991**, *35*, 1499–1522. [CrossRef]
29. Advani, S.G. *Flow and Rheology in Polymer Composites Manufacturing: Processing Short-Fiber Systems*, 1st ed.; Composite Materials Series; Elsevier Science: Amsterdam, The Netherlands, 1994; Volume 10.

30. Wang, J.; O'Gara, J.F.; Tucker, C.L. An objective model for slow orientation kinetics in concentrated fiber suspensions: Theory and rheological evidence. *J. Rheol.* **2008**, *52*, 1179–1200. [CrossRef]
31. Tseng, H.C.; Chang, R.Y.; Hsu, C.H. Phenomenological improvements to predictive models of fiber orientation in concentrated suspensions. *J. Rheol.* **2013**, *57*, 1597–1631. [CrossRef]
32. Phelps, J.H.; Tucker, C.L. An anisotropic rotary diffusion model for fiber orientation in short- and long-fiber thermoplastics. *J. Non-Newton. Fluid Mech.* **2009**, *156*, 165–176. [CrossRef]
33. Tseng, H.C.; Chang, R.Y.; Hsu, C.H. Comparison of recent fiber orientation models in injection molding simulation of fiber-reinforced composites. *J. Thermoplast. Compos. Mater.* **2020**, *33*, 35–52. [CrossRef]
34. Favaloro, A.J.; Tucker, C.L. Analysis of anisotropic rotary diffusion models for fiber orientation. *Compos. Part A Appl. Sci. Manuf.* **2019**, *126*, 105605. [CrossRef]
35. Tseng, H.C.; Chang, R.Y.; Hsu, C.H. The use of principal spatial tensor to predict anisotropic fiber orientation in concentrated fiber suspensions. *J. Rheol.* **2018**, *62*, 313–320. [CrossRef]
36. Bakharev, A., Ed. *Using New Anisotropic Rotational Diffusion Model to Improve Prediction of Short Fibers in Thermoplastic Injection Molding;* Society of Plastics Engineers: Lubbock, TX, USA, 2018.
37. Dietemann, B.; Bosna, F.; Kraft, T.; Kruggel-Emden, H.; Bierwisch, C. Folgar Tucker constants fitted against simulations of various suspensions under shear and/or elongation. *Fraunhofer Fordatis* **2021**. [CrossRef]
38. Dietemann, B.; Bosna, F.; Lorenz, M.; Travitzky, N.; Kruggel-Emden, H.; Kraft, T.; Bierwisch, C. Modeling Robocasting with Smoothed Particle Hydrodynamics: Printing gap-spanning filaments. *Addit. Manuf.* **2020**, *36*, 101488. [CrossRef]
39. Dietemann, B.; Bosna, F.; Lorenz, M.; Travitzky, N.; Kruggel-Emden, H.; Kraft, T.; Bierwisch, C. Numerical study of texture in material extrusion: Orientation in a multicomponent system of spheres and ellipsoids. *J. Non-Newton. Fluid Mech.* **2021**, *291*, 104532. [CrossRef]
40. Cintra, J.S.; Tucker, C.L. Orthotropic closure approximations for flow-induced fiber orientation. *J. Rheol.* **1995**, *39*, 1095–1122. [CrossRef]
41. Du Chung, H.; Kwon, T.H. Improved model of orthotropic closure approximation for flow induced fiber orientation. *Polym. Compos.* **2001**, *22*, 636–649. [CrossRef]
42. Takano, M. *Viscosity Effect on Flow Orientation of Short Fibers;* Monsanto Research Corporation: Springfield, MO, USA, 1973.
43. Yamane, Y.; Kaneda, Y.; Dio, M. Numerical simulation of semi-dilute suspensions of rodlike particles in shear flow. *J. Non-Newton. Fluid Mech.* **1994**, *54*, 405–421. [CrossRef]
44. Férec, J.; Ausias, G.; Heuzey, M.C.; Carreau, P.J. Modeling fiber interactions in semiconcentrated fiber suspensions. *J. Rheol.* **2009**, *53*, 49–72. [CrossRef]
45. Bertevas, E.; Férec, J.; Khoo, B.C.; Ausias, G.; Phan-Thien, N. Smoothed particle hydrodynamics (SPH) modeling of fiber orientation in a 3D printing process. *Phys. Fluids* **2018**, *30*, 103103. [CrossRef]
46. Jezek, J.; Schulmann, K.; Paterson, S. Modified Jeffery model: Influence of particle concentration on mineral fabric in moderately concentrated suspensions. *J. Geophys. Res. Solid Earth* **2013**, *118*, 852–861. [CrossRef]
47. Cieslinski, M.J.; Wapperom, P.; Baird, D.G. Fiber orientation evolution in simple shear flow from a repeatable initial fiber orientation. *J. Non-Newton. Fluid Mech.* **2016**, *237*, 65–75. [CrossRef]
48. Willems, F.; Reitinger, P.; Bonten, C. Calibration of Fiber Orientation Simulations for LFT—A New Approach. *J. Compos. Sci.* **2020**, *4*, 163. [CrossRef]
49. Favaloro, A.J.; Tseng, H.C.; Pipes, R.B. A new anisotropic viscous constitutive model for composites molding simulation. *Compos. Part A Appl. Sci. Manuf.* **2018**, *115*, 112–122. [CrossRef]
50. Tseng, H.C.; Favaloro, A.J. The use of informed isotropic constitutive equation to simulate anisotropic rheological behaviors in fiber suspensions. *J. Rheol.* **2019**, *63*, 263–274. [CrossRef]
51. Li, T.; Luyé, J.F. Flow-Fiber Coupled Viscosity in Injection Molding Simulations of Short Fiber Reinforced Thermoplastics. *Int. Polym. Process.* **2019**, *34*, 158–171. [CrossRef]

Journal of
Composites Science

MDPI

Article

Multiscale Characterization of E-Glass/Epoxy Composite Exposed to Extreme Environmental Conditions

George Youssef *, Scott Newacheck, Nha Uyen Huynh and Carlos Gamez

Experimental Mechanics Laboratory, Mechanical Engineering Department, San Diego State University, 5500 Campanile Drive, San Diego, CA 92182, USA; snewache@gmail.com (S.N.); nhuynh2@sdsu.edu (N.U.H.); cgamezvillegas@sdsu.edu (C.G.)

* Correspondence: gyoussef@sdsu.edu; Tel.: +1-619-594-6649

Abstract: Fiber-reinforced polymer matrix composites continue to attract scientific and industrial interest since they offer superior strength-, stiffness-, and toughness-to-weight ratios. The research herein characterizes two sets of E-Glass/Epoxy composite skins: stressed and unstressed. The stressed samples were previously installed in an underground power distribution vault and were exposed to fire while the unstressed composite skins were newly fabricated and never-deployed samples. The mechanical, morphological, and elemental composition of the samples were methodically studied using a dynamic mechanical analyzer, a scanning electron microscope (SEM), and an x-ray diffractometer, respectively. Sandwich composite panels consisting of E-glass/Epoxy skin and balsa wood core were originally received, and the balsa wood was removed before any further investigations. Skin-only specimens with dimensions of ~12.5 mm wide, ~70 mm long, and ~6 mm thick were tested in a Dynamic Mechanical Analyzer in a dual-cantilever beam configuration at 5 Hz and 10 Hz from room temperature to 210 °C. Micrographic analysis using the SEM indicated a slight change in morphology due to the fire event but confirmed the effectiveness of the fire-retardant agents in quickly suppressing the fire. Accompanying Fourier transform infrared and energy dispersive X-ray spectroscopy studies corroborated the mechanical and morphological results. Finally, X-ray diffraction showed that the fire event consumed the surface level fire-retardant and the structural attributes of the E-Glass/Epoxy remained mainly intact. The results suggest the panels can continue field deployment, even after short fire incident.

Keywords: glass/epoxy composite; fire-hazard; fiber-reinforced polymer composite; multiscale characterization

Citation: Youssef, G.; Newacheck, S.; Huynh, N.U.; Gamez, C. Multiscale Characterization of E-Glass/Epoxy Composite Exposed to Extreme Environmental Conditions. *J. Compos. Sci.* **2021**, *5*, 80. https://doi.org/10.3390/jcs5030080

Academic Editor: Francesco Tornabene

Received: 16 February 2021
Accepted: 10 March 2021
Published: 12 March 2021

1. Introduction and Background

The general class of composite materials offers a multitude of desirable attributes that can be readily tuned on-demand during the design analysis phase by adjusting the type and ratio of the constituents. Composite materials encompass a wide range of material types, including ceramics, metals, and polymers. Polymers can be used interchangeably as the reinforcement and the matrix phases, resulting in polymer-, ceramic-, and metal-matrix subclasses of materials. Such broadness of choices and the ability to develop novel subclasses based on characteristically different design requirements have motivated the integration into a plethora of applications spanning from aerospace and automotive to biomedical and construction industries. From a mechanical behavior perspective, the specific strength (strength-to-weight-ratio) and specific stiffness (modulus-to-weight ratio) are eagerly sought for many practical applications, including the positive influence on survivability and longevity of composite-based parts, components, and structures.

Without discounting the advances in other subclasses of composite materials, fiber-reinforced polymer matrix composite (PMCs) is the most prolific subclass of these types of materials. PMCs consist of a reinforcement phase made of continuous or discontinuous fibers (e.g., carbon, glass, Kevlar®, or natural materials) and a polymer matrix (e.g., thermosets, thermoplastic, or even biodegradable). Analogous to the major class of composite materials, PMCs offer superior strength-to-weight, stiffness-to-weight, and toughness-to-weight ratios, compared to their single-phase constituents. The construction industry, among others, found refuge in composite materials to combat several challenges, such as (1) installation time and cost since composite technologies could produce near-net-shape parts [1] and (2) the deterioration of infrastructures due to its potential field life longevity [2–5]. Infrastructures and structures are continually enduring harsh and demanding loading, operating, and environmental conditions that are somewhat distinctive from other industries [6]. Due to the cost and access constraints, glass fibers are common used in construction applications, including installation and drywall tapes, to name a few examples. In addition, reinforcement using natural materials continues to take center stage in scientific and engineering communities. However, most composite-based deployed infrastructures are made using glass fibers, which exhibit remarkable strength in simple tension but suffer under shear and bending loading scenarios [5,7,8]. Therefore, the polymer matrix plays a significant role in load-bearing and non-bearing applications, protecting the brittle glass fibers in non-axial loading scenarios against shearing stresses that may result in catastrophic failure. Researchers, e.g., [9,10], investigated different polymer matrix materials to encase glass fibers, noting that vinyl-ester trumps polyester in mechanical and thermal performances. Hence, the focus of this study is on glass fiber in a modified vinyl-ester polymer matrix. The latter was modified by adding a fire-retardant agent, given the potential of fire in the specific application domain, as discussed next.

A novel application of PMCs in the construction industry is replacing aging underground power distribution vaults made of reinforced concrete with vaults made of modular composite panels constructed of glass fibers and vinyl-ester polymer matrix. The modular composite panels facilitate an easy and quick installation resulting in reduced traffic interruptions, making them practically adventurous. On the other hand, the replacement of reinforced concrete vaults requires excavating aged structures and surrounding roads, rerouting incoming traffic and interrupting the local community [11]. In addition to the apparent harsh loading and operating conditions, e.g., humidity, temperature, and static and dynamic loading, the vaults may experience fire events due to power line explosions from arcing faults or power surges. The fire hazard, in turn, results in a localized increase in the surface temperature of the materials, and if the fire is not quickly extinguished, the severity of this loading condition may lead to a catastrophic failure of the vault structure irrespective of the construction material [11–14]. In the case of composite-based vaults, the good thermal insulation of polymers and the fusion of the matrix with fire-retardant reagents offer clear advantages compared to the traditional reinforced concrete technology.

The discussion above alludes to two crucial and potential loading scenarios threatening the performance of the composite panels used in modular replacement of underground power distribution vaults, namely, long-term static loading and hazardous fire events. Recently, the dynamic properties of field-deployed E-Glass/Epoxy composite panels were investigated and reported compared to as-fabricated panels [11]. The dynamic mechanical properties were quantified using a dynamic mechanical analyzer given its broad capability in elucidating the interdependence of properties of polymers and composites on the individual or combined effect of temperature, frequency, environmental conditions [15–17] as well as reporting several thermomechanical transitions [15,17,18]. It also allows the characterization of bulk properties that directly affect the material performance. It was reported that the storage modulus of the fire-exposed E-Glass/Epoxy panels was lower than their unloaded, as-fabricated panels, while the complex modulus was shown to be unchanged within the experimental error [11]. Most of the reported changes in the dynamic properties after the fire event were attributed to change in the molecular relaxation mechanisms of the polymer matrix phase, leaving the reinforcement phase unaffected by these operating conditions. The effectiveness of the fire-retardant properties of the composite panels was demonstrated by mitigating the impact of fire in the absence of dust and oil surface contamination. Additionally, the flexural creep response of the same composite skins was characterized and elucidated as a function of high static stress, orders of magnitude higher than those experienced by the panels in deployment, and different strain rates [5]. The mechanical properties of the E-Glass/Epoxy used in the production of modular composite panels for underground vault application were found to be generally insensitive to loading rate but resulted in permanent damage at a relatively low strain level. Alternatively, E-Glass/Vinyl-ester PMCs were reported to be mildly sensitive to the creep strain [5]. A major shortcoming of these previous investigations is the lack of analytical chemistry and morphological characterizations to substantiate and corroborate the reported results, especially in fire events.

Compared to its predecessors, the distinguishing aspect of the current study is the multiscale investigation of composite panels previously deployed in real-life, harsh operating, and environmental conditions, including a fire event stemming from a power line explosion. The objective of the research leading to this paper was to characterize E-Glass/Epoxy composite skins previously installed in an underground power distribution vault after a fire incident. The mechanical, morphological, and elemental compositions of the samples were methodically studied using a dynamic mechanical analyzer, scanning electron microscope, and x-ray diffractometer, respectively. The novelty of the current study lies in the comprehensive chemical analysis to supplement the mechanical characterization to pinpoint the source of performance degradation.

2. Materials and Methods

The samples used in this experimental investigation were extracted from sandwich composite panels consisting of E-Glass/Epoxy skin and balsa wood cores received from Armorcast Products Company, used for refitting underground utility vaults. The panels were manufactured using a hand layup process and cured under high pressure in a heated hydraulic press. The sandwich panels consisted of two 6.35 mm thick skin layers made of E-Glass/Epoxy and 140 mm thick balsa wood core. The E-Glass/Epoxy skins are quadriaxial (0°/45°/90°/−45°) stitched E-glass fabric (VectorPly, E-QXM 3705, Phenix City, AL, USA) soaked with an epoxy resin of brominated bisphenol A Epoxy vinyl-ester resin dissolved in styrene (AOC, Vipel K022, Collierville, TN, USA) with the addition of Alumina Trihydrate (R.J. Marshall Company, Southfield, MI, USA) as a fire-retardant agent.

Some of the received panels were previously exposed to harsh operating conditions, including oil and dust contamination and fire since they were in underground vault deployment before laboratory testing. The oil contamination is due to the cooling oil used in power distribution electrical transformers, while the dust is associated with the outdoor underground operating environment. The fire was associated with a power-line explosion, causing superficial burns [11]. The conditions surrounding the fire accident were not monitored by the utility provider given the lack of data logging capabilities in the vaults. Nonetheless, the extracted samples provide a unique scientific opportunity to examine the effect of real-life operating conditions on the performance of E-Glass/Epoxy PMCs. Virgin panels were also received from the same manufacturer, newly fabricated, and have never been exposed to any loading or operating conditions. The skin of the field-deployed panels and of virgin (never deployed) panels were extracted by carefully grinding the balsa wood cores. The skins were then polished using 100-grit sandpaper to eliminate any traces of the balsa wood. Skin-only specimens with dimensions of ~12.5 mm wide, ~70 mm long, and ~6 mm thick were extracted and tested in the Dynamic Mechanical Analyzer (DMA, TA Instruments: Q800, New Castle, DE, USA). DMA testing was done in a dual-cantilever beam configuration at 5 Hz and 10 Hz to measure the dynamic properties of the E-Glass/Epoxy skin materials as a function of temperature ranging from room temperature to 210 °C with temperature steps of 10 °C. The DMA testing protocol is discussed in detail in [11]. It is worth noting that the core was discarded and not included in this study since it has shown before to be indifferent to the aforementioned deployment conditions [11].

Table 1 summarizes the sample identification scheme and the corresponding extraction and testing conditions, which will be used throughout the article thereafter.

Table 1. Samples identification code and the corresponding testing and evaluation conditions.

Sample Label	Condition
Virgin	Specimens extracted from newly fabricated and never-deployed panels
Burnt	Specimens extracted for previously deployed panel in a fire event
DMA	Specimens (virgin or deployed) were tested in DMA from 32–210 °C

In addition to the mechanical testing, microscopy (SEM) and spectroscopy (XRD) characterizations were done on the virgin and burnt skin specimens before and after DMA testing. The X-ray diffraction (XRD) spectra were collected using a Philips XPert XRD (Amsterdam, The Netherlands) with a copper light source with a wavelength (λ) of 0.15405 nm at operating conditions of 45 kV and 40 mA. After cleaning and positioning each specimen individually on an amorphous glass substrate for measurement, the XRD spectral data were obtained from $2\theta = 2°$ to $70°$ at a scanning rate of $2°\ \mathrm{min}^{-1}$ to observe the structural changes in the E-Glass/Epoxy composite as a function of operating (virgin vs. deployed) and testing (before and after DMA testing) conditions. The wide range was chosen based on [19]. Additionally, scanning electron microscopy (SEM, FEI Quanta 450, Hillsboro, OR, USA) was used to elucidate the morphological and chemical changes on the specimens due to the aggressive operating and testing conditions. The non-conductive composite samples were coated with 6 nm of platinum to prevent the accumulation of electron charges on the surface before microscopy. SEM micrographs were captured at an accelerated voltage of 30 kV under a high vacuum, where the effective focal distance was approximately 10 mm. Simultaneously, backscattered electrons allowed the identification of the surface chemical composition of the micrographs using energy dispersive X-ray (EDS) analysis. The X-rays emitted as the beam of electrons interacted with the surface of each composite panel were collected using a backscattered electron, high contrast detector. The signals are then measured and interpreted using the Oxford INCA EDS (Abingdon, United Kingdom) analysis software to generate an elemental mapping of the scanned area. Finally, the samples were characterized using Attenuated Total Reflection Fourier-transform infrared (ATR-FTIR) spectroscopy (Fisher Scientific Nicolet IS5, Waltham, MA, USA) to elucidate the effect of fire event on the chemical structure of the polymer/fiber composite.

3. Results and Discussion

Figure 1a shows the dynamic mechanical analyzer (DMA) results, where the thermomechanical spectra of the virgin and burnt samples are plotted at two characteristic frequencies, namely, 5 Hz and 10 Hz, comparable to urban loading scenarios due to vehicular traffic. Each thermomechanical spectrum is a plot of the storage (E') and loss (E'') moduli as a function of temperature, ranging between 32 °C and 210 °C. The storage modulus is a measure of the material stiffness or its resistance to deformation, while the loss modulus quantifies the damping or energy dissipation in the composite material. The storage and loss moduli are the components of the complex modulus of polymers. The tangent delta (tanδ) is also plotted in the inset seen in Figure 1a. Therefore, the magnitude of the complex modulus ($E^* = E' + iE''$) is also plotted in Figure 1b to demonstrate the overall effect of deployment conditions on the dynamic mechanical behavior of E-Glass/Epoxy composite skins. The complex modulus magnitude is plotted to capture the changes in both the storage and the loss moduli simultaneously. The magnitude of the complex modulus is referred to as the effective modulus. Finally, the percent difference between the effective moduli is shown in Figure 1c, demonstrating the quantitative effect of the exposure to power-line fire explosion.

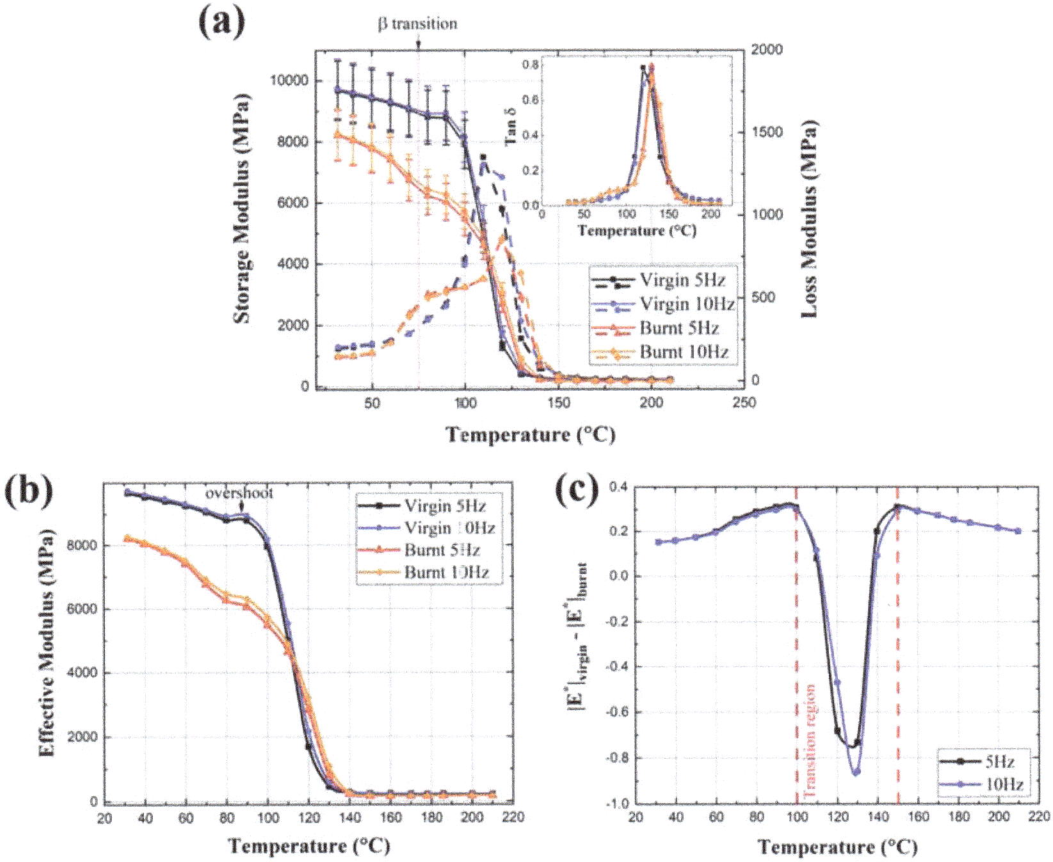

Figure 1. (**a**) Thermomechanical spectra plotting the storage and loss moduli; (**b**) magnitude of the effective moduli, and (**c**) percent change in the magnitude of the effective moduli of virgin and burnt specimens as a function of temperature at 5 and 10 Hz.

The thermomechanical spectra shown in Figure 1a are based on the average response of four virgin samples and five burnt samples. The spectra can be naturally divided into three regions. First is the glassy region, below the glass transition temperature (T_g), followed by the leathery region, where the material transitions from the glassy regime to the third and final region, the rubbery regime. As shown in Figure 1b, the storage modulus remains nearly constant in the glassy regime, followed by a sudden and significant drop in the leathery regime, and subsequently approaches another constant value in the rubbery regime. The glass transition temperature marks the thermal transition between the glassy and leathery regimes, calculated here based on the peak of the loss modulus [11]. The glass transition temperature (T_g) was found to be 112 °C for the virgin composite specimens, shifting to 120 °C for the specimens extracted from the burnt panels. Due to the low range of investigated frequencies, T_g appeared to be insensitive to the change in the testing frequency but was shown to be a frequency-dependent property if testing was done at a broader range [19]. A notable dip is observed in the storage moduli of the virgin and burnt panels before T_g, which were assigned to be the beta transition (T_β), delineating the evolution in chain mobility due to increased thermal energy from increasing temperature. The beta transition is a secondary-order transition, commonly present below the primary T_g transition, where the chains in the epoxy matrix start to slide with higher degrees of freedom than below T_β. In the latter condition, the chain motions are limited to bending and stretching mechanisms [15]. The dip in T_β is manifested in the loss modulus as a peak proceeding the T_g peak. The dip at T_β in the storage modulus corresponds to an increase in the loss modulus, indicating an uptick in the energy dissipation; therefore, a change in T_β is related to a change in toughness [11,15]. This can be clearly seen in the plot of tanδ. The burnt samples show a more pronounced peak in the loss-modulus at T_β since this transition point is associated with chain mobility due to the thermal energy, as discussed above. It is believed that the fire event had loosened the chain and alleviated some of the processing residual stresses and, hence, provided a high degree of mobility to the chains at T_β and throughout the thermomechanical spectrum. Here, T_β was found to be insensitive to the harsh deployment operating conditions since it remained constant at 75 °C for virgin and burnt panels, suggesting the overall chemical structure of the composite remained unaffected by the deployment conditions since the secondary thermal transitions are linked to different activation energies based on the microstructure [15,20].

Table 2 summarizes and collates the DMA testing results, including T_g and T_β, and the average storage and loss moduli at the two different testing frequencies (5 Hz and 10 Hz). In the following discussion, the reported flexural storage and loss moduli in the glassy and rubbery regimes are based on the average and standard deviation in the moduli values below 100 °C and above 150 °C, respectively. The effect of deployment conditions on the mechanical behavior within the leathery regime is discussed next in the context of the difference in the effective moduli. The flexural storage modulus was measured to be 9.06 ± 0.56 GPa and 6.99 ± 1.02 GPa for the virgin and burnt samples, respectively, at temperatures ≤100 °C at 5 Hz. Within the same range of temperatures, the corresponding loss moduli were 0.342 ± 0.182 GPa and 0.344 ± 0.188 GPa. Comparing these values at 5 Hz, the storage modulus exhibited a ~23% decrease while the loss modulus remained nearly unchanged, indicating a reduction in deformation resistance while maintaining toughness. The reduction in the storage modulus in the burnt samples is attributed to the localized change in temperature during the fire event, resulting in permanent loosening of the chains in the epoxy matrix or an increase in the amorphous phase. In the rubbery regime (≥150 °C), the average storage and loss moduli were found to be 0.231 ± 0.008 GPa and 0.013 ± 0.009 GPa, respectively, for the virgin panels and 0.172 ± 0.004 GPa and 0.006 ± 0.007 GPa, respectively, for the burnt samples at 5 Hz. The reduction in the moduli in the rubbery region for the burnt specimen confirms the increase in the amorphous phase of the epoxy matrix.

Table 2. Summary of the properties extracted from dynamic mechanical analyzer (DMA) testing.

Property	**Virgin**		**Burnt**	
T_g	**112 °C**		**120 °C**	
T_β	**75 °C**		**75 °C**	
Testing Freq.	5 Hz	10 Hz	5 Hz	10 Hz
E' (≤100 °C), GPa	9.062 ± 0.564	9.171 ± 0.504	6.992 ± 1.016	7.135 ± 0.935
E'' (≤100 °C), GPa	0.342 ± 0.182	0.340 ± 0.170	0.344 ± 0.188	0.340 ± 0.179
E' (≥150 °C), GPa	0.231 ± 0.008	0.233 ± 0.009	0.172 ± 0.004	0.174 ± 0.004
E'' (≥150 °C), GPa	0.013 ± 0.009	0.016 ± 0.012	0.006 ± 0.007	0.009 ± 0.012

The moduli result for the virgin and burnt specimens at 10 Hz are also listed in Table 2, showing an increase in the average moduli. For example, the average storage modulus in the glassy region for virgin samples increased to 9.17 ± 0.50 GPa when tested at 10 Hz, compared to 9.06 ± 0.56 GPa at 5 Hz. For the burnt sample, the storage modulus changed from 6.99 ± 1.02 GPa at 5 Hz to 7.13 ± 0.93 GPa at 10 Hz. Similar behavior was also reported for the loss modulus. The slight uptick in the moduli as a function of testing frequency is consistent with the expected response of the polymer matrix, where an increase in frequency corresponds to the domination of the elastic, solid-like behavior [15]. Moreover, the rubbery modulus of the burnt samples was lower than the virgin, suggesting a disruption to the network structure, as discussed before about the increase in the amorphous nature of the epoxy matrix. A closing note is warranted at this point concerning the apparent jump ("overshoot") in the storage modulus of the virgin specimens, regardless of the testing frequency. Menard postulated that this hump in the storage modulus before the onset of the leathery regime (i.e., the drop at T_g) corresponds to relief of residual stresses from the processing method [15]. As discussed above, the panels were fabricated using hand layup and hot press processes, inducing high residual stresses, and limiting the chain mobility below T_g. As the temperature increases during the DMA testing, the chain mobility increases to allow the chains to move to a lower energy state and, in turn, releasing the residual stresses. Thus, the localized increase in the temperature during the fire event and the subsequent free convection cooling resulted in the annealing of the deployed panels, relieving the residual stresses, and eliminating the pre-T_g hump. Notably, the fire-induced annealing process reduced the dynamic mechanical properties in the glassy and rubbery regimes, as discussed above.

The difference in the magnitude of the effective moduli ($|E^*|_{virgin} - |E^*|_{burnt}$) was calculated for each dataset collected at different testing frequencies and plotted in Figure 1c based on $|E^*| = \sqrt{(E'^2 + E''^2)}$, where the effective modulus ($|E^*|$) is a function of both the storage (E') and loss (E'') moduli. Excluding the leathery regime, 100 °C < T < 150 °C, the change in the effective modulus was 24 ± 5% on average, with the moduli of the specimens extracted from the burnt samples lagging from their virgin counterparts. This is in good agreement with the results discussed above and consistent with our previous investigations of the same composite panels [11]. There was a significant difference between the virgin and burnt samples in the leathery regime, reaching up to an 80% difference. The dichotomy of the moduli in the leathery regime favored the burnt specimens over their virgin counterparts, such that the moduli of the latter was consistently inferior to the former between 120 and 140 °C. That is to say, the fire-induced annealing of the composite skins was believed to result in an apparent plasticization of the polymer matrix, hence, increasing the effective modulus of the burnt samples within the leathery regime. Such behavior is also consistent with attributes of thermosets, such as intolerance to heating due to the interlocking of the chains, as will be shown later in the FTIR spectra.

Figure 2 is a collage of SEM micrographs of the exposed surface of the virgin specimen (Figure 2a), virgin specimen after DMA testing (Figure 2b), oil-contaminated area of the burnt specimen after DMA testing (Figure 2c), and a contamination-free area of the same burnt specimen also after DMA testing (Figure 2d). Before embarking on the SEM results, two clarifying notes are due. First, an emphasis is given to comparing the surface morphology after the DMA testing to delineate the effect of the gradual increase in temperature due to the flexural testing protocol in the DMA and a sudden, localized rise in temperature corresponding to the fire-event in deployment on the morphology. Second, two adjacent areas on the surface of the burnt specimen were observed under the SEM to substantiate the localized effect of the power explosion fire-event on the performance of these composite panels in field deployment, whereas the micrograph in Figure 2c was taken a few millimeters away from Figure 2d, showing a vastly different morphology. The morphology of the micrograph in Figure 2d is characteristically analogous to the micrograph in Figure 2b, where the former is of the burnt specimen while the latter is of the virgin specimen, both acquired after DMA testing. The SEM micrograph of the virgin and untested specimen, Figure 2a, explicates the rough morphology of the surface due to the curing using the heated press. Figure 2a also demonstrates the intermixing of the continuous and chopped fibers on the surface; the latter appear to have been added to the epoxy matrix during manufacturing as an additional reinforcement. Figure 2b shows an SEM micrograph of one of the DMA tested virgin specimen, subjected to a gradual increase in temperature up to 210 °C. Figure 2b reveals the continuous fibers extending across the surface and comparatively smoother morphology than the micrograph in Figure 2a. The morphology change is associated with the increase in temperature and the corresponding relief of process-induced stresses.

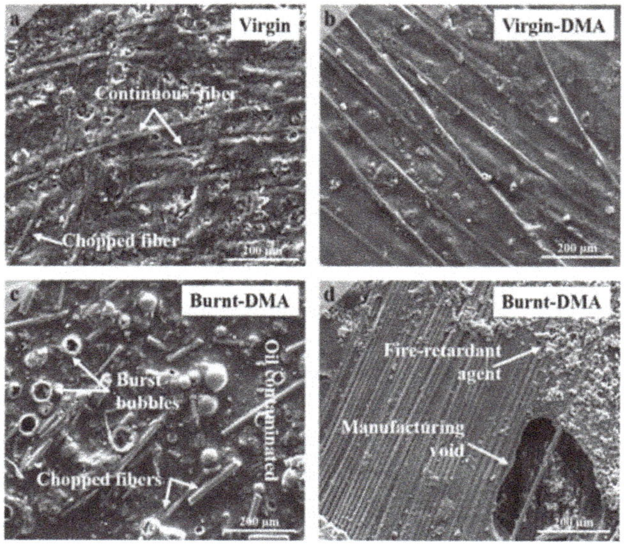

Figure 2. Collage of SEM micrographs of (**a**) virgin E-Glass/Epoxy skin surface; (**b**) the surface of a virgin sample after DMA testing, and (**c**,**d**) surface of deployed specimens after DMA testing. (Micrograph c is for portion of the sample with oil contamination while micrograph d is from the same sample but from region that was contamination-free).

The micrographs in Figure 2c,d explicate the combined effect of the power line explosion fire-event and gradual increase in temperature during DMA testing on the surface morphology. Here, one area (Figure 2c) exhibited notable oil contamination during deployment, resulting in visually noticeable burn marks on the surface. In contrast, a contamination-free area of the same sample (Figure 2d) is characteristically similar to the micrograph in Figure 2b of the virgin sample. The micrograph in Figure 2c depicts a higher presence of chopped fibers and absence of continuous fibers, which may have receded into the skin due to the significant, localized increase in temperature during the fire event before the activation of the fire-retardant agent that extinguished the fire and limited further potential damage. Notably, the covered underground vault also limited the supply of oxygen and helped in the quick extinction of the fire due to the power line explosion. Figure 2c also portrays some of the bubbles due to the oil-contamination burst because of the increase in temperature while other bubbles in the same vicinity remained intact, signifying the limited lingering time of the fire event. On the other hand, Figure 2d shows a residue of the fire-retardant agent added during the manufacturing process, which was believed to help suppress the spread of the fire and limit its effect to small regions. The manufacturing voids shown in Figure 2d were observed before using optical microscopy as reported in [11], due to the entrapped air bubble and the volatiles that escaped during the manufacturing process from the epoxy curing. The sharpness of the void edges demonstrates the limited effect of the fire since the edges resemble the character of the as-manufactured panels. Notably, these voids were also present in the virgin panels and, hence, their association with the matrix curing process. Lastly, Figure 2d shows the continuous fibers extending across the surface similar to the virgin samples, indicating that the fire event did not result in excessive sagging of the matrix and prevented fiber fracture. The continuity of the fiber and lack of evidence of delamination or interface issues corroborate the results from our previous studies [11].

Figure 3 shows the EDS results, corresponding to specific areas from the SEM micrographs shown in Figure 2, respectively, ascertaining the elemental composition of the different composite specimens studied herein. A compilation of the elemental composition of each specimen is included in Appendix A. At the onset of this section, it is important to note that the EDS results are used qualitatively to delineate the effect of the fire event while providing supporting evidence to the results from other methods. Evident from the EDS compositional spectra shown in Figure 3 and listed in Table A1 is the amorphous nature of the composite samples given the broad diversity of reported chemical elements, including aluminum, bromine, and silicon, to name a few. Generally, each spectrum, irrespective of the specimen conditions, contained all expected elements based on the constituents used in the fabrication of the composites, as discussed in the previous section. The EDS spectra reported here are also in excellent agreement with those reported in [21–23]. Since alumina trihydrate was added as a fire-retardant agent, the appearance of aluminum and oxygen traces in the EDS spectra is justified. It is important to note that despite the exposure of the burnt samples to the fire event, aluminum persisted in their corresponding EDS spectra, corroborating the previous conclusion that the fire was quickly extinguished. Similarly, bromine, used as a fire-retardant agent formulated into the epoxy resin, in all the spectra shown in Figure 3, adds additional evidence to the short lifetime of the fire event, reducing the damage potential, as discussed before. It is worth noting that the existence of aluminum and bromine even in the burnt specimen with oil-contamination after DMA testing at high temperature indicates the suitability of these composite panels for extended deployment since the fire-retardant reagents are at comparable levels as their counterparts, which were unexposed to the fire-event. The EDS characterization was performed over multiple regions on the samples, and the results were in good accordance with those reported herein.

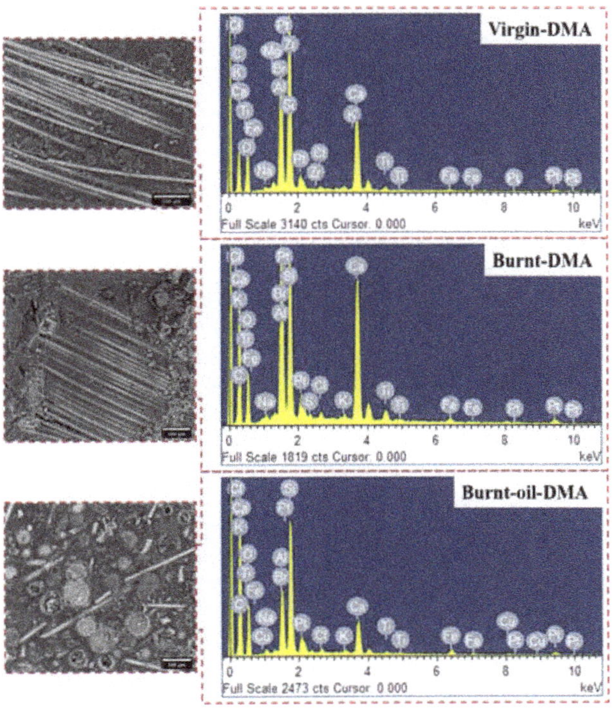

Figure 3. SEM micrographs and the accompanying elemental composition of virgin specimen (**top panel**), burnt specimen contamination-free (**middle panel**), and burnt specimen oil-contamination (**bottom panel**) after DMA testing.

The EDS spectra also revealed additional elemental compositions in all the investigated sample conditions, including C, Na, Mg, Si, and Ca. The average weight percentages of some of these elements exhibited a substantial (in some cases) change, as quantitatively reported in Table A1. For example, the weight percentage of carbon increased from 50.63 $\pm$ 9.00% in the virgin specimen after DMA testing to 62.13 $\pm$ 2.17% in the burnt specimens with oil contamination. An increase in the carbon is expected due to the exposure to the fire event, where locations with oil-contamination reporting higher percentage of carbon than elsewhere (see Table A1). The weight percentage of silicon also changed slightly from 5.9 $\pm$ 3.81% in the virgin specimen to 3.76 $\pm$ 1.02% and 4.31 $\pm$ 0.46% for the burnt specimen, indicating that the glass fibers may have experienced orientation change during the fire event due to the localized increase in temperature. The change in glass fiber orientation has been postulated above by assuming the receding of the fibers into the matrix given the relatively high temperature during the fire [24]. Hence, the EDS results confirm the potential change in orientation, corroborating with the mechanical data. In all, the study of the EDS spectra and the SEM micrographs show that the operating conditions did not significantly dismember the elemental structure of the composite panels; hence, their deployment may be extended given no other damages incurred that may compromise the structural integrity of the panels [5,25].

The chemical degradation of the epoxy matrix was also investigated using FTIR spectroscopy. Figure 4 plots the transmission of infrared light through virgin and burnt specimens. The spectra in Figure 4 are in good agreement with those reported before in the literature of similar composite material [26,27]. Most of the peaks found in the virgin sample were also present in the burnt samples except for a strong peak at 1010 cm^{-1} and a moderate peak at 1720 cm^{-1}, corresponding to the ester carboxyl group. Additionally, a few small peaks from 3620 to 3500 and from 795 to 740 cm^{-1}, corresponding to the Al(Al)OH bonds of the flame retardant, were absent in the burnt specimen, but present in the virgin samples [28,29] The FTIR spectra reveals loss of the flame retardant and some chemical degradation of the vinyl ester due to combustion. Any cleavage of the ester backbone polymer is a likely contributor to the mechanical softening observed in Figure 1.

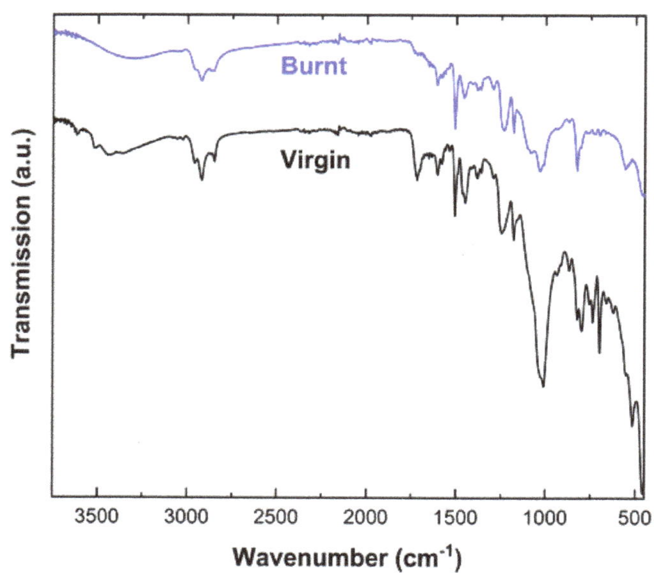

Figure 4. FTIR transmission of virgin and burnt specimens.

Finally, the characteristics of the glass fibers in the epoxy resin matrix were further analyzed using X-ray diffraction. Figure 5 is a plot of the XRD spectra of the same four composite skins samples studied above. There is an additional (010) alumina trihydrate peak at $2\theta = 18.4°$, which was truncated since it has a much higher intensity than the existing peak; hence, its inclusion would have dwarfed the remaining peaks. The broad underlying peak centered around $2\theta \approx 20°$ is characteristic of the E-Glass/Epoxy composite and is in good agreement with other XRD analyses reported in the literature [30–33]. This broad peak indicates the uniform distribution of the glass fibers within the epoxy matrix and highlights the amorphous nature of the composite [21,30]. Generally, all the major sharp peaks in the virgin specimen are due to the Gibbsite phase of the fire-retardant agent, alumina trihydrate [34]. Contrary to the EDS results, the peaks associated with alumina trihydrate disappeared from the burnt specimen XRD spectra irrespective of the presence or absence of oil contamination, suggesting the fire may have consumed the surface-level fire-retardant agent. The discrepancy in the results is associated with the difference in the penetration depth of the XRD and EDS approaches. Nonetheless, E-Glass fibers and the epoxy matrix structure appear to be mostly unaffected by the fire event since the amorphous peak persisted in all of the XRD spectra, further substantiating the results discussed above.

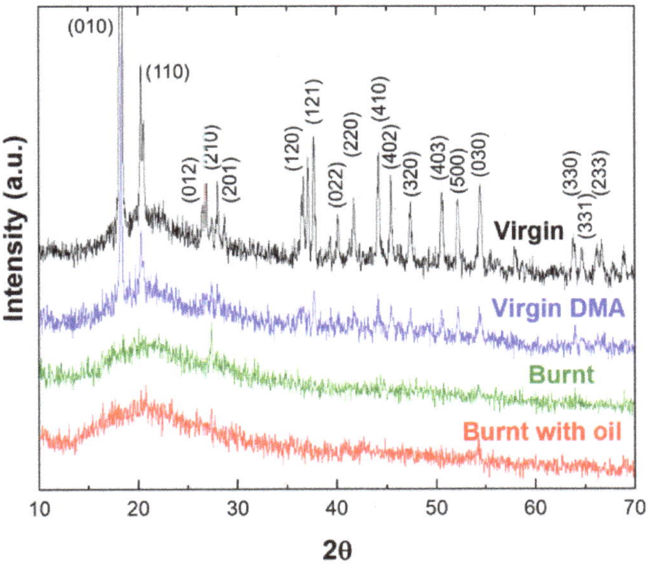

Figure 5. XRD spectral patterns of a virgin specimen and the virgin and burnt specimens after DMA testing ("Burnt" ≡ without contamination, "Burnt with oil" ≡ oil contaminated).

4. Conclusions

The results of multiscale characterization of E-Glass/Epoxy composite panels were reported herein. In doing so, the dynamic mechanical properties, surface morphology, and elemental compositions of previously deployed panels in a fire incident and as-fabricated panels were measured and compared. Deployed samples exhibited a ~23% decrease of the storage modulus (in the glassy regime at $\leq$100 °C), while the loss modulus remained nearly unchanged when tested at 5 Hz. This reduction is attributed to the localized increase in temperature during the fire event, resulting in permanent loosening of the epoxy matrix chains or an increase in the amorphous phase. These mechanical results were found to be consistent at different testing frequencies as well as in the rubbery regime at temperatures $\geq$150 °C. SEM micrographs and associated EDS spectra explicate the morphology of the surface due to the fabrication processes, deployment conditions, and the effect of DMA testing. The results exemplify the effectiveness of the fire-retardant agent additives in quickly extinguishing the fire, given the persistent presence of bromine and aluminum in the EDS spectra on burnt samples. The EDS results also showed an increase in carbon content on burnt samples with oil contamination. The chemical structural analysis using XRD confirmed the amorphous nature of the composite and the activation of the fire-retardant agent during the fire incident. In summary, the operating conditions did not significantly compromise the mechanical integrity of the E-Glass/Epoxy composite, thus, demonstrating the viability of an extended deployable lifetime of the fire-exposed composite panels. Future research will focus on the combined effect of humidity (due to the heavy raining seasons), long-term static loading, and significant change in temperature on these composite panels' long-term performance.

Author Contributions: Conceptualization, G.Y.; methodology, G.Y.; formal analysis, G.Y., S.N., N.U.H. and C.G.; investigation, G.Y., S.N. and N.U.H.; data curation, G.Y., S.N. and N.U.H.; analysis validation and verification, S.N., N.U.H. and C.G.; writing—original draft preparation, G.Y. and C.G.; writing—review and editing, S.N., N.U.H. and C.G.; visualization, S.N. and N.U.H.; supervision, G.Y.; project administration, G.Y.; funding acquisition, G.Y. All authors have read and agreed to the published version of the manuscript.

Funding: The research leading to these results was supported in part by the United States Department of Defense under Grant Agreement No. W911NF-14-1-0039 and internal funds from San Diego State University.

Institutional Review Board Statement: Not applicable.

Informed Consent Statement: Not applicable.

Data Availability Statement: The data presented in this study are available on request from the corresponding author. The data are not publicly available due to privacy.

Acknowledgments: The authors are grateful for the previous support from Armorcast Products Inc. for supplying the prefabricated panels as well as their previous financial support. The guidance and discussion with Wally Sunnaa, P.E. and Vahe Karamardian are greatly appreciated. The authors finally acknowledge the use of equipment at the San Diego State University Electron Microscopy Facility acquired by NSF instrumentation grant DBI-0959908 and the use of the X-ray diffractometer at San Diego State University funded by the Department of Geological Sciences.

Conflicts of Interest: The authors declare no conflict of interest.

Appendix A

Table A1. Elemental Compositions extracted from the EDS spectra of virgin and deployed E-Glass/Epoxy composite samples.

Element	Weight %			Atomic %		
	Virgin-DMA	Burnt-No-Oil	Burnt-Oil	Virgin-DMA	Burnt-No-Oil	Burnt-Oil
C K	50.63 ± 9	46.99 ± 7.97	62.13 ± 2.17	65.21 ± 9.53	62.14 ± 7.8	72.7 ± 1.6
O K	27.27 ± 5.21	29.14 ± 5.79	26.12 ± 1.09	26.56 ± 5.77	29.23 ± 6.84	22.96 ± 1.23
Na K	0.22 ± 0.19	0.38 ± 0.14	0.16 ± 0.14	0.15 ± 0.13	0.27 ± 0.11	0.09 ± 0.08
Mg K	0.11 ± 0.19	-	-	0.07 ± 0.13	-	-
Al K	3.1 ± 2.69	1.68 ± 0.3	1.75 ± 0.23	1.81 ± 1.57	0.99 ± 0.14	0.92 ± 0.14
Si K	5.9 ± 3.81	3.76 ± 1.02	4.31 ± 0.46	3.31 ± 2.21	2.12 ± 0.48	2.16 ± 0.26
S K	-	0.25 ± 0.12	-	-	0.12 ± 0.06	-
Cl K	0.16 ± 0.02	0.28 ± 0.06	0.15 ± 0.03	0.07 ± 0.01	0.13 ± 0.03	0.06 ± 0.01
K K	0.11 ± 0.09	0.15 ± 0.05	0.08 ± 0.07	0.04 ± 0.04	0.06 ± 0.02	0.03 ± 0.03
Ca K	2.62 ± 1.78	7.75 ± 2.68	1.36 ± 0.19	1.03 ± 0.72	3.11 ± 1.17	0.48 ± 0.07
Ti K	0.32 ± 0.1	0.5 ± 0.18	0.13 ± 0.02	0.1 ± 0.02	0.17 ± 0.06	0.04 ± 0.01
Fe K	0.06 ± 0.1	0.5 ± 0.19	0.36 ± 0.09	0.02 ± 0.03	0.14 ± 0.06	0.09 ± 0.02
Cu K	-	0.21 ± 0.18	0.19 ± 0.09	-	0.05 ± 0.05	0.04 ± 0.02
Br K	7.66 ± 4.06	6.51 ± 1.06	1.94 ± 0.46	1.47 ± 0.74	1.3 ± 0.23	0.34 ± 0.09
Pt L	1.83 ± 0.36	1.9 ± 0.07	1.33 ± 0.15	0.14 ± 0.03	0.16 ± 0.02	0.1 ± 0.01

References

1. Wang, R.M.; Zheng, S.R.; Zheng, Y.P. *Polymer Matrix Composites and Technology*; Woodhead Publishing: Cambridge, UK, 2011; ISBN 9780857092212.
2. Bank, L.C. *Composites for Construction: Structural Design with FRP Materials*; John Wiley & Sons, Inc.: Hoboken, NJ, USA, 2006; ISBN 0471681261.
3. Guedes, R.M. Durability of polymer matrix composites: Viscoelastic effect on static and fatigue loading. *Compos. Sci. Technol.* **2007**, *67*, 2574–2583. [CrossRef]
4. Guedes, R.M. Lifetime predictions of polymer matrix composites under constant or monotonic load. *Compos. Part A Appl. Sci. Manuf.* **2006**, *37*, 703–715. [CrossRef]
5. Youssef, G. Creep analysis of E-glass/vinyl-ester laminated composites for underground vault structures. *Polym. Compos.* **2020**, *41*, 3564–3574. [CrossRef]
6. Correia, J.R.; Branco, F.; Gonilha, J.; Silva, N.; Camotim, D. Glass fibre reinforced polymer pultruded flexural members: Assessment of existing design methods. *Struct. Eng. Int. J. Int. Assoc. Bridg. Struct. Eng.* **2010**, *20*, 362–369. [CrossRef]
7. Sa, M.F.; Gomes, A.M.; Correia, J.R.; Silvestre, N. Creep behavior of pultruded GRFP elements-Part 1: Literature review and experimental study. *Compos. Struct.* **2011**, *93*, 2450–2459. [CrossRef]
8. Karataş, M.A.; Gökkaya, H. A review on machinability of carbon fiber reinforced polymer (CFRP) and glass fiber reinforced polymer (GFRP) composite materials. *Def. Technol.* **2018**, *14*, 318–326. [CrossRef]
9. Gopinath, A.; Senthilkumar, M.; Babu, A. Evaluation of Mechanical Properties and Microstructure of Polyester and Epoxy Resin Matrices Reinforced with Jute, E-glass and coconut Fiber. *Mater. Today Proc.* **2018**, *5*, 20092–20103. [CrossRef]
10. Sharada Prabhakar, C.; Ramesh Babu, P. Characterization of mechanical and thermal properties of high strength glass epoxy and rayon carbon phenolic composites. *Mater. Today Proc.* **2018**, *5*, 26898–26903. [CrossRef]
11. Youssef, G.; Pessoa, G.; Nacy, S. Effect of elevated operating temperature on the dynamic mechanical performance of E-glass/epoxy composite. *Compos. Part B Eng.* **2019**, *173*, 106937. [CrossRef]
12. Im, K.-H.; Cha, C.-S.; Park, J.-W.; Cha, Y.-H.; Jung, I.-Y.Y.-A. Effect of temperatures on impact damage and residual strength of CFRP composite laminates. In *AIP Conference Proceedings*; American Institute of Physics: College Park, MD, USA, 2000; pp. 1247–1254.
13. Idicula, M.; Boudenne, A.; Umadevi, L.; Ibos, L.; Candau, Y.; Thomas, S. Thermophysical properties of natural fibre reinforced polyester composites. *Compos. Sci. Technol.* **2006**, *66*, 2719–2725. [CrossRef]
14. Choy, C.L. Thermal Conductivity of Polymers. *Polymer* **1977**, *18*, 984–1004. [CrossRef]
15. Menard, K.P. *Dynamic Mechanical Analysis: A Practical Introduction*; CRC Press: Boca Raton, FL, USA, 1999; ISBN 0849386888.
16. Youssef, G.; Whitten, I. Dynamic properties of ultraviolet-exposed polyurea. *Mech. Time-Depend. Mater.* **2017**, *21*, 351–363. [CrossRef]
17. Menczel, J.D.; Prime, R.B. (Eds.) *Thermal Analysis of Polymers: Fundamentals and Applications*; John Wiley & Sons, Inc.: Hoboken, NJ, USA, 2009; ISBN 9781787284395.
18. Turi, E.A. (Ed.) *Thermal Characterization of Polymeric Materials*; Academic Press: New York, NY, USA, 1981.
19. Feng, J.; Guo, Z. Effects of temperature and frequency on dynamic mechanical properties of glass/epoxy composites. *J. Mater. Sci.* **2016**, *51*, 2747–2758. [CrossRef]
20. Bershtein, V.A.; Egorov, V.M.; Kemp, T.J. *Differential Scanning Calorimetry of Polymers: Physics, Chemistry, Analysis, Technology*; Ellis Horwood Ltd.: New York, NY, USA, 1994.
21. Adekomaya, O.; Adediran, A.A.; Adama, K. Characterization and morphological properties of glass fiber reinforced epoxy composites fabricated under varying degrees of hand lay-up techniques. *J. Appl. Sci. Environ. Manag.* **2018**, *22*, 110–114. [CrossRef]
22. Laura, D.M.; Keskkula, H.; Barlow, J.W.; Paul, D.R. Effect of glass fiber surface chemistry on the mechanical properties of glass fiber reinforced, rubber-toughened nylon 6. *Polymer* **2002**, *43*, 4673–4687. [CrossRef]
23. Swetha, C.; Kumar, R. Quasi-static uni-axial compression behaviour of hollow glass microspheres/epoxy based syntactic foams. *Mater. Des.* **2011**, *32*, 4152–4163. [CrossRef]
24. Allred, R.E.; Wesson, S.P.; Shin, E.E.; Inghram, L.; McCorkle, L.; Papadopoulos, D.; Wheeler, D.; Sutter, J.K. The influence of sizings on the durability of high-temperature polymer composites. *High Perform. Polym.* **2003**, *15*, 395–419. [CrossRef]
25. Bledzki, A.K.; Mamun, A.A.; Volk, J. Physical, chemical and surface properties of wheat husk, rye husk and soft wood and their polypropylene composites. *Compos. Part A Appl. Sci. Manuf.* **2010**, *41*, 480–488. [CrossRef]
26. Prabhakar, M.N.; Song, J.I. Influence of chitosan-centered additives on flammable properties of vinyl ester matrix composites. *Cellulose* **2020**, *27*, 8087–8103. [CrossRef]
27. Wu, Z.; Li, S.; Liu, M.; Wang, Z.; Liu, X. Liquid oxygen compatible epoxy resin: Modification and characterization. *RSC Adv.* **2015**, *5*, 11325–11333. [CrossRef]
28. Mittal, G.; Dhand, V.; Ryu, J.I.; Rhee, K.Y.; Kim, H.J.; Jung, D.H. Fabrication of modified MMT/Glass/Vinylester multiscale composites and their mechanical properties. *J. Nanomater.* **2015**, *2015*. [CrossRef]
29. Schroeder, P. Infrared spectroscopy in clay science. *Teach. Clay Sci.* **2002**, *11*, 181–206.
30. Hao, Y.; Liu, F.; Shi, H.; Han, E.; Wang, Z. The influence of ultra-fine glass fibers on the mechanical and anticorrosion properties of epoxy coatings. *Prog. Org. Coat.* **2011**, *71*, 188–197. [CrossRef]

31. Kanny, K.; Mohan, T.P. Resin infusion analysis of nanoclay filled glass fiber laminates. *Compos. Part B Eng.* **2014**, *58*, 328–334. [CrossRef]
32. Quaresimin, M.; Salviato, M.; Zappalorto, M. Fracture and interlaminar properties of clay-modified epoxies and their glass reinforced laminates. *Eng. Fract. Mech.* **2012**, *81*, 80–93. [CrossRef]
33. Tjong, S.C. Structural and mechanical properties of polymer nanocomposites. *Mater. Sci. Eng. R Rep.* **2006**, *53*, 73–197. [CrossRef]
34. Reddy, T.R.; Thyagarajan, K.; Montero, O.A.; Reddy, S.R.L.; Endo, T. X-Ray Diffraction, Electron Paramagnetic Resonance and Optical Absorption Study of Bauxite. *J. Miner. Mater. Charact. Eng.* **2014**, *2*, 114–120. [CrossRef]

Journal of
Composites Science

MDPI

Article

Additive Manufacturing for Effective Smart Structures: The Idea of 6D Printing

Stelios K. Georgantzinos [1,*], Georgios I. Giannopoulos [2] and Panteleimon A. Bakalis [3]

1 Department of Aerospace Science and Technology, National and Kapodistrian University of Athens, 34400 Psachna, Greece
2 Department of Mechanical Engineering, University of Peloponnese, 1 Megalou Alexandrou Street, 26334 Patras, Greece; ggiannopoulos@uop.gr
3 General Department, National and Kapodistrian University of Athens, 34400 Psachna, Greece; pbakalis@uoa.gr
* Correspondence: sgeor@uoa.gr

Abstract: This paper aims to establish six-dimensional (6D) printing as a new branch of additive manufacturing investigating its benefits, advantages as well as possible limitations concerning the design and manufacturing of effective smart structures. The concept of 6D printing, to the authors' best knowledge, is introduced for the first time. The new method combines the four-dimensional (4D) and five-dimensional (5D) printing techniques. This means that the printing process is going to use five degrees of freedom for creating the final object while the final produced material component will be a smart/intelligent one (i.e., will be capable of changing its shape or properties due to its interaction with an environmental stimulus). A 6D printed structure can be stronger and more effective than a corresponding 4D printed structure, can be manufactured using less material, can perform movements by being exposed to an external stimulus through an interaction mechanism, and it may learn how to reconfigure itself suitably, based on predictions via mathematical modeling and simulations.

Keywords: additive manufacturing; 6D printing; 4D printing; 3D printing; smart materials; nanocomposites

Citation: Georgantzinos, S.K.; Giannopoulos, G.I.; Bakalis, P.A. Additive Manufacturing for Effective Smart Structures: The Idea of 6D Printing. *J. Compos. Sci.* **2021**, *5*, 119. https://doi.org/10.3390/jcs5050119

Academic Editors: Francesco Tornabene and Thanasis Triantafillou

Received: 8 April 2021
Accepted: 29 April 2021
Published: 1 May 2021

1. Introduction

Additive manufacturing (AM) or 3D printing, discovered in the 1980s as a rapid prototyping method and is now considered as a manufacturing process, practically in the same way as the conventional manufacturing processes. AM has generated an impact in all industry areas, including aerospace [1,2], automobile [3], construction [4], medical and military [5]. It offers flexibility in the product design process, a considerable reduction in material consumption, as well as makes product personalization affordable [6]. ASTM International categorizes AM processes in seven groups [7] (i.e., material extrusion (e.g., fused deposition modeling—FDM), powder bed fusion, vat photopolymerization (stereolithography—SLA), material jetting (e.g., PolyJet), sheet lamination, directed energy deposition, and binder jetting), each typified by the principle according to which matter is formed.

In terms of advances in technologies and applications concerning 3D printing, one of the extraordinary reported developments so far is 4D printing in which 3D printing technology is combined with smart materials (SMs) to produce structures that change shape over time [8–10]. More specifically, in 4D printing, the developed 3D object can self-transform or respond to external stimuli over time [11]. In other words, time could be considered as the 4th printing dimension. Here, time is not utilized to describe for how long the object is printed but expresses the fact that it can continue to evolve [12,13] after being printed or even the ability to self-regenerate [14]. 4D printing is expected to play a crucial role in future designs due to its obvious advantages. For instance, 4D printing may allow scientists to develop, micro/nano-actuators and/or smart devices, without additional equipment, i.e., wires, motors, and batteries. Applications of 4D printing have been already referred to in several fields. Characteristically, 4D printing has already been successfully applied for manufacturing various products including structures with multi-directional properties [15], soft actuators [16], and others.

The production of products with material intelligence (i.e., SMs), via the 4D printing technique, requires the achievement of the self-transformation skill. The final printed smart objects should possess expandable, adaptable, and/or deformable characteristics in response to an employed stimulus (i.e., temperature, humidity, or solvents), as well as pH or light [17]. The materials utilized for structure evolution exhibit an abundance of properties (i.e., self-healing [18], self-assembly [19], self-sensing, decision making, and self-adaptability) [20]. Shape morphing following printing is the key feature of 4D printing. Shape changes in 4D printed structures can be caused by various external stimuli, to produce folding, expansion, or shrinkage of the printed objects. These evolutions are characterized by different behaviors for objects comprised of one SM and bilayer structures with different properties and inhomogeneity [21,22].

If a 3D printed object, beyond its morphing into a new shape (4D printing), can further change its function, this transformation could be called, also as 5D printing. Nevertheless, a comprehensive definition of 4D printing is more appropriate to include the evolution of the 3D printed structure concerning changes in shape, property, as well as in functionality. On the other hand, "5D printing" is more commonly referred to as a name for five-axis AM [23,24]. The term "5D" is used to describe the five-axis printing method, firstly proposed by Mitsubishi Electric Research Laboratories (MERL), which is accomplished by the utilization of a print bed that rocks back and forth on two-axis, creating the fourth and fifth processing axis [23]. The competitive advantage of 5D printing compared with 3D printing is that it can produce stronger objects using less material [24].

Here, to our best knowledge, an attempt will be made for the first time in the open literature to introduce the concept of 6D printing followed by the necessary theoretical background, potential areas of application, and possible arisen benefits. The idea is simple, novel, worth researching, and can easily be described as follows: A 6D printed structure could be regarded as a child born from the marriage between a five-axis printer of FDM technology and SMs. It can be stronger than a corresponding 4D printed structure, can be manufactured using less material, can change shape due to an external stimulus via an interaction mechanism, while its structural quality and intelligence may be improved by introducing computational optimization methods and simulations.

2. The Ground Idea

Today, the benefits of 3D printing are well established: Design freedom, part consolidation, no tooling necessary, just-in-time inventory, faster production, easily accessible, tangible design, cost-effectiveness, creative designs, unlimited shapes and geometry, variety of raw materials, less waste production, risk reduction and others [25]. From an engineer's point of view, the list of practical advantages is boundless. Taking into account all of these benefits, one would consider that 3D printing would be a very common manufacturing procedure in our times. However, this is not true: The industry is struggling to incorporate efficient technology into its production. However, AM is still less than 2% of the entire manufacturing market. This is due to the fact that some challenges do not allow the extensive adoption of the specific AM. Common problems associated with 3D printing are imperfections in the final product, post-processing requirements, longer production timelines, limitation to ordinary material components, etc.

Any current limitation associated with the 3D printing process is inevitably transferred to the 4D Printing technology. The 4D restrictions may be divided into two main categories. The first category is related to the manufacturing process itself, while the second one has to do with the functionality, characteristics, and capabilities of the final smart structure.

Regarding the manufacturing limitations, the need for supporting material as well as the post-process requirement for its removal should be highlighted. In addition, the lack of additional printing degrees of freedom, evidently, leads to the need for extended processing routes and, thus, to longer processing times, while the achievable structural complexity and coherence, that governs the SM strength, is rather restricted. Finally, analogous limitations arise regarding the SM stimulus-response capabilities due to the lack of the extra printing directions. It should be noted that despite the fact the 4D printing may allow the production of SM components with a complex orientation by using more convoluted programming, does not have the high manufacturing flexibility regarding the layer-by-layer orientation that is offered by the 6D printing method.

To improve and optimize some features and aspects of the AM, a novel concept is proposed here. The main objective of the idea is the scientific establishment of the six-dimensional (6D) printing concept, which is schematically explained in Figure 1.

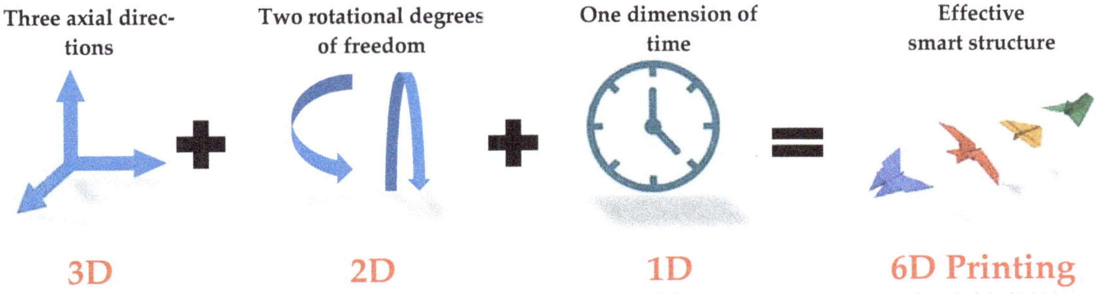

Figure 1. The 6D Printing method: Dimensionality clarification.

The 4D and 5D printing concepts should be clarified beforehand in order to enlighten the aim of the current exploration.

The 4D printing technology [26] is grounded on the 3D one, nonetheless, additionally requires the use of stimulus and stimulus-responsive raw materials. Thus, such a manufacturing process leads to the production of the famous SMs [27]. In other words, the alteration of the 4th dimension of the part is indirectly achieved via a stimulus condition which is applied after its printing finalization. On the other hand, the 5D technology [26] refers to common, non-intelligent final material component products, which; however, are manufactured by the use of a five-axis printing technique, instead of the three-axis one used in 3D printing, a fact that enables the process of objects in multiple dimensions.

The concept of 6D printing, to the authors' best knowledge, is introduced for the first time. The new method combines the 4D and the 5D printing techniques. This means that the printing process is going to use five degrees of freedom for creating the final object while the final produced material component will be a smart/intelligent one (i.e., will be capable of changing its shape due to its interaction with an environmental stimulus).

The expected arisen benefit due to the realization of such a novel manufacturing process is, first and above all, the capability of printing structures that exhibit intelligent behavior that has unique "response to stimulus" characteristics due to the arisen distinctive structural features. The new multi-axis printing capabilities are expected to lead to a new family of SMs which will also be more easily manufactured to tighter tolerances and specifications. Apart from achieving a unique material intelligence by incorporating two more printing axis, the proposed technology is additionally expected to be highly capable of producing stronger products, of better quality at minimized processing times.

Finally, yet importantly, the method may support effectively not only single-phase polymers but also nano-reinforced polymers [28] as raw materials, in a more specialized manner. The use of suitable nano-inclusions and nano-reinforcements [29] is expected to provide even more specialized final properties of the final 6D printed SM. The addition of nanomaterials to the polymeric basic raw material increases the complexity of the method design (i.e., the calibration for the printer and the whole printing setup as well as the computational modeling and simulation predictions required for the proper design of the whole manufacturing technique and final product) [30].

Overall, this work aims to highlight the benefits from the establishment of a new field in prosthetic technology, which can give new possibilities and designs whereas it has not yet been introduced in the literature. The presented concept needs to be distinguished from existing available technologies, as it will not only offer more degrees of freedom in construction and make manufacturing more efficient in terms of material use and production time but can give new designs and novel smart functionalities, which are difficult to be realized via the existing additive technology. The capabilities which arise from the idea of the 6D Printing method are inherently based on the printing production processes and its settings and not just on the properties of the material. According to the proposed definition, the single intelligence or multiple intelligences of the SM products that lead to the extra dimension (i.e., the multidirectional evolution or simultaneous change of several material attributes in time due to excitation) are all integrated into one dimension, avoiding, in this manner, any possible definition vagueness. To better differentiate each category of additive manufacturing over the proposed one, Table 1 is presented which contains the main aspects and characteristics of each relevant technique. Note that, on the whole, the proposed idea aims at improving the quality, programmability, and adjustability of the dynamic after-print multi-intelligent response of the SM object. The main advantage of the method is the fact that enables the printing in more directions and along more complicated or even curved paths, thus, leading to more well-designed products regarding both the achieved structural integrity as well as the level of intelligence. Finally, the 6D printing method accompanied with an appropriate setup may allow the use of less raw material and may simultaneously provide shorter processing times due to its inherent processing flexibility.

Table 1. Categories of additive manufacturing techniques with their basic features.

Addictive Manufacturing	Required Equipment	Basic Characteristics	Limitations
3D Printing	3D printers	Layer by layer development in the vertical axis	Inflexibility on the layer orientationProduct with no intelligence
4D Printing	3D printers Raw material with intelligences	Creation of Smart Materials (SM) Products	Limits regarding the layer orientation Intelligences of restricted programmability and strong directivity
5D Printing	5D printers or robotic arms	Layering in every direction Less material and process time	Product with no intelligence Higher setup cost
6D Printing	5D printers or robotic arms Raw material with intelligences	Layering in every direction Less material and process time Creation of SM Products of flexible and adjustable programmability	Additional calibration and need for standardization Higher setup cost

The setup cost regarding the 6D printing process is expected to be high. However, the positive economic influence of adopting the proposed method is highly dependent on the production lot size and the level of integrated automation. The more massive the production via 6D printing, the higher the overall cost reduction in comparison with the 4D printing method, due to the minimization of the raw material loss.

3. Methodology

The 6D printing process is briefly outlined diagrammatically in the flow chart of Figure 2. Each basic step of this process corresponds to the objectives that are presented in this section.

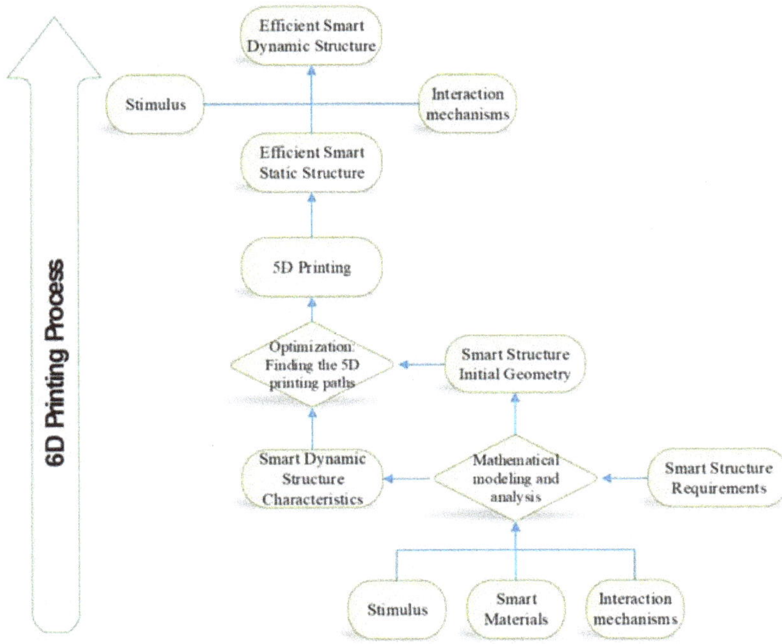

Figure 2. The concept of 6D printing.

3.1. Database of SMs—Stimuli Interaction Mechanisms

One category of printing materials that is under the spotlight in the last years is the SMs [31]. When a product is printed by using suitable raw materials, it may be transformed into an alternative structure that is intelligently sensitive to the influence of external energy that is input in the form of pressure, temperature, electric field, magnetic field, light, or other environmental stimuli. As a first step, for the establishment of the proposed process, an extended literature review is required regarding all the existing SMs and their properties, stimuli, and interaction mechanisms in order to identify the most suitable SMs for the application of the 6D AM.

Further investigations should be conducted to identify all the available polymers [32] as well as nano-reinforced polymers [33] which could be successfully melted and processed by the headed end of the 5D printer extruder, thus used as candidate printing raw materials. The development of a database concerning the aforementioned relevant raw material records may allow the development of simple and effective computational and analytical methods, which are originally required for modeling and designing the final products as well as finding the ideal manufacturing procedures.

3.2. Multi-Scale Modeling and Analysis of SMs and Structures

The final structure of the investigated printed SMs will be accomplished (a) after depositing small drops of melted polymeric media, which on occasion will be nano-reinforced, layer-by-layer; and (b) after cooling the printed smart component at room temperature. Furthermore, it is evident that the response of the SMs by a stimulus, such as environmental temperature, stress, strain, electrical current, magnetic field, etc., is strongly dependent not only on the final formation and molecular structure of the involved polymeric chains but also by the position and physical properties of the nano-reinforcement alone. Moreover, the investigation and accurate representation of the molecular kinetics, the relevant glass transition phenomena, the interphase interactions, the electrostatic and vdW forces is of high importance for developing the necessary stimulus interaction mechanisms and; therefore, for selecting the after-print SM proper design.

At the stage of the product and process optimal design, extensive comparisons between the novel 6D and the 4D printing techniques should initially be made at a theoretical level, to reach rigid conclusions about the superiority of the proposed process. Given the nanoscale significant effects governing the adopted smart materials, special formulations based on the combination of the molecular dynamics (MD) [34] and the finite element method (FEM) [35,36] must be developed to simulate the final printed material components, in an effort evaluate, control and characterize the design aspects and parameters of the investigated products. Generally, the effects at the nanoscale of both the polymeric chains as well as the involved nano-fillers under the stimulus conditions require the combination of both atomistic as well as continuum numerical techniques. Thus, appropriate MD, FEM, as well as combined multi-scale computational methods [37] should be introduced, developed, validated, and parametrically applied in order to capture the SM behavior at the nano, micro, and physical scale. The aim is the efficient modeling and simulation of the final printed products under the stimulus conditions, with low computational cost and high accuracy.

3.3. *Optimal Paths of 5D Printed Smart Structures*

In 5D printing, the printed part moves and rotates while the printer head is printing, so as for the printing to undertake the curve path of the part being printed rather than moving through a flat layer as in the case of 3D printers. This is expected to provide stronger structures using less material [35] in comparison with classical 3D printing. Besides that, both 3D and 5D printing techniques use the same technologies such as the input of 3D CAD files and the same printing materials.

The evaluation and calibration of the process require extensive experimental investigation. Numerous parts using different arrangements and printing paths should be produced and then tested. The experimental examination should be realized by using both standard 3D printing equipment as well as a five-axis printer capable of processing a variety of pure and nano-reinforced raw materials at different temperature levels.

In the conventional 3D printing process, between the CAD model that expresses the final design of the structure and the 3D printer, there is the 3D printing slicing software (slicer) essentially acting as the middleman. This software slices the 3D CAD model into 2.5D layers along the transverse axis (z-axis), generating the optimal paths of the extruder in the FDM technology through g-code. Settings in parameters like infill pattern, infill density, layer height, etc. can affect the behavior of the printed structure. Given that the printing process in a 5D printer can be performed in more axes, the presence of more parameters and combinations of parameters is inevitable. For example, the infill pattern and density as well as the presence of multiple materials can be developed in more directions than in 3D printing. The relations between the printing parameters and the resulting physical properties (mass, tensile strength, glass temperature, electrical conductivity, etc.) may influence the SM behavior, and, thus, should be systematically examined using standardized samples and a bottom-up approach, starting from simple to more complex geometries. The behaviors must be recorded, and semi-analytical equations may be extracted performing appropriate regression analysis. By determining the optimization problem, defining the objective function, variables, constraints, etc., as well as using multi-criteria analysis and/or artificial intelligence algorithms, optimum solutions could be obtained. The goal is to find the best process parameters that lead to increased SM quality as well as high productivity.

3.4. Efficient Smart Structures in Action

At the final stage, the assessment of the 6D printing method should focus on the achieved intelligence of the printed objects. Firstly, the experimental measurements should be conducted in order to determine the response of the 6D printed part under the action of static stimulus conditions. The efficiency and accuracy of the intelligent 6D printed objects in changing shape or/and properties under the corresponding stimulus is to be evaluated. Obviously, the accurate response and quick reaction to the stimulus are strongly dependent on the structural quality of the final part and, consequently, related to the followed 6D printing process. Thus, carefully designed after-print experiments and measurements may give valuable information for further improvements. After validating the proposed manufacturing procedure for producing SMs with an excellent response under a static interaction with the environment, the investigation should proceed to the characterization of the dynamic stimuli-response of the 6D printed SMs.

4. Potential Benchmarks

For convenience, the performance of the 6D concept should initially be examined through simple problems and applications, some of which are proposed in this section.

The investigation may focus on specific component geometries and material constituents. In order to evaluate the benefits of the 6D printing over the 4D one, an experimental investigation may be performed regarding the overall strength, surface quality, stimulus-response, and production efficiency of the final produced SM components (parts). The final specimens should be manufactured via both 6D and 4D methods and the processing times as well as the achieved physical properties may be set into contrast through statistical methods in order to reach confident conclusions.

The expected superiority of the 6D printing may be demonstrated via the indicative example and graphical comparison illustrated in Figure 3. The left and right material component is manufactured by using the 4D and the proposed 6D printing method, respectively, via the use of a combination of layers made of different polymer media, for example, Polylactic acid (PLA) layers [9,10] may be combined in an alternating manner with a different polymeric medium such as thermoplastic polyurethane (TPU). Here, the glass transition temperature (Tg) difference, causing internal residual stress mismatches, will be the key. Both flat specimens are expected to bend into a U-shape when environmental temperatures become higher than Tg. It should be mentioned that some or all of the plastic layers may be reinforced with nanomaterials such as graphene nanoribbons [29], using appropriate row material available in the international market.

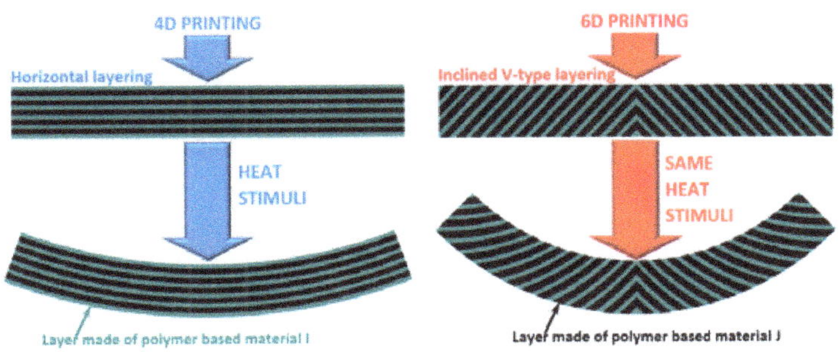

Figure 3. An indicative comparison between 4D and 6D printing: Bending deformation of a printed component is increased due to a heat stimulus when the proposed 6D printing process is utilized instead of the 4D one.

Since both investigated 4D and 6D printed SMs are composed of alternating layers of two different polymeric materials, they evidently will obtain bending deformations under the stimulus of heat/temperature [9]. The left specimen which is manufactured via

3D printing only consists of flat polymeric layers. In contrast, the right product contains V-type layering which is possible only by the use of 6D printing. Thus, it becomes evident that for this specimen, a more intensive, efficient, and controllable self-bending behavior is expected. Furthermore, utilizing the proper processing parameters, the 6D printing may require less material, lower manufacturing time for similar results.

Another potential application may be inspired by the field of magneto-sensitive smart materials, also known as magnetorheological materials. Such materials are a class of smart composites prepared by dispersing nanometer- or micrometer-sized ferromagnetic fillers into different matrix phases [38]. Their rheological properties and deformations may be effectively controlled by an external magnetic field. A possible relevant 6D printed product is illustrated in Figure 4. As depicted, the specimen has a complex structure consisting of inclined layers made of pure polymer and nanoreinforced polymer. Each nanocomposite layer is constituted by a polymer matrix material filled with magnetic nanoparticles. Due to the application of a strong magnetic field and because of the aforementioned structural attributes achieved via 6D printing, the specimen becomes highly stretched along its one direction because of the arisen common orientation of the nanoparticles as well as the shape of layering. In addition, taking into advantage the two extra available fabrication axes, it is expected that the intelligence of the final products, apart from enhanced will also be more controllable to meet the requirements of the manufacturer. Finally, the process setup may be adjusted to produce more deformable but simultaneously stronger smart material components.

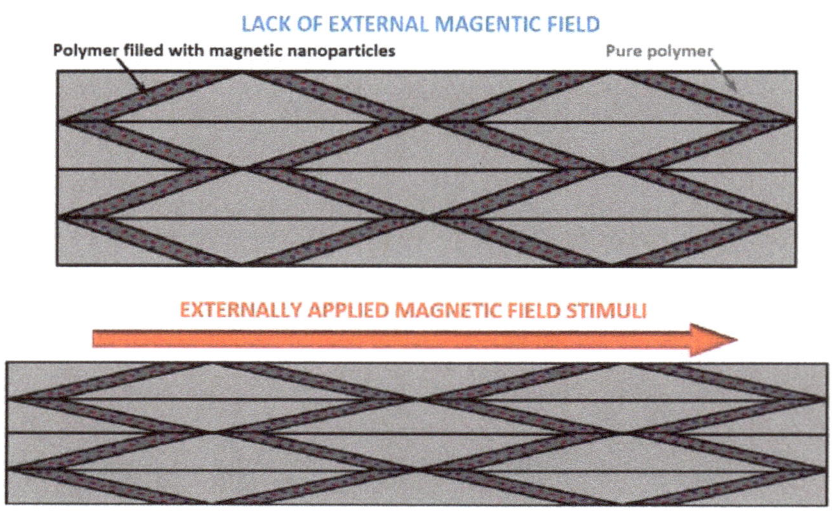

Figure 4. An indicative application of 6D printing. The inclined complex layering by using combinations of pure and nano-reinforced polymeric layers allows intensive response under the action of a magnetic stimulus.

5. The Technological Impact

As it was already mentioned, the proposed manufacturing procedure combines the 5D printing concept (which involves the motion of the printing head along with the rotation of the print bed at defined angles) with the 4DP (which leads to SMs modified with respect to the time) into a new concept called 6D printing. Evidently, the developed AM process will lead to the creation of a new generation of advanced SMs that can also contain nanoparticles in their structure. It is evident, that such a development in the field of multi-dimensional printing technology will drastically boost the whole societal impact of AM [30].

First of all, the new method is expected to provide solutions that will increase the efficiency and reduce the material cost and lead time of additive manufacturing procedures, thus leading to lighter and stronger parts, having a higher sensitivity to the relevant stimuli.

The 6D printing concept, due to its advantages, is expected to be quickly adopted by the industry which is involved in manufacturing the relevant materials and printing equipment and by other end-users that use AM. The use of 6D printing may be introduced in the same technological and manufacturing SM application field of 3D printing. However, due to the expected higher processing accuracy, new structural and functional capabilities of the proposed method, its application would be more effective in areas where precision is crucial, such as in the aerospace industry, in medical devices and electronics, in operational components for energy applications, etc. More analytically, in the future, 6D printing technology could create disruptive innovation in several fields such as:

- Medical field: Concerning the production of medical parts for prosthetics, orthopedics, or dental implants with better properties and reduced cost. Bones are not flat but have a curved surface, thus the rotation of the bed in order to follow the curve is crucial and, additionally, the created SMs could be used for smart medical implants for complex medical cases [24].
- Constructions: The proposed technique could be easily customized in order to be suitable for the construction of concrete-based structures such as buildings and other infrastructures [39] where the additional printing degrees of freedom would offer a distinct benefit Used for self-repair and self-adapting systems and products, for example, piping that could dynamically change its diameter and properties in response to the water pressure using smart materials and follow the curves of the pipe using five-axis printing.
- Aerospace engineering: Aimed at the production of curved materials that can reshape according to the temperature.
- Manufacturing industry: For the manufacturing of advanced on-demand products with reduced raw material usage and energy consumption in comparison to previous printing techniques (thus contributing to environmental sustainability and reduced environmental impact).

The proposed AM technique may allow the development of complex products that would be difficult to be manufactured by using standard 3D printing. For example, the 6D printing could permit the production of self-folding furniture that could be printed on flat boards, easily transferred and transported, and then curl up into the final shape after being heated or getting wet. In addition, five-axis movement could allow the creation of objects with curved surfaces that may change properties or geometry according to the weather or other environmental conditions.

6. Conclusions

In summary, a novel AM technique is proposed here named 6D printing. The main idea is to combine the 5D printing concept, which involves the motion of the printing head along with the rotation of the print bed at defined angles, with the 4D printing one, which leads to SMs modified with respect to the time, into a new concept called 6D printing. The realization of the proposed manufacturing procedure is expected to lead to new developments that will increase the efficiency and reduce the material cost and lead time of AM procedures, thus providing lighter and stronger parts, which also have a higher sensitivity to the relevant stimuli. The procedure requires further validation via extensive numerical and experimental investigations and for this reason, some realistic and simple benchmark applications have been also proposed. The rigid establishment of the method only requires the exploration of the SMs recent technological achievements, some previous experience on 3D printing, a five-axis printing machine, and common raw materials. In the near future, relevant research on the optimization of the proposed manufacturing technique is to be announced.

Author Contributions: Conceptualization, S.K.G. and G.I.G.; methodology, S.K.G. and G.I.G.; formal analysis, S.K.G. and G.I.G.; investigation, S.K.G., G.I.G. and P.A.B.; resources, S.K.G., G.I.G. and P.A.B.; data curation, S.K.G., G.I.G. and P.A.B.; writing—original draft preparation, S.K.G., G.I.G. and P.A.B.; writing—review and editing, S.K.G. and G.I.G.; visualization, S.K.G. and G.I.G.; supervision, S.K.G. and G.I.G. All authors have read and agreed to the published version of the manuscript.

Funding: This research received no external funding.

Institutional Review Board Statement: Not applicable.

Informed Consent Statement: Not applicable.

Conflicts of Interest: The authors declare no conflict of interest.

References

1. Moon, S.K.; Tan, Y.E.; Hwang, J.; Yoon, Y.J. Application of 3D printing technology for designing light-weight unmanned aerial vehicle wing structures. *Int. J. Precis. Eng. Manuf. Green Technol.* **2014**, *1*, 223–228. [CrossRef]
2. Ntouanoglou, K.; Stavropoulos, P.; Mourtzis, D. 4D Printing Prospects for the Aerospace Industry: A critical review. *Procedia Manuf.* **2018**, *18*, 120–129. [CrossRef]
3. Manghnani, R. An exploratory study: The impact of additive manufacturing on the automobile industry. *Int. J. Curr. Eng. Technol.* **2015**, *5*, 3407–3410.
4. Tay, Y.W.D.; Panda, B.; Paul, S.C.; Noor Mohamed, N.A.; Tan, M.J.; Leong, K.F. 3D printing trends in building and construction industry: A review. *Virtual Phys. Prototyp.* **2017**, *12*, 261–276. [CrossRef]
5. Ventola, C.L. Medical applications for 3D printing: Current and projected uses. *Pharm. Ther.* **2014**, *39*, 704.
6. Ko, H.; Moon, S.K.; Hwang, J. Design for additive manufacturing in customized products. *Int. J. Precis. Eng. Manuf.* **2015**, *16*, 2369–2375. [CrossRef]
7. Astm International. ASTM committee F42 on additive manufacturing technologies, & ASTM committee F42 on additive manufacturing technologies. Subcommittee F42. 91 on terminology. In *Standard Terminology for Additive Manufacturing Technologies*; ASTM International: West Conshohocken, PA, USA, 2012.
8. Momeni, F.; Mehdi, M.; Hassani, N.S.; Liu, X.; Ni, J. A review of 4D printing. *Mater. Des.* **2017**, *122*, 42–79. [CrossRef]
9. Kačergis, L.; Mitkus, R.; Sinapius, M. Influence of fused deposition modeling process parameters on the transformation of 4D printed morphing structures. *Smart Mater. Struct.* **2019**, *28*, 105042. [CrossRef]
10. Noroozi, R.; Bodaghi, M.; Jafari, H.; Zolfagharian, A.; Fotouhi, M. Shape-adaptive metastructures with variable bandgap regions by 4D printing. *Polymers* **2020**, *12*, 519. [CrossRef]
11. Chung, S.; Song, S.E.; Cho, Y.T. Effective software solutions for 4D printing: A review and proposal. *Int. J. Precis. Eng. Manuf. Green Technol.* **2017**, *4*, 359–371. [CrossRef]
12. Ge, Q.; Dunn, C.K.; Qi, H.J.; Dunn, M.L. Active origami by 4D printing. *Smart Mater. Struct.* **2014**, *23*, 094007. [CrossRef]
13. Pei, E. 4D printing–revolution or fad? *Assem. Autom.* **2014**, *34*, 123–127. [CrossRef]
14. Tibbits, S.; Cheung, K. Programmable materials for architectural assembly and automation. *Assem. Autom.* **2012**, *32*, 216–225. [CrossRef]
15. Quan, Z.; Wu, A.; Keefe, M.; Qin, X.; Yu, J.; Suhr, J.; Byun, J.-H.; Kim, B.-S.; Chou, T.-W. Additive manufacturing of multi-directional preforms for composites: Opportunities and challenges. *Mater. Today* **2015**, *18*, 503–512. [CrossRef]
16. Zolfagharian, A.; Kouzani, A.Z.; Khoo, S.Y.; Moghadam, A.A.A.; Gibson, I.; Kaynak, A. Evolution of 3D printed soft actuators. *Sens. Actuators A Phys.* **2016**, *250*, 258–272. [CrossRef]
17. Leist, S.K.; Zhou, J. Current status of 4D printing technology and the potential of light-reactive smart materials as 4D printable materials. *Virtual Phys. Prototyp.* **2016**, *11*, 249–262. [CrossRef]
18. Wang, Y.; Adokoh, C.K.; Narain, R. Recent development and biomedical applications of self-healing hydrogels. *Expert Opin. Drug Deliv.* **2018**, *15*, 77–91. [CrossRef]
19. Li, J.; Li, W.; Huang, W.; Zhang, G.; Sun, R.; Wong, C.P. Fabrication of highly reinforced and compressible graphene/carbon nanotube hybrid foams via a facile self-assembly process for application as strain sensors and beyond. *J. Mater. Chem. C* **2017**, *5*, 2723–2730. [CrossRef]
20. Khoo, Z.X.; Teoh, J.E.M.; Liu, Y.; Chua, C.K.; Yang, S.; An, J.; Leong, K.F.; Yeong, W.Y. 3D printing of smart materials: A review on recent progresses in 4D printing. *Virtual Phys. Prototyp.* **2015**, *10*, 103–122. [CrossRef]
21. Kwok, T.H.; Wang, C.C.; Deng, D.; Zhang, Y.; Chen, Y. Four-dimensional printing for freeform surfaces: Design optimization of origami and kirigami structures. *J. Mech. Des.* **2015**, *137*, 111712. [CrossRef]
22. Zhou, Y.; Huang, W.M.; Kang, S.F.; Wu, X.L.; Lu, H.B.; Fu, J.; Cui, H. From 3D to 4D printing: Approaches and typical applications. *J. Mech. Sci. Technol.* **2015**, *29*, 4281–4288. [CrossRef]
23. 3D Printing is So Last Week ... Say Hello to "5D Printing" at Mitsubishi Electric Research Labs! Available online: https://3dprint.com/139742/mitsubishi-electric-5d-printing/ (accessed on 15 March 2021).
24. Haleem, A.; Javaid, M.; Vaishya, R. 5D printing and its expected applications in Orthopaedics. *J. Clin. Orthop. Trauma* **2018**, *10*, 809–810. [CrossRef]

25. El-Sayegh, S.; Romdhane, L.; Manjikian, S. A critical review of 3D printing in construction: Benefits, challenges, and risks. *Arch. Civ. Mech. Eng.* **2021**, *20*, 34. [CrossRef]
26. Ravinder Reddy, P.; Anjani Devi, P. Review on the advancements to additive manufacturing-4D and 5D printing. *Int. J. Mech. Prod. Eng. Res. Dev.* **2018**, *8*, 397–402. [CrossRef]
27. Tzou, H.S.; Lee, H.-J.; Arnold, S.M. Smart materials, precision sensors/actuators, smart structures, and structronic systems. *Mech. Adv. Mater. Struct.* **2004**, *11*, 367–393. [CrossRef]
28. Yu, X.; Cheng, H.; Zhang, M.; Zhao, Y.; Qu, L.; Shi, G. Graphene-based smart materials. *Nat. Rev. Mater.* **2017**, *2*, 17046. [CrossRef]
29. Kaleem, M.A. Manufacturing graphene based polymer matrix composites (GPMCs) via 3D printing (additive manufacturing): A review. *Compos. Theory Pract.* **2019**, *19*, 126–132.
30. Huang, S.H.; Liu, P.; Mokasdar, A.; Hou, L. Additive manufacturing and its societal impact: A literature review. *Int. J. Adv. Manuf. Technol.* **2013**, *67*, 1191–1203. [CrossRef]
31. Liu, C.; Qin, H.; Mather, P.T. Review of progress in shape-memory polymers. *J. Mater. Chem.* **2007**, *17*, 1543–1558. [CrossRef]
32. Cock, F.; Cuadri, A.A.; García-Morales, M.; Partal, P. Thermal, rheological and microstructural characterisation of commercial biodegradable polyesters. *Polym. Test.* **2013**, *32*, 716–723. [CrossRef]
33. Schulz, M.J.; Kelkar, A.D.; Sundaresan, M.J. *Nanoengineering of Structural, Functional, and Smart Materials*; CRC Press: Boca Raton, FL, USA, 2005; pp. 1–712.
34. Zhang, J.; Koo, B.; Subramanian, N.; Liu, Y.; Chattopadhyay, A. An optimized cross-linked network model to simulate the linear elastic material response of a smart polymer. *J. Intell. Mater. Syst. Struct.* **2016**, *27*, 1461–1475. [CrossRef]
35. Mackerle, J. Smart materials and structures: FEM and BEM simulations A bibliography (1997–1999). *Finite Elem. Anal. Des.* **2001**, *37*, 71–83. [CrossRef]
36. Cheng, X.; Chen, Y.; Dai, S.; Bilek, M.M.M.; Bao, S.; Ye, L. Bending shape memory behaviours of carbon fibre reinforced polyurethane-type shape memory polymer composites under relatively small deformation: Characterisation and computational simulation. *J. Mech. Behav. Biomed. Mater.* **2019**, *100*, 103372. [CrossRef] [PubMed]
37. Mehar, K.; Panda, S.K.; Dewangan, H.C. Multiscale finite element prediction of thermomechanical flexural strength of nanotube-reinforced hybrid smart composite panel bonded with SMA fibre. *Structures* **2020**, *28*, 2300–2310. [CrossRef]
38. Xu, Y.; Gong, X.; Wan, Q.; Liu, T.; Xuan, S. Magneto-sensitive smart soft material and magnetorheological mechanism. *Adv. Mech.* **2015**, *45*, 461–495.
39. Bos, F.; Wolfs, R.; Ahmed, Z.; Salet, T. Additive manufacturing of concrete in construction: Potentials and challenges of 3D concrete printing. *Virtual Phys. Prototyp.* **2016**, *11*, 209–225. [CrossRef]

MDPI
St. Alban-Anlage 66
4052 Basel
Switzerland
Tel. +41 61 683 77 34
Fax +41 61 302 89 18
www.mdpi.com

Journal of Composites Science Editorial Office
E-mail: jcs@mdpi.com
www.mdpi.com/journal/jcs

www.ingramcontent.com/pod-product-compliance
Lightning Source LLC
Chambersburg PA
CBHW041932260926
51850CB00018B/68

9783036555072